QC 794.6 .S85 K35 1988
Kaku, Michio.
Introduction to superstrings

3/20/92

**Graduate Texts in Contemporary Physics**

*Series Editors*:

Joseph L. Birman
Helmut Faissner
Jeffrey W. Lynn

## Graduate Texts in Contemporary Physics

R. N. Mohapatra: **Unification and Supersymmetry: The Frontiers of Quark–Lepton Physics**

R. E. Prange and S. M. Girvin (eds.): **The Quantum Hall Effect, 2nd ed.**

M. Kaku: **Introduction to Superstrings**

J. W. Lynn (ed.): **High Temperature Superconductivity**

H. V. Klapdor (ed.): **Neutrinos**

J. H. Hinken: **Superconductor Electronics: Fundamentals and Microwave Application**

Michio Kaku

# Introduction to Superstrings

With 48 Illustrations

Springer-Verlag
New York Berlin Heidelberg
London Paris Tokyo Hong Kong

D. HIDEN RAMSEY LIBRARY
U.N.C. AT ASHEVILLE
ASHEVILLE, N. C. 28814

Michio Kaku
Department of Physics
City College of the City University of New York
New York, NY 10031, USA

*Series Editors*

Joseph L. Birman
Department of Physics
The City College of the
City University of New York
New York, NY 10031, USA

H. Faissner
Physikalisches Institut
RWTH Aachen
5100 Aachen
Federal Republic of Germany

Jeffrey W. Lynn
Department of Physics and Astronomy
University of Maryland
College Park, MD 20742, USA

Printed on acid-free paper.

Library of Congress Cataloging-in-Publication Data
Kaku, Michio.
Introduction to superstrings/Michio Kaku.
p. cm.—(Graduate texts in contemporary physics)
Bibliography: p.
1. Superstring theories. I. Title. II. Title: Introduction to superstrings. III. Series.
QC794.6.S85K35 1988
530.1'4—dc19 87-36080

© 1988 by Springer-Verlag New York Inc.
All rights reserved. This work may not be translated or copied in whole or in part without the written permission of the publisher (Springer-Verlag, 175 Fifth Avenue, New York, NY 10010, USA), except for brief excerpts in connection with reviews or scholarly analysis. Use in connection with any form of information storage and retrieval, electronic adaptation, computer software, or by similar or dissimilar methodology now known or hereafter developed is forbidden.
The use of general descriptive names, trade names, trademarks, etc. in this publication, even if the former are not especially identified, is not to be taken as a sign that such names, as understood by the Trade Marks and Merchandise Marks Act, may accordingly be used freely by anyone.

Typeset by Asco Trade Typesetting Ltd., Hong Kong.
Printed and bound by R. R. Donnelley and Sons, Harrisonburg, Virginia.
Printed in the United States of America.

9 8 7 6 5 4 3 2 1 (Corrected Second Printing, 1990)

ISBN 0-387-96700-1 Springer-Verlag New York Berlin Heidelberg
ISBN 3-540-96700-1 Springer-Verlag Berlin Heidelberg New York

*This Book is Dedicated to My Parents*

# Preface

> We are all agreed that your theory is crazy. The question which divides us is whether it is crazy enough.
>
> Niels Bohr

Superstring theory has emerged as the most promising candidate for a quantum theory of all known interactions. Superstrings apparently solve a problem that has defied solution for the past 50 years, namely the unification of the two great fundamental physical theories of the century, quantum field theory and general relativity. Superstring theory introduces an entirely new physical picture into theoretical physics and a new mathematics that has startled even the mathematicians.

Ironically, although superstring theory is supposed to provide a unified field theory of the universe, the theory itself often seems like a confused jumble of folklore, random rules of thumb, and intuition. This is because the development of superstring theory has been unlike that of any other theory, such as general relativity, which began with a geometry and an action and later evolved into a quantum theory. Superstring theory, by contrast, has been evolving backward for the past 20 years. It has a bizarre history, beginning with the purely accidental discovery of the quantum theory in 1968 by G. Veneziano and M. Suzuki.

Thumbing through old math books, they stumbled by chance on the Beta function, written down in the last century by mathematician Leonhard Euler. To their amazement, they discovered that the Beta function satisfied almost all the stringent requirements of the scattering matrix describing particle interactions. Never in the history of physics has an important scientific discovery been made in quite this random fashion.

Because of this accident of history, physicists have ever since been trying to work backward to fathom the physical principles and symmetries that underlie the theory. Unlike Einstein's theory of general relativity, which began with a geometric principle, the equivalence principle, from which the action could be derived, the fundamental physical and geometric principles that lie at the foundation of superstring theory are still unknown.

To reduce the amount of hand-waving and confusion this has caused, two themes have been stressed throughout this book. To provide the student with a solid foundation in superstring theory, we have first stressed the method of *Feynman path integrals*, which provides by far the most powerful formalism in which to discuss the model. Path integrals have become an indispensable tool for theoretical physicists, especially when quantizing gauge theories. Therefore, we have devoted Chapter 1 of this book to introducing the student to the methods of path integrals for point particles.

The second theme of this book is the method of *second quantization.* Although traditionally field theory is formulated as a second quantized theory, the bulk of superstring theory is formulated as a first quantized theory, presenting numerous conceptual problems for the beginner. Unlike the method of second quantization, where all the rules can be derived from a single action, the method of first quantization must be supplemented with numerous other rules and coventions. The hope is that the second quantized theory will reveal the underlying geometry on which the entire model is based. Thus, we have tried to stress the importance of second quantization and string field theory throughout this book. This is not just for pedagogical reasons. Ultimately, the second quantized formulation may solve the outstanding problem of superstring theory: dynamically breaking a 10-dimensional theory down to 4 dimensions.

In addition to providing the student with a firm foundation in path integrals and field theory, the other purpose of this book is to introduce students to the latest developments in superstring theory, that is, to acquaint them with the fast-paced areas that are currently the most active in theoretical research, such as:

String field theory
Conformal field theory
Kac–Moody algebras
Multiloop amplitudes and Teichmüller spaces
Calabi–Yau phenomenology
Orbifolds and four-dimensional superstrings

The goal of this book is to provide students with an overview by which to evaluate the research areas of string theory and perhaps even engage in original research. The only prerequisite for this book is a familiarity with advanced quantum mechanics. However, the mathematics of superstring theory has soared to dizzying heights. In order to provide an introduction to more advanced mathematical concepts, such as Lie groups, general relativity, supersymmetry, and supergravity, we have included a short introduction to them in the Appendix, which we hope will fill the gaps that may exist in the students' preparation. Finally, terms that may be unfamiliar to graduate students are included in the Glossary of Terms, also in the Appendix.

For the student, we should mention how to approach this book. Chapters 1–5 represent Part I, the results of first quantization. They form an essential foundation for the next chapters and cannot be skipped. Chapter 1, however, may be skipped by one who is relatively fluent in the methods of ordinary quantum field theory, such as gauge invariance and Faddeev–Popov quantization. (But we emphasize that the method of path integrals forms the foundation for this book, and hence even an advanced student may profit from reviewing Chapter 1.)

Chapters 2 and 3 form the heart of an elementary introduction to string theory. Chapter 4, however, can be omitted by one who only wants an overview of string theory. With the exception of the fermion vertex function and ghosts, most of the results of string theory can be developed using Chapters 2 and 3 without conformal field theory, and hence a beginner may overlook this chapter. (However, we emphasize that most modern approaches to first quantized string theory use the results of conformal field theory because it is the most versatile. A serious student of string theory, therefore, should be thoroughly familiar with the results of Chapter 4.)

Chapter 5 is essential to understand the miraculous cancellation of divergences of the theory, which separates string theory from all other field theories. Because the theory of automorphic functions gets increasingly difficult as one describes multiloop amplitudes, the beginner may skip the discussion of higher loops. The serious student, though, will find that multiloop amplitudes form an area of active research.

Part II begins a discussion of the field theory of strings, and Part III examines phenomenology. The order of these two parts can be interchanged without difficulty. Each part was written to be relatively independent of the other, so the more phenomenologically inclined student may skip directly to Part III without suffering any loss.

Chapters 6–8 in Part II present the evolution of three approaches to string field theory. Chapter 6 discusses the original light cone theory and how to quantize multiloop theories based on strings. However, Chapter 7 was written in a relatively self-contained fashion, so the serious student may skip Chapter 6 and delve directly into the covariant theory.

Ultimately, all of string field theory will be written geometrically, and one promising geometric formalism is presented in Chapter 8. The beginning student may find Chapter 8 a bit difficult, because it is highly mathematical. Therefore, the beginner may omit this chapter.

In Part III, the beginner may skip Chapter 9. The discussion of anomalies is rather technical and mainly based on point particles, and overlaps the discussions found in other books. Chapter 10 cannot be omitted, as it represents one of the most promising of the various superstring theories. Likewise, Chapter 11 forms an essential part of our understanding of how the superstring theory may eventually make contact with experimental data.

The author hopes that this will help both the beginner and the more advanced student to decide how to approach this book.

# Acknowledgments

I would like to thank Dr. L. Alvarez-Gaumé for making many extensive and valuable comments throughout the entire book, which have strengthened the presentation and content in several key places. Drs. M. Dine, S. Samuel, J. Lykken, O. Lechtenfeld, I. Ichinose, D.-X. Li, and A. Das also read the manuscript and contributed numerous helpful criticisms that have been incorporated into the book. Dr. D. O. Vona carefully read the entire manuscript and made important suggestions that greatly improved the draft.

I would especially like to express my sincerest appreciation to Dr. B. Sakita, Dr. J. Birman, and the faculty and students in the Physics Department of the City College of the City University of New York for constant encouragement and support throughout the writing of this manuscript, without which this book would not have been possible. I would also like to acknowledge support from the National Science Foundation and the CUNY-FRAP Program.

New York, New York

MICHIO KAKU

# Contents

Preface ........................................................ vii
Acknowledgments ........................................................ xi

## Part I. First Quantization and Path Integrals

CHAPTER 1
Path Integrals and Point Particles ........................................ 3
1.1. Why Strings? ........................................................ 3
1.2. Historical Review of Gauge Theory ........................................ 7
1.3. Path Integrals and Point Particles ........................................ 18
1.4. Relativistic Point Particles ........................................ 25
1.5. First and Second Quantization ........................................ 28
1.6. Faddeev–Popov Quantization ........................................ 30
1.7. Second Quantization ........................................ 34
1.8. Harmonic Oscillators ........................................ 37
1.9. Currents and Second Quantization ........................................ 40
1.10. Summary ........................................ 44
References ........................................ 47

CHAPTER 2
Nambu–Goto Strings ........................................ 49
2.1. Bosonic Strings ........................................ 49
2.2. Gupta–Bleuler Quantization ........................................ 59
2.3. Light Cone Quantization ........................................ 66
2.4. BRST Quantization ........................................ 69
2.5. Trees ........................................ 71
2.6. From Path Integrals to Operators ........................................ 77
2.7. Projective Invariance and Twists ........................................ 83

2.8. Closed Strings ........ 86
2.9. Ghost Elimination ........ 89
2.10. Summary ........ 94
References ........ 98

CHAPTER 3
Superstrings ........ 99
3.1. Supersymmetric Point Particles ........ 99
3.2. Two-Dimensional Supersymmetry ........ 102
3.3. Trees ........ 109
3.4. Local 2D Supersymmetry ........ 115
3.5. Quantization ........ 117
3.6. GSO Projection ........ 121
3.7. Superstrings ........ 124
3.8. Light Cone Quantization of the GS Action ........ 126
3.9. Vertices and Trees ........ 132
3.10. Summary ........ 134
References ........ 137

CHAPTER 4
Conformal Field Theory and Kac–Moody Algebras ........ 139
4.1. Conformal Field Theory ........ 139
4.2. Superconformal Field Theory ........ 148
4.3. Spin Fields ........ 153
4.4. Superconformal Ghosts ........ 156
4.5. Fermion Vertex ........ 163
4.6. Spinors and Trees ........ 166
4.7. Kac–Moody Algebras ........ 169
4.8. Supersymmetry ........ 172
4.9. Summary ........ 173
References ........ 175

CHAPTER 5
Multiloops and Teichmüller Spaces ........ 177
5.1. Unitarity ........ 177
5.2. Single-Loop Amplitude ........ 181
5.3. Harmonic Oscillators ........ 184
5.4. Single-Loop Superstring Amplitudes ........ 192
5.5. Closed Loops ........ 194
5.6. Multiloop Amplitudes ........ 199
5.7. Riemann Surfaces and Teichmüller Spaces ........ 209
5.8. Conformal Anomaly ........ 216
5.9. Superstrings ........ 219
5.10. Determinants and Singularities ........ 223
5.11. Moduli Space and Grassmannians ........ 224
5.12. Summary ........ 236
References ........ 240

# Part II. Second Quantization and the Search for Geometry

CHAPTER 6
Light Cone Field Theory ........ 245
6.1. Why String Field Theory? ........ 245
6.2. Deriving Point Particle Field Theory ........ 248
6.3. Light Cone Field Theory ........ 252
6.4. Interactions ........ 259
6.5. Neumann Function Method ........ 265
6.6. Equivalence of the Scattering Amplitudes ........ 270
6.7. Four-String Interaction ........ 272
6.8. Superstring Field Theory ........ 277
6.9. Summary ........ 283
References ........ 288

CHAPTER 7
BRST Field Theory ........ 289
7.1. Covariant String Field Theory ........ 289
7.2. BRST Field Theory ........ 295
7.3. Gauge Fixing ........ 298
7.4. Interactions ........ 301
7.5. Axiomatic Formulation ........ 306
7.6. Proof of Equivalence ........ 309
7.7. Closed Strings and Superstrings ........ 315
7.8. Summary ........ 325
References ........ 328

CHAPTER 8
Geometric String Field Theory ........ 331
8.1. Why Geometry? ........ 331
8.2. The String Group ........ 336
8.3. Unified String Group ........ 341
8.4. Representations of the USG ........ 343
8.5. Ghost Sector and the Tangent Space ........ 348
8.6. Connections and Covariant Derivatives ........ 352
8.7. Geometric Derivation of the Action ........ 357
8.8. The Interpolating Gauge ........ 360
8.9. Closed Strings and Superstrings ........ 364
8.10. Summary ........ 368
References ........ 372

# Part III. Phenomenology and Model Building

CHAPTER 9
Anomalies and the Atiyah–Singer Theorem ........ 375
9.1. Beyond GUT Phenomenology ........ 375
9.2. Anomalies and Feynman Diagrams ........ 379

9.3. Anomalies in the Functional Formalism ........ 384
9.4. Anomalies and Characteristic Classes ........ 386
9.5. Dirac Index ........ 391
9.6. Gravitational and Gauge Anomalies ........ 394
9.7. Anomaly Cancellation in Strings ........ 404
9.8. A Simple Proof of the Atiyah–Singer Index Theorem ........ 406
9.9. Summary ........ 412
References ........ 416

CHAPTER 10
Heterotic Strings and Compactification ........ 417
10.1. Compactification ........ 417
10.2. The Heterotic String ........ 422
10.3. Spectrum ........ 427
10.4. Covariant and Fermionic Formulations ........ 430
10.5. Trees ........ 432
10.6. Single-Loop Amplitude ........ 435
10.7. $E_8$ and Kac–Moody Algebras ........ 439
10.8. 10D Without Supersymmetry ........ 441
10.9. Lorentzian Lattices ........ 446
10.10. Summary ........ 448
References ........ 451

CHAPTER 11
Calabi–Yau Spaces and Orbifolds ........ 452
11.1. Calabi–Yau Spaces ........ 452
11.2. Review of de Rahm Cohomology ........ 457
11.3. Cohomology and Homology ........ 461
11.4. Kähler Manifolds ........ 467
11.5. Embedding the Spin Connection ........ 474
11.6. Fermion Generations ........ 476
11.7. Wilson Lines ........ 481
11.8. Orbifolds ........ 482
11.9. Four-Dimensional Superstrings ........ 487
11.10. Summary ........ 503
11.11. Conclusion ........ 507
References ........ 509

APPENDIX ........ 511
A.1. A Brief Introduction to Group Theory ........ 511
A.2. A Brief Introduction to General Relativity ........ 523
A.3. A Brief Introduction to the Theory of Forms ........ 528
A.4. A Brief Introduction to Supersymmetry ........ 532
A.5. A Brief Introduction to Supergravity ........ 539
A.6. Glossary of Terms ........ 544
A.7. Notation ........ 560
References ........ 562

Index ........ 563

**Part I**

# First Quantization and Path Integrals

CHAPTER 1

# Path Integrals and Point Particles

## §1.1. Why Strings?

One of the greatest scientific challenges of our time is the struggle to unite the two fundamental theories of modern physics, quantum field theory and general relativity, into one theoretical framework. Remarkably, these two theories together embody the sum total of all human knowledge concerning the most fundamental forces of nature. Quantum field theory, for example, has had phenomenal success in explaining the physics of the microcosm, down to distances less than $10^{-15}$ cm. General relativity, on the other hand, is unrivaled in explaining the large-scale behavior of the cosmos, providing a fascinating and compelling description of the origin of the universe itself. The astonishing success of these two theories is that together they can explain the behavior of matter and energy over a staggering 40 orders of magnitude, from the subnuclear to the cosmic domain.

The great mystery of the past five decades, however, has been the total incompatibility of these two theories. It's as if nature had two minds, each working independently of the other in its own particular domain, operating in total isolation of the other. Why should nature, at its deepest and most fundamental level, require two totally distinct frameworks, with two sets of mathematics, two sets of assumptions, and two sets of physical principles?

Ideally, we would want a unified field theory to unite these two fundamental theories:

$$\left.\begin{array}{l}\text{Quantum field theory}\\ \text{General relativity}\end{array}\right\}\text{Unified field theory}$$

However, the history of attempts over the past decades to unite these two theories has been dismal. They have inevitably been riddled with infinities or have violated some of the cherished principles of physics, such as causality. The powerful techniques of renormalization theory developed in quantum field theory over the past decades have failed to eliminate the infinities of quantum gravity. Apparently, a fundamental piece of the jigsaw puzzle is still missing.

Although quantum field theory and general relativity seem totally incompatible, the past two decades of intense theoretical research have made it increasingly clear that the secret to this mystery most likely lies in the power of *gauge symmetry*. One of the most remarkable features of nature is that its basic laws have great unity and symmetry when expressed in terms of group theory. Unification through gauge symmetry, apparently, is one of the great lessons of physics. In particular, the use of local symmetries in Yang–Mills theories has had enormous success in banishing the infinities of quantum field theory and in unifying the laws of elementary particle physics into an elegant and comprehensive framework. Nature, it seems, does not simply incorporate symmetry into physical laws for aesthetic reasons. Nature *demands* symmetry.

The problem has been, however, that even the powerful gauge symmetries of Yang–Mills theory and the general covariance of Einstein's equations are insufficient to yield a finite quantum theory of gravity.

At present, the most promising hope for a truly unified and finite description of these two fundamental theories is the superstring theory [1–12]. Superstrings possess by far the largest set of gauge symmetries ever found in physics, perhaps even large enough to eliminate all divergences of quantum gravity. Not only does the superstring's symmetry include that of Einstein's theory of general relativity and the Yang–Mills theory, it also includes supergravity and the Grand Unified Theories (GUTs) [13] as subsets.

Roughly speaking the way in which superstring theory solves the riddle of infinities can be visualized as in Fig. 1.1, where we calculate the scattering of two point particles by summing over an infinite set of Feynman diagrams with loops. These diagrams, in general, have singularities that correspond to "pinching" one of the internal lines until the topology of the graph is altered. By contrast, in Fig. 1.2 we have the single-loop contribution to the scattering

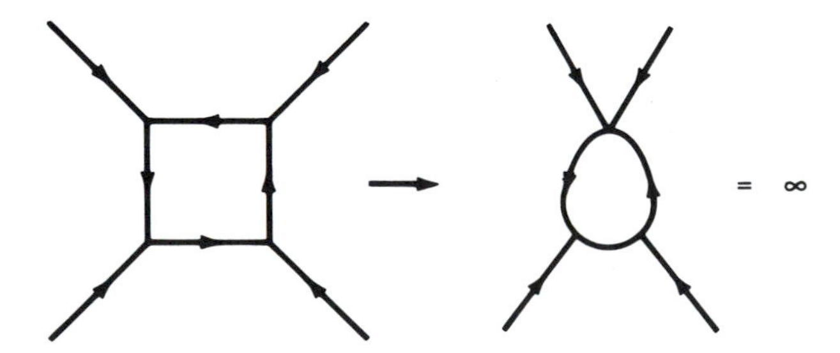


Figure 1.1. Single-loop Feynman diagram for four-particle scattering. The ultraviolet divergence of this diagram corresponds to the pinching of one internal leg, i.e., when one internal line shrinks to a point.

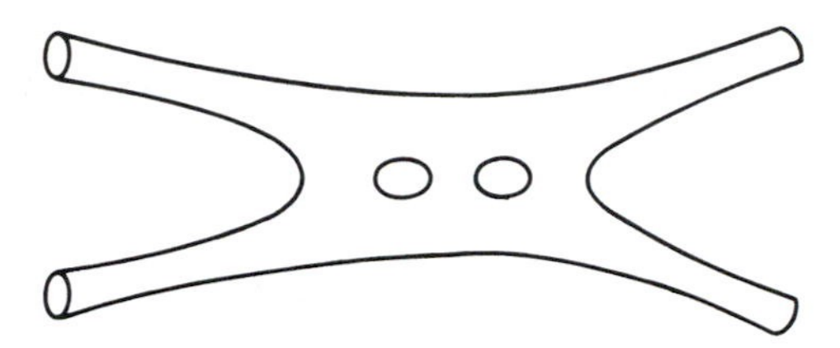

Figure 1.2. Two-loop Feynman diagram for closed string scattering. The diagram is ultraviolet finite because it cannot be pinched as in the point particle case. From topological arguments alone, we can see that string theory is less divergent than point particle theory. Infrared divergences, however, may still exist.

of two closed string states. Notice that we cannot "pinch" one of the internal lines as in the point particle case. Thus, we naively expect that the superstring theory is less divergent or even finite because of the symmetries that forbid this topological deformation.

Any theory that can simultaneously eliminate the infinities of the $S$-matrix and incorporate quantum mechanics, the general theory of relativity, GUT theory, and supergravity obviously possesses mathematics of breathtaking beauty and complexity. In fact, even the mathematicians have been startled at the mathematics emerging from the superstring theory, which links together some of the most dissimilar, far-ranging fields of mathematics, such as Kac–Moody algebras, Riemann surfaces and Teichmüller spaces, modular groups, and even Monster group theory.

*The great irony of string theory, however, is that the theory itself is not unified.* To someone learning the theory for the first time, it is often a frustrating collection of folklore, rules of thumb, and intuition. At times, there seems to be no rhyme or reason for many of the conventions of the model. For a theory that makes the claim of providing a unifying framework for all physical laws, it is the supreme irony that the theory itself appears so disunited! The secrets of the model, at its most fundamental level, are still being pried loose.

Usually, when we write down a quantum theory, we start with the geometry or symmetry of the theory and then write down the action. From the action, in turn, we derive all the predictions of the model, including the unitary $S$-matrix. Thus, a *second quantized* action is the proper way in which to formulate a quantum field theory. The fundamental reason why superstring theory seems, at times, to be a loose collection of apparently random conventions is that it is usually formulated as a first quantized theory. Because of this, we must appeal to intuition and folklore in order to construct all the Feynman diagrams for a unitary theory.

Unfortunately, the second quantized action and the geometry of the superstring are some of the last features of the model to be developed. In fact, as seen from this perspective, the model has been developing *backward* for the past 20 years, beginning with the accidental discovery of its quantum theory in 1968!

By contrast, when Einstein first discovered general relativity, he started with physical principles, such as the equivalence principle, and formulated it in the language of general covariance. Once the geometry was established, he then wrote down the action as the unique solution to the problem. Later, classical solutions to the equations were discovered in terms of curved manifolds, which provided the first successful theoretical models for the large-scale behavior of the universe. Finally, the last step in the evolution of general relativity is the development of a quantum theory of gravity. The crucial steps in the historical evolution of general relativity can thus be represented as

$$\text{Geometry} \rightarrow \text{Action} \rightarrow \text{Classical theory} \rightarrow \text{Quantum theory}$$

Furthermore, both general relativity and Yang–Mills theory are mature theories: they both can be formulated from first principles, which stresses the geometry and the physical assumptions underlying the theory. Superstring theory is just beginning to reach that stage of development.

Remarkably, Yang–Mills theory and gravity theory are the *unique* solution to two simple geometric statements:

(1) Global Symmetry
The free theory must propagate pure ghost-free spin-one and spin-two fields transforming as irreducible representations of SU($N$) and the Lorentz group.

(2) Local Symmetry
The theory must be locally SU($N$) and generally covariant.

What is remarkable is that the coupled Yang–Mills–gravity action is the unique solution of these two simple principles:

$$L = -\tfrac{1}{4}\sqrt{-g}F_{\mu\nu}F^{\mu\nu} - \frac{1}{2\kappa^2}\sqrt{-g}R_{\mu\nu}g^{\mu\nu} \tag{1.1.1}$$

(The first principle contains the real physics of the theory. It cannot be included as a subset of the second principle. There is an infinite number of generally covariant and SU($N$) symmetric invariants, so we need the first principle to input the physics and select the irreducible representations of the basic fields. By "pure" fields, we mean ghost-free fields that have at most two derivatives, which rules out $R^2$ and $F^4$ higher derivative theories.)

The question remains: *What is the counterpart to these two simple principles for superstring theory*? Much work still has to be done to formulate a truly geometric theory of strings, but the most promising candidate is presented in Chapter 8, where I discuss the geometric formulation of string field theory.

The plan of this book, of course, must reflect the fact that the theory has been evolving backward. For pedagogical reasons, we will mostly follow the historical development of the theory. Thus, Part I of this book, which introduces the first quantized theory, will at times appear to be a loose collection of conventions without any guiding principle. That is why we have chosen, in Part I, to emphasize the *path integral or functional* approach to string theory.

Only with Feynman path integrals do we have a formalism in which we can derive the other formalisms, such as the harmonic oscillator formalism. Although the path integral formulation of a first quantized theory is still woefully inadequate compared to a genuine second quantized theory, it is the most convenient formalism in which to tie together the loose ends of the first quantized theory.

In Part II of the book we will discuss the field theory itself, from which we can derive all the results of the theory from one action. However, once again we have followed historical order and presented the field theory backward. We begin with the broken theory, and then present a candidate for the geometric one.

Finally, in Part III we present the "phenomenology" of strings. Although it may be presumptuous to do phenomenology starting at $10^{19}$ GeV, it is important to establish the kinds of predictions that the theory makes.

However, to really appreciate the successes and possible defects of the superstring theory, we must first try to understand the historical problems that have plagued physicists for the past five decades. Let us now turn to a quick review of the development of gauge theories in order to appreciate the difficulty of constructing a finite theory of gravity. We will also briefly sketch the historical development of the superstring theory.

## §1.2. Historical Review of Gauge Theory

In the 1960s, elementary particle physics seemed hopelessly mired in confusion. The weak, electromagnetic, strong, and gravitational forces were each studied separately, largely in isolation of the others. Moreover, investigations into each force had reached a fundamental roadblock:

(1) The weak interactions. Theoretical models of the weak interactions had progressed embarrassingly little beyond the Fermi theory first proposed three decades earlier in the 1930s:

$$L_{\text{Fermi}} \sim \bar{\psi}_p \Gamma^A \psi_n \bar{\psi}_e \Gamma_A \psi_\nu \tag{1.2.1}$$

where the $\Gamma^A$ represents various combinations of Dirac matrices. The next major step, a theory of $W$ bosons, was plagued with the problem of infinities. Furthermore, no one knew the underlying symmetry among the leptons, or whether there was any.

(2) The strong interactions. In contrast to weak interactions, the Yukawa meson theory provided a renormalizable theory of the strong interactions:

$$L_{\text{Yukawa}} \sim g\bar{\psi}\psi\phi \tag{1.2.2}$$

However, the Yukawa theory could not explain the avalanche of "elementary" particles that were being discovered in particle accelerators. J. Robert Oppenheimer even suggested that the Nobel Prize in physics should go to the physicist who *didn't* discover a particle that year. Furthermore, the quark model, which seemed to fit the data much better than it had any right to, was plagued with the fact that quarks were never seen experimentally.

(3) The gravitational force. Gravity research was totally uncoupled from research in the other interactions. Classical relativists continued to find more and more classical solutions in isolation from particle research. Attempts to canonically quantize the theory were frustrated by the presence of the tremendous redundancy of the theory. There was also the discouraging realization that even if the theory could be successfully quantized, it would still be nonrenormalizable.

This bleak landscape changed dramatically in the early 1970s with the coming of the gauge revolution. One of the great achievements of the past 15 years has been the development of a fully renormalizable theory of spin-1 gauge particles in which, for the first time, physicists could actually calculate realistic $S$-matrix elements. Thus, it took fully 100 years to advance beyond the original gauge theory first proposed by Maxwell in the 1860s! (See the Appendix for an elementary introduction to gauge theories and group theory.)

Apparently the key to eliminating the divergences of relativistic quantum mechanics is to go to larger and more sophisticated gauge groups. Symmetry, instead of being a purely aesthetic feature of a particular model, now becomes its most important feature.

For example, Maxwell's equations, which provided the first unification of the electric force with the magnetic force, has a gauge group given by U(1). The unification of the weak and electromagnetic forces into the electro-weak force requires $SU(2) \otimes U(1)$. The forces that bind the quarks together into the hadrons, or quantum chromodynamics (QCD), are based on SU(3). All of elementary particle physics, in fact, is compatible with the minimal theory of $SU(3) \otimes SU(2) \otimes U(1)$.

Although the verdict is still not in on the GUTs, which are supposed to unite the electroweak force with the strong force, once again the unifying theme is gauge symmetry, with such proposals as SU(5), O(10), etc. symmetry.

Although the gauge revolution is perhaps one of the most important developments in decades, it is still not enough. There is a growing realization that the Yang–Mills theory by itself cannot push our understanding of the physical universe beyond the present level. Not only do the GUTs fail to explain important physical phenomena, but also there is the crucially important problem of formulating a quantum theory of gravity.

Grand Unified Theories, first of all, cannot be the final word on the unification of all forces. There are several features of GUTs that are still unresolved:

(1) GUTs cannot resolve the problem of why there are three nearly exact copies or families of elementary particles. We still cannot answer Rabi's question, "Who ordered the muon?"

(2) GUTs still have 20 or so arbitrary parameters. They cannot, for example, calculate the masses of the quarks, or the various Yukawa couplings. A truly unified field theory should have at most one arbitrary parameter.
(3) GUTs have difficulty solving the hierarchy problem. Unless we appeal to supersymmetry, it is hard to keep the physics of incredibly massive particles from mixing with everyday energies and destroying the hierarchy.
(4) The unification of particle forces occurs around $10^{-28}$ cm which is very close to the Planck length of $10^{-33}$ cm, where we expect gravitational effects to become dominant. Yet GUTs say nothing whatsoever about gravitation.
(5) So far, proton decay has not been conclusively observed, which already rules out minimal SU(5). There is, therefore, still no compelling experimental reason for introducing the theory.
(6) It is difficult to believe that no new interactions will be found between present-day energies and the unification scale. The "desert" may very well bloom with new interactions yet unknown.

The most perplexing and the most challenging of these problems, from a foundational point of view, has been to find a way of quantizing Einstein's theory of general relativity. Although Yang–Mills theories have had spectacular successes in unifying the known laws of particle physics, the laws of gravity are curiously different at a fundamental level. Clearly, Yang–Mills theory and conventional gauge theory are incapable of dealing with this problem. Thus, GUTs are faced with formidable experimental and theoretical problems when pushed to their limits.

General relativity is also plagued with similar difficulties when pushed to its limits:

(1) Classically, it has been established that Einstein's equations necessarily exhibit pointlike singularities, where we expect the laws of general relativity to collapse. Quantum corrections must dominate over the classical theory in this domain.
(2) The action is not bounded from below, because it is linear in the curvature tensor. Thus, it may not be stable quantum mechanically.
(3) General relativity is not renormalizable. Computer calculations, for example, have now conclusively shown that there is a nonzero counterterm in Einstein's theory at the two-loop level.

Naive attempts to quantize Einstein's theory of gravitation have met with disappointing failure. One of the first to point out that general relativity would be incompatible with quantum mechanics was Heisenberg, who noted that the presence of a dimensional coupling constant would ruin the usual renormalization program.

If we set

$$\frac{h}{2\pi} = 1; \qquad c = 1 \tag{1.2.3}$$

there still remains a dimensional constant even in the Newtonian theory of gravity, the gravitational constant $G$:

$$F = G\frac{m_1 m_2}{r^2} \tag{1.2.4}$$

which has dimensions of centimeters squared. When we power expand the metric tensor $g_{\mu\nu}$ around flat space with the metric $\eta_{\mu\nu} = (-+++)$, we introduce the coupling constant $\kappa$, which has dimensions of centimeters:

$$g_{\mu\nu} = \eta_{\mu\nu} + \kappa h_{\mu\nu} \tag{1.2.5}$$

Therefore:

$$G \sim \kappa^2 \tag{1.2.6}$$

In this system of units, where the only unit is the centimeter, this coupling constant $\kappa$ becomes the Planck length, $10^{-33}$ cm or $10^{19}$ GeV, which is far beyond the reach of experimentation!

Renormalization theory, however, is founded on the fundamental premise that we can eliminate all divergences with an infinite redefinition of certain constants. Having a dimensional coupling constant means that this complicated reshuffling and resumming of graphs is impossible. Unlike standard renormalizable theories, in quantum gravity we cannot add diagrams that have different powers of the coupling constant. *This means that general relativity cannot be a renormalizable theory.* The amplitude for graviton–graviton scattering, for example, is now a power expansion in a dimensional parameter (see Fig. 1.3):

$$A = \sum_{n=2}^{\infty} \kappa^n A_n \tag{1.2.7}$$

$= \kappa^2$ + $\kappa^2$ + $\kappa^4$ + $\kappa^4$ + ⋯ + $\kappa^6$ + $\kappa^6$ + ⋯

Figure 1.3. Scattering amplitude for graviton–graviton scattering. Because the coupling constant has dimensions, graphs of different order cannot be added to renormalize the theory. Thus, theories containing quantum gravity must be either divergent or completely finite order-by-order. Pure quantum gravity has been shown on computer to diverge at the two-loop level. Counterterms have also been found for quantum gravity coupled to lower-spin particles. Thus, superstring theory is the only candidate for a finite theory.

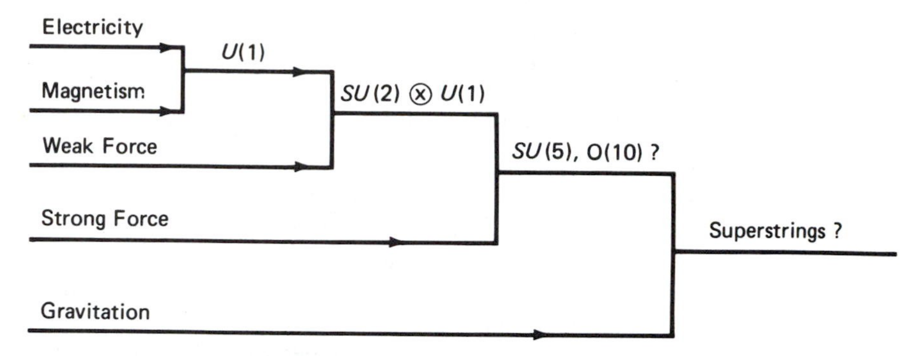


Figure 1.4. Chart showing how gauge theories based on Lie groups have united the fundamental forces of nature. Maxwell's theory, based on U(1), unites electricity and magnetism. The Weinberg–Salam model, based on SU(2) ⊗ U(1), unites the weak force with the electromagnetic force. GUT theories (based on SU(5), O(10), or larger groups) are the best candidate to unite the strong force with the electro-weak force. Superstring theory is the only candidate for a gauge theory that can unite gravity with the rest of the particle forces.

where we are no longer able to shuffle graphs with different values of $n$ to cancel the infinities, which is the heart of renormalization theory. Thus, renormalization theory breaks down.

Because general relativity is hopelessly outside the domain of conventional renormalization theory, one must reconsider Dirac's fundamental objection. It was Dirac who said that the success of quantum mechanics was based on approximation schemes where each correction term was increasingly small. But renormalization theory is flawed because it maximally violates this principle and manipulates infinite quantities and discards them at the end.

One solution might be to construct a theory of gravity that is finite to every order in the coupling constant, with no need for renormalization at all. For a while, one bright hope was supergravity [14, 15], based on the local gauge group Osp($N$/4) (see Appendix), which was the first nontrivial extension of Einstein's equations in 60 years. The hope was that this gauge group would offer us a large enough set of Ward–Takahashi identities to cancel a large class of divergent diagrams. The larger the gauge group, the more likely troublesome infinites would cancel (see Fig. 1.4):

| Theory | Gauge group |
|---|---|
| Electromagnetism | U(1) |
| Electro-weak | SU(2) ⊗ U(1) |
| Strong | SU(3) |
| GUT(?) | SU(5), O(10) |
| Gravity (?) | GL(4), O(3, 1) |
| Supergravity (?) | Osp($N$/4) |

The basic strategy being pursued was

$$\text{Gauge symmetry} \rightarrow \text{Ward} - \text{Takahashi identities}$$
$$\rightarrow \text{Cancellation of graphs} \rightarrow \text{Renormalizable theory}$$

For example, even Einstein's theory of gravity can be shown to be trivially finite at the first loop level. There exists a remarkable identity, called the Gauss–Bonnet identity, which immediately shows that all one-loop graphs in general relativity (which would take a computer to write down) sum to zero. In fact, the super-Gauss–Bonnet identities eliminate many of the divergences of supergravity, but probably not enough to make the theory finite.

The largest and most promising of the supergravities, the O(8) supergravity, is probably divergent. Unfortunately, it is possible to write down locally supersymmetric counterterms at the seventh loop level. It is highly unlikely that the coefficients of this and probably an infinite number of other counterterms can all vanish without appealing to an even higher symmetry. This is discouraging, because it means that the gauge group of the largest supergravity theory, Osp(8/4), is still too small to eliminate the divergences of general relativity.

Furthermore, the O(8) gauge group is too small to accommodate the minimal SU(3) ⊗ SU(2) ⊗ U(1) of particle physics. If we go to higher groups beyond O(8), we find that we must incorporate higher and higher spins into the theory. However, an interacting spin-3 theory is probably not consistent, making one suspect that O(8) is the limit to supergravity theories.

In conclusion, supergravity must be ruled out for two fundamental reasons:

(1) It is probably not a finite theory because the gauge group is not large enough to eliminate all possible supersymmetric counterterms. There is a possible counterterm at the seventh loop level.
(2) Its gauge group O(8) is not large enough to accommodate the minimal symmetry of particle physics, namely SU(3) ⊗ SU(2) ⊗ U(1).

Physicists, faced with these and other stumbling blocks over the years, have concluded that perhaps one or more of our cherished assumptions about our universe must be abandoned. Because general relativity and quantum mechanics can be derived from a small set of postulates, one or more of these postulates must be wrong. The key must be to drop one of our commonsense assumptions about nature on which we have constructed general relativity and quantum mechanics. Over the years, several proposals have been made to drop some of our commonsense notions about the universe:

(1) Continuity
This approach assumes that space–time must be granular. The size of these grains would provide a natural cutoff for the Feynman integrals, allowing us to have a finite *S*-matrix. Integrals like

$$\int_{\varepsilon}^{\infty} d^4x \tag{1.2.8}$$

would then diverge as $\varepsilon^{-n}$, but we would never take the limit as $\varepsilon$ goes to zero. Lattice gravity theories are of this type. In Regge calculus [16], for example, we latticize Riemannian space with discrete four-simplexes and replace the curvature tensor by the angular deficit calculated when moving in a circle around a simplex:

$$-\frac{1}{2\kappa^2}\sqrt{-g}R \rightarrow \text{angular deficit}$$

(In flat space, there is no angular deficit when walking around a closed path, and the action collapses.) Usually, in lattice theories, we take the limit as the lattice length goes to zero. Here, however, we keep it fixed at a small number [17]. At present, however, there is no experimental evidence to support the idea that space–time is granular. Although we can never rule out this approach, it seems to run counter to the natural progression of particle physics, which has been to postulate larger and more elegant groups.

(2) Causality

This approach allows small violations in causality. Theories that incorporate the Lee–Wick mechanism [18] are actually renormalizable, but permit small deviations from causality. These theories make the Feynman diagrams converge by adding a fictitious Pauli–Villars field of mass $M$ that changes the ultraviolet behavior of the propagator. Usually, the Feynman propagator converges as $p^{-2}$ in the ultraviolet limit. However, by adding a fictitious particle, we can make the propagator converge even faster, like $p^{-4}$:

$$\frac{1}{p^2 + m^2} - \frac{1}{p^2 + M^2} \rightarrow \frac{1}{p^4} \tag{1.2.9}$$

Notice that the Pauli–Villars field is a ghost because of the $-1$ that appears in the propagator. (This means that the theory will be riddled with negative probabilities.) Usually, we let the mass of the Pauli–Villars field tend to infinity. However, here we keep it finite, letting the pole go onto the unphysical sheet. Investigations of the structure of the resulting Feynman diagrams show, however, that causality is violated; that is, you can meet your parents before you are born.

(3) Unitarity

We can replace Einstein's theory, which is based on the curvature tensor, with a conformal theory based on the Weyl tensor:

$$\sqrt{-g}R_{\mu\nu}g^{\mu\nu} \rightarrow \sqrt{-g}C^2_{\mu\nu\rho\sigma} \tag{1.2.10}$$

where the Weyl tensor is defined as

$$C_{\mu\nu\rho\sigma} = R_{\mu\nu\rho\sigma} + g_{\mu[\sigma}R_{\rho]\nu} + g_{\nu[\rho}R_{\sigma]\mu} + \tfrac{1}{3}Rg_{\mu[\rho}g_{\sigma]\nu} \tag{1.2.11}$$

where the brackets represent antisymmetrization. The conformal tensor possesses a larger symmetry group than the curvature tensor, that is, invariance under local conformal transformations:

$$\begin{cases} g_{\mu\nu} \to e^{\sigma} g_{\mu\nu} \\ C^{\mu}_{\nu\rho\sigma} \to C^{\mu}_{\nu\rho\sigma} \end{cases} \tag{1.2.12}$$

The Weyl theory converges because the propagators go as $p^{-4}$; that is, it is a higher derivative theory. However, there is a "unitary ghost" that also appears with a $-1$ in the propagator, for the same reasons cited above. The most optimistic scenario would be to have these unitary ghosts "confined" by a mechanism similar to quark confinement [19, 20].

(4) Locality
Over the years, there have also been proposals to abandon some of the important postulates of quantum mechanics, such as locality. After all, there is no guarantee that the laws of quantum mechanics should hold down to distances of $10^{-33}$ cm. However, there have always been problems whenever physicists tried to deviate from the laws of quantum mechanics, such as causality. At present, there is no successful alternative to quantum mechanics.

(5) Point Particles
Finally, there is the approach of superstrings, which abandons the concept of idealized point particles, first introduced 2000 years ago by the Greeks.

The superstring theory, because it abandons only the assumption that the fundamental constituents of matter must be point particles, does the least amount of damage to cherished physical principles and continues the tradition of increasing the complexity and sophistication of the gauge group. Superstring theory does not violate any of the laws of quantum mechanics, yet manages to eliminate most, if not all, of the divergences of the Feynman diagrams. The symmetry group of the superstring model, the largest ever encountered in the history of physics, is probably large enough to make the theory finite to all orders. Once again, it is symmetry, and not the breakdown of quantum mechanics, that is the fundamental key to rendering a theory finite.

In Fig. 1.5 we see diagrammatically the evolution of various theories of gravity. First, there was Newton's theory of action at a distance, where gravitational interactions travel faster than the speed of light. Einstein replaced this with the classical interpretation of curved manifolds. Quantum gravity, in turn, makes quantum corrections to Einstein's theory by adding in loops. Finally, the superstring theory makes further corrections to the point particle quantum theory by summing over all possible topological configurations of interacting strings.

Superstring theory, however, is quite unlike its predecessors in its historical development. Unlike other physical theories, superstring theory has perhaps one of the strangest histories in science, with more twists and turns than a roller coaster.

First, two young physicists, Veneziano and Suzuki [21, 22], independently

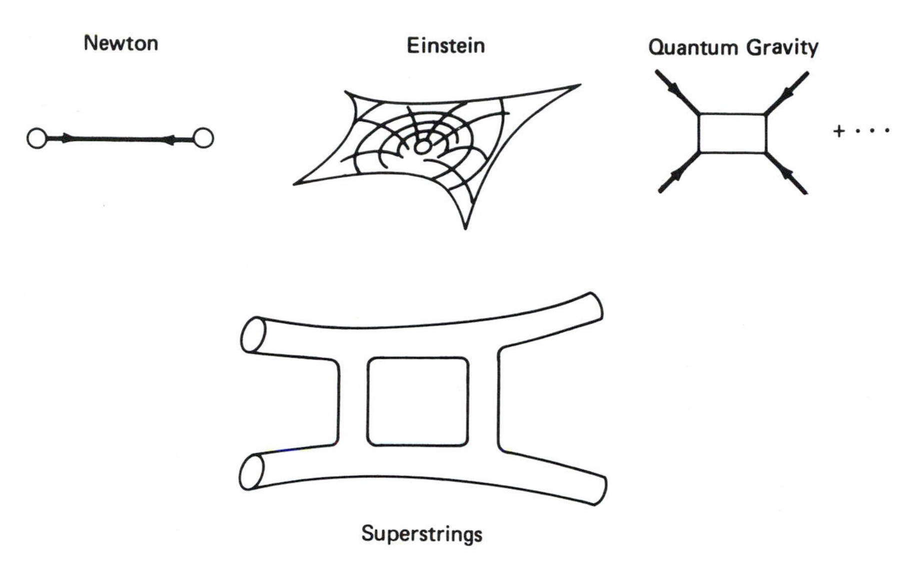


Figure 1.5. Steps in the evolution of the theory of gravitation. Each step in this chart builds on the successes of the previous step. Newton thought gravity was a force that acted instantly over a distance. Einstein proposed that gravitation was caused by the curvature of space–time. The naive merger of general relativity and quantum mechanics produces a divergent theory, quantum gravity, which assumes that gravitation is caused by the exchange of particle-like gravitons. Superstring theory proposes that gravitation is caused by the exchange of closed strings.

discovered its quantum theory when they were thumbing through a math book and accidentally noted that the Euler Beta function satisfied all the postulates of the $S$-matrix for hadronic interactions (except unitarity). Neveu, Schwarz, and Ramond [23–25] quickly generalized the theory to include spinning particles. To solve the problem of unitarity, Kikkawa, Sakita, and Virasoro [26] proposed that the Euler Beta function be treated as the Born term to a perturbation series. Finally, Kaku, Yu, Lovelace, and Alessandrini [27–33] completed the quantum theory by calculating bosonic multiloop diagrams. The theory, however, was still formulated entirely in terms of on-shell $S$-matrix amplitudes.

Next, Nambu and Goto [34–35] realized that lurking behind these scattering amplitudes was a classical relativistic string. In one sweep, they revolutionized the entire theory by revealing the unifying, classical picture behind the theory. The relationship between the classical theory and the quantum theory was quickly made by Goldstone, Goddard, Rebbi, and Thorn [36] and further developed by Mandelstam [37]. The theory, however, was still formulated as a first quantized theory, so that the measure, the vertices, the counting of graphs, etc. all had to be postulated ad hoc and not deduced from first principles.

The action (in a particular gauge) was finally written down by Kaku and

Kikkawa [38]. At last, the model could be derived from one action strictly in terms of physical variables, although the action did not have any symmetries left. However, when it was discovered that the theory was defined only in 10 and 26 dimensions, the model quickly died. Furthermore, the rapid development of QCD as a theory of hadronic interactions seemingly put the last nail in the coffin of the superstring.

For 10 years, the model languished because no one could believe that a 10- or 26-dimensional theory had any relevance to 4-dimensional physics. When Scherk and Schwarz [39] made the outrageous (for its time) suggestion that the dual model was actually a theory of all known interactions, no one took the idea very seriously. The idea fell like a lead balloon.

Finally, the discovery in 1984 by Green and Schwarz [40] that the superstring theory is anomaly-free and probably finite to all orders in perturbation theory has revived the theory. The $E_8 \otimes E_8$ "heterotic string" of Gross, Harvey, Martinec, and Rohm [41] at present seems to be the best candidate for unifying gravity with physically reasonable models of particle interactions.

One of the active areas of research now is to complete the evolution of the theory, to discover why all the "miracles" occur in the model. There has been a flurry of activity in the direction of writing down the covariant action using methods discovered in the intervening 10 years, such as BRST. However, there is now a growing realization that the covariant BRST formalism itself is a gauge-fixed formalism, much like the light cone formalism. Recently, however, there has been work on a truly *geometric field theory* where all the features of the theory can be deduced from simple physical principles. This is explained in Chapter 8. This would truly complete the evolution of the theory, which has been progressing backward for the past 20 years

$$\text{Quantum theory} \rightarrow \text{Classical theory} \rightarrow \text{Action} \rightarrow \text{Geometry}$$

Let us summarize some of the promising positive features of the superstring model:

(1) The gauge group includes $E_8 \otimes E_8$, which is much larger than the minimal group $SU(3) \otimes SU(2) \otimes U(1)$. There is plenty of room for phenomenology in this theory.
(2) The theory has no anomalies. These small but important defects in a quantum field theory place enormous restrictions on what kinds of theories are self-consistent. The symmetries of the superstring theory, by a series of "miracles," can cancel all of its potential anomalies.
(3) Powerful arguments from the theory of Riemann surfaces indicate that the theory is finite to all orders in perturbation theory (although a rigorous proof is still lacking).
(4) There is very little freedom to play with. Superstring models are notoriously difficult to tinker with without destroying their miraculous properties. Thus, we do not have the problem of 20 or more arbitrary coupling constants.

(5) The theory includes GUTs, super-Yang–Mills, supergravity, and Kaluza–Klein theories as subsets. Thus, many of the features of the phenomenology developed for these theories carry over into the string theory.

Superstring theory, crudely speaking, unites the various forces and particles in the same way that a violin string provides a unifying description of the musical tones. By themselves, the notes *A*, *B*, *C*, etc. are not fundamental. However, the violin string is fundamental; one physical object can explain the varieties of musical notes and even the harmonies one can construct from them. In much the same way, the superstring provides a unifying description of elementary particles and forces. In fact, the "music" created by the superstring is the forces and particles of nature.

Although superstring theory, because of its fabulously large set of symmetries, has "miraculous" cancellations of anomalies and divergences, we must also present a balanced picture and point out its shortcomings. To be fair we must also list the potential problems of the theory that have been pointed out by critics of the model:

(1) It is impossible experimentally to reach the tremendous energies found at the Planck scale. Therefore, the theory is in some sense untestable. A theory that is untestable is not an acceptable physical theory.
(2) Not one shred of experimental evidence has been found to confirm the existence of supersymmetry, let alone superstrings.
(3) It is presumptuous to assume that there will be no surprises in the "desert" between 100 and $10^{19}$ GeV. New, totally unexpected phenomena have always cropped up when we have pushed the energy scale of our accelerators. Superstring theory, however, makes predictions over the next 17 orders of magnitude, which is unheard of in the history of science.
(4) The theory does not explain why the cosmological constant is zero. Any theory that claims to be a "theory of everything" must surely explain the puzzle of a vanishing cosmological constant, but it is not clear how superstrings solves this problem.
(5) The theory has an embarrassment of riches. There are apparently *thousands* of ways to break down the theory to low energies. Which is the correct vacuum? Although the superstring theory can produce the minimal theory of $SU(3) \otimes SU(2) \otimes U(1)$, it also predicts many other interactions that have not yet been seen.
(6) No one really knows how to break a 10-dimensional theory down to 4 dimensions.

Of these six objections to the model, the most fundamental is the last, the inability to calculate dimensional breaking. The reason for this is simple: to every order in perturbation theory, the dimension of space–time is stable. Thus, in order to have the theory spontaneously curl up into 4- and 6-dimensional universes, we must appeal to nonperturbative, dynamical effects, which are notoriously difficult to calculate. This is why the search for the geometry underlying the theory is so important. The geometric formulation of the model may give us the key insight into the model that will allow us to make nonperturbative calculations and make definite predictions with the theory.

Thus, the criticism that the model cannot be tested at the Planck length is actually slightly deceptive. *The superstring theory, if it could be successfully broken dynamically, should be able to make predictions down to the level of everyday energies.* For example, it should be able to predict the masses of the quarks. Therefore, we do not have to wait for several centuries until we have accelerators that can reach the Planck length.

*Thus, the fundamental problem facing superstrings is not necessarily an experimental one. It is mainly theoretical. The outstanding problem of the theory is to calculate dynamical symmetry breaking, so that its predictions can be compared with experimental data at ordinary energies.*

A fundamental theory at Planck energies is also a fundamental theory at ordinary energies. Thus, the main stumbling block to the development of the theory is an understanding of its nonperturbative behavior. And the key to this understanding probably lies in a second quantized, geometric formulation of the model.

In Part I of this book, however, we will follow historical precedent and present the first quantized formulation of the model. As we will stress throughout this book, the first quantized theory seems to be a loose collection of random facts. As a consequence, we have emphasized the path integral formulation (first written down for the Veneziano model by Hsue, Sakita, and Virasoro [42, 43]) as the most powerful method of formulating the first quantized theory. Although the path integral approach cannot reveal the underlying geometric formulation of the model, it provides the most comprehensive formulation of the first quantized theory.

We will now turn to the functional formulation [44] of point particle theory, which can be incorporated almost directly into the string theory.

## §1.3. Path Integrals and Point Particles

Let us begin our discussion by analyzing the simplest of all possible systems, the classical nonrelativistic point particle. Surprisingly, much of the analysis of this simple dynamical system carries over directly to the superstring theory. The language we will use is the formalism of path integrals, which is so versatile that it can accommodate both first quantized point particles and second quantized gauge fields with equal ease.

As in classical mechanics, the starting point is the Lagrangian for a point particle:

$$L = \tfrac{1}{2}m\dot{x}_i^2 - V(x) \tag{1.3.1}$$

where the particle is moving in an external potential. The real physics is contained in the statement that the action $S$ must be minimized. The equations of motion can be derived by minimizing the action:

$$S = \int L(x_i, \dot{x}_i, t)\, dt$$
$$\delta S = 0 \tag{1.3.2}$$

To calculate the equations of motion, let us make a small variation in the path of the particle given by

$$\delta x_i; \qquad \delta \dot{x}_i \tag{1.3.3}$$

Under this small variation, the action varies as follows:

$$\int dt \left\{ \frac{\delta L}{\delta x_i} \delta x_i + \frac{\delta L}{\delta \dot{x}_i} \delta \dot{x}_i \right\} = 0 \tag{1.3.4}$$

Integrating by parts, we arrive at the Euler–Lagrange equations:

$$\frac{\delta L}{\delta x_i} - \frac{d}{dt} \frac{\delta L}{\delta \dot{x}_i} = 0 \tag{1.3.5}$$

For our point particle, the equations of motion become

$$m \frac{d^2 x_i}{dt^2} = -\frac{\partial V(x)}{\partial x_i} \tag{1.3.6}$$

which correspond to the usual classical Newtonian equations of motion.

In addition to the Lagrangian formulation of classical mechanics, there is also the Hamiltonian form. Instead of introducing the position and the velocities as fundamental objects, we now introduce the position and the momentum:

$$p_i = \frac{\delta L}{\delta \dot{x}_i} \tag{1.3.7}$$

With this definition of the conjugate variable, we have

$$H = p_i \dot{x}_i - L$$
$$H(p_i, x_i) = \frac{p_i^2}{2m} + V(x) \tag{1.3.8}$$

Finally, the Poisson brackets between the momenta and the coordinates are given by

$$[p_i, x_j]_{\mathrm{PB}} = -\delta_{ij} \tag{1.3.9}$$

A celebrated theorem in classical mechanics states that the equations of motion of Newton and the action principle method can be shown to be identical. Beginning with the action principle, we can derive Newton's laws of motion, and vice versa.

Equations of motion $\leftrightarrow$ Action principle

*This equivalence, however, breaks down at the quantum level.* Quantum mechanically, there is a fundamental difference between the two, with the equations of motion being only an approximation to the actual quantum behavior of matter. *Thus, the action principle is the only acceptable framework for quantum mechanics.*

Let us now reformulate the principles of quantum mechanics in terms of Feynman path integrals [44]:

(1) The probability $P(a, b)$ of a particle moving from point $a$ to point $b$ is the square of the absolute value of a complex number, the transition function $K(a, b)$:

$$P(a, b) = |K(a, b)|^2 \tag{1.3.10}$$

(2) The transition function is given by the sum of a certain phase factor, which is a function of the action $S$, taken over all possible paths from $a$ to $b$:

$$K(a, b) = \sum_{\text{paths}} k e^{i2\pi S/h} \tag{1.3.11}$$

where the constant $k$ can be fixed by

$$K(a, c) = \sum_{\text{paths}} K(a, b)K(b, c) \tag{1.3.12}$$

and the intermediate sum is taken over paths that go through all possible intermediate points $b$.

The second principle says that a particle "sniffs out" all possible paths from point $a$ to point $b$, no matter how complicated the paths may be. We calculate this phase factor for each of this infinite number of paths. Then the transition factor for the path between $a$ and $b$ is calculated by summing over all possible phase factors (see Fig. 1.6).

Remarkably, *the essence of quantum mechanics is captured in these two principles.* All the profoundly important implications of quantum mechanics, which represent a startling departure from classical mechanics, can be derived from these two innocent-sounding principles! In particular, these two principles summarize the essence of the quantum interpretation of the double-slit experiment, which, in turn, summarizes the essence of quantum mechanics itself.

It is apparent at this point that the results of classical mechanics can be reproduced from our two assumptions in a certain approximation. Notice that, for values of $S$ that are large compared to Planck's constant, the phase factor fluctuates rapidly, canceling out these contributions:

$$\delta S \gg \frac{h}{2\pi}: \sum_{\text{paths}} e^{i2\pi S/h} \to 0 \tag{1.3.13}$$

Thus, the only contributions to the path integral that survive are those for which the deviations in the action from the classical path are on the order of

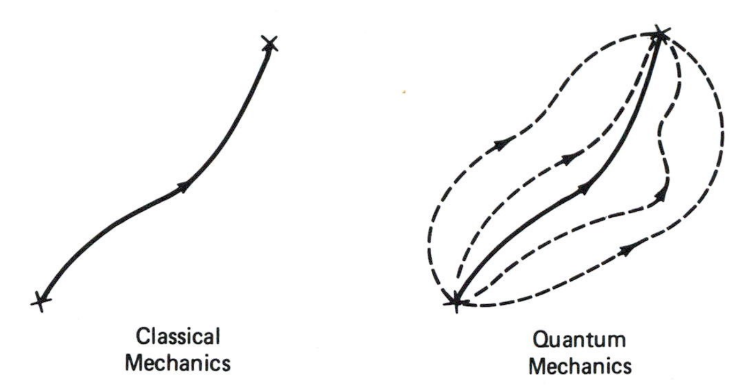


Figure 1.6. The essential difference between classical mechanics and quantum mechanics. Classical mechanics assumes that a particle executes just one path between two points based either on the equations of motion or on the minimization of the action. By constrast, quantum mechanics sums the contributions of probability functions (based on an action) for all possible paths between two points. Although the classical path is the one most favored, in principle all possible paths contribute to the path integral. Thus, the action principle is more fundamental than the equations of motion at the quantum level.

Planck's constant:

$$\delta S \sim \frac{h}{2\pi} \tag{1.3.14}$$

We see that the Euler–Lagrange equations of motion are reproduced only in a certain classical limit, that is, when Planck's constant goes to zero. Therefore the size of Planck's constant ultimately determines the probability that a particle will execute trajectories that are forbidden classically. We see the origin of Heisenberg's uncertainty principle embodied in these two principles.

Now let us try to reformulate this more precisely in terms of path integrals. The second principle now reads

$$K(a, b) = \int_a^b Dx\, e^{i2\pi S/h} \tag{1.3.15}$$

where

$$K(a, c) = \int K(a, b)K(b, c)\, Dx_b \tag{1.3.16}$$

and

$$\sum_{\text{paths}} \to \int Dx = \lim_{N\to\infty} \int \prod_{i=1}^{3} \prod_{n=1}^{N} dx_{i,n} \tag{1.3.17}$$

where the index $n$ labels $N$ intermediate points that divide the interval between the initial and the final coordinate. We will take the limit when $N$ approaches infinity.

It is absolutely essential to understand that the integration $Dx$ is not the ordinary integration over $x$. In fact, it is the product of all possible integrations over all intermediate points $x_{i,n}$ between points $a$ and $b$. This crucial difference between ordinary integration and functional integration goes to the heart of the path integral formalism.

This infinite series of integrations, in turn, is equivalent to summing over all possible paths between $a$ and $b$. Thus, we will have to be careful to include normalization factors when performing an integration over an infinite number of intermediate points.

If we take the simple case where $L = \frac{1}{2}m\dot{x}_i^2$, all functional integrations can actually be performed exactly. The integral in question is a Gaussian, which is fortunately one of the small number of functional integrals that can actually be performed. One of the great embarrassments of the method of path integrals is that one of the few integrals that can actually be performed is

$$\int_{-\infty}^{\infty} dx\, x^{2n} e^{-r^2x^2} = \frac{\Gamma(n+\frac{1}{2})}{r^{2n+1}} \tag{1.3.18}$$

We will be using this formula throughout the entire book.

Let us now break up the path into an infinite number of intermediate points, $x_{i,n}$. (Notice that the functional expression integrates over all possible values of the intermediate point $x_{i,n}$, so we cannot expect that $x_{i,n}$ and $x_{i,n+1}$ are close to each other even for small time separations.) Let us write

$$\begin{gathered} dt \to \varepsilon \\ \tfrac{1}{2}m\dot{x}_i^2\, dt \to \tfrac{1}{2}m(x_n - x_{n+1})_i^2 \varepsilon^{-1} \end{gathered} \tag{1.3.19}$$

In order to perform the functional integral over an infinite number of intermediate points, we will repeatedly use the following Gaussian integration:

$$\begin{aligned} &\int_{-\infty}^{\infty} dx_2 \exp[-a(x_1 - x_2)^2 - a(x_2 - x_3)^2] \\ &\quad = \sqrt{\frac{\pi}{2a}} \exp[-\tfrac{1}{2}a(x_1 - x_3)^2] \end{aligned} \tag{1.3.20}$$

One of the crucial points to observe here is that the integration over a Gaussian in one of the intermediate points yields another Gaussian with that intermediate point removed. This is the fundamental reason why we can perform the functional integration over an infinite number of intermediate points.

Finally, the path integral that we wish to perform is given by

$$\begin{aligned} K(a, b) = \lim_{\varepsilon \to 0} &\int\int \cdots \int dx_1\, dx_2 \cdots dx_{N-1} \\ &\times \left(\frac{2\pi i\varepsilon}{m}\right)^{-(1/2)N} \exp\left\{\frac{im}{2\varepsilon} \sum_{n=1}^{N} (x_n - x_{n-1})^2\right\} \end{aligned} \tag{1.3.21}$$

(where we have suppressed the vector index $i$). Using the previous relation (1.3.20), the final result is equal to

$$K(a, b) = \left| \frac{m}{2\pi(t_b - t_a)} \right|^{1/2} \exp \frac{\frac{1}{2} i m (x_b - x_a)^2}{t_b - t_a} \tag{1.3.22}$$

The transition probability function $K$ has some very interesting properties. For example, it solves the wave equation:

$$\frac{-1}{2m} \frac{\partial^2}{\partial x_a^2} K(a, b) = i \frac{\partial}{\partial t_a} K(a, b) \tag{1.3.23}$$

when $t_a$ is greater than $t_b$.

Later, we will generalize these expressions for the case of freely propagating strings, and we will find that these expressions for the Green's functions carry over with only small, but important, changes.

To show the relationship between the Hamiltonian and Lagrangian formalisms in the path integral approach , it is helpful to insert a complete set of intermediate states when we divide up the path from $a$ to $b$. Let us treat the variable $x$ as an operator $\hat{x}$ acting on a set of eigenstates:

$$\hat{x}|x\rangle = x|x\rangle \tag{1.3.24}$$

The $|x\rangle$ represents an eigenstate of the position operator, treating $\hat{x}$ as an operator whose eigenvalue is equal to the number $x$. Then completeness over eigenstates for coordinates and for momenta can be represented as

$$\begin{aligned} 1 &= \int |x\rangle \, dx \, \langle x| \\ 1 &= \int |p\rangle \, dp \, \langle p| \end{aligned} \tag{1.3.25}$$

We normalize our states as follows:

$$\begin{aligned} \langle x|y\rangle &= \delta(x - y) \\ \langle p|x\rangle &= \frac{e^{ipx}}{\sqrt{2\pi}} \end{aligned} \tag{1.3.26}$$

(Because of the infinite number of normalization constants that constantly appear in the path integral formalism, we will often delete them for the sake of clarity in this book. We do not lose any generality, because we can, of course, reinsert them into the path integral if we desire.)

With these eigenstates, we can now rewrite the expression for the Green's function for going from point $x_1$ to $x_N$:

$$K(1, N) = \langle x_1, t_1 | x_N, t_N \rangle \tag{1.3.27}$$

In order to derive the previous expression (1.3.22) for the transition amplitude, let us insert a complete set of intermediate states at every intermediate point between $x_1$ and $x_N$:

$$\langle x_1, t_1 | x_N, t_N \rangle = \langle x_1, t_1 | x_2, t_2 \rangle \int dx_2 \langle x_2, t_2 | \int dx_2 \cdots | x_{N-1}, t_{N-1} \rangle \int dx_{N-1} \langle x_{N-1}, t_{N-1} | x_N, t_N \rangle \tag{1.3.28}$$

Now let us examine each infinitesimal propagator in terms of the Hamiltonian, which we write as a function of the coordinates and derivatives:

$$H = H(x, \partial_x) \tag{1.3.29}$$

Then the transition for an infinitesimal interval is given by

$$\begin{aligned}
\langle x_1, t_1 | x_2, t_2 \rangle &= \langle x_1 | e^{-iH(x,\partial_x)\delta t} | x_2 \rangle \\
&= e^{-iH(x,\partial_x)\delta t} \langle x_1 | x_2 \rangle \\
&= e^{-iH(x,\partial_x)\delta t} \langle x_1 | p \rangle \int dp \langle p | x_2 \rangle \\
&= e^{-iH(x,p)\delta t} \int \frac{dp}{2\pi} e^{ip(x_2 - x_1)} \\
&= e^{-iH(x,p)\delta t} \int \frac{dp}{2\pi} e^{ip\dot{x}\delta t}
\end{aligned} \tag{1.3.30}$$

It is very important to notice that path integrals have made it possible to make the transition from classical to quantum commutators. The Hamiltonian can be expressed either as a function of derivatives with respect to the position or as a function of the canonical momenta because of the identity:

$$\partial_x e^{ipx} = ipe^{ipx} \tag{1.3.31}$$

This allows us to make the important identification:

$$\begin{cases} H(x, p) \leftrightarrow H(x, \partial_x) \\ p \leftrightarrow -i\dfrac{\delta}{\delta x} \end{cases} \tag{1.3.32}$$

In the functional formalism, the important correspondence between momenta and partial derivatives arises because of this identity.

Putting everything together, we can now write the complete transition amplitude as

$$\langle x_1, t_1 | x_N, t_N \rangle = \int_{x_1}^{x_N} Dp\, Dx \exp i \int_{t_1}^{t_N} [p\dot{x} - H(p, x)]\, dt \tag{1.3.33}$$

where

$$H = \frac{p_i^2}{2m} + V(x) \tag{1.3.34}$$

(As usual, we have dropped all the intermediate normalizations, which are just factors of $2\pi$.) Notice that the functional integral, which was once only a function of the coordinates, is now a function of both the momenta and the coordinates.

In order to retrieve the original Lagrangian, we can perform the $p$ integration exactly, because it is a simple Gaussian integral, and we arrive at

$$\langle x_1, t_1 | x_N, t_N \rangle = \int_{x_1}^{x_N} Dx \exp i \int_{t_1}^{t_N} [\tfrac{1}{2} m \dot{x}_i^2 - V(x)] \, dt \tag{1.3.35}$$

We have thus made the transition between the Lagrangian and the Hamiltonian formalism using functional methods. We can use either:

$$L = \tfrac{1}{2} m \dot{x}_i^2 - V(x) \leftrightarrow H = \frac{p_i^2}{2m} + V(x) \tag{1.3.36}$$

Functionally, the only difference between these two expressions is whether we integrate over the coordinates or a combination of the coordinates and the momenta. The transition probability can be represented as

$$\begin{aligned} K(a, b) &= \int_{x_a}^{x_b} Dx \exp i \int_{t_a}^{t_b} dt [\tfrac{1}{2} m x_i^2 - V(x)] \\ &= \int_{x_a}^{x_b} Dx \, Dp \exp i \int_{t_a}^{t_b} dt \left[ p\dot{x}_i - \frac{p_i^2}{2m} - V(x) \right] \end{aligned} \tag{1.3.37}$$

## §1.4. Relativistic Point Particles

So far, our discussion has been limited to nonrelativistic particles, where all degrees of freedom are physical. However, nontrivial complications occur when we generalize our previous discussion to the case of relativistic particles. In particular, the $(-1)$ appearing in the Lorentz metric will, in general, cause nonphysical states to propagate in the theory. These nonphysical "ghost" states, which have negative probability, must be eliminated carefully to ensure a sensible causal theory free of negative norm states.

For the relativistic case, let us assume that the location of a point particle is given by a four-vector:

$$x_\mu(\tau) \tag{1.4.1}$$

where parametrization $\tau$ does *not* necessarily refer to the time. The action is particularly simple, being proportional to the four-dimensional path length:

$$S = -m \int ds = -m(\text{length}) \tag{1.4.2}$$

The path length $ds$ can be written in terms of the coordinates:

$$ds = \sqrt{-\dot{x}_\mu^2} \, d\tau \tag{1.4.3}$$

where the dot refers to differentiation with respect to the parameter $\tau$. This action, unlike the previous nonrelativistic action, is invariant under reparametrizations of the fictitious parameter $\tau$. Let us make a change of coordinates from $\tau$ to $\tilde{\tau}$:

$$\tau \to \tilde{\tau}(\tau) \tag{1.4.4}$$

Then we find

$$\begin{aligned} d\tau &= \frac{d\tau}{d\tilde{\tau}}\, d\tilde{\tau} \\ \frac{dx}{d\tau} &= \frac{dx}{d\tilde{\tau}} \frac{d\tilde{\tau}}{d\tau} \\ \left\{\left(\frac{dx_\mu}{d\tau}\right)^2\right\}^{1/2} d\tau &= \left\{\left(\frac{dx_\mu}{d\tilde{\tau}}\right)^2\right\}^{1/2} d\tilde{\tau} \end{aligned} \tag{1.4.5}$$

Thus, the action is invariant under an arbitrary reparametrization of the variable $\tau$.

This can be written infinitesimally as

$$\begin{cases} \tau \to \tau + \delta\tau \\ \delta x_\mu = \dot{x}_\mu \delta\tau \end{cases} \tag{1.4.6}$$

As before, we can now introduce canonical conjugates:

$$p_\mu = \frac{\delta L}{\delta \dot{x}^\mu} = \frac{m\dot{x}_\mu}{\sqrt{-\dot{x}_\mu^2}} \tag{1.4.7}$$

The crucial difference, however, from our previous discussion of the nonrelativistic point particle is that not all the canonical momenta are independent. In fact, we find a constraint among them:

$$p_\mu^2 + m^2 \equiv 0 \tag{1.4.8}$$

Thus, the mass shell condition arises as an exact constraint among the momenta. If we calculate the Hamiltonian associated with this system, we find that

$$H = p^\mu \dot{x}_\mu - L \equiv 0 \tag{1.4.9}$$

The Hamiltonian vanishes identically.

These unusual features, the vanishing of the Hamiltonian and the constraints among the momenta, are typical of systems with redundant gauge degrees of freedom. The invariance under reparametrization, for example, tells us that the path integral that we wrote earlier actually diverges:

$$\int Dx\, e^{iS} = \infty \tag{1.4.10}$$

This is because there is a separate contribution from each particular parametrization. But since $Dx$ is parametrization invariant, this means that we are summing over an infinite number of copies of the same thing. Thus, the integral must diverge.

Dirac, however, explained how to quantize systems with redundant gauge degrees of freedom. For example, let us introduce canonical momenta $p$ and impose the constraint condition via a Lagrange multiplier as follows:

$$L = p_\mu \dot{x}^\mu - \tfrac{1}{2}e(p_\mu^2 + m^2) \tag{1.4.11}$$

The constraint equation (1.4.8) is imposed here as a classical equation of motion. By varying $e$, we recover the constraint on the momenta. Quantum mechanically, however, this constraint is imposed by functionally integrating out over $e$. In the path integral, we have

$$\int De \exp\left[-i \int d\tau\, \tfrac{1}{2}e(p^2 + m^2)\right] \sim \delta(p^2 + m^2) \tag{1.4.12}$$

where we have used the fact that the integral over $e^{ikx}$ (or the Fourier transform of the number 1) is equal to $\delta(x)$. Notice that new Lagrangian (1.4.11) still possesses the gauge degree of freedom. It is invariant under

$$\begin{aligned} \delta x_\mu &= \varepsilon \dot{x}_\mu \\ \delta p_\mu &= \varepsilon \dot{p}_\mu \\ \delta e &= \frac{d(\varepsilon e)}{d\tau} \end{aligned} \tag{1.4.13}$$

The advantage that this action has over the previous one is that all variables occur linearly. We do not have to worry about complications caused by the square root. (The field $e$ that we have introduced will become the metric tensor $g_{ab}$ when we generalize this action to the string.)

Let us now functionally integrate over the $p$ variable. Because the integration is again a Gaussian, we have no problem in performing the $p$ integration:

$$\begin{aligned} &\int Dp \exp\left\{i \int d\tau [p\dot{x} - \tfrac{1}{2}e(p^2 + m^2)]\right\} \\ &\quad \sim \exp\left\{i \int d\tau\, \tfrac{1}{2}(e^{-1}\dot{x}^2 - em^2)\right\} \end{aligned} \tag{1.4.14}$$

Thus, we have now obtained a third version of the point particle action. The advantage of this action is that it is linear in the coordinates and is invariant under

$$\begin{cases} \delta x_\mu = \varepsilon \dot{x}_\mu \\ \delta e = \dfrac{d(\varepsilon e)}{d\tau} \end{cases} \tag{1.4.15}$$

In summary, we have found three equivalent ways to express the relativistic point particle. The "second-order" Lagrangian (1.4.14) is expressed in terms of second-order derivatives in the variable $x_\mu(\tau)$ and the field $e$. The "non-linear" Lagrangian (1.4.3) is expressed only in terms of $x_\mu(\tau)$. It can be derived from the second-order form by functionally integrating over the $e$ field. And finally, the "Hamiltonian" form contains both $x_\mu(\tau)$ and the canonical conjugate $p_\mu(\tau)$ (it is first-order in derivatives):

$$\begin{aligned} \text{1st-order (Hamiltonian) form:} \quad & L = p_\mu \dot{x}^\mu - \tfrac{1}{2}e(p_\mu^2 + m^2) \\ \text{2nd-order form:} \quad & L = \tfrac{1}{2}(e^{-1}\dot{x}_\mu^2 - em^2) \\ \text{Non-linear form:} \quad & L = -m\sqrt{-\dot{x}_\mu^2} \end{aligned} \tag{1.4.16}$$

All three are invariant under reparametrization. Each of them has its own distinct advantages and disadvantages. This exercise in writing the action of the free relativistic particle in three different ways is an important one because it will carry over directly into the string formalism. Expressed in terms of path integrals, the point particle theory and the string theory are remarkably similar.

## §1.5. First and Second Quantization

In this section, we will quantize the classical point particle and then show the relationship to the more conventional second quantized formalism of field theory. The first quantization program, we shall see, is rather clumsy compared to the second quantized formalism that most physicists are familiar with, but historically the string theory evolved as a first quantized theory. The great advantage of the second quantized formalism is that the entire theory can be derived from a single action, whereas the first quantized theory requires many additional assumptions.

The transition from the classical to the quantum system is intimately linked with the question of eliminating redundant infinites. As we said before, the path integral is formally ill-defined because we are summing over an infinite number of copies of the same thing. The trick is to single out just one copy.

There are at least three basic ways in which the first quantized point particle may be quantized: the Coulomb gauge, the Gupta–Bleuler formalism, and the BRST formalism.

### Coulomb Quantization

Here, we choose the gauge

$$x_0 = t = \tau \tag{1.5.1}$$

In other words, we set the time component of the $x$ variable equal to the real time $t$, which now parametrizes the evolution of the string. In this gauge, the action reduces to

$$L = -m \int \sqrt{1 - v_i^2}\, dt \tag{1.5.2}$$

In the limit of velocities small compared to the velocity of light, we have

$$L \sim \tfrac{1}{2} m \dot{x}_i^2 \tag{1.5.3}$$

as before, so that the functional integral is modified to

$$\int Dx_\mu\, \delta(x_0 - t) e^{iS} = \int Dx_i \exp i \int \tfrac{1}{2} m \dot{x}_i^2\, dt \tag{1.5.4}$$

For the case of the string, this simple example will lay the basis for the light cone quantization. The advantage of the Coulomb gauge is that all ghosts have been explicitly removed from the theory, so we are dealing only with physical quantities. The other advantage is that the zeroth component of the position vector is now explicitly defined to be the time variable. The parametrization of the point particle is now given in terms of the physical time.

The disadvantage of the Coulomb formalism, however, is that manifest Lorentz symmetry is broken and we have to check explicitly that the quantized Lorentz generators close correctly. Athough this is trivial for the point particle, surprising features will emerge for the quantum string, fixing the dimension of space–time to be 26.

## Gupta–Bleuler Quantization

This approach tries to maintain Lorentz invariance. This means, of course, that particular care must be taken to prevent the negative norm states from spoiling the physical properties of the $S$-matrix. The Gupta–Bleuler method keeps the action totally relativistic, but imposes the constraint (1.4.8) on state vectors:

$$[p_\mu^2 + m^2] |\phi\rangle = 0 \tag{1.5.5}$$

(Notice that the above equation is a ghost-killing constraint, because we can use it to eliminate $p_0$.) This formalism allows us to keep the commutators fully relativistic:

$$[p_\mu, x_\nu] = -i\eta_{\mu\nu} \tag{1.5.6}$$

where we choose $\eta_{\mu\nu} = (-+++\cdots)$. Notice that this gauge constraint naturally generalizes to the Klein–Gordon equation:

$$[\Box - m^2] \phi(x) = 0 \tag{1.5.7}$$

The Gupta–Bleuler formalism is an important one because most of the calculations in string theory have been carried out in this formalism.

### BRST Quantization

The advantage of the BRST formalism [45, 46] is that it is manifestly Lorentz-invariant. But instead of regaining unitarity by applying the gauge constraints on the Hilbert space, which may be quite difficult in practice, the BRST formulation uses the Faddeev–Popov ghosts to cancel the negative metric particles. Thus, although the Green's functions are not unitary because of the propagation of negative metric states and ghosts, the final $S$-matrix is unitary because all the unwanted particles cancel among each other. Thus, the BRST formalism manages to incorporate the best features of both formalisms, i.e. the manifest Lorentz invariance of the Gupta–Bleuler formalism and the unitarity of the Coulomb or light cone formalism. In order to study the BRST formalism, however, we must first understand Faddeev–Popov quantization.

## §1.6. Faddeev–Popov Quantization

Before we discuss the BRST method, it is essential to make a digression and review the formalism developed by Faddeev and Popov [47]. As we said earlier, the path integral measure $Dx_\mu$ is ill-defined because it possesses a gauge degree of freedom, so we are integrating over an infinite number of copies of the same thing. Naively, one might insert the gauge constraint directly into the path integral. If the constraint is given by some function $F$ of the fields being set to zero:

$$F(x_\mu) = 0 \tag{1.6.1}$$

then we insert this delta functional directly into the path integral:

$$Z = \int Dx \prod_x \delta[F(x_\mu)] e^{iS} \tag{1.6.2}$$

However, this naive approach is actually incorrect because the delta functional contributes a *nontrivial measure* to the functional integral.

The key to the Faddeev–Popov method is to insert the number 1 into the functional, which obviously has the correct measure. For our purposes, the most convenient formulation of the number 1 is given by

$$1 = \Delta_{\text{FP}} \int D\varepsilon \delta[F(x_\mu^\varepsilon)] \tag{1.6.3}$$

where $\varepsilon$ is the parametrization of the gauge symmetry of the coordinate, in (1.4.6), $x_\mu^\varepsilon$ is the variation of the field with respect to this symmetry, and the Faddeev–Popov determinant $\Delta_{\text{FP}}$ is defined by the previous equation.

Notice that the integral appearing in the previous equation is an integration over all possible parametrizations of the field. Since we are integrating out over all parametrizations, then, by construction, the Faddeev–Popov determinant is gauge independent of any particular parametrization:

$$\Delta_{\text{FP}}(x) = \Delta_{\text{FP}}(x^\varepsilon) \tag{1.6.4}$$

Let us now insert the number 1 into the functional integral and make a gauge transformation to reabsorb the $\varepsilon$ dependence in $x$:

$$\begin{aligned} Z &= \int Dx\, \Delta_{\mathrm{FP}}(x) \int D\varepsilon\, \delta[F(x^{\varepsilon})]e^{iS} \\ &= \int Dx\, \Delta_{\mathrm{FP}}(x) \int D\varepsilon\, \delta[F(x)]e^{iS} \end{aligned} \tag{1.6.5}$$

Notice here that $x^{\varepsilon}$ was gauge rotated back into the original variable $x$. Since all other parts of the functional integral were already gauge independent, we now have

$$Z = \left[\int D\varepsilon\right] \int Dx\, \Delta_{\mathrm{FP}} \delta[F(x)]e^{iS} \tag{1.6.6}$$

We can now extract out the integral over the gauge parameter, which measures the infinite volume of the group space:

$$\text{volume} = \int D\varepsilon \tag{1.6.7}$$

and obtain a new expression for the functional which no longer has this infinite redundancy:

$$Z = \int Dx\, \Delta_{\mathrm{FP}} \delta[F(x)]e^{iS} \tag{1.6.8}$$

Notice that a naive quantization of the path integral would simply insert the $F$ constraint and would omit the Faddeev–Popov determinant, which is a new feature that makes the measure come out correctly.

Now let us calculate the Faddeev–Popov determinant, which carries all the information concerning the ghosts of the theory. The trick is to change variables from $\varepsilon$ to $F$. We can do this because both $\varepsilon$ and $F$ have the same number of degrees of freedom. Thus, the Jacobian can be calculated:

$$\det\left[\frac{\delta F}{\delta \varepsilon}\right] D\varepsilon = DF \tag{1.6.9}$$

We can therefore write

$$\begin{aligned} \Delta_{\mathrm{FP}} &= \left\{\int D\varepsilon\, \delta(F)\right\}^{-1} \\ &= \left\{\int DF \det\left[\frac{\delta \varepsilon}{\delta F}\right] \delta(F)\right\}^{-1} \\ &= \left\{\det\left[\frac{\delta \varepsilon}{\delta F}\right]_{F=0}\right\}^{-1} \\ &= \det\left[\frac{\delta F}{\delta \varepsilon}\right]_{F=0} \end{aligned} \tag{1.6.10}$$

Thus, the Faddeev–Popov factor can be expressed as a simple determinant of the variation of the gauge constraint. It is more convenient to introduce this factor directly into the action by exponentiating it. We use the following trick:

$$\Delta_{\mathrm{FP}} = \int D\theta \, D\bar{\theta} \, e^{iS_{\mathrm{gh}}} \tag{1.6.11}$$

where the new ghost contribution to the action is given by

$$S_{\mathrm{gh}} = \int d\tau \, \bar{\theta} \left[ \frac{\delta F}{\delta \varepsilon} \right]_{F=0} \theta \tag{1.6.12}$$

where the $\theta$ variables are anticommuting $c$-numbers called *Grassmann numbers* (see Appendix). Normally, when performing functional integrations, we expect to find the determinant of the inverse of a matrix. With functional integration over Grassmann numbers, the determinant occurs in the *numerator*, not the denominator. Grassmann numbers have the strange property that

$$\theta_i \theta_j = -\theta_j \theta_i \tag{1.6.13}$$

In particular, this means

$$\theta^2 = 0 \tag{1.6.14}$$

Normally, this would mean that $\theta$ vanishes. However, this is not the case for a Grassmann number. Thus, we also have the strange identity

$$e^{\theta} = 1 + \theta \tag{1.6.15}$$

This identity makes the integration over exponentials of Grassmann-valued fields in the functional integral rather easy, because they are simply polynomials. More identities on Grassmann numbers are presented in the Appendix, where we show that

$$\int \prod_{i=1}^{N} d\theta_i \, d\bar{\theta}_i \exp\left[ \sum_{i,j=1}^{N} \bar{\theta}_i A_{ij} \theta_j \right] = \det(A_{ij}) \tag{1.6.16}$$

This identity verifies that integration over Grassmann variables yields determinant factors in the numerator, not the denominator, so that we can express the Faddeev–Popov determinant in (1.6.11) as a Grassmann integral.

Now that we have developed the apparatus of Faddeev–Popov quantization, let us return to the BRST approach, where we wish to impose the gauge condition

$$e = 1 \tag{1.6.17}$$

(we omit some subtleties with respect to this gauge). In this gauge, we should be able to recover the usual covariant Feynman propagator. To show this, notice that our action (1.4.14) becomes

$$L = \tfrac{1}{2}(\dot{x}_\mu^2 - m^2) \tag{1.6.18}$$

Given this Lagrangian, our Green's function for the propagation of a point particle from one point to another is now given by

$$\begin{aligned}\Delta_F(x_1, x_2) &= \langle x_1 | \frac{1}{\Box - m^2} | x_2 \rangle \\ &= \langle x_1 | \int_0^\infty d\tau\, e^{-\tau(\Box - m^2)} | x_2 \rangle \\ &= \int_0^\infty d\tau \int_{x_1}^{x_2} Dx \exp\left( -\frac{1}{2} \int_0^\tau d\bar{\tau} (\dot{x}_\mu^2 - m^2) \right) \end{aligned} \tag{1.6.19}$$

Notice that this is the usual covariant Feynman propagator rewritten in first quantized path integral language.

Originally, before gauge fixing, our action was invariant under

$$\delta e = \frac{d(\varepsilon e)}{d\tau} \tag{1.6.20}$$

Thus, the Faddeev–Popov determinant associated with the gauge choice $e = 1$ is the determinant of the derivative. We now use a Gaussian integral over Grassmann states to represent the determinant, using eqn. (1.6.10):

$$\Delta_{\mathrm{FP}} = \det|\partial_\tau| = \int D\theta\, D\bar{\theta} \exp i \int d\tau\, \bar{\theta} \partial_\tau \theta \tag{1.6.21}$$

(If we had used ordinary real fields instead of Grassmann-valued fields, the determinant would have come out with the wrong power.)

Putting everything together, we find that our final action can be represented as

$$L = p_\mu \dot{x}^\mu - \tfrac{1}{2}(p_\mu^2 + m^2) - i\bar{\theta}\partial_\tau \theta \tag{1.6.22}$$

*The essence of the BRST approach is to notice that this gauge-fixed action has the additional symmetry*:

$$\begin{aligned} \delta x_\mu &= i\varepsilon\theta \dot{x}_\mu \\ \delta p_\mu &= i\varepsilon\theta \dot{p}_\mu \\ \delta\theta &= i\varepsilon\theta\dot{\theta} \\ \delta\bar{\theta} &= i\varepsilon\theta\dot{\bar{\theta}} + \tfrac{1}{2}\varepsilon(p_\mu^2 + m^2) \end{aligned} \tag{1.6.23}$$

At first, we may wonder why yet another symmetry appears after we have already fixed the gauge degree of freedom. However, this extra symmetry is *global* and hence does not allow us to impose any constraints on the theory. This symmetry, therefore, is different from the ones found earlier and cannot be used to eliminate gauge fields from the action.

We can summarize the BRST approach by extracting an operator $Q$ that will generate the symmetry found earlier:

$$\begin{aligned} \delta\phi &= [\varepsilon Q, \phi] \\ Q &= \theta(\Box - m^2) \\ Q^2 &= 0 \end{aligned} \tag{1.6.24}$$

The physical states satisfy

$$Q|\phi\rangle = 0 \tag{1.6.25}$$

Notice that enforcing this constraint recovers the Klein–Gordon equation for on-shell particles:

$$(\Box - m^2)\phi = 0 \tag{1.6.26}$$

## §1.7. Second Quantization

So far, we have been analyzing only the first quantized approach to quantum particles. We have quantized only the position and momentum vectors:

$$\text{First quantization:} \quad [p_i, x_j] = -i\delta_{ij} \tag{1.7.1}$$

The limitations of the first quantized approach, however, will soon become apparent when we introduce interactions. Let us say that we wish to describe point particles that can bump into each other and split apart, rather than introduce an external potential. We must now modify the generating functional to include summing over Feynman graphs:

$$Z = \sum_{\text{topologies}} \int Dx\, e^{-\text{length}} \tag{1.7.2}$$

(Notice that we have Wick rotated the $\tau$ integration so that the exponential converges. It will be clear from the context when the Wick-rotated theory is being used in this book because the exponential becomes real. We will not discuss the delicate question of the convergence of path integrals.)

In other words, we must, by hand, sum over the various particle topologies where point particles can split and reform. Each topology represents the history of the trajectories of the various point particles as they interact. The amplitude for $N$-particle scattering, with momenta given by $k_1, k_2, \ldots, k_N$, can now be represented as

$$\begin{aligned} A(k_1, k_2, \ldots, k_N) = &\sum_{\text{topologies}} g^n \int Dx\, \Delta_{\text{FP}} \\ &\times \exp\left\{ -\int dt\, L(t) + i \sum_{i=1}^{N} k_\mu x_i^\mu \right\} \end{aligned} \tag{1.7.3}$$

Notice that we are taking the Fourier transform of the Green's function, so that the amplitude is a function of the external momenta. This formula can be more conveniently represented as

$$A_N = \sum_{\text{topologies}} \left\langle \exp i \sum_{i=1}^{N} k_\mu x_i^\mu \right\rangle \tag{1.7.4}$$

Thus, we associate a factor $e^{ikx}$ for each external particle coming from the Fourier transform term. This path integral formula for the scattering amplitude is important because it will carry over almost exactly into the string formalism.

Notice how clumsy this description is. We must fix the set of all topologically allowed configurations and their weights *by hand*. Furthermore, unitarity of the $S$-matrix is not at all obvious.

In a second quantized description, however, we introduce a field $\psi(x)$ and quantization relations between the fields themselves, not between the coordinates:

$$\text{Second quantization:} \quad [\pi(x), \psi(y)]_{x_0=y_0} = -i\delta^{(3)}(x_i - y_i) \tag{1.7.5}$$

The advantage of the second quantized approach is that the interacting Hamiltonian can be written explicitly, without having to introduce sums over topologies. Showing that the Hamiltonian is Hermitian is sufficient to fix the weights of all diagrams and to demonstrate the unitarity of the $S$-matrix.

In summary, the pros and cons of first and second quantizations are as follows:

First Quantization

(1) Interactions must be added in by hand, order by order in the coupling constant.
(2) Unitarity of the final $S$-matrix is not obvious. This must be explicitly checked order by order.
(3) The formalism is necessarily a perturbative one, since the expansion in topologies is intimately tied to the expansion in terms of the coupling constant.
(4) It is difficult to describe the theory off-shell.

Second Quantization

(1) The interactions are explicit in the action itself.
(2) Unitarity is guaranteed if the Hamiltonian is Hermitian.
(3) The theory can be formally written nonperturbatively as well as perturbatively.
(4) The theory is necessarily off-shell.

The transition from the first to the second quantized theory can also be performed most easily in the path integral formalism in the Coulomb gauge. Earlier, we showed that the Green's function for a propagating free point particle can be explicitly evaluated:

$$K(a, b) = \left\{ \frac{m}{2\pi i(t_b - t_a)} \right\}^{1/2} \exp i \frac{m(x_b - x_a)^2}{2(t_b - t_a)} \tag{1.7.6}$$

This Green's function can also be written in a second quantized fashion. Let us start with the Hamiltonian:

$$H = -\frac{1}{2m}\nabla^2 \tag{1.7.7}$$

The Green's function satisfies

$$(i\partial_t - H)K(x, t; x', t') = \delta^{(3)}(x - x')\delta(t - t') \tag{1.7.8}$$

Solving for this Green's function, we find

$$K(a, b) = [i\partial_t - H]^{-1}_{x_a, t_a; x_b, t_b} \tag{1.7.9}$$

where we are treating the inverse Green's function as if it were a discrete matrix in $(x, t)$ space, and we have dropped trivial normalization factors. This allows us to write the integral in second quantized language. To demonstrate this, we will use the following identities throughout this book:

$$\begin{aligned}\int \prod_{i=1}^{N} dx_i \exp\left\{\sum_{i,j=1}^{N} - x_i A_{ij} x_j + \sum_{i=1}^{N} J_i x_i\right\} \\ = \frac{\pi^{(1/2)N}}{\det|A_{ij}|} \exp\left\{\tfrac{1}{4}\sum_{i,j=1}^{N} J_i (A^{-1})_{ij} J_j\right\}\end{aligned} \tag{1.7.10}$$

(This integral can easily be derived using our earlier formula for the Gaussian integral (1.3.18). We simply diagonalize the $A$ matrix by making a change of variables in $x$. Thus, the quadratic term in the integral becomes a function of the eigenvalues of the $A$ matrix. Because all the modes have now decoupled, the Gaussian integral can be performed exactly by completing the square. Finally, we make another similarity transformation to convert the eigenvalues of $A$ back into the $A$ matrix itself.)

From this, we can also derive the following:

$$\begin{aligned}\int x_n x_m \prod_{i=1}^{N} dx_i \exp\left\{\sum_{i=1}^{N} - x_i A_{ij} x_j + \sum_{i=1}^{N} J_i x_i\right\} \\ \sim \left[\frac{\delta}{\delta J_n}\frac{\delta}{\delta J_m} \exp\left\{\tfrac{1}{4} J_i (A^{-1})_{ij} J_j\right\}\right]_{J=0} \det|A_{ij}|^{-1} \\ \sim (A^{-1})_{nm} (\det|A_{ij}|)^{-1}\end{aligned} \tag{1.7.11}$$

These are some of the most important integrals in this book. Using these equations, we can now write the Green's function totally in terms of second quantized fields:

$$K(a, b) = \int \psi^*(x_a, t_a)\psi(x_b, t_b)\, D\psi^*\, D\psi \exp i \int dx\, dt\, L(\psi) \tag{1.7.12}$$

where

$$L(\psi) = \psi^*(i\partial_t - H)\psi \tag{1.7.13}$$

where we are again treating $K(a, b)$ as if it were a matrix in discretized $(x, t)$ space.

In summary, we now have two complementary descriptions of the point particle. We can write the theory either in terms of the particle's coordinates $x_i$ or in terms of its fields $\psi(x)$.

At the free level, both descriptions are totally equivalent, both in ease of description and also in mathematics. However, at the interacting level, distinct differences appear. For example, it is easy to write

$$L_I \sim \phi^3; \quad \sim \phi^4 \tag{1.7.14}$$

and we are guaranteed to get a unitary description of an interacting field. However, in the first quantized approach, the sum over topologies:

$$\sum_{\text{topologies}} \tag{1.7.15}$$

is a clumsy way in which to describe a unitary theory. We must check unitarity order by order in increasingly complicated diagrams. Furthermore, we are forced to adopt a totally perturbative description for the first quantized description. The sum over topologies in the first quantized path integral is a sum over perturbative Feynman diagrams, so the formulation is necessarily perturbative from the very beginning. That is the fundamental reason why we have divided this book into first quantization and second quantization.

## §1.8. Harmonic Oscillators

One example that will illustrate the relationship between first and second quantizations is the harmonic oscillator problem. This example will prove helpful in introducing the harmonic oscillator representation, which we will use extensively for the string model. Let us begin with a point particle governed by the following Hamiltonian:

$$H = \frac{p^2}{2m} + \tfrac{1}{2}kx^2 \tag{1.8.1}$$

where $k$ is the spring constant. Because the momenta and coordinates are conjugates, we can use the same arguments presented earlier in our discussion of path integrals to set

$$[p, x] = -i \tag{1.8.2}$$

We can now redefine our coordinates and momenta in terms of harmonic oscillators:

$$\begin{aligned} p &= (\tfrac{1}{2}m\omega)^{1/2}(a + a^\dagger) \\ x &= i(2m\omega)^{-1/2}(a - a^\dagger) \end{aligned} \tag{1.8.3}$$

where

$$k = m\omega^2 \tag{1.8.4}$$

In order to satisfy the canonical commutation relation (1.8.2), we must have

$$[a, a^\dagger] = 1 \tag{1.8.5}$$

If we insert this expression back into the Hamiltonian, we find

$$H = \tfrac{1}{2}\omega(aa^\dagger + a^\dagger a) \tag{1.8.6}$$

By extracting a $c$-number term, we can write this in normal ordered fashion:

$$H = \omega(a^\dagger a + h_0) \tag{1.8.7}$$

where $h_0$ is the zero point energy. We can now introduce the Hilbert space of harmonic oscillators. Let us define the vacuum as

$$a|0\rangle = 0 \tag{1.8.8}$$

Then an element of the Fock space of the harmonic oscillator Hamiltonian is given by

$$|n\rangle = \frac{(a^\dagger)^n}{\sqrt{n!}}|0\rangle \tag{1.8.9}$$

such that the states form an orthonormal basis:

$$\langle n|m\rangle = \delta_{nm} \tag{1.8.10}$$

The energy of the system is quantized and given by

$$E_n = (n + \tfrac{1}{2})\omega \tag{1.8.11}$$

So far, the system has been presented only in a first quantized formalism. We are quantizing only a single point particle at any time. We would now like to make the transition to the second quantized wave function by introducing

$$|\Phi\rangle = \sum_{n=0}^{\infty} \phi_n|n\rangle \tag{1.8.12}$$

where we power expand in the basis states of the harmonic oscillator. Thus, instead of describing a single excited state of a point particle, we are now introducing the wave function, which will be a superposition of an arbitrary number of excited states.

Let us make the important definition

$$\langle x|\Phi\rangle = \Phi(x) \tag{1.8.13}$$

This can be calculated explicitly. Notice that we now have two independent basis states, the harmonic oscillator basis $|n\rangle$ and the position eigenvectors $|x\rangle$. We must now calculate how to go back and forth between these two bases.

Let us first analyze the simplest matrix element:

$$\sigma_0(x) = \langle x|0\rangle \tag{1.8.14}$$

This matrix element satisfies the equation

$$\begin{aligned}0 &= \langle x|a|0\rangle \\ &= \langle x|\frac{p - im\omega x}{\sqrt{2m\omega}}|0\rangle \\ &= (2m\omega)^{-1/2}\left(-\frac{i\partial}{\partial x} - im\omega x\right)\langle x|0\rangle \\ &= -i(2m\omega)^{-1/2}\left(\frac{\partial}{\partial x} + m\omega x\right)\sigma_0(x)\end{aligned} \tag{1.8.15}$$

This last equation can be solved exactly:

$$\sigma_0(x) = (m\omega/\pi)^{1/4}e^{-1/2\xi^2} \tag{1.8.16}$$

where

$$\xi = (m\omega)^{1/2}x \tag{1.8.17}$$

It is now a straightforward step to calculate all such matrix elements. Let

$$\begin{aligned}\sigma_n(x) &= \langle x|n\rangle \\ &= \langle x|(n!)^{-1/2}a^{\dagger n}|0\rangle \\ &= (n!)^{-1/2}(2m\omega)^{-n(1/2)}\langle x|[p + im\omega x]^n|0\rangle \\ &= (n!)^{-1/2}(2m\omega)^{-(1/2)n}\left(-i\frac{\partial}{\partial x} + im\omega x\right)^n\sigma_0(x)\end{aligned} \tag{1.8.18}$$

The solution is therefore

$$\sigma_n(x) = i^n(2^n n!)^{-1/2}(m\omega/\pi)^{1/4}\left(\xi - \frac{\partial}{\partial\xi}\right)^n e^{-(1/2)\xi^2} \tag{1.8.19}$$

In general, these are nothing but Hermite polynomials $H_n$. In terms of these polynomials, we can express the eigenstate $|x\rangle$ and $|n\rangle$ in terms of each other:

$$\begin{cases}|x\rangle = \sum_{n=1}^{\infty}|n\rangle\langle n|x\rangle = \sum_{n=1}^{\infty}|n\rangle\sigma_n(x) \\ |n\rangle = |x\rangle\int dx\langle x|n\rangle = \int dx\,\sigma_n(x)|x\rangle\end{cases} \tag{1.8.20}$$

Thus, using (1.8.12) and (1.8.20), we have the power expansion of the wave function in terms of a complete set of orthogonal polynomials, the Hermite polynomials:

$$\Phi(x) = \langle x|\Phi\rangle = \langle x|\sum_{n=1}^{\infty}\phi_n|n\rangle = \sum_{n=1}^{\infty}\phi_n H_n(\xi)e^{-(1/2)\xi^2} \tag{1.8.21}$$

Similarly, it is not difficult to calculate the Green's function for the propagation of a point particle in a harmonic oscillator potential. The Green's function would be the same as if we had started with the second quantized formalism with the action:

$$L = \Phi(x)^* \left( i\partial_t + \frac{1}{2m}\nabla^2 - \tfrac{1}{2}kx^2 \right) \Phi(x) \tag{1.8.22}$$

From this second quantized action, we can therefore derive the equations of motion:

$$\begin{aligned} i\partial_t \Phi(x, t) &= \left[ \frac{-1}{2m}\nabla^2 + \tfrac{1}{2}kx^2 \right] \Phi(x, t) \\ &= H\Phi(x, t) \end{aligned} \tag{1.8.23}$$

From this, we can define the canonical momenta conjugate to $\Phi(x, t)$ such that the canonical quantization relations are satisfied:

$$[\Pi(x, t), \Phi(x', t)] = -i\delta(x - x') \tag{1.8.24}$$

## §1.9. Currents and Second Quantization

Let us begin with a discussion of the relativistic second quantized theory, which, as we have seen, is equivalent perturbatively to the first quantized theory. When quantizing the point particle in the Gupta–Bleuler formalism, we were led to the equations of motion:

$$[\Box - m^2]\phi = 0 \tag{1.9.1}$$

which can be derived from the second quantized action:

$$L = \tfrac{1}{2}[\partial_\mu \phi \partial^\mu \phi + m^2 \phi^2] \tag{1.9.2}$$

One of the most powerful techniques we used to explore the first quantized theory was symmetry. We would now like to study the question of symmetries within the second quantized formalism.

First, let us calculate the equations of motion by making a small variation in the field and requiring that the action be invariant under this variation:

$$\delta S = 0 = \int d^D x \left( \frac{\delta L}{\delta \phi} \delta\phi + \frac{\delta L}{\delta \partial_\mu \phi} \delta \partial_\mu \phi \right) \tag{1.9.3}$$

Let us now integrate by parts, using $\delta \partial_\mu \phi = \partial_\mu \delta\phi$:

$$\delta S = \int d^D x \left( \frac{\delta L}{\delta \phi} - \partial_\mu \frac{\delta L}{\delta \partial_\mu \phi} \right) \delta\phi + \int d^D x \partial_\mu \left( \frac{\partial L}{\delta \partial_\mu \phi} \delta\phi \right) \tag{1.9.4}$$

If we temporarily ignore the surface term, the action is stationary if we have the following equation of motion:

$$\partial_\mu \frac{\delta L}{\delta \partial_\mu \phi} - \frac{\delta L}{\delta \phi} = 0 \tag{1.9.5}$$

If we insert the Lagrangian into this equation, we obtain the equations of motion, which reproduce the constraint found earlier in the first quantized formalism.

Let us now make a small change in the fields, parametrized by a small, as yet unspecified, number $\varepsilon^\alpha$:

$$\delta\phi = \frac{\delta\phi}{\delta\varepsilon^\alpha}\delta\varepsilon^\alpha \tag{1.9.6}$$

If we insert this into the previous equation for the variation of the action, keep the surface term intact, and assume that the equations of motion are satisfied, then we have the following:

$$\delta S = \int d^D x \partial_\mu \left( \frac{\partial L}{\delta \partial_\mu \phi} \frac{\delta\phi}{\delta\varepsilon^\alpha} \right) \delta\varepsilon^\alpha \tag{1.9.7}$$

Let us define the tensor in the parentheses as the *current*:

$$J_\alpha^\mu = \frac{\delta L}{\delta \partial_\mu \phi} \frac{\delta\phi}{\delta\varepsilon^\alpha} \tag{1.9.8}$$

Then we have the important equation

$$\delta S = \int d^D x \partial_\mu J_\alpha^\mu \delta\varepsilon^\alpha \tag{1.9.9}$$

Thus, if the action $S$ is stationary under this variation, we have a conserved current $J^{\mu\alpha}$:

$$\partial_\mu J^{\mu\alpha} = 0 \tag{1.9.10}$$

We will use this equation over and over again in the discussion of strings when we want to extract the current for supersymmetry and conformal invariance. Finally, we note that the integrated charge $Q^\alpha$ associated with the current is constant in time:

$$\int d^D x \partial_\mu J^{\mu\alpha} = \int d^{D-1} x \partial_0 J^{0\alpha} + \text{surface term} \tag{1.9.11}$$

Thus,

$$\begin{gathered} Q^\alpha = \int d^{D-1} x J_0^\alpha \\ \partial_\mu J^{\mu\alpha} = 0 \to \frac{dQ^\alpha}{dt} = 0 \end{gathered} \tag{1.9.12}$$

Finally, we wish to construct yet another conserved current associated with the action. Let us make a small variation in the space–time variable:

$$\delta x^\mu = \varepsilon^\mu \tag{1.9.13}$$

Under this change, the volume element of the integral changes as

$$\delta d^D x = d^D x \partial_\mu \delta x^\mu \tag{1.9.14}$$

Therefore, the variation of the action under this change is

$$\begin{aligned} \delta S &= \int d^D x [L \partial_\mu \delta x^\mu + \delta L] \\ \delta L &= \delta x^\mu \partial_\mu L + \frac{\delta L}{\delta \phi} \delta \phi + \frac{\delta L}{\delta \partial_\mu \phi} \delta \partial_\mu \phi \end{aligned} \tag{1.9.15}$$

Now, if we assume that the equations of motion are satisfied, we have

$$\delta S = \int d^D x \partial_\mu \left\{ \left( + L \delta^\mu_\nu - \frac{\delta L}{\delta \partial_\mu \phi} \partial_\nu \phi \right) \delta x^\nu \right\} \tag{1.9.16}$$

If we now define the *energy–momentum tensor* as

$$T_{\mu\nu} = \frac{\delta L}{\delta \partial^\mu \phi} \partial_\nu \phi - \eta_{\mu\nu} L \tag{1.9.17}$$

then we have the equation

$$\delta S = \int d^D x \partial_\mu (T^{\mu\nu} \delta x_\nu) \tag{1.9.18}$$

So if the action is invariant under this change, then the energy–momentum tensor is conserved:

$$\partial_\mu T^{\mu\nu} = 0 \tag{1.9.19}$$

For example, for the scalar particle action, the energy–momentum tensor becomes

$$T_{\mu\nu} = \partial_\mu \phi \partial_\nu \phi - \eta_{\mu\nu} L \tag{1.9.20}$$

which is conserved if the equations of motion hold.

Lastly, it is instructive to investigate how the various quantization procedures treat the Yang–Mills field (see Appendix). Let us begin with the SU($N$) invariant action:

$$L = -\tfrac{1}{4} [F^a_{\mu\nu}]^2 \tag{1.9.21}$$

where

$$F^a_{\mu\nu} = \partial_\mu A^a_\nu - \partial_\nu A^a_\mu - f^{abc} A^b_\mu A^c_\nu \tag{1.9.22}$$

The action is invariant under

$$\delta A^a_\mu = \partial_\mu \Lambda^a - f^{abc} A^b_\mu \Lambda^c \tag{1.9.23}$$

where $\Lambda^a$ is a gauge parameter.

The path integral method begins with the functional

$$Z = \int \prod_{\mu, x} dA_\mu(x) e^{i \int d^4x - (1/4) F^2_{\mu\nu}} \tag{1.9.24}$$

Now we consider the three methods of quantization:

## Coulomb Quantization

The gauge invariance permits us to take the gauge:

$$\nabla_i A_i^a = 0$$

We can integrate over the $A_0$ component because it has no time derivatives, so the Coulomb formulation is explicitly ghost-free. (The price we pay for this, of course, is the loss of manifest Lorentz invariance, which must be checked by hand.) In this gauge, the action becomes

$$L = +\tfrac{1}{2}(\partial_0 A_i^a)^2 - \tfrac{1}{4}(F_{ij}^a)^2 + \cdots \tag{1.9.25}$$

where all fields are transverse. This is the canonical form for the Lagrangian.

## Gupta–Bleuler Quantization

The advantage of the Gupta–Bleuler formulation is that we can keep manifest Lorentz symmetry without violating unitarity. For example, let us take the gauge

$$\partial_\mu A^{\mu a} = 0 \tag{1.9.26}$$

In this gauge, the propagator for massless vector particles becomes

$$\frac{\eta_{\mu\nu}}{p^2} \tag{1.9.27}$$

Notice that the propagator explicitly contains a ghost. The timelike excitation has a coefficient of $-1$ in the propagator, which represents a ghost. However, we are free to quantize in this covariant approach beause we will impose the ghost-killing constraint on the Hilbert space:

$$\langle \phi | \partial_\mu A^{\mu a} | \psi \rangle = 0 \tag{1.9.28}$$

This constraint allows us to solve for and hence eliminate the ghost modes. Thus, although the free propagator will allow ghosts to propagate, the Hilbert space is ghost-free, so the theory itself is both Lorentz invariant and ghost-free.

## BRST Quantization

The BRST approach begins by calculating the Faddeev–Popov determinant (1.6.10). Let us calculate the determinant of the matrix:

$$\begin{aligned} M^{ab}(x, y) &= \frac{\delta(\partial_\mu A^{\mu a}(x))}{\delta \Lambda^b(y)} \\ &= \partial_\mu D^\mu \frac{\delta \Lambda^a(x)}{\delta \Lambda^b(y)} \\ &= \partial_\mu D^\mu (\delta^4(x - y)\delta^{ab}) \end{aligned} \tag{1.9.29}$$

As before, we can write the determinant of $M^{ab}$ by exponentiating it into the action using (1.6.10):

$$L = -\tfrac{1}{4}F_{\mu\nu}^{a2} + \frac{1}{2\alpha}(\partial_\mu A^{\mu a})^2 + \bar{c}^a M^{ab} c^b \tag{1.9.30}$$

where the anticommuting Faddev–Popov ghost fields are represented by $c$ and $\bar{c}$. This action is invariant under the following BRST transformation:

$$\text{BRST:} \quad \begin{cases} \delta A_\mu^a = (\nabla_\mu c)^a \varepsilon \\ \delta c^a = -\tfrac{1}{2} f^{abd} c^b c^d \varepsilon \\ \delta \bar{c}^a = \dfrac{1}{\alpha}(\partial_\mu A^{\mu,a})\varepsilon \end{cases} \tag{1.9.31}$$

Once again, it is important to notice that the BRST transformation is nilpotent. The BRST symmetry is not connected to the conservation of any observable quantity. From the previous invariance, we can extract out the generator of this transformation $Q$ such that

$$Q^2 = 0 \tag{1.9.32}$$

The physical states of the theory then satisfy

$$Q|\text{phy}\rangle = 0 \tag{1.9.33}$$

## §1.10. Summary

The great irony of string theory, which is supposed to provide a unifying framework for all known interactions, is that the theory itself is so disorganized. String theory is often frustrating to the beginner because it is full of folklore, conventions, and arbitrary rules of thumb. The fundamental reason for this is that string theory has historically evolved backward as a first quantized theory, rather than as a second quantized theory, where the entire theory is defined in terms of a fundamental action. The disadvantages of the first quantized approach are that

(1) The interactions of the theory must be introduced by hand. They cannot be derived from a single action.
(2) Unitarity is not obvious in this approach. The counting of graphs must be checked tediously.
(3) The formulation is perturbative, so that crucial nonperturbative calculations, such as dimensional breaking, are beyond its scope.
(4) The formulation is basically on-shell, rather than off-shell.

By contrast, the advantage of the second quantized approach is precisely that everything can be derived from a single off-shell action, where unitarity is manifest and nonperturbative calculations can, in principle, be performed.

Unfortunately, string theory evolved historically as a first quantized theory. Thus, string theory has been evolving backward, with the second quantized geometric theory still in its infancy. For pedagogical reasons, we have introduced string theory from a semihistorical point of view, beginning with the first quantized theory and later developing the second quantized theory and the underlying geometry. We hope that future accounts of string theory will reverse this sequence.

To reduce the level of arbitrariness in the first quantized theory as much as possible, in this chapter we have tried to lay the groundwork for string theory in the formalism of path integrals. This functional formalism has the great advantage that we can express the first and second quantized gauge theories with equal ease. We find, in fact, that large portions of the path integral formulation of point particles can be incorporated wholesale into string theory.

The path integral method postulates two fundamental principle that express the essence of quantum mechanics:

(1) The probability $P(a, b)$ of a particle going from point $a$ to point $b$ is given by the absolute value squared of a transition function $K(a, b)$:

$$P(a, b) = |K(a, b)|^2$$

(2) The transition function is given by the sum of a phase factor $e^{iS}$, where $S$ is the action, taken over all possible paths from $a$ to $b$:

$$K(a, b) = \sum_{\text{paths}} k e^{iS}$$

In the limit of continuous paths, we have

$$K(a, b) = \int_a^b Dx\, e^{iS}$$

where

$$Dx = \lim_{N\to\infty} \prod_{i=1}^{3} \prod_{n=1}^{N} dx_{i,n}$$

The action $S$ of the first quantized point particle is given by the length of the path that the particle sweeps out in space-time. We can represent the Lagrangian for the point particle in three ways:

$$\begin{aligned} \text{1st-order (Hamiltonian) form:} &\quad L = p_\mu \dot{x}^\mu - \tfrac{1}{2}e(p_\mu^2 + m^2) \\ \text{2nd-order form:} &\quad L = \tfrac{1}{2}(e^{-1}\dot{x}_\mu^2 - em^2) \\ \text{Non-linear form:} &\quad L = -m\sqrt{-\dot{x}_\mu^2} \end{aligned} \tag{1.10.1}$$

Unfortunately, because all three forms of the action are parametrization-invariant, the path integral diverges. Thus, the quantization procedure must break this gauge symmetry and yield the correct measure in the functional.

These actions can be quantized in three ways, each with its own advantages and disadvantages:

(1) Coulomb Quantization
By explicitly fixing the value of some of the fields, such as

$$x_0 = t = \tau$$

we can eliminate the troublesome negative metric states and the Lagrangian becomes $\frac{1}{2}mv_i^2$. The Coulomb quantization method is therefore manifestly ghost-free. However, the disadvantage of this method is that it is very awkward because manifest Lorentz symmetry is broken and must be checked at every level.

(2) Gupta–Bleuler Quantization
The advantage of the Gupta–Bleuler quantization method is that we have a manifestly covariant quantization program. Of course, negative metric ghosts are now allowed to circulate in the theory, but they are eventually eliminated by imposing the gauge constraints directly onto the Hilbert space:

$$[p_\mu^2 + m^2]|\phi\rangle = 0$$

Thus, the $S$-matrix is ultimately ghost-free. The disadvantage of this approach, however, is that the imposition of these gauge constraints, especially at the interacting level, is frequently quite difficult.

(3) BRST Quantization
This method of quantization keeps the good features of both approaches. The theory is manifestly covariant, but the $S$-matrix is still unitary because the addition of ghost fields in the theory cancels precisely against the negative metric states. The BRST method imposes the gauge $e = 1$ in the first-order form and then inserts the Faddeev–Popov term $\Delta_{FP}$ into the functional to get the correct measure. We can exponentiate this determinant into the action by using Grassman variables:

$$\Delta_{FP} = \det|\partial_\tau| = \int d\theta \, d\bar{\theta} e^{i\int d\tau\, \bar{\theta}\partial_\tau\theta}$$

The resulting gauge-fixed action has a residual symmetry, called the BRST symmetry, which is generated by $Q$, the BRST charge. (This new symmetry does not result in the elimination of more fields).

When we generalize these methods to the interacting case, the path integral formulation begins with the fundamental formula for the transition function for $N$-particle scattering:

$$\begin{aligned} A(k_1, k_2, \ldots, k_N) &= \sum_{\text{topologies}} g^n \int Dx \, \Delta_{FP} \\ &\quad \times \exp\left\{ i \int dt \, L(t) + i \sum_{i=1}^{N} k_\mu x_i^\mu \right\} \\ &= \sum_{\text{topologies}} \int Dx \langle e^{i\sum_{i=1}^N k_\mu x_i^\mu} \rangle \end{aligned} \tag{1.10.2}$$

The first quantized description for the $N$-particle scattering amplitude is clumsy because one must explicitly sum over certain topologies, which must be put in by hand. This means that unitarity is not obvious in the first quantized formalism. Later, we will see that this problem in the first quantized point particle theory carries over directly into the first quantized string theory. In the second quantized description, however, all topologies can be derived explicitly from a single action.

The transition from a first to a second quantized description is straightforward in the path integral formulation. For example, the propagator can be written in either first or second quantized language:

$$\begin{aligned}\Delta_{ab} &= \int_{x_a}^{x_b} Dx\, e^{i\int_{t_a}^{t_b} dt\, L(t)} \\ &= \int D\psi\, D\psi^*\, \psi(x_a)\psi^*(x_b)e^{i\int Dx\, L(\psi)}\end{aligned} \tag{1.10.3}$$

where

$$\begin{aligned}\langle x|\psi\rangle &= \psi(x) \\ L(t) &= \tfrac{1}{2}m\dot{x}_i^2 \\ L(\psi) &= \psi^*(i\partial_t - H)\psi\end{aligned} \tag{1.10.4}$$

The last equation is the Lagrangian for the Schrödinger wave equation, which can be derived beginning with the postulates of path integrals and $L = \frac{1}{2}mv_i^2$.

For the interactions, it is also possible to extract the second quantized vertices from the first quantized theory in exactly the same fashion. We simply write down the functional integral over a world sheet where the point particle splits into other point particles, and then write this Green's function as a functional integral over second quantized fields.

We will shortly see the advantage of carefully working out the details of point particle path integrals. We will find that almost all of this formalism carries over directly into the string formalism!

## References


[1] For reviews of string theory, see J. H. Schwarz, ed., *Superstrings*, Vols. 1 and 2, World Scientific, Singapore, 1985; J. H. Schwarz, *Phys. Rep.* **89**, 223 (1982).
[2] M. B. Green, *Surv. High Energy Phy.* **3**, 127 (1983).
[3] V. Alessandrini, D. Amati, M. LeBellac, and D. I. Olive, *Phys. Rep.* **1C**, 170 (1971).
[4] S. Mandelstam, *Phys. Rep.* **13C**, 259 (1974).
[5] C. Rebbi, *Phys. Rep.* **12C**, 1 (1974).
[6] P. Frampton, *Dual Resonance Models*, Benjamin, New York, 1974.
[7] J. Scherk, *Rev. Mod. Phys.* **47**, 1213 (1975).
[8] G. Veneziano, *Phys. Rep.* **9C**, 199 (1974).
[9] M. Jacob, ed., *Dual Theory*, North-Holland, Amsterdam, 1974.
[10] J. H. Schwarz, *Phys. Rep.* **8C**, 269 (1973).

[11] M. B. Green and D. Gross, eds., *Unified String Theories*, World Scientific, Singapore, 1986.
[12] M. B. Green, J. H. Schwarz, and E. Witten, *Superstring Theory*, Vols. 1 and 2, Cambridge University Press, Cambridge, 1986.
[13] H. Georgi and S. L. Glashow, *Phys. Rev. Lett.* **32**, 438 (1974).
[14] D. Z. Freedman, P. Van Nieuwenhuizen, and S. Ferrara, *Phys. Rev.* **D13**, 32 (1976).
[15] S., Deser and B. Zumino, *Phys. Lett.* **62B**, 335 (1976).
[16] T. Regge, *Nuovo Cim.* **19** 558 (1961).
[17] N. H. Christ, R. Friedberg, and T. D. Lee, *Nucl. Phys.* **B202**, 89 (1982).
[18] G. C. Wick and T. D. Lee, *Nucl. Phys.* **B9**, 209 (1969).
[19] M. Kaku, *Nucl. Phys.* **B203**, 285 (1982).
[20] M. Kaku, *Phys. Rev. Lett.* **50**, 1895 (1983); *Phys. Rev.* **D27**, 2809, 2819 (1983).
[21] G. Veneziano, *Nuovo Cim.* **57A**, 190 (1968).
[22] M. Suzuki, unpublished.
[23] P. Ramond, *Phys. Rev.* **D3**, 2415 (1971).
[24] A. Neveu and J. H. Schwarz, *Nucl. Phys.* **B31**, 816 (1971).
[25] A. Neveu and J. H. Schwarz, *Phys. Rev.* **D4**, 1109 (1971).
[26] K. Kikkawa, B. Sakita, and M. A. Virasoro, *Phys. Rev.* **184**, 1701 (1969).
[27] M. Kaku and L. P. Yu, *Phys. Lett.* **33B**, 166 (1970).
[28] M. Kaku and L. P. Yu, *Phys. Rev.* **D3**, 2992, 3007, 3020 (1971).
[29] M. Kaku and J. Scherk, *Phys. Rev.* **D3**, 430 (1971).
[30] M. Kaku and J. Scherk, *Phys. Rev.* **D3**, 2000 (1971).
[31] V. Alessandrini, *Nuovo Cim.* **2A**, 321 (1971).
[32] C. Lovelace, *Phys. Lett.* **32B**, 703 (1970).
[33] C. Lovelace, *Phys. Lett.* **34B**, 500 (1971).
[34] Y. Nambu, Lectures at the Copenhagen Summer Symposium (1970).
[35] T. Goto, *Prog. Theor. Phys.* **46**, 1560 (1971).
[36] P. Goddard, J. Goldstone, C. Rebbi, and C. B. Thorn, *Nucl. Phys.* **B56**, 109 (1973).
[37] S. Mandelstam, *Nucl. Phys.* **B64**, 205 (1973); **B69**, 77 (1974).
[38] M. Kaku and K. Kikkawa, *Phys. Rev.* **D10**, 1110, 1823 (1974).
[39] J. Scherk and J. H. Schwarz, *Nucl. Phys.* **B81**, 118 (1974). See also T. Yonea, *Prog. Theor. Phys.* **51**, 1907 (1974).
[40] M. B. Green and J. H. Schwarz, *Phys. Lett.* **149B**, 117 (1984); **151B**, 21 (1985).
[41] D. J. Gross, J. A. Harvey, E. Martinec, and R. Rohm, *Nucl. Phys.* **B256**, 253 (1986); **B267**, 75 (1986).
[42] C. S. Hsue, B. Sakita, and M. A. Virasoro, *Phys. Rev.* **D2**, 2857 (1970).
[43] J. L. Gervais and B. Sakita, *Nucl. Phys.* **B34**, 632 (1971); *Phys. Rev.* **D4**, 2291 (1971); *Phys. Rev. Lett.* **30**, 716 (1973); see also D. B. Fairlie and H. B. Nielsen, *Nucl. Phys.* **B20**, 637 (1970).
[44] R. P. Feynman and A. R. Hibbs, *Quantum Mechanics and Path Integrals*, McGraw-Hill, New York, 1965.
[45] C. Becchi, A. Rouet, and R. Stora, *Ann. Phys.* **98**, 287 (1976).
[46] I. V. Tyupin, Lebedev preprint FIAN No. 39 (1975), unpublished.
[47] L. D. Faddeev and V. N. Popov, *Phys. Lett.* **25B**, 29 (1967).

CHAPTER 2

# Nambu–Goto Strings

## §2.1. Bosonic Strings

String theory, at first glance, seems divorced from the standard techniques developed over the past 40 years for second quantized field theories. This is because string theory was first historically discovered as a *first quantized theory*. This is the reason why string theory at times appears to be a random collection of arbitrary conventions. Although a second quantized field theory can be derived completely from a single action, a first quantized theory requires additional assumptions. In particular, the vertices, the choice of interactions, and the weights of these perturbation diagrams must be postulated by hand and checked to be unitary later.

Fortunately, the path integral formalism for the first quantized point particle has been generalized for the string by J. L. Gervais and B. Sakita, which enables us to write down the dynamics of interacting strings with remarkable ease.

In the previous chapter, we laid the crucial mathematical groundwork for a discussion of the first quantized point particle theory. Surprisingly, almost all of the main features of the Nambu–Goto string have some form of analogue in the first quantized point particle theory. Of course, entirely new features are found in the string theory, such as the existence of powerful symmetries on the world sheet, but the basic methods of quantization can be carried over directly from the point particle case studied in the previous chapter.

We saw that the usual formulation of second quantized field theory can be rewritten in first quantized form. Thus, the traditional covariant Feynman propagator (1.6.19) can be written via (1.3.28), (1.3.30), (1.3.37) as

$$\begin{aligned}\Delta_F(x_1, x_2) &= \langle x_1 | \frac{1}{\Box - m^2} | x_2 \rangle \\ &= \langle x_1 | \int_0^\infty d\tau\, e^{-\tau(\Box - m^2)} | x_2 \rangle \\ &= \int_0^\infty d\tau \int_{x_1}^{x_2} Dx\, e^{-(1/2)\int_0^\tau d\bar{\tau}(\dot{x}_\mu^2 - m^2)} \end{aligned} \tag{2.1.1}$$

where we integrate over all possible trajectories of a particle located at $x_\mu(\tau)$ which start at $x_1$ and end at point $x_2$. The interactions, we saw, were introduced by hand into the theory by postulating a particular set of topologies over which this particle can roam. The scattering amplitude, for example, is

$$\begin{aligned}A(k_1, k_2, \ldots, k_N) &= \sum_{\text{topologies}} \int Dx\, e^{-\int L\, d\tau + i\sum_{i=1}^N k_i \cdot x_i} \Delta_{FP} \\ &= \sum_{\text{topologies}} \left\langle \prod_{i=1}^N e^{ik_i \cdot x^i} \right\rangle \end{aligned} \tag{2.1.2}$$

where we integrate over topologies that form the familiar Feynman diagrams for $\phi^3$ or $\phi^4$ theory.

It is important to notice that the resulting Feynman diagram is a graph, *not* a manifold. At the interaction point, the local topology is not $\mathbf{R}^n$, so it cannot be a manifold. There is no correlation between the internal lines and the interaction points. This means that we can introduce arbitrarily high spins at the interaction point of the first quantized relativistic point particle. Thus, the first quantized point particle theory has an infinite degree of arbitrariness, corresponding to the different spins and masses we can place at the interaction point. Furthermore, the ultraviolet singularities of each Feynman diagram correspond to the number of ways we can "pinch" the diagram by shrinking an internal line to zero, thus deforming the local topology.

This picture, however, totally changes with the string. Although the path integral formalism looks almost identical, there are profoundly important differences. In particular, the sum over histories becomes a sum over all possible tubes or sheets that one can draw between two different strings (see Fig. 2.1). This world sheet, in turn, is a genuine manifold, a Riemann surface, so the set of interactions consistent with the propagator is severely limited. *Thus, we expect to find a very small number of string theories, in contrast to the infinite number of point particle theories one can write.* Furthermore, the superstring theory does not suffer from ultraviolet divergences caused by shrinking one of the internal lines to zero. You cannot "pinch" the string world sheet to obtain an ultraviolet divergence. Thus, string theory is free of ultraviolet divergences from strictly topological arguments. (We must be careful to point out that this pinched diagram, however, can be reinterpreted as an *infrared* divergence representing the emission of massless, spin-zero par-

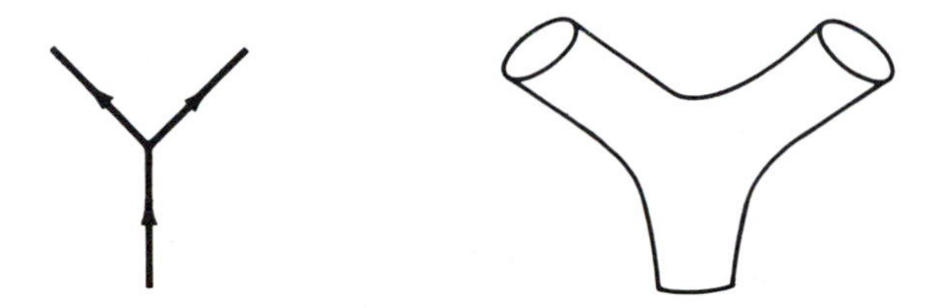

Figure 2.1. Vertex functions for point particles and strings. A large number of point particle theories are possible, based on different spins and isospins, because the Feynman diagrams are graphs. Only a few string theories are known, however, because the interactions are restricted to be manifolds, not graphs. Conformal symmetry, modular invariance, and supersymmetry place enormous restrictions on the manifolds one may use to construct superstring theories that have no counterpart in point particle theory.

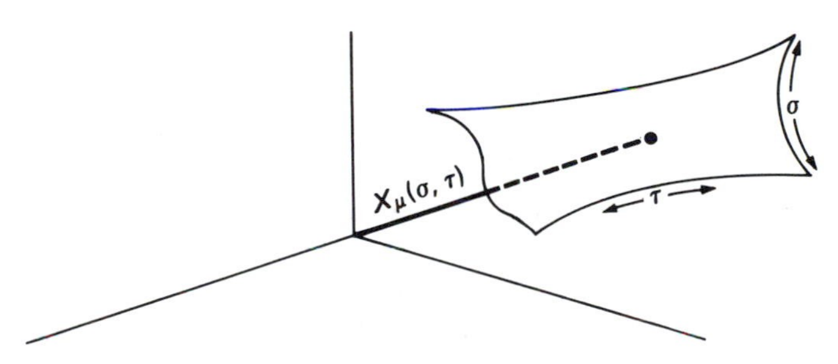


Figure 2.2. The two-dimensional world sheet swept out by a string. When a string, which is parametrized by $\sigma$, moves in space–time, it sweeps out a two-dimensional surface parametrized by $\sigma$ and $\tau$. The string variable $X_\mu(\sigma, \tau)$ is just a vector that extends from the origin to a point on this two-dimensional manifold.

ticle into the vacuum. Fortunately, supersymmetry eliminates these infrared divergences.)

In summary, although the path integral formalism can treat both the first quantized point particle and the first quantized string theory with relative ease, there are profoundly important physical differences between the two theories that arise from strictly topological arguments.

We begin our discussion of strings by first introducing the coordinate of a string vibrating in physical space–time. Let the points along the string be parametrized by the variable $\sigma$, and then let the string propagate in time. Let the vector

$$X_\mu(\sigma, \tau) \tag{2.1.3}$$

represent the space–time coordinates of this string (see Fig. 2.2) parametrized by two variables. When the string moves, it sweeps out a two-dimensional surface, which we call the "world sheet." We will parametrize this world sheet with two variables, $\sigma$ and $\tau$. The vectors that are tangent to the surface are given by the derivatives of the coordinate:

$$\text{Tangent vectors} = \frac{\partial X_\mu}{\partial \tau}; \qquad \frac{\partial X_\mu}{\partial \sigma} \tag{2.1.4}$$

The contraction of two of these tangent vectors yields a metric:

$$g_{ab} = \partial_a X_\mu \partial_b X^\mu \tag{2.1.5}$$

where we have now replaced the two variables $(\tau, \sigma)$ with the set $(a, b)$, where $a, b$ can equal either 0 or 1. The infinitesimal area on this surface can be written simply as

$$d\text{ Area} \sim \sqrt{\det |g_{ab}|}\, d\sigma\, d\tau \tag{2.1.6}$$

In analogy to the point particle case, where the action is the length swept out by the point, we now define our action to be the surface area of this world sheet. Our Lagrangian is therefore [1–4]:

$$L = \frac{1}{2\pi\alpha'}\sqrt{\dot{X}_\mu^2 X'^{\mu 2} - (\dot{X}_\mu X'^\mu)^2} \tag{2.1.7}$$

where the prime represents $\sigma$ differentiation and the dot represents $\tau$ differentiation. The action is just the Lagrangian integrated over the world sheet, which is the total area of the two-dimensional surface:

$$S = \int d\sigma\, d\tau\, L(\sigma, \tau) \tag{2.1.8}$$

The Green's function for the propagation of a string from configuration $X_a$ at "time" $\tau_a$ to configuration $X_b$ at "time" $\tau_b$, as well as the path integral over a surface that expresses the topology of several interacting strings, can be represented as

$$\begin{aligned} K(X_a, X_b) &= \int_{X_a}^{X_b} DX\, e^{-\int_{\tau_a}^{\tau_b} d\tau \int_0^\pi d\sigma\, L} \\ Z &= \sum_{\text{Topologies}} \int d\mu\, DX\, e^{-\text{area}} \end{aligned} \tag{2.1.9}$$

where $DX = \prod_{\mu,\sigma,\tau} dX_\mu(\sigma, \tau)$, $d\mu$ represents the measure of integeration over the location of the external legs, and where we have made a Wick rotation in the $\tau$ variable $(\tau \to -i\tau)$ so the integral converges.

The correspondence between the point particle path integral formalism that we carefully developed in the previous chapter and the string formalism is quite remarkable. We find that almost the entire point particle formalism can be imported into the string formalism:

$$\left\{ \begin{array}{c} x_\mu(\tau) \\ \text{length} \\ \prod_{\mu,\tau} dx_\mu(\tau) \end{array} \right\} \to \left\{ \begin{array}{c} X_\mu(\sigma, \tau) \\ \text{area} \\ \prod_{\mu,\sigma,\tau} dX_\mu(\sigma, \tau) \end{array} \right\}$$

Similarly, the path integral for the point particle and the string theory have surprising similarities. The $N$-point function for the $N$-string scattering amplitude can also be written as a Fourier transform, similar to point particle path integrals:

$$\left\{ \begin{array}{c} \displaystyle\int_{x_i}^{x_j} Dx\, e^{-\int_{\tau_1}^{\tau_2} L(x)\, d\tau} \\ \left\langle \displaystyle\prod_{i=1}^{N} e^{ik_i x^i} \right\rangle \end{array} \right\} \rightarrow \left\{ \begin{array}{c} \displaystyle\int_{X_i}^{X_j} DX\, e^{-\int_0^\pi d\sigma \int_{\tau_1}^{\tau_2} L(X)\, d\tau} \\ \left\langle \displaystyle\prod_{i=1}^{N} e^{ik_i X^i} \right\rangle \end{array} \right\}$$

Although there are remarkable similarities between point particle and string theories when expressed in the language of path integrals, the crucial difference between them emerges when we analyze the topologies over which the objects can move. For the point particle case, the topologies are *graphs*, as in Feynman graphs, whereas the topologies for string theories are *manifolds*:

$$\text{Graphs} \rightarrow \text{Manifolds}$$

One of the crucial reasons why there are so many point particle actions (and so few string actions) is the difference between graphs and manifolds. The nontrivial restrictions placed on manifolds severely restrict the number of consistent string theories.

As in the case of the point particle, the choice of parametrization was totally arbitrary. Thus, our action must be reparametrization invariant. To see this, let us make an arbitrary change of variables:

$$\begin{aligned} \tilde{\sigma} &= \tilde{\sigma}(\sigma, \tau) \\ \tilde{\tau} &= \tilde{\tau}(\sigma, \tau) \end{aligned} \tag{2.1.10}$$

Under this reparametrization, the string variable changes as

$$\delta X^\mu = X'^\mu \delta\sigma + \dot{X}^\mu \delta\tau \tag{2.1.11}$$

Because the area of a surface is independent of the parametrization, the action is manifestly reparametrization invariant, which is easily checked.

As before, let us now write down the canonical conjugates of the theory:

$$P_\mu = \frac{\delta L}{\delta \dot{X}^\mu} = \frac{1}{2\pi\alpha'} \frac{X'^2 \dot{X}_\mu - (\dot{X}_\nu X'^\nu) X'_\mu}{\sqrt{\det |\partial_a X^\nu \partial_b X_\nu|}}$$

As in the point particle case, these momenta are not all independent. In fact, we find two identities that are satisfied by the canonical momenta:

$$\text{Constraints:} \quad \begin{cases} P_\mu^2 + \dfrac{1}{(2\pi\alpha')^2} X_\mu'^2 \equiv 0 \\ P_\mu X'^\mu \equiv 0 \end{cases} \tag{2.1.12}$$

Thus, the canonical momenta are constrained by these two conditions. If we calculate the Hamiltonian of the system, we find that it vanishes identically as in (1.4.9):

$$H = P_\mu \dot{X}^\mu - L \equiv 0 \tag{2.1.13}$$

The vanishing of the Hamiltonian and the presence of constraints among the various momenta are indications that the system is a gauge system with an infinite redundancy. The reparametrization invariance of the system is the origin of this redundancy. We can therefore write down the close correspondence between the constraints of the point particle theory and the string theory:

$$\{p^2 + m^2 = 0\} \to \left\{ \begin{array}{c} P_\mu^2 + \dfrac{1}{(2\pi\alpha')^2} X_\mu'^2 = 0 \\ P_\mu X'^\mu = 0 \end{array} \right\}$$

Before we begin a detailed discussion of the quantization of the string, it is instructive to investigate the purely classical motions of this string. Let us first classically set the parameter $\tau$ equal to time, so that

$$\begin{aligned} \dot{X}_\mu &= (1, v_i) \\ X'_\mu &= (0, X'_i) \end{aligned} \tag{2.1.14}$$

Then let us factor out $X'^2$ from the action (2.1.7):

$$L = -\frac{1}{2\pi\alpha'} |X_i'^2|^{1/2} (1 - \tilde{v}_i^2)^{1/2} \tag{2.1.15}$$

where $\tilde{v}$ is the velocity component perpendicular to the string:

$$\tilde{v}_i = v_i - \frac{v_k X'_k}{X_j'^2} X'_i \tag{2.1.16}$$

The boundary conditions that we derive from this gauge-fixed action include

$$\tilde{v}_i^2 = 1 \tag{2.1.17}$$

This means that the ends of the classical string travel at the speed of light.

We can also calculate the energy of the classical string. Let us assume that the string is in a configuration that maximizes its angular momentum, i.e., it is a rigid rod that rotates with angular velocity $\omega$ around an axis labeled by the unit vector $\mathbf{r}$. The string can be parametrized as

$$\mathbf{X} = \sigma \mathbf{r} \tag{2.1.18}$$

where

$$\begin{aligned} \dot{\mathbf{r}} &= \boldsymbol{\omega} \times \mathbf{r} \\ \mathbf{r} \cdot \boldsymbol{\omega} &= 0 \end{aligned} \tag{2.1.19}$$

and $-l \leq \sigma \leq l$.

To calculate the energy and the angular momentum of the system, we must first write down the Lorentz generators associated with the string:

$$M^{\mu\nu} = \int d\sigma (P^\nu X^\mu - P^\mu X^\nu) \tag{2.1.20}$$

Notice that this generates the algebra of the Lorentz group if we impose the Poisson brackets:

$$[X_\mu(\sigma), P_\nu(\sigma')] = \eta_{\mu\nu}\delta(\sigma - \sigma') \tag{2.1.21}$$

We can now calculate the energy and the angular momentum from the components of the Lorentz generator [5]:

$$\begin{aligned} E &= \frac{1}{2\pi\alpha'}\int_{-l}^{l} d\sigma(1 - \omega^2\sigma^2)^{-1/2} \sim \frac{l}{2\alpha'} \\ \mathbf{J} &= \frac{1}{2\pi\alpha'}\int_{-l}^{l} d\sigma \frac{\sigma^2}{(1 - \omega^2\sigma^2)^{1/2}}[\mathbf{r} \times (\boldsymbol{\omega} \times \mathbf{r})] \\ &= \boldsymbol{\omega}(4\alpha'\omega^3)^{-1} \\ &\sim \boldsymbol{\omega}E^2\alpha' \end{aligned} \tag{2.1.22}$$

Thus, the angular momentum of the rotating string is proportional to the square of the energy of the system:

$$|\mathbf{J}| \sim E^2 \tag{2.1.23}$$

If we plot the energy squared on the $x$-axis and the angular momentum on the $y$-axis, then we obtain a curve called the *Regge trajectory*. The slope of the Regge trajectory is given by $\alpha'$ and the curve is linear. Thus, we have obtained the leading Regge trajectory for the classical motion of a rigid rotator. We will, throughout this book, take the normalization $\alpha' = \frac{1}{2}$. This is an arbitrary convention. However, we will see later that the intercept $a_0$ of the leading trajectory must be set equal to one, which is fixed by conformal invariance once we quantize the theory. Thus, we set

$$\begin{aligned} \alpha' &= \tfrac{1}{2} \\ a_0 &= 1 \end{aligned} \tag{2.1.24}$$

When we quantize the system, we will find that there is an infinite number of such parallel Regge trajectories, but with increasingly negative $y$-intercepts.

As we have stressed, there is a crucial difference between the point particle case and the string, which is that the string system has a larger set of constraints that generate the gauge group of reparametrizations. For example, if we introduce the canonical Poisson brackets, then we can explicitly show that the constraints generate an algebra.

To calculate this algebra, we decompose the $X$ and $P$ in terms of normal modes for the open string:

$$\begin{aligned} X^\mu(\sigma) &= x^\mu + 2\sum_{n=1}^{\infty} \frac{1}{\sqrt{n}} X_n^\mu \cos n\sigma \\ P^\mu(\sigma) &= \frac{1}{\pi}\left\{p^\mu + \sum_{n=1}^{\infty} \sqrt{n} P_n^\mu \cos n\sigma\right\} \end{aligned} \tag{2.1.25}$$

Notice that the $X$ decomposition is given strictly in terms of cosines. This is because, when we calculate the equations of motion of the string, we must integrate by parts and hence obtain unwanted surface terms at $\sigma = \pi$ and $\sigma = 0$. In order to eliminate these surface terms, we must impose

$$X'_\mu = 0 \tag{2.1.26}$$

at the boundary. This boundary condition eliminates all sine modes of the string.

We will sometimes find it convenient to take advantage of this form of the expansion. In particular, it means that

$$\begin{aligned} X_\mu(\sigma) &= X_\mu(-\sigma) \\ X'_\mu(\sigma) &= -X'_\mu(-\sigma) \end{aligned} \tag{2.1.27}$$

The same applies for the modes of the canonical conjugate $P_\mu$. This, in turn, allows us to combine both constraints into one, using the properties of the string under reflection from $\sigma$ into $-\sigma$. If we let the string parametrization length be $\pi$ we can define:

$$L_f = \frac{1}{4\pi} \int_{-\pi}^{\pi} d\sigma\, f(\sigma) \left( \sqrt{2\alpha'}\pi P_\mu + \frac{X'_\mu}{\sqrt{2\alpha'}} \right)^2 \tag{2.1.28}$$

where $f(\sigma)$ is an arbitrary function defined from $-\pi$ to $\pi$. Notice that both constraints are now combined into one equation because of this reflection symmetry. Using (2.1.21), we can show that these generators form a closed algebra:

$$[L_f, L_g] = L_{f \times g} \tag{2.1.29}$$

where

$$f \times g = fg' - gf' \tag{2.1.30}$$

It is also possible to show that this algebra satisfies the Jacobi identities:

$$[L_{[f}, [L_g, L_{h]}] = 0 \tag{2.1.31}$$

where the brackets represent all possible cyclic symmetrizations. This algebra is called the *Virasoro algebra* [6], which will turn out to be one of the most powerful tools we have in constructing the string theory.

As in (1.4.11), we can elevate the constraints into the action with Lagrange multipliers $\lambda(\sigma, \tau)$ and $\rho(\sigma, \tau)$:

$$L = P_\mu \dot{X}^\mu + \pi\alpha'\lambda \left[ P_\mu^2 + \frac{X'^2_\mu}{(2\pi\alpha')^2} \right] + \rho P_\mu X'^\mu \tag{2.1.32}$$

By functionally integrating out over these Lagrange multipliers, we arrive at the previous set of constraints. Not surprisingly, this new action has its own reparametrization group parametrized by $\eta$ and $\varepsilon$:

$$\begin{aligned} \delta X_\mu &= 2\pi\alpha'\varepsilon P_\mu + \eta X'_\mu \\ \delta P_\mu &= \left[\frac{\varepsilon X'_\mu}{2\pi\alpha'} + \eta P_\mu\right]' \\ \delta\lambda &= -\dot{\varepsilon} + \lambda'\eta - \eta'\lambda + \rho'\varepsilon - \rho\varepsilon' \\ \delta\rho &= -\dot{\eta} + \lambda'\varepsilon - \lambda\varepsilon' + \rho'\eta - \eta'\rho \end{aligned} \tag{2.1.33}$$

The advantage of this form for the action is that it is first-order and does not have the bothersome square roots of the original action. As in the point particle case, this indicates that there exists yet one more form for the action, expressed in terms of an auxiliary field. To find this third formulation of the action, let us introduce a new independent field

$$g_{ab}(\sigma, \tau) \tag{2.1.34}$$

which represents a metric on a two-dimensional surface. Unlike our previous discussion, this metric is now totally independent of the string variable. Let us write down the Polyakov form of the action [7] ($g = |\det g_{ab}|$):

$$L = -\frac{1}{4\pi\alpha'}\sqrt{g}g^{ab}\partial_a X_\mu \partial_b X^\mu \tag{2.1.35}$$

This is a generalization of the second-order point particle action (1.4.14). Notice that the Polyakov action resembles an action with scalar fields interacting with an external two-dimensional gravitational field. This action, too, possesses manifest reparametrization invariance:

$$\begin{aligned} \delta X^\mu &= \varepsilon^a \partial_a X^\mu \\ \delta g^{ab} &= \varepsilon^c \partial_c g^{ab} - g^{ac}\partial_c \varepsilon^b - g^{bc}\partial_c \varepsilon^a \\ \delta\sqrt{g} &= \partial_a(\varepsilon^a \sqrt{g}) \end{aligned} \tag{2.1.36}$$

The action is also trivially invariant under Weyl rescaling:

$$\delta g^{ab} = \Lambda g^{ab} \tag{2.1.37}$$

The Polyakov action is entirely equivalent at the classical level to the earlier Nambu–Goto action. As in the Nambu–Goto formalism, we can derive the Virasoro algebra. By varying with respect to the metric tensor, we obtain the energy–momentum tensor, which we can set to zero:

$$T_{ab} = -4\pi\alpha' \frac{1}{\sqrt{g}} \frac{\delta L}{\delta g^{ab}} \tag{2.1.38}$$

Working this out explicitly, we find

$$T_{ab} = \partial_a X_\mu \partial_b X^\mu - \tfrac{1}{2} g_{ab} g^{cd} \partial_c X^\mu \partial_d X_\mu \tag{2.1.39}$$

The moments of the energy–momentum tensor will correspond to the Virasoro generators. Thus, we have another way of deriving the Virasoro generators from this new but equivalent formalism.

Notice that the metric field $g_{ab}$ is *not* a propagating field. The metric tensor does not have any derivatives acting on it. Thus, we can eliminate it via its own equations of motion. This leads us to

$$\frac{\delta L}{\delta g_{ab}} = 0 \to g_{ab} = \frac{2\partial_a X_\mu \partial_b X^\mu}{g^{cd}\partial_c X_\nu \partial_d X^\nu} \tag{2.1.40}$$

Substituting this value of the metric tensor back into the action, we rederive the original Nambu–Goto action. Thus, at the classical level, the two actions are identical.

In summary, as in the point particle case, we now have three different ways in which to write down the action, all of which are equivalent classically. Each has its own particular advantages and disadvantages when we make the transition to the quantum system. These string equations are direct generalizations of the three point particle Lagrangians found in (1.4.16). As before, we have the second-order formalism, which is expressed in terms of the string variable $X_\mu$ as well as the metric tensor $g_{ab}$; the non-linear formalism, which is expressed entirely in terms of $X_\mu$; and the Hamiltonian formalism, where we have $X_\mu$ and its canonical conjugate $P_\mu$ (or the pair $\partial_a X_\mu$ and $P^{a\mu}$):

$$\begin{aligned}
\text{1st-order (Hamiltonian) form: } L &= P_\mu \dot{X}^\mu + \pi\alpha'\lambda\left[P_\mu^2 + \frac{X_\mu'^2}{(2\pi\alpha')^2}\right] + \rho P_\mu X'^\mu \\
&\sim \sqrt{g} P_\mu^a g_{ab} P^{b\mu} + P^{a\mu}\partial_a X_\mu \sqrt{g}/\pi\alpha' \\
\text{2nd-order form: } L &= \frac{-1}{4\pi\alpha'}\sqrt{g} g^{ab}\partial_a X_\mu \partial_b X^\mu \\
\text{Non-linear form: } L &= \frac{1}{2\pi\alpha'}(\dot{X}_\mu^2 X_\nu'^2 - (\dot{X}_\mu X'^\mu)^2)^{1/2}
\end{aligned} \tag{2.1.41}$$

At first, we suspect that these actions are totally equivalent, so that we can choose one and drop the others. This is apparently not so, for two subtle reasons:

(1) Because we are dealing with a first quantized theory, we have to take the sum over all interacting topologies that are swept out by the string. For the Nambu–Goto string, the precise nature of these topologies is ambiguous and must be specified by hand. However, for the Polyakov form of the action, which contains an independent metric tensor, we can eliminate most of this ambiguity by specifying that we sum over all *conformally and modular inequivalent configurations*. (These terms will be defined later.) This will become a powerful constraint once we start to derive loops and will determine the functional measure uniquely. The measure and the topologies in the Nambu–Goto action, however, are not well defined. (We must point out, however, that this rule of integrating over inequivalent

surfaces does not automatically satisfy unitarity. This still must be checked by hand.)

(2) The gauge fixing of Weyl invariance for the Polykov action, although trivial classically, poses problems when we make the transition to quantum mechanics. An anomaly appears when we carefully begin the quantization process. In fact, this conformal anomaly will disappear only in 26 dimensions!

Let us now discuss the quantization of the string action. The strategy we will take in quantizing the free theory to obtain the physical Hilbert space will be first to extract the symmetry of the action, then the currents, and then the algebra formed by the generators of this symmetry. (For the string, the symmetry will be reparametrization invariance and the algebra will be the Virasoro algebra.) Then we must apply the constraints onto the Hilbert space, which eliminates the ghosts and creates a unitary theory. It is important to keep this strategy in mind as we begin the quantization of the string:

Action → Symmetry → Current → Algebra → Constraints → Unitarity

As in the point particle case, we can begin the quantization program in several ways. There are three formalisms in which to fix the gauge of the theory: (1) Gupta–Bleuler (conformal gauge), (2) light cone gauge, and (3) BRST formalism. The advantages and disadvantages of each are as follows:

(1) The Gupta–Bleuler is perhaps the simplest of the three formalisms. We allow ghosts to appear in the action, which permits us to maintain manifest Lorentz invariance. The price we must pay, however, is that we must impose ghost-killing constraints on the Hilbert space. Projection operators must be inserted in all propagators. For trees, this is trivial. For higher loops, however, this is exceedingly difficult.
(2) The advantage of the light cone gauge formalism is that it is explicitly ghost-free in the action as well as the Hilbert space. There are no complications when going to loops. However, the formalism is very awkward and Lorentz invariance must be checked at each step of the way.
(3) The BRST formalism combines the best features of the previous two formalisms. It is manifestly covariant, like the Gupta–Bleuler formalism, and it is unitary, like the light cone formalism, because the negative metric ghosts cancel against the Faddeev–Popov ghosts.

Let us now discuss each quantization scheme separately.

## §2.2. Gupta–Bleuler Quantization

The Gupta–Bleuler formalism will maintain Lorentz invariance by imposing the Virasoro constraints on the state vectors of the theory:

$$\langle \phi | L_f | \psi \rangle = 0 \tag{2.2.1}$$

where $\langle\phi|$ and $|\psi\rangle$ represent states of the theory. This constraint will eliminate ghosts in the state vectors, allowing us to keep nonphysical negative metric ghosts intact in the action.

Classically, the metric tensor has three degrees of freedom that we can gauge away, two arising from reparametrization invariance and one from Weyl invariance. Since the metric tensor has three degrees of freedom, we can gauge all of its components away:

$$g_{ab} = \delta_{ab} = \begin{pmatrix} -1 & 0 \\ 0 & 1 \end{pmatrix} \tag{2.2.2}$$

which we call the *conformal gauge*. (There are complications, as we have said, in taking the conformal gauge for the quantum theory and for higher loops.) Our action reduces to

$$S = \frac{1}{4\pi\alpha'} \int_0^\pi d\sigma \int d\tau (\dot{X}_\mu^2 - X_\mu'^2) \tag{2.2.3}$$

This is exceptionally simple because the action now corresponds to an uncoupled free string. This action yields the free equations of motion:

$$\left(\frac{\partial^2}{\partial\sigma^2} - \frac{\partial^2}{\partial\tau^2}\right) X_\mu(\sigma, \tau) = 0 \tag{2.2.4}$$

with the boundary condition:

$$X_\mu'(0, \tau) = X_\mu'(\pi, \tau) = 0 \tag{2.2.5}$$

which we need to enforce when we integrate by parts and eliminate the surface term. The solutions of the equations of motion are arbitrary functions of $\sigma + \tau$ and $\sigma - \tau$:

$$X^\mu(\sigma, \tau) = X_1^\mu(\sigma + \tau) + X_2^\mu(\sigma - \tau) \tag{2.2.6}$$

The canonical commutation relations are now

$$[P_\mu(\sigma), X_\nu(\sigma')] = -i\eta_{\mu\nu}\delta(\sigma - \sigma') \tag{2.2.7}$$

where

$$\delta(\sigma - \sigma') = \frac{1}{\pi}\left(1 + 2\sum_{n=1}^{\infty} \cos n\sigma \cos n\sigma'\right) \tag{2.2.8}$$

There is, of course, an infinite number of possible representations of the path integral. However, as in the point particle case, we can always choose the simplest one, the harmonic oscillator basis [8], where the Hamiltonian becomes diagonal. Unlike the point particle case, however, we now have an infinite number of oscillators, one set for each normal mode:

$$\begin{aligned} X_n^\mu &= \tfrac{1}{2} i\sqrt{2\alpha'}(a_n^\mu - a_{-n}^\mu) \\ P_n^\mu &= \frac{1}{\sqrt{2\alpha'}}(a_n^\mu + a_{-n}^\mu) \end{aligned} \tag{2.2.9}$$

where we can satisfy the canonical commutation relations if we set

$$[a_{n\mu}, a^{\dagger}_{m\nu}] = \delta_{nm}\eta_{\mu\nu} \tag{2.2.10}$$

It is also conventional to introduce an equivalent set of oscillators:

$$\alpha^{\mu}_{m} = \sqrt{m}a^{\mu}_{m}; \qquad m > 0$$

$$\alpha^{\mu}_{-m} = \sqrt{m}a^{\dagger\mu}_{m}; \qquad m > 0$$

$$X^{\mu}(\sigma, \tau) = x^{\mu} + 2\alpha' p^{\mu}\tau + i\sqrt{2\alpha'} \sum_{n\neq 0} \frac{\alpha^{\mu}_{n}}{n} e^{-in\tau} \cos n\sigma \tag{2.2.11}$$

Written in this basis, the Hamiltonian takes on an especially simple form (see (1.3.37)):

$$\begin{aligned} H &= \int_0^{\pi} d\sigma (P_{\mu}\dot{X}^{\mu} - L) \\ &= \pi\alpha' \int_0^{\pi} \left( P_{\mu}^2 + \frac{1}{(2\alpha'\pi)^2} X'^2_{\mu} \right) d\sigma \\ &= \sum_{n=1}^{\infty} n a^{\dagger}_{n\mu} a^{\mu}_{n} + \alpha' p^2_{\mu}; \qquad \alpha_{0\mu} = \sqrt{2\alpha'} p_{\mu} \end{aligned} \tag{2.2.12}$$

where we have made an infinite shift in the zero point energy. At this point, the mass of the lowest order particle is not well defined because we made this infinite shift, but we will later show that this lowest particle is actually a tachyon. We will show that the intercept of the model is fixed at 1.

Notice that each oscillator mode is basically uncoupled from the other oscillator modes. In fact, the Hamiltonian is diagonal in the Fock space of harmonic oscillator excitations. Taking this specific representation of the string function from the infinite number of possibilities is a great advantage because the allowed eigenstates of our Hamiltonian are now simply the products of the Fock spaces of all possible harmonic oscillators:

$$\text{Eigenstates:} \quad \prod_{n,\mu} \{a^{\dagger}_{n,\mu}\} |0\rangle \tag{2.2.13}$$

where the vacuum is defined as

$$a_{n\mu}|0\rangle = 0; \qquad n \geq 0 \tag{2.2.14}$$

The spectrum of the lower lying states can be categorized as (see Fig. 2.3):

$$\begin{aligned} \text{Tachyon} &\rightarrow |0\rangle \\ \text{Massless vector} &\rightarrow a_1^{\dagger\mu}|0\rangle \\ \text{Massless scalar} &\rightarrow k_{\mu}a_1^{\dagger\mu}|0\rangle \\ \text{Massive spin two} &\rightarrow a_1^{\dagger\mu}a_1^{\dagger\nu}|0\rangle \\ \text{Massive vector} &\rightarrow a_2^{\dagger\mu}|0\rangle \end{aligned} \tag{2.2.15}$$

As expected, we recover the leading Regge trajectory that we obtained earlier

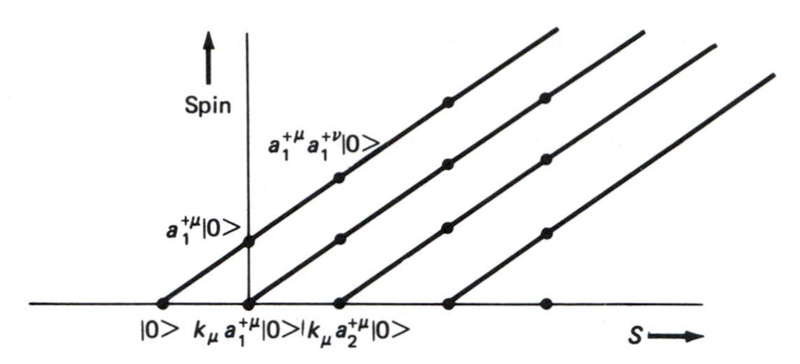


Figure 2.3. Regge trajectories for the open string. The $x$-axis corresponds to the energy squared and the $y$-axis to the spin. The particle farthest to the left is the tachyon, which corresponds to the vacuum of the Fock space. The massless spin-1 particle is the Maxwell or Yang–Mills field, which corresponds to a single creation operator acting on the vacuum. There is an infinite number of Regge trajectories, corresponding to the infinite excitations of a relativistic string or the infinite number of states in the Fock space.

from classical arguments, and also an infinite number of daughter trajectories that have increasingly negative $y$-intercept. In this gauge, the propagator or Green's function takes on a simple form:

$$K(a, b) = \langle X_a | e^{-H\delta\tau} | X_b \rangle$$
$$= \int_{X_a}^{X_b} DX \exp - \int d\sigma\, d\tau \left\{ \frac{1}{4\pi\alpha'} (\dot{X}_\mu^2 - X_\mu'^2) \right\} \tag{2.2.16}$$

where

$$DX = \prod_\mu \prod_\sigma dX^\mu(\sigma) = \prod_{\mu, n} dX_n^\mu \tag{2.2.17}$$

We must be careful to note that the functional integral over $DX$ is an infinite product of integrals over each point along the string, or each Fourier mode of the string.

Fortunately, this integration can be carried out explicitly. Let the particle move from $\tau = 0$ to $\tau = \infty$. Because the Hamiltonian is diagonal on the space of harmonic oscillators, we can perform the integration over $\tau$ exactly. We find

$$\langle X_a | \int_0^\infty d\tau\, e^{-H\tau} | X_b \rangle = \langle X_a | \frac{1}{L_0 - 1} | X_b \rangle \tag{2.2.18}$$

where $L_0$ is the zeroth Fourier component of the Virasoro generators (2.1.28). In other words, our propagator for the free theory is

$$D \equiv \frac{1}{L_0 - 1} \tag{2.2.19}$$

sandwiched between states of the string. However, because we have the identity

$$|X\rangle \int DX \, \langle X| = 1 \tag{2.2.20}$$

we can remove the path integrals explicitly at every intermediate point between the initial and the final state. In fact, because of the simplicity of the $N$-point function, we will be able to remove all functional integrations over string states explicitly, leaving only the harmonic oscillators. It is important to stress again that the harmonic oscillator formalism is nothing but a specific representation of the path integral. Its simplicity is a consequence of the Hamiltonian being diagonal in the Fock space of harmonic oscillator states.

In much the same way, the closed string can also be written in terms of harmonic oscillators. For the closed string, there is no $X' = 0$ boundary condition, and we can use both sine and cosine functions to decompose its normal modes. Thus, we expect twice the number of oscillators for the closed string. The oscillator decomposition is given by

$$X_\mu(\sigma) = x_\mu + (\tfrac{1}{2}\alpha')^{1/2} \sum_{n=1}^{\infty} \frac{1}{\sqrt{n}}(a_n e^{-in\sigma} + \tilde{a}_n e^{in\sigma} + a_n^\dagger a^{in\sigma} + \tilde{a}_n^\dagger e^{-in\sigma})_\mu$$

$$P_\mu(\sigma) = \frac{p_\mu}{2\pi} + \frac{1}{2\pi\sqrt{2\alpha'}} \sum_{n=1}^{\infty} \sqrt{n}( - ia_n e^{-in\sigma} - i\tilde{a}_n e^{in\sigma} + ia_n^\dagger e^{in\sigma} + i\tilde{a}_n^\dagger e^{-in\sigma})_\mu \tag{2.2.21}$$

The Hamiltonian for the closed string is

$$\begin{aligned} H &= \pi \int_0^{2\pi} d\sigma \left( \alpha' P_\mu^2 + \frac{X_\mu'^2}{4\pi^2\alpha'} \right) \\ &= \sum_{n=1}^{\infty} (n a_n^\dagger a_n + n \tilde{a}_n^\dagger \tilde{a}_n) + \alpha' p_\mu^2 \end{aligned} \tag{2.2.22}$$

Again, the Fock space consists of all elements created out of harmonic oscillators, but this time there is an extra constraint that is not found for the open string:

$$(L_0 - \tilde{L}_0)|\phi\rangle = 0 \tag{2.2.23}$$

(The interpretation of this constraint is that the closed string must be independent of the origin of the $\sigma$ coordinate. For example, the operator $\int d\sigma \, e^{i\sigma(L_0 - \tilde{L}_0)}$ can be interpreted in two ways. First, it generates rotations in $\sigma$ space, so we average over a rotation of $2\pi$ in $\sigma$ space. Second, if we perform the integral, we have $\delta(L_0 - \tilde{L}_0)$, which is constraint (2.2.23) when applied to the Hilbert space. We will return to this constraint later.)

The Fock space consists of (see Fig. 2.4):

$$\begin{aligned} &\text{Tachyon} \to |0\rangle \\ &\text{Massless spin two} \to a_1^{\dagger\mu}\tilde{a}_1^{\dagger\nu}|0\rangle \\ &\text{Massless scalar} \to k_\mu k_\nu a_1^{\dagger\mu}\tilde{a}_1^{\dagger\nu}|0\rangle \end{aligned} \tag{2.2.24}$$

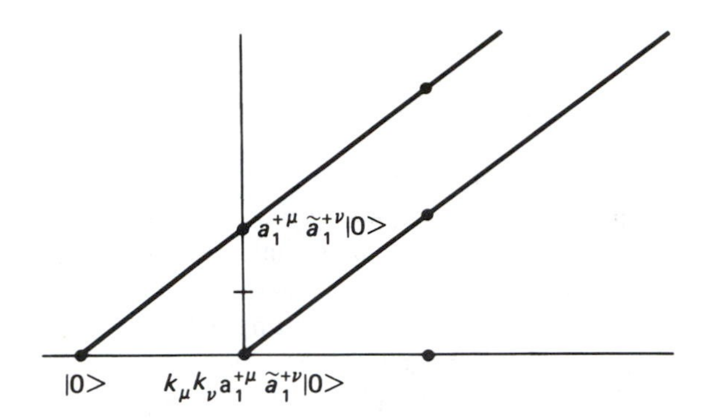


Figure 2.4. Regge trajectories for the closed string. The Fock space is built out of two commuting sets of harmonic oscillators. The massless spin-2 particle is the graviton, which corresponds to the product of both types of operators acting on the vacuum.

Notice that a massless spin-two particle occurs in the spectrum of the closed string. When the string model was first being interpreted as a model of hadrons, the presence of this gravitonlike spin-two particle was a great embarrassment. Attempts were made to associate this particle with the Pomeron trajectory found in $S$-matrix theory. It can be shown that, when we generalize to trees and loops, this massless spin-two particle has a gauge invariance equivalent to the graviton of Einstein's theory. By dropping the earlier interpretation of the string theory as a model of hadrons, we find a natural place for this gravitonlike object as the graviton itself.

In summary, the Gupta–Bleuler formalism in the conformal gauge appears simple and elegant mainly because we are allowing ghosts to appear throughout the action. The theory reduces to the simplest possible string theory, a free propagating string.

The price we must pay for this simplicity, however, is the imposition of the constraints on the Fock space. We can write the Virasoro generators as [6]:

$$\begin{aligned} L_n &= \frac{\pi}{4}\int_{-\pi}^{\pi} d\sigma\, e^{in\sigma}\left[P_\mu + \frac{X'_\mu}{2\pi\alpha'}\right]^2 \\ &= \tfrac{1}{2}\sum_{m=-\infty}^{\infty} \alpha_{n-m,\mu}\alpha_m^\mu \\ L_0 &= \sum_{n=1}^{\infty} \alpha_{-n\mu}\alpha_n^\mu + \tfrac{1}{2}\alpha_0^2 \end{aligned} \tag{2.2.25}$$

The *physical states* of the theory must therefore satisfy

$$\begin{aligned} L_n|\phi\rangle &= 0; \qquad n > 0 \\ (L_0 - 1)|\phi\rangle &= 0 \end{aligned} \tag{2.2.26}$$

The algebra generated by these opertors is

$$[L_n, L_m] = (n-m)L_{n+m} + \frac{D}{12}(n^3 - n)\delta_{n,-m} \tag{2.2.27}$$

where $D$ is the dimension of space–time. The fact that there is a $c$-number central term appearing in this equation at first sounds surprising, but this can be calculated explicitly by taking the vacuum expectation product of a simple commutator:

$$\langle 0|[L_2, L_{-2}]|0\rangle = \frac{D}{2} \tag{2.2.28}$$

The origin of this central term is that we *normal ordered* the generator $L_0$ to obtain finite matrix elements. This normal ordering was even found in the point particle case in (1.8.7), except now the energy shift is infinite. *The price we pay for finite matrix elements is that locality in $\sigma$ is lost.* The normal ordering spoils the fact that the generators $L_f$ were originally local functions in $\sigma$. The process of quantization necessarily destroys the locality of Virasoro generators in the variable $\sigma$, and hence the $c$-number central term occurs. Thus, the quantization scheme and regularization scheme used to extract finite information from the model are actually inconsistent with conformal symmetry. Fortunately, this inconsistency can be eliminated if we fix the dimension of space–time to be 26.

That there are ghosts in the theory is due to the fact that the zeroth component of the harmonic oscillator has negative metric:

$$\text{Ghost} = \{a_{n,0}^{\dagger}\}|0\rangle \tag{2.2.29}$$

Thus, the coefficient of the Green's function occurs with a negative sign. In addition, there are zero norm states and negative norm states that must be taken into account.

To analyze the spectrum, let us define a *spurious state* $|S\rangle$ to be one that is orthogonal to all physical real states $|R\rangle$. Spurious states can be written as

$$|S\rangle = L_{-n}|\chi\rangle; \qquad n > 0$$

$$\langle R|S\rangle = 0$$

for some integer $n$ and some state $|\chi\rangle$. (If we take the matrix element of this state with a physical state, the scalar product always vanishes because $L_n$ destroys a physical state.) Now let us construct the spurious state:

$$|\psi\rangle = [L_{-2} + aL_{-1}^2]|\phi\rangle \tag{2.2.30}$$

We do not want this state to be part of the physical Hilbert space. However, let us see the conditions under which it might be part of the physical spectrum. Let us set

$$L_1|\psi\rangle = L_2|\psi\rangle = 0 \tag{2.2.31}$$

This fixes the following:

$$3 - 2a = 0; \qquad \tfrac{1}{2}D - 4 - 6a = 0 \tag{2.2.32}$$

which, in turn, fixes

$$D = 26; \qquad a = \tfrac{3}{2} \tag{2.2.33}$$

Thus, this spurious state satisfies (2.2.26) and hence is part of the physical Fock space. At first, this seems disastrous. We want our physical Hilbert space to be ghost-free. But notice that in 26 dimensions this state has zero norm (not negative norm). Since $|\phi\rangle$ was arbitrary, we have constructed an infinite class of states $|\psi\rangle$ that are simultaneously spurious and physical. If we take the norm of this higher order state, we find that it also vanishes in 26 dimensions, making it a null spurious state. This state is still acceptable because the norm of the state is nonnegative. Thus, in 26 dimensions we have an acceptable spectrum for this set of states.

Similar analyses of the state vectors of the theory show that a physical state $|\theta\rangle$ can be constructed which actually has negative norm if $D$ is greater than 26:

$$|\theta\rangle = (a\alpha_{-1}^2 + bk\cdot\alpha_{-2} + c(k\cdot\alpha_{-1})^2)|0\rangle$$

Imposing $L_1|\theta\rangle = L_2|\theta\rangle = 0$, we find that $b = a(D-1)/5$ and $c = a(D+4)/10$. We find that the norm of the state is

$$\langle\theta|\theta\rangle = \tfrac{2}{25}a^2(D-1)(26-D)$$

Thus, the dimension of space–time cannot exceed 26 or else negative norm states exist as part of the physical states. In general, we find that the spectrum is ghost-free if the dimension of space–time is less than or equal to 26:

$$\text{Ghost-free:} \quad \begin{Bmatrix} a = 1, & D = 26 \\ a \leq 1, & D \leq 25 \end{Bmatrix} \tag{2.2.34}$$

This exercise was done only for a piece of the Fock space. But can ghosts be eliminated to all orders in the string model? We will return to the difficult problem of ghost elimination in the Gupta–Bleuler formalism at the end of this chapter, when we actually construct the physical Hilbert space and show that it has no negative norm states in 26 dimensions.

## §2.3. Light Cone Quantization

Choosing the light cone gauge, where all unphysical degrees of freedom are explicitly removed from the very beginning, is possible because we have two gauge degrees of freedom, and hence two gauge-fixing conditions can be inserted into our path integral. One of these gauge-fixing constraints can be the elimination of nonphysical modes from the Hilbert space, as in the Coulomb gauge. Thus the elimination of ghost states, which is quite involved for the Gupta–Bleuler formalism (as we shall see at the end of this chapter), becomes trivial in the light cone gauge.

Let us choose the notation

$$\begin{aligned} X^+ &= \frac{1}{\sqrt{2}}[X^0 + X^{D-1}] \\ X^- &= \frac{1}{\sqrt{2}}[X^0 - X^{D-1}] \end{aligned} \tag{2.3.1}$$

then

$$A_\mu B^\mu = A_i B_i - A^+ B^- - A^- B^+ \tag{2.3.2}$$

Depending on which version of the action in (2.1.41) we use, we will have different gauge constraints. If we start with the original Nambu–Goto action, for example, the gauge conditions in the path integral are

$$Z = \int DX\, M\Delta_{\text{FP}} \prod_\sigma \delta(X^+(\sigma) - p^+\tau)\delta(\dot{X}_\mu^2 + X_\mu'^2 - 2\dot{X}_\mu X'^\mu)e^{-S} \tag{2.3.3}$$

where $M$ is a measure term that must be added to have a unitary theory, and the two delta functions represent the gauge-fixing constraints. The remarkable feature of the second constraint is that the Nambu–Goto action, which is expressed as a highly nonlinear square root, completely linearizes [12]

$$\sqrt{\dot{X}_\mu^2 X_\nu'^2 - (\dot{X}_\mu X'^\mu)^2} \sim \tfrac{1}{2}(\dot{X}_\mu^2 - X_\mu'^2) \tag{2.3.4}$$

(Because the light cone action is no longer a square root, we have a well-behaved action that can be canonically quantized.)

The constraint $X^+ = p^+\tau$ means that the $\sigma$ dependence within $X^+$ has completely disappeared and that the "time" $\tau$ now beats in synchronism with $X^+$. We can use the second constraint, in turn, to eliminate the $X^-$ modes, and hence all longitudinal modes have completely disappeared. The action can now be expressed totally in terms of transverse modes.

Next, we will solve the constraints in the first action in (2.1.41). Let us integrate over the Lagrange multipliers $\rho$ and $\lambda$ in the Hamiltonian form of the action, and then impose the gauge-fixing constraints:

$$Z = \int DX\, DP \prod_\sigma \delta(X^+(\sigma) - p^+\tau)\delta\left(P^+(\sigma) - \frac{p^+}{\pi}\right)\delta\left(P_\mu^2 + \frac{X_\mu'^2}{\pi^2}\right)\delta(P_\mu X'^\mu)e^{-S} \tag{2.3.5}$$

Because the covariant Hamiltonian (2.1.13) is equal to zero, the only term in the Lagrangian is $P_\mu \dot{X}^\mu$ (the $X^-$ term drops out):

$$L = \int_0^\pi d\sigma\, P_\mu \dot{X}^\mu = \int_0^\pi d\sigma (P_i \dot{X}^i - p^+ P^-(\sigma)) \tag{2.3.6}$$

There are several remarkable features to this formalism. First, we can apply four, not two, constraints onto our Hilbert space, two from gauge fixing and two by integrating over $\lambda$ and $\rho$. Second, because the covariant Hamiltonian is equal to zero, the action only consists of $P_\mu \dot{X}^\mu$, but the light cone Hamiltonian emerges out of the decomposition of (2.3.6):

$$H = p^+ \int_0^\pi d\sigma\, P^-(\sigma) \tag{2.3.7}$$

On the other hand, we can solve the constraint for $P^-$:

$$P^-(\sigma) = \frac{\pi}{2p^+}\left(P_i^2 + \frac{X_i'^2}{\pi^2}\right) \tag{2.3.8}$$

Plugging the value for $P^-$ into the definition of the light cone Hamiltonian (2.3.7), we now have

$$H = \frac{\pi}{2}\int_0^\pi \left(P_i^2 + \frac{X_i'^2}{\pi^2}\right) d\sigma \tag{2.3.9}$$

which is the just the Hamiltonian (2.2.12) defined over physical transverse states.

Similarly, we can also eliminate the $X^-$ modes by solving another constraint:

$$P_\mu X'^\mu = 0 \tag{2.3.10}$$

which can be solved for $X^-$, yielding

$$X^-(\sigma) = \int_0^\sigma d\sigma' \frac{\pi}{p^+}[P_i X_i'] \tag{2.3.11}$$

Putting everything together, our functional now becomes

$$Z = \int DX_i\, DP_i\, e^{i\int d\tau \int d\sigma (P_i \dot{X}_i - H)} \tag{2.3.12}$$

where $H$ is the light cone Hamiltonian density. The great advantage of the light cone gauge is that the Virasoro constraints have been explicitly solved, so there is no need to impose them on states. All $+$ modes have been gauged away from the start, and the $-$ modes have been eliminated in terms of transverse states because the Virasoro conditions have been solved exactly through (2.3.8) and (2.3.11). Instead of imposing the Virasoro constraints on the Hilbert space, we simply solve them exactly and eliminate the $-$ modes.

However, the great disadvantage of the formalism is that we must tediously check for Lorentz invariance at each step of the calculation. Normally, the generators of the Lorentz group are given by

$$\begin{aligned} M^{\mu\nu} &= \int_0^\pi d\sigma [X^\mu P^\nu - X^\nu P^\mu] \\ &= x^\mu p^\nu - x^\nu p^\mu - i \sum_{n=1}^\infty \frac{1}{n}[\alpha_{-n}^\mu \alpha_n^\nu - \alpha_{-n}^\nu \alpha_n^\mu] \end{aligned} \tag{2.3.13}$$

It is easy to check from (2.2.7) that this satisfies the correct commutation relations for the Lorentz group:

$$[M^{\mu\nu}, M^{\alpha\beta}] = i\eta^{\mu\alpha}M^{\nu\beta} + \cdots \tag{2.3.14}$$

However, Lorentz invariance has to be checked once again in light of the fact that we have explicitly eliminated all ghost modes. Most of the commutators are trivial to check, because they are linear. The troublesome term comes from $X^-$, which is highly nonlinear and is written as

$$X^- = x^- + p^-\tau + i \sum_{n\neq 0}^{\infty} \frac{1}{n}\alpha_n^- e^{-in\tau} \cos n\sigma \tag{2.3.15}$$

where

$$\alpha_n^- = \frac{1}{2p^+} \sum_{i=1}^{D-2} \sum_{m=-\infty}^{\infty} [:\alpha_{n-m}^i \alpha_m^i: - 2a\delta_{n,0}] \tag{2.3.16}$$

and $a$ is the intercept. All commutators are easily calculated, except for the one involving $X^-$. The calculation, which is lengthy and must be performed carefully, yields the final answer:

$$[M^{-i}, M^{-j}] = \frac{-1}{p^{+2}} \sum_{n=1}^{\infty} [\alpha_{-n}^i \alpha_n^j - \alpha_{-n}^j \alpha_n^i] \Delta_n \tag{2.3.17}$$

where

$$\Delta_n = \frac{n}{12}(26 - D) + \frac{1}{n}\left[\frac{D - 26}{12} + 2 - 2a\right] \tag{2.3.18}$$

For this commutator to vanish, we must have

$$D = 26; \qquad a = 1 \tag{2.3.19}$$

This fixes both the dimension of space–time $D$ and the intercept $a$ for the model.

## §2.4. BRST Quantization

As in the point particle case, the **BRST** quantization method starts with the Faddeev–Popov quantization prescription and then extracts out a new nilpotent symmetry operator. The action is invariant under a reparametrization (2.1.36):

$$\begin{aligned} \delta g_{ab} &= g_{ac}\partial_b \delta v^c + \partial_a \delta v^c g_{cb} - g_{ab}\partial_c \delta v^c \\ &= \nabla_a \delta v_b + \nabla_b \delta v_a \end{aligned} \tag{2.4.1}$$

(the second line has been written covariantly and contains two-dimensional Christoffel symbols; see Appendix). This allows us to impose the gauge constraint:

$$g_{ab} = \delta_{ab} \tag{2.4.2}$$

This, in turn, generates a Faddev–Popov determinant:

$$\Delta_{\mathrm{FP}} = \det(\nabla_a) = \det(\nabla_z)\det(\nabla_{\bar{z}}) \tag{2.4.3}$$

This determinant can be calculated by exponentiating it into the action. To do this, we must introduce anticommuting ghost fields $b$ and $c$ (see (1.6.22)). As before, these ghost fields make it possible to place the determinant of a derivative in the exponential. Let us define $z = \tau + i\sigma$; then:

$$L = \frac{1}{\pi}\left(\tfrac{1}{2}\partial_z X_\mu \partial_{\bar{z}} X^\mu + b\partial_{\bar{z}} c + \bar{b}\partial_z \bar{c}\right) \tag{2.4.4}$$

Notice that this action is invariant under

$$\begin{aligned}
\delta X_\mu &= \varepsilon[c\partial_z X_\mu + \bar{c}\partial_{\bar{z}} X_\mu] \\
\delta c &= \varepsilon[c\partial_z c] \\
\delta \bar{c} &= \varepsilon[\bar{c}\partial_{\bar{z}} \bar{c}] \\
\delta b &= \varepsilon[c\partial_z b + 2\partial_z c b - {}^1\!/\!_2\partial_z X_\mu \partial_z X^\mu] \\
\delta \bar{b} &= \varepsilon[\bar{c}\partial_{\bar{z}} \bar{b} + 2\partial_{\bar{z}} \bar{c} \bar{b} - {}^1\!/\!_2\partial_{\bar{z}} X_\mu \partial_{\bar{z}} X^\mu]
\end{aligned} \tag{2.4.5}$$

From this variation and (1.9.8), we can extract out the nilpotent BRST operator $Q$. However, it is also important to note that in general, given *any* Lie algebra with commutation relations $[\lambda_m, \lambda_n] = f_{mn}^p \lambda_p$, it is possible to construct a nilpotent operator $Q$ [10] out of anticommuting operators $c_n$ and $b_m$:

$$Q = \sum_{n=-\infty}^{\infty} c_{-n}[\lambda_n - \tfrac{1}{2} f_{nm}^p c_{-m} b_p] \tag{2.4.6}$$

where

$$\{c_n, b_m\} = \delta_{n,-m} \tag{2.4.7}$$

Thus, our nilpotent BRST operator can be written in this form:

$$\begin{aligned}
Q &= \sum_{n=-\infty}^{\infty} :c_{-n}(L_n^X + \tfrac{1}{2} L_n^{\mathrm{gh}} - a\delta_{n,0}): \\
&= c_0(L_0 - a) + \sum_{n=1}^{\infty} [c_{-n}L_n + L_{-n}c_n] - \tfrac{1}{2} \sum_{n,m=-\infty}^{\infty} :c_{-m}c_{-n}b_{n+m}:(m-n)
\end{aligned} \tag{2.4.8}$$

where $L_n^X$ equals the $X$-dependent Virasoro generator, and $L_n^{\mathrm{gh}}$ is the ghost contribution to the generator. At this point, there are two unspecified parameters in the above equation, the value of the intercept $a$ and the dimension of space–time. Let us calculate the square of $Q$, which should be zero:

$$Q^2 = \tfrac{1}{2} \sum_{m=-\infty}^{\infty} \left(\frac{D}{12}(m^3 - m) + \tfrac{1}{6}(m - 13m^3) + 2am\right) c_m c_{-m} \tag{2.4.9}$$

For this to vanish, we must fix the dimension of space–time to be 26 and the intercept to be equal to 1.

As in the point particle case, we find that the physical states of the theory are given by

$$Q|\text{phy}\rangle = 0 \tag{2.4.10}$$

When we separate out the modes, we find that the lowest state satisfies

$$\begin{aligned}(L_0 - 1)|\phi\rangle &= 0\\ L_n|\phi\rangle &= 0\end{aligned} \tag{2.4.11}$$

as before.

## §2.5. Trees

So far we have quantized only free strings. The interaction of strings can be determined either through functional methods or through harmonic oscillator methods. Of the two, functional methods are by far the more powerful. In fact, we can view the harmonic oscillator method as just one specific representation of the functional method. For trees and even the first loop, oscillator methods provide a fast and convenient method of calculating the amplitudes, but for higher loops the oscillator method rapidly becomes impractical. The functional method, because of its vast versatility, can always be used either to derive the oscillator method for trees and the first loop, or to derive all loops.

Historically, the dual model was first discovered accidentally as a quantum amplitude for the scattering of tachyons. These amplitudes possessed a remarkable property called *duality*, meaning that they could be factorized in terms of $s$-channel poles *or* $t$-channel poles. Usually, in the theory of point particles, the Feynman diagrams sum the $s$- and $t$-channel poles separately (see Fig. 2.5), but the dual amplitudes already possessed poles in both channels simultaneously. Thus, it would be overcounting to include dual amplitudes for the $s$- and $t$-channel poles separately. For more complicated diagrams, we see that the same $N$-point amplitude could be factorized in any number of ways. Because the counting of these diagrams was not the usual counting of Feynman diagrams, for years it was erroneously believed that a true field theory interpretation of the dual model was not possible.

Following the analogy with the point particle case, let us begin by defining the $N$-point scattering amplitude for tachyon scattering [11–13]:

$$\begin{aligned}A_N(k_1, k_2, \ldots, k_N) &= \sum_{\text{topologies}} \int d\mu_N \int Dg_{ab} \int DX\\ &\quad \times \Delta_{\text{FP}} \exp i \int L\, d\sigma\, d\tau \left\{ \prod_{i=1}^{N} \sqrt{g} \exp ik_{i,\mu} X^{\mu i} \right\}\\ &= \sum_{\text{topologies}} \int Dg_{ab} \int d\mu \left\langle \prod_{i=1}^{N} \sqrt{g} \exp ik_{i,\mu} X^{\mu i} \right\rangle\end{aligned} \tag{2.5.1}$$

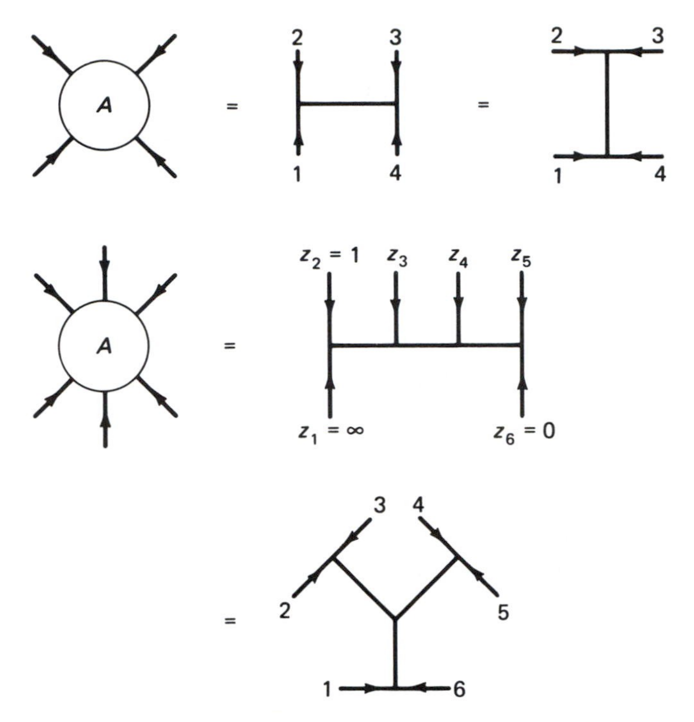


Figure 2.5. Duality of the Veneziano model. The four-point amplitude can be decomposed in terms of either $s$-channel or $t$-channel poles. This is in contrast to standard point particle field theory, which sums over both $s$- and $t$-channel poles. This property of duality extends to the $N$-point function. For this reason, it was once thought that a field theory of strings was impossible. A field theory of strings would be plagued by overcounting, especially at higher orders.

This expression will be the fundamental path integral from which we will derive the theory of interacting strings. *It will be the single most important formula in the first quantized formulation.*

In the expression, momentum $k_i$ from the $i$th tachyon flows into the boundary of the surface at a point $z_i$. The string variable $X^{\mu i}$ is defined at that $i$th point $z_i$ on the boundary where momentum $k_i$ flows into the diagram. The measure of integration $d\mu$ is an integral over the various $z_i$, which we will determine shortly.

This expression simplifies considerably if we take the conformal gauge. In particular, we get

$$\begin{aligned} A_N &= \int d\mu \int DX \exp\left\{\frac{-1}{4\pi\alpha'}\int (\partial_z X_\mu \partial_{\bar{z}} X^\mu)\, d^2z + i\sum_{i=1}^{N} k_{i\mu} X^\mu\right\} \\ &= \sum_{\text{Topologies}} \int d\mu \left\langle \prod_{i=1}^{N} e^{ik_i\cdot X}\right\rangle \end{aligned} \tag{2.5.2}$$

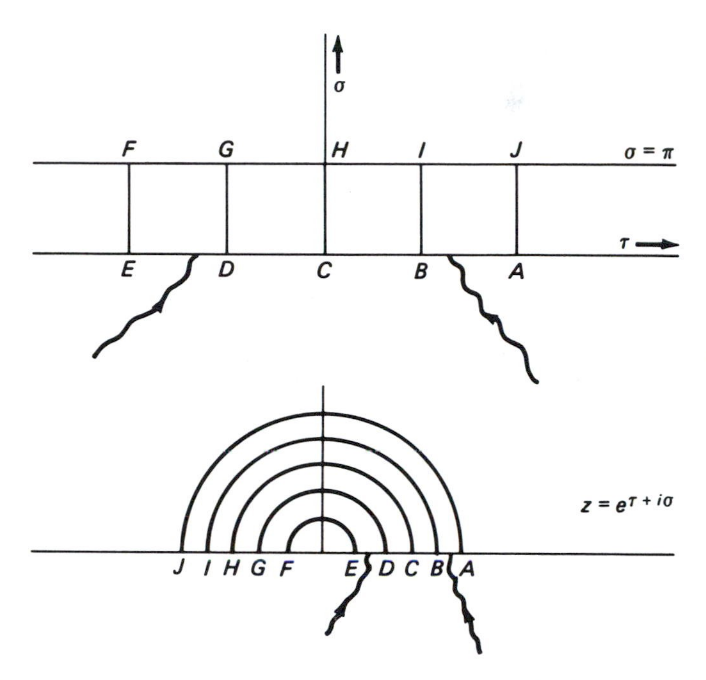


Figure 2.6. Conformal surfaces for open string propagation. In the $\rho$ plane, the surface over which a string propagates is a horizontal strip of width $\pi$. The wavy lines at the bottom corresponds to "zero width" strings or external tachyons. In the $z$-plane, the surface becomes the upper half complex plane. The mapping from one surface to the other is given by the exponential.

In Fig. 2.6 we show how to simplify the string interaction diagram. By letting the string interaction length to go to zero for the external tachyons, we see that the string interaction surface can be reduced to an infinite horizontal strip in the complex plane, extending from $\sigma = 0$ to $\sigma = \pi$ and $\tau = \pm\infty$.

Notice that the action, although it is no longer reparametrization invariant because we have fixed $g_{ab} = \delta_{ab}$, is still conformally invariant. Thus, to avoid overcounting of conformally equivalent surfaces, we will take the set of topologies over which we must sum to be the set of all *conformally inequivalent* two-dimensional complex surfaces.

Let the world sheet of the $N$-point tree be a horizontal strip of width $\pi$ that extends horizontally in the complex plane. The $x$-axis corresponds to $\tau$ and the $y$-axis corresponds to $\sigma$. Let us now change coordinates to complex variables:

$$z = e^{\tau + i\sigma} \tag{2.5.3}$$

This mapping takes this infinite strip, which describes the world sheet of the interacting string (with zero width tachyons), into the upper half of the complex plane.

Fortunately, the functional integral is a Gaussian that can be evaluated with the identities presented in the previous chapter. Let us define $X_{\text{classical}}$ as the solution to the classical equations of motion:

$$\nabla^2 X_{\mu,\text{cl}} = -2i\pi\alpha' J_\mu \tag{2.5.4}$$

where

$$J_\mu(z) = \sum_{i=1}^{N} k_{i\mu}\delta(z - z_i) \tag{2.5.5}$$

where $z_i$ are points on the real axis of the complex plane that correspond to the external zero width tachyons interacting with the string. After a Wick rotation in the $\tau$ variable, this is just Poisson's equation for electrostatics. To solve this, we need the Green's function:

$$\nabla^2 G(z, z') = 2\pi\delta(z - z') \tag{2.5.6}$$

We must calculate the Green's function, with Neumann boundary conditions, for the upper half-plane. The easiest way to calculate this is to borrow a trick from the theory of electrostatics, namely the method of images. Let us place a point charge at the point $z'$ in the upper half-plane. Consider another point charge at the point $\bar{z}'$ that is symmetrically reflected through the $x$-axis; $\bar{z}'$ is in the lower half-plane. If we are sitting on a point charge $z$ in the upper half-plane, then the potential at that point is proportional to

$$G(z, z') = \ln|z - z'| + \ln|z - \bar{z}'| \tag{2.5.7}$$

Notice that if we are sitting on the $x$-axis so that $z$ is real, then the derivative of the Green's function normal to the $x$-axis is zero. Thus, these boundary conditions are precisely what we want, so (by the uniqueness theorem) this is the Green's function for the upper half-plane.

We can now insert this Green's function back into the integral. The classical value of $X$ that solves (2.5.4) is

$$X_{\text{cl}} = -i\alpha' \int G(z, z')J(z')\, dz' \tag{2.5.8}$$

Let us now make a shift in the integration variable:

$$X_\mu \to X_{\mu,\text{cl}} + X_\mu \tag{2.5.9}$$

We find, therefore, that the functional integrals can be performed using (1.7.10):

$$\begin{aligned} &\int DX \exp\left\{-\frac{1}{4\pi\alpha'}\int \partial_z X_\mu \partial_{\bar{z}} X^\mu\, d^2z + i\int J_\mu X^\mu\, d^2z\right\} \\ &= \exp\left\{\frac{\alpha'}{2}\int J_\mu(z)G(z, z')J^\mu(z')\, dz\, dz'\right\} \\ &= \prod_{i \neq j} \exp\{\alpha' k_i \cdot k_j \ln|z_i - z_j|\} \\ &= \prod_{i<j} |z_i - z_j|^{2\alpha' k_i \cdot k_j} \end{aligned} \tag{2.5.10}$$

where:

$$J^\mu(z) = \sum_j \delta(z - z_j) k_j^\mu$$

Putting everything together in (2.5.2), we find

$$A_N = \int d\mu \prod_{1 \le i < j \le N} |z_i - z_j|^{2\alpha' k_i \cdot k_j} \tag{2.5.11}$$

(Notice that we have explicitly removed the "self-energy" term for $i = j$, which would be divergent. We can truncate the integral and still maintain the conformal properties of the theory. We will find that this truncation will also have to be performed in the harmonic oscillator method.)

Now we must complete the last step, which is to fix the measure $d\mu$.

Our first guess is that, when the amplitude is expressed in terms of $z_i$, the measure is simply equal to one. This is the correct choice that is compatible with conformal invariance. To prove it, recall that earlier we said we must sum over *all conformally inequivalent surfaces.* Consider the set of conformal transformations that map the upper half-plane into itself, such that the real axis is mapped into itself. In general, the points on the real axis that are mapped into each other transform under a subset of the conformal transformations, the *projective* or Möbius transformations:

$$y' = \frac{ay + b}{cy + d} \tag{2.5.12}$$

for real $a$, $b$, $c$, and $d$ such that $ad - bc = 1$. This set of four parameters defines a real matrix with unit determinant:

$$\begin{pmatrix} a & b \\ c & d \end{pmatrix} \tag{2.5.13}$$

In general, the group defined by the set of all real $2 \times 2$ matrices that have unit determinant is SL(2, $R$) (see Appendix). Notice that this group of transformations can be generated by making successive transformations:

$$\begin{cases} y \to y + b \\ y \to ay \\ y \to \dfrac{1}{y} \end{cases} \tag{2.5.14}$$

Thus, we wish the amplitude, including the contribution of the measure, to be projectively invariant.

Let us make a projective transformation on the integrand to see how it transforms:

$$\prod_{i<j} (z_i - z_j)^{2\alpha' k_i \cdot k_j} = \prod_{i<j} (z_i' - z_j')^{2\alpha' k_i \cdot k_j} \prod_k (a - cz_k')^2 \tag{2.5.15}$$

We want our measure to cancel out the noninvariant term in the above expression. Let us take our measure to be the number 1 and the integration region to be fixed by $z_i \geq z_{i+1}$. Then there is one last complication. We must still "fix the gauge" for projective transformations or else we will have overcounting. We must integrate once and only once over each projectively distinct configuration of the $z_i$ variables. If the external momentum flows into the upper half-plane at points given by $z_i$, then we are allowed to fix three of these points at random. This corresponds to "gauge fixing" the projective invariance, which selects out only projecitvely inequivalent parametrizations. Our final result is

$$d\mu = \frac{\theta(z_i - z_{i+1})}{dV_{abc}} \prod_{i=1}^{N} dz_i \tag{2.5.16}$$

where we have explicitly removed the contribution from the three fixed points $z_{a,b,c}$:

$$dV_{abc} = dz_a\, dz_b\, dz_c (z_a - z_b)^{-1} (z_b - z_c)^{-1} (z_c - z_a)^{-1} \tag{2.5.17}$$

where $z_{a,b,c}$ are three points that can be *randomly* chosen on the real axis. Notice that we have fixed three variables so that we only integrate over projectively inequivalent configurations. (If we integrated over these three variables, we would have overcounting of the integration region. We would be integrating over an infinite number of copies of the same thing.)

The simplest choices for the three fixed points are

$$\begin{aligned} z_1 &= \infty \\ z_2 &= 1 \\ z_N &= 0 \end{aligned} \tag{2.5.18}$$

In this configuration, our final result for the $N$-point amplitude becomes [14–19]:

$$A_N = \int \prod_{i=3}^{N-1} dz_i \prod_{2 \leq i < j \leq N} (z_i - z_j)^{2\alpha' k_i \cdot k_j} \tag{2.5.19}$$

where the region of integration is

$$\infty = z_1 \geq z_2 = 1 \geq z_3 \cdots z_{N-1} \geq z_N = 0 \tag{2.5.20}$$

This is our final result for the $N$-point amplitude, which we derived using only functional techniques.

Let us summarize what we have done:

(1) We have taken the string lengths to be zero for the external tachyons, so the world sheet of the $N$-point amplitude was a horizontal strip in the complex plane that has width $\pi$. Momentum from the external tachyons flowed into this strip at designated points along the real axis $z_i$ (see Fig. 2.6).
(2) By making a conformal transformation from the strip to the upper half complex plane, we explicitly solved for the Neumann function, which we calculated using a trick from electrostatics, the method of images.
(3) Projective invariance fixed the measure of integration to be the number 1. We ordered the external points $z_i \geq z_{i+1}$.

(4) We "fixed the gauge" left over from projective transformations; i.e., we fixed three of the $N$ points of the integrand.

The previous expression for the $N$-point amplitude (2.5.19) considered only the case with external tachyons entering the world sheet at selected points on the boundary. In principle, we can have particles with arbitrary spin on the boundary. For higher spins, we have the same factor $e^{ik\cdot X}$, which represents part of the Fourier transform, multiplied by the polarization tensor of the higher spins. Thus, the vertex function for the tachyon is not actually $e^{ik\cdot X}$, which is universal for all spins because it is part of the Fourier transform. The true vertex for the tachyon is actually just the number one.

Spin-two vertices can be represented as

$$V = \sqrt{g}g^{ab}\partial_a X^\mu \partial_b X^\nu \varepsilon_{\mu\nu} e^{ik_{i\mu}X^\mu} \tag{2.5.21}$$

where $\varepsilon_{\mu\nu}$ is the polarization tensor. For the general case:

$$V = \sqrt{g}g^{[a_1 a_2}\cdots g^{a_{2m-1}a_{2m}]}\partial_{a_1} X^{\mu_1}, \ldots, \partial_{a_{2m}} X^{\mu_{2m}} \varepsilon_{\mu_1,\ldots,\mu_{2m}} e^{ik_\mu X^\mu} \tag{2.5.22}$$

## §2.6. From Path Integrals to Operators

We used functional methods to calculate the $N$-point scattering amplitude. The only nontrivial part of the calculation was the integration over conformally inequivalent two-dimensional complex surfaces, which determined the measure of integration of the amplitude.

The same calculation can also be performed using the harmonic oscillator formalism, which, as we have stressed, is nothing but a particular representation of the path integral for which the Hamiltonian is diagonal. For trees and the first loop, the harmonic oscillator method is quite simple because the Hamiltonian is diagonal on the Fock space of the harmonic oscillators. However, for higher loops this is no longer true: the harmonic oscillator method becomes increasingly difficult and impractical. The path integral method thus provides the only systematic way in which to analyze the higher loop amplitudes with relative ease. (The calculation of anomalies, however, is easier in the harmonic oscillator formalism, where the integer index $n$ serves as a cutoff for the theory. In the path integral formalism, one must use point-splitting techniques and other regulators, as we shall see in Chapter 5.)

In the functional formalism, we know that the propagator for free strings is given by

$$\langle X_a| \int_0^\infty d\tau\, e^{-\tau H} |X_b\rangle = \langle X_a| \frac{1}{L_0 - 1} |X_b\rangle \tag{2.6.1}$$

Similarly, we also know that, by inserting a complete set of intermediate states within the path integral, the vertex function for the $i$th tachyon is

$$\langle X_a| e^{ik_{i\mu}X_i^\mu} |X_b\rangle \tag{2.6.2}$$

where

$$X_\mu^i = X_\mu(\sigma = 0, \tau = \tau_i) \tag{2.6.3}$$

Because of the identity

$$|X\rangle \int DX\,\langle X| = 1$$

we can remove all path integrals from the $N$-point function and make the transition from the path integral formalism to the harmonic oscillator formalism. For example, let us begin with the path integral expression for the $N$-point tachyon amplitude and make the substitution to harmonic oscillators:

$$\begin{aligned}
A_N &= \int DX\, d\mu\, e^{-S} \prod_{i=1}^{N} e^{ik_i X(\tau_i)} \\
&= \int d\mu\, e^{-S} \left\{ \cdots e^{ik_j X(\tau_j)} |X_j\rangle \int DX_j\, \langle X_j| e^{ik_{j+1} X(\tau_{j+1})} \cdots \right\} \\
&= \int d\mu \left\{ \cdots e^{ik_j X(0)} |X_j\rangle \int DX_j\, \langle X_j| e^{-(\tau_{j+1}-\tau_j)H} e^{ik_{j+1} X(0)} \cdots \right\} \\
&= \int \prod_i d\bar{\tau}_i \langle 0, k_1 | e^{ik_2 X(0)} e^{-\bar{\tau}_1 H} e^{ik_3 X(0)} \cdots |0, k_N\rangle \\
&= \langle 0, k_1 | V(k_2) D V(k_3) \cdots V(k_{N-1}) |0, k_N\rangle
\end{aligned}$$

Several points in this derivation must be clarified. First, in the path integral formalism, we found that we had to omit the $i = j$ contribution to the integrand in (2.5.10). Similarly, we also have to truncate the harmonic oscillator formalism. The expression for the vertex function (2.6.2) is formally infinite when we convert to harmonic oscillators. For example, if we naively take the vacuum expectation value of the exponential, we arrive at a divergent sum. The normal ordered expression is the one we desire, which has finite matrix elements. Let us exponentiate (2.2.11):

$$:\exp ik_{i\mu} X^{\mu i} (\sigma = 0, \tau_i): = \exp\left\{ k \cdot \sum_{n=1}^{\infty} \frac{\alpha_{-n}}{n} e^{in\tau_i} \right\} e^{ik \cdot x(\tau_i)} \exp\left\{ -k \cdot \sum_{n=1}^{\infty} \frac{\alpha_n}{n} e^{-in\tau_i} \right\} \tag{2.6.4}$$

(Normal ordering consists of moving all creation (destruction) operators to the left (right) so that the resulting operator has finite matrix elements.)

Second, in the transition from path integrals to operators we used the fact that the Hamiltonian is the generator of $\tau$ shifts. The expression $L_0 - 1$ acts as the effective Hamiltonian for horizontal displacements in the complex plane. We will find the following formula useful:

$$y^{L_0} f(\alpha_n) y^{-L_0} = f(\alpha_n y^{-n}) \tag{2.6.5}$$

for *any* function $f$. Thus, we can define

$$y^{L_0} V_0(k_i) y^{-L_0} = V(k_i, y_i = e^{-\tau_i}) \tag{2.6.6}$$

We used this expression in the transition from path integrals to harmonic oscillators when we converted the vertex function at $\tau_i$ to the vertex function at the origin.

Third, in our derivation of the harmonic oscillator formalism we introduced the vacuum state $|0, k\rangle$. Notice that the functional formalism started with the world sheet of the interacting strings being an infinite horizontal strip of width $\pi$ in the complex plane. The external lines were placed along the real axis. Therefore the effect of a string coming in from negative (positive) infinity corresponds to the functional integral over a semi-infinite strip. Thus, the vacuum state $|0\rangle$ is the functional integral of the string theory over a semi-infinite strip. We can represent the tachyon vacuum with momentum $k$ as:

$$|0; k\rangle = e^{ik\cdot x}|0; 0\rangle \tag{2.6.7}$$

where

$$\alpha_n|0; 0\rangle = 0; \qquad n \geq 0 \tag{2.6.8}$$

Putting everything together, we can now convert the path integral to harmonic oscillator form:

$$\begin{aligned} A_N &= \int d\mu \int DX\, e^{-s} \prod_{i=3}^{N-1} dz_i \exp ik_i X^{\mu i}(\sigma = 0; \tau_i) \\ &= \langle 0; k_1 | V_0(k_2) D V_0(k_3) D \cdots D V_0(k_{N-1})|0; k_N\rangle \end{aligned} \tag{2.6.9}$$

It is important to realize that the harmonic oscillator expression is a direct consequence of the functional formalism.

Fortunately, this amplitude is easy to calculate. As an exercise, let us first calculate the four-point function explicitly:

$$\begin{aligned} A_4 &= \langle 0; k_1 | V(k_2) D V(k_3) | 0; k_4\rangle \\ &= \int_0^1 dx\, x^{k_3\cdot k_4 - 1} \langle 0| \exp\left\{-k_2 \cdot \sum_{n=1}^{\infty} \frac{\alpha_n}{n}\right\} x^R \exp\left\{k_3 \cdot \sum_{n=1}^{\infty} \frac{\alpha_{-n}}{n}\right\}|0\rangle \\ &= \int_0^1 dx\, x^{-(1/2)s-2}(1-x)^{-(1/2)t-2} \\ &= \frac{\Gamma(-\alpha(s))\Gamma(-\alpha(t))}{\Gamma(-\alpha(s) - \alpha(t))} \\ &= B(-\alpha(s), -\alpha(t)) \end{aligned} \tag{2.6.10}$$

where

$$\begin{aligned} \alpha(s) &= 1 + \tfrac{1}{2}s \\ s &= -(k_1 + k_2)^2 \\ t &= -(k_2 + k_3)^2 \\ u &= -(k_1 + k_3)^2 \end{aligned} \tag{2.6.11}$$

are the Mandelstam variables and $R$ is the number operator: $R = \sum_{n=1}^{\infty} n a_n^\dagger a_n$.

Thus, the four-point scattering amplitude has an elegant form. In fact, it can be represented as the Euler Beta function. This Beta function has fascinating physical properties that first excited the imagination of high-energy theorists in the late 1960s. It was discovered by G. Veneziano and M. Suzuki quite by accident as they were looking for a way in which to satisfy finite energy sum rules for the hadronic $S$-matrix. At that time, theorists wanted to construct a scattering amplitude for hadrons that had the following of G. Chew's criteria for an $S$-matrix:

(1) Unitarity
(2) Lorentz symmetry
(3) *CPT* invariance
(4) Analyticity
(5) Crossing symmetry

To this list, theorists sometimes added:

(6) Regge behavior, i.e.,

$$A(s, t) \sim s^{\alpha(t)} \tag{2.6.12}$$

for large $s$ and fixed $t$, and also

(7) Duality, i.e.,

$$A(s, t) = \sum_J \frac{c_J(t)}{s - M_J^2} = \sum_J \frac{c_J(s)}{t - M_J^2} \tag{2.6.13}$$

The list of "axioms" for the hadronic $S$-matrix was so large that physicists were pessimistic that they could ever be satisfied. Then, when the Euler Beta function was accidentally discovered while thumbing through a math book, these two young physicists were surprised that this analytic function satisfied *all but one* of these axioms! For example, Regge behavior can be shown by using Sterling's approximation on the $\Gamma$ function:

$$\Gamma(x) \approx \sqrt{2\pi}\, \frac{x^x}{\sqrt{x}} e^{-x} \tag{2.6.14}$$

Duality can also be shown by evaluating the poles of the integrand near the limits of integration $x = 0$ and $x = 1$:

$$A(s, t) \sim \frac{1}{n!} \frac{(\alpha(t) + 1)(\alpha(t) + 2) \cdots (\alpha(t) + n)}{\alpha(s) - n} \tag{2.6.15}$$

A similar expression exists for the $t$-channel poles. In fact, all the postulates but the first, unitarity, can be satisfied. (Unitarity is violated for the simple reason that the analytic structure of the amplitude is a series of poles in the $s$ and $t$ plane. A true unitary amplitude, however, should also have imaginary parts instead of poles and cuts along the real axis, as we shall see in Chapter 5.)

Using operator methods, we can now solve for the $N$-point function.

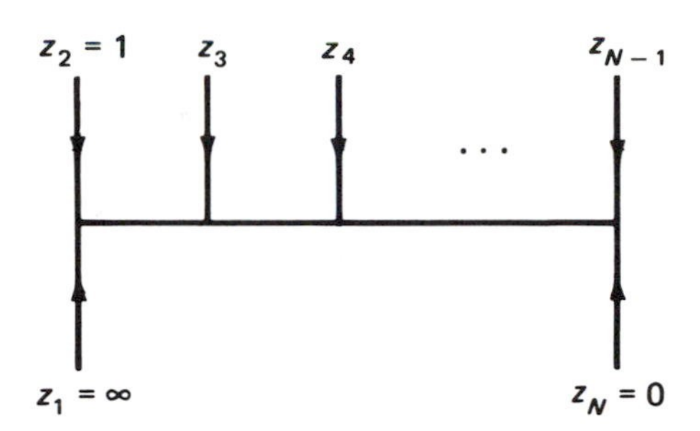


Figure 2.7. The $N$-point function. Projective invariance allows us to fix three of the $N$ variables. The most convenient parametrization is to take the first variable to be infinity, the second to be 1, the last to be 0, and all the others to be ordered and between 1 and 0.

The $N$-point amplitude can be written as (see Fig. 2.7):

$$A_N = \langle 0; k_1 | V(k_2) D \cdots V(k_{N-1}) | 0; k_N \rangle$$
$$= \int_0^1 \prod_{i=3}^{N-1} \frac{dx_i}{x_i} \langle 0; k_1 | V(k_2, 1) V(k_3, y_3) V(k_4, y_4) \cdots V(k_{N-1}, y_{N-1}) | 0; k_N \rangle \tag{2.6.16}$$

where

$$\prod_{i=3}^{N-1} \frac{dx_i}{x_i} = \prod_{i=3}^{N-1} \frac{dy_i}{y_i} \tag{2.6.17}$$
$$y_i = x_3 x_4 \cdots x_i$$

(Notice that we commuted the factors of $x^R$ to the right, until they annihilated on the vacuum.) This expression can most easily be simplified using the coherent state formalism. Let an arbitrary state of the Fock space be represented by

$$|\lambda\rangle = \sum_{n=0}^{\infty} \frac{\lambda^n}{n!} (a^\dagger)^n |0\rangle = e^{\lambda a^\dagger} |0\rangle \tag{2.6.18}$$

Then we have

$$\langle \mu | \lambda \rangle = e^{\mu^* \lambda}$$
$$x^{a^\dagger a} |\lambda\rangle = |x\lambda\rangle \tag{2.6.19}$$
$$e^{\mu a^\dagger} |\lambda\rangle = |\lambda + \mu\rangle$$

By using these identities in succession, we find

$$\langle 0; k_1 | \prod_{i=2}^{N-1} \frac{V(k_i, y_i)}{y_i} | 0; k_N \rangle = \prod_{i<j} (y_i - y_j)^{2\alpha' k_i \cdot k_j} \tag{2.6.20}$$

Thus, we find the same expression for the $N$-point function (2.5.19) that we derived using functional methods:

$$A_N = \int \prod_{i=3}^{N-1} dy_i \prod_{2 \le i < j \le N} (y_i - y_j)^{2\alpha' k_i \cdot k_j} \tag{2.6.21}$$

where the $y$'s are ordered along the real axis as before. Notice that we have already fixed the values of three points along the real axis to be 0, 1, and $\infty$, as expected.

Using oscillator methods, we can show that the $N$-point function is cyclically symmetric, which is not obvious when we write the amplitude as the sequence $VDVDVDVDV$. We use the following identity, which expresses what happens when two vertices are pushed past each other:

$$V(k_1, y_1)V(k_2, y_2) = V(k_2, y_2)V(k_1, y_1)\exp[2\alpha' \pi i k_i \cdot k_j \varepsilon(y_1 - y_2)] \tag{2.6.22}$$

where $\varepsilon(x) = 1$ if $x > 0$ and $\varepsilon(x) = -1$ if $x < 0$.

We also have to rewrite the vacuum $|0; k_N\rangle$ in a form where it resembles the other vertices, so we can calculate the properties of the amplitude under a cyclic rearrangement of the vertices. We will use the following identities:

$$\begin{aligned} \lim_{y=0} \frac{V(k, y)}{y} |0; 0\rangle &= |0; k\rangle \\ \lim_{y=\infty} \langle 0; 0| yV(k, y) &= \langle k; 0| \end{aligned} \tag{2.6.23}$$

Written in this fashion, the tachyon states at the extreme right and left are no longer treated unsymmetrically from the others. The vertex on the right is now defined at 0, and the vertex on the left is defined at $\infty$, and all other vertices occur within that interval.

Thus, when we push the $N$th vertex completely around the amplitude, we get

$$\begin{aligned} &\langle V(k_1, y_1) \cdots V(k_N, y_N)\rangle \\ &\quad = \langle V(k_N, y_N)V(k_1, y_1) \cdots V(k_{N-1}, y_{N-1})\rangle \exp\left[2\alpha' i \pi k_N \cdot \sum_{i=1}^{N-1} k_i \varepsilon\right] \end{aligned} \tag{2.6.24}$$

Notice that the last factor vanishes if we have conservation of momentum:

$$\sum_{i=1}^{N} k_i = 0 \tag{2.6.25}$$

where we have also used the on-shell condition $\alpha' k_N^2 = 1$. We see that only *on-shell* do we have an amplitude that is cyclically symmetric. (Historically, this was another reason why physicists believed that a field theory of strings was probably not possible. A field theory, by definition, is an off-shell formulation, while the marvelous features of the Veneziano model worked only on-shell.)

## §2.7. Projective Invariance and Twists

In addition to cyclic symmetry, we say that the integrand of the $N$-point function was also Möbius invariant; i.e., the function remains the same if we make the following change of variables:

$$y' = \frac{ay + b}{cy + d} \tag{2.7.1}$$

where

$$ad - bc = 1 \tag{2.7.2}$$

We showed Möbius invariance in the functional formalism. Now let us make a Möbius transformation directly on the harmonic oscillator matrix elements:

$$\begin{aligned} &\langle 0; 0| \frac{V(k_1, y_1)}{y_1} \cdots \frac{V(k_N, y_N)}{y_N} |0; 0\rangle \\ &= \langle 0; 0| \frac{V(k_1, y_1')}{y_1'} \cdots \frac{V(k_N, y_N')}{y_N'} |0; 0\rangle \prod_{i=1}^{N} (a - cy_i')^2 \end{aligned} \tag{2.7.3}$$

As we saw earlier, this factor, in turn, can be canceled by the transformation in the measure:

$$d\mu \to d\mu \prod_i (a - cy_i')^{-2} \tag{2.7.4}$$

It turns out that we can actually calculate the generator of these infinitesimal transformations on the vertex function. Not surprisingly, the generator of these Möbius transformations will be the Virasoro generators. Let us define

$$\begin{aligned} T &= 1 - \varepsilon \sum_n \alpha_n L_n \\ z &= z' + \varepsilon \sum_n \alpha_n z^{n+1} \end{aligned} \tag{2.7.5}$$

Then

$$\begin{aligned} TV(k_i, z_i)T^{-1} &= \left(1 - \varepsilon\alpha' k_i^2 \sum_n n\alpha_n z^n\right) V(k_i, z_i') \\ [L_n, V(k_i, z_i)] &= z^n \left[ z\frac{d}{dz} + n\alpha' k_i^2 \right] V(k_i, z_i) \end{aligned} \tag{2.7.6}$$

We say that the vertex $V$ has *conformal weight* equal to $\alpha' k_i^2$. (The conformal weight will play an important part in the proof that the string model is ghost-free. In Chapter 4, we will explain the origin of the conformal weight, which labels irreducible representations of the conformal group generated by the Virasoro $L_n$. For a vertex function to be properly defined, it must have weight 1 on-shell, which is true for the tachyon.)

Clearly, the $L_n$ generate conformal transformations when acting on the vertex functions. In fact, we can show that a specific representation of the Virasoro algebra on conformal fields is given by

$$L_n = -z^{n+1}\partial_z$$

This representation satisfies the definition of the conformal algebra in (2.2.27) (minus the central term) and generates conformal transformations in functions of the complex variable $z$.

We are interested in the subgroup of the conformal transformations that maps the upper half-plane into itself and the real axis into itself, i.e., the projective subgroup. This subgroup is generated by only $L_1$, $L_{-1}$, $L_0$, which generates SL(2, $R$):

$$\text{SL}(2, R): \quad \begin{cases} [L_1, L_0] = L_1 \\ [L_{-1}, L_0] = -L_{-1} \\ [L_1, L_{-1}] = 2L_0 \end{cases} \tag{2.7.7}$$

We can easily calculate the transformation of a vertex function induced by these generators:

$$\begin{aligned} e^{aL_1}V(y)e^{-aL_1} &= V[y(1-ay)^{-1}] \\ e^{bL_0}V(y)e^{-bL_0} &= V[e^b y] \\ e^{cL_{-1}}V(y)e^{-cL_{-1}} &= V(y+c) \end{aligned} \tag{2.7.8}$$

If $U$ is an element of SL(2, $R$), then we also have:

$$U|0; 0\rangle = |0; 0\rangle \tag{2.7.9}$$

(At first this might seem surprising, because we know that real states are annihilated by $L_n$ for positive, not negative, integers $n$. However, $|0; 0\rangle$ is not a real state. It is annihilated by $L_{-1}$ because

$$L_{-1}|0; 0\rangle \sim \alpha_0 \cdot \alpha_{-1}|0; 0\rangle = 0 \tag{2.7.10}$$

because $\alpha_0|0; 0\rangle = 0$. Thus, $|0; 0\rangle$ corresponds to the true vacuum of the SL(2, $R$) group, which is not a real state of the theory. When we multiply $|0; 0\rangle$ by $e^{ikx}$, then it becomes a real state (because $x$ has commutation relations with $\alpha_0$) and is annihilated by $L_n$ for positive $n$).

Putting everything together, we have

$$\begin{aligned} &\langle 0; 0| U^{-1}U \frac{V(k_1, y_1)}{y_1} \cdots \frac{V(k_N, y_N)}{y_N} U^{-1}U|0; 0\rangle \\ &= \langle 0; 0| \frac{V(k_1, y_1')}{y_1'} \cdots \frac{V(k_N, y_N')}{y_N'} |0; 0\rangle \prod_{i=1}^{N} (a - cy_i')^2 \end{aligned} \tag{2.7.11}$$

which reproduces (2.7.3).

Besides the Virasoro gauge group, one other feature of string theory is totally missing in the point particle case, and that is the "twist" operator.

Remember that a string sweeps out a two-dimensional world sheet, not just a single line. Thus, if we were to twist the world sheet, a topologically inequivalent world sheet would be swept out. At the first loop level, for example, this is the crucial topological difference between a disk with a hole and a Möbius strip.

A simple expression for the twist operator $\Omega$ [20] can be derived from the observation that a vertex function located on the real axis should convert into a vertex function located on the top of the strip:

$$\Omega V(\sigma = \pi)\, \Omega^{-1} = V(\sigma = 0) \tag{2.7.12}$$

Notice that the only change in the vertex function under this transformation is that every oscillator at the $n$th level is multiplied by $(-1)^n$. Thus, the twist operator must be

$$\Omega = (-1)^{N+1} \tag{2.7.13}$$

where

$$N = \sum_{n=1}^{\infty} \alpha_{-n\mu}\alpha_n^\mu \tag{2.7.14}$$

Notice that this satisfies

$$\Omega^2 = 1 \tag{2.7.15}$$

as it should. This establishes the twist operator up to an overall sign. However, since $N$ even states are charge conjugation even and $N$ odd states are odd under $C$, this fixes the value of $\Omega$ that we have chosen in the expression.

There is an equivalent method of obtaining the form of the twist operator. Notice that the action of the twist operator on a tree is to reverse the orientation of the external lines (see Fig. 2.8)

$$\Omega V_0(k_1) D V_0(k_2) \cdots V_0(k_{N-1})|0; k_N\rangle = V_0(k_N) D V_0(k_{N-1}) \cdots V_0(k_2)|0; k_1\rangle \tag{2.7.16}$$

Notice that the cyclic ordering of the tree has been reversed by the twist operator. To extract the operator that will perform this reversal, let us first

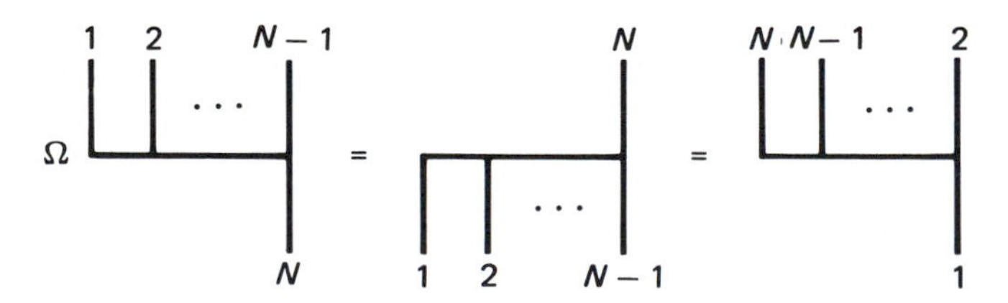


Figure 2.8. Action of the twist operator. The twist operator twists the factorized leg, which is equivalent to flipping the entire diagram upside-down. By duality, however, we can always rewrite the diagram in the original configuration (with external legs renumbered).

write the expression in terms of $y$ variables:

$$\Omega \int \prod_{i=2}^{N-1} dy_i \, V_0(k_1, y_1) \cdots V_0(k_N, y_N)|0; 0\rangle \tag{2.7.17}$$

where we take the limit as $y_1 \to 1$ and $y_N \to 0$.

When written in this fashion, we now see that we can reverse the cyclic ordering of the amplitude and interchange $y_1$ and $y_N$ by a change of variables:

$$y_i' = 1 - y_i$$

Now we wish to write down an operator that will perform this change of variables. Previously, we wrote down the generators of SL(2, $R$) that will perform a projective transformation on a vertex function. By examining this change of variables, we easily find that the twist operator must be given by

$$\Omega = (-1)^R e^{-L_{-1}} \tag{2.7.18}$$

Although the two forms of the twist operator seem totally different, they are actually identical on-shell. Since the Veneziano amplitude is strictly on-shell, we have the freedom of choosing either form of the twist operator.

## §2.8. Closed Strings

So far, our discussion has been specific to open strings, where the external tachyons attach themselves to the endpoints of the conformal strip swept out by the string. Poles emerge in the Veneziano model when two points $z_i$ and $z_j$ along the edge of the strip come close together. Let us now analyze the Shapiro–Virasoro model, which is based on closed rather than open strings and corresponds to a path integral taken on the tube (or sphere) swept out by a closed string. The pole structure of this model is much larger than the original Veneziano function, because external states can attach themselves anywhere along the surface of the tube as it moves in space–time [21]:

$$A(s, t, u) = \frac{\Gamma(-\frac{1}{2}\alpha(s))\Gamma(-\frac{1}{2}\alpha(t))\Gamma(-\frac{1}{2}\alpha(u))}{\Gamma[-\frac{1}{2}(\alpha(t) + \alpha(u))]\Gamma[-\frac{1}{2}(\alpha(u) + \alpha(s))]\Gamma[-\frac{1}{2}(\alpha(s) + \alpha(t))]} \tag{2.8.1}$$

This function, unlike the earlier Veneziano function, has poles simultaneously in all *three* channels, rather than just two channels. This was quickly generalized to the $N$-point function:

$$A_N = \int d\mu \prod_{2 \le i < j \le N} |z_i - z_j|^{(1/2)k_i \cdot k_j} \tag{2.8.2}$$

where

$$d\mu = |z_a - z_b|^2 |z_b - z_c|^2 |z_c - z_a|^2 \frac{\prod_{i=1}^N d^2 z_i}{d^2 z_a \, d^2 z_b \, d^2 z_c} \tag{2.8.3}$$

As in the case of the Veneziano function, the poles in this amplitude emerge when two variables $z_i$ and $z_j$ come close together, except now the poles can occur anywhere in the complex plane, not just on the real axis.

Our starting point for the quantization of the closed string is (2.2.21), where we decompose the string $X_\mu(\sigma)$ and its conjugate into normal modes. The canonical commutation relations remain the same as (2.2.7), which leads to the Hamiltonian given in (2.2.22). As in the open string case, we can factorize the amplitude into vertices and propagators in the harmonic oscillator formalism, but now several important differences emerge:

(1) We now have two sets of mutually commuting harmonic oscillators $\alpha_n$ and $\tilde{\alpha}_n$ to sum over, rather than just one set as in the open string case.
(2) The Virasoro conditions now consist of two sets of conformal generators $L_n$ and $\tilde{L}_n$ acting on the physical states:

$$\begin{gathered} L_n|\phi\rangle = \tilde{L}_n|\phi\rangle = 0 \\ (L_0 - 1)|\phi\rangle = (\tilde{L}_0 - 1)|\phi\rangle = 0 \end{gathered} \tag{2.8.4}$$

(3) We must integrate over all shifts in $\sigma$, because the closed string states should be independent of where we chose the origin of the $\sigma$ coordinate.
(4) The amplitude is not just the sequential product of vertices and propagators. Because external lines can occur anywhere in the complex plane, we must sum over all different orderings of the external lines.
(5) Fixing the vertex function to have weight 1 and using the anomaly cancellation argument show that the intercept for the closed string model must be 2 and $\alpha' k^2 = 2$, meaning that the theory necessarily contains the massless graviton. In fact, the linearized general covariance gauge symmetry of general relativity simply emerges as the lowest order Virasoro gauge symmetry (we will expand on this in Chapter 7).

Let us begin by discussing the propagator:

$$\begin{aligned} D &= \frac{1}{2\pi} \int_{|z| \leq 1} z^{L_0 - 2} \bar{z}^{\tilde{L}_0 - 2}\, d^2 z \\ &= \frac{\sin \pi(L_0 - \tilde{L}_0)}{\pi(L_0 - \tilde{L}_0)} \frac{1}{L_0 + \tilde{L}_0 - 2} \end{aligned} \tag{2.8.5}$$

This actually has a simple physical interpretation. Notice that the factor of the inverse $L$'s simply is the usual propagator of a closed string. The function involving the sines, however, is equal to zero unless

$$L_0 = \tilde{L}_0 \tag{2.8.6}$$

This, in turn, can be represented as

$$\int d\sigma \exp i2\pi\sigma(L_0 - \tilde{L}_0) \tag{2.8.7}$$

There are two ways to interpret this operator. If we explicitly perform the integral, we have the operator $\delta(L_0 - \tilde{L}_0)$, which is a projection operator acting on the full Hilbert space that eliminates states $|\phi\rangle$ which do not satisfy $(L_0 - \tilde{L}_0)|\phi\rangle = 0$. Second, we notice that this operator is the generator of $\sigma$ rotations by one full cycle, so that the propagator simply expresses the fact that, as a closed string moves, one must integrate over one full cycle. The closed string amplitude is therefore independent of where we chose the origin of the $\sigma$ parametrization. (This constraint will have important consequences when we discuss the compactification of the closed string and the heterotic string in later chapters.)

In Fig. 2.9, we see that the world sheet originally consists of a horizontal strip in the complex plane that is $2\pi$ in width, such that the top and bottom sides are identified with each other (creating a long horizontal tube). External tachyon lines can enter this tube internally. We see that by exponentiating these coordinates, we can transform this horizontal tube into the entire complex plane. External lines then couple to all point within the plane.

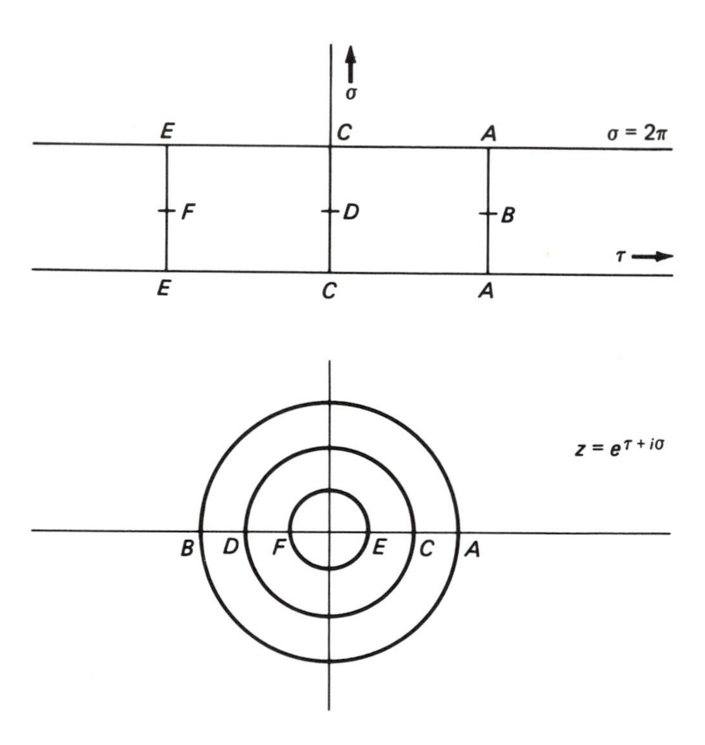


Figure 2.9. Conformal surfaces for closed string propagation. In the $\rho$ plane, the string propagates on a horizontal strip of width $2\pi$, such that the upper and lower horizontal lines are identified with each other, which is topologically equivalent to a tube. In contrast to the open string case, the external lines attach themselves to the interior of surface, not the boundary. In the $z$-plane, this surface maps to the entire complex plane by the exponential map.

As before, the vertex function is once again given by $:e^{ikX}:$, where the string $X$ is defined at the point in the complex plane where the external moment is entering. Because the two sets of harmonic oscillators commute, the vertex function factorizes into the product of two open string vertex functions. The final expression for the $N$-point scattering amplitude of tachyons is given by [22, 23]:

$$A_N = \sum_{\text{perm}} \langle 0; k_1 | V(k_2) D \cdots V(k_{N-1}) | 0; k_N \rangle \tag{2.8.8}$$

The summation over all possible permutations of the ordering of the external lines guarantees that the $z_i$ variables can roam freely in the complex plane.

## §2.9. Ghost Elimination

We have developed the harmonic oscillator formalism using Gupta–Bleuler quantization. In the conformal gauge, the theory maintains manifest Lorentz invariance and, in fact, becomes a theory based on free fields. This accounts for the fact that the theory at the tree level is very easy to write down.

The price we have to pay for this simplicity, however, is that we must impose the Virasoro constraints directly onto the Hilbert space to eliminate ghosts.

In general, the proof that ghost states do not couple in tree diagrams is quite simple. (The proof breaks down in loops, where extreme care must be exercised to properly eliminate ghost states.) Let us define a real physical state as one that satisfies the Gupta–Bleuler constraints:

$$\begin{aligned} L_n |R\rangle &= 0; \qquad n > 0 \\ [L_0 - 1] |R\rangle &= 0 \end{aligned} \tag{2.9.1}$$

Let us define a spurious state as one that does not couple to real states:

$$\langle S | R \rangle = 0 \tag{2.9.2}$$

We can conveniently represent such a state as

$$|S\rangle = L_{-n} |\chi\rangle \tag{2.9.3}$$

for some state $\chi$. We now wish to show that spurious states do not couple to trees, i.e.,

$$\langle S | \text{Tree} \rangle = 0 \tag{2.9.4}$$

To show this, we need two more identities:

$$\begin{aligned} [L_n - L_0 - n + 1] V_0 &= V_0 [L_n - L_0 + 1] \\ [L_n - L_0 + 1] \frac{1}{L_0 - 1} &= \frac{1}{L_0 + n - 1} [L_n - L_0 - n + 1] \end{aligned} \tag{2.9.5}$$

(This can easily be proved from (2.7.6), which in turn crucially depends on the vertex function having conformal weight 1 with external tachyons satisfying

$\alpha' k^2 = 1$. Thus, the restriction on the conformal weight of the vertex is crucial for the elimination of all ghosts in the theory.)

From these two identities, we can now easily show

$$[L_n - L_0 - n + 1] V_0 D V_0 D \cdots V_0 |0\rangle = 0 \tag{2.9.6}$$

This is the desired result; it shows that the $L$'s can be pushed to the right until they finally annihilate on the vacuum.

In summary, we have shown that spurious states do not couple to trees:

$$L_n |\text{Tree}\rangle = 0 \to \langle S | \text{Tree}\rangle = 0 \tag{2.9.7}$$

This means that we do not have to make any special change in the functional for tree amplitudes because of the presence of ghosts. The ghosts states propagating within the tree amplitude automatically cancel by themselves. However, we find that the loop functions do indeed have problems with ghosts propagating internally. This is because

$$\langle S | \text{Tree} | S \rangle \neq 0 \tag{2.9.8}$$

When we naively trace over the trees to obtain loops, we thus inadvertently include the presence of ghosts, which must be factored out explicitly:

$$A_{\text{loop}} = \sum_n \langle n | \text{Tree} | n \rangle \tag{2.9.9}$$

The sum over $|n\rangle$ explicitly contains ghost states.

The fact that ghosts decouple from trees and only contribute to loop diagrams is precisely equivalent to what happens in the Yang–Mills case. Faddeev–Popov ghost contributions to the Yang–Mills theory do not contribute to the trees of the theory, but they do contribute to the loops. The Faddeev–Popov ghost contribution is (see (1.9.30)):

$$\bar{c} \partial_\mu D^\mu(A) c \tag{2.9.10}$$

Notice that this produces the following interaction of the gauge field $A$ with the ghost field $c$:

$$L_I \sim \bar{c} A c \tag{2.9.11}$$

This coupling means that a single ghost cannot couple to a tree diagram consisting of gauge fields. They can only contribute to loops, where the $c$ fields can circulate internally. Thus, particular attention was required to guarantee that the Faddeev–Popov ghosts canceled against the negative metric states in the loops.

The proof that the Virasoro conditions completely eliminate all possible ghost states is, however, an extremely complicated task. In principle, the ability to go to the light cone gauge is normally sufficient to prove that the theory is ghost-free. However, one can never be sure that the quantum theory doesn't introduce anomalies that destroy this fact. Therefore, it becomes important to check explicitly that the Fock space is ghost-free in the Gupta–Bleuler formalism. There are two independent proofs of this theorem, neither of which is very simple, due to Brower, Thorn, and Goddard [24, 25]. The reader may skip the following discussion.

Because of the numerous details involved, let us first sketch our strategy for the elimination of ghosts. We would like to construct a set of physical operators $V_m^i$ and $\bar{V}_m^-$ such that

(1) they commute with the Virasoro generators:

$$\begin{aligned}[L_n, V_m^i] &= 0 \\ [L_n, \bar{V}_m^-] &= 0\end{aligned} \tag{2.9.12}$$

(2) the $V_n^i$ and $\bar{V}_n^-$ generate the physical Hilbert space:

$$|\text{phy}\rangle = V_{-n_1}^{\mu_1} V_{-n_2}^{\mu_2} \cdots V_{-n_N}^{\mu_N} \bar{V}_{-m_1}^- \cdots \bar{V}_{-m_P}^- |0; p_0\rangle \tag{2.9.13}$$

such that the states have either positive or zero norm, but never negative norm.

The original Fock space of $D$-dimensional harmonic oscillators should be equivalent to the set of states generated by these $D$-2-dimensional transverse operators, the $-$ component of these oscillators, and the original Virasoro operators:

$$\{V_{-n}^i, \bar{V}_{-n}^-, L_{-n}\} \tag{2.9.14}$$

The $L_{-n}$ generate ghost states, so that by only taking the states $V_{-n}^i$ and $V_{-n}^-$ we will generate the correct ghost-free states. This is the desired ghost-killing result that we now want to prove.

Let us begin by first defining

$$A_n^i = \frac{1}{2\pi} \int_0^{2\pi} V^i(nk_0, \tau)\, d\tau \tag{2.9.15}$$

where

$$V^i(nk_0, \tau) = \dot{X}^i e^{ink_{0\mu}X^\mu(\tau)} \tag{2.9.16}$$

where $k_0$ is a null vector:

$$\begin{cases} k_0^2 = 0 \\ k_0^- = -1 \\ k_0^+ = k_0^i = 0 \end{cases} \tag{2.9.17}$$

We can view this vertex operator as the vertex for the insertion of a massless vector particle. This particular vertex function was chosen for the following reason. Notice that, because $k_0$ is a null vector, the Virasoro generators have the following commutation relations with this vertex operator:

$$[L_m, V(\tau)] = -i\frac{d}{d\tau}(e^{im\tau}V(\tau)) \tag{2.9.18}$$

Notice that this is a total derivative. When integrated over a circle, this means immediately that

$$[L_m, A_n^i] = 0 \tag{2.9.19}$$

Thus, we chose the form of the vertex function such that we fulfill the first criterion: the $A$'s commute with the Virasoro generators. So far, however, we have no constraints on $k_0$ other than that it is a null vector. We will need to place more constraints so that we fulfill the second condition.

Let us start with a ground state vector of momentum $p_0$, such that $p_0$ and $k_0$ satisfy

$$\begin{aligned} \alpha' p_0^2 &= 1 \\ k_0 \cdot p_0 &= 1 \end{aligned} \tag{2.9.20}$$

Now let us apply the vertex operator $A_n^i$ on the state vector $|0; p_0\rangle$. The resulting state has momentum $p_0 + nk_0$. We want this resulting state to satisfy the mass shell condition as well. As a consequence, we demand

$$\alpha' M^2 = -\tfrac{1}{2}(p_0 + nk_0)^2 = -1 - n \tag{2.9.21}$$

Thus, we demand that the momentum vector $k_0$ be constrained to be a null state so that the vertex operator commutes with the Virasoro generators, and we demand that $k_0 \cdot p_0 = 1$ in order to generate additional transverse mass shell states when acting on the ground state.

Now, we must fully implement the last condition, namely that these operators, in fact, generate the entire physical space of states. There are some problems here. At first, one might suspect that the $-$ modes can be created by simply covariantizing the value of $k_0$. Normally, since the conformal spin of a product of operators is just their sum, we expect this to be true. However, the normal ordering destroys this. In fact, we find that the $-$ components do not commute with the Virasoro generators:

$$[L_n, :\dot{X}^- e^{inX^+}:] = e^{im\tau}\left(-i\frac{d}{d\tau} + m\right):\dot{X}^- e^{inX^+}: + \tfrac{1}{2}nm^2 e^{im\tau + inX^+} \tag{2.9.22}$$

In order to remedy this difficulty, we must add an extra term to the definition. The complete definition is now

$$V^\mu(k, \tau) = :\dot{X}^\mu: e^{ik\cdot X} + \tfrac{1}{2}ik^\mu \frac{d}{d\tau}(\log k\cdot \dot{X})e^{ik\cdot X} \tag{2.9.23}$$

It can be shown that the above combination has conformal spin one and reduces, when we take the transverse components, to the previous expression when we take its integral.

Next, let us take the commutator between these fields:

$$[V_m^\mu, V_n^\nu] = nk_0^\mu V_{m+n}^\nu - mk_0^\nu V_{m+n}^\mu + C_m^{\mu\nu}\delta_{m,-n} \tag{2.9.24}$$

where

$$C_m^{\mu\nu} = 2m^3 k_0^\mu k_0^\nu + m\eta^{\mu\nu} \tag{2.9.25}$$

This allows us to write

$$\begin{aligned} V_m^+ &= \delta_{m,0} \\ [V_m^i, V_n^j] &= m\delta_{ij}\delta_{m,-n} \\ [V_m^-, V_n^i] &= -nV_{m+n}^i \\ [V_m^-, V_n^-] &= (m-n)V_{m+n}^- + 2m^3\delta_{m,-n} \end{aligned} \tag{2.9.26}$$

Notice that the + components are trivial and that the commutation relations of the transverse operators are exactly the same as for the usual harmonic oscillators. Now let us redefine the − modes once again:

$$\bar{V}_n^- = V_n^- - \tfrac{1}{2}\sum_{m=1}^{\infty}\sum_{i=1}^{D-2} :V_m^i V_{n-m}^i: \tag{2.9.27}$$

Putting everything together, the final commutation relations are

$$\begin{aligned} [V_m^i, V_n^j] &= m\delta^{ij}\delta_{m,-n} \\ [\bar{V}_m^-, V_n^i] &= 0 \\ [\bar{V}_m^-, \bar{V}_n^-] &= (m-n)\bar{V}_{m+n}^- + \frac{26-D}{12}m^3\delta_{m,-n} \\ [L_n, \bar{V}_m^-] &= [L_n, V_m^i] = 0 \end{aligned} \tag{2.9.28}$$

This is the final set of commutation relations. Notice that new $\bar{V}_m^-$ operators commute with the original $V_m^i$ operators and that they both commute with the Virasoro generators. Thus, we have now created a new Hilbert space of linearly independent operators that we can use to replace the original Fock space:

$$\{a_{-n}^\mu\} \to \{V_{-n}^i, \bar{V}_{-n}^-, L_{-n}\} \tag{2.9.29}$$

*The physical Hilbert space is thus generated by the operators* $\{V_{-n}^i, \bar{V}_{-n}^-\}$, which are equivalent to the DDF operators [26] which generate the physical space in the light cone gauge. Furthermore, we know that the following state has zero norm:

$$\bar{V}_{-n_1}^- \bar{V}_{-n_2}^- \cdots \bar{V}_{-n_N}^- |0\rangle \tag{2.9.30}$$

This can be checked explicitly by taking the norm of the state and using the commutation relations of the $\bar{V}_n^-$. Thus, we have the final statement, that the set of states generated by

$$\{V_{-n}^i, \bar{V}_{-n}^-\} \tag{2.9.31}$$

have either positive norm or zero norm, but never negative norm. This completes the proof that the Virasoro conditions completely remove the ghosts from the Hilbert space.

## §2.10. Summary

The first quantized theory of strings is remarkably similar to the theory of point particles, except for the nontrivial addition of gauge symmetries representing the reparametrization invariance of the world sheet. As in point particle theory (see (1.4.16)), we have three equivalent actions for the string:

$$\text{1st-order (Hamiltonian) form:}\quad L = P_\mu \dot{X}^\mu + \pi\alpha'\lambda\left[P_\mu^2 + \frac{X'^2}{(2\pi\alpha')^2}\right] + \rho P_\mu X'^\mu$$

$$\sim \sqrt{g} P^a_\mu g_{ab} P^{b\mu} + P^{a\mu}\partial_a X_\mu \sqrt{g}/\pi\alpha' \tag{2.10.1}$$

$$\text{2nd-order form:}\quad L = \frac{-1}{4\pi\alpha'}\sqrt{g} g^{ab}\partial_a X_\mu \partial_b X^\mu$$

$$\text{Non-linear form:}\quad L = \frac{1}{2\pi\alpha'}(\dot{X}_\mu^2 X_\nu'^2 - (\dot{X}_\mu X'^\mu)^2)^{1/2}$$

Let us summarize the similarities found in the point particle case and the string theory when expressed in path integral language:

$$x_\mu(\tau) \to X_\mu(\sigma, \tau)$$

$$\text{length} \to \text{area}$$

$$Dx = \prod_{\mu,\tau} dx_\mu(\tau) \to DX = \prod_{\mu,\tau,\sigma} dX_\mu(\sigma, \tau)$$

$$\int_{x_i}^{x_j} Dx\, e^{i\int L(x)\,d\tau} \to \int_{X_i}^{X_j} DX\, e^{i\int L(X)\,d\sigma\,d\tau}$$

$$\left\langle \prod_{i=1}^N e^{ik_i x^i} \right\rangle \to \left\langle \prod_{i=1}^N e^{ik_i X^i} \right\rangle$$

$$\text{Graph} \to \text{Manifold}$$

$$p^2 + m^2 = 0 \to \left(P_\mu + \frac{X'_\mu}{2\pi\alpha'}\right)^2 = 0$$

Our strategy for quantizing the string action is to write down the symmetry of the theory, extract the currents from this symmetry, calculate the algebra satisfied by these currents, and apply these currents onto the Hilbert space in order to eliminate ghosts. The strategy we follow throughout the book is

Action → Symmetry → Current → Algebra → Constraints → Unitarity

All three actions possess reparametrization symmetry, which generates the Virasoro algebra:

$$L_n = \frac{\pi}{4}\int_{-\pi}^{\pi} d\sigma\, e^{in\sigma}\left[P_\mu + \frac{X'_\mu}{2\pi\alpha'}\right]^2 \tag{2.10.2}$$

The algebra generated by these operators is

$$[L_n, L_m] = (n - m)L_{n+m} + \frac{D}{12}(n^3 - n)\delta_{n,-m} \tag{2.10.3}$$

As in the point particle case, there are three ways in which to quantize the theory:

## Gupta–Bleuler Quantization

In the Gupta–Bleuler formalism, we fix the gauge:

$$g_{ab} = \delta_{ab} \tag{2.10.4}$$

and the resulting action breaks reparametrization invariance but maintains the subgroup of conformal transformations:

$$L = \frac{1}{2\pi}(\dot{X}_\mu^2 - X_\mu'^2) \tag{2.10.5}$$

This Lagrangian, of course, propagates negative metric ghosts associated with the timelike mode of $X$. To eliminate them, the Gupta–Bleuler quantization method states that the state vectors of the theory must satisfy

$$\begin{aligned} L_n|\phi\rangle &= 0; \qquad n > 0 \\ (L_0 - 1)|\phi\rangle &= 0 \end{aligned} \tag{2.10.6}$$

This means that the constraints are constructed to vanish on the Hilbert space.

Although the action in this formalism is quite elegant, the price we pay is that the elimination of ghosts on state vectors is quite difficult, especially for loops. In fact, a large mathematical apparatus is necessary to show that all ghosts can be eliminated in this gauge.

## Light Cone Quantization

In the light cone quantization program, we set

$$X^+ = 2\alpha' p^+ \tau \tag{2.10.7}$$

The advantage of this approach is that we can eliminate all the redundant ghost modes of the theory from the very start in terms of only the transverse components. The disadvantage is that the formalism is awkward and we must reestablish Lorentz symmetry at every step. The surprising feature is that the Lorentz generators of the broken theory close only in 26 dimensions. Specifically, the problem occurs with the commutator:

$$[M^{-i}, M^{-j}] = \frac{-1}{p^{+2}} \sum_{n=1}^{\infty} [\alpha_{-n}^i \alpha_n^j - \alpha_{-n}^j \alpha_n^i] \Delta_n \tag{2.10.8}$$

where

$$\Delta_n = \frac{n}{12}(26 - D) + \frac{1}{n}\left[\frac{D-26}{12} + 2 - 2a\right] \tag{2.10.9}$$

For this commutator to vanish, we must have

$$D = 26; \qquad a = 1 \tag{2.10.10}$$

## BRST Quantization

The BRST method combines the best features of both approaches. We still maintain Lorentz covariance of the action, but the negative metric states do not worry us because they cancel against the ghosts circulating in the theory because of the Faddeev–Popov determinant. When we exponentiate the Faddeev–Popov determinant, we must introduce two anticommuting ghost fields $b$ and $c$. The resulting gauge-fixed action still has a residual symmetry generated by the BRST charge:

$$Q = c_0(L_0 - a) + \sum_{n=1}^{\infty} [c_{-n}L_n + L_{-n}c_n] - \tfrac{1}{2} \sum_{n,m=-\infty}^{\infty} :c_{-m}c_{-n}b_{n+m}:(m - n) \tag{2.10.11}$$

Fixing

$$Q^2 = 0 \tag{2.10.12}$$

sets the intercept to be one and the dimension of space–time to be 26. The physical states of the theory are defined by

$$Q|\text{phy}\rangle = 0 \tag{2.10.13}$$

As in point particle theory, interactions are introduced by summing over different topological configurations in the path integral. The fundamental functional equation for the interacting theory, upon which this entire chapter rests, is given by

$$\begin{aligned} A_N(k_1, k_2, \ldots, k_N) &= \sum_{\text{topologies}} \int d\mu_N \int Dg_{ab} \int DX \\ &\quad \times \Delta_{\text{FP}} \exp i \int L \, d\sigma \, d\tau \left\{ \prod_{i=1}^{N} \sqrt{g} \exp ik_{i,\mu}X^{\mu i} \right\} \\ &= \sum_{\text{topologies}} \int Dg_{ab} \int d\mu \left\langle \prod_{i=1}^{N} \sqrt{g} \exp ik_{i,\mu}X^{\mu i} \right\rangle \end{aligned} \tag{2.10.14}$$

The amplitude can be computed exactly in the conformal gauge by using the identity

$$\begin{aligned} &\int DX \exp\left\{ -\frac{1}{4\pi\alpha'} \int \partial_z X_\mu \partial_{\bar{z}} X^\mu \, d^2z + i \int J_\mu X^\mu \, d^2z \right\} \\ &= \exp\left\{ \frac{\alpha'}{2} \int J_\mu(z) G(z, z') J^\mu(z') \, dz \, dz' \right\} \\ &= \prod_{i<j} |z_i - z_j|^{2\alpha' k_i \cdot k_j} \end{aligned} \tag{2.10.15}$$

In the conformal gauge, the sum over topologies is given by the sum over all *conformally inequivalent* configurations. If we consider conformal transformations that map the upper half-plane into itself and the real axis into itself, then the points along the real axis transform according to a projective transformation SL(2, $R$):

$$y' = \frac{ay + b}{cy + d} \tag{2.10.16}$$

where the coefficients are real and satisfy $ad - bc = 1$. This fixes the measure $d\mu$, so the formula for the $N$-point function becomes

$$A_N = \int \prod_{i=3}^{N-1} dz_i \prod_{2 \le i < j \le N} (z_i - z_j)^{2\alpha' k_i \cdot k_j} \tag{2.10.17}$$

where the $z$'s are ordered along the real axis.

The transition to the harmonic oscillator formalism is straightforward because, in the conformal gauge, the Hamiltonian is diagonal on the Fock space of oscillator modes. The propagator from the configuration $X_a$ to $X_b$ is

$$\langle X_a | \int_0^\infty e^{-\tau H} \, d\tau \, | X_b \rangle = \langle X_a | \frac{1}{L_0 - 1} | X_b \rangle \tag{2.10.18}$$

The vertex is equal to

$$\langle X_c | e^{i k_{i\mu} X^{\mu i}(0, \tau_i)} | X_d \rangle \tag{2.10.19}$$

We can remove all the string eigenstates because

$$| X \rangle \int DX \, \langle X | = 1 \tag{2.10.20}$$

Thus, the $N$-point function is equal to

$$A_N = \langle 0; k_1 | V_0(k_2) D V_0(k_3) \cdots V_0(k_{N-1}) | 0; k_N \rangle \tag{2.10.21}$$

In the operator formalism, projective invariance can be reestablished by using the fact that the operators $L_{\pm 1}$ and $L_0$ generate the projective group SL(2, $R$). In fact, under an arbitrary conformal transformation, the vertices transform as

$$T V(k_i, z_i) T^{-1} = \left( 1 - \varepsilon \alpha' k_i^2 \sum_{n=1}^{\infty} n \alpha_n z^n \right) V(k_i, z_i') \tag{2.10.22}$$

or:

$$[L_n, V(k_i, z_i)] = z^n \left[ z \frac{d}{dz} + n \alpha' k_i^2 \right] V(k_i, z_i) \tag{2.10.23}$$

We say that the vertex $V$ has conformal weight equal to $\alpha' k_i^2 = 1$.

Finally, we can explicitly construct the operators that will generate only physical states with zero or positive norm. The previous identity shows that operators with conformal spin equal to 1 automatically commute with the Virasoro generators. This allows us to create operators, based on the vertex for a massless spin-one particle, that will generate the physical space. We can construct three sets of mutually commuting operators which together generate the entire Fock space of harmonic oscillators:

$$\{V_{-n}^i, \bar{V}_{-n}^-, L_{-n}\}|0\rangle \tag{2.10.24}$$

By taking only those states built out of $V_{-n}^i$ and $\bar{V}_{-n}^-$ and dropping the $L_{-n}$, we obtain a new Fock space that has only states which have positive and zero norm. Thus, the set of states that satisfy:

$$\begin{aligned} L_n|\phi\rangle &= 0 \\ (L_0 - 1)|\phi\rangle &= 0 \end{aligned} \tag{2.10.25}$$

is free of negative norm states in 26 dimensions.

## References

[1] Y. Nambu, Lectures at the Copenhagen Summer Symposium (1970).
[2] T. Goto, *Prog. Theor. Phys.* **46**, 1560 (1971).
[3] For earlier formulations, see also H. B. Nielsen, *15th International Conference of High Energy Physics* (Kiev) 1970.
[4] L. Susskind, *Nuovo Cim.* **69A**, 457 (1970).
[5] See C. B. Thorn, in *Unified String Theory* (edited by M. B. Green and D. Gross), World Scientific, Singapore, 1985.
[6] M. A. Virasoro, *Phys. Rev.* **D1**, 2933 (1970).
[7] A. M. Polyakov, *Phys. Lett.* **103B**, 207, 211 (1981).
[8] S. Fubini, D. Gordon, and G. Veneziano, *Phys. Lett.* **29B**, 679 (1969).
[9] P. Goddard, J. Goldstone, C. Rebbi, and C. B. Thorn, *Nucl Phys.* **B56**, 109 (1973).
[10] M. Kato and K. Ogawa, *Nucl. Phys.* **B212**, 443 (1983).
[11] C. S. Hsue, B. Sakita, and M. A. Virasoro, *Phys. Rev.* **D2**, 2857 (1970).
[12] J. L. Gervais and B. Sakita, *Nucl. Phys.* **B34**, 632 (1971); *Phys. Rev.* **D4**, 2291 (1971); *Phys. Rev. Lett.* **30**, 716 (1973).
[13] D. B. Fairlie and H. B. Nielsen, *Nucl. Phys.* **B20**, 637 (1970).
[14] K. Bardakçi and H. Ruegg, *Phys. Rev.* **181**, 1884 (1969).
[15] M. A. Virasoro, *Phys. Rev. Lett.* **22**, 37 (1969).
[16] C. J. Goebel and B. Sakita, *Phys. Rev. Lett.* **22**, 257 (1969).
[17] H. M. Chan, *Phys. Lett.* **28B**, 425 (1969).
[18] H. M. Chan and S. T. Tsou, *Phys. Lett.* **28B**, 485 (1969).
[19] Z. J. Koba and H. B. Nielsen, *Nucl. Phys.* **B12**, 517 (1969); **B10**, 633 (1969).
[20] I. Caneschi, A. Schwimmer, and G. Veneziano, *Phys. Lett.* **30B**, 351 (1969).
[21] M. A. Virasoro, *Phys. Rev.* **177**, 2309 (1969).
[22] J. Shapiro, *Phys. Lett.* **33B**, 361 (1970).
[23] M. Yoshimura, *Phys. Lett.* **34B**, 79 (1971).
[24] P. Goddard and C. B. Thorn, *Phys. Lett.* **40B**, 235 (1972).
[25] R. C. Brower and K. A. Friedman, *Phys. Rev.* **D7**, 535 (1973).
[26] E. Del Giudice, P. Di Vecchia, and S. Fubini, *Ann. Phys.* **70**, 378 (1972).

CHAPTER 3

# Superstrings

## §3.1. Supersymmetric Point Particles

Supersymmetry is one of the most elegant of all symmetries, uniting bosons and fermions into a single multiplet:

$$\text{Fermions} \leftrightarrow \text{Bosons}$$

By uniting fields of differing statistics, supersymmetry and supergroups have also opened up an entirely new area of mathematics.

However, the irony is that there is not a single shred of experimental evidence in its favor. For example, physicists have tried to fit the electron or neutrino into supersymmetric multiplets, but the scalar partners of these leptons have never been seen. In fact, none of the presently known particles has a supersymmetric partner. Some critics have called supersymmetry a "solution looking for a problem."

Although there are absolutely no empirical data to support the notion of supersymmetry, it is undeniable that supersymmetry provides a wealth of highly desirable theoretical mechanisms that hold tremendous promise. Supersymmetry is more than just an elegant way in which to unite elementary particles into aesthetically pleasing multiplets; it also has definite practical applications to quantum field theory:

(1) Supersymmetry generates super-Ward–Takahashi identities that cancel many normally divergent Feynman graphs. For example, Feynman loop diagrams with bosons and fermions circulating internally in the loop differ by a factor of $-1$. Because of supersymmetry, the boson loop can cancel against the fermion loop, leaving us with a much milder divergence. We see, therefore, that Yang–Mills theories with supersymmetry have better

renormalization properties than ordinary gauge theories. In fact, certain "nonrenormalization theorems" can be proved to all orders in perturbation theory.

(2) Supersymmetry may solve the "hierarchy problem" that plagues ordinary Grand Unified-type theories. In GUTs, there are two widely separated energy scales, the energy scale of ordinary particle physics in the billion electron volt range and also the GUT energy scale at $10^{15}$ or so billion electron volts. In between these two energy scales is a vast "desert" where no new phenomena are found. However, when renormalization effects are calculated, the two energy scales inevitably begin to mix. Loop correctings, for example, to the quark masses can push them up near the GUT scale, which is unacceptable. "Fine-tuning" one's coupling constants and masses by hand can, in principle, solve the hierarchy problem, but this is very contrived and artificial. Fortunately, the Ward–Takahashi identities of supersymmetry are strong enough to enforce "nonrenormalization theorems" to all orders in perturbation theory. Thus, supersymmetry is necessary to stabilize these two mass scales in perturbation theory and prevent mixing.

(3) Supersymmetry may help shed light on the "cosmological constant" problem. Empirically, the cosmological constant term $\lambda\sqrt{-g}$, which is a correction to the Einstein–Hilbert action, is exceedingly small on an astronomical scale. The problem is to explain the near vanishing of the cosmological constant without "fine-tuning." Supersymmetry is probably strong enough to force the cosmological constant to be zero to all orders in perturbation theory (because this term breaks supersymmetry). This doesn't completely solve the cosmological constant problem, however, because we must inevitably break supersymmetry to reach ordinary energies. (The problem, therefore, is to explain the vanishing of the cosmological constant after supersymmetry breaking has occurred.)

(4) Supersymmetry eliminates many undesirable particles. The tachyon that appears in the bosonic string model, for example, is eliminated because it violates supersymmetry. By eliminating these particles, supersymmetry also reduces the divergence of the higher loop graphs. In Chapter 5 we will show that the potential divergences of the superstring theory are associated with the infrared emission of tachyons and dilatons. Thus, by eliminating these particles, we simultaneously eliminate the source of potential divergences.

(5) Lastly, when supersymmetry is elevated into a local guage theory, it naturally reduces the divergences of quantum gravity. This is because local supersymmetry can be defined only in the presence of gravitons (see Appendix). Local supersymmetry is thus intimately tied up with general relativity. In fact, local supersymmetry successfully eliminates the lower loop divergences of supergravity. However, the largest of the supergravity theories, O(8) supergravity, probably has a divergence at the seventh loop level, which most likely rules out supergravity as an acceptable

quantum field theory. Only when we combine local supersymmetry with the conformal invariance of the string theory do we have a large enough gauge group to eliminate perhaps all the divergences of quantum gravity.

Supersymmetry, as an invariance of an action, was first discovered in the string theory. Gervais and Sakita [1] showed that an extension of the usual bosonic action possessed a symmetry that converted bosons into fermions. Unfortunately, it languished for many years because the supersymmetry of the early string model was a two-dimensional supersymmetry on the world sheet. It wasn't until relatively recently that it was finally proved that the string model possessed *both* two-dimensional and 10-dimensional space–time supersymmetry.

Let us begin our discussion by considering the simplest possible spinning action, the spinning point particle. In addition to the variable $x_\mu$, which locates the position of the point particle, let us introduce Dirac spinors $\theta^A$, where $A$ is arbitrary, and the Dirac matrix in $D$ dimensions $\Gamma^\mu$. In general, a Dirac spinor in $D$ dimensions has $2^{(1/2)D}$ complex components. We can write [2, 3]

$$S = \frac{1}{2}\int e^{-1}(\dot{x}_\mu - i\bar{\theta}^A\Gamma_\mu\dot{\theta}^A)^2\,d\tau \tag{3.1.1}$$

This point particle action is invariant under

$$\begin{cases} \delta\theta^A = \varepsilon^A \\ \delta\bar{\theta}^A = \bar{\varepsilon}^A \\ \delta x^\mu = i\bar{\varepsilon}^A\Gamma^\mu\theta^A \\ \delta e = 0 \end{cases} \tag{3.1.2}$$

Notice that the combination

$$\Pi^\mu = \dot{x}^\mu - i\bar{\theta}^A\Gamma^\mu\dot{\theta}^A \tag{3.1.3}$$

is an invariant all by itself under this transformation. Thus, any expression involving this combination will be invariant under this symmetry. The strange feature of this action, however, is that half of the components of the fermionic field cancel from the action by itself.

By varying the $e$, $x_\mu$, and $\theta$ fields, we can derive several equations of motion:

$$\begin{aligned} \Pi^2 &= 0 \\ \dot{\Pi}^\mu &= 0 \\ \Gamma\cdot\Pi\dot{\theta} &= 0 \end{aligned} \tag{3.1.4}$$

We also have

$$(\Gamma\cdot\Pi)^2 = -\Pi^2 = 0 \tag{3.1.5}$$

Thus, half the eigenvalues of the matrix $\Gamma\cdot\Pi$ vanish. But since $\theta$ always appears in the combination $\bar{\theta}^A\Pi_\mu\Gamma^\mu\theta^A$, half the components of the Dirac spinor are actually missing from the action. Thus $\theta$ is not an independent spinor but satisfies a constraint that reduces its components by half.

The reason for this decoupling is that the action is invariant under yet another local symmetry [2]:

$$\begin{cases} \delta\theta^A = i\Gamma\cdot\Pi\kappa^A \\ \delta x^\mu = i\bar{\theta}^A\Gamma^\mu\delta\theta^A \\ \delta e = 4e\dot{\bar{\theta}}^A\kappa^A \end{cases} \tag{3.1.6}$$

Furthermore, there is also another bosonic invariance of the action:

$$\begin{cases} \delta\theta^A = \lambda\dot{\theta}^A \\ \delta x^\mu = i\bar{\theta}^A\Gamma^\mu\delta\theta^A \\ \delta e = 0 \end{cases} \tag{3.1.7}$$

When we try to commute two supersymmetry operations given in (3.1.6), we find that the algebra does not close unless we use the equations of motion:

$$[\delta_1, \delta_2]\theta^A = (2i\Gamma_\mu\kappa_2^A\dot{\bar{\theta}}^B\Gamma\cdot\Pi\Gamma^\mu\kappa_1^B + 4i\Gamma\cdot\Pi\kappa_2^A\dot{\bar{\theta}}^B\kappa_1^B) - (1\leftrightarrow 2) \tag{3.1.8}$$

(This is actually typical of supersymmetric actions. Notice that the numbers of components of $x_\mu$ and $\theta^A$ do not necessarily match off-shell, which means that auxiliary fields are in general necessary to close the algebra.)

Lastly, if we calculate the canonical conjugate to the coordinates, we find

$$\pi_\theta^A = \frac{\delta L}{\delta\dot{\theta}^A} = i\Gamma^\mu\Pi_\mu\theta^A \tag{3.1.9}$$

This poses vast complications for the covariant quantization. Because there is an explicit dependence on $x$ within the canonical conjugate, the quantization relations become nonlinear and hence exceedingly difficult to solve. Worse, it turns out that the term $\Gamma^\mu\Pi_\mu$ becomes a projection operator in the theory, meaning that we cannot invert the transformation and solve for $\theta^A$. In other words, a naive covariant quantization of the spinning point particle does not seem to exist at all! This is a warning that the supersymmetric string is not going to be as simple as super-Yang–Mills or supergravity theory. As a consequence, we will discuss the simpler two-dimensional world supersymmetry of the Neveu–Schwarz–Ramond model first, and then the more complicated 10-dimensional space–time supersymmetry of the Green–Schwarz model.

## §3.2. Two-Dimensional Supersymmetry

Given the problems associated with superparticles (such as lack of a convariant quantization and lack of an algebra that closes off-shell), let us temporarily abandon space–time supersymmetry and discuss the two-dimensional world sheet symmetry of the simplest possible action involving free strings and free fermions. This action will already be gauge fixed in the conformal gauge, but it will reveal all the essential features of two-dimensional supersymmetry. In this gauge-fixed formulation, we will impose the gauge constraints on the Fock space by hand. Later, we will present the complete action, where we will be able to derive these constraints starting from a locally symmetric action.

In the conformal gauge, we have [1]

$$L = -\frac{1}{2\pi}(\partial_a X_\mu \partial^a X^\mu - i\bar{\psi}^\mu \rho^a \partial_a \psi_\mu) \tag{3.2.1}$$

where $a = 1, 2$ and labels two-dimensional vectors, and $\mu$ is a space–time index.

Notice that $\psi$ is a bizarre object, an anticommuting Majorana spinor in two dimensions and a *vector* in real space–time. Let us define

$$\begin{aligned}
\psi^\mu &= \begin{pmatrix} \psi_0^\mu \\ \psi_1^\mu \end{pmatrix}; \qquad \bar{\psi}^\mu = \psi^\mu \rho^0 \\
\rho^0 &= \begin{pmatrix} 0 & -i \\ i & 0 \end{pmatrix} \\
\rho^1 &= \begin{pmatrix} 0 & i \\ i & 0 \end{pmatrix} \\
\{\rho^a, \rho^b\} &= -2\eta^{ab}
\end{aligned} \tag{3.2.2}$$

If we write this out explicitly in components, we have

$$L = \frac{1}{2\pi}(\dot{X}\cdot\dot{X} - X'\cdot X' + i\psi_0(\partial_\tau + \partial_\sigma)\psi_0 + i\psi_1(\partial_\tau - \partial_\sigma)\psi_1) \tag{3.2.3}$$

Although this action is gauge fixed, it is still invariant under the global transformation:

$$\begin{aligned}
\delta X^\mu &= \bar{\varepsilon}\psi^\mu \\
\delta\psi^\mu &= -i\rho^a \partial_a X^\mu \varepsilon
\end{aligned} \tag{3.2.4}$$

Thus, by temporarily abandoning our attempt to produce a theory of strings with genuine space–time spinors, we have achieved a two-dimensional supersymmetric theory that is quite simple, involving free bosonic and anti-commuting fields.

Historically, the NS–R theory [4–6] was the first successful attempt at introducing spin into the dual model. It was also the first example of a linear supersymmetric action [1] and was soon followed by four-dimensional super-symmetric point particle actions [7, 8].

Now that we have written down our 2D supersymmetric action, let us retrace the steps we used in the previous chapter to solve the system. The next step is to extract the currents associated with these symmetries, then establish the algebra that these currents generate, and finally apply these constraints on the Hilbert space. The sequence we will follow in this section is a straightforward generalization of the steps we developed in the previous chapter:

$$\text{Action} \to \text{Symmetries} \to \text{Current} \to \text{Algebra} \to \text{Constraints} \to \text{Unitarity} \tag{3.2.5}$$

Following this strategy, let us now calculate the supersymmetric current associated with this symmetry. As we saw in the last chapter, the existence of conformal symmetry is sufficient to generate a conserved current. In (1.9.8), we saw that this conserved current could be written as

$$J_\mu^\alpha = \frac{\delta L}{\delta \partial_\mu \phi} \frac{\delta \phi}{\delta \varepsilon^\alpha} \tag{3.2.6}$$

By inserting (3.2.4) into (3.2.6), we find

$$J_a = \tfrac{1}{2} \rho^b \rho_a \psi^\mu \partial_b X_\mu \tag{3.2.7}$$

To check that this is conserved, we must first write down the equations of motion for the system, which is especially easy since it is a free action:

$$\begin{aligned} [\partial_\tau + \partial_\sigma] \psi_0^\mu &= 0 \\ [\partial_\tau - \partial_\sigma] \psi_1^\mu &= 0 \end{aligned} \tag{3.2.8}$$

It is now easy to show

$$\partial_a J^a = 0 \tag{3.2.9}$$

Written out in components, this equals ($J_\pm \equiv \tfrac{1}{2}(J_0 \pm J_1)$)

$$\begin{aligned} J_- &= \tfrac{1}{2} \psi_0^\mu (\partial_\tau - \partial_\sigma) X_\mu \\ J_+ &= \tfrac{1}{2} \psi_1^\mu (\partial_\tau + \partial_\sigma) X_\mu \end{aligned} \tag{3.2.10}$$

Using the equations of motion, one can show that

$$\begin{aligned} (\partial_\tau + \partial_\sigma) J_- &= 0 \\ (\partial_\tau - \partial_\sigma) J_+ &= 0 \end{aligned} \tag{3.2.11}$$

In addition to the supersymmetric current, we also have the energy–momentum tensor, which as we saw earlier in (1.9.17) can be written as

$$T_\nu^\mu = \frac{\delta L}{\delta \partial_\mu \phi} \partial_\nu \phi - L \delta_\nu^\mu \tag{3.2.12}$$

This can easily be adapted for our purposes. Inserting (3.2.1) into (3.2.12), we find

$$\begin{aligned} T_{ab} &= \partial_a X^\mu \partial_b X_\mu + \frac{i}{4} \bar{\psi}^\mu \rho_a \partial_b \psi_\mu + \frac{i}{4} \bar{\psi}^\mu \rho_b \partial_a \psi_\mu - (\text{Trace}) \\ \partial_a T^{ab} &= 0 \end{aligned} \tag{3.2.13}$$

where the $T_{ab}$ is traceless because we have explicitly subtracted out the trace. Written out explicitly, this is

$$\begin{aligned} T_{00} &= T_{11} \\ &= \tfrac{1}{2}(\dot{X}^2 + X'^2) - \frac{i}{4}\psi_0^\mu(\partial_\tau - \partial_\sigma)\psi_{0\mu} - \frac{i}{4}\psi_1^\mu(\partial_\tau + \partial_\sigma)\psi_{1\mu} \\ T_{01} &= T_{10} \\ &= \dot{X}\cdot X' + \frac{i}{4}\psi_0^\mu(\partial_\tau - \partial_\sigma)\psi_{0\mu} - \frac{i}{4}\psi_1^\mu(\partial_\tau + \partial_\sigma)\psi_{1\mu} \end{aligned} \tag{3.2.14}$$

Our basic strategy (3.2.5) is to use these currents to restrict the Fock space so that all ghosts are eliminated:

$$\begin{aligned} \langle T_{ab} \rangle &= 0 \\ \langle J_a \rangle &= 0 \end{aligned} \tag{3.2.15}$$

(We must stress again, however, that we are simply imposing this constraint ad hoc on the Fock space. Because the action is not locally supersymmetric, we cannot derive this constraint from first principles. Later, when we present the fully symmetric action, we will see that these constraints emerge because of a local symmetry.)

Unlike the bosonic string, we actually have a choice of two distinct boundary conditions that we can place on the $\psi_0$ and $\psi_1$ fields, either periodic or antiperiodic. At $\sigma = 0$, we can always choose $\psi_0 = \psi_1$. However, at $\sigma = \pi$, we have the choice of two distinct boundary conditions. If the field is periodic (antiperiodic), we have the Ramond (Neveu–Schwarz) boundary conditions [9, 10]:

$$\begin{cases} \text{Ramond} & \psi_0(\pi, \tau) = \psi_1(\pi, \tau) \\ \text{Neveu–Schwarz} & \psi_0(\pi, \tau) = -\psi_1(\pi, \tau) \end{cases} \tag{3.2.16}$$

Given two different types of boundary conditions for our spinor, we naturally have two different ways in which to power expand these two fields in Fourier components with integer (half-integer) modes [8, 9]:

$$\text{R:} \quad \begin{cases} \psi_0^\mu(\sigma, \tau) = 2^{-1/2} \displaystyle\sum_{n=-\infty}^{\infty} d_n^\mu \, e^{-in(\tau-\sigma)} \\ \psi_1^\mu(\sigma, \tau) = 2^{-1/2} \displaystyle\sum_{n=-\infty}^{\infty} d_n^\mu \, e^{-in(\tau+\sigma)} \end{cases} \tag{3.2.17}$$

$$\text{NS:} \quad \begin{cases} \psi_0^\mu(\sigma, \tau) = 2^{-1/2} \displaystyle\sum_{r \in Z+1/2} b_r^\mu \, e^{-ir(\tau-\sigma)} \\ \psi_1^\mu(\sigma, \tau) = 2^{-1/2} \displaystyle\sum_{r \in Z+1/2} b_r^\mu \, e^{-ir(\tau+\sigma)} \end{cases} \tag{3.2.18}$$

The next step in our strategy (3.2.5) is to isolate the algebra that these currents generate. In the previous chapter, we saw that the symmetries yielded the energy–momentum tensor, which in turn generated the Virasoro or conformal algebra. For the NS–R action, we will find that the superconformal algebra is generated.

With this decomposition of the fields, the moments $L_n$ of the current $T_{ab}$ can now be written as

$$L_n = \frac{1}{2\pi}\int_0^\pi d\sigma\{e^{in\sigma}(T_{00} + T_{01}) + e^{-in\sigma}(T_{00} - T_{01})\} \tag{3.2.19}$$

Notice that the definition of the conformal generators now includes the contributions from the anticommuting sector.

We can repeat this Fourier expansion for the supersymmetric current as well. For the R sector, we have

$$F_n = \frac{\sqrt{2}}{\pi}\int_0^\pi d\sigma\{e^{in\sigma}J_+ + e^{-in\sigma}J_-\} \tag{3.2.20}$$

For the NS sector, we have

$$G_r = \frac{\sqrt{2}}{\pi}\int_0^\pi d\sigma\{e^{ir\sigma}J_+ + e^{-ir\sigma}J_-\} \tag{3.2.21}$$

To extract the algebra from these moments of the currents, we need to construct the canonical conjugates of the fields. If we define

$$\pi_A^\mu = \frac{\delta L}{\delta\psi_{A,\mu}} \tag{3.2.22}$$

then we find that the fermionic field is self-conjugate. Thus, we impose

$$\{\psi_A^\mu(\sigma,\tau), \psi_B^\nu(\sigma',\tau)\} = \pi\delta_{AB}\delta(\sigma-\sigma')\eta^{\mu\nu} \tag{3.2.23}$$

Therefore the oscillators in (3.2.17) and (3.2.18) obey

$$\begin{cases} \text{R:} & \{d_n^\mu, d_m^\nu\} = \eta^{\mu\nu}\delta_{n,-m} \\ \text{NS:} & \{b_r^\mu, b_s^\nu\} = \eta^{\mu\nu}\delta_{r,-s} \end{cases} \tag{3.2.24}$$

It is important to notice that the zero component of the Ramond sector is proportional to the Dirac gamma matrices:

$$\{d_0^\mu, d_0^\nu\} = \eta^{\mu\nu} \tag{3.2.25}$$

Hence, we will see that the Ramond sector corresponds to the fermionic sector, and that the NS sector, even though it contains anticommuting operators, still remains a bosonic sector. The fact that boundary conditions like (3.2.16) play such an important role in the development of the fermionic and bosonic sectors of the theory is quite novel and apparently unique to the string theory. (This apparently obscure fact will play an important role when we discuss multiloop amplitudes defined over surfaces with holes. Then these boundary conditions will determine what are called the "spin structures" of the manifold.)

Once the commutation relations are established, we can construct the algebra that these tensors generate. As before, the moments of these currents witll generate a closed algebra. The algebra for the NS sector is

$$\begin{aligned}
[L_m, L_n] &= (m-n)L_{m+n} + \frac{D}{8}(m^3 - m)\delta_{m,-n} \\
[L_m, G_r] &= (\tfrac{1}{2}m - r)G_{m+r} \\
\{G_r, G_s\} &= 2L_{r+s} + \tfrac{1}{2}D(r^2 - \tfrac{1}{4})\delta_{r,-s}
\end{aligned} \tag{3.2.26}$$

Explicitly, we have, in terms of oscillators in the NS sector:

$$\begin{aligned}
L_m &= \tfrac{1}{2}\sum_{n=-\infty}^{\infty} :\alpha_{-n}\alpha_{m+n}: + \tfrac{1}{2}\sum_{r=-\infty}^{\infty}(r + \tfrac{1}{2}m):b_{-r}b_{m+r}: \\
G_r &= \sum_{n=-\infty}^{\infty} \alpha_{-n}b_{r+n}
\end{aligned} \tag{3.2.27}$$

Repeating the same steps for the R sector, we have

$$\begin{aligned}
[L_m, L_n] &= (m-n)L_{m+n} + \frac{D}{8}m^3\delta_{m,-n} \\
[L_m, F_n] &= (\tfrac{1}{2}m - n)F_{m+n} \\
\{F_m, F_n\} &= 2L_{m+n} + \tfrac{1}{2}Dm^2\delta_{m,-n}
\end{aligned} \tag{3.2.28}$$

Written out explicitly, we have

$$\begin{aligned}
L_m &= \tfrac{1}{2}\sum_{n=-\infty}^{\infty} :\alpha_{-n}\alpha_{m+n}: + \tfrac{1}{2}\sum_{n=-\infty}^{\infty}(n + \tfrac{1}{2}m):d_{-n}d_{m+n}: \\
F_m &= \sum_{n=-\infty}^{\infty} \alpha_{-n}d_{m+n}
\end{aligned} \tag{3.2.29}$$

We can read off the Hamiltonian directly from the components of the energy–momentum tensor. In the conformal gauge, the Hamiltonian for the NS sector is

$$H = \sum_{n=1}^{\infty} na_{-n}^{\mu}a_{n\mu} + \sum_{r=1/2}^{\infty} rb_{-r}^{\mu}b_{r\mu} + \alpha' p_0^2 \tag{3.2.30}$$

For the R sector, it is

$$H = \sum_{n=1}^{\infty} na_{-n}^{\mu}a_{n\mu} + \sum_{m=1}^{\infty} md_{-m}^{\mu}d_{m\mu} + \alpha' p_0^2 \tag{3.2.31}$$

Notice that this Hamiltonian is diagonal on the Fock space of harmonic oscillators. This is extremely important, because this means that we can easily make the transition from path integrals to the harmonic oscillator formalism. The propagator for the theory becomes a simple integral over the exponential, which is easily performed.

The states of the theory, as in the bosonic sector, are simply given by the

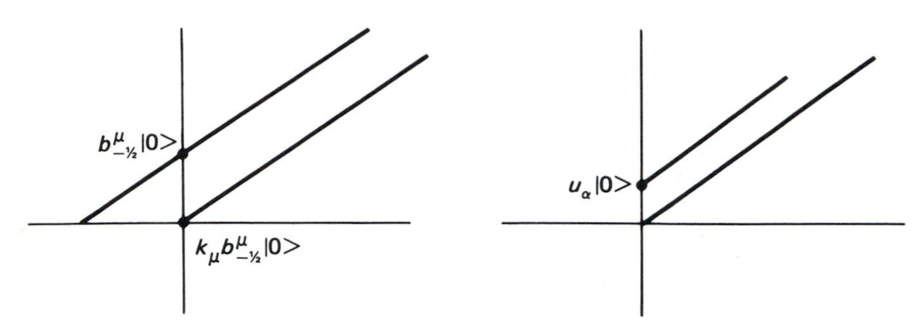


Figure 3.1. Regge trajectories for the NS–R model for open strings. The boson states correspond to the Fock space generated by all possible products of $a^+$ and $b^+$ harmonic oscillators acting on the vacuum. The fermion states correspond to the Fock space generated by all possible products of $a^+$ and $d^+$ oscillators acting on the vacuum, which is a space–time spinor, not a scalar. On the left, the massless spin-1 particle corresponds to the Maxwell or Yang–Mills field (all tachyon states, including the vacuum, can be eliminated when we impose the GSO projection). On the right, the massless spin-$\frac{1}{2}$ fermion is the supersymmetric partner of the massless vector field.

Fock space of the product of these oscillators:

$$\text{NS eigenstates:} \quad \prod_{n,\mu}\prod_{m,\nu}\{a^\dagger_{n,\mu}\}\{b^\dagger_{m,\nu}\}|0\rangle \tag{3.2.32}$$

$$\text{R eigenstates:} \quad \prod_{n,\mu}\prod_{m,\nu}\{a^\dagger_{n,\mu}\}\{d^\dagger_{m,\mu}\}|0\rangle u_\alpha \tag{3.2.33}$$

where $u_\alpha$ corresponds to an arbitrary spinor, which as yet has no restrictions placed on it. Some of the lowest lying states in the NS sector are as follows (see Fig. 3.1):

$$\text{NS states} \quad \begin{cases} \text{vacuum} & |0\rangle; \qquad k^2 = 2 \\ \text{Tachyon} & k_\mu b^\mu_{-1/2}|0\rangle; \qquad k^2 = 1 \\ \text{Massless vector} & a^\mu_{-1}|0\rangle \end{cases} \tag{3.2.34}$$

$$\text{R states} \quad \begin{cases} \text{Spin } \tfrac{1}{2} \text{ fermion} & |0\rangle u_\alpha \\ \text{Spin } \tfrac{3}{2} \text{ fermion} & d^\mu_{-1}|0\rangle u_\alpha \\ \text{Spin } \tfrac{3}{2} \text{ fermion} & a^\mu_{-1}|0\rangle u_\alpha \end{cases} \tag{3.2.35}$$

In summary, we find a remarkable similarity between the NS–R model and the original bosonic string. In each case, we begin with an action, define its symmetries, generate its currents from the symmetries, construct the algebra from the currents, and then apply the constraints onto the Hilbert space. The main difference is the addition of the supercurrent, which generates the superconformal algebra (3.2.26) and (3.2.28).

The last step in our strategy is then to apply this superconformal algebra onto the Hilbert space of the NS–R model. The ghost-killing conditions are

$$F_n \text{ or } G_r|\phi\rangle = 0$$

for positive $n$ and $r$. To establish the mass–shell conditions, however, we will now have to investigate the tree amplitudes for the NS–R model.

## §3.3. Trees

Once again, our starting point for the interacting string theory is the functional integral. Unfortunately, we have no guiding principle for the construction of the interacting theory except for intuition. As a guess, let us multiply the usual vertex term $V_0$ with a factor of $k_\mu \psi^\mu$ at the point at which a spin-zero particle enters the diagram. Then a reasonable assumption is that the $N$-point scattering amplitude for this scalar particle is a generalization of (2.5.2) and is given by

$$A(1, 2, 3, \ldots, N) = \sum_{\text{Topologies}} \int d\mu \, DX \, D\psi \prod_{i=1}^{N} k_{i\mu} \psi^\mu e^{ik_{i\mu} \cdot X^{i\mu}}$$

$$= \sum_{\text{Topologies}} \int d\mu \left\langle \prod_{i=1}^{N} k_{i\mu} \psi^{i\mu} e^{ik_{i\mu} \cdot X^{i\mu}} \right\rangle \tag{3.3.1}$$

where

$$D\psi = \prod_\mu \prod_{\sigma,\tau} d\psi_\mu(\sigma, \tau) \tag{3.3.2}$$

where we are now functionally integrating over an infinite sequence of Grassman variables (see Appendix).

As in the bosonic functional integral, we can remove the functional integrals at each intermediate point along the string sheet because the Hamiltonian is diagonal on this space. Thus, by using

$$1 = |X\rangle|\psi\rangle \int DX \, D\psi \langle\psi|\langle X| \tag{3.3.3}$$

at each intermediate point, we can remove all the functional integrals, leaving only the harmonic oscillators. Thus, once again the functional integral permits a derivation of the harmonic oscillator formalism, which we view as only one particular representation of the functional integral.

In oscillator language, the vertex for the emission of a scalar particle with momentum $k_\mu$ becomes

$$V = k_\mu \psi^\mu V_0 \tag{3.3.4}$$

The underlying motivation for this choice is the rule that vertex functions must have conformal weight 1, in order to preserve the ghost-killing conditions (2.9.5). From (3.2.27) and (3.2.28), we can calculate the commutator of $L_n$ with $\psi_\mu$, and we find that it has conformal weight $\frac{1}{2}$. Since $V_0$ has conformal weight $\alpha' k^2$ and $\psi$ has weight $\frac{1}{2}$, the weight of $V$ is the sum of the two:

$$\alpha' k^2 + \tfrac{1}{2} = 1 \rightarrow k^2 = \frac{1}{2\alpha'} \tag{3.3.5}$$

where $\alpha'$ is equal to $\frac{1}{2}$. This vertex function, which corresponds to the emission or absorption of a tachyon, is guaranteed to satisfy the correct ghost-killing conditions.

The propagator is easy to calculate. Notice that the Hamiltonian is diagonal in the space of harmonic oscillators. Therefore, we will use an explicit representation of the functional based on the normal modes of the harmonic oscillator. Then we have

$$D = \int_0^\infty e^{-\tau(L_0-1)}\,d\tau = \frac{1}{L_0-1} \tag{3.3.6}$$

Let us construct an $N$-point function of tachyons:

$$\langle 0; k_1 | k_1 \cdot b_{1/2} V(k_2) D \cdots V(k_{N-1}) k_N \cdot b_{-1/2} | 0; k_N \rangle \tag{3.3.7}$$

where we have placed the tachyon states to the left and the right of all the vertices and propagators. In this formalism, we can prove cyclic symmetry in the same way it was shown for the bosonic string. We first note:

$$\begin{aligned} &\lim_{y\to 0} \frac{V(k, y)}{y} |0; 0\rangle = k \cdot b_{-1/2} |0; k\rangle \\ &\lim_{y\to\infty} \langle 0; 0| y V(k, y) = \langle 0; -k | k \cdot b_{1/2} \end{aligned} \tag{3.3.8}$$

This permits us to treat all external tachyons alike. By commuting the last vertex to the left, we can show, in exact parallel to the bosonic case studied in (2.6.24), that the amplitude is cyclic symmetric.

At this point, however, we have a problem. The tachyon state that we have just constructed corresponds to $k_\mu b^\mu_{-1/2} |0; k\rangle$, which satisfies

$$[L_0 - 1] k_\mu b^\mu_{-1/2} |0; k\rangle = 0 \to \alpha' k^2 = \tfrac{1}{2} \tag{3.3.9}$$

However, this means that the vacuum state $|0; k\rangle$ has even lower mass because it satisfies the following condition:

$$[L_0 - 1] |0; k\rangle = 0 \to \alpha' k^2 = 1 \tag{3.3.10}$$

We are faced, therefore, with the unusual problem that the true vacuum of the theory (3.3.10) does not correspond to the tachyon that we have just constructed (3.3.9). In other words, the Hilbert space seems to be too large; the true vacuum is an unnecessary state. We seem to have two lowest lying states.

The resolution to this puzzle comes from the fact that the vacuum state, indeed, is a redundant state that can be removed from the theory. This process is accomplished by redefining the Hilbert space of the theory.

The formalism that we have been working with is called the $F_1$ formalism and is a clumsy one. The vacuum of the theory is not equal to the tachyon, so the Fock space is actually larger than necessary. Although cyclic symmetry is easy to prove in this formalism, we would like to introduce a more streamlined formalism, called $F_2$ [10], which works with the reduced Fock space by removing the vacuum state at $k^2 = 1/\alpha'$.

To accomplish this, let us rewrite the tachyon state as

$$k_\mu b^\mu_{-1/2}|0; k\rangle = G_{-1/2}|0; k\rangle \tag{3.3.11}$$

$$\langle 0; k|k_\mu b^\mu_{1/2} = \langle 0; k|G_{1/2} \tag{3.3.12}$$

Now we rewrite the $N$-point tachyon scattering amplitude (3.3.7) using (3.3.12):

$$A_N = \langle 0; k_1|G_{1/2}V(k_2)D\cdots V(k_{N-1})G_{-1/2}|0; k_N\rangle \tag{3.3.13}$$

Now comes the crucial step. We will push the $G_{1/2}$ to the right successively through various vertices and propagators. We need the formulas

$$\begin{gathered} \{G_r, V\} = [L_{2r}, V_0] = [L_0 + r - 1]V_0 - V_0(L_0 - 1) \\ G_{1/2}\frac{1}{L_0 - 1} = \frac{1}{L_0 - \frac{1}{2}}G_{1/2} \end{gathered} \tag{3.3.14}$$

It is essential to notice that when we shove $G_{1/2}$ to the right, we change the intercept of the propagator

$$\frac{1}{L_0 - 1} \to \frac{1}{L_0 - \frac{1}{2}} \tag{3.3.15}$$

while all the other terms involving the $L$'s vanish as in the bosonic case. Finally, we have pushed $G_{1/2}$ all the way to the right, where it vanishes on the tachyon state:

$$G_{1/2}G_{-1/2}|0\rangle = (2L_0 - G_{-1/2}G_{1/2})|0\rangle = 0 \tag{3.3.16}$$

Thus, putting all terms together, we finally have

$$\begin{aligned} A_N &= \langle 0; k_1|k_1\cdot b_{1/2}V(k_2)\frac{1}{L_0 - 1}\cdots V(k_{N-1})k_N\cdot b_{-1/2}|0; k_N\rangle \\ &= \langle 0; k_1|V(k_2)\frac{1}{L_0 - \frac{1}{2}}\cdots V(k_{N-1})|0; k_N\rangle \end{aligned} \tag{3.3.17}$$

Surprisingly, we have now completely rewritten the original amplitude such that the set of resonances has been shifted. In particular, the old vacuum state in the $F_1$ formalism at $\alpha' k^2 = 1$ has vanished from the Hilbert space. It has decoupled completely in the new formalism, which we call $F_2$ [10]. Instead, we are left with the tachyon state being represented by $|0; k\rangle$, which now satisfies the new condition $[L_0 - \frac{1}{2}]|0; k\rangle = 0$ with $\alpha' k^2 = \frac{1}{2}$.

This is quite remarkable. The state $|0; k\rangle$, which used to represent the old vacuum state at $\alpha' k^2 = 1$, has now suddenly transformed into the tachyon state at $\alpha' k^2 = \frac{1}{2}$. Thus, in the $F_2$ formalism, the tachyon state and the new vacuum are the same particle. (One should be careful in noting that the *same* symbol $|0; k\rangle$ can represent either the old vacuum in the $F_1$ formalism or the tachyon in the new $F_2$ formalism.)

This shifting of the Hilbert space is a novel feature not found in point particle field theories. In fact, when we discuss the conformal field theory in the next chapter, we will find that this strange "picture changing" phenomenon arises whenever we construct irreducible representations of the superconformal group. Hence, it is not a trick, but an essential feature of the group theory. (We should also mention that when we make the model space–time supersymmetric, we will eliminate the tachyon state as well. Thus, the tachyon decouples from the true superstring Hilbert space, leaving a unitary theory.)

Let us summarize the differences between these two formalisms:

$$F_1: \quad \begin{cases} \text{Vertex} = V \\ \text{Propagator} = (L_0 - 1)^{-1} \\ \text{Tachyon:} \quad k_\mu \cdot b^\mu_{-1/2}|0; k\rangle \ (\alpha' k^2 = \tfrac{1}{2}); \qquad \text{Vacuum: } |0; k\rangle \ (\alpha' k^2 = 1) \end{cases} \tag{3.3.18}$$

$$F_2: \quad \begin{cases} \text{Vertex} = V \\ \text{Propagator} = (L_0 - \tfrac{1}{2})^{-1} \\ \text{Tachyon} = \text{Vacuum:} \quad |0; k\rangle; \ (\alpha' k^2 = \tfrac{1}{2}) \end{cases} \tag{3.3.19}$$

The advantages and disadvantages of the two formalisms are as follows:

(1) In the $F_1$ formalism, manifest cyclic symmetry is much easier to prove. However, we must carry along the excess baggage of the vacuum state, which decouples completely.
(2) In the $F_2$ formalism, cyclic symmetry is obscure, but ghost elimination and gauge transformations are easy to perform. The advantage is that we are now working in a smaller Fock space.

Using the $F_2$ formalism, it is not hard to evaluate the four-point function exactly:

$$\langle 0, -k_1 | V(k_2) \frac{1}{L_0 - \frac{1}{2}} V(k_3) |0; k_4\rangle = \frac{\Gamma(1 - \alpha(s))\Gamma(1 - \alpha(t))}{\Gamma(1 - \alpha(s) - \alpha(t))} \tag{3.3.20}$$

where $\alpha(s) = 1 + \alpha' s$.

There is also a straightforward generalization to the $N$-point function. The $N$-point function is represented as

$$A_N = \int d\mu_N \langle 0; 0| \frac{V(k_1, y_1)}{y_1}, \ldots, \frac{V(k_N, y_N)}{y_N} |0; 0\rangle \tag{3.3.21}$$

In general, this expression contains many factors that are tedious to work out. A simple method of obtaining the entire result is to use the relation

$$\frac{V(y)}{\sqrt{y}} = \int d\theta \exp\{ik \cdot X + \theta k \cdot \psi / \sqrt{y}\} \tag{3.3.22}$$

Notice that if we power expand the exponential, only the linear term survives the $\theta$ integration, and we wind up with the previous expression for the vertex.

We have, at this point, done nothing. Now, let us use two identities:

$$\langle 0|\frac{\psi^\mu(y_1)}{\sqrt{y_1}}\frac{\psi^\nu(y_2)}{\sqrt{y_2}}|0\rangle = \frac{\eta^{\mu\nu}}{y_1 - y_2}$$

$$\langle 0|\frac{V(y_1)}{\sqrt{y_1}}\frac{V(y_2)}{\sqrt{y_2}}|0\rangle = \int d\theta_1\, d\theta_2 \exp[k_1\cdot k_2 \ln(y_1 - y_2 - \theta_1\theta_2)] \tag{3.3.23}$$

It is not hard to generalize the above relation to the $N$-point tachyon amplitude:

$$\langle 0|\prod_{i=1}^{N}\frac{V(k_i, y_i)}{y_i}|0\rangle = \int \prod \frac{d\theta_i}{\sqrt{y_i}} \prod_{i<j}(y_i - y_j - \theta_i\theta_j)^{k_i\cdot k_j} \tag{3.3.24}$$

The advantage of this expression is that we can now read off the various terms that appear in the formula by successively power expanding and integrating over the Grassman variables.

In addition to tachyon–tachyon scattering amplitudes, we can also calculate the scattering of massless vector particles (which correspond to Maxwell and Yang–Mills particles). The choice of the vertex function for the massless vector particle is constrained by the fact that it must have conformal weight 1 and the correct spin. A natural choice for this vertex is

$$V(\zeta, k) = \{G_r, \zeta\cdot\psi e^{ikX}\} = (\zeta\cdot\dot{X} - \zeta\cdot\psi k\cdot\psi)e^{ikX}$$

where $\zeta$ is the polarization vector of the vector particle and $k^2 = \zeta\cdot k = 0$. The conformal weight of this vertex is equal to the sum of the conformal weights of its individual factors. Since $\psi$ has conformal weight $\frac{1}{2}$ and $e^{ikX}$ has conformal weight $\alpha' k^2$, this means that the conformal weight of this vertex is given by

$$\tfrac{1}{2} + \tfrac{1}{2} + 0 = 1$$

which is the desired weight. This vertex is guaranteed to satisfy the $G$ and $L$ ghost-killing conditions. It is not difficult to calculate the $N$-point scattering amplitude for this massless gauge particle using the same formalism developed for the tachyon. For example, the scattering amplitude for four massless gauge particles is given by

$$A_4 = K\frac{\Gamma(-\frac{1}{2}s)\Gamma(-\frac{1}{2}t)}{\Gamma(1 - \frac{1}{2}s - \frac{1}{2}t)}$$

where the kinematic factor $K$ is given by

$$\begin{aligned} K = &-\tfrac{1}{4}(st\zeta_{13;24} + su\zeta_{23;14} + tu\zeta_{12;34}) + \tfrac{1}{2}s(k_{14}k_{32}\zeta_{24} + k_{23}k_{41}\zeta_{13} \\ &+ k_{13}k_{42}\zeta_{23} + k_{24}k_{31}\zeta_{14}) + \tfrac{1}{2}t(k_{21}k_{43}\zeta_{31} + k_{34}k_{12}\zeta_{24} \\ &+ k_{24}k_{13}\zeta_{34} + k_{31}k_{42}\zeta_{12}) + \tfrac{1}{2}u(k_{12}k_{43}\zeta_{32} + k_{34}k_{21}\zeta_{14} \\ &+ k_{14}k_{23}\zeta_{34} + k_{32}k_{41}\zeta_{12}) \end{aligned}$$

where

$$k_{ij} = \zeta_i \cdot k_j; \qquad \zeta_{ij} = \zeta_i \cdot \zeta_j; \qquad \zeta_{ij;kl} = \zeta_{ij}\zeta_{kl}$$

(The scattering involving fermions can also be calculated; they will also have the basic form, except that the kinematic factor $K$ will depend on the external spinors.)

We should add that another advantage of working with the $F_2$ formalism is that we can explicitly show the invariance of the amplitude under the following:

$$\begin{aligned} \delta y_i &= \theta_i \varepsilon \\ \delta \theta_i &= \varepsilon \end{aligned} \tag{3.3.25}$$

This generates the group Osp(1, 2) (see Appendix), which is the supersymmetric generalization of the projective group SL(2, $R$). This group is generated by the algebra formed by the set

$$G_{\pm 1/2}, \quad L_{\pm 1}, \quad L_0 \tag{3.3.26}$$

Let $\Omega$ be an element of the group Osp(1, 2). Then the proof of super-projective invariance can be shown by noting:

$$\Omega |0; 0\rangle = |0; 0\rangle \tag{3.3.27}$$

(Notice that, as in the bosonic case, the vacuum state $|0; 0\rangle$ is not a physical state. Thus $\Omega$, which does not in general kill physical states, can annihilate that vacuum state.) The vertex rotates as

$$\Omega V(y, \theta)\Omega^{-1} = V(y', \theta') \tag{3.3.28}$$

where $V(y, \theta)$ is the vertex function before we integrate out over $\theta$.

Now that we have established the properties of the three-boson vertex function in the Neveu–Schwarz formalism, the fermion–fermion–boson coupling (with an external boson line) for the Ramond model can also be calculated. We choose

$$V(k) = \Gamma :e^{ik \cdot X}: \tag{3.3.29}$$

where

$$\Gamma = \gamma_{11}(-1)^{\sum_n d_{-n} \cdot d_n} \tag{3.3.30}$$

where $\gamma_{11}$ is the product of the Dirac matrices. In addition to this vertex function, we also need the propagator:

$$D = \frac{1}{F_0} = \frac{F_0}{L_0} \tag{3.3.31}$$

where $F_0$ is given in (3.2.20). We have chosen the intercept to be zero (which we will see is the proper choice in order to guarantee conformal invariance; at this early stage, however, we cannot yet motivate the choice of zero intercept).

Finally, the vacuum state is now a spin-$\frac{1}{2}$ fermion, given by

$$\sum_{\alpha} |0; q\rangle_\alpha u_\alpha \tag{3.3.32}$$

where we sum over the spinor indices $\alpha$ of the spinor $u_\alpha$. The scattering amplitude for a fermion interacting with several bosons is now given by

$$\bar{u}(q_1)\langle 0; q_1 | V(k_2) D \cdots D V(k_{N-1}) | 0; q_N \rangle u(q_N) \tag{3.3.33}$$

So far, we have only exhibited amplitudes with two external fermion lines. Surprisingly, our first guesses, based on simple intuition, have all been successful. Because the action in the conformal gauge was that of a free theory, the NS–R model has been exceedingly simple. There is a price to be paid for this, however.

In principle, because the string model can be factorized in any channel, it should be possible to factorize the R model in the meson channel and rederive the NS model or to extract out the multifermion vertex function. Actually, this is quite difficult. In particular, the fermion vertex function (with an external fermion leg coupled to internal meson and fermion lines) is an exceedingly difficult object to work with. This, in turn, makes it difficult to calculate multifermion amplitudes in the NS–R formalism.

Although it can be shown that the R model can be factorized in the various channels to derive the NS model, the NS–R formalism is actually quite clumsy when calculating multifermion amplitudes. In the next chapter, we will see that the techniques of conformal field theory make the covariant calculation of multifermion amplitudes possible.

## §3.4. Local 2D Supersymmetry

Notice that we imposed the following conditions by hand:

$$\begin{aligned} \langle T_{ab} \rangle &= 0 \\ \langle J_a \rangle &= 0 \end{aligned} \tag{3.4.1}$$

No justification was given. We merely appealed to the fact that there must exists a higher action from which these constraints can be derived ab initio.

We will now describe the local generalization of the previous theory. The key to the construction of this locally supersymmetric action is to add more fields to the theory. In addition to the supersymmetric pair

$$(X_\mu, \psi_\mu) \tag{3.4.2}$$

we introduce a two-dimensional *vierbein* into the theory and also its supersymmetric partner:

$$(e_\alpha^a, \chi_\alpha) \tag{3.4.3}$$

where Greek letters such as $\alpha$ and $\beta$ label two-dimensional vectors in curved space, Greek letters like $\mu$ and $\nu$ continue to label 10-dimensional space–time vectors, Roman letters such as $a$, $b$, $c$ label two-dimensional vectors in flat space, where $\chi_\alpha$ is a two-dimensional spinor as well as a two-dimensional vector, and where we have suppressed two-dimensional spinor indices. Thus, both the vierbein and the spinor $\chi_\alpha$ have four components, as required by supersymmetry. Let us now construct the complete action, first written down by Brink, Di Vecchia, Howe, Deser, and Zumino [11–13]:

$$L = -\frac{1}{4\pi\alpha'}\sqrt{g}[g^{\alpha\beta}\partial_\alpha X^\mu \partial_\beta X_\mu - i\bar{\psi}^\mu \rho^\alpha \nabla_\alpha \psi_\mu + 2\bar{\chi}_\alpha \rho^\beta \rho^\alpha \psi^\mu \partial_\beta X_\mu + \tfrac{1}{2}\bar{\psi}_\mu \psi^\mu \bar{\chi}_\alpha \rho^\beta \rho^\alpha \chi_\beta] \tag{3.4.4}$$

Notice that the $\rho$ matrices used above are actually defined in two-dimensional curved space because they are multiplied by the vierbein:

$$\rho^\alpha = e_a^\alpha \rho^a$$

This action is invariant under

$$\text{2D SUSY:} \quad \begin{cases} \delta X^\mu = \bar{\varepsilon}\psi^\mu \\ \delta\psi^\mu = -i\rho^\alpha \varepsilon(\partial_\alpha X^\mu - \bar{\psi}^\mu \chi_\alpha) \\ \delta e_\beta^a = -2i\bar{\varepsilon}\rho^a \chi_\beta \\ \delta\chi_\alpha = \nabla_\alpha \varepsilon \end{cases} \tag{3.4.5}$$

It is also invariant under Weyl (scale) transformations:

$$\text{Weyl:} \quad \begin{cases} \delta X_\mu = 0 \\ \delta\psi^\mu = -\tfrac{1}{2}\sigma\psi^\mu \\ \delta e_\beta^a = \sigma e_\beta^a \\ \delta\chi_\alpha = \tfrac{1}{2}\sigma\chi_\alpha \end{cases} \tag{3.4.6}$$

It is also invariant, due to certain identities in two dimensions, under

$$\begin{aligned} \delta\chi_\alpha &= i\rho_\alpha \eta \\ \delta e_\beta^a = \delta\psi_\mu &= \delta X_\mu = 0 \end{aligned} \tag{3.4.7}$$

Finally, by construction, it is also manifestly invariant under local two-dimensional Lorentz transformations and reparametrizations because of the presence of the vierbein. (The presence of ordinary derivatives like $\partial_\alpha$ in the action instead of curved space covariant derivatives $D_\alpha$ is due to the fact that the two-dimensional connection field vanishes in the action.)

Now we are in a position to derive the constraints that we imposed on the Fock space by hand. Notice that the variation of the action with respect to the vierbein yields

$$T_\alpha^a = 0 \tag{3.4.8}$$

while variation of the vierbein's supersymmetric partner yields

$$\rho^\alpha \rho^\beta \partial_\alpha X_\mu \psi^\mu - \tfrac{1}{4}(\bar{\psi}\psi)\rho^\alpha \rho^\beta \chi_\alpha = 0 \tag{3.4.9}$$

These are the constraints that we promised earlier, which are now derived as a consequence of the variation of fields.

We can now take the gauge:

$$\begin{aligned} e_\alpha^a &= \delta_\alpha^a \\ \chi_\alpha &= 0 \end{aligned} \tag{3.4.10}$$

Notice that there are enough symmetries on the $\chi_\alpha$ field to eliminate it entirely, a combination of two-dimensional supersymmetry (3.4.5) and the symmetry in (3.4.7). Then the constraints reduce to

$$T_{\alpha\beta} = \partial_\alpha X_\mu \partial_\beta X^\mu + \tfrac{1}{2} i\bar{\psi}\rho_{(\alpha}\partial_{\beta)}\psi - (\text{trace}) = 0 \tag{3.4.11}$$

$$J^\alpha = \tfrac{1}{2}\rho^\beta \rho^\alpha \partial_\beta X_\mu \psi^\mu = 0 \tag{3.4.12}$$

(where the parentheses symbolize taking one-half of the symmetrized sum). These, of course, are the constraints we originally introduced at the beginning of our discussion in (3.2.7) and (3.2.13). Thus, we now have derived these constraints from an invariant action, rather than simply imposing them on the states of the system.

In summary, this action is invariant under several local symmetries:

(1) Local 2D Lorentz symmetry
(2) Reparametrization
(3) Weyl rescaling
(4) 2D supersymmetry

but *not* 10D space–time supersymmetry.

## §3.5. Quantization

Quantization of the NS–R action is a straightforward extension of the quantization of the bosonic string. The surprising feature is that the highly coupled NS–R action becomes a free theory after choosing the gauge.

Again, we will use three methods: (1) Gupta–Bleuler, (2) light cone, (3) BRST.

### Gupta–Bleuler

The Gupta–Bleuler formalism is the simplest of the three and is the formalism we have actually been using all this time. Using the large set of symmetries in the model, we can choose the conformal gauge:

$$\begin{cases} \chi_\alpha = 0 \\ e_\alpha^b = \delta_\alpha^b \end{cases} \tag{3.5.1}$$

Notice that in this gauge the action reduces to the one that we have been using, with the provision that we impose

$$\text{NS:}\quad \begin{cases} L_n|\Phi\rangle = 0 \\ G_r|\Phi\rangle = 0 \end{cases} \tag{3.5.2}$$

$$\text{R:}\quad \begin{cases} L_n|\Psi\rangle = 0 \\ F_m|\Psi\rangle = 0 \end{cases} \tag{3.5.3}$$

for positive $n$, $m$, and $r$. Thus, we have been quantizing the NS–R action in the Gupta–Bleuler formalism.

## Light Cone Quantization

The essence of the light cone formalism is that we eliminate all unphysical ghost states by solving the constraints (3.4.11) and (3.4.12) explicitly in terms of the transverse or physical modes, rather than applying them onto states. We can impose:

$$\begin{cases} \psi^1 = 0 \\ \psi^\mu(\partial_\tau + \partial_\sigma)X_\mu = 0 \end{cases} \tag{3.5.4}$$

Most of the generators are only slightly changed. The real difference occurs in the $-$ components of the fields, which contain the transverse components of the conformal generators:

$$\alpha_n^- = \alpha_n^-(X) + \frac{1}{2p^+}\sum_{r=-\infty}^{\infty}(r - \tfrac{1}{2}n):b_{n-r}^i b_r^i: - \tfrac{1}{2}\frac{a\delta_{n,0}}{p^+} \tag{3.5.5}$$

where the first term on the right refers to the bosonic generator we found earlier in (2.3.16), and

$$b_r^- = \frac{1}{p^+}\sum_{i=1}^{D-2}\sum_{r=-\infty}^{\infty}\alpha_{r-s}^i b_s^i \tag{3.5.6}$$

The troublesome Lorentz generator is the one that contains the $-$ components:

$$\begin{aligned} M^{-i} &= M_0^{-i} + K^{-i} \\ K^{-i} &= \frac{-i}{2p^+}\sum_{m=-\infty}^{\infty}\sum_{j=1}^{D-2}\sum_{r=-\infty}^{\infty}\alpha_{-m}^j(b_{m-r}^i b_r^j - b_{m-r}^j b_r^i) \end{aligned} \tag{3.5.7}$$

We find, finally:

$$[M^{-i}, M^{-j}] = \frac{-1}{(p^+)^2}\sum_{n=1}^{\infty}(\alpha_{-n}^i\alpha_n^j - \alpha_{-n}^j\alpha_n^i)(\Delta_n - n) \tag{3.5.8}$$

where

$$\Delta_n = n\left(\frac{D-2}{8}\right) + \frac{1}{n}\left(2a - \frac{D-2}{8}\right) \tag{3.5.9}$$

Thus, Lorentz invariance is achieved only with $D = 10$ and $a = \frac{1}{2}$.

## BRST Quantization

BRST quantization begins with the calculation of the Faddeev–Popov determinant associated with the conformal gauge. Because of the large set of symmetries of the action, we can enforce:

$$
\begin{aligned}
e_\alpha^a &= \delta_\alpha^a \\
\chi_\alpha &= 0
\end{aligned}
\tag{3.5.10}
$$

As before, we must analyze the constraint when we make a variation of the fields in (3.4.5) and (3.4.7)

$$
\delta\chi_\alpha = \nabla_\alpha \varepsilon + i\rho_\alpha \eta \tag{3.5.11}
$$

As before, the Faddeev–Popov determinant (1.6.10) associated with the gauge constraint is given by

$$
\Delta_{\mathrm{FP}} = \det\left[\frac{\delta}{\delta\varepsilon}\nabla_\alpha \varepsilon'\right] \tag{3.5.12}
$$

(The $\eta$ field does not contribute to the Faddeev–Popov ghost determinant.) The Faddeev–Popov determinant associated with the anticommuting sector is thus

$$
\det[\nabla_\alpha] = \det[\nabla_z]\det[\nabla_{\bar z}] \tag{3.5.13}
$$

Analyzing this is a bit tricky because of the number of degrees of freedom within $\chi_a$, which is both a two-dimensional spinor and a two-dimensional vector, and thus has four components.

As before, the easiest way to calculate the determinant of an operator is to exponentiate it into the action:

$$
\begin{aligned}
\det[\nabla_\alpha] &= \int D\beta\, D\gamma\, e^{-S_{\mathrm{gh}}} \\
S_{\mathrm{gh}} &= \frac{1}{\pi}\int d^2 z\, \beta\bar\partial\gamma + \text{c.c.}
\end{aligned}
\tag{3.5.14}
$$

where now the ghost fields $\beta$ and $\gamma$ are *commuting* variables. (Although this action looks similar to the previous ghost action (2.4.4) found for the bosonic string, it is essential to point out that these new ghost fields $\beta$ and $\gamma$ transform differently under the conformal group than the $b$, $c$ ghosts because they were derived from variations of a spinor $\chi_\alpha$ rather than the tensor $g_{ab}$. We will discuss this further in the next chapter.)

The complete action, therefore, is now the sum of the original action with the sum over the anticommuting ghosts $b$, $c$ and the commuting ghosts $\beta$, $\gamma$.

We can easily calculate the energy–momentum tensor and the supersymmetric current for the superconformal ghosts. Taking their moments, we arrive at the super-Virasoro algebra:

$$L_m^{\mathrm{gh}} = \sum_{m,n=-\infty}^{\infty} (m+n):b_{m-n}c_n: + \sum_{m,n=-\infty}^{\infty} (\tfrac{1}{2}m+n):\beta_{m-n}\gamma_n:$$
$$F_m^{\mathrm{gh}} = -2\sum_{m,n=-\infty}^{\infty} b_{-n}\gamma_{m+n} + \sum_{m,n=-\infty}^{\infty} (\tfrac{1}{2}n-m)c_{-n}\beta_{m+n} \tag{3.5.15}$$

Similarly, the combined action, including the ghost contribution, is still invariant under a nilpotent transformation. The generator of this nilpotent transformation is given by $Q$, where [14–16]

$$Q = \sum_{n=-\infty}^{\infty} (L_{-n}c_n + F_{-n}\gamma_n) - \tfrac{1}{2}\sum_{m,n=-\infty}^{\infty} (m-n):c_{-m}c_{-n}b_{m+n}:$$
$$+ \sum_{m,n=-\infty}^{\infty}\left(\frac{3n}{2}+m\right)c_{-n}\beta_{-m}\gamma_{m+n} + \sum_{m,n=-\infty}^{\infty}\gamma_{-m}\gamma_{-n}b_{m+n} - ac_0 \tag{3.5.16}$$

where the $L$ and $F$ generators are functions only of the $\alpha$ and $d$ oscillators. If we let $Q^2 = 0$, then we arrive at

$$D = 10$$
$$a = \begin{cases} \tfrac{1}{2} & \text{(NS)} \\ 0 & \text{(R)} \end{cases} \tag{3.5.17}$$

The physical condition is now

$$Q|\mathrm{phy}\rangle = 0 \tag{3.5.18}$$

Solving this, we arrive at the usual physical state conditions:

$$\text{NS:} \quad \begin{cases} (L_0 - \tfrac{1}{2})|\phi\rangle = 0 \\ G_r|\phi\rangle = 0; \qquad r > 0 \end{cases} \tag{3.5.19}$$

$$\text{R:} \quad \begin{cases} (L_0)|\phi\rangle = 0 \\ F_n|\phi\rangle = 0; \qquad n > 0 \end{cases} \tag{3.5.20}$$

Let us now calculate the anomaly contribution of these oscillators. We will find that the two contributions add to yield a vanishing anomaly, but only in 10 dimensions. The ghost contribution yields

$$\{F_m^{\mathrm{gh}}, F_n^{\mathrm{gh}}\} = 2L_{m+n}^{\mathrm{gh}} - 5m^2 \tag{3.5.21}$$

From the Ramond sector, the anomaly contribution was $\tfrac{1}{2}Dm^2 + 2a$. Thus, we have

$$\tfrac{1}{2}Dm^2 + 2a - 5m^2 = 0 \tag{3.5.22}$$

which fixes

$$D = 10$$
$$a = 0 \tag{3.5.23}$$

Similarly, for the NS sector, the anomaly is equal to

$$\{G_r, G_s\} = 2L_{r+s} + \tfrac{1}{4} - 5r^2 \tag{3.5.24}$$

This must cancel against the anomaly coming from the bosonic sector: $\frac{1}{2}D(r^2 - \frac{1}{4}) + 2a$. Thus

$$\tfrac{1}{2}D(r^2 - \tfrac{1}{4}) + 2a + \tfrac{1}{4} - 5r^2 = 0 \tag{3.5.25}$$

which vanishes if

$$\begin{aligned} D &= 10 \\ a &= \tfrac{1}{2} \end{aligned} \tag{3.5.26}$$

## §3.6. GSO Projection

The action (3.4.4) for the NS–R model was not invariant under 10D space–time supersymmetry. Although it is not obvious, the NS–R theory is also invariant under 10-dimensional space–time supersymmetry if we make a certain truncation of the Fock space. To see this, let us first analyze the space of states of the string.

The total number of bosonic states at the $n$th level is equal to the partition of the integer $n$, that is, $p(n)$. To see this, let us first take the example of the fourth level. The total number of states that are eigenfunctions of the Hamiltonian at that level is five

$$\begin{gathered} (a_{-1})^4|0\rangle; \qquad (a_{-2})^2|0\rangle; \qquad a_{-1}a_{-3}|0\rangle; \\ (a_{-1})^2a_{-2}|0\rangle; \qquad a_{-4}|0\rangle \end{gathered} \tag{3.6.1}$$

Notice that this is precisely the total number of ways in which the integer four can be broken up as the sum of other integers, i.e., the partition of four. In general, each state at the $N$th level can be represented as one of the various ways we can partition an integer:

$$|\phi\rangle = a_{-n_1}^{\lambda_1} a_{-n_2}^{\lambda_2} a_{-n_3}^{\lambda_3} \cdots a_{-n_M}^{\lambda_M} |0\rangle \tag{3.6.2}$$

where

$$\begin{aligned} R|\phi\rangle &= N|\phi\rangle \\ N &= \sum_{i=1}^{M} n_i^{\lambda_i} \end{aligned} \tag{3.6.3}$$

and $R$ is the number operator for this space.

In statistical mechanics, we often introduce the partition function that counts how many states there are at a given energy:

$$Z = \sum_n \langle n| e^{-H\tau} |n\rangle \tag{3.6.4}$$

We will find the partition function useful because it will enable us to count the number of states of the string:

$$Z = \mathrm{Tr}(x^R) \tag{3.6.5}$$

$$x = e^{-\tau} \tag{3.6.6}$$

The coefficient of $x^N$ in this power expansion is nothing but the partition of $N$, that is, $p(N)$. This trace can be performed in a number of ways. The simplest is to introduce coherent states:

$$1 = \int d\lambda\, d\lambda^*\, e^{-|\lambda|^2} |\lambda\rangle\langle\lambda| \tag{3.6.7}$$

Thus, the trace can be represented as

$$Z = \int d\lambda\, d\lambda^*\, e^{-|\lambda|^2} \langle\lambda| x^R |\lambda\rangle \tag{3.6.8}$$

Performing the integral, we find

$$Z = \prod_{n=1}^{\infty} (1 - x^n)^{-26} \tag{3.6.9}$$

As a check, we can power expand this function (in one dimension) and show that each coefficient of $x^N$ is equal to $p(N)$, i.e.,

$$Z = \sum_{i=1}^{\infty} p(i) x^i \tag{3.6.10}$$

This function is related to a famous function in mathematics, the Hardy–Ramanujan function. In fact, some of the miraculous properties of the string model are related to the identities of the Hardy–Ramanujan function.

Next, we would like to compute the states of the NS and the R sector to see if they are supersymmetric. At the very least, we want the number of on-shell states to be equal in the two sectors.

Notice that the NS–R Fock space actually divides up into two smaller Fock spaces, depending on whether a given state has an even or odd number of $b$ oscillators. Because the $b$ oscillators are anticommuting, the total number $N$ external lines must be even, because $b$ oscillators must be contracted in pairs. Thus, we can define a "$G$-parity" operator that simply counts the number of $b$ oscillators in a given state. It is $-1$ for an odd number and $+1$ for an even number:

$$G = (-1)^{\sum_{r=1/2} b_{-r} b_r} \tag{3.6.11}$$

If we take the even-$G$ parity sector of the theory, *this eliminates the tachyon immediately*, leaving the massless vector particle as the lowest lying state, with $10 - 2 = 8$ physical states.

Meanwhile, the Ramond sector can also be reduced. If we demand that our lowest lying massless fermion is Majorana–Weyl, then it has only 16 surviving modes. The counting goes like this. A Dirac spinor in $D$ dimensions has

$2^{(1/2)D} = 32$ complex components. Demanding that it be Majorana (real) reduces this number by half, and demanding that it be Weyl reduces this number again by half. Let us now calculate the partition function that counts the total number of states at each level. For the NS sector, we have

$$f_{\rm NS} = \sum_{n=0}^{\infty} \dim V_n x^n$$
$$= x^{-1/2}\,\mathrm{Tr}[\tfrac{1}{2}(1+G)x^R] \tag{3.6.12}$$

where

$$R = \sum_{n=1}^{\infty} n a_n^\dagger a_n + \sum_{r=1/2}^{\infty} r b_r^\dagger b_r \tag{3.6.13}$$

The trace can be evaluated exactly, leaving us with

$$f_{\rm NS} = \tfrac{1}{2}x^{-1/2}\prod_{n=1}^{\infty}(1-x^n)^{-8}\left[\prod_{n=1}^{\infty}(1+x^{n-1/2})^8 - \prod_{n=1}^{\infty}(1-x^{n-1/2})^8\right] \tag{3.6.14}$$

By contrast, the trace over the Ramond sector becomes

$$f_{\rm R} = 8\,\mathrm{Tr}(x^R) \tag{3.6.15}$$

where

$$R = \sum_{n=1}^{\infty} n(a_n^\dagger a_n + d_n^\dagger d_n) \tag{3.6.16}$$

We find

$$f_{\rm R} = 8\prod_{n=1}^{\infty}(1-x^n)^{-8}(1+x^n)^8 \tag{3.6.17}$$

At first, one might suspect that there is no relationship between these two sectors. However, it was known to Jacobi as early as 1829 that these two expressions are precisely equivalent:

$$f_{\rm NS} = f_{\rm R} \tag{3.6.18}$$

This, of course, is not a proof that the reduced NS–R model is supersymmetric, but it is a necessary condition. This projection is called the GSO projection (after Gliozzi, Scherk, and Olive [17]) and it has a profoundly important role to play in the superstring. (In Chapter 5, we will find that the GSO projection is equivalent to modular invariance of the closed superstring one-loop amplitude.)

Thus, one of the great disadvantages of the NS–R approach is that space–time supersymmetry is very obscure. A second great disadvantage of the NS–R model is that the vertex function for the emission of a fermion is very difficult to work with and does not have nice properties under the Virasoro gauges. As a result, it took many years before two-fermion scattering amplitudes could be worked out. In the NS–R theory, the vertices for boson and fermion couplings are [18–22]

$$\begin{cases} V_{\mathrm{BB}} = k_\mu \psi^\mu V_0 \\ V_{\mathrm{FF}} = \gamma_{11}(-1)^{\sum_n d_{-n}\cdot d_n} V_0 \\ V_{\mathrm{FB}} = e^{-L_1^d} W V_0 \end{cases} \tag{3.6.19}$$

where $L_1^d$ is a Virasoro generator and

$$W = \langle 0|_{\mathrm{B}} \exp\left[\frac{1}{i\sqrt{2}} \sum_{m=0,r=1/2}^{\infty} (-1)^{m-r} \binom{m-\frac{1}{2}}{r-\frac{1}{2}} d_m^{\mu\dagger} b_{\mu r}\right]$$
$$\times \exp\left\{-\tfrac{1}{4} \sum_{r,s=1/2}^{\infty} \frac{r-s}{r+s}(-1)^{r+s} \binom{-\frac{1}{2}}{r-\frac{1}{2}} \binom{-\frac{1}{2}}{s-\frac{1}{2}} b_r^\mu b_{s\mu}\right\} \gamma_{11}|u\rangle_{\mathrm{F}} \tag{3.6.20}$$

where this vertex takes us from the bosonic Fock space to the fermionic Fock space because we have a bosonic vacuum on the left and a fermionic vacuum on the right. Unfortunately, this vertex has conformal spin $\frac{3}{8}$, making it difficult to construct an amplitude that is conformally invariant.

Thus, although the NS–R model is appealing because it is essentially a free theory in the conformal gauge, the problems with the fermionic sector force us to go to higher versions of the theory. Next, we will investigate the GS action which is equivalent to the GSO projected NS–R theory and is manifestly space–time supersymmetric.

## §3.7. Superstrings

Although the NS–R was simple and elegant, it displayed only 2-dimensional world sheet supersymmetry and not genuine 10-dimensional supersymmetry. Let us now present the Green–Schwarz action, where the 10-dimensional supersymmetry is manifest [23, 24]:

$$S = \frac{-1}{4\alpha'\pi} \int d\sigma\, d\tau \{\sqrt{g} g^{\alpha\beta} \Pi_\alpha \cdot \Pi_\beta + 2i\varepsilon^{\alpha\beta} \partial_\alpha X^\mu (\bar\theta^1 \Gamma_\mu \partial_\beta \theta^1 - \bar\theta^2 \Gamma_\mu \partial_\beta \theta^2)$$
$$- 2\varepsilon^{\alpha\beta} \bar\theta^1 \Gamma^\mu \partial_\alpha \theta^1 \bar\theta^2 \Gamma_\mu \partial_\beta \theta^2\} \tag{3.7.1}$$

where

$$\Pi_\alpha^\mu = \partial_\alpha X^\mu - i\bar\theta^A \Gamma^\mu \partial_\alpha \theta^A \tag{3.7.2}$$

where the two $\theta^A$ fields ($A = 1, 2$) are genuine space–time fermion fields (rather than space–time vectors as in the NS–R model). These 10D spinors, however, transform as scalars in two dimensions, not two-component spinors. This action, like the supersymmetric point particle action (3.1.1), is invariant under global supersymmetry:

$$\begin{aligned} \delta\theta^A &= \varepsilon^A \\ \delta X^\mu &= i\bar\varepsilon^A \Gamma^\mu \theta^A \end{aligned} \tag{3.7.3}$$

However, in the proof, we have to make use of the identity

$$\Gamma_\mu \psi_{[1} \bar{\psi}_2 \Gamma^\mu \psi_{3]} = 0 \tag{3.7.4}$$

To prove this identity, we must make use of a Fierz transformation. It turns out that this identity holds only under the following conditions:

(1) $D = 3$ and the fermions are Majorana.
(2) $D = 4$ and the fermions are Majornan or Weyl.
(3) $D = 6$ and the fermions are Weyl.
(4) $D = 10$ and the fermions are both Majorana and Weyl.

The standard Dirac representation is complex and exists in any space–time dimension. The Majorana representation of the Dirac matrices is one where they are all real (or imaginary). Thus, they have half as many components as the Dirac representation. They exist only in $D = 2, 3, 4 \bmod 8$ dimensions. A Weyl representation of the Dirac matrices is one where we have projected out half the components by the Weyl projection operator:

$$\tfrac{1}{2}(1 \pm \Gamma_{D+1}); \qquad \Gamma_{D+1} = \Gamma_0 \Gamma_1 \cdots \Gamma_{D-1}$$

The Weyl representation exists only in even dimensions. And lastly, spinors that are simultaneously Majorana and Weyl are defined only in 2 mod 8 dimensions:

$$\text{Majorana–Weyl spinor:} \quad D = 2 \bmod 8$$

For our present case, we will choose Majorana–Weyl spinors in 10 dimensions.

To prove that the action (3.7.1) is locally supersymmetric in 10 dimensions, let us first construct the following projection operators, which project onto self-dual and anti-self-dual pieces of two-dimensional vectors:

$$P_\pm^{\alpha\beta} = \tfrac{1}{2}(g^{\alpha\beta} \pm \varepsilon^{\alpha\beta}/\sqrt{g}) \tag{3.7.5}$$

This projection operator has the following properties:

$$\begin{aligned}
P_\pm^{\alpha\beta} g_{\beta\gamma} P_\pm^{\gamma\delta} &= P_\pm^{\alpha\delta} \\
P_\pm^{\alpha\beta} g_{\beta\gamma} P_\mp^{\gamma\delta} &= 0 \\
\kappa^{1\alpha} &= P_-^{\alpha\beta} \kappa_\beta^1 \\
\kappa^{2\alpha} &= P_+^{\alpha\beta} \kappa_\beta^2
\end{aligned}$$

where the supersymmetric parameter is given by $\kappa^{A\alpha a}$, where $A = 1$ or 2, $\alpha$ represents a two-dimensional vector index, and $a$ is a spinorial index in 10 dimensions. We will usually suppress the spinorial index. Notice that the last two equations state that $\kappa$ is self-dual (anti-self-dual) for $A = 2(1)$.

Then the action can be shown to be invariant under

$$\begin{aligned}
\delta\theta^A &= 2i\Gamma \cdot \Pi_\alpha \kappa^{A\alpha} \\
\delta X^\mu &= i\bar{\theta}^A \Gamma^\mu \delta\theta^A \\
\delta(\sqrt{g} g^{\alpha\beta}) &= -16\sqrt{g}(P_-^{\alpha\gamma} \bar{\kappa}^{1\beta} \partial_\gamma \theta^1 + P_+^{\alpha\gamma} \kappa^{2\beta} \partial_\gamma \theta^2)
\end{aligned} \tag{3.7.6}$$

The great achievement of the GS action is that it explicitly contains 10-dimensional supersymmetry. Because the fundamental field is now a genuine anticommuting space–time spinor (rather than the anticommuting vector field in the NS–R theory), we can explicitly construct the space–time supersymmetric operator $Q_\alpha$.

The great defect of the presentation, however, is that naive covariant quantization, as in the point particle case, does not exist. Once again, as in (3.1.9) for the point particle case, we must face the fact that the quantization relations are nonlinear because

$$\pi_A = \frac{\delta L}{\delta \psi_A} \sim i\Gamma^\mu \partial_\sigma X_\mu \theta^A + \cdots \tag{3.7.7}$$

As a consequence, the commutation relations between fields are highly nonlinear. Worse, we find that the commutation relations actually are proportional to the inverse of the constraints; i.e., they probably don't exist [2, 24–32].

As a result, *we will be forced to use light cone quantization.*

## §3.8. Light Cone Quantization of the GS Action

The counting of the number of independent degrees of freedom within the spinors is important when we fix the gauge. Normally, Dirac spinors in $D$ dimensions ($D$ even) have $2^{(1/2)D}$ complex components. Thus, the spinors $\theta^{1,2}$ in 10 dimensions have $32 + 32$ complex components. However, when we restrict them to being Majorana spinors, we are left with half as many: $32 + 32$ real components. Restricting them to being Weyl spinors further reduces them down by half once again, to $16 + 16$ real components. Choosing the light cone gauge will reduce the number of independent components down further to $8 + 8$ real components. Lastly, when we go on-shell and impose the Dirac equation on these spinors, the number of independent components goes down by half again, to 8. But this is precisely the number of components necessary to form a supermultiplet with the 8 bosonic components of the string $X_i$. Thus, we have precisely the correct number of components necessary to satisfy supersymmetry on-shell. If $N$ is the number of components in the spinors $\theta^1$ and $\theta^2$, then we have

$$\begin{aligned}
\text{Dirac:} \quad & N = 32 + 32 \text{ complex components} \\
\text{Majorana:} \quad & N = 32 + 32 \text{ real components} \\
\text{Majorana–Weyl:} \quad & N = 16 + 16 \text{ real components} \\
\text{Light cone:} \quad & N = 8 + 8 \text{ real components} \\
\text{On-shell:} \quad & N = 8 \text{ real components}
\end{aligned}$$

Let us now begin this reduction down to the light cone gauge. We will choose the gauge constraints:

$$\begin{aligned} \Gamma^+\theta^{1,2} &= 0 \\ \Gamma^\pm &= 2^{-1/2}(\Gamma^0 \pm \Gamma^9) \end{aligned} \tag{3.8.1}$$

Because these gamma matrices satisfy

$$(\Gamma^+)^2 = (\Gamma^-)^2 = 0 \tag{3.8.2}$$

this means that exactly half of the original 16 + 16 components of the spinors are eliminated. The final light cone action becomes remarkably simple in this gauge:

$$S = \frac{-1}{4\pi\alpha'} \int d\sigma\, d\tau (\partial_a X^i \partial^a X^i - i\bar{S}\rho^b \partial_b S) \tag{3.8.3}$$

where we make the substitution

$$\sqrt{p^+}\theta \to S \tag{3.8.4}$$

The essential point is to observe that all the complicated nonlinear terms in (3.7.1) that prevented a simple covariant quantization of the superstring have now disappeared. (Notice that a curious phenomenon has also occurred. In the covariant GS action, $\theta^1$ and $\theta^2$ in two dimensions were scalars, independent of each other. Now, in the light cone gauge, these two independent scalars have merged into a single two-dimensional spinor.)

Quantization is exceptionally simple because the system reduces to that of free particles (while the covariant theory was intractibly coupled). The equations of motion are those of free strings:

$$\begin{aligned} (\partial_\tau + \partial_\sigma) S^{1a} &= 0 \\ (\partial_\tau - \partial_\sigma) S^{2a} &= 0 \end{aligned} \tag{3.8.5}$$

The commutation relations are

$$\{S^{Aa}(\sigma, \tau), S^{Bb}(\sigma', \tau)\} = \pi\delta^{ab}\delta^{AB}\delta(\sigma - \sigma') \tag{3.8.6}$$

However, we can have, as before, several boundary conditions on these fields. For an open string (Type I), we have the following boundary conditions:

$$\begin{cases} S^{1a}(0, \tau) = S^{2a}(0, \tau) \\ S^{1a}(\pi, \tau) = S^{2a}(\pi, \tau) \end{cases} \tag{3.8.7}$$

Notice that these two string fields have the same SO(8) chirality. This is because only these relations are compatible with a global supersymmetry transformation that equates the two supersymmetry parameters $\varepsilon^A$. This means that the open Type I superstring has only $N = 1$ supersymmetry. (Reversing the signs in the previous equation, yielding opposite chiralities, would make supersymmetry totally impossible. This actually is chosen in another type of string theory, the $O(16) \otimes O(16)$ theory.) The normal mode expansion thus reads

$$
\begin{aligned}
S^{1a}(\sigma, \tau) &= 2^{-1/2} \sum_{n=-\infty}^{\infty} S_n^a e^{-in(\tau-\sigma)} \\
S^{2a}(\sigma, \tau) &= 2^{-1/2} \sum_{n=-\infty}^{\infty} S_n^a e^{-in(\tau+\sigma)}
\end{aligned}
\tag{3.8.8}
$$

where

$$
\{S_m^a, S_n^b\} = \delta^{ab}\delta_{n,-m}
\tag{3.8.9}
$$

For the closed string (Type II), however, we actually have two types of options; the fields can either be chiral or not. Closed strings are, by definition, periodic in sigma, which yields the following normal mode expansion:

$$
\begin{cases}
S^{1a}(\sigma, \tau) = \sum_{n=-\infty}^{\infty} S_n^a e^{-2in(\tau-\sigma)} \\
S^{2a}(\sigma, \tau) = \sum_{n=-\infty}^{\infty} \tilde{S}_n^a e^{-2in(\tau+\sigma)}
\end{cases}
\tag{3.8.10}
$$

If these two fields have different chiralities, then they are called Type IIA. If they have the same chiralities as in Type I strings, they are called Type IIB, i.e.,

$$
\begin{aligned}
\text{Type I} &= \begin{cases} \text{open and closed string} \\ \text{same chiralities} \end{cases} \\
\text{Type IIA} &= \begin{cases} \text{closed string} \\ \text{opposite chiralities} \end{cases} \\
\text{Type IIB} &= \begin{cases} \text{closed string} \\ \text{same chiralities} \end{cases}
\end{aligned}
\tag{3.8.11}
$$

The spectrum of the light cone superstring is especially appealing because the theory is on-shell supersymmetric, meaning that we should immediately see all particle states arranged such that the helicity states of the bosons match the number of fermions. The mass of each particle is determined by the Hamiltonian:

$$
N = \alpha'(\text{mass})^2 = \sum_{n=1}^{\infty} (\alpha_{-n}^i \alpha_n^i + n S_{-n}^a S_n^a)
\tag{3.8.12}
$$

For Type I, the ground state of the theory consists of the massless vector particle and its spinor partner. Let $|i\rangle$ represent the eight physical transverse polarizations of the massless vector field. Then the spinor partner $|a\rangle$ can be represented as

$$
|a\rangle = \frac{i}{8}(\gamma_i S_0)^a |i\rangle
\tag{3.8.13}
$$

We can normalize our states as

$$\begin{aligned} \langle i|j\rangle &= \delta_{ij} \\ \langle a|b\rangle &= \tfrac{1}{2}(h\gamma^+)^{ab} \\ |i\rangle &= \frac{i}{8}(\bar{S}_0\gamma_{i+})^a|a\rangle \end{aligned} \tag{3.8.14}$$

where $h$ is the Weyl projection operator. (We will use the notation that $\gamma^{\mu_1,\mu_2,\ldots,\mu_n}$ is equal to the product of gamma matrices summed over all permutations in the indices. The normalization is such that $\gamma^{12} = \gamma^1\gamma^2$).

If we were to quantize the 10-dimensional super Maxwell theory in the light cone gauge, then the supersymmetric pair $(A_\mu, \psi^\alpha)$ would correspond to

$$\text{Supersymmetric multiplet:} \quad \begin{cases} A_i \to |i\rangle \\ \psi^a \to |a\rangle \end{cases} \tag{3.8.15}$$

Thus, the light cone theory reproduces the super-Maxwell theory at the lowest level.

At the next level, we have 128 boson and 128 fermion states:

$$128 \text{ Bosons} \quad \begin{cases} \alpha^i_{-1}|j\rangle \to 64 \text{ states} \\ S^a_{-1}|b\rangle \to 64 \text{ states} \end{cases} \tag{3.8.16}$$

$$128 \text{ Fermions} \quad \begin{cases} \alpha^i_{-1}|a\rangle \to 64 \text{ states} \\ S^a_{-1}|i\rangle \to 64 \text{ states} \end{cases} \tag{3.8.17}$$

At the $N = 2$ level, we have 1152 bosons and an equal number of fermions:

$$1152 \text{ Bosons} \quad \begin{cases} \alpha^i_{-1}\alpha^j_{-1}|k\rangle \to 288 \text{ states} \\ S^a_{-1}S^b_{-1}|i\rangle \to 224 \text{ states} \\ \alpha^i_{-2}|j\rangle \to 64 \text{ states} \\ \alpha^i_{-1}S^a_{-1}|b\rangle \to 512 \text{ states} \\ S^a_{-2}|b\rangle \to 64 \text{ states} \end{cases}$$

$$1152 \text{ Fermions} \quad \begin{cases} \alpha^i_{-1}S^a_{-1}|j\rangle \to 512 \text{ states} \\ S^a_{-2}|i\rangle \to 64 \text{ states} \\ \alpha^i_{-1}\alpha^j_{-1}|a\rangle \to 288 \text{ states} \\ \alpha^i_{-2}|a\rangle \to 64 \text{ states} \\ S^a_{-1}S^b_{-1}|c\rangle \to 224 \text{ states} \end{cases} \tag{3.8.18}$$

This procedure can be repeated at the next level, where we have 15,360 states.

Not surprisingly, we can also regroup these massive multiplets according to O(9). This is because *massless* $D$-dimensional supergravity, when compactified, yields $D - 1$ *massive* states. Thus, the $N = 1$ sector, with 128 bosons and 128 fermions, can be regrouped as 44 + 84 bosons in irreducible O(9) representations, and a single spin-$\frac{3}{2}$ 128 multiplet for the fermions. At the $N = 2$ level, we have the bosons being regrouped as the 9 + 36 + 126 + 156 + 231 + 594 representations of O(9), while the fermions can be reassembled in 16 + 128 + 432 + 576.

For the Type II closed string, the spectrum becomes even more interesting, because we obtain supergravity at the massless level, with 128 bosons and 128 fermions:

$$\text{SUSY multiplet} \quad \begin{cases} 128 \text{ Bosons } |i\rangle|j\rangle; |a\rangle|b\rangle \\ 128 \text{ Fermions } |i\rangle|a\rangle; |a\rangle|i\rangle \end{cases} \tag{3.8.19}$$

where the first state refers to the $S$ oscillators and the second to the $\tilde{S}$ oscillators. If the fermions have the opposite handedness (Type IIA), then this represents the $N = 2$, $D = 10$-dimensional reduction of ordinary $N = 1$, $D = 11$ supergravity.

However, if the fermions have the same handedness, then there are complications. This theory contains a fourth-rank antisymmetric tensor in 10 dimensions, with 35 independent components in 8 dimensions. It has been shown that no covariant action for such a particle exists! Thus, we have the unusual situation where a supersymmetric action does not exist for this sector. The theory, of course, is still well defined. The light cone theory and the $S$-matrix elements are explicitly calculable. However, the $S$-matrix, strictly speaking, does not seem to arise from a covariant action [33].

Finally, if we take the restriction to symmetrized states of the Type I theory, we obtain $N = 1, D = 10$ supergravity. Let us summarize the zero slope limits of these theories:

$$\begin{aligned} \text{Type I} &\to N = 1, D = 10 \text{ super-Yang–Mills} \\ \text{Type I} &\to N = 1, D = 10 \text{ supergravity} \\ \text{Type IIA} &\to N = 2, D = 10 \text{ supergravity} \\ \text{Type IIB} &\to \text{nonexistent} \end{aligned} \tag{3.8.20}$$

At higher and higher levels, it becomes prohibitive to continue this analysis of the spectrum to show supersymmetry. We can prove, however, that the entire spectrum, at arbitrarily high orders, is supersymmetric simply by showing that supersymmetry generators exist with commutation relations that commute with the Hamiltonian. We will now show this by explicitly calculating the supersymmetry generators to all orders.

In the light cone gauge, the two supersymmetry transformations (3.7.6) and (3.7.3) become quite simple:

$$\begin{gathered} \begin{cases} \delta S^a = (2p^+)^{1/2} \eta^a \\ \delta X^i = 0 \end{cases} \\ \begin{cases} \delta S^a = -i\sqrt{p^+}\rho^\alpha \partial_\alpha X^i \gamma^i_{a\dot{a}} \varepsilon^{\dot{a}} \\ \delta X^i = p^{+-1/2} \gamma^i_{a\dot{a}} \bar{\varepsilon}^{\dot{a}} S^a \end{cases} \end{gathered} \tag{3.8.21}$$

(See the Appendix for a representation of the $\gamma$ matrices. There are three 8 representations of SO(8). The first is a vector representation $8_v$ which we denote by the index $i$. The other two representations are both spinorial. We use the index $a$ for one representation and $\dot{a}$ for the other. Thus, the $\gamma$ matrix has indices that transform as: $\gamma^i_{a\dot{a}}$.) We can find an explicit form for the two supersymmetry generators in terms of the fields that reproduces the transformations given in (3.8.21):

$$\begin{aligned} Q^a &= (2p^+)^{1/2} S_0^a \\ Q^{\dot{a}} &= (p^+)^{-1/2} \gamma^i_{\dot{a}a} \sum_{n=-\infty}^{\infty} S^a_{-n} \alpha^i_n \end{aligned} \tag{3.8.22}$$

It is now a straightforward task to calculate the anticommutation relations between these generators:

$$\begin{cases} \{Q^a, Q^b\} = 2p^+ \delta^{ab} \\ \{Q^a, Q^{\dot{a}}\} = \sqrt{2}\gamma^i_{a\dot{a}} p^i \\ \{Q^{\dot{a}}, Q^{\dot{b}}\} = 2H\delta^{\dot{a}\dot{b}} \end{cases} \tag{3.8.23}$$

where

$$H = \frac{1}{p^+}\left\{\sum_{n=1}^{\infty} [\alpha^i_{-n}\alpha^i_n + nS^a_{-n}S^a_n] + \tfrac{1}{2}p_i^2\right\} \tag{3.8.24}$$

The proof that the light cone theory is Lorentz invariant is now expanded to the proof that the superstring is super-Poincaré invariant. In addition to the usual commutators, we must show that

$$[M^{\mu\nu}, Q_\alpha] \sim -\tfrac{1}{2}i(\gamma^{\mu\nu})^\beta_\alpha Q_\beta \tag{3.8.25}$$

where $\alpha$, $\beta$ represent spinor indices in 10 dimensions. Again, the only difficult commutator involves the $-$ component of the oscillators:

$$\alpha_n^- = \frac{1}{2p^+} \sum_{m=-\infty}^{\infty} (\alpha^i_{n-m}\alpha^i_m + (m - \tfrac{1}{2}n)S^a_{n-m}S^a_m) \tag{3.8.26}$$

The difficult commutator has an extra piece:

$$M^{-i} = M_0^{-i} + \frac{-i}{4p^+} \sum_{n,m=-\infty}^{\infty} S^a_{n-m}\gamma^{ij}_{ab}S^b_m\alpha^j_{-n} \tag{3.8.27}$$

Because the theory is already uniquely defined to be in 10 dimensions, we find

$$[M^{-i}, M^{-j}] = 0 \tag{3.8.28}$$

Furthermore, we also find the commutator for the super-Poincaré group:

$$[M^{-i}, Q^a] = \frac{-i}{\sqrt{2}}\gamma^i_{a\dot{a}} Q^{\dot{a}} \tag{3.8.29}$$

This completes the proof that the spectrum of the GS model is 10D supersymmetric to all levels.

## §3.9. Vertices and Trees

We now turn to the question of the interactions. We will have to rely on several educated guesses as to the nature of the vertex functions, but demanding that the vertices transform under supersymmetry places stringent conditions on their final form. Having established space–time supersymmetry for the superstring, let us now show that this restriction is sufficient to allow us to construct the trees. We will demand that the supersymmetry operators take us from fermion vertices to boson ones and vice versa. This will impose an enormous number of constraints on our theory, which will, in effect, uniquely determine the theory itself.

We demand that, under a supersymmetric transformation generated by the $Q$'s, the vertices $V_F$ and $V_B$ transform into each other

$$\begin{aligned} [\eta^a Q^a, V_F(u, k)] &\sim V_B(\tilde{\zeta}, k) \\ [\eta^a Q^a, V_B(\zeta, k)] &\sim V_F(\tilde{u}, k) \\ [\varepsilon^{\dot{a}} Q^{\dot{a}}, V_F(\hat{u}, k)] &\sim V_B(\hat{\zeta}, k) \\ [\varepsilon^{\dot{a}} Q^{\dot{a}}, V_B(\hat{\zeta}, k)] &\sim V_F(\hat{u}, k) \end{aligned} \tag{3.9.1}$$

where $\zeta$ stands for the polarization tensor of the massless vector particle, and $u$ stands for its Majorana–Weyl spinor partner. Notice that these rules are complicated by the fact that a supersymmetric transformation necessarily generates rotations in space–time, so that the $\zeta$ and $u$ must also transform properly into $\tilde{\zeta}$, $\hat{\zeta}$ and $\tilde{u}$, $\hat{u}$ as we make this supersymmetric transformation.

A natural assumption is that these vertices can be expressed as

$$\begin{cases} V_B(\zeta, k) = \zeta \cdot B e^{ik \cdot X} \\ V_F(u, k) = u F e^{ik \cdot X} \end{cases} \tag{3.9.2}$$

for some vector $B$ and spinor $F$. Remarkably, when all the details are worked out in the light cone gauge, we can satisfy all these conditions for simple values of $B$ and $F$.

We begin by first calculating the rotated spinors and polarization tensors. To calculate how the polarization and the spinor transform, we take the zeroth component of the supersymmetric generators and see how they transform states. At the zeroth level, we find

$$\begin{aligned} Q^a &\to (2p^+)^{1/2} S_0^a \\ Q^{\dot{a}} &\to (p^+)^{-1/2} (\gamma^i p^i)_{a\dot{a}} S_0^a \end{aligned} \tag{3.9.3}$$

These zeroth components transform the polarization tensor into Majorana–Weyl spinors. Let us define

$$\begin{cases} \eta^a Q^a |u\rangle = |\tilde{\zeta}\rangle \\ \eta^a Q^a |\zeta\rangle = |\tilde{u}\rangle \\ \varepsilon^{\dot{a}} Q^{\dot{a}} |\zeta\rangle = |\hat{u}\rangle \\ \varepsilon^{\dot{a}} Q^{\dot{a}} |u\rangle = |\hat{\zeta}\rangle \end{cases} \tag{3.9.4}$$

Now insert (3.9.3) into (3.9.4) and solve for the transformed spinors and vectors. It is now a simple matter to show that the rotated spinors and polarization tensors are

$$\begin{cases} \tilde{\zeta}^i = (\gamma^i)_{a\dot{a}}\eta^a u^{\dot{a}} \\ \tilde{u}^{\dot{a}} = \eta^a k^+(\gamma^i\zeta^i)_{a\dot{a}} \\ \hat{u}^{\dot{a}} = \dfrac{-1}{\sqrt{2}}(\varepsilon\gamma_{ij})^{\dot{a}}k^i\zeta^j \\ \hat{\zeta}^i = \dfrac{1}{\sqrt{2}}(\varepsilon^{\dot{a}}\gamma^i_{a\dot{a}}u) + \dfrac{\sqrt{2}}{k^+}\varepsilon^{\dot{a}}u^{\dot{a}}k^j \end{cases} \tag{3.9.5}$$

When we insert (3.9.5) into (3.9.1), we find the solution for the $B$ and the $F$ fields:

$$\begin{cases} B^+ = p^+ \\ B^i = \dot{X}^i - R^{ij}k^j \\ F^{\dot{a}} = (2p^+)^{-1/2}[(\gamma\cdot\dot{X}S)^{\dot{a}} + \frac{1}{3}:(\gamma^i S)^{\dot{a}}R^{ij}: k^j] \\ F^a = (p^+/2)^{1/2}S^a \end{cases} \tag{3.9.6}$$

where

$$R^{ij} = \tfrac{1}{4}\gamma^{ij}_{ab}S^aS^b \tag{3.9.7}$$

We can now contract over vertices and propagators to obtain trees. The great advantage of this approach is that we can calculate multifermion amplitudes with almost the same ease with which we calculate multiboson amplitudes. For example, the propagator and the amplitude for the scattering of fermions and bosons are given by

$$D = \left\{\sum_{n=1}^{\infty} \alpha^i_{-n}\alpha^i_n + nS^a_{-n}S^a_n + \tfrac{1}{2}p^2\right\}^{-1} \tag{3.9.8}$$

$$A_N = \langle 0; k_1|V(k_2)D\cdots V(k_{N-1})|0; k_N\rangle \tag{3.9.9}$$

where the various vertices can be either fermionic or bosonic. The four-point amplitude, for example, is equal to

$$A_4 = K\frac{\Gamma(-\frac{1}{2}s)\Gamma(-\frac{1}{2}t)}{\Gamma(1-\frac{1}{2}s-\frac{1}{2}t)} \tag{3.9.10}$$

where $K$ is a matrix that gives us the appropriate spin structure for the amplitude. If we let $u$ represent a spinor and $\zeta$ the polarization of massless fields, then:

$$\begin{aligned} K(u_1,\zeta_2,u_3,\zeta_4) &= -\tfrac{1}{2}t(\bar{u}_1\Gamma^\mu\Gamma^\nu\Gamma^\lambda u_3)\zeta_{2\mu}(k_3+k_4)_\nu\zeta_{4\lambda} \\ &\quad + \tfrac{1}{2}s(\bar{u}_1\Gamma^\mu\Gamma^\nu\Gamma^\lambda u_3)\zeta_{4\mu}(k_2+k_3)_\nu\zeta_{2\lambda} \\ K(u_1,u_2,u_3,u_4) &= -\tfrac{1}{2}s(\bar{u}_2\Gamma^\mu u_3)(\bar{u}_1\Gamma_\mu u_4) + \tfrac{1}{2}t(\bar{u}_1\Gamma^\mu u_2)(\bar{u}_4\Gamma_\mu u_3) \end{aligned} \tag{3.9.11}$$

As we mentioned before, the basic form (3.9.10) for the massless spinor and vector scattering amplitudes is the same. The only difference comes in the factor $K$.

## §3.10. Summary

In summary, there are two equivalent versions of the superstring theory. The relative advantages and disadvantages of the NS–R and the GS actions are as follows:

(1) The NS–R action is linear and easily quantized, both covariantly and in the light cone gauge. Trees and loops for bosons are easily constructed. It is manifestly invariant under two-dimensional supersymmetry. However, these great advantages must be balanced against the fact that 10-dimensional space–time supersymmetry is very obscure, to say the least. Because it is based on anticommuting vector fields $\psi_\mu$, there are problems with spin statistics, which requires a GSO projection. Furthermore, the vertex for fermion emission is prohibitively difficult to use, thus putting multifermion amplitude calculations beyond practical reach.

(2) The GS action is manifestly supersymmetric in 10 dimensions at all mass levels. It is based on genuine space–time spinors, not anticommuting vector fields. However, the action is highly nonlinear, and a covariant quantization does not seem possible at present. Presently, it is only quantizable in the light cone gauge. In this noncovariant gauge, all amplitudes, including multifermion trees, can be calculated.

Let us review a few facts about the NS–R theory. The complete NS–R action is given by

$$L = \frac{-1}{4\pi\alpha'}\sqrt{g}[g^{\alpha\beta}\partial_\alpha X^\mu \partial_\beta X_\mu - i\bar{\psi}^\mu \rho^\alpha \nabla_\alpha \psi_\mu + 2\bar{\chi}_\alpha \rho^\beta \rho^\alpha \psi^\mu \partial_\beta X_\mu$$

$$+ \tfrac{1}{2}\bar{\psi}_\mu \psi^\mu \bar{\chi}_\alpha \rho^\beta \rho^\alpha \chi_\beta] \tag{3.10.1}$$

This action is invariant under

$$\text{2D SUSY} \quad \begin{cases} \delta X^\mu = \bar{\varepsilon}\psi^\mu \\ \delta\psi^\mu = -i\rho^\alpha \varepsilon(\partial_\alpha X^\mu - \bar{\psi}^\mu \chi_\alpha) \\ \delta e^a_\beta = -2i\bar{\varepsilon}\rho^a \chi_\beta \\ \delta\chi_\alpha = \nabla_\alpha \varepsilon \end{cases} \tag{3.10.2}$$

as well as Weyl, reparametrization, and local two-dimensional Lorentz invariance. The advantage of this formalism is that the energy–momentum tensor and the supersymmetric current generate an algebra, the super-Virasoro algebra, that allows us to explicitly eliminate all ghosts in the theory. We do not have to impose these gauge constraints by hand, as in the case of the superconformally invariant action.

Our strategy for quantization will be the same one used for the bosonic string. We first isolate the symmetries of the action, extract the algebra from these symmetries, then use this algebra to eliminate the ghost states and obtain a unitary theory:

Action → Symmetries → Current → Algebra → Constraints → Unitarity

If we eliminate the vierbein and the $\chi$ field, we get the NS–R theory in the superconformal gauge:

$$L = -\frac{1}{2\pi}(\partial_a X_\mu \partial^a X^\mu - i\bar{\psi}^\mu \rho^a \partial_a \psi_\mu) \tag{3.10.3}$$

Because of the invariance of the original action, which is locally supersymmetric, we can construct the following currents and set them equal to zero:

$$J_a = \tfrac{1}{2}\rho^b \rho_a \psi^\mu \partial_b X_\mu \tag{3.10.4}$$

$$T_{ab} = \partial_a X^\mu \partial_b X_\mu + \frac{i}{4}\bar{\psi}^\mu \rho_a \partial_b \psi_\mu + \frac{i}{4}\bar{\psi}^\mu \rho_b \partial_a \psi_\mu - (\text{Trace}) \tag{3.10.5}$$

If we calculate the moments of these constraints, they form the algebra:

$$\text{NS:}\quad \begin{cases} [L_m, L_n] = (m-n)L_{m+n} + \dfrac{D}{8}(m^3 - m)\delta_{m,-n} \\ [L_m, G_r] = (\tfrac{1}{2}m - r)G_{m+r} \\ \{G_r, G_s\} = 2L_{r+s} + \tfrac{1}{2}D(r^2 - \tfrac{1}{4})\delta_{r,-s} \end{cases}$$

$$\text{R:}\quad \begin{cases} [L_m, L_n] = (m-n)L_{m+n} + \dfrac{D}{8}m^3\delta_{m,-n} \\ [L_m, F_n] = (\tfrac{1}{2}m - n)F_{m+n} \\ \{F_m, F_n\} = 2L_{m+n} + \tfrac{1}{2}Dm^2\delta_{m,-n} \end{cases} \tag{3.10.6}$$

As in the bosonic case, it is straightforward to make the transition from path integrals to the harmonic oscillator formalism. The Hamiltonian is diagonal in the Fock space of harmonic oscillators, so we can remove all intermediate functional integrations.

We find that the tachyon vertex function with weight one can be written as

$$k_\mu \psi^\mu V_0 \tag{3.10.7}$$

However, we find that there are two distinct ways or "pictures" in which we can write down the $N$-point amplitude, called the $F_1$ and $F_2$ formalisms. The rules for forming the $N$-point functions are

$$F_1:\quad \begin{cases} \text{Vertex} = V \\ \text{Propagator} = \dfrac{1}{L_0 - 1} \\ \text{Tachyon} = k_\mu \cdot b_{-1/2}|0; k\rangle; \qquad \alpha' k^2 = \tfrac{1}{2} \\ \text{Vacuum} = |0; k\rangle; \qquad \alpha' k^2 = 1 \end{cases} \tag{3.10.8}$$

$$F_2: \begin{cases} \text{Vertex} = V \\ \text{Propagator} = \dfrac{1}{L_0 - \frac{1}{2}} \\ \text{Tachyon} = \text{Vacuum} = |0; k\rangle; \qquad \alpha' k^2 = \frac{1}{2} \end{cases} \tag{3.10.9}$$

The advantages and disadvantages of the two formalisms are as follows:

(1) In the $F_1$ formalism, manifest cyclic symmetry is much easier to prove. However, we must carry along the excess baggage of the vacuum state, which decouples completely in the theory.
(2) In the $F_2$ formalism, cyclic symmetry is obscure, but ghost elimination is easy to perform. The advantage of the $F_2$ picture is that we are now working in a smaller Fock space.

The $N$-point function is represented as

$$A_N = \int d\mu_N \langle 0; 0| \frac{V(k_1, y_1)}{y_1}, \ldots, \frac{V(k_N, y_N)}{y_N} |0; 0\rangle \tag{3.10.10}$$

Although the NS–R model with GSO projection can be shown to have the same number of fermion and boson states, the proof that the theory is actually space–time supersymmetric is prohibitively difficult. As a result, we will show space–time supersymmetry by postulating the new GS action, where we have genuine space–time spinors. The GS action is

$$S = \frac{-1}{4\alpha'\pi} \int d\sigma\, d\tau \{ \sqrt{g} g^{\alpha\beta} \Pi_\alpha \cdot \Pi_\beta + 2i\varepsilon^{\alpha\beta} \partial_\alpha X^\mu (\bar{\theta}^1 \Gamma_\mu \partial_\beta \theta^1 - \bar{\theta}^2 \Gamma_\mu \partial_\beta \theta^2)$$
$$- 2\varepsilon^{\alpha\beta} \bar{\theta}^1 \Gamma^\mu \partial_\alpha \theta^1 \bar{\theta}^2 \Gamma_\mu \partial_\beta \theta^2 \} \tag{3.10.11}$$

where

$$\Pi_\alpha^\mu = \partial_\alpha X^\mu - i\bar{\theta}^A \Gamma^\mu \partial_\alpha \theta^A \tag{3.10.12}$$

This action is invariant under

$$\begin{aligned} \delta\theta^A &= \varepsilon^A \\ \delta X^\mu &= i\bar{\varepsilon}^A \Gamma^\mu \theta^A \end{aligned} \tag{3.10.13}$$

as well as

$$\begin{aligned} \delta\theta^A &= 2i\Gamma \cdot \Pi_\alpha \kappa^{A\alpha} \\ \delta X^\mu &= i\bar{\theta}^A \Gamma^\mu \delta\theta^A \\ \delta(\sqrt{g} g^{\alpha\beta}) &= -16\sqrt{g}(P_-^{\alpha\gamma} \bar{\kappa}^{1\beta} \partial_\gamma \theta^1 + P_+^{\alpha\gamma} \kappa^{2\beta} \partial_\gamma \theta^2) \end{aligned} \tag{3.10.14}$$

The advantage of this GS formulation is that it is space–time supersymmetric to all levels. However, the price we pay for this is that covariant quantization of the theory is notoriously difficult, because it is a coupled system even at the free level. Therefore, we will go to the light cone gauge, where the theory becomes a linear theory:

$$S = \frac{-1}{4\pi\alpha'} \int d\sigma\, d\tau (\partial_a X^i \partial^a X^i - i\bar{S}\rho^b \partial_b S) \tag{3.10.15}$$

This action is also space–time supersymmetric, with generators given by

$$\begin{aligned} Q^a &= (2p^+)^{1/2} S_0^a \\ Q^{\dot{a}} &= (p^+)^{-1/2} \gamma^i_{a\dot{a}} \sum_{n=-\infty}^{\infty} S^a_{-n} \alpha^i_n \end{aligned} \tag{3.10.16}$$

The anticommutation relations between these generators are given by

$$\begin{cases} \{Q^a, Q^b\} = 2p^+ \delta^{ab} \\ \{Q^a, Q^{\dot{a}}\} = \sqrt{2}\gamma^i_{a\dot{a}} p^i \\ \{Q^{\dot{a}}, Q^{\dot{b}}\} = 2H\delta^{\dot{a}\dot{b}} \end{cases} \tag{3.10.17}$$

where

$$H = \frac{1}{p^+} \left\{ \sum_{n=1}^{\infty} [\alpha^i_{-n}\alpha^i_n + nS^a_{-n}S^a_n] + \frac{p_i^2}{2} \right\} \tag{3.10.18}$$

Next, we will investigate the conformal field theory, which combines the best features of both the NS–R and GS formalisms.

## References


[1] J. L. Gervais and B. Sakita, *Nucl. Phys.* **B34**, 632 (1971).
[2] W. Siegel, *Phys. Lett.* **128B**, 397 (1983); *Nucl. Phys.* **B263**, 93 (1985); *Class. Quant. Grav.* **2**, L95 (1985).
[3] L. Brink and J. H. Schwarz, *Phys. Lett.* **100B**, 310 (1981).
[4] P. Ramond, *Phys. Rev.* **D3**, 2415 (1971).
[5] A. Neveu and J. H. Schwarz, *Nucl. Phys.* **B31**, 86 (1971).
[6] A. Neveu, J. H. Schwarz, and C. B. Thorn, *Nucl. Phys.* **B31**, 529 (1971).
[7] J. Wess and B. Zumino, *Nucl. Phys.* **B70**, 39 (1974).
[8] See also Y. A. Gol'fand and E. P. Likhtman, *JETP Lett.* **13**, 323 (1971); D. V. Volkov and V. P. Akulov, *JETP Lett.* **16**, 621 (1972) (English, p. 438).
[9] M. A. Virasoro, unpublished.
[10] Y. Aharonov, A. Casher, and L. Susskind, *Phys. Rev.* **D5**, 988 (1972).
[11] L. Brink, P. Di Vecchia, and P. Howe, *Phys. Lett.* **65B**, 471 (1976).
[12] S. Deser and B. Zumino, *Phys. Lett.* **65B**, 369 (1976).
[13] For an earlier attempt, see Y. Iwasaki and K. Kikkawa, *Phys. Rev.* **D8**, 440 (1973).
[14] N. Ohta, *Phys. Rev.* **D33**, 1681 (1986).
[15] M. Ito, T. Morozumi, S. Nojiri, and S. Uehara, *Prog. Theor. Phys.* **75**, 934 (1986).
[16] J. Schwarz, *Suppl. Prog. Theor. Phys.* **86**, 70 (1986).
[17] F. Gliozzi, J. Scherk, and D. Olive, *Nucl. Phys.* **B122**, 253 (1977).
[18] C. B. Thorn, *Phys. Rev.* **D4**, 1112 (1971).
[19] L. Brink, D. Olive, C. Rebbi, and J. Scherk, *Phys. Lett.* **45B**, 379 (1973).
[20] S. Mandelstam, *Phys. Lett.* **46B**, 447 (1973).

[21] J. H. Schwarz and C. C. Wu, *Phys. Lett.* **47B**, 453 (1973).
[22] D. Bruce, E. Corrigan, and D. Olive, *Nucl. Phys.* **B95**, 427 (1975).
[23] M. Green and J. H. Schwarz, *Phys. Lett.* **136B**, 367 (1984).
[24] M. Green and J. H. Schwarz, *Nucl. Phys.* **B198**, 252, 441 (1982).
[25] T. Hori and K. Kamimura, *Prog. Theor. Phys.* **73**, 476 (1985).
[26] M. Kaku and J. Lykken, in *Symposium on Anomalies, Geometry, and Topology* (edited by W. A. Bardeen and A. R. White) World Scientific, Singapore, 1986.
[27] See also I. Bengtsson and M. Cederwall, Goteborg preprint 84-21.
[28] T. J. Allen, Cal Tech preprint CALT-68-1373.
[29] T. Hori, K. Kamimura, and M. Tatewaki, *Phys. Lett.* **185B**, 367 (1987).
[30] C. Crnkovic, *Phys. Lett.* **173B**, 429 (1986).
[31] R. E. Kallosh, *Pis'ma JETPh* **45**, 365 (1987); *Phys. Lett.* **195B**, 369 (1987).
[32] I. A. Batalin, R. E. Kallosh, and A. Van Proeyen, KUL-TF-87/17.
[33] N. Marcus and J. H. Schwarz, *Phys. Lett.* **115B**, 111 (1982).

CHAPTER 4

# Conformal Field Theory and Kac–Moody Algebras

## §4.1. Conformal Field Theory

One of the mysteries of the superstring is the existence of two ways of formulating the theory: the first is the NS–R model (after the GSO projection) with anticommuting vectors, and the second is the GS model with genuine anticommuting spinors. Each formulation has its own distinct advantages and disadvantages. In this chapter, we will discuss the conformal field theory, which allows us to see the dynamic link between the two formalisms. The conformal field theory of Friedan and Shenker [1] combines the best features of both theories. Conformal field theory allows us to

(1) Introduce covariant anticommuting spinor fields based entirely on free fields. The GS formalism, on the other hand, is based on complicated interacting fields that make covariant quantization exceedingly difficult.
(2) Construct explicit covariant tree graphs for multifermion scattering. In the NS–R formalism, however, this is prohibitively difficult because it is necessary to introduce complicated projection operators that extract out the ghosts. The conformal field theory replaces these awkward projection operators with free Faddeev–Popov ghosts, which are easy to manipulate.
(3) Interpolate between the GS and NS–R formulations and see their relationship. It provides the bridge by which one can interpret the results of one formulation in terms of the other:

$$\text{Conformal field theory} \leftrightarrow \begin{cases} \text{GS model} \\ \text{NS–R model} \end{cases}$$

(4) Construct the covariant supersymmetry generators. This is impossible in the NS–R formalism, and is possible in the GS formalism only in the light cone gauge.

(5) Describe both the NS and R sectors of the theory with the same vacuum, rather than using the clumsy formalism of two distinct Hilbert spaces based on the vacua $|0\rangle_{\text{NS}}$ and $|0\rangle_{\text{R}} u_\alpha$. This is accomplished by a process called bosonization, i.e., constructing fermions out of bosons in two dimensions.

There is a small price we must pay, however, for conformal field theory. Ghosts and anti-ghosts proliferate quite rapidly in this formalism, especially for superstrings, and a strange phenomenon called "picture changing" (see [3.3.18–9]) must be introduced. Fortunately, these ghosts and anti-ghosts are free fields, and hence easy to manipulate. Furthermore, conformal field theory does not seem to be derived from any action, as in the GS and NS–R formalisms. Conformal field theory stresses the group theoretic behavior of the fields, rather than proceeding from an action. Therefore, we suspect that there is a higher, yet undiscovered first quantized action that exists beyond the GS and NS–R actions.

(We should stress that conformal field theory is not a field theory in the sense of second quantization, i.e., where we start with the formalism of Schwinger, Tomonaga, and Feynman. The second quantized field theory of strings will be discussed in Part II of this book.)

The essence of conformal field theory is that it stresses the use of conformal invariance alone to calculate correlation functions between different fields [2–6]. It is quite remarkable that conformal invariance by itself is sufficient to determine almost completely the structure of $N$-point scattering amplitudes. For example, in (2.6.9) we have encountered operators that have the following matrix elements:

$$\langle \phi(z_1)\phi(z_2) \rangle = f(z_1 - z_2) \tag{4.1.1}$$

where the function $f$ can be a power or a logarithm. For example, the matrix element for tree amplitudes between two bosonic strings is

$$\langle X(w)X(z) \rangle \sim \log(w - z)$$

while the matrix element of two normal ordered vertices is

$$\langle e^{ikX(w)} e^{-ikX(z)} \rangle \sim (w - z)^{-k^2}$$

So far, we have used explicit representations of the fields in order to calculate the matrix elements. However, the process can also be reversed. If we knew all about a field's group properties, we should be able to calculate its matrix elements and even reconstruct the field.

The essential idea behind conformal field theory is to use the conformal properties of the field $\phi$ to completely determine all of its matrix elements and even reconstruct the field. In conformal field theory, this is accomplished by knowing the short-distance behavior of left-moving or right-moving fields

$$\phi(z_1)\phi(z_2) \sim \frac{1}{(z_1 - z_2)^n} + \text{less singular terms} \tag{4.1.2}$$

*Thus, in principle, all possible matrix elements can be calculated from the conformal properties of the fields themselves.*

In conformal field theory, we also construct a *spin field* $S_\alpha$ that transforms as a genuine spinor under space–time Lorentz transformations and has conformal weight $\frac{5}{8}$. However, coming from the Faddeev–Popov ghost sector, we find yet another field with conformal weight $\frac{3}{8}$. It is the product of these two fields, one being a spin field and the other being a ghost field, that allows us to construct the complete fermion vertex with the proper weight one. This vertex, although it involves exponentials of fields, is defined totally in terms of free fields and thus solves the problem of constructing a simple fermion vertex function.

The way in which this spin field is actually introduced is through the process called "bosonization" [7, 8], i.e., creating a fermion out of a boson. It is easy, of course, to create a boson out of two fermions. However (in two dimensions) we have the option of being able to create fermions out of bosons, which was once thought to be impossible. Actually, the hint of how to "bosonize" was already given in the previous chapter, when we introduced the vertex function

$$V = :e^{ik\cdot X}: \tag{4.1.3}$$

In (2.6.22) we saw that

$$V_1 V_2 = V_2 V_1 e^{i\pi k_1 \cdot k_2 \varepsilon} \tag{4.1.4}$$

Notice that, by choosing various values of the momenta in the exponential, we can actually create operators that satisfy the relation

$$V_1 V_2 = (\pm 1) V_2 V_1 \tag{4.1.5}$$

In other words, we can create a field with fermionic commutation relations out of bosonic harmonic oscillators. The key to this construction is the normal ordering of oscillator modes in two dimensions. This feature does not carry over to four-dimensional field theories. (In hindsight, it is possible to see why fermions and bosons are so closely linked in two dimensions but not in higher dimensions. For the Lorentz group O(1, 1), which has only one generator, the concept called "spin" in one spatial dimension does not have much meaning.)

This method of bosonization via normal ordering of fields will also be the key to constructing the conformal field theory. We will construct the fermion spin field $S_\alpha$ via normal ordering of exponentials of bosonic fields. From this, we will construct the covariant supersymmetry operator. The advantage of this is that we can now discuss space–time spinors using one common vacuum for both bosons and fermions and that the entire construction is based on free fields.

Before we discuss the superconformal case, let us begin our discussion by analyzing conformal field theory. Let us make the most general conformal transformation on the world sheet variable $z$:

$$z \to \tilde{z}(z) \tag{4.1.6}$$

Under this conformal transformation, we say that a primary analytic function transforms with *conformal weight h* if it transforms as

$$\phi(z) = \tilde{\phi}(\tilde{z})\left(\frac{d\tilde{z}}{dz}\right)^h \tag{4.1.7}$$

(Secondary fields transform as derivatives of $\phi(z)$.) This conformal weight is the same concept we introduced back in (2.7.6) when we were discussing the conformal properties of vertices. We see now that mathematically the conformal weight is an index to label irreducible representations of the conformal group generated by the Virasoro algebra.

We can now construct objects that are invariant under a conformal transformation:

$$\phi(z)\, dz^h = \tilde{\phi}(\tilde{z})\, d\tilde{z}^h \tag{4.1.8}$$

Once we have defined how fields transform under a conformal transformation, we must then check that we have closure of the group under two such conformal operations. Let us say that we make two successive conformal transformations:

$$z \to z_1(z) \to z_2[z_1(z)] \tag{4.1.9}$$

Then the field transforms as

$$U_1 \phi(z) U_1^{-1} = \phi(z_1)\left(\frac{dz_1}{dz}\right)^h \tag{4.1.10}$$

$$U_2 U_1 \phi(z) U_1^{-1} U_2^{-2} = \phi[z_2(z_1(z))]\left(\frac{dz_2}{dz_1}\frac{dz_1}{dz}\right)^h \tag{4.1.11}$$

Thus, the conformal transformations form a group with the composition law

$$\begin{aligned} U_3 &= U_2 U_1 \\ z_3 &= z_2(z_1(z)) \end{aligned} \tag{4.1.12}$$

The closure of the algebra can most easily be seen if we take an infinitesimal conformal transformation. Consider an infinitesimal variation of the coordinate:

$$\delta z = \varepsilon(z) \tag{4.1.13}$$

Under the transformation (4.1.7), we have the infinitesimal transformation

$$\delta\phi_\varepsilon = [\varepsilon\partial + h(\partial\varepsilon)]\phi \tag{4.1.14}$$

(We will abbreviate $\partial_z$ by $\partial$.) If we commute two such variations, then we arrive at

$$\delta_{[\varepsilon_1, \varepsilon_2]} = [\delta_{\varepsilon_1}, \delta_{\varepsilon_2}] \tag{4.1.15}$$

where

$$[\varepsilon_1, \varepsilon_2] = \varepsilon_1 \partial\varepsilon_2 - \varepsilon_2 \partial\varepsilon_1 \tag{4.1.16}$$

(Compare this relation with (2.1.30).)

We have shown once again that the group closes with arbitrary weight $h$. To get a better understanding of the meaning of the conformal weight $h$, let us calculate the conformal weight of the string. Under the infinitesimal variation of the coordinate, we use the chain rule to show that the string transforms as

$$\delta X_\mu(z) = \varepsilon \partial X_\mu(z) \tag{4.1.17}$$

Thus, the string field was weight 0. Likewise, it is easy to show that the derivative of the string field has weight 1:

$$\delta \partial X_\mu(z) = [\varepsilon \partial + (\partial \varepsilon)] \partial X_\mu(z) \tag{4.1.18}$$

Let us summarize the weights of some common string fields:

| Field | Weight |
|---|---|
| $X$ | 0 |
| $\partial X$ | 1 |
| $\partial X \cdot \partial X$ | 2 |

(4.1.19)

In particular, this means that the energy–momentum tensor or the Virasoro generators have weight 2.

Weights are additive. The product of two fields of weights $h_1$ and $h_2$ produces a field of weight $h_1 + h_2$ at the same point:

$$\phi^{(h_1)}(z)\phi^{(h_2)}(z) = \phi^{(h_1+h_2)}(z) \tag{4.1.20}$$

Notice the important fact, which will be used throughout this book, that the integral of an object of weight 1 is an invariant:

$$\begin{aligned} \delta \oint \phi^{(1)}(z)\, dz &= \oint [v\partial\phi + (\partial v)\phi]\, dz \\ &= \oint \partial[v\phi]\, dz \\ &= 0 \end{aligned} \tag{4.1.21}$$

We will use this fact in constructing vertex operators and also the action of the second quantized theory.

Now let us investigate in detail how the energy–momentum tensor acts on the basic fields of the theory.

If the action is written in terms of $X_\mu(z, \bar{z})$,

$$S = \frac{1}{2\pi} \int d^2 z\, \bar{\partial} X^\mu \partial X_\mu \tag{4.1.22}$$

then the energy–momentum tensor in (1.9.17) associated with this action is

$$T_B = -\tfrac{1}{2}\partial X^\mu \partial X_\mu \tag{4.1.23}$$

The transformation of fields under conformal transformations is generated by the integrated energy–momentum tensor parametrized by a small function $\varepsilon$:

$$T_\varepsilon = \frac{1}{2\pi i}\oint_{C_0} dz\, \varepsilon(z) T(z) \tag{4.1.24}$$

where we take the line integral that encircles the origin in $z$ space. The importance of the energy–momentum tensor $T$ is that it is the generator of the conformal transformations we have been studying. To see this, write

$$\begin{aligned}
\delta\phi(z_2) &= [T_\varepsilon, \phi(z_2)] \\
&= \left[\frac{1}{2\pi i}\oint_{C_0} dz\, \varepsilon(z) T(z), \phi(z_2)\right] \\
&= \frac{1}{2\pi i}\oint_{C_{0,2}-C_0} dz\, P\varepsilon(z) T(z)\phi(z_2) \\
&= \frac{1}{2\pi i}\oint_{C_2} P\varepsilon(z) T(z)\phi(z_2) \\
&= \frac{1}{2\pi i}\oint_{C_2} dz\, \varepsilon(z) T(z)\phi(z_2)
\end{aligned} \tag{4.1.25}$$

Notice that the curve $C_{0,2}$ encircles both the origin and the point $z_2$, whereas the curve $C_2$ encircles only the point $z_2$, because

$$C_2 = C_{0,2} - C_0 \tag{4.1.26}$$

(see Fig. 4.1). Notice also that we have adopted *radial* ordering $P$ of our operators in the complex plane, where the operators are ordered according to their distance from the origin:

$$|z| \le |z_1| \tag{4.1.27}$$

In the last step, notice that we have dropped radial ordering, because the radial-ordered products are analytic.

From before, we also know that

$$\delta\phi = [\varepsilon\partial + h(\partial\varepsilon)]\phi \tag{4.1.28}$$

By comparing the two expressions (4.1.25) and (4.1.28) for $\delta\phi$, we can now read off the small-distance behavior of the field $\phi$:

$$T(z)\phi(w, \bar{w}) \sim \frac{h}{(z-w)^2}\phi(w, \bar{w}) + \frac{1}{z-w}\partial_w \phi + \cdots \tag{4.1.29}$$

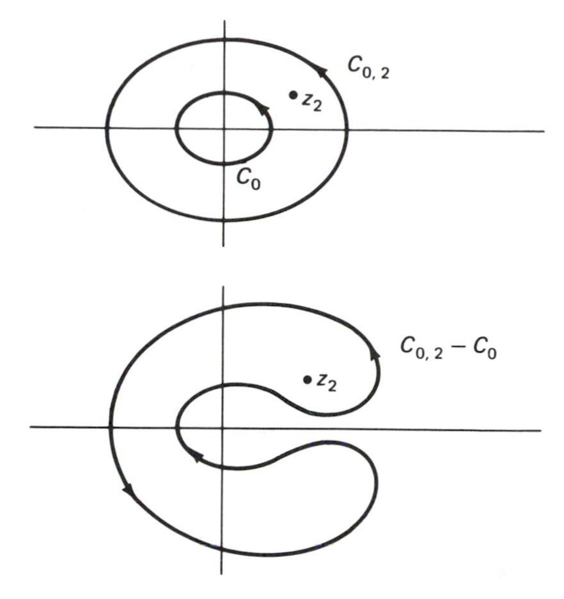


Figure 4.1. Contours for conformal field theory integrations. In the first diagram, the point $z$ lies between two concentric circles. In the second diagram, the direction of the inner contour has been reversed and merged with the outer contour, forming a closed path. The point $z$ lies within the closed path, but the origin does not.

Since the energy–momentum tensor itself has weight 2, we can also insert it into the short-distance equation. Thus, we derive the short-distance behavior of the generators themselves:

$$T(z)T(w) = \frac{1}{2}\frac{c}{(z-w)^4} + \frac{2}{(z-w)^2}T(w) + \frac{1}{(z-w)}\partial_w T(w) + \cdots \tag{4.1.30}$$

The second term on the right-hand side of the equation shows that the generators themselves transform as weight 2 conformal fields.

Let us now introduce normal modes and make the link with the oscillator formalism. We can always decompose any field $\psi$ of weight $h$ as follows:

$$\psi(z) = \sum_{n=-\infty}^{\infty} z^{-n-h}\psi_n \tag{4.1.31}$$

This allows us to decompose the energy–momentum into normal modes. We write

$$\begin{aligned} L_n &= \oint \frac{dz}{2\pi i} z^{n+1} T(z) \\ T(z) &= \sum_{n=-\infty}^{\infty} z^{-n-2} L_n \end{aligned} \tag{4.1.32}$$

It will be instructive to actually derive, step by step, the Virasoro algebra from these abstract expressions to see the equivalence of commutators and operator product expansions. Using (4.1.30) and (4.1.32), we find

$$\begin{aligned}
[L_n, L_m] &= \oint \frac{dw}{2\pi i} w^{n+1} \oint \frac{dz}{2\pi i} z^{n+1} \\
&\quad \times \left[ \frac{\frac{1}{2}c}{(z-w)^4} + \frac{2}{(z-w)^2} T(w) + \frac{1}{(z-w)} \partial_w T \right] \\
&= \oint \frac{dw}{2\pi i} w^{n+1} \Big\{ (n+1) w^n 2T(w) + w^{n+1} \partial_w T \\
&\quad + \frac{\frac{1}{2}c}{3!} w^{n-2} (n+1)n(n-1) \Big\} \\
&= \oint \frac{dw}{2\pi i} \Big\{ (2n+2) w^{n+m+1} T(w) - (m+n+2) w^{n+m+1} T(w) \\
&\quad + \frac{c}{12} n(n+1)(n-1) w^{m+n-1} \Big\}
\end{aligned}$$

This finally reduces down to the usual Virasoro algebra:

$$[L_m, L_n] = (m-n) L_{m+n} + \frac{c}{12}(m^3 - m)\delta_{m,-n} \tag{4.1.33}$$

Inserting the expression for $L_n$ into the line integral, we also obtain

$$[L_n, \phi(z)] = \{z^{n+1}\partial + h(n+1)z^n\}\phi(z) \tag{4.1.34}$$

(This can also be derived from (4.1.28).)

We have shown that this description is equivalent to the usual one in terms of commutators that we studied in previous chapters. In fact, the previous equation is nothing but (2.7.6), which we used to calculate the conformal weight of a vertex function.

Not only can we express the usual conformal commutators in this fashion, we can also express the Faddeev–Popov ghost contribution in the conformal gauge in this equivalent language. The Faddeev–Popov ghosts will give us a different representation of the Virasoro algebra based on $b$ and $c$ ghost fields rather than the string variable $X$. If we rewrite (2.4.1), we have

$$\begin{aligned}
\delta g_{zz} &= \nabla_z \delta\varepsilon_z \\
\delta g_{\bar{z}\bar{z}} &= \nabla_{\bar{z}} \delta\varepsilon_{\bar{z}}
\end{aligned} \tag{4.1.35}$$

These constraints, in turn, allow us to calculate the change in the functional measure. A simple calculation of the Jacobian of the transformation yields

$$Dg_{zz} Dg_{\bar{z}\bar{z}} = (\det \nabla_z)(\det \nabla_{\bar{z}}) D\varepsilon_z D\varepsilon_{\bar{z}} \tag{4.1.36}$$

To exponentiate the determinant into the action, let us introduce two ghost fields, $b$ and $c$. Then the ghost action (2.4.4) can be written as

$$S_{\mathrm{gh}} = \frac{1}{\pi} \int d^2 z \, (b_{zz} \partial_{\bar{z}} c^z + \bar{b}_{\bar{z}\bar{z}} \partial_z \bar{c}^{\bar{z}}) \tag{4.1.37}$$

(Note that we have written the ghost action such that the tensorial nature of the fields is manifest under the conformal group. In this way, the invariance of the action under conformal transformations is transparent. Thus, the field $b_{zz}$ is a second-rank tensor, while $c^z$ is a first-rank tensor.)

Given this action, we can construct the energy–momentum tensor associated with it:

$$T_{\mathrm{gh}} = c\partial_z b + 2(\partial_z c)b \tag{4.1.38}$$

These ghost fields have the following matrix element:

$$\langle b_{zz}(z) c^z(w) \rangle = \frac{1}{z - w}$$

If we now carefully calculate the short-distance behavior of these fields, we find

$$T_{\mathrm{gh}}(z) T_{\mathrm{gh}}(w) = \frac{-13}{(z-w)^4} + \frac{2}{(z-w)^2} T_{\mathrm{gh}}(z) + \frac{1}{z-w} \partial T_{\mathrm{gh}}(z) + \cdots \tag{4.1.39}$$

Notice the factor of $-13$ in the anomaly calculation for the ghost energy–momentum tensor. If we take the sum of these two energy–momentum tensors, (4.1.23) and (4.1.38), we find that the combined tensor has a vanishing anomaly in 26 dimensions:

$$\begin{aligned} T(z) &= T_X(z) + T_{\mathrm{gh}}(z) \\ T(z)T(w) &\sim \tfrac{1}{2} \frac{D - 26}{(z-w)^4} + \cdots \end{aligned} \tag{4.1.40}$$

Notice that the sum of the two energy–momentum tensors has zero central term only in 26 dimensions. *This, in fact, fixes the dimension of space–time to be* 26. To summarize, the bosonic representation of the energy–momentum tensor in terms of string fields $X$ yields $D$ for the anomaly term, while the ghost representation of $T$ yields $-26$. The sum of these two tensors is the true energy–momentum tensor, which has vanishing central term only if $D = 26$.

Whenever we have a Lie group, the first question to ask is: What are its representations? Let us now say a few words about the representations of the Virasoro algebra.

For SU(2) or SU(3), for example, we know that the representations can be formed by using the familiar "ladder operators" to construct triplets, octets, and higher representations. For a general Lie algebra, we can construct representations by taking all possible products of "raising operators" acting on a "highest weight" vacuum. The set of all such states created by the raising operators is called the "universal enveloping algebra." The same process can be used for the Virasoro algebra, treating $L_{-n}$ as ladder or raising operators.

We will define the *highest weight vector* $|h\rangle$ as follows:

$$\begin{aligned} L_0|h\rangle &= h|h\rangle \\ L_n|h\rangle &= 0; \quad \text{where } n > 0 \end{aligned} \tag{4.1.41}$$

Notice that any physical state $|R\rangle$ with $h = 1$ is a highest weight vector, since it satisfies $(L_0 - 1)|R\rangle = 0$. Notice that the vacuum state $|0; k\rangle$ is also highest weight vector, but that $|0; 0\rangle$ is not.

Now define the set of states generated by all "raising" operators $L_{-n}$ acting on the highest weight vector:

$$|\omega\rangle = L_{-n_1}^{\lambda_1} L_{-n_2}^{\lambda_2} \cdots L_{-n_N}^{\lambda_N}|h\rangle \tag{4.1.42}$$

The set of all such states is called a *Verma module*. If $|h\rangle$ is considered a physical state, then the Verma module associated with it is the set of all spurious states that we can generate from that real state.

Notice that, under an arbitrary conformal transformation, this state simply transforms into another member of the Verma module:

$$\Omega|\omega\rangle = |\omega'\rangle \tag{4.1.43}$$

where

$$\Omega = e^{\sum_n c^n L_{-n}} \tag{4.1.44}$$

This is because a Virasoro generator that hits an element of a Verma module simply creates another state within the same Verma module. The Verma module therefore forms a representation of the group.

In addition to the highest weight vector, let us also define the SL(2, $R$) invariant vacuum, which we introduced earlier:

$$\begin{aligned} L_{\pm 1}|0; 0\rangle &= 0 \\ L_0|0; 0\rangle &= 0 \end{aligned} \tag{4.1.45}$$

The relationship between these two different types of vacua is

$$|h\rangle = \phi(0)|0; 0\rangle \tag{4.1.46}$$

where $\phi(0)$ is a primary field of weight $h$. This is easy to check if we hit both sides of the equation with the Virasoro generators.

## §4.2. Superconformal Field Theory

Now let us discuss the more complex question of writing the superconformal field theory. Instead of writing down the transformation of a field as a function of a complex variable, we write down the most arbitrary transformation of a pair of variables:

$$\mathbf{z} = \mathbf{z}(z, \theta) \tag{4.2.1}$$

where $\theta$ is a Grassman variable. The most arbitrary transformation of this pair is given by

$$\tilde{\mathbf{z}}(\mathbf{z}) = (\tilde{z}(z, \theta), \tilde{\theta}(z, \theta)) \tag{4.2.2}$$

It is disappointing that this transformation is much too general. In fact, it doesn't close properly if we take the product of two such transformations. We need to *impose a constraint* on the system in order to make the group of transformations close properly.

We define the supersymmetric derivative as (see Appendix)

$$D \equiv \frac{\partial}{\partial \theta} + \theta \frac{\partial}{\partial z} \tag{4.2.3}$$

so that

$$D^2 = \frac{\partial}{\partial z} \tag{4.2.4}$$

Now let us calculate how this supersymmetric derivative transforms under a reparametrization:

$$D = (D\tilde{\theta})\tilde{D} + [D\tilde{z} - \tilde{\theta}D\tilde{\theta}]\tilde{D}^2 \tag{4.2.5}$$

This composition law is also disappointing because it is highly nonlinear. We want, as in the conformal case, a transformation of the derivative to be linear. Therefore, we simply eliminate the nonlinear terms by imposing the constraints

$$\begin{aligned} D &= (D\tilde{\theta})\tilde{D} \\ D\tilde{z} - \tilde{\theta}D\tilde{\theta} &= 0 \end{aligned} \tag{4.2.6}$$

These constraints, which linearize the transformation of the reparametrization, are precisely the constraints we want. By imposing them, we can show that two distinct superconformal transformations close properly:

$$\mathbf{z} \to \tilde{\mathbf{z}} \to \hat{\mathbf{z}} \tag{4.2.7}$$

In summary, we say a field transformation is superconformal if it satisfies

$$\begin{cases} \dfrac{d\tilde{\mathbf{z}}}{d\mathbf{z}} = D\tilde{\theta} \\ \phi(\mathbf{z}) = \tilde{\phi}(\tilde{\mathbf{z}})(D\tilde{\theta})^{2h} \end{cases} \tag{4.2.8}$$

It is easy to check that if we eliminate the $\theta$ components, this expression reduces back to the conformal equation written earlier in (4.1.7).

As in the previous section, we can write this down as an infinitesimal variation on the fields:

$$\delta_\varepsilon \phi = [\varepsilon \partial + \tfrac{1}{2}(D\varepsilon)D + h(\partial \varepsilon)]\phi \tag{4.2.9}$$

where $\varepsilon$ parametrizes the superconformal transformations. Notice that this expression is exactly identical to the bosonic transformation (4.1.28) except for the term that has two $D$'s in it. The closure of the algebra can now be written

$$\delta_{[\varepsilon_1, \varepsilon_2]} = [\delta_{\varepsilon_1}, \delta_{\varepsilon_2}] \tag{4.2.10}$$

where

$$[\varepsilon_1, \varepsilon_2] = \varepsilon_1 \partial \varepsilon_2 - \varepsilon_2 \partial \varepsilon_1 + \tfrac{1}{2}(D\varepsilon_1)(D\varepsilon_2) \tag{4.2.11}$$

Notice that we reduce back to the usual Virasoro case (4.1.16) when we set $D\varepsilon_1$ and $D\varepsilon_2$ equal to zero.

We find that the Grassman parametrizations can be written in terms of two parameters, $\partial z = \varepsilon_a$ and $\partial\theta = \varepsilon_b$:

$$\begin{aligned} \varepsilon(\mathbf{z}) &= \delta z + \theta\delta\theta = \varepsilon_a + \theta\varepsilon_b \\ \delta z &= \varepsilon - \theta\delta\theta = \varepsilon_a + \tfrac{1}{2}\theta\varepsilon_b \\ \delta\theta &= \tfrac{1}{2}D\varepsilon = \tfrac{1}{2}(\varepsilon_b + \theta\partial\varepsilon_a) \end{aligned} \tag{4.2.12}$$

The previous equation is a specific solution of (4.2.6), so we know that the group closes properly when we make successive transformations on the fields. This solution will be the basis on which we will construct the superconformal currents and then the superconformal algebra itself.

So far, we have made only very general remarks about the superconformal group without referring to any specific model. Let us now rewrite the NS–R model in terms of conformal fields. The action is

$$S = \frac{1}{2\pi} \int d^2z \, d\theta \, d\bar{\theta} \, \bar{D}X^\mu \, DX_\mu \tag{4.2.13}$$

The equations of motion are

$$\bar{D} \, DX_\mu = 0 \tag{4.2.14}$$

whose solution is

$$X^\mu(z, \theta, \bar{z}, \bar{\theta}) = X^\mu(z, \theta) + X^\mu(\bar{z}, \bar{\theta}) \tag{4.2.15}$$

Thus we choose

$$X^\mu(z, \theta) = X^\mu(z) + \theta\psi^\mu(z) \tag{4.2.16}$$

so that the action can be written as

$$S = \frac{1}{2\pi} \int d^2z (\bar{\partial}X^\mu \partial X_\mu - \psi\bar{\partial}\psi - \bar{\psi}\partial\bar{\psi}) \tag{4.2.17}$$

Notice that this is nothing but the original NS–R (3.2.1) action written in conformal language. From this action, we can read off the super-energy–momentum tensor: with a slightly different normalization.

$$T = T_F + \theta T_B = -\tfrac{1}{2}DX^\mu \partial X_\mu \tag{4.2.18}$$

Written out explicitly, this is

$$T_F = -\tfrac{1}{2}\psi_\mu \partial X^\mu$$
$$T_B = -\tfrac{1}{2}\partial X^\mu \partial X_\mu - \tfrac{1}{2}\partial\psi^\mu \psi_\mu \tag{4.2.19}$$

This is nothing but the currents (3.2.7) and (3.2.13) written in conformal language. Note that the supercurrent $J_a$ is now written as part of the same tensor as the energy–momentum tensor $T_{ab}$, which means that they are supersymmetric partners of each other.

As in (4.1.25), the variation of a superfield is given by

$$\delta\phi(\mathbf{z}_2) = \frac{1}{2\pi i}\oint_{c_2} d\mathbf{z}_1\, \varepsilon(\mathbf{z}_1) T(\mathbf{z}_1)\phi(\mathbf{z}_2) \tag{4.2.20}$$

If the field $\phi$ has weight $h$ and the energy–momentum tensor has weight $\frac{3}{2}$, then $\phi$ and $T$ transform under a superconformal transformation as

$$T(\mathbf{z}_1)\phi(\mathbf{z}_2) \sim h\frac{\theta_{12}}{z_{12}^2}\phi(\mathbf{z}_2) + \frac{\frac{1}{2}}{z_{12}}D_2\phi + \frac{\theta_{12}}{z_{12}}\partial_2\phi + \cdots$$
$$T(\mathbf{z}_1)T(\mathbf{z}_2) \sim \frac{\hat{c}}{4}\frac{1}{z_{12}^3} + \frac{3}{2}\frac{\theta_{12}}{z_{12}^2}T(\mathbf{z}_2) \tag{4.2.21}$$
$$+ \frac{1}{2}\frac{1}{z_{12}}D_2 T(\mathbf{z}_2) + \frac{\theta_{12}}{z_{12}}\partial_2 T(\mathbf{z}_2) + \cdots$$

where

$$\theta_{12} = \theta_1 - \theta_2$$
$$z_{12} = z_1 - z_2 - \theta_1\theta_2 \tag{4.2.22}$$

Actually, these relations are a deceptively compact way of expressing a large number of equations. To see this, let us expand out the previous expression. Inserting (4.2.18) into (4.2.21), we find

$$T_B(z_1)T_B(z_2) \sim \frac{3\hat{c}/4}{(z_1 - z_2)^4} + \frac{2}{(z_1 - z_2)^2}T_B(z_2) + \frac{1}{z_1 - z_2}\partial_2 T_B + \cdots$$
$$T_B(z_1)T_F(z_2) \sim \frac{\frac{3}{2}}{(z_1 - z_2)^2}T_F(z_2) + \frac{1}{z_1 - z_2}\partial_2 T_F + \cdots \tag{4.2.23}$$
$$T_F(z_1)T_F(z_2) \sim \frac{\hat{c}/4}{(z_1 - z_2)^3} + \frac{\frac{1}{2}}{z_1 - z_2}T_B(z_2) + \cdots$$

These relations, when expanded out, yield the superconformal algebra that we wrote out explicitly in (3.2.26) and (3.2.28). Thus, these equations contain a considerable amount of information.

The field transformation in (4.2.21) can be written out in detail if $\phi =$

$\phi_0 + \theta\phi_1$

$$\begin{aligned} T_B(z_1)\phi_0(z_2) &\sim \frac{h}{(z_1-z_2)^2}\phi_0(z_2) + \frac{1}{z_1-z_2}\partial_2\phi_0 + \cdots \\ T_B(z_1)\phi_1(z_2) &\sim \frac{h+\frac{1}{2}}{(z_1-z_2)^2}\phi_1(z_2) + \frac{1}{z_1-z_2}\partial_2\phi_1 + \cdots \\ T_F(z_1)\phi_0(z_2) &\sim \frac{\frac{1}{2}}{z_1-z_2}\phi_1(z_2) + \cdots \\ T_F(z_1)\phi_1(z_2) &\sim \frac{h}{(z_1-z_2)^2}\phi_0(z_2) + \frac{\frac{1}{2}}{z_1-z_2}\partial_2\phi_0 + \cdots \end{aligned} \tag{4.2.24}$$

If we power expand as

$$\begin{aligned} T_F(z) &= \tfrac{1}{2}\sum_n z^{-n-3/2}G_n \\ T_B(z) &= \sum_n z^{-n-2}L_n \\ \phi_0(z) &= \sum_n z^{-n-h}\phi_{0,n} \\ \phi_1(z) &= \sum_n z^{-n-h-1/2}\phi_{1,n} \end{aligned} \tag{4.2.25}$$

then we obtain the transformation of $\varphi(z,\theta)$ under a superconformal transformation:

$$\begin{aligned} [L_m, \phi_0(z)] &= z^{m+1}\partial\phi_0 + h(m+1)z^m\phi_0(z) \\ [L_m, \phi_1(z)] &= z^{m+1}\partial\phi_1 + (h+\tfrac{1}{2})(m+1)z^m\phi_1(z) \\ [\varepsilon G_m, \phi_0(z)] &= \varepsilon z^{m+1/2}\phi_1(z) \\ [\varepsilon G_m, \phi_1(z)] &= \varepsilon[z^{m+1/2}\partial\phi_0 + 2(m+\tfrac{1}{2})hz^{m-1/2}\phi_0(z)] \end{aligned} \tag{4.2.26}$$

$$\begin{aligned} [L_m, \phi_{0,n}] &= [(h-1)m-n]\phi_{0,m+n} \\ [L_m, \phi_{1,n}] &= [(h-\tfrac{1}{2})m-n]\phi_{1,m+n} \\ [\varepsilon G_m, \phi_{0,n}] &= \varepsilon\phi_{1,m+n} \\ [\varepsilon G_m, \phi_{1,n}] &= \varepsilon[(2h-1)m-n]\phi_{0,m+n} \end{aligned} \tag{4.2.27}$$

Thus, relations (4.2.21) contain the entire information concerning superconformal transformations.

Now let us calculate the contribution to the anomaly. Comparing (4.1.29) and (4.2.23), we see that a boson $X_\mu$ contributes $D$ to the anomaly, whereas a fermion $\psi_\mu$ contributes only $\frac{1}{2}D$. We can also show that the contributions of the $b$, $c$ ghosts and $\beta$, $\gamma$ ghosts yield factors of $-26$ and $+11$ for the anomaly. Later in this chapter, in fact, we will show that the anomaly contribution by a ghost of conformal weight $\lambda$ is

$$c = -2\varepsilon(6\lambda(\lambda-1)+1) \tag{4.2.28}$$

where $\varepsilon$ equals $+1$ $(-1)$ for Fermi (Bose) statistics. Summarizing this, we find

| Field | Anomaly |
|---|---|
| $X_\mu$ | $D$ |
| $\psi_\mu$ | $\frac{1}{2}D$ |
| $b, c$ | $-26$ |
| $\beta, \gamma$ | $11$ |

(4.2.29)

Adding everything together, we find that the total anomaly contribution is

$$D + \tfrac{1}{2}D - 26 + 11 = \tfrac{3}{2}(D - 10) \tag{4.2.30}$$

so $D = 10$ in order to cancel this term.

## §4.3. Spin Fields

So far, almost nothing is new. We have only rederived old results that could have been obtained with the harmonic oscillator formalism of the previous chapters. We have simply chosen to rewrite the generators of the NS–R algebra in a language that stresses the conformal weight and the $z$-plane singularity structure of the commutators, rather than expanding them out in terms of their Fourier moments. What is new, however, will be the introduction of a new field $S_\alpha$ that transforms as a genuine spinor under the Lorentz group. This will allow us to rewrite the NS–R model with only one vacuum, rather than two, by the process called bosonization.

Let us first write down the generators $j^{\mu\nu}$ of the 10-dimensional Lorentz group SO(10) in terms of the anticommuting vector field $\psi$ found in the NS–R theory:

$$j^{\mu\nu}(z) = \psi^\mu\psi^\nu(z) \tag{4.3.1}$$

We demand that they have the commutation relations of the usual Lorentz algebra, but reexpressed in terms of fields defined in the $z$-plane. Given the known anticommutation relations of the $\psi$ field given in (3.2.23), we can readily show

$$j^{\mu\nu}(z)j^{\sigma\tau}(w) \sim \frac{1}{z-w}\eta^{\mu\sigma}j^{\nu\tau}(w)(1 - \mu\leftrightarrow\nu)(1 - \sigma\leftrightarrow\tau)$$

$$+ \frac{1}{(z-w)^2}(\eta^{\mu\tau}\eta^{\nu\sigma} - \mu\leftrightarrow\nu) + \cdots \tag{4.3.2}$$

Notice that the generators of this algebra are functions of $z$ and hence form a larger algebra than that of SO(10). In fact, this algebra is called a *Kac–Moody algebra* and we will use such algebras extensively in this book.

When studying the group theory behind Lorentz symmetry, we know that the representations of the group can be grouped into tensors and spinors. The transformation properties of these fields can be uniquely determined from group theory alone. Similarly, we will define tensor and spinor representations of the Kac–Moody algebra associated with SO(10), whose transformation properties are uniquely determined by group theory alone. These transformation properties, it turns out, are so powerful that we can determine the matrix elements from them.

In particular, a vector, by definition, transforms under the SO(10) Kac–Moody algebra as follows:

$$j^{\mu\nu}(z)\psi^{a}(w) \sim \frac{1}{z-w}(\eta^{\mu\sigma}\psi^{\nu} - \eta^{\nu\sigma}\psi^{\mu})(w) + \cdots \tag{4.3.3}$$

A spin field $S_\alpha$, because it transforms as a spinor under SO(10), must satisfy, by definition, the following transformation property:

$$j^{\mu\nu}(z)S_\alpha(w) \sim \tfrac{1}{4}\frac{1}{z-w}(\gamma^{[\mu}\gamma^{\nu]})^{\beta}_{\alpha}S_\beta(w) + \cdots \tag{4.3.4}$$

The remarkable claim of conformal field theory is that the two previous identities, which show how vectors and spinors transform under SO(10), are sufficient to determine practically *all* of the correlation functions of the theory!

The energy–momentum tensor, in turn, can be written as the normal-ordered square of the Lorentz generators:

$$T_B^{\psi} = \frac{-\frac{1}{4}}{D-1}j^{\mu\nu}j_{\mu\nu}(z) \tag{4.3.5}$$

Explicitly calculating the commutators of the previous equation verifies that it is the Virasoro generator.

Unfortunately, the spin field $S_\alpha$ has conformal weight $\frac{5}{8}$, which we will prove later in this chapter when we construct an explicit representation of these fields. We can see this intuitively, however, for the following reason. We saw earlier in (4.1.46) that the highest weight vector $|h\rangle$ of a Verma module can be written as

$$|h\rangle = \phi(0)|0;0\rangle \tag{4.3.6}$$

where $\phi$ is a field of weight $h$ and $|0;0\rangle$ is the SL(2, $R$) vacuum. For the superconformal case, we actually have two highest weight vacuum vectors:

$$\begin{aligned} &|h\rangle = S(0)|0;0\rangle \\ &G_0|h\rangle \end{aligned} \tag{4.3.7}$$

We wish to eliminate the second highest weight vacuum vector in order to preserve 2D supersymmetry. In the superstring, we know that we only have one spinor in our theory at the lowest level, not two. In order to eliminate the second state, use the identity

$$G_0^2 = L_0 - \frac{\hat{c}}{16} \tag{4.3.8}$$

(We are taking a slightly different form for the $c$-number anomaly in the Ramond algebra than (3.2.28). This choice also satisfies the Jacobi identities.) If we hit the state $G_0|h\rangle$ with another $G_0$, then we want the state to vanish. This means

$$G_0 G_0|h\rangle = \left(h - \frac{\hat{c}}{16}\right)|h\rangle = 0 \tag{4.3.9}$$

To satisfy this, the weight $h$ of the spin field must be $\frac{10}{16} = \frac{5}{8}$, as expected.

The key to calculating the matrix elements of any conformal field is to calculate its short-distance behavior with the superconformal group and other fields. Therefore let us now calculate the short-distance behavior of the spin field, using invariance arguments alone to determine its short-distance structure and its matrix elements.

First of all, the previous identity (4.3.4) tells us that the operator product of two $S$'s must contain at least a $\psi$ field:

$$S_\alpha(z)S^\beta(w) = \frac{1}{(z-w)^{1/4}}(\gamma^{\mu\nu})_\alpha^\beta \psi_\mu(z)\psi_\nu(w) + \text{other terms} \tag{4.3.10}$$

To calculate the operator product of $\psi$ and $S$, we consider the three-point function

$$\langle 0|_{\text{NS}} S_\alpha(z_1)\psi^\mu(z_2)S_\beta(z_3)|0\rangle_{\text{NS}} \tag{4.3.11}$$

where the vacuum is the NS vacuum. In the limit of $z_1 \to \infty$ and $z_3 \to 0$, the spin field changes the NS vacuum into a vacuum with spinor quantum numbers, i.e., the Ramond vacuum $|0\rangle_{\text{R}}$:

$$S_\alpha(0)|0\rangle_{\text{NS}} = |0\rangle_{\text{R}} u_\alpha; \qquad \langle 0|_{\text{NS}} S_\alpha(\infty) = u_\alpha \langle 0|_{\text{R}} \tag{4.3.12}$$

(Notice that the spin operator allows us to go from the NS vacuum to the R vacuum, which was not possible in the earlier NS–R theory.) This means that (4.3.11) can be written as

$$u_\alpha \langle 0|_{\text{R}} \psi^\mu(z_2)|0\rangle_{\text{R}} u_\beta \tag{4.3.13}$$

But only the zero modes of the $\psi$ field survive this vacuum expectation product, so we are left with the matrix element of a Dirac matrix:

$$\psi^\mu(z)S_\alpha(w) \sim \frac{1}{\sqrt{2}}\frac{1}{(z-w)^{1/2}}(\gamma^\mu)_\alpha^\beta S_\beta(z) + \cdots \tag{4.3.14}$$

This, in turn, means that $\psi$ occurs in the operator product of two $S$'s:

$$S_\alpha(z)S^\beta(w) \sim \frac{1}{\sqrt{2}(z-w)^{3/4}}(\gamma_\mu)_\alpha^\beta \psi^\mu(z) + \cdots \tag{4.3.15}$$

Finally, we need to know the short-distance behavior between two spin fields. We saw that the conformal weight of the spin field is $\frac{5}{8}$. Thus,

$$S_\alpha(z)S^\beta(w) \sim -\delta_\alpha^\beta(z-w)^{-5/4} + \cdots \tag{4.3.16}$$

where $\frac{5}{4}$ is twice the dimension of the spin field.

Finally, we can bring (4.3.10), (4.3.14), and (4.3.16) together, which yields the short-distance behavior of two spin fields. In summary, invariance arguments have established the short-distance behavior of the spin field, which can be represented as

$$\begin{aligned} S_\alpha(z)S^\beta(w) \sim \frac{-1}{(z-w)^{5/4}}\delta_\alpha^\beta + \frac{1}{\sqrt{2}(z-w)^{3/4}}(\gamma_\mu)_\alpha^\beta\psi^\mu(z) \\ + \frac{1}{\sqrt{2}(z-w)^{1/4}}(\gamma^{\mu\nu})_\beta^\alpha\psi_\mu(z)\psi_\nu(z) + \cdots \end{aligned} \tag{4.3.17}$$

We saw that the spin field has dimension $\frac{5}{8}$. What we want is a fermion vertex of dimension 1, suitable for use in a multifermion amplitude. To find the missing factor of $\frac{3}{8}$, let us now turn to the ghost sector of the theory. The ghost sector will give us the final missing piece.

## §4.4. Superconformal Ghosts

Using conformal language, let us now rewrite the Faddeev–Popov determinant arising from the superconformal gauge fixing. Using (3.4.5), we find

$$\begin{aligned} \delta g_{zz} &= \nabla_z \delta\xi_z \\ \delta\chi_z &= \nabla_z \delta\varepsilon \end{aligned} \tag{4.4.1}$$

and their complex conjugates. Thus, the Faddeev–Popov determinant is

$$\det{}_g(\nabla_z)\det{}_\chi(\nabla_z) \tag{4.4.2}$$

and their complex conjugates. At first, this looks very much like the determinant we found in (2.4.3) for the Nambu–Goto string, i.e., the determinant of $\partial_z$ and $\partial_{\bar{z}}$. There are, however, several important differences. The Hilbert space over which these operators act has now changed. If we expand the determinant in terms of basis states, we find that the conformal weights have changed, the tensor transformations under the conformal group have changed, and the statistics of the fields is reversed.

When we exponentiate these determinants into the action (see (1.6.16)), we arrive at the ghost action:

$$S_{\text{gh}} = \frac{1}{\pi}\int d^2z\, d\theta\, d\bar{\theta}\, B\bar{D}C + \text{c.c.} \tag{4.4.3}$$

where

$$\begin{cases} B(z) = \beta(z) + \theta b(z) \\ C(z) = c(z) + \theta\gamma(z) \end{cases} \tag{4.4.4}$$

where $\beta$ and $\gamma$ are commuting operaters. Performing the integration over $\theta$, we have

$$S_{\mathrm{gh}} = \frac{1}{\pi}\int d^2z(b\bar{\partial}c + \beta\bar{\partial}\gamma) + \text{c.c.} \tag{4.4.5}$$

$$\begin{aligned} \bar{D}B &= 0 \\ \bar{D}C &= 0 \end{aligned} \tag{4.4.6}$$

where $(b, c)$ are the ghost fields arising from gauge fixing the metric, and the $\beta, \gamma$ are ghosts arising from fixing the $\chi_\alpha$ field (either in the NS or the R sector). The superconformal ghosts (which have commutation relations among them) were first encountered in (3.5.14) when we quantized the NS–R model. The difference now is that we wish to stress the conformal properties of these fields, i.e., their weights.

If we summarize the weights of these fields, we find

| Field | Weight | Statistics |
|---|---|---|
| $b$ | 2 | Fermi |
| $c$ | $-1$ | Fermi |
| $\beta$ | 3/2 | Bose |
| $\gamma$ | $-\frac{1}{2}$ | Bose |

(4.4.7)

Given the action (4.4.5), we can extract out the energy–momentum tensor, which is given by the sum of two pieces:

$$\begin{cases} T_X = -\frac{1}{2}DX^\mu\partial X_\mu = T_F + \theta T_B \\ T_{\mathrm{gh}} = -C\partial B + \frac{1}{2}DC\,DB - \frac{3}{2}\partial CB \end{cases} \tag{4.4.8}$$

Written out explicitly, we have

$$\begin{aligned} T_F^{\mathrm{gh}}(z) &= -c\partial\beta - \tfrac{3}{2}\partial c\beta + \tfrac{1}{2}\gamma b \\ T_B^{\mathrm{gh}}(z) &= c\partial b + 2\partial cb - \tfrac{1}{2}\gamma\partial\beta - \tfrac{3}{2}\partial\gamma\beta \end{aligned} \tag{4.4.9}$$

It is awkward always having to write down $b$ and $c$ and $\beta$ and $\gamma$, especially when the equations for the bosonic and fermionic fields are so similar. We will, therefore, adopt a system where we can describe all ghosts fields at once, with arbitrary weights. Let us write down the *generic* ghost action, using boldface fields to represent either the commuting or anticommuting ghosts:

$$S = \frac{1}{\pi} \int d^2 z \, \mathbf{b} \bar{\partial} \mathbf{c} + \text{c.c.} \tag{4.4.10}$$

$$\bar{\partial} \mathbf{b} = \bar{\partial} \mathbf{c} = 0 \tag{4.4.11}$$

where we define the field $\mathbf{b}$ to have arbitrary weight $\lambda$ and the field $\mathbf{c}$ to have weight $1 - \lambda$. Remember that the anticommuting $b$, $c$ ghost system has $\lambda = 2$ and that the commuting $\beta$, $\gamma$ ghost system has weight $\lambda = \frac{3}{2}$. This action is a compact expression for (4.4.5), except now we have generalized our discussion to all possible conformal weights.

Starting from this generic action, we can easily write down the energy–momentum tensor:

$$T^{bc}(z) = -\lambda \mathbf{b} \partial \mathbf{c} + (1 - \lambda) \partial \mathbf{b} \mathbf{c} \tag{4.4.12}$$

If we take the decompositions:

$$\begin{aligned} \mathbf{b}(z) &= \sum_{n \in \delta - \lambda + Z} z^{-n-\lambda} \mathbf{b}_n \\ \mathbf{c}(z) &= \sum_{n \in \delta + \lambda + Z} z^{-n-1+\lambda} \mathbf{c}_n \end{aligned} \tag{4.4.13}$$

where $\delta$ equals 0 for the NS sector and $\frac{1}{2}$ for the R sector. With this representation, the Virasoro generators are

$$L_n^{bc} = \sum_k (k - (1 - \lambda)n) b_{n-k} c_k \tag{4.4.14}$$

One can check that these generate the usual commutation relations for arbitrary $\lambda$.

In addition to the energy–momentum tensor, we can construct two other currents from the action, the BRST current and the ghost number current.

Following (1.9.12), the BRST current is due to the fact that the original gauge action, plus its Faddeev–Popov ghosts, has a residual gauge symmetry that is nilpotent (and hence cannot be used to eliminate any more fields). Any gauge symmetry has a current associated with it, so the BRST current can be derived directly from the action:

$$J_{\mathrm{BRST}}(\mathbf{z}) = DC(CDB - \tfrac{3}{4} DCB) \tag{4.4.15}$$

Following (1.9.12), the BRST charge is the superintegral of the BRST current:

$$Q_{\mathrm{BRST}} = \frac{1}{2\pi i} \oint dz \, d\theta \, J_{\mathrm{BRST}} \tag{4.4.16}$$

This charge can be divided into three pieces:

$$Q_{\mathrm{BRST}} = Q^{(0)} + Q^{(1)} + Q^{(2)} \tag{4.4.17}$$

where

$$Q^{(0)} = \frac{1}{2\pi i} \oint dz (cT_B(X, \psi, \beta, \gamma) - c\partial cb)$$
$$Q^{(1)} = \frac{1}{2\pi i} \oint dz \, \tfrac{1}{2}\gamma\psi_\mu \partial X^\mu \tag{4.4.18}$$
$$Q^{(2)} = \frac{1}{2\pi i} \oint dz \, \tfrac{1}{4}\gamma^2 b$$

By carefully analyzing each piece, we can show that the sum is nilpotent:

$$Q_{\text{BRST}}^2 = 0 \tag{4.4.19}$$

In addition to the BRST current, there is also the $U(1)$ current, called the "ghost number current." Naively, because the ghosts always occur in pairs in the action, we expect that they must conserve a certain quantum number, somewhat analogous to the situation with baryon number. However, surprisingly enough, we find that there is a correction term to the ghost current. Let us write down the ghost number current, which is just bilinear in the ghost fields:

$$\mathbf{j}(z) = -\mathbf{bc} = \sum_n z^{-n-1} \mathbf{j}_n \tag{4.4.20}$$

where

$$\mathbf{j}_n = \sum_k \varepsilon \mathbf{c}_{n-k} \mathbf{b}_k \tag{4.4.21}$$

$\varepsilon = +1$ (Fermi) or $\varepsilon = -1$ (Bose). We have all the identities necessary to calculate the short-distance behavior between the ghost current and the energy–momentum tensor:

$$T(z)\mathbf{j}(w) \sim \frac{Q}{(z-w)^3} + \frac{\mathbf{j}(z)}{(z-w)^2} + \cdots \tag{4.4.22}$$

where $Q = \varepsilon(1 - 2\lambda)$. (Notice that the presence of this $Q$ factor is anomalous.) This implies

$$[L_m, \mathbf{j}_n] = -n\mathbf{j}_{m+n} + \tfrac{1}{2}Qm(m+1)\delta_{m,-n} \tag{4.4.23}$$

This ghost number current assigns a ghost number to each of the ghost fields. This ghost number current has unusual properties under complex conjugation:

$$\mathbf{j}_m^\dagger = -\mathbf{j}_{-m} - Q\delta_{m,0} \tag{4.4.24}$$

where

$$Q = \varepsilon(1 - 2\lambda)$$
$$\lambda = \tfrac{1}{2}(1 - \varepsilon Q) \tag{4.4.25}$$
$$\varepsilon = +1 \text{ (Fermi statistics);} \quad -1 \text{ (Bose statistics)}$$

(If the $Q$ term were missing in (4.4.24), it would have normal complex conjugation properties.) The quantum numbers of the ghost fields are

$$\begin{aligned} &b, c: \quad \varepsilon = 1; \qquad \lambda = 2; \qquad Q = -3; \qquad c = -26 \\ &\beta, \gamma: \quad \varepsilon = -1; \qquad \lambda = \tfrac{3}{2}; \qquad Q = 2; \qquad c = 11 \end{aligned} \tag{4.4.26}$$

and the anomaly contribution of the ghost is $c = -2\varepsilon(6\lambda(\lambda - 1) + 1)$. One of the unusual features of this ghost structure is the existence of an *infinite number of vacua*, which arises because of (4.4.23) and (4.4.24). Let us define the vacua by

$$\begin{aligned} \mathbf{b}_n|q\rangle = 0; \qquad n > \varepsilon q - \lambda \\ \mathbf{c}_n|q\rangle = 0; \qquad n \geq -\varepsilon q + \lambda \end{aligned} \tag{4.4.27}$$

The zeroth component of the ghost number current and the $L_0$ have the following action on these states:

$$\begin{aligned} \mathbf{j}_0|q\rangle &= q|q\rangle \\ L_0^{bc}|q\rangle &= \tfrac{1}{2}\varepsilon q(Q + q)|q\rangle \end{aligned} \tag{4.4.28}$$

In particular, the last identities show that the only nonvanishing matrix elements are

$$\langle -q - Q|q\rangle = 1 \tag{4.4.29}$$

The easiest way to show this is to take the matrix elements of the currents between different vacua, where each vacuum is labeled by the number $q$:

$$\langle q'|L_0|q\rangle; \qquad \langle q'|j_0|q\rangle \tag{4.4.30}$$

All the matrix elements are zero, except when $q' = -q - Q$.

This is highly unusual. In the usual Veneziano model, there was only one unique vacuum. Now, there appears to be an *infinite number of vacua* labeled by $q$ for the ghost sector of the NS and R models! *The existence of an infinite number of Fermi and Bose vacua is one of the unusual features of conformal field theory.*

This means that there is an anomaly in the ghost current. The problem lies in (4.4.22) and (4.4.23), i.e., the fact that the U(1) ghost current has anomalous commutation relations with the energy–momentum tensor. The anomalous term corresponds to the violation of the conservation of ghost number. In fact, the divergence of the current is given by $\partial_{\bar{z}} j_z = \frac{1}{8}Q\sqrt{g}R^{(2)}$, i.e., the two-dimensional curvature density. The ultimate origin of all these difficulties actually lies back in (4.1.36), when we calculated the Faddeev–Popov determinant over $\nabla_z$ and $\nabla_{\bar{z}}$. A careful analysis of the eigenvalues of these operators shows that we have to remove the zero modes, or else the determinants make no sense. When we exponentiate these determinants by expressing them in terms of Faddeev–Popov ghosts, these zero modes, in turn, correspond to nontrivial solutions to the equations $\partial_{\bar{z}}c^z = \partial_{\bar{z}}b_{zz} = 0$. We cannot give a full discussion of these zero modes, unfortunately, because it requires new information that will not be discussed until Chapter 9, when we analyze anomalies in detail. (Briefly, this anomaly can be shown to be related to the topology of the Riemann surface swept out by the string. By integrating the divergence equation for the U(1) ghost current, we can use the Gauss–Bonnet theorem to show that $\int d^2z\,\sqrt{g}R^{(2)} = -8\pi(g-1)$, where $g$ is the number of holes or handles in a closed Riemann surface. Then we use the Riemann–Roch theorem, which tells us that the number of zero modes of $c$ minus the zero modes of $b$ equals $(1-2\lambda)(1-g)$. The zero modes of $c$ correspond to conformal Killing vectors, and the zero modes of $b$ correspond to moduli. This is a useful result, because it tells us that the number of complex moduli for a sphere with $g$ handles is $3g-3$, which we will use extensively in Chapter 5. It also tells us that the number of supermoduli is $4(g-1)$, which is difficult to prove using other methods.)

It can be shown that the Fermi vacua (for the ghosts of the usual Nambu–Goto bosonic string) are actually equivalent vacua; i.e., by multiplying Fermi vacua with various monomials of $b$ and $c$, one can reproduce the other vacua. Thus, the various vacua yield equivalent representations. However, the situation with the Bose vacua (for the ghosts of the NS–R model) is different. We find that the various vacua are actually inequivalent vacua. No monomial in the fields $\beta$ and $\gamma$ can take us from one vacuum to another.

For the NS sector, the Bose vacua are labeled by integers, while for the R sector the Bose vacua are labeled by half-integers. The vacuum states that come closest to the usual definition of the vacua (i.e., they are annihilated by all positive frequency parts of the oscillators) are

$$\begin{aligned} \text{R:} \quad & \begin{cases} |-\tfrac{1}{2}\rangle \\ |-\tfrac{3}{2}\rangle \end{cases} \\ \text{NS:} \quad & \{|-1\rangle \end{aligned} \tag{4.4.31}$$

They are normalized as follows:

$$\begin{aligned} \langle -\tfrac{3}{2}|-\tfrac{1}{2}\rangle &= 1 \\ \langle -1|-1\rangle &= 1 \end{aligned} \tag{4.4.32}$$

Although the appearance of an infinite number of inequivalent vacua appears at first to be a disaster, we will show later on that this yields a perfectly acceptable theory. In particular, we will show that, on-shell, it really makes no difference which of the various vacua we choose. All on-shell matrix elements for the same physical process, with arbitrary choices of vacua, will yield the same numbers. In fact, we will be able to create "picture changing" operators that will take us from one vacuum to another. The situation is exactly analogous to the one found earlier with the $F_1$ and $F_2$ pictures discussed in (3.3.18) and (3.3.19).

Having worked out the structure of the ghost sector of the NS–R model, our next step is to find the field that gives us conformal weight $\frac{3}{8}$, the missing piece in the fermion vertex function.

Let us now bosonize the ghost current by introducing a new field $\phi$ into the theory:

$$\mathbf{j}(z) = \varepsilon\partial\phi(z) \tag{4.4.33}$$

The new object we wish to investigate is

$$:e^{q\phi}: \tag{4.4.34}$$

Its short-distance behavior is given by

$$j(z)e^{q\phi(w)} \sim \frac{q}{z-w}e^{q\phi(w)} + \cdots \tag{4.4.35}$$

and

$$T(z)e^{q\phi(w)} \sim [\tfrac{1}{2}\varepsilon q(q+Q)(z-w)^{-2} + (z-w)^{-1}\partial_w]e^{q\phi(w)} + \cdots \tag{4.4.36}$$

This means that

$$e^{q\phi(0)}|0\rangle = |q\rangle \tag{4.4.37}$$

$$\text{weight:}\quad \tfrac{1}{2}\varepsilon q(q+Q) \tag{4.4.38}$$

This means that multiplying by the bosonized field $e^{q\phi}$ allows us to go from one ghost vacuum to another. Notice that the NS ghost vacua are integral, while the R ghost vacua are fractional. Because $q$ can be fractional, *this allows us to go back and forth between the various NS and R vacua by multiplying with* $e^{q\phi}$.

Let us use this bosonization technique to write the anticommuting $b$ and $c$ fields in terms of a new scalar boson field $\sigma$:

$$\begin{aligned} b(z) &= e^{-\sigma(z)} \\ c(z) &= e^{\sigma(z)} \end{aligned} \tag{4.4.39}$$

(We can check that the fields have the correct conformal weight. From (4.4.38), a field $e^{q\sigma}$ has conformal weight $\frac{1}{2}q(q-3)$. Then for $q = -1$, the field $e^{-\sigma}$ has weight 2, while for $q = 1$ the field $e^{\sigma}$ has weight $-1$. They therefore have the correct weight.) Notice that both the left and the right sides are anticommuting fields, even though $\sigma$ itself is a commuting field. It is now easy to show:

$$\langle\sigma(z)\sigma(w)\rangle = \log(z-w); \qquad e^{-\sigma(z)}e^{\sigma(w)} \sim \frac{1}{z-w}$$

For the NS–R ghost sector, however, this is more subtle. These ghosts are already commuting, so bosonization does not seem possible. We can, however, use a trick:

$$\begin{aligned} \beta &= e^{-\phi}\partial\xi \\ \gamma &= e^{\phi}\eta \end{aligned} \tag{4.4.40}$$

where the left-hand sides are commuting fields, and the right-hand side is the product of two anticommuting fields; i.e., $\xi$ and $\eta$ are also anticommuting fields (which, in turn, can be bosonized). Thus, we have written commuting fields in terms of anticommuting fields. We can also reverse this procedure:

$$\begin{aligned} \eta &= \partial\gamma e^{-\phi} \\ \partial\xi &= \partial\beta e^{\phi} \end{aligned} \tag{4.4.41}$$

Notice that the $\xi$ and $\eta$ fields are themselves anticommuting, so we can bosonize them. Let us express these two anticommuting fields in terms of the boson field $\chi$:

$$\begin{aligned} \xi &= e^{\chi} \\ \eta &= e^{-\chi} \end{aligned} \tag{4.4.42}$$

Although conformal field theory has the great advantage that all fields are free fields, one small price to pay is that we must keep track of all these different free fields. Because it is crucial to keep all these boson fields clearly defined, let us tabulate their quantum numbers:

| Boson | Charge | Anomaly | Weight |
|---|---|---|---|
| $\phi$ | $Q = 2$ | $c = 13$ | $\text{wt}(\exp q\phi) = -\frac{1}{2}q(q+2)$ |
| $\chi$ | $Q = -1$ | $c = -2$ | $\text{wt}(\exp q\chi) = \frac{1}{2}q(q-1)$ |
| $\sigma$ | $Q = -3$ | $c = -26$ | $\text{wt}(\exp q\sigma) = \frac{1}{2}q(q-3)$ |

(4.4.43)

There is a crucial subtlety in the definition of the Bose NS–R ghosts. Notice that $\beta$ is defined in terms of the derivative of $\xi$, so that it is independent of the zero mode of the $\xi$ field. Thus, the usual Fock space is independent of the zero mode of the $\xi$ field. Thus, we have two possible Fock spaces. The "small" Fock space is missing the zero mode of $\xi$. The "large" Fock space contains the zero mode of $\xi$ and is reducible. Because

$$[\eta_0, \xi_0] = 1 \tag{4.4.44}$$

this means that the vacuum of the $\eta$, $\xi$ system is degenerate.

The purpose of this construction is that we can now write down the missing piece of the fermion vertex operator.

## §4.5. Fermion Vertex

Let us calculate the conformal weights of the bosonized fields:

$$\begin{aligned} \text{Wt}(e^{-(1/2)\phi}) &= \tfrac{1}{2}\varepsilon q(q+Q) = \tfrac{3}{8} \\ \text{Wt}(e^{(1/2)\phi}) &= \tfrac{1}{2}\varepsilon q(q+Q) = -\tfrac{5}{8} \end{aligned} \tag{4.5.1}$$

We now have found the missing piece. Since the bosonized field $e^{-(1/2)\phi}$ has weight $\frac{3}{8}$, *we can now construct the true fermion vertex of the theory*:

$$V_{-1/2} = u^\alpha e^{-(1/2)\phi} S_\alpha e^{ik\cdot X} \tag{4.5.2}$$

which has conformal weight

$$\tfrac{5}{8} + \tfrac{3}{8} + \alpha' k^2 \tag{4.5.3}$$

If we place the external fermion line on-shell

$$\begin{aligned} k^2 &= 0 \\ \gamma^\mu k_\mu u_\alpha &= 0 \end{aligned} \tag{4.5.4}$$

then we have a vertex of conformal weight 1. This vertex has the correct properties under the BRST charge:

$$\{Q_{\mathrm{BRST}}, V_{-1/2}\}_+ = 0 \tag{4.5.5}$$

up to terms that vanish on-shell. This is one of the great accomplishments of conformal field theory, the creation of a genuine fermion vertex function with conformal weight 1 based entirely on free fields. The key observation was to use the ghost sector to supply the missing conformal weight of $\frac{3}{8}$.

Although we have now attained our goal, we are not yet out of the woods. We mentioned earlier in the last section that there is a problem with respect to the infinite Bose sea. It turns out that if we insert this fermion vertex within a boson scattering matrix element, we get 0:

$$\langle \cdots V_{-1/2} \cdots V_{-1/2}, \ldots \rangle = 0 \tag{4.5.6}$$

What is needed, of course, is a vertex with ghost charge $+\frac{1}{2}$ to cancel the $-\frac{1}{2}$ coming from the fermion vertex. This new vertex $V_{1/2}$ must anticommute with the BRST charge up to terms that vanish on-shell. It is easy to show that any vertex

$$V = [Q_{\mathrm{BRST}}, \Phi] \tag{4.5.7}$$

for arbitrary $\Phi$ has a vanishing anticommutation relation with the BRST charge because $Q$ is nilpotent. However, all these vertices are spurious. These states are null and do not couple to real states $|R\rangle$, which satisfy the relation

$$Q_{\mathrm{BRST}}|R\rangle = 0 \tag{4.5.8}$$

so they cannot be used as vertex functions. They simply produce zero matrix elements with the physical sector of the theory. However, there is a vertex function for which this reasoning does not apply:

$$V_{1/2} = 2[Q_{\mathrm{BRST}}, \xi V_{-1/2}] \tag{4.5.9}$$

Normally, one would expect that such a vertex is also spurious and does not couple to real states of the theory. However, $\xi V_{1/2}$, as we said earlier, is not part of the irreducible Fock space of the theory, and thus we cannot simply say that this commutator vanishes on contraction with real states. This vertex does not necessarily vanish because

$$Q_{\text{BRST}}\xi|0\rangle \neq 0 \tag{4.5.10}$$

After working out the details, we find that this vertex is equal to

$$V_{1/2} = u^\alpha(k)e^{ik\cdot X}[e^{(1/2)\phi}(\partial X^\mu + \tfrac{1}{4}ik\cdot\psi\psi^\mu)(\gamma_\mu)_{\alpha\beta}S^\beta + \tfrac{1}{2}e^{3\phi/2}\eta b S_\alpha] \tag{4.5.11}$$

This, it turns out, is the correct fermion vertex that allows us to contract onto $V_{-1/2}$.

However, at this point we realize a rather disturbing problem. We now have too many possible vertices! For example, we could also have written

$$\begin{aligned} V_{3/2} &= [Q_{\text{BRST}}, \xi V_{1/2}] \\ V_{5/2} &= [Q_{\text{BRST}}, \xi V_{3/2}] \\ V_{7/2} &= \cdots \end{aligned} \tag{4.5.12}$$

In fact, there is an infinite number of such vertices, each one linking inequivalent Bose sea vacua. This is certainly an embarrassment of riches. However, it is possible to show that we need only use $V_{1/2}$ and $V_{-1/2}$, and that the other vertices do not yield any new matrix elements. We wish to show the following identity:

$$\begin{aligned} &\langle\cdots V_{-1/2}(u_1, k_1, z_1), \ldots, V_{1/2}(u_2, k_2, z_2)\cdots\rangle \\ &= \langle\cdots V_{1/2}(u_1, k_1, z_1), \ldots, V_{-1/2}(u_2, k_2, z_2)\cdots\rangle \end{aligned} \tag{4.5.13}$$

The proof that we can simply switch the various $\frac{1}{2}$ and $-\frac{1}{2}$ ghost indices on the vertices at will involves some rather subtle arguments that take us between the irreducible small Fock space (which does not include the zero mode of $\xi$) and the reducible large Fock space (which does include the zero mode of $\xi$).

We begin by rewriting the vertex $V_{1/2}$ in the equivalent form

$$V_{1/2}(z_2) = \frac{1}{2\pi i}\oint dw\, j_{\text{BRST}}(w)\xi(z_2)V_{-1/2}(z_2) \tag{4.5.14}$$

Notice that the vertex is now written in the large Fock space, so we must insert another $\xi(z)$ to redefine the vacua so that we have a nonvanishing matrix element. However, because the position of $\xi(z)$ is irrelevant (because we want only its zero mode), we can now rewrite the matrix element as

$$\langle\cdots\xi(z_1)V_{-1/2}\cdots\frac{1}{2\pi i}\oint dw\, j_{\text{BRST}}(w)\xi(z_2)V_{-1/2}(z_2)\cdots\rangle \tag{4.5.15}$$

So far we have done nothing. We have simply gone from the small Fock space, where the zero mode of $\xi$ never appears, to the larger Fock space, where it does. However, the value of the matrix element remains precisely the same.

Now we make the following observation. The contour integral that encircles $z_2$ can be changed at will, so let us enlarge it until it goes around the back of the Riemann surface (a sphere) and finally comes around and encircles the point $z_1$. Of course, the enlarging of the contour cuts across other bosonic vertices, but notice that $j_{\mathrm{BRST}}$ commutes with all of them on-shell, so that we can move both the contour and the current $j_{\mathrm{BRST}}$ at will until they encircle $z_1$.

At this point, notice that the contour integral is now around $\xi V_{-1/2}$, which can now be written as $V_{+1/2}$. Thus, we have totally reversed the position of the contour integral, and we can eliminate all the $\xi$ and go back to the small Fock space. Symbolically, the steps can be represented as follows:

$$\begin{aligned} \langle \cdots V_{-1/2} \cdots V_{1/2} \cdots \rangle &\to \left\langle \cdots (\xi V_{-1/2}) \cdots \left[ \oint dw \; \xi j_{\mathrm{BRST}} V_{-1/2} \right] \cdots \right\rangle \\ &\to \left\langle \cdots \left[ \oint dw \; \xi j_{\mathrm{BRST}} V_{-1/2} \right] \cdots (\xi V_{-1/2}) \cdots \right\rangle \\ &\to \langle \cdots V_{1/2} \cdots V_{-1/2} \cdots \rangle \end{aligned} \tag{4.5.16}$$

The purpose of this exercise was to show that we can successfully reverse the position of the $\frac{1}{2}$ and $-\frac{1}{2}$ in the matrix element. This means that, although there is an infinite number of fermion vertices, they will all yield the same on-shell matrix element.

## §4.6. Spinors and Trees

In order to calculate tree amplitudes, it is necessary to construct an explicit representation of the spin fields in terms of NS–R operators. Although these fields were highly coupled operators in the GS formalism, the great advantage of the conformal field theory is the fact that the interacting spin field can be composed of free fields, making the calculation of correlation functions explicitly possible.

We will construct the spin field out of the generators of the SO(10) algebra, which in turn are composed of the $\psi(z)$. The SO(10) algebra is a rank 5 Lie algebra (see Appendix). This means that, out of the $10(10-1)/2 = 45$ generators of the SO(10) algebra, 5 mutually commute among themselves, forming the Cartan subalgebra. The commutation relations among these 5 commuting generators and 40 noncommuting elements are

$$\begin{aligned} [H_i, H_j] &= 0 \\ [H_i, E_{\boldsymbol{\alpha}}] &= \boldsymbol{\alpha}_i E_{\boldsymbol{\alpha}} \\ [E_{\boldsymbol{\alpha}}, E_{\boldsymbol{\beta}}] &= \varepsilon(\boldsymbol{\alpha}, \boldsymbol{\beta}) E_{\boldsymbol{\alpha}+\boldsymbol{\beta}} \end{aligned} \tag{4.6.1}$$

The last identity is true if $E_{\boldsymbol{\alpha}+\boldsymbol{\beta}}$ is a raising or lowering operator. Each $\boldsymbol{\alpha}$ is a root vector of the group SO(10). The $\varepsilon$ terms are the structure constants of the group, which obey various symmetry conditions and associativity.

Let us now introduce five mutually commuting fields $\phi_j$ and express the 45 generators of SO(10) in terms of them. We can represent the five mutually commuting members of the SO(10) algebra as

$$\partial\phi_j = H_j \tag{4.6.2}$$

To represent the remaining 40 generators, write

$$E_{\boldsymbol{\alpha}} = :e^{i\beta^j_{\boldsymbol{\alpha}}\phi_j}: c_{\boldsymbol{\alpha}} \tag{4.6.3}$$

where $\beta$ is a matrix that takes us from the root vector $\boldsymbol{\alpha}$ into a linear combination of the $\phi_j$ vectors:

$$\beta_{\boldsymbol{\alpha}} = (\pm 1, \pm 1, 0, 0, 0) \tag{4.6.4}$$

We also include its permutations. Notice that there are four possible arrangements of the + and − signs and that there are 10 possible ways in which these signs can be placed into five slots. Thus, this matrix has 40 elements in it. Each slot corresponds to one of the five $\phi_j$.

The purpose of the factor $c_{\boldsymbol{\alpha}}$ in (4.6.3) is to make the commutation relations come out correctly. If we insert these expressions back into the definition of the algebra, we find that we must set

$$c_{\boldsymbol{\alpha}}c_{\boldsymbol{\beta}} = \varepsilon(\boldsymbol{\alpha}, \boldsymbol{\beta})c_{\boldsymbol{\alpha}+\boldsymbol{\beta}} \tag{4.6.5}$$

If we demand associativity of the commutators, then we also have

$$\varepsilon(\boldsymbol{\alpha}, \boldsymbol{\beta})\varepsilon(\boldsymbol{\alpha} + \boldsymbol{\beta}, \gamma) = \varepsilon(\boldsymbol{\alpha}, \boldsymbol{\beta} + \gamma)\varepsilon(\boldsymbol{\beta}, \gamma) \tag{4.6.6}$$

One of many possible representations of these two cocycles is

$$\varepsilon(\boldsymbol{\alpha}, \boldsymbol{\beta}) = (-1)^{\sigma(\boldsymbol{\alpha}, \boldsymbol{\beta})} \tag{4.6.7}$$

where

$$\sigma(\boldsymbol{\alpha}, \boldsymbol{\alpha}) = \tfrac{1}{2}\langle\boldsymbol{\alpha}, \boldsymbol{\alpha}\rangle \tag{4.6.8}$$

and

$$\sigma(\boldsymbol{\alpha}, \boldsymbol{\beta}) + \sigma(\boldsymbol{\beta}, \boldsymbol{\alpha}) = \langle\boldsymbol{\alpha}, \boldsymbol{\beta}\rangle \bmod 2 \tag{4.6.9}$$

(see [9, 10] for other conventions.)

We can also represent the NS anticommuting vector field in terms of this bosonization. Let us write

$$\psi^{\rho_\alpha} = :e^{i\rho^j_\alpha\phi_j}: c_{\rho_\alpha} \tag{4.6.10}$$

where

$$\rho_{\boldsymbol{\alpha}} = (\pm, 0, 0, 0, 0) \tag{4.6.11}$$

with permutations. There are $2 \times 5$ elements within $\rho$, which is the number of elements within $\psi^\mu$.

Let us now represent the spin field in terms of this bosonized picture. Let us define the matrix

$$\lambda_{\boldsymbol{\alpha}} = \tfrac{1}{2}(\pm, \pm, \pm, \pm, \pm) \tag{4.6.12}$$

Notice that there are $2^5 = 32$ terms in this matrix. Let us define

$$S_\alpha = :e^{i\lambda_\alpha^j\phi_j}: c_\alpha \tag{4.6.13}$$

This field $S_\alpha$ has the correct number of components for a spinor in 10 dimensions. Furthermore, it has weight $\frac{5}{8}$. This is because each individual factor has weight $\frac{1}{8}$, and there are five of them, for a total of $\frac{5}{8}$. This confirms the earlier observation made in (4.3.9), which was based purely on group theoretic arguments, that the spin field $S_\alpha$ has weight $\frac{5}{8}$.

The field $e^{-(1/2)\phi}$ has weight $\frac{3}{8}$, and $e^{(1/2)\phi}$ has weight $-\frac{5}{8}$. Thus, our vertex with weight 1 is given by

$$V_{-1/2,\alpha} = S_\alpha e^{-(1/2)\phi} :e^{ik\cdot X}: \tag{4.6.14}$$

Now that we have an explicit representation of the spin field, we can calculate matrix elements for fermion–fermion scattering.

Let us calculate the four-fermion scattering amplitude, represented by the product of three independent factors involving $X$, $S_\alpha$, and $e^{-(1/2)\phi}$:

$$\langle V_{\alpha_1}(z_1)V_{\alpha_2}(z_2)V_{\alpha_3}(z_3)V_{\alpha_4}(z_4)\rangle \tag{4.6.15}$$

we set $z_1 = \infty$, $z_2 = 1$, $z_3 = z$, $z_4 = 0$.

Let us calculate each factor separately. Using the formula derived earlier in Chapter 2, we find that the $X$-dependent factors are equal to

$$\langle V(k_1, z_1)V(k_2, z_2)\cdots V(k_N, z_N)\rangle = \prod_{i\neq j}(z_i - z_j)^{k_i k_j} \tag{4.6.16}$$

where

$$V(k, z) = :e^{ikX(z)}: \tag{4.6.17}$$

and the sum of all $k_i$ equals zero.

Let us now specialize to the case of the ghost contribution, which equals

$$\langle e^{-(1/2)\phi(\infty)}e^{-(1/2)\phi(1)}e^{-(1/2)\phi(z)}e^{-(1/2)\phi(0)}\rangle = [z(1-z)]^{-1/4} \tag{4.6.18}$$

Lastly, the spin field contribution is

$$\begin{aligned}&\langle S_\alpha(\infty)S_\beta(1)S_\gamma(z)S_\delta(0)\rangle\\ &\quad = [z(1-z)]^{-3/4}\{(1-z)(\gamma^\mu)_{\alpha\beta}(\gamma_\mu)_{\alpha\delta} - z(\gamma^\mu)_{\alpha\delta}(\gamma_\mu)_{\beta\gamma}\}\end{aligned} \tag{4.6.19}$$

Putting all three pieces together, we arrive at

$$\begin{aligned}A_4 = g^2 &\int_0^1 dz\, z^{k_3\cdot k_2 - 1}(1-z)^{k_3\cdot k_4 - 1}\\ &\times\{(1-z)(\gamma^\mu)_{\alpha\beta}(\gamma_\mu)_{\alpha\delta} - z(\gamma^\mu)_{\alpha\delta}(\gamma_\mu)_{\beta\gamma}\}\end{aligned} \tag{4.6.20}$$

Rewriting this amplitude, we finally arrive at

$$\begin{aligned}A_4 = g^2\{&B(1-\tfrac{1}{2}t, -\tfrac{1}{2}s)(\gamma^\mu)_{\alpha\beta}(\gamma_\mu)_{\gamma\delta}\\ &- B(-\tfrac{1}{2}t, 1-\tfrac{1}{2}s)(\gamma^\mu)_{\alpha\delta}(\gamma_\mu)_{\beta\gamma}\}\end{aligned} \tag{4.6.21}$$

This calculation and others involving $N$-point fermion scattering amplitudes are not that difficult with conformal field theory [9, 10], but would have been extremely difficult in the older covariant NS–R and GS formalisms.

## §4.7. Kac–Moody Algebras

Although the results of conformal field theory are powerful, to the beginner the various rules and conventions may seem a bit arbitrary and random. It appears, at first glance, that conformal field theory depends on clever tricks and accidents rather than anything fundamental.

In reality, the underlying consistency and elegance of conformal field theory are due to new infinite-dimensional Lie algebras, called Kac–Moody algebras [11–22], which are powerful extensions of the usual finite-dimensional Lie algebras. They were discovered by mathematicians V. G. Kac and R. V. Moody in 1967, although one form of them was already known to physicists in the mid-1960s as current algebras. Together with the superconformal group in two dimensions, the SO(10) Kac–Moody algebra provides the mathematical framework for conformal field theory. Many of the matrix elements for conformal field theory, in fact, can be viewed as Clebsch–Gordan coefficients for Kac–Moody algebras.

We define a Kac–Moody algebra to be a generalization of an ordinary Lie algebra such that its generators obey

$$[T_m^i, T_n^j] = if^{ijl}T_{m+n}^l + km\delta^{ij}\delta_{m,-n} \tag{4.7.1}$$

The algebra looks very much like an ordinary Lie algebra, except for the infinite integer index $m$ on each generator and the constant $k$, which is called the level. The zeroth component of the $T$'s is nothing but the algebra of a finite Lie Algebra. We will often find it convenient to rewrite the generators of the Kac–Moody algebra as the Fourier components of a single function defined over a circle:

$$T^i(\theta) = \sum_n T_n^i e^{-in\theta} \tag{4.7.2}$$

We can also break down the generators of the Kac–Moody algebra into the Cartan subalgebra and its eigenvectors:

$$H_i(\theta); \qquad E_{\boldsymbol{\alpha}}(\theta) \tag{4.7.3}$$

Thus, a Kac–Moody algebra looks like an ordinary Lie algebra smeared over a circle.

Let us now construct what is called the "basic representation" of the Kac–Moody algebra using vertex operators. This representation holds only for simply laced groups (i.e., groups with roots of equal length, which are $A$, $D$, and $E$ Lie groups) with level one ($k = 1$). Let us first define the string variable

$$\phi_i(\theta) = q_i + p_i\theta + i \sum_{n \neq 0} \frac{\alpha_n^i e^{-in\theta}}{n} \tag{4.7.4}$$

Now introduce the basis vectors for the lattice of our Lie algebra such that

$$\mathbf{e}_i \cdot \mathbf{e}_j = \delta_{ij} \tag{4.7.5}$$

This allows us to write the string variable and the vertex function as vectors on the lattice:

$$\begin{aligned} \boldsymbol{\phi}(\theta) &= \sum \mathbf{e}_i \phi_i \\ q_i(\theta) &= :e^{i\phi_i(\theta)}: c_i \end{aligned} \tag{4.7.6}$$

where the $c_i$ are the familiar cocycles introduced in (4.6.5) in order to get the commutation relations with the correct sign. They have many possible representations, one of which is

$$c_i = \exp\left\{\tfrac{1}{2} i\pi \left(\sum_{k<i} - \sum_{k>i}\right) \mathbf{e}_k \cdot \mathbf{p}\right\} \tag{4.7.7}$$

Finally, this allows us to introduce the generators of the Kac–Moody algebra:

$$\begin{aligned} H_i(\theta) &= :q_i^*(\theta) q_i(\theta): = -\frac{d\phi_i}{d\theta} \\ E_{\boldsymbol{\alpha}} &= :q_j^*(\theta) q_k(\theta): = \pm i c_j^* c_k :e^{i\boldsymbol{\alpha}\boldsymbol{\phi}(\theta)}: \end{aligned} \tag{4.7.8}$$

where $\boldsymbol{\alpha} = \mathbf{e}_k - \mathbf{e}_j$ and we use the $+$ sign for $k > j$ and the $-$ sign for $k < j$. Because the $i$th and $j$th elements commute, the $H$'s are generalizations of the Cartan subalgebra, i.e., the set of mutually commuting elements of the Lie algebra.

Let us now calculate the commutator between these generators. In much the same way as before, we find that the product of two vertex functions is

$$\begin{aligned} V(\boldsymbol{\alpha}, \theta) &= :e^{i\boldsymbol{\alpha}\boldsymbol{\phi}(\theta)}: \\ V(\boldsymbol{\alpha}, \theta) V(\boldsymbol{\beta}, \theta') &= \Delta(\theta - \theta')^{-\boldsymbol{\alpha}\cdot\boldsymbol{\beta}} :e^{i\boldsymbol{\alpha}\boldsymbol{\phi}(\theta) + i\boldsymbol{\beta}\boldsymbol{\phi}(\theta')}: \end{aligned} \tag{4.7.9}$$

where

$$\Delta(\theta - \theta') = e^{-i(1/2)(\theta-\theta')}(1 - (1 - \varepsilon)e^{-i(\theta-\theta')})^{-1} \sim \frac{-i}{\theta - \theta' + i\varepsilon} + \cdots \tag{4.7.10}$$

We now have all the identities necessary to compute the commutator between all the generators. We find, by direct computation,

$$\begin{aligned} [H_i(\theta), H_j(\theta')] &= 0 \\ [H_i(\theta), E_{\boldsymbol{\alpha}}(\theta')] &= -2\pi\delta(\theta - \theta')\boldsymbol{\alpha}_i E_{\boldsymbol{\alpha}}(\theta) \end{aligned} \tag{4.7.11}$$

Notice that the $H$'s are mutually commuting, as they are in the ordinary Lie algebra, and that the $\boldsymbol{\alpha}_i$ are eigenvalues of the $H$'s. The remaining commutators between the various $E$'s are given by

$$[E_{\boldsymbol{\alpha}}(\theta), E_{-\boldsymbol{\alpha}}(\theta)] = 2\pi\delta(\theta - \theta') \sum_i \boldsymbol{\alpha}_i H_i(\theta) + 2\pi i\delta'(\theta - \theta') \tag{4.7.12}$$

and

$$[E_{\boldsymbol{\alpha}}(\theta), E_{\boldsymbol{\beta}}(\theta')] = \begin{cases} 2\pi\delta(\theta - \theta')E_{\boldsymbol{\alpha}+\boldsymbol{\beta}}(\theta) & \text{if } \boldsymbol{\alpha} + \boldsymbol{\beta} \in \Gamma \\ 0 & \text{otherwise} \end{cases} \tag{4.7.13}$$

($\Gamma$ is the root lattice.) Thus, we see that the commutators of a Kac–Moody algebra are very similar to the ordinary ones for a Lie algebra, except that the generators are smeared over a circle.

It is also crucial to notice that the semidirect product between the Virasoro algebra and the Kac–Moody algebra is possible. The commutators are

$$\begin{aligned} [L_m, T_n^i] &= -nT_{m+n}^i \\ [L_n, L_m] &= (n - m)L_{n+m} + \frac{c}{12}n(n^2 - 1)\delta_{n,-m} \end{aligned} \tag{4.7.14}$$

Depending on the representation of the algebra that we choose, we can get a correlation between the level $k$ of the Kac–Moody algebra and the central term $c$ of the Virasoro algebra. One remarkable property of the Kac–Moody algebra is that we can construct a representation of the Virasoro algebra entirely in terms of the Kac–Moody algebra. Let us write the Virasoro generators smeared over a circle:

$$\begin{aligned} L(\theta) &= \tfrac{1}{2}N\left(\sum_i :H_i(\theta)^2: + \sum_\alpha :E_{\boldsymbol{\alpha}}(\theta)E_{-\boldsymbol{\alpha}}(\theta):\right) \\ &= \tfrac{1}{2}N\left(\sum_{i=1}^d :T_i(\theta)^2:\right) \end{aligned} \tag{4.7.15}$$

This is called the Sugawara form.

We have all the necessary identities to commute these past each other, and we find

$$\begin{aligned} [L(\theta), L(\theta')] &= 2\pi i\delta'(\theta - \theta')(L(\theta) + L(\theta')) \\ &\quad - \frac{c}{12}2\pi i(\delta'''(\theta - \theta') + \delta'(\theta - \theta')) \end{aligned} \tag{4.7.16}$$

where

$$\begin{aligned} N &= \frac{1}{1 + \frac{1}{2}c_2(G)} \\ c &= \frac{d}{1 + \frac{1}{2}c_2(G)} \end{aligned} \tag{4.7.17}$$

where $d$ is the dimensionality of the group and $c_2(G)$ is the value of the quadratic Casimir operator for the adjoint representation.

The last commutator yields the semidirect product between the Kac–Moody and Virasoro algebras:

$$[L(\theta), E_{\boldsymbol{\alpha}}(\theta')] = \delta'(\theta - \theta')E_{\boldsymbol{\alpha}}(\theta) \tag{4.7.18}$$

We compute the value of the central charge for a few groups:

| $c$ | Group |
|---|---|
| $n - 1$ | $\mathrm{SU}(n)$ |
| $n + \frac{1}{2}$ | $\mathrm{SO}(2n + 1)$ |
| $n$ | $\mathrm{SO}(2n)$ |
| $n$ | $E_n$ |
| $\dfrac{n(2n + 1)}{n + 2}$ | $\mathrm{Sp}(n)$ |

From the perspective of Kac–Moody algebras, let us now reanalyze conformal field theory, reinterpreting the results of conformal field theory from group theory. We see that the spin fields given in (4.6.13), under the SO(10) Kac–Moody Lie algebra, transform as a 32-component spinor. Likewise, the NS anticommuting vector fields given in (4.6.10), under the SO(10) Kac–Moody algebra, transform as 10-component vectors. We can also view the energy–momentum tensor (4.3.5) as just the Sugawara construction of the Virasoro algebra from an SO(10) Kac–Moody algebra. We can also view the $b$, $c$, $\beta$, $\gamma$ ghosts of conformal field theory as forming representations of the superconformal group with different values of the central term given in (4.4.43). Finally, we can also view the correlation functions, which are the heart of conformal field theory, as the Clebsch–Gordon coefficients found in the different tensor products of various representations of the SO(10) Kac–Moody algebra coupled with the superconformal group.

## §4.8. Supersymmetry

Finally, we can now construct the supersymmetry operator in this theory that allows us to go back and forth between the fermion and the boson sectors of the theory. Because $V_{\pm 1/2}$ has conformal weight 1, this means that its integral is an invariant under the conformal group. Thus, if we take $k = 0$, we find the supersymmetry operator

$$Q_{\pm 1/2, \alpha} = \frac{1}{2\pi i} \oint dz \, V_{\pm 1/2, \alpha} \tag{4.8.1}$$

Depending on whether we take the + or − fermion vertex, we find

$$\begin{aligned} Q_{-1/2,\alpha} &= S_\alpha e^{-(1/2)\phi} \\ Q_{1/2,\alpha} &= e^{(1/2)\phi} S^\beta (\gamma^\mu)_{\beta\alpha} \partial_z X_\mu \end{aligned} \tag{4.8.2}$$

These operators indeed have the property of converting the fermion sector into a boson sector, and vice versa:

$$\begin{aligned} [Q_\alpha, V_F(u, k, z)]_+ &= V_B(\zeta^\mu = u^\beta (\gamma^\mu)_{\beta\alpha}, k, z) \\ [Q_\alpha, V_B(\zeta, k, z)] &= V_F(u^\beta = ik^\mu \zeta^\nu (\gamma_{\mu\nu})^\beta_\alpha, k, z) \end{aligned} \tag{4.8.3}$$

Once again, however, we are faced with the problem of having an infinite number of such supersymmetry operators, which link the various inequivalent Bose sea states together. However, because of arguments given earlier, we do not have to worry about this. Although there is an infinite number of inequivalent Bose vacua, the matrix elements between them yield the same answer.

## §4.9. Summary

The great advantage of the superconformal field theory is that we have the best features of the NS–R and GS models. The superconformal field theory has the advantage that it is manifestly covariant, based entirely on free fields, possesses a covariant supersymmetric generator, has spin fields of weight $\frac{5}{8}$ and fermion vertices of weight 1, and is easy to manipulate.

There are two drawbacks, however, to the conformal field theory. First, we must carefully keep track of the many bosonized ghosts. Fortunately, these ghosts are free fields and operate on different Hilbert spaces. Second, we have an infinite number of "pictures." Fortunately, the final $S$-matrix is independent of whatever choice of "pictures" we take.

The essence of conformal field theory is that we can calculate the correlation functions of various fields simply by knowing their transformation properties and their short-distance behavior. For example, the correlation function

$$\langle e^{ikX(w)} e^{-ikX(z)} \rangle \sim (w - z)^{-k^2}$$

can be calculated once we know the short-distance behavior of two string fields.

One of the achievements of the conformal field theory is the construction of a fermion vertex function with conformal weight 1. This vertex function is dependent on a spin field $S_\alpha$, which must transform as a genuine spinor under the SO(10) Lorentz group:

$$j^{\mu\nu}(z) S_\alpha(w) \sim \frac{1}{4} \frac{1}{z - w} (\gamma^{[\mu} \gamma^{\nu]})^\beta_\alpha S_\beta(w) + \cdots \tag{4.9.1}$$

Other considerations, such as the field having dimension $\frac{5}{8}$, allow us to fix the relations

$$\psi^\mu(z)S_\alpha(w) \sim \frac{1}{\sqrt{2}(z-w)^2}(\gamma^\mu)_\alpha^\beta S_\beta(z) + \cdots \tag{4.9.2}$$

$$\begin{aligned} S_\alpha(z)S^\beta(w) \sim \frac{-1}{(z-w)^{5/4}}\delta_\alpha^\beta + \frac{1}{\sqrt{2}(z-w)^{3/4}}(\gamma_\mu)_\alpha^\beta\psi^\mu(z) \\ + \frac{1}{\sqrt{2}(z-w)^{1/4}}(\gamma^{\mu\nu})_\alpha^\beta\psi^\mu(z)\psi^\nu(z) + \cdots \end{aligned} \tag{4.9.3}$$

Once we know all the short-distance behavior of the spin field, we can construct the vertex function that has weight 1. We need an extra piece that has weight $\frac{3}{8}$, which comes from the ghost sector:

$$S_{\text{gh}} = \frac{1}{\pi}\int d^2z\, d\theta\, d\bar{\theta}\, B\bar{D}C + \text{c.c.} \tag{4.9.4}$$

where

$$\begin{cases} B(z) = \beta(z) + \theta b(z) \\ C(z) = c(z) + \theta\gamma(z) \end{cases} \tag{4.9.5}$$

The ghost sector can be bosonized according to

$$\begin{aligned} \beta &= e^{-\phi}\partial\xi \\ \gamma &= e^{\phi}\eta \end{aligned} \tag{4.9.6}$$

This, in turn, finally allows us to write down the vertex function with conformal weight 1 on-shell:

$$V_{-1/2} = u^\alpha e^{-(1/2)\phi}S_\alpha e^{ik\cdot X} \tag{4.9.7}$$

which has conformal weight

$$\tfrac{5}{8} + \tfrac{3}{8} + \alpha' k^2 \tag{4.9.8}$$

In order to write down amplitudes that can contract with the $-\frac{1}{2}$ vertex operator, we need a corresponding vertex function with $+\frac{1}{2}$, which is given by

$$\begin{aligned} V_{1/2} = u^\alpha(k)e^{ik\cdot X}[e^{(1/2)\phi}(\partial X^\mu + \tfrac{1}{4}ik\cdot\psi\psi^\mu)(\gamma_\mu)_{\alpha\beta}S^\beta \\ + \tfrac{1}{2}e^{3\phi/2}\eta b S_\alpha] \end{aligned} \tag{4.9.9}$$

With these two vertices, we can now construct multifermion scattering amplitudes.

One drawback of this formalism, however, is that we have an infinite number of vacua and hence an infinite number of vertex functions. These vacua can be constructed as

$$e^{q\phi(0)}|0\rangle = |q\rangle \tag{4.9.10}$$

$$\text{weight:}\quad \tfrac{1}{2}\varepsilon q(q+Q) \tag{4.9.11}$$

This, in turn, allows us to construct an infinite number of fermion vertex functions. For example,

$$\begin{aligned} V_{3/2} &= [Q_{\mathrm{BRST}}, \xi V_{1/2}] \\ V_{5/2} &= [Q_{\mathrm{BRST}}, \xi V_{3/2}] \\ V_{7/2} &= \cdots \end{aligned} \tag{4.9.12}$$

are all acceptable vertex functions with weight 1. This is an embarrassment of riches. However, it can be shown that, at least on-shell, all these vertices are equivalent to each other. Thus, we can take any one of these at will and construct the fermion–fermion scattering amplitudes.

One of the great advantages of this formalism is that we can now construct the supersymmetry generator, which is nothing but the vertex function integrated over $z$. In fact, we actually now have an infinite number of inequivalent supersymmetry operators, all of them being equivalent on-shell:

$$\begin{aligned} Q_{-1/2,\alpha} &= S_\alpha e^{-(1/2)\phi} \\ Q_{1/2,\alpha} &= e^{(1/2)\phi} S^\beta (\gamma^\mu)_{\beta\alpha} \partial_z X_\mu, \text{ etc.} \end{aligned} \tag{4.9.13}$$

Finally, it is possible to obtain a coherent overall picture of conformal field theory if we view it as one way of calculating matrix elements for infinite-dimensional Kac–Moody algebras and the superconformal algebra. A Kac–Moody algebra has commutation relations

$$[T^i_m, T^j_n] = if^{ijl} T^l_{m+n} + km\delta^{ij}\delta_{m,-n}$$

where the $f$'s are the structure constants of a finite-dimensional Lie algebra. For level one simply laced algebras, we can construct the basic representation of the Kac–Moody algebra. The basic representation is given in terms of vertex operators first found in string theory.

In this formalism, the spin fields $S_\alpha$ and NS–R $\psi$ fields can be viewed as spinorial and vector representations of SO(10) Kac–Moody, and the various $(b, c)$ and $(\beta, \gamma)$ ghosts can be viewed as forming different representations of the superconformal algebra. Correlation functions can then be viewed as Clebsch–Gordon coefficients for the combined SO(10) Kac–Moody and superconformal algebra.

## References


[1] See D. Friedan, in *Unified String Theories* (edited by M. B. Green and D. Gross), World Scientific, Singapore, 1986.

[2] V. G. Knizhnik and A. B. Zamolodchikov, *Nucl. Phys.* **B247**, 83 (1984).

[3] A. A. Belavin, A. M. Polyakov, and A. B. Zamolodchikov, *Nucl. Phys.* **B241**, 333 (1984).

[4] D. Friedan, E. Martinec, and S. Shenker, *Nucl. Phys.* **B271**, 93 (1986).

[5] D. Friedan, E. Martinec, and S. Shenker, *Phys. Lett.* **160B**, 55 (1985).

[6] V. G. Knizhnik, *Phys. Lett.* **160B**, 403 (1985).

[7] S. Mandelstam, *Phys. Rev.* **D11**, 3026 (1975).
[8] S. Coleman, *Phys. Rev.* **D11**, 2088 (1975).
[9] J. Cohn, D. Friedan, Z. Qiu, and S. Shenker, *Nucl. Phys.* **B278**, 577 (1986).
[10] V. A. Kostelecky, O. Lechtenfeld, W. Lerche, and S. Samuel, *Nucl. Phys.* **B288**, 173 (1987).
[11] V. Kac, *Functional Anal. Appl.* **1**, 328 (1967).
[12] R. V. Moody, *J. Algebra* **10**, 211 (1968).
[13] P. Goddard and D. Olive, *Int. J. Mod. Phys.* **A1**, 303 (1986).
[14] P. Goddard, W. Nahm, and D. Olive, *Phys. Lett.* **160B**, 111 (1985).
[15] P. Goddard, A. Kent, and D. Olive, *Comm. Math. Phys.* **103**, 105 (1986).
[16] J. Lepowsky and S. Mandelstam, in *Vertex Operators in Mathematics and Physics*, Springer-Verlag, New York, 1985.
[17] K. Bardakçi and M. B. Halpern, *Phys. Rev.* **D3**, 2493 (1971).
[18] M. B. Halpern, *Phys. Rev.* **D12**, 1684 (1975).
[19] P. Goddard, W. Nahm, and D. Olive, *Phys. Lett.* **160B**, 111 (1985).
[20] P. Goddard, A. Kent, and D. Olive, *Comm. Math. Phys.* **103**, 105 (1986).
[21] V. Kac, *Infinite Dimensional Lie Algebras*, Birkhauser, Boston, 1983.
[22] R. V. Moody, *Bull. Am. Math. Soc.* **73**, 217 (1974).

CHAPTER 5

# Multiloops and Teichmüller Spaces

## §5.1. Unitarity

One of the most exciting aspects of string theory is the possiblity of a theory of gravity that is totally finite and therefore independent of conventional renormalization theory. String theory may provide a framework in which, for the first time, a finite theory of quantum gravity may emerge. What is particularly fascinating is the mechanism by which the cancellation of all potential divergences takes place, namely the use of *topological arguments* to eliminate certain divergences. Once again, we see the enormous power of symmetry that is built into the string model. We will show, for example, that the diagrams that are potentially divergent are topologically equivalent to the emission of an effective dilaton. Therefore, by eliminating the dilaton from the theory, we obtain a theory without any apparent divergences. Thus, mechanisms that never before appeared in point particle quantum field theory are responsible for the elimination of potentially harmful graphs.

So far, we have developed only the first quantized theory of interacting strings without loops. This, of course, cannot yield a unitary theory. The Euler Beta function, we saw earlier, has poles on the real $s$-plane axis, without imaginary parts or cuts, and therefore the theory describes only tree diagrams. Early attempts were made to modify the original Beta function by adding an imaginary part to the mass of the resonances:

$$A(s, t) \to \sum_J \frac{\alpha_J}{s - M_J^2 + i\Gamma_J} \tag{5.1.1}$$

but then the marvelous properties of the Beta function were inevitably destroyed.

The correct proposal to unitarize the model was finally made by Kikkawa, Sakita, and Virasoro [1] in 1969 by adding loops, treating the Beta function as the Born term in a perturbative approach to defining the $S$-matrix. The actual multiloop amplitudes were calculated by Kaku, Yu, Lovelace, and Alessandrini [2–8].

To understand how the perturbation series is set up, let us begin with the time evolution operator $U$, which transforms an initial state at $t = -\infty$ into the final state at $t = \infty$. The $S$-matrix is the matrix element of $U$:

$$S_{if} = \langle i | U(-\infty, \infty) | f \rangle \tag{5.1.2}$$

Because the time evolution operator is unitary, the $S$-matrix itself is also unitary:

$$S^\dagger S = S S^\dagger = 1 \tag{5.1.3}$$

In matrix form, this reads

$$\sum_n \langle i | S | n \rangle \langle n | S^\dagger | j \rangle = \delta_{ij} \tag{5.1.4}$$

where the $n$'s are a complete set of intermediate states. If we separate out the state that corresponds to no scattering, we get the $T$ matrix:

$$S = 1 - iT \tag{5.1.5}$$

Then:

$$i(T - T^\dagger) = TT^\dagger \tag{5.1.6}$$

If we take matrix elements for the scattering of the multiparticle initial state $\langle i |$ going to the multiparticle final state $| j \rangle$, then we have

$$\mathrm{Im}\, T_{ij} = -\tfrac{1}{2} \sum_n \langle i | T | n \rangle \langle n | T^\dagger | j \rangle \tag{5.1.7}$$

(see Fig. 5.1) If we represent the four-string scattering amplitude as $\langle i | T | k \rangle$, then clearly we must combine various four-point functions to obtain the next order in the perturbation series. Because strings can "twist" as they sweep out a two-dimensional surface, the set of Feynman diagrams for the loops is larger

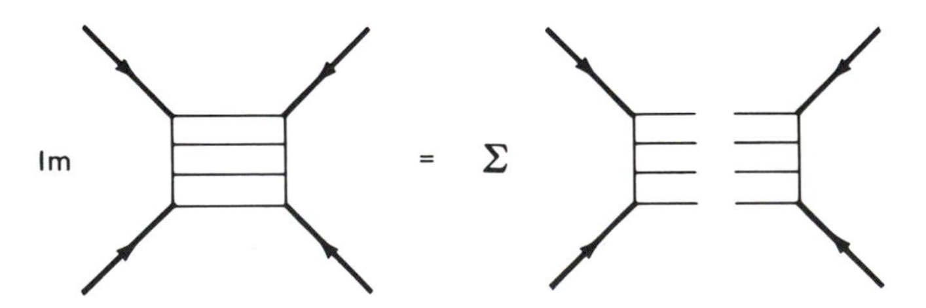


Figure 5.1. Unitarity of the $S$-matrix. The imaginary part of the scattering amplitude is proportional to the square of its absolute value. In this fashion, we can construct higher loops from lower-order tree diagrams. This was the original unitarization scheme advocated by Kikkawa, Sakita, and Virasoro.

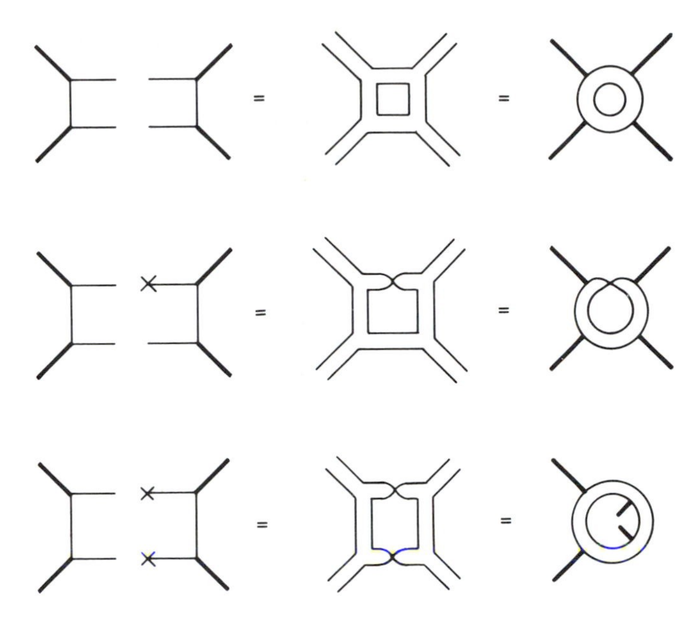


Figure 5.2. Planar, nonplanar, and nonorientable single-loop open string diagrams. Unitarity forces us to sew together three types of graphs. (The × on the string corresponds to twisting the leg.) The nonorientable diagram corresponds to a Möbius strip. The nonplanar diagram corresponds to placing some external tachyon lines on the boundary of the interior loop.

than for simple planar diagrams. In fact, as Fig. 5.2 shows, there are three types of diagrams that one can construct using the optical theorem.

For the open string, the interactions sweep out a world surface that is topologically equivalent to a disk with holes. In addition, we can also have "twists" in the disk. (A twist is created when we cut a line between two holes and then rejoin the cut by reversing the orientation of the points along the cut.) The three types of open string diagrams are

(1) Planar diagrams, which are topologically equivalent to a disk with $N$ holes punched in the interior with the external lines located on the exterior edge.
(2) Nonplanar orientable diagrams, where the external lines can be located on some of the internal holes as well as the external edge, or where the holes overlap.
(3) Nonorientable diagrams, where we have an odd number of twists in the surface of the disk. The Möbius strip is an example of a nonorientable diagram.

For the closed string, the tree diagram swept out by the interacting string is topologically equivalent to a sphere. There are two types of loop diagrams we can create out of the closed string. Let us cut $2N$ holes in the sphere and carefully mark the orientation of points along the circular edge of each hole.

Then rejoin $N$ pairs of holes to obtain a sphere with $N$ handles. The two types of diagrams are:

(1) Planar diagrams, where the orientation of the circular edge of each pair of holes is preserved when we resew the diagram. A doughnut, for example, is a planar single loop diagram.
(2) Nonorientable diagrams, where the pairs of points along the circular edges of the holes are joined by reversing the orientation of the holes. A Klein bottle, for example, is a nonorientable diagram (in Fig. 5.3, we see how Klein bottles can be assembled from flat two-dimensional surfaces that have their boundaries identified). (We note that only Type I strings, which carry no orientation or direction, have Klein bottles in their perturbation

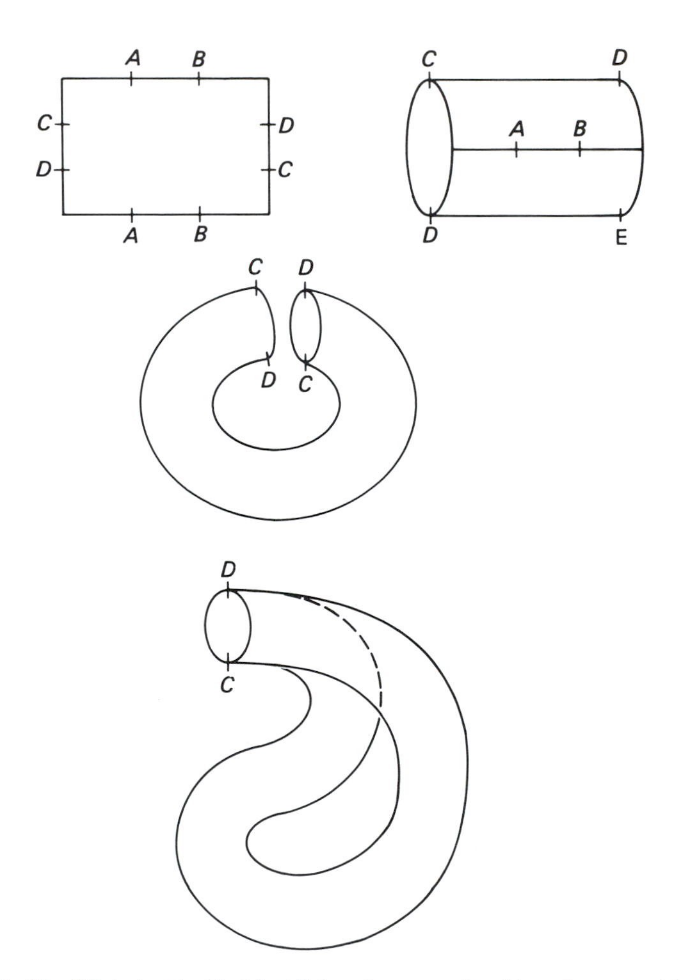


Figure 5.3. The Klein bottle. By identifying the opposite sides of a rectangle, we can obtain either a torus or (if the orientations of the sides are reversed) a Klein bottle, shown here. The Klein bottle is a two-dimensional closed surface with only one side.

series. Type II strings carry an intrinsic orientation and cannot produce Klein bottles.)

In the functional formalism, we saw earlier that all tree diagrams were constructed by calculating the Neumann function for the disk or the sphere [9–11]. The simplest method of calculating this Neumann function was to conformally map the disk or sphere to the upper half-plane or the entire complex plane. We then borrowed a method developed in electrostatics, the method of images, to write down the Neumann function:

$$G(z, z) = \ln|z - z'| + \ln|z - \bar{z}'| \tag{5.1.8}$$

Now, we will generalize this discussion to Riemann surfaces with holes. Fortunately, mathematicians long ago wrote down the Neumann function for the disk and sphere with $N$ holes. In fact, Burnside [12] solved this problem in 1891! The solutions to this classical problem are given in terms of *automorphic functions*, which we will now analyze.

## §5.2. Single-Loop Amplitude

First, let us set up the functional integral for the single-loop diagram:

$$A = \int_S DX\, d\mu\, e^{iS} \prod_{i=1}^{N} e^{ik_i X_i} \tag{5.2.1}$$

where we functionally integrate over a horizontal strip in the complex plane (see Fig. 5.4) that has finite length and then identify the left and right edges. In this way, we construct a surface topologically equivalent to a disk with a hole. The functional integral, of course, can be calculated explicitly, leaving us with the factor

$$\exp \sum k_i N(z_i, z_j') k_j \tag{5.2.2}$$

where $N$ is the Neumann function.

Now, let us topologically deform the horizontal strip into the following surface. Consider an annulus, defined as the region in the upper half-plane that has an outer radius of $r_b$ and an inner radius of $r_a$ and a ratio $w = r_b/r_a$. Now impose the fact that the outer perimeter is to be identified with the inner perimeter. This means that a point $z$ on the surface of the outer perimeter is to be identified with the point on the inner perimeter with the same polar angle:

$$z \to wz \tag{5.2.3}$$

This identification creates a semicircular tube in the upper half-plane. The conformal mapping from the horizontal strip to this tube is just the exponential.

A tube, in turn, can be mapped into a disk with a hole by stretching one end of the tube until it becomes a large circle and then shrinking the other end. When we construct Neumann functions in this annulus, we obviously are

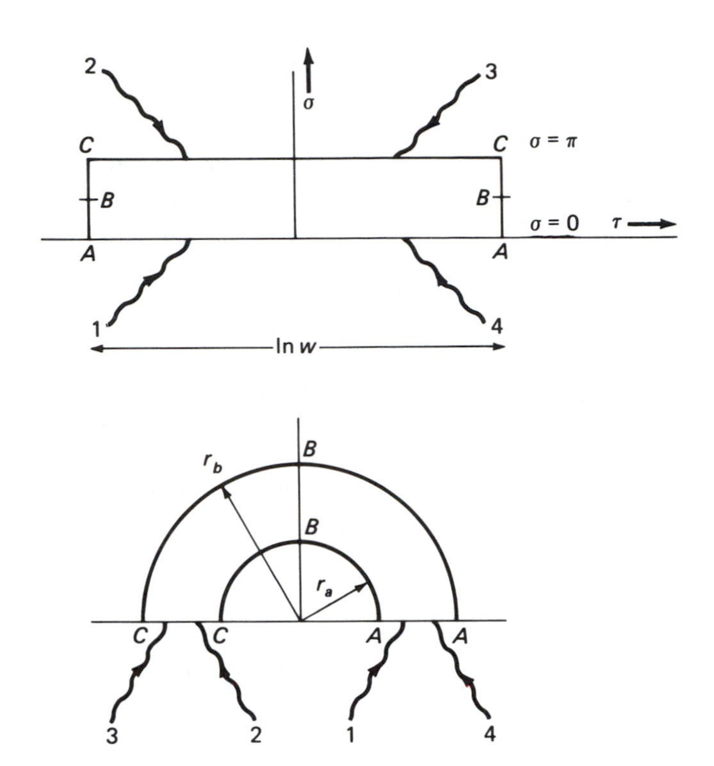


Figure 5.4. Conformal surface of the single-loop open string diagram. In the $\rho$ plane, the surface corresponds to a rectangle of width $\pi$ and of arbitrary length, such that we identify the ends. Rectangles of constant width with varying lengths are all conformally inequivalent, so we must integrate over all lengths in the path integral. The wavy lines correspond to "zero width" strings or tachyons, which may attach to the surface on either the upper or lower boundary. In the $z$-plane, the surface corresponds to a narrow tube that is bent into a half-circle, with external lines emanating from the ends.

interested in functions that have the property

$$\psi(z) = \psi(wz) \tag{5.2.4}$$

If we take arbitrary powers of $w$, this identification actually divides the upper half-plane into an infinite number of concentric circles. Each concentric circle has radius $w^n$. By identifying the outer perimeter of one annulus with its inner perimeter, we can create an infinite succession of tubes. *Thus, the entire upper half-plane can be decomposed into an infinite sequence of these tubes, but we are interested in only one of them.*

Thus, we have divided up the upper half-plane by making an identification of concentric circles generated by multiplication by the number $w$. This number $w$ thus parametrized the disk with a hole. This is called the *Teichmüller parameter* for the single-loop diagram. These parameters have a natural generalization to surfaces with $N$ holes.

In general, we can also divide up the upper half-plane by using an arbitrary projective transformation, which, as we saw earlier, maps the real axis into the real axis. (In general, circles are mapped into circles under a projective transformation.) We define an *automorphic function* as one that has the property

$$\psi(z) = \psi(z') \tag{5.2.5}$$

where we make a projective or SL(2, $R$) transformation:

$$\text{SL}(2, R): \quad \begin{cases} z' = \dfrac{az + b}{cz + d} \\ ad - bc = 1 \end{cases} \tag{5.2.6}$$

for real $a$, $b$, $c$, and $d$. (Phases can sometimes enter into the periodicity properties of these functions.) Fortunately, mathematicians have calculated this function. The Neumann function is

$$N(z, z') = \ln|\psi(z'/z, w)| + \ln|\psi(\bar{z}'/z, w)| \tag{5.2.7}$$

where we demand, up to a phase, that the function be periodic:

$$\psi(z, w) = \psi(wz, w) \tag{5.2.8}$$

Explicitly, this periodic function is

$$\ln \psi(x, w) = \ln(1 - x) - \tfrac{1}{2}\ln x + \frac{\ln^2 x}{2 \ln w}$$
$$+ \sum_{n=1}^{\infty} [\ln(1 - w^n x) + \ln(1 - w^n/x) - 2\ln(1 - w^n)] \tag{5.2.9}$$

Notice that this function explicitly has the required periodicity property mentioned earlier. Thus, when we exponentiate the Neumann function, the momentum factor in the integrand contains factors like

$$\prod_n \prod_{i<j}^{N} (z_i - w^n z_j)^{k_i \cdot k_j} \tag{5.2.10}$$

This function can also be rewritten in terms of Jacobi theta functions:

$$\psi(x, w) = -2\pi i \exp(i\pi\xi^2/\tau)\frac{\Theta_1(\xi|\tau)}{\Theta_1'(0|\tau)} \tag{5.2.11}$$

where

$$\xi = \frac{\ln x}{2\pi i}$$
$$\tau = \frac{\ln w}{2\pi i} \tag{5.2.12}$$

There are four Jacobi theta functions, given by

$$\begin{aligned}
\Theta_1(v|\tau) &= i \sum_{n=-\infty}^{\infty} (-1)^n q^{(n-1/2)^2} e^{i\pi(2n-1)v} \\
&= 2f(q^2)q^{1/4} \sin \pi v \prod_{n=1}^{\infty} (1 - 2q^{2n} \cos 2\pi v + q^{4n}) \\
\Theta_2(v|\tau) &= \Theta_1(v + \tfrac{1}{2}|\tau) \\
\Theta_3(v|\tau) &= \sum_{n=-\infty}^{\infty} q^{n^2} e^{2i\pi n v} \\
&= f(q^2) \prod_{n=1}^{\infty} (1 + 2q^{2n-1} \cos 2\pi v + q^{4n-2}) \\
\Theta_4(v|\tau) &= \Theta_3(v + \tfrac{1}{2}|\tau) = -ie^{i\pi(v+(1/4)\tau)}\Theta_1(v + \tfrac{1}{2}\tau|\tau)
\end{aligned} \tag{5.2.13}$$

where

$$f(q^2) = \prod_{n=1}^{\infty} (1 - q^{2n}); \qquad q = e^{i\pi\tau} \tag{5.2.14}$$

is the partition function.

The amplitude is thus

$$A_{\text{loop}} = \int d\mu \int d^{26}p\, \Delta \prod_n \prod_{i<j} (z_i - w^n z_j)^{k_i \cdot k_j}$$

Using conformal invariance, we can also determine the factors $\Delta$ and $d\mu$. However, we will find it convenient simply to summarize the results of the harmonic oscillator approach, which also arrives at the correct integration measure.

## §5.3. Harmonic Oscillators

The most convenient language in which to discuss the planar, nonplanar, and nonorientable single-loop diagrams is the operator language, which, of course, is a specific representation of the functional integral. Let us begin with a strip in the complex plane $\tau + i\sigma$, the set of all points from $\sigma = 0$ to $\pi$. Now place the interaction with external lines along the $x$-axis. As before, if we take this to be our surface for the functional integral, then we can insert a complete set of intermediate states everywhere:

$$|X\rangle \int DX_\mu \langle X| = 1 \tag{5.3.1}$$

and we obtain for the functional integral

$$\cdots \int DX_1 \langle X_1|V_0|X_2\rangle \int DX_2 \langle X_2| \frac{1}{L_0 - 1} \cdots \tag{5.3.2}$$

As before, we note that the above expression for the functional integral is possible because the Hamiltonian is diagonal in the harmonic oscillator basis. By eliminating the functional integration at all intermediate points along the strip, we obtain the first planar loop amplitude, in operator language:

$$A_P = \int d^D P \, \mathrm{Tr}[V_0 D V_0 D \cdots V_0 D] \tag{5.3.3}$$

Carefully choosing the gauge for the multiloop calculation is crucial. For the tree amplitudes, we saw in (2.9.5) that ghost states do not couple to trees, so they can essentially be ignored. However, ghost states do couple to factorized trees because of (2.9.9), so they can propagate internally in a loop unless they are carefully eliminated. An identical problem occurs when quantizing Yang-Mills theory (which is not surprising, since both Yang-Mills and strings are gauge theories).

There are three standard ways of eliminating the ghosts in loop amplitudes. First, we can insert projection operators which explicitly remove the ghosts from the Hilbert space. However, this technique is rather cumbersome and prohibitively difficult for fermion loops. Second, we can allow Feddeev-Popov ghost states to propagate and cancel the ghosts states. Using BRST and conformal field theory, even higher fermion loop amplitudes can be calculated. And third, we can use the light cone gauge, which we will use in this chapter.

The advantage of choosing the light cone condition is that it vastly simplifies the calculation. Ghost states, for the most part, can simply be ignored.

In addition to the light cone gauge, we will also choose a specific kinematic frame for the external legs of a multiparticle amplitude:

$$k^+ = 0 \tag{5.3.4}$$

(We should caution the reader that certain complications exist with respect to the kinematics of enforcing this for an arbitrary number of external particles with arbitrary spin. For example, we may have to perform analytic continuations of the external components of momenta in order to preserve this frame. A careful analysis shows that we can always choose this frame for fewer than 26 external boson legs and less than 10 supersymmetric legs. There are no inherent problems, however, with the light cone approach because one can always perform a Lorentz rotation on the vertex functions to a frame where the + components are nonvanishing. The light cone gauge is compatible with arbitrary values of the + components of the momenta. As a result, we will omit further discussion of this delicate point in our discussion.)

Fortunately, this trace is easily evaluated using the coherent state formalism (2.6.18, 19) that we used for the tree diagram:

$$A_N = \int \prod_{i=1}^{N} dx_i \int d^D p \, \mathrm{Tr}[V_0(k_1, x_1) V_0(k_2, x_1 x_2) \cdots V_0(k_N, x_1 x_2 \cdots x_N) w^{L_0 - 2}] \tag{5.3.5}$$

where

$$w = x_1 x_2 \cdots x_N \tag{5.3.6}$$

The trace can be explicitly calculated using coherent state methods. We use the identity

$$\operatorname{Tr}(M) = \frac{1}{\pi} \int d^2\lambda \, e^{-|\lambda|^2} \langle \lambda | M | \lambda \rangle \tag{5.3.7}$$

Carefully working out the different factors, we have

$$A_N = \int d^D p \prod_{i=1}^{N} dx_i \, x_i^{(1/2)p_i^2 - 2} T \tag{5.3.8}$$

$$T = f(w)^{2-D} \prod_{i<j} \prod_{m=1}^{\infty} \left\{ \frac{(1 - w^{m-1} c_{ji})(1 - w^m c_{ji}^{-1})}{(1 - w^m)^2} \right\}^{k_i \cdot k_j} \tag{5.3.9}$$

where

$$\begin{aligned} \rho_i &= x_1 x_2 \cdots x_i \\ f(w) &= \prod_{n=1}^{\infty} (1 - w^n) \\ c_{ji} &= \rho_j / \rho_i \end{aligned} \tag{5.3.10}$$

We can also explicitly perform the $p$ integration:

$$\int d^D p \prod_{i=1}^{D} x_i^{(1/2)p_i^2} = \left( \frac{-2\pi}{\ln w} \right)^{(1/2)D} \prod_{1 \le i < j \le N} \left[ c_{ji}^{-1/2} \exp \frac{\ln^2 c_{ji}}{2 \ln w} \right]^{k_i \cdot k_j} \tag{5.3.11}$$

Combining everything together, we have [13, 14]

$$A_N = \pi^{-1} \int_0^1 \prod_{i=1}^{N-1} \theta(v_{i+1} - v_i) \, dv_i \int_0^1 \frac{dq}{q^3} \left( \frac{-2\pi^2}{\ln q} \right)^N f(q^2)^{-24} \prod_{i<j} (\psi_{P,ij})^{k_i \cdot k_j} \tag{5.3.12}$$

where

$$\begin{aligned} \psi_{P,ij} &= \psi(c_{ji}, w) \\ v_i &= \ln \rho_i / \ln w \\ q &= e^{2\pi^2/\ln w} \end{aligned} \tag{5.3.13}$$

and the $\theta$ function here simply orders the various $v_i$ factors along the real axis.

This is the final result for the planar single-loop amplitude. Notice several features about the result:

(1) As predicted by the path integral method, the integrand is an automorphic function.
(2) The measure is easily evaluated in the harmonic oscillator approach. (The calculation is a bit harder in the path integral approach.)
(3) The integral diverges at $q = 0$, which corresponds to the inner hole shrinking to zero radius. The divergence is a mild one and can be eliminated once we add superpartners to the strings.

For future reference, we will define the planar (P), nonplanar (NP), and nonorientable (NO) integrands at the same time:

$$\psi_{\mathrm{P}}(x, w) = \frac{1-x}{\sqrt{x}} \exp\left(\frac{\ln^2 x}{2 \ln w}\right) \prod_{n=1}^{\infty} \left\{\frac{(1-w^n x)(1-w^n/x)}{(1-w^n)^2}\right\} \tag{5.3.14}$$

$$\psi_{\mathrm{NP}}(x, w) = \frac{1+x}{\sqrt{x}} \exp\left(\frac{\ln^2 x}{2 \ln w}\right) \prod_{n=1}^{\infty} \left\{\frac{(1-w^n x)(1+w^n/x)}{(1-w^n)^2}\right\} \tag{5.3.15}$$

$$\psi_{\mathrm{NO}}(x, w) = \frac{1-x}{\sqrt{x}} \exp\left(\frac{\ln^2 x}{2 \ln w}\right) \prod_{n=1}^{\infty} \left\{\frac{(1-(-w)^n x)(1-(-w)^n/x)}{(1-(-w)^n)^2}\right\} \tag{5.3.16}$$

These functions, in turn, can be reexpressed in a form in which their link with Jacobi theta functions is more apparent:

$$\psi_{\mathrm{P}}(x, w) = \frac{-2\pi}{\ln q} \sin \pi v \prod_{n=1}^{\infty} \left\{\frac{(1-2q^{2n} \cos 2\pi v + q^{4n})}{(1-q^{2n})^2}\right\} \tag{5.3.17}$$

$$\psi_{\mathrm{NO}}(x, w) = \frac{-4\pi}{\ln q} \sin \tfrac{1}{2}\pi v \prod_{n=1}^{\infty} \left\{\frac{(1-2(-\sqrt{q})^n \cos \pi v + q^n)}{(1-(-\sqrt{q})^n)^2}\right\} \tag{5.3.18}$$

$$\psi_{\mathrm{NP}}(x, w) = -\frac{\pi}{\ln q} q^{-1/4} \prod_{n=1}^{\infty} \left\{\frac{(1-2q^{2n-1} \cos 2\pi v + q^{4n-2})}{(1-q^{2n})^2}\right\} \tag{5.3.19}$$

Written explicitly in terms of theta functions, we have

$$\begin{aligned} \psi_{\mathrm{P}}(x, w) = {} & -2\pi i \exp \frac{\ln^2 x}{2 \ln w} \\ & \times \Theta_1\left(\frac{\ln x}{2\pi i}\middle|\frac{\ln w}{2\pi i}\right) \Theta_1'^{-1}\left(0\middle|\frac{\ln w}{2\pi i}\right) \end{aligned} \tag{5.3.20}$$

$$\begin{aligned} \psi_{\mathrm{NP}}(x, w) = {} & 2\pi \exp \frac{\ln^2 x}{2 \ln w} \\ & \times \Theta_2\left(\frac{\ln x}{2\pi i}\middle|\frac{\ln w}{2\pi i}\right) \Theta_1'^{-1}\left(0\middle|\frac{\ln w}{2\pi i}\right) \end{aligned} \tag{5.3.21}$$

$$\begin{aligned} \psi_{\mathrm{NO}}(x, w) = {} & -2\pi i \exp \frac{\ln^2 x}{2 \ln w} \\ & \times \Theta_1\left(\frac{\ln x}{2\pi i}\middle|\frac{\ln w}{2\pi i} + \tfrac{1}{2}\right) \Theta_1'^{-1}\left(0\middle|\frac{\ln w}{2\pi i} + \tfrac{1}{2}\right) \end{aligned} \tag{5.3.22}$$

Now that we have calculated the planar single-loop graph, let us calculate the nonorientable single-loop function. The trace we want to evaluate is

$$A_{\mathrm{NO}} = \int d^D p \, \mathrm{Tr}(\Omega V_0 D V_0 \cdots V_0 D) \tag{5.3.23}$$

where $\Omega$ is the twist operator in (2.7.13). Notice that the twist can be placed

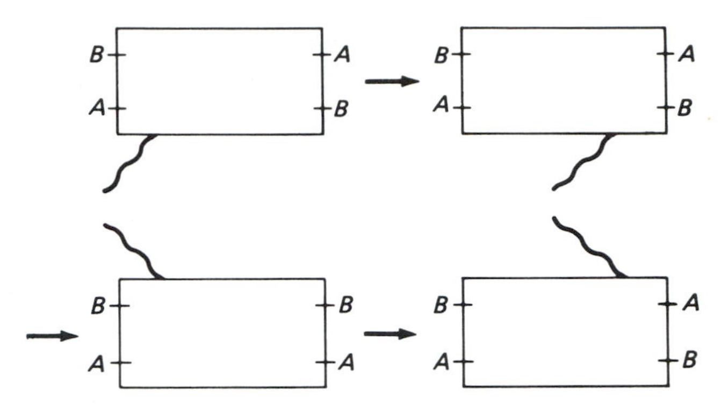


Figure 5.5. Integrating around a Möbius strip. For the Möbius strip, an external leg must travel twice the length of the strip in order to make a complete path around the boundary. Thus, the nonorientable strip has an integration region that has twice the range of the planar one.

anywhere along the chain, and nothing changes. The trace can be evaluated using the same coherent state techniques, and the major change is that $w$ turns into $-w$. The final result is

$$A_{\mathrm{NO}} = \int_0^1 \prod_{i=1}^{N} dx_i\, w^{-2} f(-w)^{-24} \left(\frac{-2\pi}{\ln w}\right)^{13} \prod_{i<j} (\psi_{\mathrm{NO},ij})^{k_i \cdot k_j}$$
$$= \int_0^2 \prod_{t-1}^{N} \theta(v_{i+1} - v_i)\, dv_i \int_0^1 \frac{dq}{q^3}$$
$$\times (-\ln q)^{-N} f(-q^2)^{-24} \prod_{i<j} (\psi_{\mathrm{NO},ij})^{k_i \cdot k_j} \tag{5.3.24}$$

Notice that the integration region is *twice* the usual one form 0 to 1. This is because, for the external lines to go completely around the Möbius strip, the lines must go around the edge twice (see Fig. 5.5). This fact will eventually play a crucial role in the cancellation of anomalies in Chapter 9.

The nonplanar diagram can also be evaluated. Notice that placing an even number of twists on the loop separates the external lines into two classes, those that revolve around the outer edge and those that revolve around the inner. The final answer depends on which lines are on the inner and outer edges. Let us place the twists so that the first to the $L$th lines are split apart from the others:

$$A_{\mathrm{NP}} = \int d^D p\, \mathrm{Tr}(\Omega D V_1 D \cdots V_L \Omega D V_{L+1} \cdots D V_N) \tag{5.3.25}$$

The integral can again be performed, with the answer [15]

$$A_{\mathrm{NP}} = \int_R \prod_{i=1}^{N-1} dv_i \int_0^1 \frac{dq}{q^3} \left(\frac{-2\pi^2}{\ln q}\right)^N [f(q^2)]^{-24} \prod_{i<j} (\psi_{\mathrm{NP},ij} \text{ or } \psi_{\mathrm{P},ij})^{k_i \cdot k_j} \tag{5.3.26}$$

where we use $\psi_{NP}$ if the $ij$ lines are on opposite sides of the disk and $\psi_P$ if they lie on the same side. The integration region for the lines reflects the fact that there are two disjoint regions of the disk. The external lines are integrated sequentially, except that there are now two disjoint regions.

For convenience, let us now put all three amplitudes in the same general expression:

$$A_J \sim \int_{R_J} \prod_{i=1}^{N-1} dv_i \int_0^1 \frac{dq_J}{q_J^3} \left( \frac{-2\pi^2}{\ln q_J} \right)^N f_J(q)^{-24} \prod_{i<j} (\psi_{J,ij})^{k_i \cdot k_j} \tag{5.3.27}$$

where

$$\begin{aligned} J &= \text{P, NO, NP} \\ R_P &= \{0 \le v_1 \cdots \le v_N = 1\} \\ R_{NO} &= \{0 \le v_1 \cdots \le v_N = 2\} \\ f_P(q^2) &= f_{NP}(q^2) \\ f_{NO}(q^2) &= f(-q^2) \\ q &= q_P = q_{NP} = q_{NO}^4 \end{aligned} \tag{5.3.28}$$

and where the nonplanar region of integration reflects the fact that there are two disjoint regions of integration.

Now that we have written down explicit representations for the various open string amplitudes, let us draw some rather remarkable conclusions from these amplitudes:

## (1) Closed Strings from Open Strings

One of the strange features of the nonplanar single-loop diagram is that it has more poles than those found by factorizing on the usual open string channels [16]. By examining the factor

$$\int_0^1 dq\, q^{-3-(1/4)s} \tag{5.3.29}$$

where the $\frac{1}{4}s$ factor comes from the momentum-dependent integrand) we find that there are extra poles at $s/4 = -2, 0, 2, 4, 6, \ldots$, which are precisely the locations of the poles of the closed string sector. Thus, the open string sector automatically contains the closed string sector. This can most easily be shown with dual diagrams by thinking of the nonplanar diagram being a cylinder, with two sets of external lines that can rotate around the top and bottom edges of the cylinder. However, by factorization, we can slice the cylinder horizontally, such that the intermediate state is a closed loop. *Thus, the closed string emerges as a "bound state" of the open string sector* (see Fig. 5.6). The closed string sector, by itself, is an entirely unitary theory. However, the open string sector, by itself, is not. We see that the presence of open strings

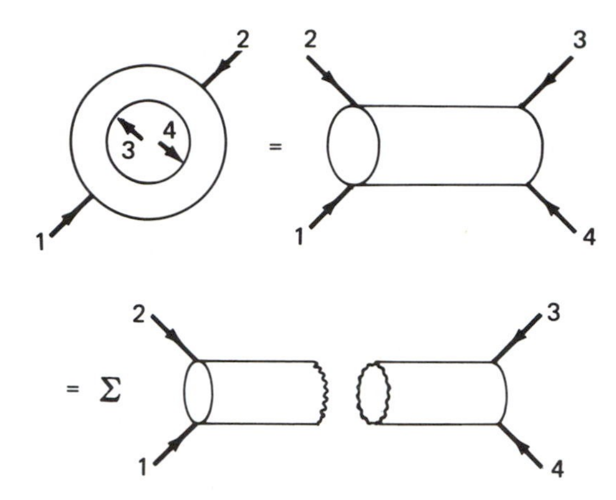


Figure 5.6. Emergence of the closed string theory from open strings. A curious feature of the open string theory is that, at the first-loop level, it already contains the closed string theory as a "bound state." The nonplanar diagram for open strings can be stretched until it becomes a cylinder, which in turn can be factorized into two smaller cylinders. Thus, the intermediate state must be a closed string.

demands the existence of closed strings as intermediate states, or else the theory is not unitary. In fact, as noticed by Lovelace [17], this unwanted singularity in the complex plane for the nonplanar diagram is actually a cut (which would be disastrous), but becomes a pole only in 26 dimensions. In fact, this was the first indication that the string model was consistent only in 26 dimensions.

## (2) Slope Renormalization

Notice that the divergence of the planar diagram arises from

$$\int_0^1 \frac{dq}{q^3}; \qquad \int_0^1 \frac{dq}{q} \tag{5.3.30}$$

This divergence at $q \to 0$ corresponds to the hole in the disk shrinking to zero. This divergence arises from the fact that we are summing over an infinite number of intermediate states propagating in the interior of the loop.

This is not, however, the ultraviolet divergence that we usually associate with Feynman diagrams. For point particles, the divergences of the various amplitudes arise when we deform the local topology of a particular diagram, such that a propagator shrinks to a point. *Thus, divergences of Feynman graphs are associated with deformations of the local topology of the graphs.*

In string theory, however, because of conformal invariance, we cannot pinch a propagator to a point. *Thus, conformal invariance on the world sheet rules out ultraviolet divergences.* However, we still have the infrared divergence of the interior points shrinking to zero.

Again, conformal invariance tells us that we can always map the shrinking hole into a dilaton or tachyon vanishing into the vacuum. We can always "pinch" this shrinking hole and extract a closed string resonance with vacuum quantum numbers vanishing into the vacuum. Thus, conformal invariance gives us an entirely new interpretation of the divergences of string theory. This new interpretation associates with each divergence a closed string state "pinching" off from the hole, with zero momentum, corresponding to tachyons for the $q^{-3}$ divergence or dilatons for the $q^{-1}$ divergence vanishing into the vacuum (see Fig. 5.7).

When we pinch the shrinking hole and extract a closed string state with vanishing momentum, we notice that the remaining diagram looks just like a tree without the hole. Thus, by extracting out the divergent pole contribution, what we have left is a finite tree. This, in turn, allows us to treat the divergences as a redefinition of the free parameter, the Regge slope $\alpha'$. This process, called "slope renormalization," works fine for the dilaton $q^{-1}$ pole, but it is not clear how to handle the tachyon pole contribution $q^{-3}$. Thus, the bosonic theory might not be totally free of divergences.

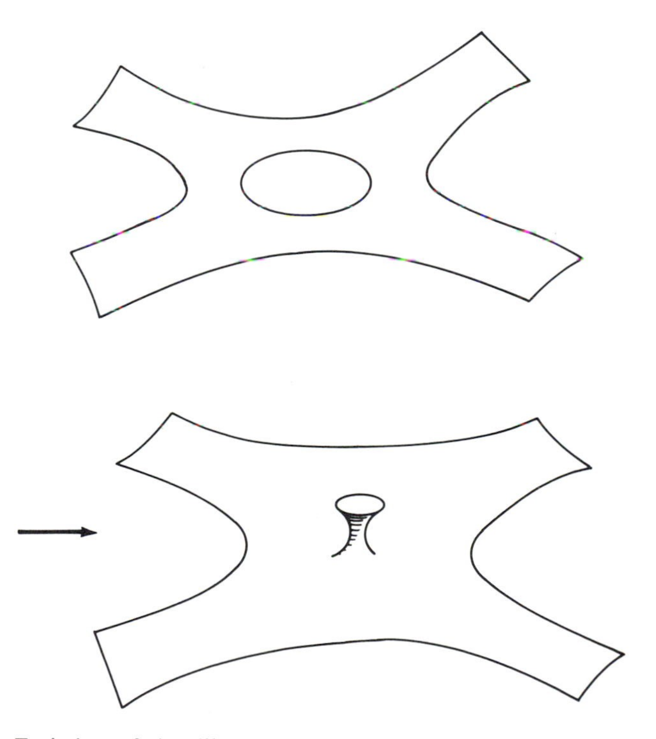

Figure 5.7. Emission of the dilaton. By conformal invariance, we can deform the single-loop open string diagram and pinch the interior loop. The resulting diagram corresponds to the emission of a tachyon or dilaton into the vacuum, which results in an infrared divergence. Superstring theory is constructed so that these poles do not exist, so the theory is formally finite.

### (3) Finite Superstrings

Experience with supersymmetric loop calculations for point particle theories has shown that the internal bosonic line cancels against the internal fermion line, yielding amplitudes that are much less divergent than expected. The same thing occurs for superstrings. Let us now turn to the superstring graphs, where we will be able to perform this "slope renormalization" [18, 19] explicitly. We will find that the $q^{-3}$ divergence in the open string Type I theory cancels by itself (this is also expected because the theory has no tachyons). This leaves only the $q^{-1}$ pole, so slope renormalization is possible. The truly remarkable thing, however, is that the Type II theory is actually *finite* by itself, without any slope renormalization!

## §5.4. Single-Loop Superstring Amplitudes

The single-loop open superstring can be calculated in either the GS or the NS–R formalism. In the GS formalism, we must use the light cone formulation, because a satisfactory covariant one does not exist. The advantage of this formalism, however, is that the graphs are manifestly space-time supersymmetric. In the NS–R formalism, we must either use a projection operator to extract out the ghosts or use the BRST techniques that allow the ghost to propagate and cancel the negative metric ghosts. Unfortunately, supersymmetry is not manifest until we add bosonic and fermionic loop contributions separately and insert the GSO projection operator into each loop.

Let us use GS formalism developed in Section 3.9, where supersymmetry is manifest. As before, the massless vector boson vertex is

$$V_B(\zeta, k, \tau) = \zeta^i B^i(\tau) :e^{ik\cdot X}: \tag{5.4.1}$$

where

$$\begin{aligned} B^i(\tau) &= P^i(\tau) + k^j R^{ij}(\tau) \\ R^{ij}(\tau) &= \tfrac{1}{8}\bar{S}(\tau)\gamma^{ij-}S(\tau) \\ S^a(\tau) &= \sum_{n=-\infty}^{\infty} S_n^a e^{-in\tau} \end{aligned} \tag{5.4.2}$$

where we have also set $\zeta^+ = 0$ for the polarization vector of a massless boson.

The fermion vertex is

$$V_F(u, k, \tau) = \bar{F}(\tau)u :e^{ik\cdot X}: \tag{5.4.3}$$

where

$$\bar{F}^a(\tau) = i(p^+)^{-1/2}(\bar{S}(\tau)\gamma\cdot P(\tau) - \tfrac{1}{3}:R^{ij}(\tau)k^i\bar{S}(\tau)\gamma^j:)^a \tag{5.4.4}$$

Let us consider only the trace over external bosons. There are a great many simplifications. For example, the trace over the $S_0$ operators requires at least eight of these operators. Thus, amplitudes with two and three external legs vanish all by themselves. As a result, *there are no self-energy and vertex corrections at all in the theory.*

The first nonzero amplitude occurs for four external legs. Even then, the amplitude actually vanishes, except for the contribution from the term

$$\zeta^i k^j R_0^{ij} :e^{ik\cdot X}: \tag{5.4.5}$$

In fact, the only diverging contribution to the trace is

$$\mathrm{Tr}(w^{\sum(1/2)n\bar{S}_{-n}\gamma^- S_n}) = f(w)^8 \tag{5.4.6}$$

As expected, this cancels against the other contribution coming from the boson loop. Thus, the final single-loop open superstring amplitude is

$$A_{\text{loop}} = K \int_0^1 \prod_{I=1}^{3} \theta(v_{I+1} - v_I)\, dv_I \int_0^1 \frac{dq}{q} \prod_{I<J} (\psi_{IJ})^{k_I\cdot k_J} \tag{5.4.7}$$

where $K$ is the same kinematic factor (3.9.11) found in the tree graph with external massless boson legs. As expected, the graph diverges only as $q^{-1}$ because there are no tachyons in the theory to give us a $q^{-3}$ divergence.

Let us now extract out the infinite piece of this graph. By carefully extracting the finite piece near $q = 0$, we find that the $\psi$ function reduces to an ordinary sine function, so that the finite piece $\bar{A}$ becomes

$$\bar{A}_{\text{loop}} = K \int_0^1 \prod_{I=1}^{3} \theta(v_{I+1} - v_I) \prod_{I<J} [\sin \pi(v_J - v_I)]^{k_I\cdot k_J} \tag{5.4.8}$$

To show that we can perform slope renormalization on this graph, let us actually perform the $v_2$ and $v_3$ integrations. If we define

$$x = \frac{\sin \pi(v_2 - v_1) \sin \pi v_3}{\sin \pi(v_3 - v_1) \sin \pi v_2} \tag{5.4.9}$$

then, after the integrations, we find

$$\bar{A}_{\text{loop}} \sim K \int_0^1 \left(\frac{\ln x}{1 - x} + \frac{\ln(1 - x)}{x}\right) x^{-\alpha' s}(1 - x)^{-\alpha' t}\, dx \tag{5.4.10}$$

It is easy checked that this, in turn, is precisely the derivative of the Born term with respect to the slope. Thus

$$\bar{A}_{\text{loop}} \sim \frac{1}{(\alpha')^2} \frac{\partial}{\partial \alpha'} A_{\text{tree}} \tag{5.4.11}$$

Notice that we can absorb this divergence into a renomalization of the slope. This is our desired result, *that a redefinition of the Regge slope can render the Type I superstring theory renormalizable at the first loop.*

Now, let us consider the case of the closed string amplitude, where even more surprises take place.

## §5.5. Closed Loops

It is again a straightforward process to calculate the single-loop amplitude of the closed bosonic string. Let us stress the differences:

(1) The world sheet of the string, which was topologically equivalent to the upper half-plane for the open string case, now becomes the entire complex plane.
(2) The propagator must contain an integration over $\sigma$ so that it is independent of the origin of $\sigma$ space.
(3) We must sum over different orderings of the vertex functions.
(4) The external lines, which were once attached to the boundary of the strip, are now attached to the interior of the complex surface.

In Fig. 5.8, we see the horizontal strip defined from $\sigma = 0$ to $\sigma = 2\pi$, such that the top and bottom edges are identitied with each other. For the single loop, we must now also identity the left and right edges. By the exponential mapping, we then map the finite horizontal strip to the entire complex plane.

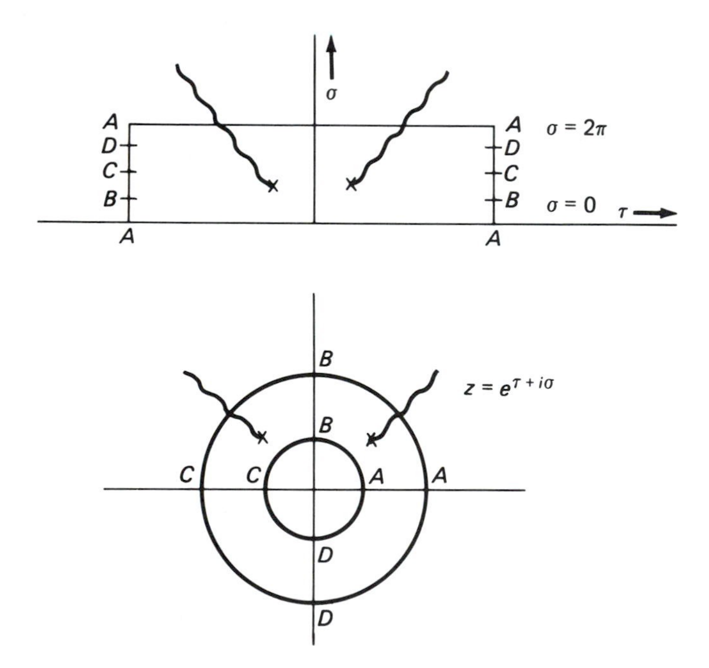


Figure 5.8. Conformal surface for the single-loop closed string diagram. In the $\rho$ plane, the surface is a rectangle of width $2\pi$ and arbitrary length, with the opposite edges identified. In the $z$-plane, the surface corresponds to a doughnut. External legs can attach themselves to any point within the surface.

By the usual coherent state methods, we find [20] (taking the slope $\alpha' = \frac{1}{4}$)

$$\begin{aligned} A &= \int d^D p \operatorname{Tr}(V_1 D V_2 \cdots V_N D) \\ &= \int \prod_{i=1}^{N} d^2 z_i \, |w|^{-4} |f(w)|^{-48} \left( \frac{-4\pi}{\ln|w|} \right)^{13} \prod_{i<j} \chi_{ij}^{(1/2)k_i \cdot k_j} \end{aligned} \tag{5.5.1}$$

where

$$\begin{aligned} v_j &= (2\pi i)^{-1} \ln z_1 z_2 \cdots z_j \\ v_{ji} &= v_j - v_i \\ \tau &= v_M = (2\pi i)^{-1} \ln w \\ w &= z_1 z_2 \cdots z_N \\ c_{ji} &= z_{i+1} z_{i+2} \cdots z_j \end{aligned} \tag{5.5.2}$$

and

$$\begin{aligned} \chi(z, w) &= \exp \frac{(\ln^2|z|)}{2\ln|w|} \left| z^{-1/2}(1-z) \prod_{m=1}^{\infty} \left\{ \frac{(1-w^m z)(1-w^m/z)}{(1-w^m)^2} \right\} \right| \\ \chi_{ij} &= \chi(c_{ji}, w) \\ &= 2\pi \exp\left( \frac{-\pi(\operatorname{Im} v_{ji})^2}{\operatorname{Im}\tau} \right) \left| \frac{\Theta_1(v_{ji}|\tau)}{\Theta_1'(0|\tau)} \right| \end{aligned} \tag{5.5.3}$$

Let us rewrite this as

$$A = \int d^2\tau (\operatorname{Im}\tau)^{-2} C(\tau) F(\tau) \tag{5.5.4}$$

where

$$\begin{aligned} C(\tau) &= 4(\tfrac{1}{2}\operatorname{Im}\tau)^{-12} e^{4\pi \operatorname{Im}\tau} |f(e^{2\pi i \tau})|^{-48} \\ F(\tau) &= \pi^N \operatorname{Im}\tau \int \prod_{i=1}^{N-1} d^2 v_i \prod_{i<j} (\chi_{ij})^{(1/2)k_i \cdot k_j} \end{aligned}$$

Notice that the integrand is doubly periodic:

$$\chi(v+1, \tau) = \chi(v+\tau, \tau) = \chi(v, \tau) \tag{5.5.5}$$

This is easy to see because the $\Theta$ function has the following properties:

$$\begin{aligned} \Theta_1(v+1|\tau) &= \Theta_1(v|\tau) \\ \Theta_1(v+\tau|\tau) &= -e^{-i\pi(2v+\tau)} \Theta_1(v|\tau) \end{aligned} \tag{5.5.6}$$

This is important, because it shows that a naive integration over the $v$ variables will drastically overcount the proper region of integration. Consider the parallelogram formed by the origin and the points 0, 1, $\tau$, and $1+\tau$. When opposite sides are identified, this becomes topologically equivalent to a torus or doughnut. Notice that the double periodicity divides up the complex plane into an infinite number of these parallelograms. Thus, we want to integrate over only one parallelogram, or else we will have infinite overcounting. Thus, for fixed $\tau$, we must restrict the integration over the $v$ variables or else we will be integrating over an infinite number of copies of the same thing. We will choose the following truncation:

$$\begin{gathered} 0 \leq \operatorname{Im} v_i \leq \operatorname{Im} \tau \\ -\tfrac{1}{2} \leq \operatorname{Re} v_i \leq \tfrac{1}{2} \end{gathered} \tag{5.5.7}$$

Surprisingly, in addition to this truncation in $v$ space, we must perform yet another truncation in $\tau$ space. The integrand of the closed loop diagram is actually invariant under yet another transformation, given by

$$\tau' = \frac{a\tau + b}{c\tau + d} \tag{5.5.8}$$

where $a, b, c, d$ are all *integers* and $ad - bc = 1$. This generates what is called the *modular group* SL(2, $Z$). Let us show that the integral is invariant under this transformation:

$$\begin{gathered} d^2\tau \to |c\tau + d|^{-4}\, d^2\tau \\ \operatorname{Im} \tau \to |c\tau + d|^{-2} \operatorname{Im} \tau \end{gathered} \tag{5.5.9}$$

Therefore the following are actually invariant under this modular transformation:

$$\frac{d^2\tau}{\operatorname{Im}^2 \tau} = \frac{d^2\tilde{\tau}}{\operatorname{Im}^2 \tilde{\tau}} \tag{5.5.10}$$

Now let us calculate how the other terms transform under a modular transformation:

$$e^{-(\pi/4)\operatorname{Im}\tau}|f(e^{2\pi i\tau})|^3 \to |c\tau + d|^{3/2} e^{-(\pi/4)\operatorname{Im}\tau}|f(e^{2\pi i\tau})|^3 \tag{5.5.11}$$

Thus,

$$C(\tau) = C(\tilde{\tau}) \tag{5.5.12}$$

Finally, we also have

$$\chi\left(\frac{v}{c\tau + d}, \frac{a\tau + b}{c\tau + d}\right) = |c\tau + d|^{-1}\chi(v, \tau) \tag{5.5.13}$$

Thus,

$$F(\tau) = F(\tilde{\tau}) \tag{5.5.14}$$

The integrand is therefore invariant under a modular transformation, which was first pointed out by Shapiro [20]. But what is the intuitive meaning of this symmetry?

The Polykov action tells us that we must integrate over all conformally inequivalent surfaces. We first observe that two parallelograms with different values of $\tau$ are conformally inequivalent when we identify opposite sides. Thus, naively we expect that the integration over $\tau$ automatically integrates over all conformally inequivalent surfaces. However, this is not the case.

There are actually two kinds of reparametrizations of a surface that must be carefully distinguished. The first are the reparametrizations that can smoothly be deformed back to the identity, i.e., the set of smooth reparametrizations that contains the identity map. The second is the set of reparametrizations that cannot be smoothly deformed back to the identity. Global diffeomorphisms are of this category. For example, take the parallelogram and identify only one set of opposite sides. This creates a tube. Normally, we would bring the two ends of the tube together to make a torus. However, now twist one of the open ends of the tube by $2\pi$ and then resew the ends together. By examing the resulting surface, we find that this has caused a genuine reparametrization of the surface, but that the identity map cannot be represented in this way. This twist is called a "Dehn twist" and it generates a discrete group. For the torus, it can be shown that the group generated by Dehn twists is the modular group $\mathrm{Sp}(2, Z)$.

Concretely, the integral is invariant under $\tau \to -1/\tau$ and $\tau \to \tau + 1$. By successive applications of these two transformations, we can show that we generate the entire modular group. But carefully examining the effect of these two transformations shows that they simply interchange the boundary of the parallelogram, generating Dehn twists.

In summary, this second symmetry, called modular invariance, results from the fact that we must gauge-fix not only reparametrizations that can smoothly reach the identity map but also global diffeomorphisms that are not connected to the identity map. Thus, we must divide up the complex $\tau$ plane so that we only integrate over one surface invariant under the transformations $\tau \to \tau + 1$ and $\tau \to -1/\tau$. Thus, the complex $\tau$ plane is divided up into an infinite set of redundant copies. To eliminate this infinite overcounting, we will take the following fundamental region of integration:

$$\text{Fundamental region} = \begin{cases} -\frac{1}{2} \leq \operatorname{Re} \tau \leq \frac{1}{2} \\ \operatorname{Im} \tau \geq 0 \\ |\tau| \geq 1 \end{cases} \tag{5.5.15}$$

(see Fig. 5.9).

We will shortly see that modular invariance is perhaps one of the most powerful tools we have in checking for the self-consistency of new string compactifications.

As we said, the effect of a modular transformation is to perform a Dehn twist, i.e., shuffle the boundary conditions on the parallelogram. For example, if we have a string defined on the parallelogram labeled by $X(\sigma_1, \sigma_2)$, then a modular transformation changes the boundary conditions by

$$X(\sigma_1, \sigma_2) \to X(a\sigma_1 + b\sigma_2, c\sigma_1 + d\sigma_2) \tag{5.5.16}$$

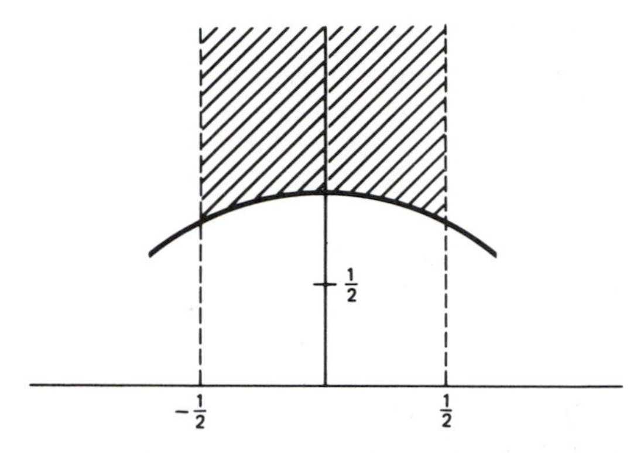


Figure 5.9. Fundamental region for the single-loop closed string amplitude. Modular invariance of the amplitude divides the complex plane into an infinite number of equivalent regions. Thus, we must choose only one such region, or else the amplitude is infinite. The most convenient region lies between Re $\tau = -\frac{1}{2}$ and $+\frac{1}{2}$ with $|\tau|$ being greater than one.

Specifically, we can check that the transformations $\tau \to \tau + 1$ and $\tau \to -1/\tau$ change the boundary conditions in the following fashion:

$$\begin{cases} \tau \to \tau + 1 \\ \tau \to -1/\tau \end{cases} \to \begin{cases} X(\sigma_1, \sigma_2) \to X(\sigma_1 + \sigma_2, \sigma_2) \\ X(\sigma_1, \sigma_2) \to X(\sigma_2, -\sigma_1) \end{cases} \tag{5.5.17}$$

Now let us analyze the divergence structure of the closed string amplitude. We first note that

$$\chi(v, \tau) \to 2\pi|v| \tag{5.5.18}$$

as $v \to 0$. Thus, poles occur, as expected, in the amplitude when external lines coincide. The poles occur at $s = -8, 0, 8, 16$, etc. This divergence can be interpreted as coming from the self-energy diagram on the external leg, which, unfortunately, is on mass shell.

Now let us generalize this calculation to the Type II superstring, which can easily be evaluated using the same techniques. The vertices all have the form

$$W = V\tilde{V} \tag{5.5.19}$$

where the $V$'s correspond to open string vertices (with half the momentum) of the two sets of oscillators. Once again, the two and three point loops vanish because of the trace over the $S_0$ modes. The trace can be evaluated exactly as in the open string case, except we now have double the number of oscillators. The answer for the boson scattering amplitude is

$$A_{\text{loop}} = K \int_F d^2\tau (\text{Im } \tau)^{-2} F_S(\tau) \tag{5.5.20}$$

where

$$F_S(\tau) = (\text{Im } \tau)^{-3} \int \prod_{I=1}^{3} d^2 v_I \prod_{I<J} (\chi_{IJ})^{(1/2)k_I \cdot k_J} \tag{5.5.21}$$

Notice that the factor $C(\tau)$ is missing and that the power of Im $\tau$ has changed. The remarkable thing is that *this amplitude is totally finite*! This finiteness is due to several factors:

(1) The absence of two and three point amplitudes makes it impossible to place tadpole or self-energy insertions on the external lines. Thus, we do not have the poles found earlier for the closed bosonic string.
(2) We can reduce the divergence of the graph by taking a fundamental domain free of divergences.
(3) There are no contributions from tachyons vanishing into the vacuum, since there are no tachyons.
(4) Fermion internal lines cancel with boson internal lines to reduce the divergence of the graph.

## §5.6. Multiloop Amplitudes

The multiloop function can also be written down explicitly in terms of path integrals over Riemann surfaces with holes. The main problem with constructing these amplitudes is the choice of parametrization of the Riemann surface. Four types of parametrizations have been developed for the multiloop amplitudes:

(1) *Schottky groups*. Multiloop amplitudes, which were originally calculated in this formalism [2–8], are discussed in this section. There are several advantages to this parametrization of a Riemann surface. First, it is explicit. There is no guesswork in the choice of integration variables, which are known exactly. The integration variables, in fact, are intuitively related to the topological structure of the Riemann surface. Second, the amplitudes factorize (because this representation was originally calculated by sewing together multiresonance vertex functions). Thus, unitarity can be shown. The disadvantage of this formalism, like other formalisms, is that modular invariance is not obvious at all. The integration region must be truncated by hand.
(2) *Constant curvature metrics*. This formalism, which will be discussed in the next section, is based on Riemann surfaces with constant Gaussian curvature. The advantage of this formalism is that it arises naturally when quantizing the Polyakov action. The other advantage is that a considerable mathematical literature exists for Riemann surfaces with constant curvature. The disadvantage of this approach, as in the Schottky representation, is that modular invariance of the higher loop amplitudes is still obscure. The integration region must be truncated by hand. Unlike the Schottky representation, however, explicit representations of the $6N - 6$ modular parameters for arbitrary surfaces with constant metrics are rare. Furthermore, because this formalism is not derived by sewing together three-resonance vertex functions, factorization and hence unitarity are not obvious.

(3) *Theta functions.* This is perhaps the most natural formalism, because modular invariance is built-in from the very start. We will discuss it in section 11. This method is based on generalizing the theta functions introduced in (5.2.11) for the single-loop amplitude to include functions which are quasi-periodic in several variables. The natural integration variable is the period matrix $\Omega_{ij}$ itself, which can be defined for any Riemann surface. This formalism can also be easily extended to include theta functions defined on various spin structures. Although this formalism holds much promise and is the subject of much research, there are also severe drawbacks. For example, beyond three loops, the period matrix becomes an extremely awkward method of parametrizing moduli space. (This difficulty of parametrizing moduli space by the period matrix beyond three loops is called the "Schottky problem." Only recently have mathematicians solved this problem. Unfortunately, the solution is highly nonlinear, and much more work has to be done to develop the formalism beyond three loops.) As in the previous formalism, factorization (and hence unitarity) are obscure. Higher loop theta functions are constructed by making educated guesses and appealing to the uniqueness of the final result, not by sewing together vertex functions.

(4) *Light cone formalism.* Because the light cone method is based strictly on physical variables, without any ghosts, this formalism is manifestly unitary and hence one intuitively expects that it automatically provides one cover of moduli space. This conjecture, which was never considered before by mathematicians in their study of moduli space, has recently been proven to all orders. We will only briefly discuss the light cone formalism at the end of this chapter because this formalism will be further developed in the next chapter in the context of developing the field theory of strings. The advantage of the light cone formalism is that it is manifestly unitary, factorizable, modular invariant, and easily generalized to a genuine second quantized field theory. The disadvantage is that it is obviously gauge-fixed.

All four formalisms, of course, must eventually yield equivalent results.

Let us first discuss the Schottky groups and rewrite the single-loop amplitude in a form that can most easily be generalized to the multiloop case. (We will discuss the $\Theta$ function method in Section 5.11.) Let us define a projective transformation (2.7.1) that forms the group SL(2, $R$), in a way that emphasizes its geometric properties. Let us define the *invariant points* $x_1$ and $x_2$ of the projective transformations to be the points that are left invariant under the transformation:

$$\text{Invariant points:} \quad \begin{cases} P(x_1) = x_1 \\ P(x_2) = x_2 \end{cases} \tag{5.6.1}$$

Then we can rewrite an arbitrary projective transformation, which has three arbitrary parameters, in terms of the two invariant points and the *multiplier* $X$:

$$P(z) = \frac{z(x_2 - Xx_1) - x_1 x_2(1 - X)}{z(1 - X) + x_2 X - x_1} \tag{5.6.2}$$

Another convenient form for the projective transformation is

$$\frac{P(z) - x_2}{P(z) - x_1} = X \frac{z - x_2}{z - x_1} \tag{5.6.3}$$

The advantage of writing the projective transformation in terms of the multiplier is that products of projective transformations have simple multipliers:

$$\begin{aligned} (X_P)^n &= X_{(P^n)} \\ X_{PQ} &= X_{QP} \\ (X_P)^{-1} &= X_{(P)^{-1}} \end{aligned} \tag{5.6.4}$$

Under a projective transformation, any $P$ can be brought to the form

$$P \to \begin{pmatrix} \sqrt{X} & 0 \\ 0 & \sqrt{X}^{-1} \end{pmatrix} \tag{5.6.5}$$

Notice that the trace of $P$ can be written as

$$\text{Tr } P(z) = \sqrt{X} + \frac{1}{\sqrt{X}} \tag{5.6.6}$$

This, in turn, allows us to define "conjugacy classes." Two projective transformations $P_1$ and $P_2$ belong to the same conjugacy class if they have the same multiplier.

Depending on the multiplier, we can define several types of projective transformations:

(1) $P$ is hyperbolic if $X$ is real, positive, and not equal to one.
(2) $P$ is parabolic if $X$ is real and equals one.
(3) $P$ is elliptic if $|X|$ equals one and $X$ does not equal one.
(4) $P$ is loxodromic if $X$ is complex and none of the above.

For a real projective transformation, the multiplier can be greater than or less than one, depending on whether $x_1 < x_2$ or vice versa. We thus have the freedom to choose all our projective transformations to be hyperbolic, so the multiplier is less than one. For the single loop, we now recognize the projective transformation to be

$$\begin{aligned} P(z) &= wz \\ P(z) &= \begin{pmatrix} \sqrt{w} & 0 \\ 0 & \dfrac{1}{\sqrt{w}} \end{pmatrix} \end{aligned} \tag{5.6.7}$$

Thus, the multiplier of the single-loop transformation is simply $w$ itself.

Notice that we have $3N$ parameters associated with $N$ projective transformations. However, we noted earlier that 3 points along the real axis can always be fixed because of projective invariance. Thus, the open string $N$-loop amplitude will have $3N - 3$ parameters that describe the surface. These are called *Teichmüller parameters* and are the minimum number of parameters needed to characterize inequivalent Riemann surfaces with a boundary. For the closed string, we have spheres and complex projective operators, so we have $6N - 6$ parameters. In summary:

$$\text{Teichmüller parameters} \begin{cases} \text{closed string:} & 6N - 6 \\ \text{open string:} & 3N - 3 \end{cases} \tag{5.6.8}$$

(Intuitively, we may say that it takes two parameters to locate the position of the center of a hole and one parameter to label its radius. Thus, $3N$ parameters are needed to describe a surface with $N$ holes. We must subtract three to eliminate conformally equivalent ways of putting the external lines on the real axis. This makes a total of $3N - 3$ parameters.)

Let us now rewrite the integrand of the single loop function in terms of manifestly invariant functions. From (5.6.4), we note that the partition function can be rewritten as:

$$\prod_{n=1}^{\infty} (1 - w^n) = \prod_{n=1}^{\infty} (1 - (X_P)^n) = \prod_{n=1}^{\infty} (1 - X_{P^n}) \tag{5.6.9}$$

Thus, the partition function is now written in terms of the set of projective transformations $P^n$.

Second, we can rewrite the momentum integrand in terms of projective transformations. Before, we showed that the combination

$$d\mu \prod_{i<j} (z_i - z_j)^{k_i \cdot k_j} \Delta \tag{5.6.10}$$

transformed nicely under a projective transformation because of momentum conservation and the fact that the external lines are on-shell. Now, we want to show that the single loop can be written as an invariant integrand:

$$d\mu \prod_{n=1, i<j} (z_i - w^n z_j)^{k_i \cdot k_j} \Delta = d\mu \prod_{n=1, i<j} (z_i' - P^n z_j')^{k_i \cdot k_j} \Delta' \tag{5.6.11}$$

Thus, the single-loop integrand can be rewritten as

$$\int d\mu \prod_{\substack{n=1 \\ i<j}} (z_i - P^n z_j)^{k_i \cdot k_j} (1 - X_{P^n})^{-24} \Delta \tag{5.6.12}$$

It is crucial that we identify the region of the complex plane that we are integrating over. In Fig. 5.10 we see how the upper half of the complex plane is divided into equivalent sectors by the action of the operator $P$. Notice that a projective transformation maps circles into circles. Thus, a point near one invariant point, after being repeatedly hit with $P$, gradually migrates over to the other invariant point. In the process, the upper half-plane is sliced up into disjoint regions. We can take *any* of these disjoint regions for our integral. In the figure, we have taken the region that is midway between the two invariant points. The arrows show how a point migrates under $P$, i.e., we see the identification of the points along the surface. When we add in external lines, they lie only on the $x$-axis and can only move up to the edge of these circles.

This procedure easily generalizes to the multiloop amplitude. We demand

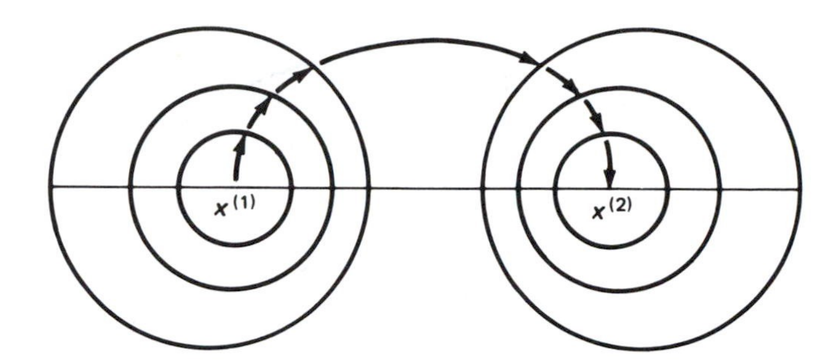


Figure 5.10. Action of a projective transformation. Under successive applications of the same projective transformation, circles surrounding one invariant point map into circles surrounding the other invariant point. After an infinite number of projective transformations, a point on one of these circles comes arbitrarily close to the invariant point.

(1) Projective invariance under SL(2, $R$) transformations. There should be $N$ projective operators for the $N$-loop amplitude.
(2) Modular invariance for the closed string, so we integrate over one fundamental domain in parameter space.

From these invariance arguments alone, we can almost write down the unique multiloop amplitude.

First, let us write down the multiloop divergence. Let

$$P_1, P_2, \ldots, P_N \tag{5.6.13}$$

represent $N$ real projective transformations, such that their invariant points are arranged sequentially on the real axis. Let $\{P\}$ represent the set of all possible distinct products of the various $P$ transformations, raised to any positive or negative power. (In Fig. 5.11 we see that the invariant points are all aligned on the real axis.) The set $\{P\}$ forms a group. Our region of integration will be the region of the real axis that exists between the limit points of the elements of $\{P\}$. If the limit points of all the elements of $\{P\}$ fills up the entire complex plane, then this is uninteresting. Then there is no region of integration for our variables.

Our interest lies in groups $\{P\}$ whose limit points form a discrete set on the real axis, so there is a finite region of integration. Groups with this property are called *Schottky groups.*

Let $\{\tilde{P}\}$ represent all distinct products of these transformations modulo cyclic permutations. Then the divergent term for the $N$-loop amplitude is a product over the set $\{\tilde{P}\}$ [4, 5]:

$$\prod_{\{\tilde{P}\}} (1 - X_{\{\tilde{P}\}})^{-24} \tag{5.6.14}$$

and the momentum-dependent term is given by a product over the set $\{P\}$:

$$\prod_{i<j,\{P\}} (z_i - \{P\} z_j)^{k_i \cdot k_j} \tag{5.6.15}$$

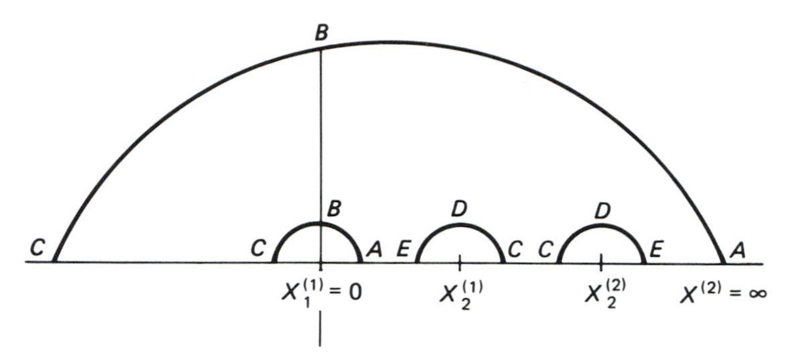


Figure 5.11. Schottky representation for the double-loop open string diagram. There are two sets of invariant points lying on the real axis, corresponding to each of the two projective transformations. The circles surrounding the invariant points are to be identified with each other.

The integrand for the $N$-loop amplitude is therefore surprisingly easy to write down. The only difficulty is determining the region of integration such that we preserve projective invariance. The final formula for the $N$-loop amplitude with $M$ external tachyon legs is [2–8]

$$\begin{aligned} A_N = \int \prod_{\alpha=1}^{N} d^{26}k_\alpha \, d\mu \prod_{\{P\}} \prod_{\{\tilde{P}\}} (1 - X_{\{\tilde{P}\}})^{-24} \\ \times \prod_{1=i<j=M} (z_i - \{P\} z_j)^{k_i \cdot k_j} \prod_{\substack{\beta,\lambda=1 \\ \beta \neq \lambda}}^{N} dX_\beta \, X_\beta^{-\alpha(k_\beta)-1} \\ \times \prod_{i=1}^{M} \left( \frac{z_i - \{P\} x_\lambda^{(2)}}{z_i - \{P\} x_\lambda^{(1)}} \right)^{k_i \cdot k_\lambda} \\ \times \left\{ \frac{x_\beta^{(1)} - \{P\} x_\lambda^{(1)}}{x_\beta^{(2)} - \{P\} x_\lambda^{(1)}} \frac{x_\beta^{(2)} - \{P\} x_\lambda^{(2)}}{x_\beta^{(2)} - \{P\} x_\lambda^{(2)}} \right\}^{(1/2) k_\beta \cdot k_\lambda} \end{aligned} \tag{5.6.16}$$

where

$$d\mu = \prod_{i=1}^{M} dz_i \, dV_{abc}^{-1} \prod_{\alpha=1}^{N} dx_\alpha^{(1)} \, dx_\alpha^{(2)} (x_\alpha^{(1)} - x_\alpha^{(2)})^{-2} \tag{5.6.17}$$

where the Roman letters $i$ and $j$ represent the external tachyon lines, and Greek letters represent only the loop variables. Each projective transformation $P_\alpha$ has two invariant points and a multiplier:

$$x_\alpha^{(1)}; \qquad x_\alpha^{(2)}; \qquad X_\alpha \tag{5.6.18}$$

The product over $\{P\}$ is taken over all inequivalent products of the projective operators $P_\alpha$. (We exclude those products where there is overcounting; i.e., if a member of $\{P\}$ is a product of transformations that ends with the transformation $P_\alpha$, then it cannot be allowed to operate on the invariant point of $P_\alpha$, or else we will overcount.)

The essential aspect of this integrand is the region of integration, which must count *each conformally inequivalent configuration once and only once.*

On the real line, we can always make a projective transformation such that all the external lines are bunched together on one side and all the invariant points are bunched together on the other side. (If a point $z$ lies to the left of the invariant points of a projective transformation $P_\alpha$, then $P_\alpha(z)$ moves $z$ across its invariant points to the other side. Thus, by successive projective transformations we can shove all external lines to one side of the real axis and all the invariant points to the other side.) We must be careful, however, to exclude the possibility of a projective transformation that whips a point all the way around the diagram. The transformation

$$P = \prod_{\alpha=1}^{N} P_\alpha \tag{5.6.19}$$

has the property that it moves points all the way around the real axis past all the invariant points. Therefore, we must be careful to extract the periodicity that enters in by circling the entire real axis an arbitrary number of times.

Let $x^{(1)}$ and $x^{(2)}$ represent the invariant points of this product. Then we have the limits of integration:

$$x^{(1)} < P(z_1) < z_M < z_{M-1} < \cdots < z_1 < x^{(2)} < x_1^{(1)} < x_1^{(2)} < x_2^{(1)} < \cdots x_N^{(2)} \tag{5.6.20}$$

with the restriction that all multipliers for $P_\alpha$, including the product $P$, remain less than or equal to one. Of course, there are other equivalent choices for the region of integration because we have taken a specific truncation of the region. In particular, we could have external lines placed between the invariant points of different projective operators. Because the set $\{P\}$ was a Schottky group, we are guaranteed that there is a finite region of integration for our variables. (We must point out that the calculation of the multiloop amplitudes was done in 1970 using projection operators that eliminated the ghosts on the leading trajectory. With BRST ghost fields, the calculation can be redone to eliminate all possible ghosts. By the uniqueness of the Neumann function over a surface, we expect the same answer.)

So far, we have discussed only open string multiloop amplitudes. Now, let us generalize these previous statements by actually writing down the Neumann function for a sphere with $N$ handles or holes for the closed string sector [21].

Let us first consider the complex plane with $2N$ holes cut out. Call this region, which lies exterior to $2N$ holes, the region $S$. Let us call these pairs of holes $a$-cycles $a_i$ and $\bar{a}_i$, where $i = 1$ to $N$. Projective transformations, we have seen, map circles into circles, so let us define $N$ projective transformations $P_i$ which take us from one $a$-cycle into their partner (see Fig. 5.12)

$$P_i\colon a_i \to \bar{a}_i \tag{5.6.21}$$

These projective transformations $P_i$, of course, can in turn be parametrized

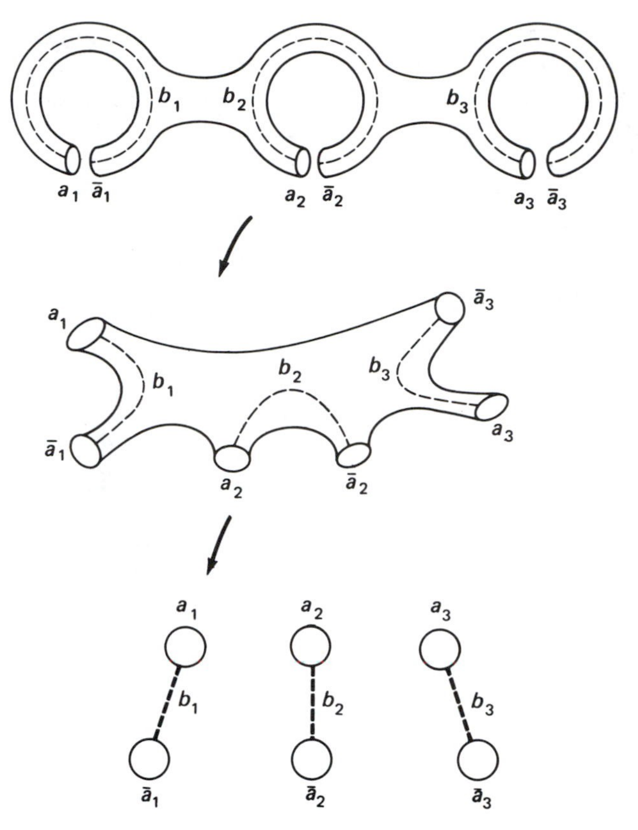


Figure 5.12. Homology cycles of an arbitrary closed Riemann surface. The $a$ and $b$ cycles are $2g$ closed lines on the surface of genus $g$ that cannot be continuously shrunk to a point. By cutting the surface along the $a$-cycles, we can flatten the surface into a plane with $2g$ holes cut out. Notice that the $b$-cycles are now lines that connect pairs of holes, which are defined by the $a$-cycles.

by two invariant points $z_i^{(1)}$ and $z_i^{(2)}$ and by the multiplier $X_i$. We will find that the invariant points lie within each of the $a$-cycles, so they do not appear in the complex plane with holes cut out. In general, a point that lies within the surface $S$ (the region exterior to all $a$-cycles) is mapped into the interior of one of the $a$-cycles by the action of a projective transformation $P_i$.

In addition to the $a$-cycles, we also have $b$-cycles, corresponding to cutting a line between $a_i$ and $\bar{a}_i$. The $b$-cycles are thus circular lines that encircle the $i$th hole when analyzing a sphere with $N$ holes. Notice that the projective transformation $z \to wz$ in the single-loop diagram took us from the interior to the exterior radii of the annulus. Thus, the projective transformation moved us along the $b$-cycle. In general, the projective transformations $P_i$ will map points in one $a$-cycle into its partner, so that we are moving along a $b$-cycle.

Let us now define $V_I$ to be the product of all distinct products of the various

$P_i$. Notice that $i$ runs from 1 to $N$, while $I$ runs over an infinite number of indices. Now let us define the function $\psi$:

$$\ln \psi(z, z') = \ln(z - z') + \sum_I{}' \ln \frac{(z - V_I z')(z' - V_I z)}{(z - V_I z)(z' - V_I z')} \tag{5.6.22}$$

where the prime in the summation means that we include either $V_I$ or $V_I^{-1}$ but not both. Normally, one would expect to define the Neumann function as

$$N(z, z') = \ln|\psi(z, z')| \tag{5.6.23}$$

However, this function actually does not have the correct properties. We want to define automorphic functions that change only up to a constant when we go around the various loops in the $z$ plane. The above expression does not have this periodicity property, which was the crucial criterion for the Neumann function. Thus, this function cannot be an automorphic function.

The single-loop function, by contrast, did have this property. Notice that the function $\ln z$, for example, changes by $2\pi i$ when moving around an $a$-cycle:

$$\ln(ze^{2\pi i}) = \ln z + 2\pi i \tag{5.6.24}$$

and by a factor of $\ln w$ when going across a $b$-cycle:

$$\ln wz = \ln w + \ln z \tag{5.6.25}$$

These factors are called *periods* of the function $\ln z$. The ratio of these two periods is

$$\frac{\ln w}{2\pi i} = \tau \tag{5.6.26}$$

which is called the *period matrix. Thus, $\tau$ is the period matrix for a single loop.*

In this way, we can show that the entire single-loop Neumann function changes only up to a constant when traveling around the $a$- or $b$-cycles.

We can show that the previous $N$-loop candidate for the Neumann function is invariant when traveling around the $r$th $a$-cycle:

$$\delta_{ar}\psi(z, z') = 0 \tag{5.6.27}$$

However, we can show that this function, when we travel around a $b$-cycle by making a transformation $z \to P_r(z)$, is not invariant. It is not periodic because

$$\delta_{br} \ln \psi(z, z') = -v_r(z) + v_r(z') - \tfrac{1}{2}i\pi\tau_{rr} + \tfrac{1}{2}i\pi - \ln(c_r z + d_r) \tag{5.6.28}$$

where

$$v_r(z) = \sum_I{}^{(r)} \ln \frac{z - V_I z_r^{(1)}}{z - V_I z_r^{(2)}} \tag{5.6.29}$$

(the summation symbol $(r)$ simply means that, to avoid double counting in the combination $V_I z_r^{(1,2)}$, we must delete any $P_r$ that might be acting on the invariant point $z_r^{(1,2)}$) and $c$ and $d$ correspond to the factors found in the original definition of the projective transformation. When we take the function

$v_r(z)$ and move $z$ around the $s$th $a$- or $b$-cycle, we find

$$\begin{aligned}\delta_{as} v_r(z) &= 2\pi i \delta_{rs} \\ \delta_{bs} v_r(z) &= 2\pi i \tau_{rs}\end{aligned} \tag{5.6.30}$$

where

$$\tau_{rs} = \frac{1}{2\pi i} \sum_I^{(rs)} \ln \frac{(z_s^{(1)} - V_I z_r^{(1)})(z_s^{(2)} - V_I z_r^{(2)})}{(z_s^{(1)} - V_I z_r^{(2)})(z_s^{(2)} - V_I z_r^{(1)})} \tag{5.6.31}$$

The matrix $\tau_{rs}$, which we get by taking $v_r$ around the $s$th $b$-cycle, is again called the *period matrix*, which is a natural generalization of the period matrix $\tau$ we found earlier for the single-loop diagram in (5.5.2) and (5.6.25).

With the generalized definition of the period matrix, we can write down the complete Neumann function that has the correct periodicity properties. We modify $\psi$ to read

$$\ln \bar{\psi}(z, z') = \ln \psi(z, z') - \frac{1}{2\pi} \sum_{r,s} \operatorname{Re}(v_r(z) - v_r(z'))((\operatorname{Im} \tau)^{-1})_{rs} \times \operatorname{Re}(v_s(z) - v_s(z')) \tag{5.6.32}$$

so that the Neumann function becomes

$$N(z, z') = \ln|\bar{\psi}(z, z')| \tag{5.6.33}$$

We can put everything together and write down the $N$-loop amplitude for the closed string. The changes that occur from the single-loop amplitude are:

(1) The single-loop partition function now becomes

$$\prod_I |1 - X_I|^{-48} \tag{5.6.34}$$

where we take the multiplier of the product over all *inequivalent* conjugacy classes of $V_I$.

(2) The factor $(\ln|w|)^{-1}$, which was related to the period matrix, is now replaced by the determinant of the $N$-loop period matrix:

$$\det|\operatorname{Im} \tau|_{rs} \tag{5.6.35}$$

(3) Another factor of

$$\prod_r |z_r^{(1)} - z_r^{(2)}|^{-4} \tag{5.6.36}$$

occurs in the amplitude.

The final result is therefore [21]

$$A_{N\text{-loop}} = \int_F \prod_{rs} \frac{d^2 z_r \, d^2 z_s^{(1)} \, d^2 z_s^{(2)} \, d^2 X_s}{dV_{abc}} \prod_s |X_s(z_s^{(1)} - z_s^{(2)})|^{-4} \times |\operatorname{Im} \tau|^{-13} \prod_I |1 - X_I|^{-48} \prod_{r>r'} |\bar{\psi}(z_r, z_r')|^{k_i \cdot k_j} \tag{5.6.37}$$

where we integrate over the fundamental region $F$ of the surface with $N$ holes.

## §5.7. Riemann Surfaces and Teichmüller Spaces

Although the previous calculation gave us an explicit formula for the multi-loop amplitude in terms of Schottky groups, we also had to be careful in truncating the region of integration so that we avoid overcounting. This seems to be an inherent problem with the Nambu–Goto formalism, where the functional integral does not uniquely fix the region of integration.

The Polyakov formalism, however, provides a way in which we can eliminate the overcounting from the very beginning, using powerful theorems for Riemann surfaces. One of the greatest advantages of this formalism is that we can bring to bear the full force of the last century's worth of mathematical research on Riemann surfaces. Of special importance will be the fact that the determinant found earlier, which contains the singularity structure of the $N$-loop diagram, can be expressed in terms of the Selberg zeta function.

Recall that the Polyakov action is given by (2.1.35)

$$L = \frac{-1}{2\pi} \int d^2z \sqrt{g} g^{ab} \partial_a X^\mu \partial_b X_\mu \tag{5.7.1}$$

The generating functional is

$$Z = \sum_{\text{topologies}} \int_{\text{metric}} Dg_{ab} \int_{\text{embeddings}} DX^\mu \, e^{-S} \tag{5.7.2}$$

where we must be careful to divide out Riemann surfaces that are equivalent to others by a conformal transformation. We will follow the derivation of Alvarez [22] for closed strings.

The functional integration over the string variable $X$ is a Gaussian and hence easy to perform (see (1.7.10)):

$$\int DX \exp\left( -\int d^2z \sqrt{g} g^{ab} \partial_a X^\mu \partial_b X_\mu \right) = \left( \frac{2\pi}{\int d^2z \sqrt{g}} \det{}'(-\nabla^2) \right)^{-(1/2)D} \tag{5.7.3}$$

where

$$\nabla^2 = \frac{-1}{\sqrt{g}} \partial_m \sqrt{g} g^{mn} \partial_n \tag{5.7.4}$$

where the prime on the determinant will always mean deleting the zero mode.

The partition function (5.2.14), which contained the divergence of the single loop, comes out of this determinant.

However, the functional integration over the metric is considerably more involved because of the presence of gauge parameters and also the presence of loops. As we saw in (2.4.1), the measure is invariant under $2D$ general covariance and rescaling:

$$\delta g_{ab} = g_{ac} \partial_b \delta v^c + \partial_a \delta v^c g_{cb} - \partial_c \delta v^c g_{ab} + 2\delta\sigma g_{ab} \tag{5.7.5}$$

If we add in the Christoffel symbols, we can rewrite this expression covariantly:

$$\delta g_{ab} = \nabla_a \delta v_b + \nabla_b \delta v_a + 2\delta\sigma g_{ab} \tag{5.7.6}$$

where $\sigma$ parametrizes a Weyl rescaling, and $\delta v_a$ parametrizes a reparametrization of the two-dimensional surface. In general, this means that the measure of integration over $g_{ab}$ is actually infinite. Notice that $g_{ab}$ has three independent components and that $\delta v_a$ and $d\sigma$ also have three components, so that naively we expect that we can set all the components of the metric to the delta function:

$$g_{ab} = \delta_{ab} \tag{5.7.7}$$

Naively, choosing the conformal gauge is equivalent to factoring out the integration over the infinite volume due to Weyl rescalings and to two-dimensional reparametrizations over the surface. Gauge fixing thus means replacing the functional integration over metrics with

$$Dg_{ab} \to Dg_{ab}(\Omega_{\text{Diff}})^{-1}(\Omega_{\text{Weyl}})^{-1} \tag{5.7.8}$$

where the infinite volume of the space can be represented as (see (1.6.7))

$$\begin{aligned} \Omega_{\text{Diff}} &= \int Dv_a \\ \Omega_{\text{Weyl}} &= \int D\sigma \end{aligned} \tag{5.7.9}$$

For spheres and disks without holes, this is actually true. However, for surfaces with larger numbers of loops, or handles, this is no longer true because of complications due to the parametrization of the loops.

In general, a sphere with $N$ holes or handles requires a number of parameters to describe the location and size of each hole. Basically, for a disk with holes we need one parameter to label the radius of each hole and two more parameters to give us the coordinates of the center of each hole. Thus, we need $3N$ parameters to describe $N$ holes. (For a sphere with handles, we need $3N$ complex parameters to describe $N$ pairs of holes.) As we saw earlier, three parameters can be fixed overall (and set equal, for example, to 0, 1, and infinity), so a disk with $N$ holes can be described by

$$3N - 3 \tag{5.7.10}$$

real parameters, or double that number if the surface is a sphere with $N$ handles. Thus, $g_{ab}$ *cannot* be set equal to $\delta_{ab}$ for surfaces with higher genus (holes). However, any two-dimensional metric is conformally equivalent to a metric of constant curvature. By a conformal transformation, we can always set the curvature to a constant. Thus, the space of metrics over which we want to integrate is the space of constant curvature metrics divided out by the diffeomorphisms on the surface $M$. *Moduli space* is the space of constant curvature metrics when we eliminate the overcounting arising from reparametrization invariance:

$$\text{Moduli space} = \frac{M_{\text{const}}}{\text{Diff}(M)} \tag{5.7.11}$$

However, as we saw in the single-loop discussion in Section 5.5, there are actually two kinds of reparametrizations, those that can be connected to the identity and those that cannot. In the previous identity, we divided out by the set of all diffeomorphisms of the surface. Thus, we also have the possibility of dividing out by the diffeomorphisms $\mathrm{Diff}_0(M)$ that are connected only to the identity. The resulting space is called *Teichmüller space*.

$$\text{Teichmüller space} = \frac{M_{\text{const}}}{\mathrm{Diff}_0(M)} \tag{5.7.12}$$

The relationship between moduli space and Teichmüller space, of course, is necessarily very close. In fact, they are actually equivalent up to the action of a discrete group, called the *mapping class group*:

$$\text{Moduli space} = \frac{\text{Teichmüller space}}{\text{MCG}} \tag{5.7.13}$$

Thus

$$\text{MCG} = \frac{\mathrm{Diff}(M)}{\mathrm{Diff}_0(M)} \tag{5.7.14}$$

In other words, the only differences between Teichmüller space and moduli space are *global* diffeomorphisms that cannot be connected to the identity. For example, think of cutting a torus as in Fig. 5.13, twisting one of the sliced edges by $2\pi$, and then reconnecting the two edges. This is called a Dehn twist. Notice that this transformation generates a diffeomorphism that cannot reach the identity. It is a global diffeomorphism. Thus, Teichmüller space is "larger" than moduli space because you have to divide out the global diffeomorphisms, or Dehn twists, to arrive at moduli space. The dimensions of both these spaces

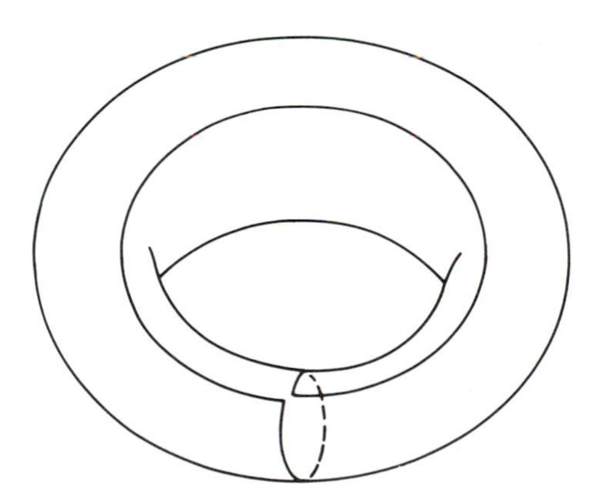

Figure 5.13. Action of a Dehn twist. The torus has been cut along its $a$-cycle. One end has been twisted one full turn and then rejoined. Notice that the original $b$-cycle has now become the sum of an $a$- and a $b$-cycle. This mapping of the torus into itself cannot be continuously deformed back to the identity map. The set of all Dehn twists generates the mapping class group.

are given by

$$\dim \text{Teich} = \dim \text{Moduli} = \begin{cases} 0, & \text{if } N = 0 \\ 2, & \text{if } N = 1 \\ 6N - 6, & \text{if } N \geq 2 \end{cases} \tag{5.7.15}$$

This, of course, is precisely the number of parameters necessary to describe a sphere with $N$ holes. Thus, this description gives us the necessary parameters to parametrize the $N$-loop diagram. (We note that the modular group and the mapping class group are identical, and we will use these two terms interchangeably.)

In practice, the volume factor due to the mapping class group can be factored out trivially, so that, for our purposes, we may treat Teichmüller space and moduli space as essentially the same.

So far, the discussion has been general. To actually extract out these extra parameters that describe the holes of a Riemann surface, we need to use the theory of Teichmüller spaces.

Some of the manipulations are a bit involved, so we must always clearly have our goal in mind, which is to rewrite the functional measure $Dg_{ab}$ in terms of the parameters $Dv_a$ $D\sigma$ and the $3N - 3$ Teichmüller parameters $Dt_i$. Thus, our goal is to establish

$$\text{Objective:} \quad Dg_{ab} = \mu \, Dv_a \, D\sigma \, Dt_i \tag{5.7.16}$$

Then, by simply dividing out by $Dv_a$ and $D\sigma$, we will have successfully eliminated the infinite redundancy introduced by reparametrization and scale invariance.

Let us now rewrite the variation of the metric tensor (5.7.6) in a more convenient form, revealing the fact that $g_{ab}$ is also a function of $t_i$, the Teichmüller parameters. Using the chain rule, we can formally represent the dependence of the metric on the Teichmüller parameters via $\delta t_i \, \delta/\delta t^i$:

$$\delta g_{ab} = [\nabla_a \delta v_b + \nabla_b \delta v_a - (\nabla_c \delta v^c) g_{ab}] + (\nabla_c \delta v^c) g_{ab} + 2\delta\sigma g_{ab} + \delta t^i T^i$$

where

$$T^i = \frac{\partial g_{ab}}{\partial t^i} - (\text{trace}) \tag{5.7.17}$$

where we have explicitly taken out the variation of the metric as a function of the $3N - 3$ parameters (Teichmüller parameters) that we label $t^i$ associated with the $N$ loops. Notice that we have subtracted out the trace in the brackets and that, at this point, we do not have to specify precisely how the metric depends on the various $t_i$.

This variation can be rewritten simply as

$$\delta g_{ab} = P_1(\delta v)_{ab} + 2\delta\sigma g_{ab} + \delta t^i \partial_i g_{ab} \tag{5.7.18}$$

where

$$P_1(\delta v)_{ab} = \nabla_a \delta v_b + \nabla_b \delta v_a - g_{ab} \nabla_c \delta v^c \tag{5.7.19}$$

The operator $P_1$ plays a crucial role in the theory of Teichmüller spaces. Notice that it is an elliptic operator that maps vectors into traceless symmetric tensors. Let ker $P_1$ represent the kernel of the operator, i.e., the set of vectors that are mapped to zero by the operator. Ker $P_1$ is called the set of "conformal Killing vectors." For surfaces with genus $N$, the dimension of the kernel of the operator is given by

$$\begin{aligned} &\dim \ker P_1 = 6 \quad \text{for genus } 0 \\ &\dim \ker P_1 = 2 \quad \text{for genus } 1 \\ &\dim \ker P_1 = 0 \quad \text{for higher genus} \end{aligned} \tag{5.7.20}$$

We also wish to define the adjoint of $P_1$, which we will call $P_1^\dagger$. To do this, of course, we first have to define how to take the inner product. Let us define

$$\|\delta g_{ab}\|^2 = \int d^2 z \sqrt{g} g^{ac} g^{bd} \delta g_{ab} \delta g_{cd} \tag{5.7.21}$$

$$\|\delta v_a\|^2 = \int d^2 z \sqrt{g} g_{ab} \delta v^a \delta v^b \tag{5.7.22}$$

Once we have defined a scalar product, this allows us to define the adjoint $P_1^\dagger$ via the definition $\langle a|Pb\rangle = \langle P^\dagger a|b\rangle$:

$$P_1^\dagger(\delta g)_a = -2\nabla^b \delta g_{ab} \tag{5.7.23}$$

Let us now analyze the space ker $P_1^\dagger$, i.e., the space of vectors that are annihilated by $P_1^\dagger$. If we rewrite everything in terms of $z$ and $\bar{z}$, then the elements of ker $P_1^\dagger$ satisfies

$$\nabla^z \delta g_{zz} = \partial_{\bar{z}} \delta g_{zz} = 0 \tag{5.7.24}$$

The kernel of $P_1^\dagger$ is spanned by what are called *quadratic differentials*. Fortunately, the dimensions of the spaces of quadratic differentials for Riemann surfaces are known:

$$\begin{aligned} &\dim \ker P_1^\dagger = 0 \quad \text{for genus } 0 \\ &\dim \ker P_1^\dagger = 2 \quad \text{for genus } 1 \\ &\dim \ker P_1^\dagger = 6N - 6 \quad \text{for genus } N \end{aligned} \tag{5.7.25}$$

Thus, the number of parameters within the kernel of $P_1^\dagger$ is equal to the number of Teichmüller parameters needed to describe $N$ loops or handles. Symbolically, we can summarize how to divide the measure of integration into its constituent parts as

$$\{\delta g_{ab}\} = \{\delta\sigma\} \oplus \{P_1 \delta v_a\} \oplus \{\ker P_1^\dagger\} \tag{5.7.26}$$

This equation has a rather simple meaning. It says that the components within $g_{ab}$ can be broken up into three parts, a dilation part, a traceless part, and also the hidden Teichmüller parameters. It also means that we are very close to attaining our goal, (5.7.16), but there are some subtle complications.

Let us explicitly make a change of variables and calculate the Jacobian of the transformation. Let us first assume that there are no Teichmüller parameters to worry about. Then we make a change of variables from $h_{ab}$ (which is the traceless part of the metric) and $\tau$ (which is the trace of $g_{ab}$) to $\delta v_a$ and $\sigma$:

$$Dg_{ab} = \det\left[\frac{\partial(\tau, h)}{\partial(\sigma, v)}\right] D\sigma D v_a \tag{5.7.27}$$

where

$$\det\left[\frac{\partial(\tau, h)}{\partial(\sigma, v)}\right] = \det\begin{bmatrix} 1 & X \\ 0 & P_1 \end{bmatrix} = \det P_1 = [\det P_1 P_1^\dagger]^{1/2} \tag{5.7.28}$$

where the value of $X$ is arbitrary, since it drops out of the determinant.

Notice that the square root of the determinant of $P_1 P_1^\dagger$ is just the Faddeev–Popov determinant for the conformal gauge, which in turn can be written in terms of the BRST ghosts. Thus, we can write the Faddeev–Popov determinant first found in (2.4.3) as

$$\det{}^{1/2}\, P_1 P_1^\dagger = \Delta_{\text{FP}} = \int Db\, Dc\, e^{-S_{gh}} \tag{5.7.29}$$

Next, we wish to calculate the Jacobian factor due to the fact that the measure actually depends on the Teichmüller parameters $t_i$. The problem is that the Teichmüller parameters are not orthogonal to $P_1\,\delta v_a$, so the Jacobian will contain *cross terms* that have to be factored out.

Let us begin our discussion by introducing a set of $3N - 3$ complex fields $\psi^a$ that will be an orthogonal basis of $\ker P_1^\dagger$. Let us now decompose the factor [22, 23]:

$$T_{ab}^i = \frac{\partial g_{ab}}{\partial t_i} - (\text{trace}) \tag{5.7.30}$$

into the following identity:

$$T^i = \left(1 - P_1 \frac{1}{P_1^\dagger P_1} P_1^\dagger\right) T^i + P_1 \frac{1}{P_1^\dagger P_1} P_1^\dagger T^i \tag{5.7.31}$$

It is important to note that we have done nothing. We have only added and subtracted the same term. However, the term that we have introduced contains the operator $P_1$ acting on another state. Thus, the above identity is useful because it allows us to extract the piece of $T^i$ that lies along the direction of $P_1$:

$$T^i \equiv \psi^a(\psi^a, T^i) + P_1 v^i \tag{5.7.32}$$

where

$$v^i = \frac{1}{P_1^\dagger P_1} P_1^\dagger T^i \tag{5.7.33}$$

$$(\psi^a, T^i) = \int d^2 z \sqrt{g} g^{bc} g^{de} T^i_{bd} \psi^a_{ce} \tag{5.7.34}$$

As expected, there is a piece of $T^i$ that lies along the direction $P_1$ that must be accounted for when we write down the Jacobian. Let us now insert this expression back into the measure for $\delta g_{ab}$:

$$\|\delta g_{ab}\|^2 = \|\delta\sigma\| + (T^i, \psi^a)(\psi_a, \psi_b)^{-1}(\psi^b, T^j)\delta t_i \delta t_j + \|P_1 \delta v_a\|^2 \tag{5.7.35}$$

This, finally, gives us the Jacobian that includes the contribution from the Teichmüller parameters. We have also explicitly taken into account the fact that $T^i$ originally contained a piece that lay in the direction of $P_1$:

$$Dg_{ab} = D\sigma Dv_a \, Dt_i \det{}^{1/2}(P_1^\dagger P_1) \frac{\det(\psi_a, T_b)}{\det^{1/2}(\psi_a, \psi_b)} \tag{5.7.36}$$

This, in turn, allows us to put the entire functional integration together into the following factor:

$$\begin{aligned} \int DX \, e^{-s} \, Dg_{ab} \, \Omega_{\text{Diff}}^{-1} \Omega_{\text{Weyl}}^{-1} = \int & Dv_a \, \Omega_{\text{Diff}}^{-1} \, D\sigma \, \Omega_{\text{Weyl}}^{-1} \\ & \times Dt_i \det{}^{1/2}(P_1^\dagger P_1) \frac{\det(\psi^a, T^i)}{\det^{1/2}(\psi^a, \psi^b)} \\ & \times \left( \frac{2\pi}{\int d^2 z \sqrt{g}} \det{}'(-\nabla^2) \right)^{-(1/2)D} \end{aligned} \tag{5.7.37}$$

We have finally attained our goal, (5.7.16), up to the question of anomalies that may spoil scale invariance.

Although this discussion might appear long and difficult, the end result is quite simple. The final answer shows that the measure of integration, including the Faddeev-Popov ghost determinant, can be written totally in terms of determinants of the operator $P_1$ and its adjoint. There are three parts to the measure: (1) first is the Faddeev-Popov term, which is written as the square root of the determinant of $P_1 P_1^\dagger$; (2) the second is the term involving $T^i$, which arises because the moduli parameters $t_i$ are not perpendicular to the basis vectors $\psi^a$ which form an orthogonal basis for ker $P_1^\dagger$; (3) the third factor is the determinant of the Laplacian, which we will also show can be written in terms of the operator $P_1$.

However, it is not yet possible to remove the integration over the scalar parameter $\sigma$ in (5.7.37). Unfortunately, the various measure terms contain the scalar parameter. We will find, in fact, that in general the scale factor cannot be eliminated from the measure terms at all, breaking conformal invariance.

In the next section, we will show that there is an obstruction, an anomaly, to scale invariance, called the conformal anomaly, which can be removed only if the dimension of space-time is 26. This fixes the dimension of space-time.

In the process, we will show how to write all determinants totally in terms of the $P_1$ operator.

## §5.8. Conformal Anomaly

We note that we can cancel the integration over the volume of the reparametrization group, because

$$\Omega_{\text{Diff}}^{-1} \int Dv_a = 1 \tag{5.8.1}$$

Next, we wish to eliminate the Weyl rescaling term:

$$\Omega_{\text{Weyl}}^{-1} \int D\sigma = 1 \tag{5.8.2}$$

However, the integration over the $\sigma$ rescaling term is considerably more complicated, because $\sigma$ terms exist in the other factors of the measure as well. Thus, we must very carefully extract all $\sigma$-dependent terms from each of the other factors before we can eliminate the rescaling term. It will turn out that the Weyl factor can also be factored out of the measure, but only if the dimension of space–time is 26! The calculation of the conformal anomaly is rather involved, so we will only sketch the highlights of the calculation.

First, we note that the term $(T^i, \psi^a)$ is already Weyl invariant. This is because if we rescale according to

$$\begin{aligned} \hat{g}_{ab} &= e^{\sigma} g_{ab} \\ \hat{T}^i &= e^{\sigma} T^i \end{aligned} \tag{5.8.3}$$

we notice that by the very definition of $(T^i, \psi^a)$, this term remains invariant. (The $\psi$ variable remains unchanged under a Weyl rescaling.)

Thus, the only term we have to worry about is

$$\left\{ \frac{\det' P_1^\dagger P_1}{\det(\psi^a, \psi^b)} \right\}^{1/2} \left\{ \frac{\det'(-\nabla^2)}{\int d^2z \sqrt{g}} \right\}^{-(1/2)D} \tag{5.8.4}$$

(The prime still means that we have extracted out the zero mode of the determinant.)

Fortunately, the extraction of the Weyl rescaling parameter can be performed simultaneously on both terms if we use a few facts about Riemann surfaces. The trick is to rewrite $P_1$, $P_1^\dagger$, and $\nabla^2$ in a way such that all three operators can be expressed in terms of the same differential operator.

Let

$$T^{ijlkmn\ldots}_{abcdef\ldots} \tag{5.8.5}$$

represent an arbitrary tensor in the two-dimensional Riemann surface. Of course, we can always rewrite this tensor in terms of complex coordinates $z$ and $\bar{z}$. In these coordinates, we denote as $K^n$ the set of all tensors $T$ that transforms as

$$K^n = \left\{T\colon T \to \left(\frac{\partial z'}{\partial z}\right)^n T\right\} \tag{5.8.6}$$

(e.g. see (2.7.6) and (4.1.7)) Clearly, the operator

$$\nabla_z^q T = (g^{z\bar{z}})^q \partial_z [(g_{\bar{z}z})^q T] \tag{5.8.7}$$

maps a tensor $T$ into another tensor. More precisely, it maps $K^q$ into $K^{q-1}$:

$$(\nabla_z^q\colon K^q \to K^{q-1}) \tag{5.8.8}$$

It is also possible to define an operator that acts in the opposite direction. The operator

$$\nabla_q^z T = g^{z\bar{z}} \partial_{\bar{z}} T \tag{5.8.9}$$

maps $K^q$ into $K^{q+1}$:

$$\nabla_q^z\colon K^q \to K^{q+1} \tag{5.8.10}$$

Now let us define the operators

$$P_q = \begin{pmatrix} \nabla_q^z & 0 \\ 0 & \nabla_z^{-q} \end{pmatrix} \tag{5.8.11}$$

and

$$P_q^\dagger = \begin{pmatrix} -\nabla_z^{q+1} & 0 \\ 0 & -\nabla_{-q-1}^z \end{pmatrix} \tag{5.8.12}$$

Using this notation, we can show

$$-P_1^\dagger P_1 = \begin{pmatrix} \nabla_z^2 \nabla_1^z & 0 \\ 0 & \nabla_{-2}^z \nabla_z^{-1} \end{pmatrix} \tag{5.8.13}$$

Thus,

$$\begin{aligned} \det P_1^\dagger P_1 &= \det(\nabla_z^2 \nabla_1^z) \det(\nabla_{-2}^z \nabla_z^{-1}) \\ &= \det \Delta_1^+ \det \Delta_{-1}^- \end{aligned} \tag{5.8.14}$$

Since the + Laplacian is the complex conjugate of the − Laplacian, we have

$$\det{}^{1/2} P_1^\dagger P_1 = \det \Delta_1^+ \tag{5.8.15}$$

Finally, we can also define

$$\Delta_0^+ \equiv \nabla_1^z \nabla_z^0 \tag{5.8.16}$$

Armed with these definitions, we have now expressed the determinant in the form

$$\left[\frac{\det' P_1^\dagger P_1}{\det(\psi^a, \psi^b)}\right]^{1/2}\left[\frac{\det'(-\nabla^2)}{\int d^2\sigma\,\sqrt{g}}\right]^{-(1/2)D} = \frac{\det \Delta_1^+}{\det^{1/2}(\psi^a, \psi^b)}(\det \Delta_0^+)^{-(1/2)D} \tag{5.8.17}$$

In other words, both pieces of the Jacobian, which at first appear dissimilar, are now represented by the *same* operator $\Delta_q^+$, for $q = 1$ and for $q = 0$. Now, it remains to calculate the variation of the previous expression when we do a Weyl rescaling. It is convenient to write the equation in terms of the heat kernel of the determinant:

$$\log \det H = -\int_\varepsilon^\infty \frac{dt}{t}\,\mathrm{Tr}'\, e^{-tH} \tag{5.8.18}$$

where $\varepsilon$ is a small number. Using this equation, we can now express the variation of the determinant due to a Weyl rescaling. The actual calculation is quite involved, so we will only quote the final result [22–24]:

$$\delta \log \frac{\det \Delta_q^+}{\det^{1/2}(\psi^a, \psi^b)} = \frac{1 + 6q(1+q)}{12\pi}\int d^2z\,\sqrt{g}R\delta\sigma \tag{5.8.19}$$

where $R$ is the contracted curvature tensor in two dimensions. Thus, we want to calculate this factor when $q = 1$ and $q = 0$. The final result is

$$\delta \log\left[\frac{\det^{1/2} P_1^\dagger P_1}{\det^{1/2}(\psi^a, \psi^b)}\left(\frac{\det'(-\nabla^2)}{\int d^2z\,\sqrt{g}}\right)^{-(1/2)D}\right]$$

$$= \frac{13 - \frac{1}{2}D}{12\pi}\int d^2z\,\sqrt{g}R\delta\sigma \tag{5.8.20}$$

Notice that for $D = 26$, the conformal anomaly disappears. Thus, we have shown

$$D = 26\!:\quad \Omega_{\mathrm{Weyl}}^{-1}\int D\sigma = 1 \tag{5.8.21}$$

Our final result for the functional is thus

$$Z = \int Dt_i \frac{\det(\psi^a, T^i)}{\det^{1/2}(\psi^a, \psi^b)}\det{}_{\hat{g}}^{1/2}\, P_1^\dagger P_1\left(\frac{\det'(-\nabla^2)}{\int d^2z\,\sqrt{\hat{g}}}\right)^{-13} \tag{5.8.22}$$

where all terms are evaluated with the metric $\hat{g}_{ab}$ where the $\sigma$ term has been completely factored out. Thus, our goal of attaining (5.7.16) has now been accomplished, at least in 26 dimensions.

Now that we have developed this powerful formalism, it is convenient to reduce this case to the problem of the single-loop amplitude and rederive our earlier result (5.3.12) in terms of our new perspective on Riemann surfaces [25].

For a single-loop graph, the region of interest is a doughnut. Our flat space metric $\hat{g}_{ab}$ corresponds to a lattice in $C$:

$$\omega_1\mathbf{Z} + \omega_2\mathbf{Z} \tag{5.8.23}$$

where **Z** corresponds to an integer. Thus, the region of interest is completely determined by the ratio

$$\tau = \omega_2/\omega_1 \tag{5.8.24}$$

and the fact that we normalize the domain to have unit area. Then, the mapping class group is simply SL(2, **Z**), and the fundamental domain, as before, can be taken to be ($\tau = \tau_1 + i\tau_2$)

$$\begin{cases} |\tau| \geq 1 \\ -\frac{1}{2} \leq \tau_1 < \frac{1}{2} \end{cases} \tag{5.8.25}$$

The measure

$$(\det{}'^{1/2}\, P_1^\dagger P_1)\left(\frac{2\pi}{\int d^2 z\, \sqrt{g}} \det{}'\, \Delta_0^+\right)^{-13} \tag{5.8.26}$$

can be further reduced. For example, the factor (5.8.26) equals

$$\tfrac{1}{2}(\det{}'\, \Delta)^{-12}(2\pi\tau_2^{-1})^{-13} \tag{5.8.27}$$

where

$$\det{}'\, \Delta = e^{-\pi\tau_2/3}\tau_2^2 \left| \prod_{n=1}^{\infty} (1 - e^{2\pi i n\tau}) \right|^4 \tag{5.8.28}$$

The invariant measure can be written, as before, as

$$dt_i = d^2\tau/\tau_2^2 \tag{5.8.29}$$

Combining everything together, we arrive at

$$Z = \int_{\text{Fund region}} \frac{d\tau_1\, d\tau_2}{\tau_2^2} (2\pi\tau_2)^{-12} e^{4\pi\tau_2} \left| \prod_{n=1}^{\infty} (1 - e^{2\pi i n\tau}) \right|^{-48} \tag{5.8.30}$$

as in (5.5.4).

## §5.9. Superstrings

Fortunately, the generalization of multiloop amplitudes to superstrings is straightforward, except for the complications due to defining spinors on Riemann surfaces of genus $g$.

Let us consider the torus in two dimensions $\sigma_1$, $\sigma_2$, constructed on a parallelogram by identifying opposite sides. A string $X$ defined on the parallelogram must satisfy periodicity properties:

$$X(\sigma_1, \sigma_2) = X(\sigma_1 + 2\pi, \sigma_2) = X(\sigma_1, \sigma_2 + 2\pi) \tag{5.9.1}$$

However, the addition of spinors to this surface increases the number of possibilities. A spinor can be either periodic (+) satisfying Ramond boundary conditions or antiperiodic (−) satisfying NS boundary conditions in *either* $\sigma_1$ or $\sigma_2$. (So far, we have only considered different boundary conditions in the

$\sigma$ direction in (3.2.16), not in the $\tau$.) Thus, the total number of different combinations for boundary conditions is four. The possible choices one can take to define a spinor on a torus correspond to $(\pm, \pm)$. We say that these four possible choices for a torus define its *spin structure.*

Furthermore, we know that modular transformations will mix up the boundary conditions, and hence the spin structures, so all four of them can contribute to the final amplitude. For example, the transformation

$$(\sigma_1, \sigma_2) \to (\sigma_1 + \sigma_2, \sigma_2) \tag{5.9.2}$$

changes $(-, -)$ into $(+, -)$. Likewise, the transformation

$$(\sigma_1, \sigma_2) \to (\sigma_2, -\sigma_1) \tag{5.9.3}$$

interchanges $(+, -)$ and $(-, +)$. Thus modular invariance, which interchanges the boundary conditions, forces us to have $(+, -)$, $(-, +)$, and $(- -)$ in the amplitude. $(+ +)$ is invariant by itself.

To calculate the contribution of each spin structure to the one-loop amplitude, let us calculate the trace of the Hamiltonian over all four possibilities. The Hamiltonians for the NS and the R sector are

$$H_{\mathrm{NS}} = \sum_{r=1/2}^{\infty} r\psi_{-r}^i \psi_r^i - \frac{1}{48} \tag{5.9.4}$$

$$H_{\mathrm{R}} = \sum_{n=1}^{\infty} n\psi_{-n}^i \psi_n^i + \frac{1}{24} \tag{5.9.5}$$

($\frac{1}{24}$ comes from zeta function regularization. See Chapter 11.) These Hamiltonians are either NS or R depending on their periodicity properties in the $\sigma$ direction. However, when we take the trace over these Hamiltonians, we are inserting a complete set of intermediate states in the $\tau$ direction. The path integral in the $\tau$ only selects out the antiperiodic boundary conditions, which is incomplete. We must modify the sum to include all possible spin structures, which is made possible by the insertion of a factor of $(-1)^F$, where $F$ is the fermion number. The final amplitude is the sum of amplitudes, $A(\pm, \pm)$ defined for all four traces, multiplied by some coefficient $C$:

$$A(\tau) = \sum_{\pm} C(\pm, \pm) A(\pm, \pm) \tag{5.9.6}$$

Explicitly, each trace can be written as follows:

$$\begin{aligned} A(-, -) &= \mathrm{Tr}\, e^{2\pi i \tau H_{\mathrm{NS}}} = [\Theta_3(0|\tau)/\eta(\tau)]^4 \\ A(+, -) &= \mathrm{Tr}\, e^{2\pi i \tau H_{\mathrm{R}}} = [\Theta_2(0|\tau)/\eta(\tau)]^4 \\ A(-, +) &= \mathrm{Tr}(e^{2\pi i \tau H_{\mathrm{NS}}}(-1)^F) = [\Theta_4(0|\tau)/\eta(\tau)]^4 \\ A(+, +) &= \mathrm{Tr}(e^{2\pi i \tau H_{\mathrm{R}}}(-1)^F) = 0 \end{aligned} \tag{5.9.7}$$

where $\eta$ is the Dedekind eta function:

$$\eta(\tau) = e^{\pi i \tau/12} \prod_{n=1}^{\infty} (1 - e^{2\pi i \tau n}) \tag{5.9.8}$$

Notice that the last trace vanishes by itself. (The first three theta functions $\Theta_{2,3,4}$ are all even under $z \to -z$, while $\Theta_1$ is odd. Thus, only the even spin structures survive in the trace.)

The significance of this is as follows. For the NS sector, for example, we include both the $(-, +)$ and $(-, -)$ sectors, so we have to add their contributions to the trace:

$$\mathrm{Tr}(1 + (-1)^F)e^{2\pi i \tau H}) \tag{5.9.9}$$

But this is precisely the GSO projection [26], which projects out states even under $(-1)^F$! We therefore have a new, physical interpretation of the GSO projection operator, which was introduced in (3.6.12) to eliminate the non-supersymmetric sectors of the NS–R theory. It's surprising that modular invariance, supersymmetry, and the GSO projection are so intimately linked.

When we calculate the vacuum amplitude, which has no external legs (and which appears in the calculation of the cosmological constant), we must add all four contributions to the spin structure. But then we have the remarkable result of Jacobi:

$$\Theta_2^4(0|\tau) + \Theta_4^4(0|\tau) - \Theta_3^4(0|\tau) = 0 \tag{5.9.10}$$

This states that the vacuum energy of the superstring, i.e., the one-loop correction to the cosmological constant, vanishes exactly!

To sum up, we have proved that modular invariance and the GSO projection operation are essentially the same for the closed superstring. Modular invariance, which mixes up the boundary conditions, demands that we add all four spin structures to the final amplitude, which in turn corresponds to adding precisely the GSO insertions of the operator $(-1)^F$ into the trace. Originally, the GSO projection was imposed on the NS–R superstring in order to have supersymmetry. We now know that the GSO projection enforces modular invariance as well [27].

As a bonus, we find that the one-loop contribution to the cosmological term is precisely equal to zero. (However, this does not mean that superstrings solves the vanishing cosmological constant problem. After supersymmetry is broken, we no longer expect that the vacuum contribution is zero, and hence superstring theory still does not explain why the cosmological constant is zero *after* supersymmetry breaking.)

These statements reinforce our conviction that the internal consistency of the superstring theory is quite remarkable.

Now let us generalize our comments to construct multiloop amplitudes with external lines.

Fortunately, the apparatus that was constructed for Riemann surfaces carries over rather simply to the superstring case if we use the NS–R formulation of the model.

Generalizing (5.7.2), the correlation functions can be computed from [28]:

$$Z = \sum_{\text{topologies}} \sum_{\substack{\text{spin}\\ \text{structures}}} \int_{\text{metrics}} Dg_{ab} \int_{\text{embeddings}} DX^\mu \int D\psi^\mu \int D\chi_\alpha \, e^{-S} \tag{5.9.11}$$

The new addition here is the integration over the two anticommuting vector fields $\psi$ and $\chi$ and also the sum over all spin structures.

We showed before in (3.4.5) that the NS–R action is invariant under

$$\delta\chi_\alpha = D_\alpha \varepsilon \tag{5.9.12}$$

Let us now, in analogy with the bosonic string, construct two operators $P_{1/2}$ and $P_{1/2}^\dagger$, such that

$$(P_{1/2}\varepsilon)_\alpha = 2D_\alpha\varepsilon - \rho_\alpha\rho^\beta D_\beta\varepsilon \tag{5.9.13}$$

In addition to the integration over moduli, we now have to integrate over *supermoduli*, which are defined as the space of traceless $\chi_\alpha$ that cannot be gauged away by a local supersymmetry transformation. Thus, they satisfy

$$P_{1/2}^\dagger\chi = -2D_\alpha\chi^\alpha = 0 \tag{5.9.14}$$

For a surface of genus $N$, the dimension of the supermoduli space is equal to the dimension of ker $P_{1/2}^\dagger$:

$$\dim\ker P_{1/2}^\dagger: \quad \begin{cases} 0 & \text{if } N = 0 \\ 2 & \text{if } N = 1 \text{ (periodic–periodic)} \\ 0 & \text{if } N = 1 \text{ (otherwise)} \\ 4N-4 & \text{if } N \geq 2 \end{cases} \tag{5.9.15}$$

(where periodic–periodic stands for boundary conditions on the NS–R field.) We wish to construct the Jacobian for

$$Dg_{ab}\, D\chi_\alpha \tag{5.9.16}$$

The Jacobian for the fermion field is calculated as before:

$$D\chi_\alpha = (\det \hat{P}_{1/2}^\dagger \hat{P}_{1/2})^{-1/2} e^{-(11/2)S(\sigma)}\, D\zeta\, D\lambda \prod_{i=1}^{4N-4} da_i \tag{5.9.17}$$

where $a_i$ are the supermoduli, the counterpart of the Teichmüller parameters, and where the integration over the fermionic $\lambda$ and $\zeta$ represents the integration over the redundancy introduced by supersymmetry and super-Weyl transformations. We define

$$S = \frac{1}{48\pi}\int d^2z\,\sqrt{\hat{g}}(\hat{g}^{ab}\partial_a\sigma\partial_b\sigma - \tfrac{1}{2}i\lambda\gamma^\alpha\hat{D}_\alpha\lambda$$
$$+ \mu^2(e^\sigma - 1) + 2^{-3/2}\mu\lambda\lambda e^{(1/2)\sigma} + \tfrac{1}{2}R_{\hat{g}}\sigma) \tag{5.9.18}$$

where $R$ is the curvature tensor. Putting everything together, we arrive at the following:

$$Dg_{ab}\, D\chi_\alpha = (\det{}^{1/2}\, \hat{P}_1^\dagger \hat{P}_1)(\det{}^{-1/2}\, \hat{P}_{1/2}^\dagger \hat{P}_{1/2})$$
$$\times e^{-15S_L} D\sigma\, Dv_a\, D\lambda\, D\zeta \prod_{i=1}^{4N-4} Da_i \prod_{j=1}^{6N-6} Dt_j \tag{5.9.19}$$

By dividing out by $D\sigma$, $Dv_a$, $D\lambda$, and $D\zeta$, we extract out the infinite redundancy introduced by the symmetry of the superstring action. In practice, however, the parametrization of the moduli, especially beyond the third loop, is quite difficult. The nature of the supermoduli, unfortunately, is even less understood. These are some of the main stumbling blocks to a definitive understanding of the multiamplitude, i.e., the choice of coordinates on a genus $g$ Riemann surface consistent with modular and supermodular invariance.

## §5.10. Determinants and Singularities

The advantage of the Riemann surface method is that, at least formally, one can obtain results for the general singularity structure of the $N$-loop amplitude. It turns out that the determinant that contains the singularity of the $N$-loop amplitude is the Selberg zeta function.

We saw earlier that projective transformations naturally fall into conjugacy classes. Two projective transformations are said to be within the same conjugacy class if they have the same multiplier. Thus, any two members of the same conjugacy class can, by a projective transformation, be brought to the form

$$z \to e^l z \tag{5.10.1}$$

This value $l$ is sometimes called the "length" of a closed geodesic. Let $z$ and $z'$ represent two points in the complex plane. Then we define the "distance" between these two points as

$$d(z, z') = 1 + \frac{|z - z'|^2}{2 \operatorname{Im} z \operatorname{Im} z'} \tag{5.10.2}$$

The advantage of this definition of the "distance" between two points is that it is the same for all elements of the same conjugacy class.

We saw, from the example of the single loop, that the multiplication by the multiplier $w$ moved across the $b$-cycle. Topologically, this simply means executing a closed geodesic on the Riemann surface. Thus, $l$ is the length of simple closed geodesics on the surface. We say that a transformation is *primitive* if it is not a power (greater than or equal to 2) of any other element of the set of projective transformations. Thus, we are interested in the set of primitive geodesics. We define the Selberg zeta function as [23]

$$Z_v(s) = \prod_{l\,\text{primitive}} \prod_{p=0}^{\infty} [1 - v(\gamma)e^{-(s+p)l}] \tag{5.10.3}$$

where we take products over the lengths of closed primitive geodesics $\gamma$ on the surface, and $v(\gamma) = 1$ for bosons and $v(\gamma) = \pm 1$ for fermions, depending on the spin structure.

Remarkably, we can write down the various determinants found earlier in the multiloop calculation in terms of the Selberg zeta function [28]:

$$\begin{aligned} \det' \Delta_0^+ &= Z'(1)e^{-c_0\chi(M)} \\ \det \Delta_1^+ &= Z_1(2)e^{-c_1\chi(M)} \end{aligned} \tag{5.10.4}$$

The fermion determinants can also be expressed as follows:

$$\begin{aligned} (\det \hat{P}_{1/2}^\dagger \hat{P}_{1/2})_v^{1/2} &= e^{-c_{(1/2)}\chi(M)} Z_v(3/2) \\ (\det' \gamma^\alpha D_\alpha \gamma^\beta D_\beta)_v^{1/2} &= \frac{e^{-c_{-(1/2)}\chi(M)}}{(2p)!} Z_v^{(2p)}\left(\frac{1}{2}\right) \end{aligned} \tag{5.10.5}$$

where $\chi$ is the Euler number ($p$ is the number of zero modes of $\gamma^\alpha D_\alpha$). The numerical constant $c$ is given by

$$\begin{aligned} 2c_n = &\sum_{0 \le m < |n| - 1/2} (2|n| - 2m - 1) \log(2|n| - m) \\ &- (|n| + \tfrac{1}{2})^2 + (n - [n])^2 + (n + \tfrac{1}{2}) \log 2\pi + 2\zeta'(-1) \end{aligned} \tag{5.10.6}$$

Thus, by analyzing the structure of one function, we obtain the singularity structure of the $N$-loop superstring!

Mathematically, it is known that the Selberg zeta function is well behaved unless the Riemann surface degenerates topologically. For example, we find a divergence as the length of the primitive geodesic approaches zero. This corresponds to the "neck" of one of the handles of the sphere stretching to infinity. By carefully analyzing the Selberg zeta function as one of the lengths of the primitive geodesics goes to zero, we find

$$\det{}^{1/2} P_1^\dagger P_1 \left( \frac{\det' (-\Delta_0^+)}{\int \sqrt{g}\, d^2 z} \right)^{-13} \sim l^{-2} e^{4\pi^2/l} \tag{5.10.7}$$

which reveals the pole corresponding to the tachyon vanishing into the vacuum.

Being able to isolate the singularities of the multiloop amplitude in terms of a known mathematical function, the Selberg zeta function, is a great step toward solving the main problem facing string perturbation theory, i.e., rigorously proving finiteness to all orders. Formally, it can be shown that the divergences of the multiloop amplitude may occur when the integration over moduli changes the topology of the Riemann surface, e.g., when two holes separate and create a long goose-like neck. However, there are many delicate points related to the question of the cancellation of these divergences that have not been totally solved. Although preliminary results are encouraging, the rigorous proof of the cancellation of divergences is still an outstanding problem.

## §5.11. Moduli Space and Grassmannians

Although enormous progress has been made in elucidating the mathematical structure of multiloop amplitudes, in some sense the tangible results have been disappointing. Back in 1970, it was already known that the divergences of the multiloop amplitude, by explicit calculation, corresponded to the deformation of the topology of a Riemann surface [4, 5]. So far, the elaborate mathematical techniques we have introduced still haven't answered the key questions: Can one rigorously show the theory is finite to all orders? If so, how do you sum the perturbation series? How do you extract nonperturbative information from the theory? Knowing that the divergences of the theory can be expressed in terms of Selberg zeta functions [23, 29–32], although an important result, still hasn't solved these mysteries. In summary, the progress has been in the area of mathematics, rather than physics.

At this point, one can take at least two diverging attitudes. One can give up on the perturbation series and proceed directly to the *field theory of strings*, in which nonperturbative information might be extracted. This is the traditional approach taken in ordinary point particle theories and is the subject of the next series of chapters. Or second, one might imagine some kind of symmetry, such as modular invariance, by which one might be able to manipulate the entire sum over all Riemann surfaces of arbitrary genus.

In ordinary point particle Feynman diagrams, the second strategy is probably impossible. The symmetries are too small, and the Feynman diagrams are graphs, not manifolds. However, the Feynman series for string theory are sums over manifolds, Riemann surfaces, where modular invariance plays a crucial role. Thus, it is conceivable that the *entire perturbation series* might be manipulated mathematically. This is the approach of Friedan and Shenker [33], who propose exploring the properties of the "universal moduli space" of all Riemann surfaces, including infinite genus.

Until recently, this program was too ambitious and complex to extract meaningful results. However, two recent developments have given some impetus to this approach:

(1) First, Belavin and Knizhnik [34] have recently shown that the measure for the multiloop bosonic amplitude is simply the absolute value of a certain holomorphic $3g - 3$ form. This is called "holomorphic factorization." In principle, it may make possible writing down the multiloop measure by inspection! In practice, however, there are certain problems in specifying the parametrization of the period matrix beyond three loops. This is called the "Schottky problem." Holomorphic factorization has made the two- and three-loop measure almost trivial, but the Schottky problem prevents the multiloop measure from being easily written down.
(2) In 1984 mathematicians finally solved the Schottky problem. Thus, a key obstruction to utilizing holomorphic factorization seems to be removed. Solving the Schottky problem, moreover, provides one with an even more powerful tool. It allows one to describe an infinite-dimensional space, called the "Grassmannian" Gr, in which all Riemann surfaces of genus $g$ are treated as single points [35–37]. Thus, by studying the properties of the points in the Grassmannian, it may be possible to manipulate the set of all perturbation diagrams at once.

(Although the Grassmannian finally provides one with the conceptual framework in which to treat the entire perturbation series as a whole, one still does not know how to sum the entire series. For example, phase space reduces the set of all possible positions and momenta of a particle to a series of points. Thus, phase space, in principle, contains all possible motions of all possible particles in the universe. However, this says everything and it says nothing. We still have to impose equations of motion and boundary conditions to extract any meaningful information from phase space. The same applies to the Grassmannian.)

This ambitious program requires fully exploiting the modular transformations of Riemann surfaces of genus $g$. Of particular importance is the mapping class group MCG, which reduces to the modular group SL(2, $Z$) for the torus. What we want is a way of studying the mapping class group for arbitrary genus. The key will be the construction of $\Theta$ functions defined on Riemann surfaces of arbitrary genus. This is in constrast to the Schottky method that we studied in Section 5.6, where modular invariance was not so obvious.

In Fig. 5.12, we have written down the $a$- and $b$-cycles for an arbitrary closed Riemann surface, which we call the "canonical homology basis." Let the antisymmetric symbol $(a, b)$ represent whether or not two cycles intersect. It is equal to 0 if they don't intersect and equal to $\pm 1$ if they do. For example, consider the torus, where we have only the $a$- and $b$-cycles. Notice that $(a, b)$ is equal to 1, because the $a$- and $b$-cycles intersect, but $(a, a)$ is equal to 0. Then the four possible combinations for the torus create a matrix:

$$\begin{pmatrix} 0 & 1 \\ -1 & 0 \end{pmatrix} \leftrightarrow \begin{cases} (a, a) = (b, b) = 0 \\ (a, b) = -(b, a) = 1 \end{cases} \tag{5.11.1}$$

The elements of the mapping class group do not change this intersection matrix. However, we know, by definition, that the group that leaves this matrix invariant is Sp(2, $Z$). (See the Appendix.) To describe the elements of Sp(2, $Z$), consider a Dehn twist $D_a$ created by slicing the surface along the $a$-cycle, twisting the cut by $2\pi$, and then resplicing the surface back again. Under this Dehn twist, the $b$-cycle converts into the sum of the $a$-cycle and the $b$-cycle (see Fig. 5.13). Thus,

$$D_a(a) = a; \qquad D_a(b) = a + b \tag{5.11.2}$$

Now represent this in terms of matrix language. Let the Dehn twist operate on the column vector $[a, b]$. Then the Dehn twist can be represented symbolically as

$$D_a = \begin{pmatrix} 1 & 0 \\ 1 & 1 \end{pmatrix}; \qquad D_b = \begin{pmatrix} 1 & 1 \\ 0 & 1 \end{pmatrix} \tag{5.11.3}$$

Similarly, we can describe the mapping class group for the two-loop surface by taking Dehn twists along $a_1, a_2, b_1, b_2$ as well as the cycle $a_1^{-1}a_2$, which is a circular line that encircles the two holes. Using the same reasoning, we can show that the Dehn twists, operating on the column vector $[a_1, b_1, a_2, b_2]$, can be represented as [38]

$$D_{a_1} = \begin{pmatrix} 1 & 0 & 0 & 0 \\ 0 & 1 & 0 & 0 \\ 1 & 0 & 1 & 0 \\ 0 & 0 & 0 & 1 \end{pmatrix}; \qquad D_{b_1} = \begin{pmatrix} 1 & 0 & 1 & 0 \\ 0 & 1 & 0 & 0 \\ 0 & 0 & 1 & 0 \\ 0 & 0 & 0 & 1 \end{pmatrix}$$

$$D_{a_2} = \begin{pmatrix} 1 & 0 & 0 & 0 \\ 0 & 1 & 0 & 0 \\ 0 & 0 & 1 & 0 \\ 0 & 1 & 0 & 1 \end{pmatrix}; \qquad D_{b_2} = \begin{pmatrix} 1 & 0 & 0 & 0 \\ 0 & 1 & 0 & 1 \\ 0 & 0 & 1 & 0 \\ 0 & 0 & 0 & 1 \end{pmatrix} \tag{5.11.4}$$

$$D_{a_1^{-1}a_2} = \begin{pmatrix} 1 & 0 & 0 & 0 \\ 0 & 1 & 0 & 0 \\ -1 & 1 & 1 & 0 \\ 1 & -1 & 0 & 1 \end{pmatrix}$$

Now let us treat the closed Riemann surface of arbitrary genus. The intersection matrix can be represented as

$$\begin{cases} (a_i, a_j) = (b_i, b_j) = 0 \\ (a_i, b_j) = -(b_i, a_j) = \delta_{ij} \end{cases} \tag{5.11.5}$$

This is nothing but a block-diagonal matrix, with eqn. (5.11.1) in each block. The group that preserves this matrix is $\mathrm{Sp}(2g, Z)$. We might suspect, therefore, that the mapping class group is just $\mathrm{Sp}(2g, Z)$. This is not quite right. There are also Dehn twists $D_c$ around cycles that are homologically trivial and do not transform the $a$- or $b$-cycles. Thus, they are represented as the unit matrix in the basis that we have been using. Nevertheless, these Dehn twists are legitimate global diffeomorphisms that must be included in the mapping class group. This subgroup is called the Torelli group $T$, and hence we finally have the desired result:

$$\frac{\mathrm{MCG}}{T} = \mathrm{Sp}(2g, Z) \tag{5.11.6}$$

(Fortunately, the effect of the Torelli group on spin structures is trivial, so we will delete further discussion of it.)

Next, we wish to describe the period matrix for the Riemann surface and how it transforms under $\mathrm{Sp}(2g, Z)$. In addition to the $2g$ cycles we can write down on the Riemann surface, we can also write down $2g$ independent harmonic one-forms $\omega_i$ and $\bar{\omega}_i$ on the surface (see Appendix). Because we have equal numbers of cycles and one-forms (due to the Hodge–de Rahm theorem), we can always normalize the integration over $a$-cycles such that

$$\int_{a_i} \omega_j = \delta_{ij} \tag{5.11.7}$$

In general, the $b$-cycle integration will produce a $g \times g$ square matrix:

$$\int_{b_i} \omega_j = \Omega_{ij} \tag{5.11.8}$$

called the period matrix, which generalizes the variable $\tau$ introduced for the single-loop amplitude in (5.5.2) and equals the period matrix introduced for the multiloop amplitude in (5.6.31). It can be shown rather easily that the period matrix is symmetric and has a positive definite imaginary part. In general, the $\frac{1}{2}g(g+1)$ elements of a symmetric $g \times g$ matrix (with positive definite imaginary part) generate a space called the *Siegel upper half-plane*.

The advantage of introducing the period matrix is that no two inequivalent Riemann surfaces can have the same $\Omega_{ij}$, which is called Torelli's theorem. Thus, up to Sp$(2g, Z)$ transformations, the period matrix gives us a convenient way in which to characterize different Riemann surfaces. Under an Sp$(2g, Z)$ transformation, the period matrix transforms as

$$\Omega' = (A\Omega + B)(C\Omega + D)^{-1} \tag{5.11.9}$$

where the $A$, $B$, $C$, $D$ are each symplectic $g \times g$ matrices. Thus, two different period matrices may describe the same Riemann surface if they are related by an Sp$(2g, Z)$ transformation.

Now that we have a mathematical description of the mapping class group in terms of Sp$(2g, Z)$ defined on Dehn twists, the next task is to write down functions defined over a Riemann surface with $g$ handles. We saw earlier that the single-loop amplitude is described in terms of the quasi-doubly periodic theta function $\Theta$. Now, we wish to write down generalized $\Theta$ functions that have certain periodicity properties defined for the surface with $g$ handles.

If we take the single-loop $\Theta$ function and then force it to have these periodicity properties over the Riemann surface with $g$ handles, we naturally add an additional set of infinite terms to the summation. Eventually, this summation approaches [39, 40]

$$\Theta(\mathbf{z}|\Omega) = \sum_{n \in Z^g} e^{i\pi \mathbf{n}\cdot\Omega\cdot\mathbf{n} + 2\pi i \mathbf{n}\cdot\mathbf{z}} \tag{5.11.10}$$

where the sum over the vector $\mathbf{n}$ is taken over a $g$ dimensional lattice. The $\Theta$ function is defined not over the complex variable $z$, but over the vector $\mathbf{z}$ defined by

$$\mathbf{z} = \int_{p_0}^{z} \boldsymbol{\omega} \tag{5.11.11}$$

where $p_0$ is an arbitrary point on the surface. Since $\boldsymbol{\omega}$ has $g$ components $\omega_i$, then $\mathbf{z}$ is also a vector with $g$ components $z_i$.

It also possible to construct $\Theta$ functions with spin structures. For the torus, we saw that a spinor parallel transported around a parallelogram picked up a phase of $+1$ $(-1)$, depending on whether the spinor was periodic (anti-periodic). Likewise, the genus $g$ Riemann surface can be represented as a

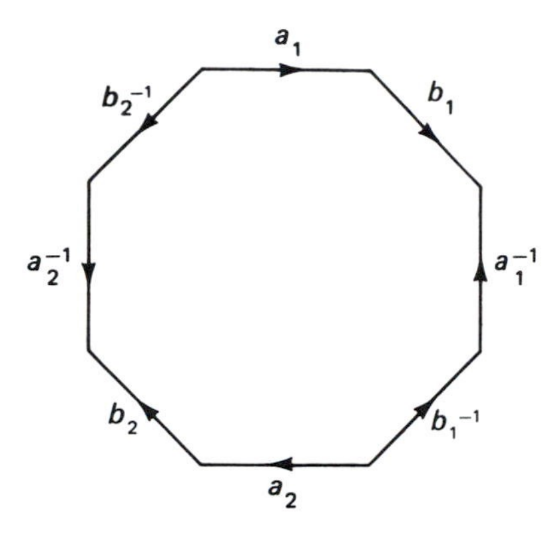


Figure 5.14. Canonical homology cycles of the Riemann surface of genus 2. This figure is formed by drawing the $a$ and $b$ homology cycles of a sphere with two handles or holes, cutting along the lines that form the cycles, and then unraveling the surface. The advantage of this basis is that the homology cycles for a surface of arbitrary genus can be represented as a polygon.

polygon with $4g$ sides (Fig. 5.14). We now have to define $2^{2g}$ phases in order to describe the different ways in which we can parallel transport a spinor around this polygon. The set of characters $[\alpha] = (\mathbf{a}, \mathbf{b})$ defines the spin structure of the manifold.

We can now define the generalized $\Theta$ function with spin structure:

$$\begin{aligned}\Theta[\alpha](\mathbf{z}|\Omega) &= \sum_{n \in Z^g} e^{i\pi(\mathbf{n}+\mathbf{a})\cdot\Omega\cdot(\mathbf{n}+\mathbf{a}) + 2\pi i(\mathbf{n}+\mathbf{a})\cdot(\mathbf{z}+\mathbf{b})} \\ &= e^{i\pi\mathbf{a}\cdot\Omega\cdot\mathbf{a} + 2\pi i\mathbf{a}(\mathbf{z}+\mathbf{b})}\Theta(\mathbf{z} + \Omega\mathbf{a} + \mathbf{b}|\Omega) \qquad (5.11.12)\end{aligned}$$

Under a shift in the lattice, we find that the generalized $\Theta$ function is periodic up to a phase:

$$\Theta[\alpha](\mathbf{z} + \Omega\cdot\mathbf{n} + \mathbf{m}|\Omega) = e^{2\pi i\mathbf{a}\cdot\mathbf{m} - i\pi\mathbf{n}\cdot\Omega\cdot\mathbf{n} - 2\pi i\mathbf{n}\cdot(\mathbf{z}+\mathbf{b})}\Theta[\alpha](\mathbf{z}|\Omega) \qquad (5.11.13)$$

Under the change $\mathbf{z} \to -\mathbf{z}$, the $\Theta$ function transforms as

$$\Theta[\alpha](-\mathbf{z}|\Omega) = (-1)^{4\mathbf{a}\cdot\mathbf{b}}\Theta[\alpha](\mathbf{z}|\Omega) \qquad (5.11.14)$$

We can thus categorize spin structures, as we did for the single-loop function in (5.9.7), according to whether or not they are even or odd under $\mathbf{z} \to -\mathbf{z}$. We find that there are $2^{g-1}(2^g - 1)$ odd and $2^{g-1}(2^g + 1)$ even spin structures on a Riemann surface of genus $g$.

It is now a straightforward task to calculate how these generalized $\Theta$ functions transform under a mapping class group transformation (5.11.9). Under $Sp(2g, Z)$, we find

$$\Theta[\alpha](\mathbf{z}'|\Omega') = \varepsilon e^{-i\pi\phi} \det(C\Omega + D)^{1/2}\Theta[\alpha](\mathbf{z}|\Omega) \qquad (5.11.15)$$

where

$$\mathbf{z}' = (C\Omega + D)^{-1}\cdot\mathbf{z} \qquad (5.11.16)$$

and

$$\begin{bmatrix} \mathbf{a}' \\ \mathbf{b}' \end{bmatrix} = \begin{pmatrix} D & -C \\ -B & A \end{pmatrix} \begin{bmatrix} \mathbf{a} \\ \mathbf{b} \end{bmatrix} = \tfrac{1}{2} \begin{bmatrix} (CD^T)_d \\ (AB^T)_d \end{bmatrix} \tag{5.11.17}$$

and the phase factor is given by

$$\phi = \mathbf{a}D^T B\mathbf{a} + \mathbf{b}C^T A\mathbf{b} - 2\mathbf{a}B^T C\mathbf{b} - (\mathbf{a}D^T - \mathbf{b}C^T)(AB^T)_d$$

where $T$ represents the transpose and $d$ represents taking the diagonal element. The symbol $\varepsilon$ represents the eighth root of unity (with a certain restriction that does not worry us here).

Now that we have defined $\Theta$ functions with the proper periodicity on the Riemann surface, our next task is to actually calculate the measure of the multiloop amplitude. Here, we are greatly aided by a remarkable result of Belavin and Knizhnik [34], which simply states that the measure of the multiloop amplitude is the square of a certain holomorphic function (up to zero modes):

$$\text{Holomorphic factorization:} \quad Z = \int_F \frac{\eta \wedge \bar{\eta}}{(\det \operatorname{Im} \Omega)^{13}} \tag{5.11.18}$$

where $F$ is the fundamental region of the surface, and $\eta$ is a $3g - 3$ form:

$$\eta = \prod_{i=1}^{3g-3} dy_i \, F(y) \tag{5.11.19}$$

where $y_i$ are a parametrization of the Teichmüller parameters, and $F(y)$ is holomorphic, has no zeros, and has second-order poles at the boundary where the surface degenerates. It is easy to check, for example, that the single-loop function (5.5.4) possesses this property (up to zero mode terms).

The statement of holomorphic factorization, on one level, is intuitively obvious. It simply says that the closed loop is a product of the left- and right-moving modes of the string, except for the zero modes. The proof of this powerful result, however, is rather involved, but can be summarized as follows. Using the techniques derived from the previous sections, we know how to formally write down the measure of the multiloop amplitude in terms of Teichmüller parameters and complicated determinants. Then, it can be shown that

$$\partial\bar{\partial} \log W = (D - 26)\cdots \tag{5.11.20}$$

where $W$ is the measure term up to the factor raised to the 13th power. Notice that the right-hand side is the conformal anomaly term, which vanishes in 26 dimensions. Thus, in 26 dimensions we have $\partial\bar{\partial} \log W = 0$, in which case it can be written as the absolute value of a holomorphic function $|F|^2$. When we are in any other dimension, there is an obstruction to holomorphic factorization. Then we say that there is an analytic anomaly.

What is remarkable is that the conditions on $F$ are probably stringent enough to uniquely fix the measure for all multiloops [41–61]. This is an enormously powerful result, which may eventually allow us to write down the multiloop measure by inspection.

For example, consider the two-loop function. Let us choose the Teichmüller parameters as the period matrix itself, since they both have three independent complex variables. Then the following combination is modular invariant:

$$\frac{d\Omega}{(\det \operatorname{Im} \Omega)^3} \tag{5.11.21}$$

We wish to find an expression for $|F|^2(\det \operatorname{Im} \Omega)^{-10}$. Using arguments from the theory of Riemann surfaces, we find that the unique answer is

$$\text{2-Loop:} \quad F = \left\{ \prod_{\mathbf{a},\mathbf{b}} \Theta[\alpha](0|\Omega) \right\}^{-2} \tag{5.11.22}$$

where the product is taken over the 10 even characters such that $4\mathbf{a} \cdot \mathbf{b} = 0 \bmod 2$.

Likewise, using similar arguments about holomorphic functions, we can represent the $F$ function for the three-loop amplitude as

$$\text{3-Loop:} \quad F = \left\{ \prod_{\mathbf{a},\mathbf{b}} \Theta[\alpha](0|\Omega) \right\}^{-1/2} \tag{5.11.23}$$

where once again the Teichmüller variables $dy_i$ are taken to be the elements of the period matrix $d\Omega_{ij}$.

This simple procedure, however, stops abruptly at the four-loop level. In general, there are $\frac{1}{2}g(g+1)$ elements in a symmetric matrix defined in the Siegel upper half-plane, while the number of Teichmüller parameters is $3g - 3$. Thus, they have the same number of elements only for $g = 2, 3$, which makes a characterization of the period matrix of higher loops exceedingly difficult. This is the origin of the Schottky problem, the fact that the set of elements in the Siegel upper half-plane agrees with the elements in the period matrix only for $g = 2, 3$. The question is: What conditions do we have to place on our functions defined in the Siegel upper half-plane such that they become functions of the period matrix?

Fortunately, mathematicians have recently solved the Schottky problem, making it possible to simply characterize period matrices for arbitrary $g$ and hence describe holomorphic functions defined over Riemann manifolds of arbitrary genus. In fact, this allows us to envision an infinite-dimensional space, the Grassmannian, where Riemann surfaces of arbitrary genus are represented as points.

What we want is a generalization of an *operator formalism for a conformal field theory defined on a Riemann manifold of arbitrary genus.* We want to construct operators acting on a generating function [35–37] that will allow us to derive all correlation functions defined on surfaces of genus $g$. Let us begin by defining fermion operators

$$\psi(z) = \sum_n \psi_n z^{-n-1/2} \tag{5.11.24}$$

with the usual anticommutation relations:

$$\{\psi_n, \psi_m^*\} = \delta_{n,m} \tag{5.11.25}$$

and the bilocal current:

$$J(z, w) = :\psi(z)\psi^*(w): \tag{5.11.26}$$

The current operator is the diagonal element

$$J(z, z) = \sum_{n \in Z} J_n z^{-n-1} \tag{5.11.27}$$

Now define the generating function

$$\tau(x) = \langle 0| e^{H(x)} g |0\rangle \tag{5.11.28}$$

where

$$H(x) = \sum_{n=0}^{\infty} x_n J_n \tag{5.11.29}$$

$g$ is an element of the Clifford group given by

$$g = \exp\left\{\oint_{z=\infty} dz \oint_{w=\infty} dw\, f(z, w) J(z, w)\right\} \tag{5.11.30}$$

where the integral is taken around infinity. It is important to know that $g$ is a function of the period matrix $\Omega$ as well as the spin structure defined over the surface, which is parametrized by the characters $\mathbf{a}$ and $\mathbf{b}$. Thus, all the information concerning the nature of the Riemann surface is encoded within the function $g$.

In ordinary point particle path integrals, we know that the action of the operator $\delta/\delta J$ on the generator of the Green's function creates the insertion of a field into the functional integral. The analog of this is given by the vertex operator

$$v(z) = e^{\sum_{n=0}^{\infty} x_n z^n} e^{\log z \partial_0 + \sum_{n=1}^{\infty} n^{-1} z^{-n} \partial_n} \tag{5.11.31}$$

where $x_n$, $\partial_m$ have the same commutation relations as the usual harmonic oscillators. By construction, we have the identity

$$\frac{\langle 0| \psi(z)\psi^*(w) g |0\rangle}{\langle 0| g |0\rangle} = \tau^{-1}(x) v(z) v^*(w) \tau(x)|_{x=0} \tag{5.11.32}$$

Thus, the action of the vertex operator is to create the correlation function for the conformal field theory, similar to the action of $\delta/\delta J$ on the generating function in ordinary field theory.

Now let us impose the conditions, called the Hirota equations:

$$\oint_{z=\infty} dz\, v(z)\tau(x)v^*(z)\tau(y) = 0 \tag{5.11.33}$$

The Hirota equations can be shown to be intimately linked to the equations of the KP (Kadomtsev–Petviasvili) hierarchy.

We now have all the tools to write down an explicit expression for $\tau(x)$ [35–37]:

$$\tau(x) = e^{\sum_{n,m\geq 0} x_n Q_{nm} x_m}\Theta[\alpha]\left(-\sum_{n=1}^{\infty} A_n x_n|\Omega\right) \tag{5.11.34}$$

where the $A_n$ appear in the power expansion of the period matrix:

$$\Omega(z) = \sum_{n=0}^{\infty} A_n z^{-n-1}\, dz \tag{5.11.35}$$

and

$$Q_{nm} = \tfrac{1}{2}[(m-1)!n(n-1)!]^{-1}\partial_t^m \partial_y^n \log \frac{E(t, y)}{(t-y)}\bigg|_{t=y=0} \tag{5.11.36}$$

where $E(t, y)$ is called the "prime form" and is straightforward generalization of the function $z - w$ found in the conformal field theory of the sphere. In fact, for $z$ near $w$, it behaves like $z - w$ even on a surface of genus $g$. Explicity, it equals

$$E(z, w) = \Theta[\alpha]\left(\int_w^z \omega|\Omega\right)(h_\alpha(z)h_\alpha(w))^{-1/2} \tag{5.11.37}$$

where

$$h_\alpha(z) = \sum_i \partial_i\Theta[\alpha](0|\Omega)\omega_i(z) \tag{5.11.38}$$

Now we come to the main result of this discussion. We find that the $\tau$ function satisfies the Hirota equations of the KP hierarchy if and only if $\Omega$ is, in fact, the period matrix of a genus $g$ Riemann surface. This is the constraint that we were looking for. $\Omega$ which has dimension $\frac{1}{2}g(g+1)$, can now be restricted to dimension $3g - 3$ and is equal to the period matrix of a Riemann surface of genus $g$. This, in principle, solves the Schottky problem (although the result is highly non-linear).

As an added bonus, because $\tau(z)$ can be viewed as the generating function for conformal field theory defined on an Riemann surface of genus $g$, we can calculate arbitrary correlation functions. The matrix element of two fermions on this Riemann surface is given by

$$\frac{\langle 0|\psi(z)\psi^*(w)g|0\rangle}{\langle 0|g|0\rangle} = \frac{\Theta[\alpha](\int_w^z \omega|\Omega)}{\Theta[\alpha](0|\Omega)E(z, w)} = P_\alpha[z, w] \tag{5.11.39}$$

Mathematicians know this as the Szego kernel, which is the unique meromorphic half-differential in $z$ and $w$ with a single pole with residue 1 for $z = w$. Physicists know this simply as the two-point function. The $N$-point generalization is equal to

$$\frac{\langle 0|\psi(z_1)\psi^*(w_1)\cdots\psi(z_n)\psi^*(w_n)g|0\rangle}{\langle 0|g|0\rangle}$$

$$= \prod_{i<j} E(z_i, z_j) \prod_{k<l} E(w_k, w_l)\Theta^{-1}[\alpha](0|\Omega)$$

$$\times \prod_{i,k} E^{-1}(z_i, w_k)\Theta[\alpha]\left(\sum_i z_i - \sum_k w_k|\Omega\right) \tag{5.11.40}$$

There is a way in which to check the consistency of these equations. We know that there are two ways in which to calculate the $N$-point function, in terms of either fermionic oscillators or, through bosonization, bosonic oscillators with the form $e^{i\phi}$. For example, by using the Wick decomposition on the $N$-point function, we can write it terms of the two-point function as

$$\langle 0| \prod_{i=1}^{N} \psi(z_i) \prod_{j=1}^{N} \psi^*(w_j)g\,|0\rangle = \det_{ij} P_\alpha(z_i, w_j) \tag{5.11.41}$$

(5.11.40) can be obtained using the bosonic representation, while (5.11.41) can be obtained using the fermionic representation. Fortunately, the equivalence of these two different expressions was proved by Fay [39]; this is called the "trisecant" addition theorem for $\Theta$ functions. The mathematical equivalence of these expressions is another check on our bosonization scheme.

In summary, what we have developed is an operator formalism for conformal field theory defined on an arbitrary Riemann surface. The ket vector $g|0\rangle$ represents a specific Riemann surface, and applying various field operators on it corresponds to taking matrix elements over that Riemann surface. It is important to note that each element $\tau(x)$, if it satisfies the Hirota conditions for the KP hierarchy, now correctly characterises a Riemann surface of arbitrary genus $g$. Thus, each $\tau(x)$ defines a point in our Grassmannian, which is the desired result.

Notice that we are still a great distance away from our final goal, which is to sum the perturbation series and extract nonperturbative information from it. However, we have made a significant step in this direction because we can now characterize every Riemann surface of arbitrary genus, including its spin structure, as a point in a Grassmannian space. With this operator formalism, we can generate points on the Grassmannian at will. It remains to be seen how useful the Grassmannian will become in the future.

We should also mention, however, some difficulties with this formalism. Although the Schottky problem is now formally solved, in practice, the Hirota equations are quite nonlinear, so it remains to be seen precisely how practical this solution will be in the explicit construction of multiloop amplitudes in terms of theta functions. Furthermore, we must mention that there exists a certain amount of confusion in the literature concerning how to define the super-moduli space for the superstring amplitudes. Although the super

Riemann-Roch theorem tells us how many super-moduli there are, there still remains the difficult problem of constructing them explicitly in a well defined way. For an introduction to super-moduli space and its problems, see [62].

In the next series of chapters, we will discuss an alternative to the Grassmanian approach, which requires no information about the genus $g$ at all. In this approach, in fact, the entire perturbation series on Riemann surfaces is only one possible approximation scheme. This is the field theory of strings.

Before closing this chapter, we should mention that, of the various multiloop formalisms discussed so far, only the light cone formalism can be naturally derived from a field theory of strings. We thus intuitively expect that the construction of light cone Feynman diagrams by assembling vertex functions in the $S$-matrix yields a specific "triangulation" of moduli space. This result, which was proven in [63–4], was never previously considered by the mathematicians. (Briefly, proof is based on constructing certain integrals on a Riemann surface of genus $g$ and calculating their periods as we move around each loop or handle. Then it can be shown that the periods are purely imaginary, which yields a modular invariant description of the surface.)

For example, in Fig. 5.15, we see how a multiloop surface is parametrized with light cone parameters, with $\tau_i$ representing the various interaction times when the string splits or joins, and $\theta_i$ representing "twisting" each string by one full revolution. To complete $6g - 6$ parameters for a surface of genus $g$, we also need to integrate over the circumference of each cylinder. Thus, all $6g - 6$ moduli parameters have a natural interpretation in terms of the physical parameters of a Feynman diagram. The light cone formalism has several advantages over the previous formalisms. First, there is no need

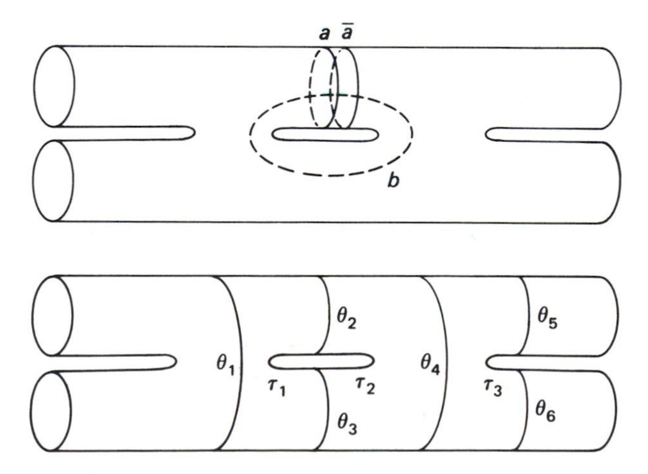


Figure 5.15. In the last diagram, we see the angles, lengths, and "times" that parametrize the light cone surface. These parameters make up the Teichmüller space, and six of them are required to parametrize each internal loop. (These parameters automatically yield one cover of the fundamental modular region.)

to truncate the integration region, which is necessary in the Schottky and the constant curvature formalisms for modular invariance. Second, it easily generalizes to an arbitrary number of loops (which cannot be said for the theta function method.) Third, it is unitary and based on physical variables. And fourth, it is naturally derived from a field theory.

Instead of discussing this formalism, we will discuss the light cone method in the context of the field theory of strings. We will devote the next series of chapters to this important topic.

## §5.12. Summary

In summary, we have shown that the unitarization scheme for the string can be implemented by using the unitarity equation, using the $N$-point amplitude as the Born term:

$$\text{Im } T_{ij} = -\tfrac{1}{2}\sum_n \langle i|T|n\rangle\langle n|T^\dagger|j\rangle \tag{5.12.1}$$

The function integral is now modified to

$$A = \int_S DX\, d\mu e^{iS} \prod_{i=1}^N e^{ik_iX_i} \tag{5.12.2}$$

where we sum over all conformally inequivalent surfaces $S$ that are disks or spheres with $N$ holes. In particular, the open string graphs come in three types: planar, nonplanar, and nonorientable (as in a Möbius strip). The closed string graphs come in only planar and nonorientable (as in the Klein bottle).

Fortunately, mathematicians have already calculated the Neumann functions on these surfaces, which are particular examples of automorphic functions.

For the first loop, the Neumann function is given in terms of

$$\ln \psi(x, w) = \ln(1 - x) - \tfrac{1}{2}\ln x + \frac{\ln^2 x}{2 \ln w}$$
$$+ \sum_{n=1}^{\infty} [\ln(1 - w^n x) + \ln(1 - w^n/x) - 2\ln(1 - w^n)] \tag{5.12.3}$$

Putting everything together, the first loop planar amplitude is equal to

$$A_N = \pi^{-1}\int_0^1 \prod_{i=1}^{N-1} \theta(v_{i+1} - v_i)\, dv_i$$
$$\times \int_0^1 \frac{dq}{q^3}\left(\frac{-2\pi^2}{\ln q}\right)^N f(q^2)^{-24} \prod_{i<j} (\psi_{P,ij})^{k_i\cdot k_j} \tag{5.1.24}$$

We found the string graphs never have ultraviolet divergences. The infinite sum over the intermediate Regge trajectories renders the graphs finite in the ultraviolet region. However, as in Feynman graph theory, the divergence of a graph is given by the local change in topology of the surface. The single-loop open string graph diverges when the interior hole vanishes. This corresponds, by conformal invariance, to extracting a closed loop with zero momentum. Thus, the divergences correspond to a tachyon or a dilaton scattering into the vacuum.

The divergence structure of these graphs are as follows:

(1) The Nambu–Goto open string diverges as $q^{-3}$ and $q^{-1}$. The latter divergence, associated with a dilaton, probably can be absorbed into a slope renormalization to all orders. The tachyon divergence is more troublesome. The nonplanar graphs, when analyzed in the complex $s$ and $t$ plane, actually contain cuts in them, which reduce to poles in 26 dimensions. These poles correspond to closed strings. Thus, the open string theory, by itself, is incomplete. Closed strings emerge as "bound states" at the first-loop level.
(2) The closed string amplitude diverges because of tadpole or self-energy insertions on external legs, which are on-shell.
(3) The Type I superstring has only $q^{-1}$ poles, and hence we can have slope renormalization to absorb this infinity. The $q^{-3}$ divergence never occurs because the boson and fermion internal legs cancel against each other.
(4) The Type II superstring is actually finite. There are no two- and three-point single-loop graphs, and hence tadpole or self-energy insertions on external legs are simply forbidden.

These results generalize nicely to the $N$-loop amplitude using the theory of automorphic functions. There are several parametrizations we can use. The first is the Schottky group method, which has the advantage that its choice of variables is explicit and it was derived in a manifestly factorizable fashion by sewing multireggeon vertices together. However, modular invariance is not so obvious. Another is the $\Theta$ function method, in which modular invariance is build in from the very start. It also easily generalizes to spin structures. The disadvantage of this method is that the choice of variables is far from clear. Also, factorization and hence unitarity are obscure. These amplitudes are simply postulated by hand to satisfy the boundary conditions, rather than being constructed by sewing vertices together. Ultimately, these methods are probably identical.

Let us start with the complex plane with $2N$ arbitrary holes cut out. Let us pair up these holes into $N$ pairs, labeled $a_i$ and $\bar{a}_i$. These are called $a$-cycles. If we cut a line between any pair of $a$-cycles, we obtain $b$-cycles. Let us define projective operators $P_i$ that map one $a$-cycle $a_i$ into its partner $\bar{a}_i$. Then these projective operators can be parametrized by two invariant points $z_1$ and $z_2$ and a multiplier $X$, such that

$$\frac{P(z) - z_1}{P(z) - z_2} = X\frac{z - z_1}{z - z_2} \tag{5.12.5}$$

For the open string, the centers of the $a$-cycles lie on the real axis, as well as the invariant point of the projective transformations. The final open string

$N$-loop amplitude is given by

$$
\begin{aligned}
A_N = \int \prod_{\alpha=1}^{N} d^{26}k_\alpha \, d\mu \prod_{\{P\}} \prod_{\{\tilde{P}\}} (1 - X_{\{\tilde{P}\}})^{-24} \\
\times \prod_{1=i<j=M} (z_i - \{P\} z_j)^{k_i \cdot k_j} \prod_{\substack{\beta, \lambda=1 \\ \beta \neq \lambda}}^{N} dX \, X_\beta^{-\alpha(k_\beta)-1} \\
\times \prod_{i=1}^{M} \left( \frac{z_i - \{P\} x_\lambda^{(2)}}{z_i - \{P\} x_\lambda^{(1)}} \right)^{k_i \cdot k_\lambda} \\
\times \left\{ \frac{x_\beta^{(1)} - \{P\} x_\lambda^{(1)}}{x_\beta^{(2)} - \{P\} x_\lambda^{(1)}} \frac{x_\beta^{(2)} - \{P\} x_\lambda^{(2)}}{x_\beta^{(2)} - \{P\} x_\lambda^{(2)}} \right\}^{(1/2)k_\beta \cdot k_\lambda}
\end{aligned}
\tag{5.12.6}
$$

where

$$
d\mu = \prod_{i=1}^{M} dz_i \, dV_{abc}^{-1} \prod_{\alpha=1}^{N} dx_\alpha^{(1)} \, dx_\alpha^{(2)} (x_\alpha^{(1)} - x_\alpha^{(2)})^{-2} \tag{5.12.7}
$$

where the Roman letters refer to the external lines and the Greek letters represent the loops. Notice that the region of integration lies between the limit points of the set $\{P\}$. We must have these limit points forming a discrete region on the real axis. Thus, we must have Schottky groups.

The $N$-loop amplitude can also be directly reformulated in the language of Riemann surfaces if we use the Polyakov action instead of the Nambu–Goto action. There are several new contributions to the integral:

(1) A factor $\Delta_{\text{FP}}$ coming from the Faddeev–Popov determinant, which can be rewritten as

$$
\det{}^{1/2} P_1^\dagger P_1 \tag{5.12.8}
$$

(2) A factor from the integration over the $X$ variable:

$$
\begin{aligned}
\int DX \exp\left( - \int d^2 z \, \sqrt{g} g^{ab} \partial_a X^\mu \partial_b X_\mu \right) \\
= \left( \frac{2\pi}{\int d^2 z \, \sqrt{g}} \det{}'(-\nabla^2) \right)^{-(1/2)D}
\end{aligned}
\tag{5.12.9}
$$

where

$$
\nabla^2 = \frac{-1}{\sqrt{g}} \partial_m \sqrt{g} g^{mn} \partial_n \tag{5.12.10}
$$

(3) The Teichmüller parameters $t_i$ must be added to the integral. To calculate this last factor, let us write

$$
\begin{aligned}
\delta g_{ab} = [\nabla_a \delta v_b + \nabla_b \delta v_a - (\nabla_c \delta v^c) g_{ab}] \\
+ (\nabla_c \delta v^c) g_{ab} + 2\delta\sigma g_{ab} + \delta t_i T^i
\end{aligned}
\tag{5.12.11}
$$

where $t_i$ are the Teichmüller parameters and

$$T^i = \frac{\partial g_{ab}}{\partial t^i} - (\text{trace}) \tag{5.12.12}$$

We can rewrite this as

$$\partial g_{ab} = P_1(\delta v)_{ab} + 2\delta\sigma g_{ab} + \delta t^i \partial_i g_{ab} \tag{5.12.13}$$

where

$$P_1(\delta v)_{ab} = \nabla_a \delta v_b + \nabla_b \delta b_a - g_{ab} \nabla_c \delta v^c \tag{5.12.14}$$

We change variables to

$$T^i = \psi^a(\psi^a, T^i) + P_1 v^i \tag{5.12.15}$$

where

$$v^i = \frac{1}{P_1^\dagger P_1} P_1^\dagger T^i \tag{5.12.16}$$

$$(\psi^a, T^i) = \int d^2z\, \sqrt{g} g^{bc} g^{de} T^i_{bd} \psi^a_{ce} \tag{5.12.17}$$

Our final result for the measure is therefore:

$$\begin{aligned} Dg_{ab}\, \Omega^{-1}_{\text{Diff}} \Omega^{-1}_{\text{Weyl}} = {} & Dv_a\, \Omega^{-1}_{\text{Diff}}\, D\sigma\, \Omega^{-1}_{\text{Weyl}} \\ & \times Dt^i \det{}^{1/2}(P_1^\dagger P_1) \frac{\det(\psi^a, T^i)}{\det^{1/2}(\psi^a, \psi^b)} \\ & \times \left( \frac{2\pi}{\int d^2z\, \sqrt{g}} \det{}'(-\nabla^2) \right)^{-(1/2)D} \end{aligned} \tag{5.12.18}$$

The great advantage of this approach to the multiloop amplitude is that we can use the power of Riemann surfaces to analyze the divergences of the amplitude. In particular, we can analyze the singularities of the Selberg zeta function:

$$Z(s) = \prod_{l \text{ primitive}} \prod_{p=0}^{\infty} [1 - v(\gamma) e^{-(s+p)l}] \tag{5.12.19}$$

Analysis of the Selberg functions confirms the fact that the measure has a singularity where the topology of the genus $g$ surface degenerates.

Fortunately, the construction of the measure on multiloop surfaces is greatly facilitated by the holomorphic factorization theorem, which states that the measure is equal to

$$Z = \int_F \frac{\eta \wedge \bar{\eta}}{(\det \operatorname{Im} \Omega)^{13}} \tag{5.12.20}$$

where $\eta$ is a holomorphic $3g - 3$ form. The formula is intuitively obvious if we consider the equal contribution of left- and right-movers to the amplitude (except for zero modes). The only complication to the proof is an anomaly (called the analytic anomaly) that vanishes in 26 dimensions. Using this result, we can basically guess the answer for the integrand of the multiloop amplitude, since the function with the correct periodicity and holomorphic singularity structure must be unique.

Another method, based on $\Theta$ functions, exploits the modular properties of these periodic functions from the very beginning. For surfaces of genus $g$, we have $2^{2g}$ spin structures, corresponding to all possible periodic (antiperiodic) boundary conditions on the surface. Our task is therefore to construct Neumann functions on this surface with spin structure that have the correct periodicity properties and singularities. The answer is given in terms of two functions, the generalized theta functions and the prime form. The theta function is given by

$$\Theta[\alpha](\mathbf{z}|\Omega) = \sum_{n \in Z^g} e^{i\pi(\mathbf{n}+\mathbf{a})\cdot\Omega\cdot(\mathbf{n}+\mathbf{a})+2\pi i(\mathbf{n}+\mathbf{a})\cdot(\mathbf{z}+\mathbf{b})} \tag{5.12.21}$$

where $\alpha$ represents the spin structure and the prime form (which is the holomorphic generalization of $z - w$ found for the sphere) is given by

$$E(z, w) = \Theta[\alpha]\left(\int_w^z \omega|\Omega\right)(h_\alpha(z)h_\alpha(w))^{-1/2} \tag{5.12.22}$$

where

$$h_\alpha = \sum_i \partial_i \Theta[\alpha](0|\Omega)\omega_i(z) \tag{5.12.23}$$

Given the theta function and the prime form, we can calculate all possible Green's functions for bosonic and fermionic operators on Riemann surfaces. Thus, we can construct a new conformal field theory on Riemann surfaces, not just spheres. For example, the two-point function of two fermions is given by the Szego kernel:

$$\frac{\Theta[\alpha](\int_w^z \omega|\Omega)}{\Theta[\alpha](0|\Omega)E(z, w)} \tag{5.12.24}$$

One difficulty in solving the multiloop problem is the question of choosing the correct coordinates. One ideal choice would be to use the period matrix $\Omega$ as the coordinate. The problem, however, is that there are $\frac{1}{2}g(g+1)$ dimensions to a square matrix like the period matrix, but only $3g - 3$ complex Teichmüller parameters. For two and three loops, they coincide, but for higher loops they do not. This is the Schottky problem, which only recently has been solved. The solution to this problem makes possible the use of Grassmannians to characterize the entire perturbation series. Each multiloop amplitude with spin structure represents a single point in the Grassmannian, so we may (at least in principle) have a way of manipulating the *entire* perturbation series over Riemann surfaces all at once. This may, in turn, eventually yield nonperturbative information about the superstring theory.

## References


[1] K. Kikkawa, B. Sakita, and M. B. Virasoro, *Phys. Rev.* **184**, 1701 (1969).
[2] M. Kaku and L. P. Yu, *Phys. Lett.* **33B**, 166 (1970).
[3] M. Kaku and L. P. Yu, *Phys. Rev.* **D3**, 2992, 3007, 3020 (1971).
[4] M. Kaku and J. Scherk, *Phys. Rev.* **D3**, 430 (1971).
[5] M. Kaku and J. Scherk, *Phys. Rev.* **D3**, 2000 (1971).
[6] V. Alessandrini, *Nuovo Cim.* **2A**, 321 (1971).
[7] C. Lovelace, *Phys. Lett.* **32B**, 703 (1970).
[8] C. Lovelace, *Phys. Lett.* **32B**, 203 (1971).
[9] C. S. Hsue, B. Sakita, and M. A. Virasoro, *Phys. Rev.* **D2**, 2857 (1970).
[10] J. L. Gervais and B. Sakita, *Phys. Rev.* **D4**, 2291 (1971); *Nucl. Phys.* **B34**, 632 (1971); *Phys. Rev. Lett.* **30**, 716 (1973).
[11] D. B. Fairlie and H. B. Nielsen, *Nucl. Phys.* **B20**, 637 (1970).
[12] W. Burnside, *Proc. London Math. Sci.* **23**, 49 (1891).
[13] D. Amati, C. Bouchiat, and J-L. Gervais, *Nuovo Cim. Lett.* **2**, 399 (1969).
[14] K. Bardakçi, M. B. Halpern, and J. A. Shapiro, *Phys. Rev.* **185**, 1910 (1969).
[15] M. Kaku and C. B. Thorn, *Phys. Rev.* **1D**, 2860 (1970).
[16] E. Cremmer and J. Scherk, *Nucl. Phys.* **B50**, 222 (1972).
[17] C. Lovelace, *Phys. Lett.* **34B**, 500 (1971).
[18] A. Neveu and J. Scherk, *Phys. Rev.* **D1**, 2355 (1970).
[19] See J. H. Schwarz, *Phys. Rep.* **13C**, 259 (1974).
[20] J. A. Shapiro, *Phys. Rev.* **D5**, 1945 (1972).
[21] S. Mandelstam, in *Unified String Theories* (edited by M. B. Green and D. Gross) World Scientific, Singapore, 1986.
[22] O. Alvarez, *Nucl. Phys.* **B216**, 125 (1983).
[23] E. D'Hoker and D. H. Phong, *Nucl. Phys.* **B269**, 205 (1986).
[24] A. M. Polyakov, *Phys. Lett.* **103B**, 207, 211 (1981).
[25] J. Polchinski, *Comm. Math. Phys.* **104**, 37 (1986).
[26] F. Gliozzi, J. Scherk, and D. Olive, *Nucl. Phys.* **B122**, 253 (1977).
[27] N. Seiberg and E. Witten, *Nucl. Phys.* **B276**, 272 (1986).
[28] E. D'Hoker and D. H. Phong, *Nucl. Phys.* **B278**, 225 (1986); *Comm. Math. Phys.* **104**, 537 (1986).
[29] G. Gilbert, *Nucl. Phys.* **B277**, 102 (1986).
[30] M. A. Namazie and S. Rajeev, *Nucl. Phys.* **B277**, 332 (1986).
[31] F. Steiner, *Phys. Lett.* **188B**, 447 (1987).
[32] A. Selberg, *J. Indian Math. Soc.* **20**, 47 (1956).
[33] D. Friedan and S. Shenker, *Phys. Lett.* **B175**, 287 (1986); *Nucl. Phys.* **B281**, 509 (1987).
[34] A. A. Belavin and V. G. Knizhnik, *Phys. Lett.* **168B**, 201 (1986).
[35] N. Ishibashi, Y. Matsuo, and H. Ooguri, *Mod. Phys. Lett.* **A2**, 119 (1987).
[36] L. Alvarez-Gaumé, C. Gomez, and C. Reina, *Phys. Lett.* **55B**, 55 (1987).
[37] C. Vafa, *Phys. Lett.* **190B**, 47 (1987).
[38] L. Alvarez-Gaumé, G. Moore, and C. Vafa, *Comm. Math. Phys.* **106**, 1 (1986).
[39] J. Fay, *Theta Functions on Riemann Surfaces*. Lecture Notes in Mathematics, Vol. 352, Springer-Verlag, Berlin, 1973.
[40] D. Mumford, *Tata Lectures on Theta*, Birkhauser, Basel, 1983.
[41] G. Moore and P. Nelson, *Nucl. Phys.* **B266**, 58 (1986).
[42] P. Nelson, Harvard preprint HUTP-86 A047 (1986).
[43] Yu. Manin, *Phys. Lett.* **172B**, 184 (1986).
[44] E. Verlinde and H. Verlinde, *Nucl. Phys.* **B288**, 357 (1987).
[45] K. Miki, *Nucl. Phys.* **B291**, 349 (1987).
[46] G. Moore, J. Harris, P. Nelson, and I. Singer, *Phys. Lett.* **178B**, 167 (1986).

[47] M. A. Namazie, K. S. Narain, and M. H. Sarmadi, *Phys. Lett.* **177B**, 329 (1986).
[48] A. Belavin, V. Knishnik, A. Morozov, and A. Perelomov, *Phys. Lett.* **177B**, 324 (1986).
[49] G. Moore, *Phys. Lett.* **176B**, 369 (1986).
[50] A. Kato, Y. Matso, and S. Odake, *Phys. Lett.* **179B**, 241 (1986).
[51] H. Sonoda, *Phys. Lett.* **178B**, 390 (1986).
[52] O. Lechtenfeld, CCNY preprint (1987).
[53] E. Verlinde and H. Verlinde, *Phys. Lett.* **192B**, 95 (1987).
[54] A. Morozov, *Phys. Lett.* **184B**, 171, 177 (1987).
[55] F. Gliozzi, *Phys. Lett.* **194B**, 30 (1987).
[56] M. Bonini and R. Iengo, *Phys. Lett.* **191B**, 56 (1987).
[57] A. Restuccia and J. G. Taylor, *Phys. Lett.* **187B**, 267, 273 (1987).
[58] F. Steiner, *Phys. Lett.* **188B**, 447 (1987).
[59] H. Sonoda, *Phys. Lett.* **184B**, 336 (1987).
[60] A. Parkes, *Phys. Lett.* **184B**, 19 (1987).
[61] A. Morozov and A. Perelomov, *Phys. Lett.* **183B**, 296 (1987).
[62] J. J. Atick, G. Moore, and A. Sen, *Nucl. Phys.* **B308**, 1 (1988).
[63] S. Giddings and S. Wolpert, *Comm. Math. Phys.* **109**, 177 (1987).
[64] E. D'Hoker and S. B. Giddings, *Nucl. Phys.* **B291**, 90 (1987).

# Part II

# Second Quantization and the Search for Geometry

CHAPTER 6

# Light Cone Field Theory

## §6.1. Why String Field Theory?

In Part I, we saw that the development of the first quantized theory often appeared disjoint and seemingly random, appealing to rules of thumb and folklore that often seemed quite arbitrary. The choice of vertex functions, the measure of integration, the counting of diagrams, etc. all were inserted by hand. In summary, the first quantized theory suffers from several important defects:

(1) The interactions must be introduced ad hoc, without any rigorous, overall motivation.
(2) Unitarity cannot easily be shown in the first quantized approach. There is no Hermitian Hamiltonian from which we can derive the interacting theory.
(3) The theory is necessarily perturbative and on-shell, which makes difficult the calculation of nonperturbative effects.

Most important, the first quantized string theory is unsuitable for a calculation of dynamical symmetry breaking. Unfortunately, to all orders in the coupling constant, the dimension of space–time seems to be stable. This means, of course, that there is little hope within the first quantized approach of calculating dynamical breaking of 26 or 10 dimensions down to 4 dimensions. Thus, it is doubtful that rigorous phenomenology can be done within the first quantized framework. The first quantized theory, although it can yield virtually tens of thousands of classical solutions, cannot select out the true vacuum from among the classical vacua that the model can exhibit.

Thus, in Part II we now turn to the *field theory of strings* [1–5], which has the greatest promise of yielding a nonperturbative formalism in which the true vacuum may be found.

At first, the second quantized approach seems to be totally redundant to the usual first quantized approach. Perturbatively, we simply reproduce the same diagrams as the first quantized approach. However, there are several important advantages to the second quantized field theory:

(1) Interactions are introduced through a new gauge group whose elements close on interacting strings. There is now a group theoretical motivation for introducing the interactions of string theory.
(2) The theory is manifestly unitary because the Hamiltonian is Hermitian. All weights of perturbative diagrams are fixed at the very beginning.
(3) Most important, we have, in principle, a method for calculating dynamical effects in the theory.

Historically, it was once thought that a field theory of strings was impossible because it would violate a host of fundamental and cherished principles of quantum mechanics. Specifically:

(1) A field theory of strings would be a nonlocal theory riddled with fatal problems, such as the violation of causality.
(2) A field theory of strings would necessarily be off-shell, yet the crucial properties of the Veneziano model, such as cyclic symmetry, only hold on-shell. Thus, a field theory of strings would not reproduce the dual model.
(3) A field theory of strings would not be both Lorentz invariant and unitary at the same time. This is because a quantization program for strings does not exist that is both Lorentz invariant and unitary off-shell. For the first quantized theory, this is not important because the theory is on-shell. But the field theory of strings is necessarily off-shell and hence must violate either Lorentz invariance or unitarity.
(4) Most important, a field theory of strings would be plagued with violations of unitarity because of severe problems with overcounting of perturbation diagrams. Field theories add the sum of $s$ and $t$-channel poles separately, which violates duality.

Fortunately, a field theory of strings that answers each of these four objections is, indeed, possible.

First, string field theory does not violate causality because the interactions, such as the breaking of a string, take place instantaneously. Furthermore, the information concerning the change in the topology of the string travels along the string at or less than the speed of light. In other words, string theory is multilocal. *Thus, string field theory is the only known nonlocal field theory consistent with the principles of quantum mechanics.*

Second, string field theory generates Green's functions that necessarily violate some of the important properties of the Veneziano model. But this is irrelevant, because these Green's functions correctly reproduce the Veneziano model on-shell, and only on-shell matrix elements can be measured.

Third, the BRST method explicitly breaks unitarity off-shell with Faddeev–

Popov ghosts. This makes no difference, however, because the Faddeev–Popov ghosts cancel with the unitary ghosts on-shell. Similarly, the light cone method maintains unitarity but breaks Lorentz invariance off-shell. The point is that on-shell, the theory is both Lorentz invariant and unitary, so the theory is still well defined.

Fourth, string field theory actually breaks duality off-shell, but this makes no difference on-shell. For example, in light cone string field theory we sum separately over $s$- and $t$-channel poles in the Feynman series:

$$A = \sum_I \frac{A_I}{s - M_I^2} + \sum_J \frac{A_J}{t - M_J^2} + \cdots \tag{6.1.1}$$

The field theory of strings solves the problem of double counting by breaking up the amplitude into its separate $t$- and $s$-channel parts, consistent with a string interpretation (see Fig. 6.1). Thus, *string field theory breaks manifest duality*, but recovers it at the level of the $S$-matrix. Although string field theory explicitly solves the problem of unitarity, it does so at the price of breaking manifest duality, which is recovered only at the end when we sum over all Feynman diagrams.

For closed strings, we also find that the individual Feynman diagrams break manifest modular invariance. Only the sum is modular invariant. Thus, the light cone field theory is manifestly unitary (because the Hamiltonian is manifestly Hermitian) but the price we pay is that we break manifest modular invariance:

First quantization (modular invariance) → Second quantization (unitarity)

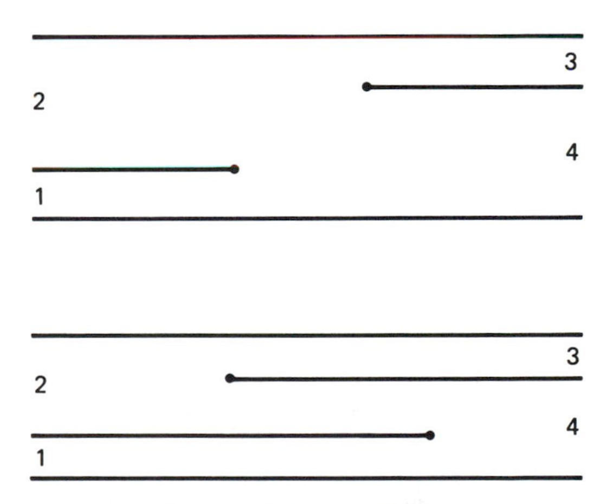


Figure 6.1. Surfaces generated by the light cone field theory. As in any field theory, we now sum over $s$- and $t$-channel graphs separately, breaking duality. Only the sum of these two graphs is dual. Thus, the light cone field theory of strings solves the problem of double counting: duality is explicitly broken for each diagram. Only the sum (the $S$-matrix) is dual.

We will begin a discussion of the second quantized theory starting with the light cone formalism [1], because this retraces the historical development of the theory and because it is the most fully developed formalism. However, because the light cone gauge is a gauge-fixed formalism, in which all the local gauge degrees of freedom have been explicitly removed, there is no trace of the elegant group theoretical formalism from which string field theory can be derived. Thus, for pedagogical reasons we begin our discussion with the light cone field theory, keeping in mind that only a true geometric theory, like the one proposed in Chapter 8, can derive the entire theory from first principles.

We begin by once again discussing the field theory of point particles. We will trace how Feynman derived the Schrödinger equation from the first quantized theory of point particles.

## §6.2. Deriving Point Particle Field Theory

In Chapter 1, we started with the path integral for a point particle moving from a point $x_i$ to $x_j$. Associated with each path connecting these two points was a phase factor $e^{is}$. The fundamental postulate of quantum mechanics is that the probability amplitude of a particle moving between these two points is the sum over all possible phases associated with each path. After a Wick rotation, we have

$$\Delta_{ij} = \int_{x_i}^{x_j} Dx\, e^{-S} \tag{6.2.1}$$

where

$$S = \int_{t_i}^{t_j} dt(\tfrac{1}{2}mv_i^2 - V(x)) \tag{6.2.2}$$

The evolution of a quantum mechanical wave, by assumption, obeys Huygens' principle:

$$\psi(x_j, t_j) = \int_{-\infty}^{\infty} \Delta(x_j, t_j; x_i, t_i)\psi(x_i, t_i)\, dx_i \tag{6.2.3}$$

Now we would like to calculate the variation of this wave functional after a small displacement $\varepsilon$ in time. Earlier, we found that the propagator was equal to

$$\Delta_{ij} = \left[\frac{m}{2\pi(t_j - t_i)}\right]^{1/2} \exp\frac{\frac{1}{2}im(x_j - x_i)^2}{t_j - t_i} \tag{6.2.4}$$

For small time intervals $\varepsilon = t_j - t_i$, we find, therefore,

$$\psi(x, t + \varepsilon) = \int_{-\infty}^{\infty} A^{-1} \exp\left\{\frac{im(x - y)^2}{2\varepsilon}\right\}\psi(y, t)\, dy \tag{6.2.5}$$

where $A$ is a normalization constant. Remember that the time interval $\varepsilon$ is very small, but the separation between $x$ and $y$ need not be small.

We want to keep only terms of order $\varepsilon$. If we let $y = x + \eta$, where $\eta$ is not necessarily a small number, the integral becomes

$$\psi(x, t + \varepsilon) = \int_{-\infty}^{\infty} A^{-1} e^{im\eta^2/2\varepsilon} \psi(x + \eta, t)\, d\eta \tag{6.2.6}$$

It is important to notice that there is no limit, in principle, to the size of $\eta$. However, in the functional integral, keeping terms of order $\varepsilon$ will necessarily restrict us to terms second order in $\eta$. Now let us power expand the left-hand side in terms of $\varepsilon$ and the right-hand side in terms of $\eta$:

$$\psi(x, t) + \varepsilon \frac{\partial \psi}{\partial t} = \int_{-\infty}^{\infty} A^{-1} e^{im\eta^2/2\varepsilon} \times \left\{ \psi(x, t) + \eta \frac{\partial \psi}{\partial x} + \tfrac{1}{2}\eta^2 \frac{\partial^2 \psi}{\partial x^2} \right\} d\eta \tag{6.2.7}$$

This integral can be explicitly performed. First, the integration constant can be determined to be

$$A = \left( \frac{2\pi i \varepsilon}{m} \right)^{1/2} \tag{6.2.8}$$

Writing out the terms in the right-hand side, we notice that the only terms that survive the Gaussian integration are those for which $\eta$ appears with an even power in the integrand. Finally, we are left with

$$i \frac{\partial \psi}{\partial t} = -\frac{1}{2m} \frac{\partial^2 \psi}{\partial x^2} \tag{6.2.9}$$

Thus, we have now *derived the Schrödinger equation*, starting only from the assumption that $L = \frac{1}{2} m \dot{x}_i^2$ and the basic assumptions of quantum mechanics. If we include the effects of a potential term and extend the expression to all three spatial directions, the derivation is basically unchanged, and we arrive at

$$i \frac{\partial \psi}{\partial t} = -\frac{1}{2m} \nabla^2 \psi + V(x) \psi \tag{6.2.10}$$

We can also dispense with external potentials entirely and introduce $\psi^3$- or $\psi^4$-type interactions directly into the action. This is the second quantized analog of summing over Y-shaped or X-shaped topologies in the first quantized point particle theory.

We have followed Feynman's original derivation of the Schrödinger equation, based on calculating the time evolution of the wave. However, there is yet another way in which we can make the transition from the first to the second quantized formalism in which the derivation is more direct. Let us now turn to the path integral action and show that we can make the transition to the second quantized formalism starting from the Green's functions themselves, rather than appealing to the equations of motion.

In Chapter 1, when we made the transition from the Hamiltonian to the Lagrangian formalism in the first quantized point particle theory, we inserted an infinite set of intermediate states, the eigenvectors of the $x$-coordinate:

$$1 = |x_i, t_i\rangle \int Dx_i \langle x_i, t_i| \tag{6.2.11}$$

into the expression

$$\Delta_{ij} = \langle x_i, t_i | x_j, t_j \rangle \tag{6.2.12}$$

at each intermediate point between $x_i$ and $x_j$. This allowed us to make the transition between the Hamiltonian and the Lagrangian approaches.

Now we wish to replace this with the integration over a complete set of *second quantized fields*:

$$1 = |\psi\rangle \int D^2\psi \, e^{-\langle\psi|\psi\rangle} \langle\psi| \tag{6.2.13}$$

Following (1.8.21), we define

$$\begin{aligned} \psi(x) &= \langle x|\psi\rangle \\ \psi^*(x) &= \langle\psi|x\rangle \\ D^2\psi &= \prod_x d\psi(x)\, d\psi^*(x) \end{aligned} \tag{6.2.14}$$

Thus, at each intermediate point between the initial and final states of a point particle, we are now going to introduce an infinite set of intermediate functional states $|\psi\rangle$ rather than $x$-eigenstates $|x\rangle$. The easiest way to check the validity of (6.2.13) is to take the following matrix element and insert a complete set of intermediate string states:

$$\begin{aligned} \delta(x-y) &= \langle x|y\rangle \\ &= \langle x|1|y\rangle \\ &= \langle x|\psi\rangle \int D^2\psi \exp\left\{-\int \langle\psi|z\rangle \int Dz \langle z|\psi\rangle\right\} \langle\psi|y\rangle \\ &= \int D^2\psi\, \psi^*(x)\psi(y) \exp - \left\{\int Dz\, \psi^*(z)\psi(z)\right\} \end{aligned} \tag{6.2.15}$$

(where as usual we drop overall normalization factors in the path integral). We treat $\psi(x)$ as an element of a column vector labeled by discrete elements $x$. Discretizing the expression, we find

$$\delta_{xy} \sim \int \prod_z d\psi_z^*\, d\psi_z\, \psi_x^* \psi_y \exp - \left\{\sum_z \psi_z^* \psi_z\right\} \tag{6.2.16}$$

This last relation is just (1.7.11). Thus, we have now shown that we can move interchangeably from first quantized basis elements $|x\rangle$ to second quantized basis elements $|\psi\rangle$, such that $\psi(x) = \langle x|\psi\rangle$.

Now let us begin our derivation of the Green's function for the Schrödinger equation entirely in terms of second quantized field functionals, without recourse to the equations of motion or Huygens' principle. We will insert the following identity at every intermediate point along the path:

$$1 = |\psi_1\rangle \int D^2\psi_1 D^2\psi_2 \exp\{-\psi_1^*(\psi_2 - \psi_1) - \psi_2^*\psi_1\} \langle\psi_2| \tag{6.2.17}$$

(We can prove this identity by functionally integrating the expression over $\psi_1^*$, which then reduces to the completeness expression written in terms of the $\psi_2$ field.)

We insert the above expression between two infinitesimally close position eigenstates in (6.2.12):

$$\begin{aligned}
\Delta_{12} &= \langle x_1|e^{-iH\delta t}|x_2\rangle \\
&= \langle x_1|x_2\rangle - i\langle x_1|H\delta t|x_2\rangle + \cdots \\
&= \int D^2\psi_{12}\langle x_1|\psi_1\rangle\langle\psi_2|x_2\rangle \exp[-\psi_1^*(\psi_2 - \psi_1) - \psi_2^*\psi_1] \\
&\quad - i\int D^2\psi_{1234}\langle x_1|\psi_1\rangle\langle\psi_2|H\delta t|\psi_3\rangle\langle\psi_4|x_2\rangle \\
&\quad \times \exp[-\psi_1^*(\psi_2 - \psi_1) - \psi_2^*\psi_1 - \psi_3^*(\psi_4 - \psi_3) - \psi_4^*\psi_3] + \cdots
\end{aligned} \tag{6.2.18}$$

where the subscript 12 or 1234 simply means the product of two or four of these functional differentials. We now take the limit of small time separations:

$$\sum_{i=1}^{N} \psi_i^*(\psi_{i+1} - \psi_i) \to \int_{t_1}^{t_N} \psi^*\dot{\psi}\, dt \tag{6.2.19}$$

After taking the limit, we find

$$\Delta_{1N} = \int D^2\psi\ \psi(x_1)\psi^*(x_N) \exp i\int dt\ \psi^*(i\partial_t - H)\psi \tag{6.2.20}$$

Thus, our new Lagrangian is

$$L = \psi^*\left\{i\frac{\partial}{\partial t} - H\right\}\psi \tag{6.2.21}$$

which is the Lagrangian for the Schrödinger wave equation. Therefore we have derived the Schrödinger equation given the postulates of quantum mechanics and the classical first quantized formalism for a point particle, without making use of the equations of motion or Huygen's principle.

The main reason why we went through this analysis for the point particle is that we will now repeat the same steps for the light cone and the BRST field theories. Surprisingly enough, we will find that this entire functional apparatus carries over directly into the string field theory for the light cone and the BRST actions. Only when we finally reach the geometric field theory of Chapter 8 will we start from first principles and postulate an entirely new gauge group.

## §6.3. Light Cone Field Theory

As in the point particle case, we now want to make the transition from the first quantized to the second quantized string formalism, using (1.8.21):

$$\langle X|\Phi\rangle = \Phi(X) \tag{6.3.1}$$

We must be careful to state that the field functional is not a local function of $X(\sigma)$ at a specific point $\sigma$ on a string. By contrast, it is actually a *multilocal* functional defined at all points along the string. If we discretize the string into a series of points:

$$\sigma_1, \sigma_2, \sigma_3, \ldots, \sigma_N \tag{6.3.2}$$

then the field functional becomes

$$\Phi(X) = \Phi[X_i(\sigma_1), X_i(\sigma_2), X_i(\sigma_3), \ldots, X_i(\sigma_N)] \tag{6.3.3}$$

and we take the limit $N \to \infty$. *Thus, the string functional is simultaneously a function of every point along the string* [1].

Let us begin by defining the Hilbert space of string excitations. It is convenient to power expand the field $|\Phi\rangle$ in terms of harmonic oscillators, in which case we have

$$|\Phi\rangle = \phi(x)|0\rangle + A_i a^i_{-1}|0\rangle + h_{ij}a^i_{-1}a^j_{-1}|0\rangle + \cdots \tag{6.3.4}$$

We see immediately the difference between the first and the second quantized formalism even at the free level. In the first quantized formalism the basic object is $X_\mu$, which represents just one possible configuration of the string. In the second quantized formalism we are dealing with the field functional $\Phi$, *which is simultaneously a composite sum of all possible string configurations.*

To make things concrete, let us now introduce a specific representation of the $|X\rangle$ eigenstates in terms of harmonic oscillators. We want the string variable $X$ to act on the string eigenvector $|X\rangle$ such that we reproduce (2.2.9):

$$X_{i,n}|X\rangle = \frac{i}{2}(a_{i,n} - a_{i,-n})|X\rangle \tag{6.3.5}$$

For the moment, let us make the assumption that the eigenvector $|X\rangle$ can be written as

$$\begin{aligned} |X\rangle &= \Omega|0\rangle \\ &= k\prod_{i,n}\exp(aX^2_{i,n} + bX_{i,n}a^\dagger_{i,n} + ca^\dagger_{i,n}a^\dagger_{i,n})|0\rangle \end{aligned} \tag{6.3.6}$$

where $a$, $b$, $c$, and $k$ are arbitrary constants.

Let us determine these constants $a$, $b$, and $c$ by acting on this state vector by the operator $X$:

$$\begin{aligned} X_{i,n}|X\rangle &= \Omega\Omega^{-1}\tfrac{1}{2}i(a_{i,n} - a^\dagger_{i,n})\Omega|0\rangle \\ &= \Omega(\tfrac{1}{2}i(bX_{i,n} + 2ca^\dagger_{i,n} - a^\dagger_{i,n}))|0\rangle \end{aligned} \tag{6.3.7}$$

Thus, we find

$$\begin{aligned} b &= -2i \\ c &= \tfrac{1}{2} \end{aligned} \tag{6.3.8}$$

Second, we want the operator $\partial_X$ to have the correct commutation relations with $X$. If we operate on the state vector, we find

$$\frac{\delta}{\delta X_{i,n}}|X\rangle = (2aX_{i,n} + ba_{i,n}^{\dagger})|X\rangle \tag{6.3.9}$$

Because we want

$$\frac{\delta}{\delta X_{i,n}}|X\rangle = -i(a_{i,n} + a_{i,n}^{\dagger})|X\rangle \tag{6.3.10}$$

this fixes the coefficient to be $a = -1$. Our final result for the state vector is therefore

$$|X\rangle = k \prod_{i,n} \exp(-X_{i,n}^2 - 2iX_{i,n}a_{i,n}^{\dagger} + \tfrac{1}{2}a_{i,n}^{\dagger}a_{i,n}^{\dagger})|0\rangle \tag{6.3.11}$$

where $k$ is a normalization constant. This expansion, in turn, allows us to calculate the field function as a power expansion in Hermite polynomials. For example, we can calculate

$$\begin{aligned} \langle 0|X\rangle &= k \prod_{i,n} e^{-X_{i,n}^2} \\ \langle 0|a_{i,n}|X\rangle &= k(-2iX_{i,n}) \prod_{j,m} e^{-X_{j,m}^2} \end{aligned} \tag{6.3.12}$$

This, in turn, allows us to rewrite the original field functional (6.3.4) in terms of Hermite polynomials, using (6.3.11):

$$\begin{aligned} \langle X|\Phi\rangle = \phi(x) \prod_{i,n} H_0(X_{i,n}) \\ + A_i(x)H_1(X_{i,1}) \prod_{\substack{i,n \\ n\neq 1}} H_0(X_{i,n}) + \cdots \end{aligned} \tag{6.3.13}$$

(where we have taken a different normalization of the Hermite polynomials than usual).

Let us now quantize this field functional, following closely the steps used in ordinary point particle field theory in eqs. (1.8.9) to (1.8.12). In general, we can power expand the field function in an arbitrary series of orthogonal polynomials. Let us choose an arbitrary element of the Fock space, which is a product of creation operators raised to some power. Let the set of integers $n_l^i$ denote the number of creation oscillators in a particular state with Lorentz index $i$ and level number $l$. Thus, an arbitrary state of the Hilbert space can be written as

$$|\{n\}\rangle = c \prod_{i,l} (a_{i,-l})^{n_l^i}|0\rangle \tag{6.3.14}$$

where this state is a product of an arbitrary sequence of creation oscillators, and we choose the normalization constant to be

$$c = (\langle\{n\}|\{n\}\rangle)^{-1/2} \tag{6.3.15}$$

such that

$$\langle\{n\}|\{m\}\rangle = \delta_{\{n\},\{m\}} \tag{6.3.16}$$

Thus, these states can be normalized so that they form an orthonormal basis. In fact, the basis is complete:

$$1 = \sum_{\{n\}} |\{n\}\rangle\langle\{n\}| \tag{6.3.17}$$

The matrix element of $|\{n\}\rangle$ in (6.3.14) with the string eigenstate $|X\rangle$ in (6.3.11) is just a product of Hermite polynomials:

$$\langle X|\{n\}\rangle = H_{\{n\}}(X)e^{-\sum_{i,n} X_{i,n}^2} \tag{6.3.18}$$

Let us now explore some useful properties of $|\{n\}\rangle$. If we power expand our field functional in terms of this orthonormal basis, we can generalize (1.8.12):

$$|\Phi\rangle = \sum_{\{n\}} \phi_{\{n\}}|\{n\}\rangle \tag{6.3.19}$$

to the following:

$$\langle X|\Phi\rangle = \sum_{\{n\}} \phi_{\{n\}} H_{\{n\}}(X)e^{-\sum_{i,n} X_{i,n}^2}$$

The inner product between two such field functionals is easily calculated:

$$\langle\Psi|\Phi\rangle = \sum_{\{n\}} \psi_{\{n\}}^* \phi_{\{n\}} \tag{6.3.20}$$

We can also show

$$\delta_{\{n\},\{m\}} = \int \phi_{\{n\}}^* \phi_{\{m\}} \exp\left(-\sum_{\{n\}} |\phi_{\{n\}}|^2\right) D^2\phi \tag{6.3.21}$$

where the measure of integration for this space is given by

$$D^2\phi = \prod_{\{n\}} d\phi_{\{n\}}\, d\phi_{\{n\}}^* \tag{6.3.22}$$

Using these identities, we can now show that the number 1 can be written in a form analogous to (6.2.13):

$$1 = \int |\Phi\rangle\, D^2\Phi \langle\Phi| \exp(-\langle\Phi|\Phi\rangle) \tag{6.3.23}$$

The above equation can be proved by power expanding each of the various field functionals via (6.3.19) and explicitly performing the integration over the coefficients. Thus,

$$
\begin{aligned}
1 &= \left[\sum_{\{n\}} \int \phi_{\{n\}} |\{n\}\rangle\right] \prod_{\{m\}} d^2\phi_{\{m\}} \left[\sum_{\{p\}} \phi^*_{\{p\}} \langle\{p\}|\right] e^{-\langle\Phi|\Phi\rangle} \\
&= \sum_{\{n\},\{p\},\{m\}} \phi_{\{n\}} |\{n\}\rangle\langle\{p\}| \phi^*_{\{p\}} \int d^2\phi_{\{m\}} e^{-\sum_{\{q\}} |\phi_{\{q\}}|^2} \\
&= \sum_{\{n\},\{p\}} |\{n\}\rangle\langle\{p\}| \delta_{\{n\},\{p\}}
\end{aligned}
\tag{6.3.24}
$$

Hence, we are justified in taking this expansion of the number 1 given by (6.3.23).

The matrix element between two string states $|X\rangle$ and $|Y\rangle$ can also be written in terms of $|\Phi\rangle$.

$$
\begin{aligned}
\langle X|Y\rangle &= \prod_{i=1}^{24} \prod_{0\le\sigma\le\pi} \delta(X_i(\sigma) - Y_i(\sigma)) \\
&= \langle X|\Phi\rangle \int D^2\Phi \langle\Phi|Y\rangle \exp\left(-\langle\Phi|Z\rangle \int DZ\langle Z|\Phi\rangle\right) \\
&= \int D^2\Phi\, \Phi^*(X)\Phi(Y) e^{-\int DZ\, \Phi^*(Z)\Phi(Z)}
\end{aligned}
\tag{6.3.25}
$$

Finally, we need to generalize (6.2.17) before deriving the Schrödinger equation for strings:

$$
1 = \int |\Phi_1\rangle\, D^2\Phi_1\, D^2\Phi_2 e^{-\langle\Phi_1|\Phi_2-\Phi_1\rangle - \langle\Phi_2|\Phi_1\rangle} \langle\Phi_2| \tag{6.3.26}
$$

As before, this identity is most easily proved by first functionally integrating over $\Phi_1^*$.

Now that we have all the identities in place, let us first consider the matrix element between two string states that are infinitesimally close:

$$
\langle X_1| e^{-iHd\tau} |X_2\rangle \tag{6.3.27}
$$

We can reexpress this matrix element either in terms of the first quantized picture, where we insert a complete set of momentum eigenstates, or in terms of the second quantized picture, where we insert a complete set of field functionals. Thus,

$$
\begin{cases}
\text{1st quantization:} \quad 1 = |P\rangle \int DP\langle P| \\
\text{2nd quantization:} \quad 1 = |\Phi\rangle \int D^2\Phi\, e^{-\langle\Phi|\Phi\rangle} \langle\Phi|
\end{cases}
\tag{6.3.28}
$$

Retracing our steps in Section 1.3, we inserted the momentum eigenstates and then derived the infinitesimal Green's function for the first quantized formalism:

$$\begin{aligned}\langle X_1|e^{-iH d\tau}|X_2\rangle &= \langle X_1|P\rangle \int dP \langle P|e^{-iH d\tau}|X_2\rangle \\ &= \int DP\, e^{-iH(X,P)d\tau} e^{i\int d\sigma P(X_1 - X_2)} \\ &= \int DP\, e^{i\int d\sigma P\dot{X}\, d\tau} e^{-iH(X,P)d\tau} \end{aligned} \tag{6.3.29}$$

where we used the fact that $\langle P|X\rangle \sim e^{i\int d\sigma P_\mu X^\mu}$. Then we inserted an infinite number of these first quantized intermediate states between every infinitesimal interval between the initial and final points:

$$\begin{aligned}\langle X_1|e^{-iH\tau}|X_N\rangle &= \int_{X_1}^{X_N} DX\, DP\, e^{i\int_{\tau_1}^{\tau_N} d\tau\, d\sigma (P\dot{X} - H)} \\ &= \int_{X_1}^{X_N} DX\, DP\, e^{i\int_{\tau_1}^{\tau_N} d\tau\, d\sigma\, L} \end{aligned} \tag{6.3.30}$$

where $L = P\dot{X} - H$.

Thus, we could go back and forth between Hamiltonian and Lagrangian formalisms in the first quantized string theory.

Now let us repeat all our steps, inserting a complete set of second quantized field functionals into our action. Let us start with an infinitesimal transition amplitude:

$$\begin{aligned}\Delta_{12} &= \langle X_1|e^{-iH\, d\tau}|X_2\rangle \\ &= \langle X_1|X_2\rangle - i\langle X_1|H\, d\tau|X_2\rangle + \cdots \\ &= \int D^2\Phi_{12}\langle X_1|\Phi_1\rangle\langle\Phi_2|X_2\rangle \exp\{-\langle\Phi_1|\Phi_2 - \Phi_1\rangle - \langle\Phi_2|\Phi_1\rangle\} \\ &\quad - i\int D^2\Phi_{1234}\langle X_1|\Phi_1\rangle\langle\Phi_2|H\, d\tau|\Phi_3\rangle\langle\Phi_4|X_2\rangle \\ &\quad \times \exp\left\{-\sum_{i=1,3}\langle\Phi_i|\Phi_{i+1} - \Phi_i\rangle - \langle\Phi_{i+1}|\Phi_i\rangle\right\} + \cdots \end{aligned} \tag{6.3.31}$$

As before, we now take the limit as

$$\sum_i \langle\Phi_i|\Phi_{i+1} - \Phi_i\rangle \to \int d\tau \langle\Phi|\dot{\Phi}\rangle$$

Next, we insert this complete set of intermediate field states between all infinitesimal intervals between the initial string configuration and the final string configuration. Then the matrix element becomes

$$\Delta_{1N} = \int \Phi^*(X_1)\Phi(X_N)\, D^2\Phi \exp i\int d\tau(\langle\Phi|i\partial_\tau - H|\Phi\rangle) \tag{6.3.32}$$

This is our main result for this section. Therefore, our second quantized action is [1]

$$L = \Phi^*(i\partial_\tau - H)\Phi \tag{6.3.33}$$

In summary, we have now shown the equivalence of the first and second quantized formalisms for the free light cone string theory. We have shown that we can write, using (6.3.28), the Green's function in either first or second quantized language:

$$\begin{aligned}\Delta_{ij} &= \int_{X_i}^{X_j} DX \exp i \int_{\tau_i}^{\tau_j} d\tau\, L(\tau) \\ &= \int D^2\Phi\, \Phi^*(X_i)\Phi(X_j) \exp i \int d\tau \int_{X_i}^{X_j} L(\Phi)\, DX\end{aligned} \tag{6.3.34}$$

where

$$L(\tau) = \frac{1}{2\pi}\int_0^\pi d\sigma\{\dot{X}_i^2 - X_i'^2\} \tag{6.3.35}$$

and where

$$H = \frac{\pi}{2}\int_0^\pi d\sigma\left(P_i^2 + \frac{X_i'^2}{\pi^2}\right) \tag{6.3.36}$$

We derived this second quantized field theory action from the requirement that it reproduce the Green's function for the propagation of a string. This shows the strong parallel between first and second quantization for the case of free strings. (This parallel, however, will be drastically broken when we discuss the interactions.)

Let us now quantize our action. The equation of motion derived from our action is

$$(i\partial_\tau - H)\Phi(X) = 0 \tag{6.3.37}$$

The energy that corresponds to a particular basis state is given by:

$$\begin{aligned}H|\{n\}\rangle &= E_{\{n\}}|\{n\}\rangle \\ E_{\{n\}} &= \frac{1}{2p^+}\sum_{i,l} l n_l^i + \alpha' p_i^2\end{aligned} \tag{6.3.38}$$

We will find it convenient to take the Fourier transform of these coefficients with respect to $x^-$:

$$\phi_{\{n\}}(x^+, x^-, X_i) \to \phi_{p^+,\{n\}}(\tau, X_i) \tag{6.3.39}$$

Thus, the coefficients must satisfy the equations of motion:

$$\phi_{p^+,\{n\}}(\tau, X_i) = \int dp_i\, e^{i(\mathbf{p}\cdot\mathbf{x} - E_{\{n\}}\tau)} A_{p^+,p_i,\{n\}} \tag{6.3.40}$$

Notice that we have introduced a new operator $A$; which is the creation or annihilation operator associated with the state $\{n\}$. It is crucial to notice that $A$ is not the same thing as $a_n^\dagger$ introduced in Chapter 1. $a_n^\dagger$ creates a single vibratory mode of the string, while $A$ creates or destroys an element of the Hilbert space spanned by all possible products of $a_n^\dagger$. $A$ creates or destroys states defined on an infinite-component field theory. It satisfies the commutation relations:

$$[A_{p^+,p_i,\{n\}}, A^\dagger_{q^+,q_j,\{m\}}] = \delta(p^+ - q^+)\delta(p_i - q_i)\delta_{\{n\},\{m\}} \tag{6.3.41}$$

Combining everything, we now have an expansion of the field functional in terms of plane waves:

$$|\Phi\rangle = \sum_{\{n\}} \int \prod_i dp_i \, A_{p^+,p_i,\{n\}} e^{i(\mathbf{p}\cdot\mathbf{x} - E_{\{n\}}\tau)} |\{n\}\rangle \tag{6.3.42}$$

We can also express this power expansion in the $X$ basis (see (1.8.21)):

$$\begin{aligned} \langle X|\Phi\rangle &= \Phi_{p^+}(X) \\ &= \sum_{\{n\}} \int \prod_i dp_i \, A_{p^+,p_i,\{n\}} H_{\{n\}}(X) e^{-\sum_{i,n} X_{i,n}^2} e^{i(\mathbf{p}\cdot\mathbf{x} - E_{\{n\}}\tau)} \end{aligned} \tag{6.3.43}$$

We are now in a position to derive the canonical commutation relations in the second quantized string field theory. Because we have power expanded the field in terms of orthonormal polynomials, it is easy to show that [1]

$$[\Phi_{p^+}(X, \tau), \Phi^\dagger_{q^+}(Y, \tau)] = \delta(p^+ - q^+) \prod_i \prod_\sigma \delta(X_i(\sigma) - Y_i(\sigma)) \tag{6.3.44}$$

Let us denote the vacuum of the $A$ oscillators by $|0\rangle\rangle$. Notice that this vacuum state is the product of the vacua of *all* the higher spin fields contained within $\Phi$. This state is the vacuum of the infinite-component field theory and hence has nothing to do with $|0\rangle$. Using the previous identities, we find an expression for the Green's function in terms of the field functionals:

$$\begin{aligned} \Delta_{12} &= \langle\langle 0|\Phi_{p^+}(X_1, \tau_1)\Phi^*_{q^+}(X_2, \tau_2)|0\rangle\rangle \\ &= \delta(p^+ - q^+) \int DX \, e^{i\int L \, d\sigma \, d\tau} \prod_\sigma \delta(\mathbf{X}(\sigma, \tau_1) - \mathbf{X}_1(\sigma)) \\ &\quad \times \prod_{\sigma'} \delta(\mathbf{X}(\sigma', \tau_2) - \mathbf{X}_2(\sigma')) \end{aligned} \tag{6.3.45}$$

where the initial and final states are labeled by $X(\sigma, \tau_1)$ and $X(\sigma', \tau_2)$. It is important to notice that we have now expressed the Green's function for the propagation of a free string in both the first and the second quantized language. Thus, at least at the free level, we can go back and forth between these two formalisms. We now have the tools with which we can write an explicit expression for the Green's function for the free string, which generalizes the expression (6.2.4) for point particles:

$$\Delta_{12} = \delta(p^+ - q^+) \prod_{k=1}^{\infty} \left[ \frac{k}{\pi \sinh(kT/\alpha)} \right]^{1/2(D-2)}$$
$$\times \left( \frac{\alpha}{4\pi T} \right)^{1/2(D-2)} \exp\left\{ -\frac{\alpha}{4T}(\mathbf{X}_1^2 - \mathbf{X}_2^2) \right\}$$
$$\times \exp\left\{ k \sinh^{-1} \frac{kT}{\alpha} \left[ \cosh \frac{kT}{\alpha} (\mathbf{X}_1^2 + \mathbf{X}_2^2) - 2\mathbf{X}_1 \cdot \mathbf{X}_2 \right] \right\} \qquad (6.3.46)$$

where $T$ is the time interval. Notice that now we have a derivation of the transition function $\Delta_{ij}$ entirely in terms of second quantized field functionals.

In summary, we have derived the free light cone string field action by directly inserting a complete set of intermediate string states at every string configuration between the initial and final string configurations.

## §6.4. Interactions

Now we turn our attention to the question of interactions. Our derivation of the free action from the first quantized theory can now be generalized to calculate the interactions of the second quantized field theory. Again, we will insert the important identity

$$1 = |\Phi\rangle \int D^2\Phi \, e^{-\langle\Phi|\Phi\rangle} \langle\Phi| \qquad (6.4.1)$$

within the path integral in order to extract the vertex function.

Historically, many of the early pioneers in quantum physics, such as Heisenberg and Yukawa, looked into the question of nonlocal field theories and found that such theories violated causality, i.e., interactions could propagate faster than the speed of light. The nonlocal interactions, which can involve two distant points $x_1$ and $x_2$, could transmit information faster than the speed of light, which is forbidden.

Miraculously, string field theory solves this problem. The solution is simple but elegant: string field theory does not violate causality and the laws of quantum mechanics because it is not a nonlocal theory, *it is actually a multilocal theory*. The interactions of the string, in which strings can break or reform, are such that strings break or reform instantaneously and then the vibrations travel down the string at or below light speed. Thus, we do not have a violation of causality.

Because all the symmetries have been extracted out in the light cone gauge, there is no overall guiding principle for the theory. We will, therefore, simply postulate the following principle:

*The only interacting string configurations that are allowed in the action are those that instantaneously change the local topology of strings.*

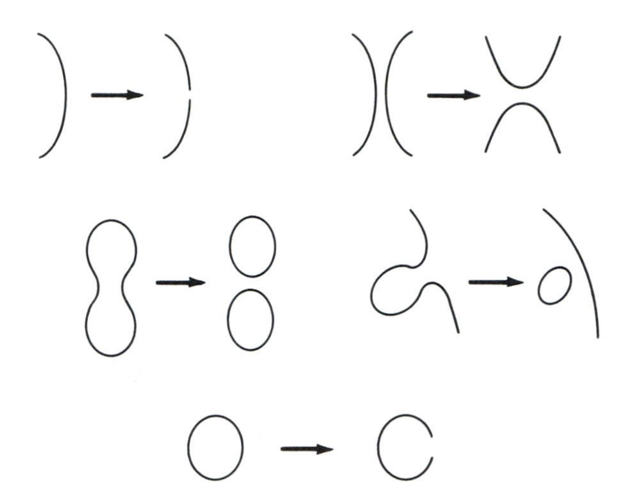

Figure 6.2. The five interactions of the light cone field theory. Open and closed strings can break, fission, and pinch. Notice that each interaction takes place locally along the string. This is the solution to the problem of causality, which is broken in all nonlocal point particle theories. Thus string field theory is the only known field theory based on extended objects that preserves causality.

Although this principle is defined only in the light cone gauge, we will find that it is sufficient to determine all the possible interactions of the field theory. There are only five such local interactions (see Fig. 6.2) that are consistent with this new definition of locality and also the conservation of momentum. (We will choose the parametrization length of the light cone string to be proportional to $p^+$, that is, $\alpha = 2p^+$, and hence momentum conservation implies that the sum over string lengths is conserved.) Thus, we postulate

$$L_I = \sum_{i=1}^{5} L_i \tag{6.4.2}$$

where each term corresponds to a specific interaction for an open string field $\Phi$ and a closed string field $\Psi$. These five interactions, symbolically speaking, can be represented as

$$L_I = \Phi^3 + \Phi^4 + \Psi^3 + \Phi^2\Psi + \Phi\Psi \tag{6.4.3}$$

We will write down explicitly what some of these interactions are, consistent with the condition of locality. Once we have written down specific representations for these five interactions, we must check that they reproduce the known results from the first quantized theory.

The simplest interaction involves the breaking of a string into two smaller pieces. To enforce the condition of locality, the string can break only at a single interior point along the string. The disturbances from this rupture should later spread along the string at velocities less than or equal to the speed of light. Thus, we are led to postulate that the points along the string are continuous across the boundary of the interactions. The unique form of the vertex func-

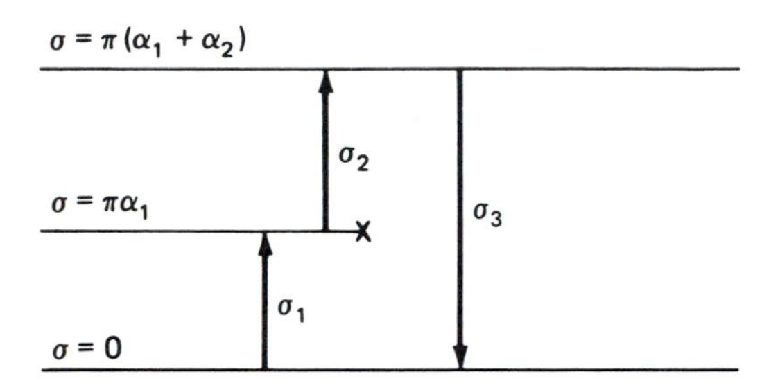


Figure 6.3. Parametrization of the three-string vertex. The parametrization length of each string is given by $\pi\alpha_i$. The sum of all three $\pi\alpha_i$ is equal to zero.

tion, consistent with momentum conservation and locality, is a series of Dirac delta functions that ensure continuity of the three strings. Our vertex is [1]

$$S_3 = \int dp_r^i \delta\left(\sum_{r=1}^{3} p^{+r}\right) \int DX_{123} \Phi^\dagger(X_3) \Phi^\dagger(X_1) \Phi(X_2) \delta_{123} + \text{h.c.} \quad (6.4.4)$$

where

$$\begin{aligned} \delta_{123} &= \prod_{\sigma_3} \delta[X_3(\sigma_3) - \theta(\pi\alpha_1 - \sigma)X_1(\sigma_1) - \theta(\sigma - \pi\alpha_1)X_2(\sigma_2)] \\ DX_{123} &= DX_1 DX_2 DX_3 \end{aligned} \quad (6.4.5)$$

(see Fig. 6.3). We will use the notation

$$\begin{gathered} 0 \leq \sigma \leq \pi(\alpha_1 + \alpha_2) \\ \sigma_1 = \sigma \quad \text{for } 0 \leq \sigma \leq \pi\alpha_1 \\ \sigma_2 = \sigma - \pi\alpha_1 \quad \text{for } \pi\alpha_1 \leq \sigma \leq \pi(\alpha_1 + \alpha_2) \\ \sigma_3 = \pi(\alpha_1 + \alpha_2) - \sigma \quad \text{for } 0 \leq \sigma \leq \pi(\alpha_1 + \alpha_2) \\ \sum_{i=1}^{3} \alpha_i = 0 \end{gathered} \quad (6.4.6)$$

where the parametrization length of each string is given by $\pi\alpha_i$ and where the $X$ variable represents only the transverse modes of the string.

Remarkably, we can actually perform the $DX$ integration in the above vertex function because it is a simple Gaussian. Let us define

$$\Phi_i(X) = \langle X | \Phi_i \rangle \quad (6.4.7)$$

Then we can rewrite (6.4.4)

$$S_3 = \int dp^{+r} \delta\left(\sum_{r=1}^{3} p^{+r}\right) \langle \Phi_1 | \langle \Phi_2 | \langle \Phi_3 | V_{123} \rangle \quad (6.4.8)$$

where we have introduced the vertex function

$$|V_{123}\rangle = \int DX_{123} |X_1\rangle |X_2\rangle |X_3\rangle \delta_{123} \quad (6.4.9)$$

Since the eigenvector $|X\rangle$ in (6.3.11) is a simple Gaussian in $X$, we can explicitly calculate the integration over $DX_{123}$ and obtain an exact formula for the vertex function in terms of harmonic oscillators.

The actual calculation will be more conveniently carried out in the $P$ representation. By taking Fourier transforms, we can easily convert from $X$ eigenstates in (6.3.11) to $P$ eigenstates:

$$|P\rangle = k \prod_{i,n} \exp(-\tfrac{1}{4}P_{i,n}^2 + P_{i,n}a_{i,n}^\dagger - \tfrac{1}{2}a_{i,n}^\dagger a_{i,n}^\dagger)|0\rangle \tag{6.4.10}$$

We easily check that this expression reproduces the correct eigenvector equation (2.2.9):

$$P_{i,n}|P\rangle = (a_{i,n} + a_{i,n}^\dagger)|P\rangle \tag{6.4.11}$$

Let us write down explicit Fourier decompositions for all three string states in the vertex, in both the $X$ and $P$ representations:

$$\begin{cases} P_i^{(r)} = \dfrac{1}{\pi|\alpha_r|}\left(p_i^r + \sum\limits_{n=1}^{\infty} \sqrt{n}p_{i,n}^r \cos\dfrac{n\sigma_r}{\alpha_r}\right)\theta_r \\ X_i^{(r)} = \left(x_i^r + \sum\limits_{n=1}^{\infty} \dfrac{1}{\sqrt{n}} x_{i,n}^r \cos\dfrac{n\sigma_r}{\alpha_r}\right)\theta_r \end{cases} \tag{6.4.12}$$

where

$$\begin{cases} \theta_1 = \theta(\pi\alpha_1 - \sigma) \\ \theta_2 = \theta(\sigma - \pi\alpha_1) \\ \theta_3 = \theta_1 + \theta_2 = 1 \end{cases} \tag{6.4.13}$$

The integral that we would like to perform is given by

$$\begin{aligned} |V_{123}\rangle &= \int DP_{123} \prod_{i=1}^{3} |P_i\rangle \delta_{123} \\ &= \int DP_{123}\delta\left(\sum_{r=1}^{\infty} \theta_r(\sigma_r)P_i^{(r)}(\sigma)\right) \prod_{r=1}^{3} \exp(-\tfrac{1}{4}P_{i,n}^{(r)^2} \\ &\quad + P_{i,n}^{(r)}a_{i,n}^{(r)^\dagger} - \tfrac{1}{2}a_{i,n}^{(r)^\dagger}a_{i,n}^{(r)^\dagger})|0\rangle \end{aligned} \tag{6.4.14}$$

The integral is simple and straightforward because it is a Gaussian. The only complication that we will face is the explicit form of the delta functional $\delta_{123}$ in (6.4.5) written in terms of harmonic oscillators. Let us now extract the Fourier components of the delta functional $\delta_{123}$ in order to perform the functional integration. Let us take the cosine transform of the constraint equation among the various $P$'s. Ordinarily, cosine and sine transforms yield delta functions. Here, because the three strings all have different lengths, we find that the cosine and sine transforms yield matrix equations, not delta functions. In particular, the statement that

$$\sum_{r=1}^{3} \theta_r(\sigma_r)P_i^{(r)}(\sigma_r) = P_i^{(3)}(\sigma_3) + P_i^{(2)}(\sigma_2)\theta_2 + P_i^{(1)}(\sigma_1)\theta_1 = 0 \tag{6.4.15}$$

now becomes, when we take the cosine transform,

$$p_{im}^{(3)} + \sum_{n=1}^{\infty} (A_{mn}^{(1)} p_{n,i}^{(1)} + A_{mn}^{(2)} p_{n,i}^{(2)}) + B_m^{(1)} p_i^{(1)} + B_m^{(2)} p_i^{(2)} = 0 \qquad (6.4.16)$$

where the $A$'s and $B$'s are the various overlap integrals between the cosine functions. Instead of enforcing this identity as an exact equation, this operator equation need only be satisfied when acting on the vertex function. We want the vertex function $|V_{123}\rangle$ to vanish when acted on by the $\delta_{123}$. This means [1, 6, 7]

$$\left(\sum_{r=1}^{3}\sum_{n=1}^{\infty} A_{mn}^{(r)}(a_n^{(r)} + a_{-n}^{(r)}) + B_m \mathbf{P}\right)|V_{123}\rangle = 0$$
$$\left(a_m^{(r)} + a_{-m}^{(r)} - \frac{\alpha_r}{\alpha_3}\sum_{n=1}^{\infty}(C^{-1}A^{(r)T}C)_{mn}(a_n^{(3)} + a_{-n}^{(3)})\right)|V_{123}\rangle = 0 \qquad (6.4.17)$$

where

$$(C)_{mn} = m\delta_{mn}$$
$$\mathbf{P} = 2p_1^+ p_2^i - 2p_2^+ p_1^i \qquad (6.4.18)$$

Thus, the Fourier coefficients are given by taking the Fourier transform of the various cosine modes. Because the three strings all have different lengths, we will in general have nontrivial Fourier coefficients. By explicit construction, we have the following Fourier coefficients:

$$A_{mn}^{(1)} = 2(n/m)^{1/2}(-1)^m \frac{1}{\pi\alpha_1}\int_0^{\pi\alpha_1} \cos\frac{n\sigma}{\alpha_1}\cos\frac{m\sigma}{\alpha_3}\,d\sigma$$
$$= -(2/\pi)(mn)^{1/2}(-1)^{m+n}\frac{\beta\sin(m\pi\beta)}{n^2 - m^2\beta^2} \qquad (6.4.19)$$

$$A_{mn}^{(2)} = 2(n/m)^{1/2}(-1)^m \frac{1}{\pi\alpha_2}\int_{\pi\alpha_1}^{\pi(\alpha_1+\alpha_2)} \cos\frac{n(\sigma - \pi\alpha_1)}{\alpha_2}\cos\frac{m\sigma}{\alpha_3}$$
$$= -(2/\pi)(mn)^{1/2}(-1)^m\frac{(\beta+1)\sin(m\pi\beta)}{n^2 - m^2(\beta+1)^2} \qquad (6.4.20)$$

$$A_{mn}^{(3)} = \delta_{mn}$$

$$B_m^{(1)} = 2(m)^{-1/2}(-1)^m\frac{1}{\pi\alpha_1}\int_0^{\pi\alpha_1}\cos\frac{m\sigma}{\alpha_3}\,d\sigma$$
$$= 2\alpha_3(\pi\alpha_1)^{-1}(-1)^m(m)^{-3/2}\sin(m\pi\beta) \qquad (6.4.21)$$

$$B_m^{(2)} = 2(m)^{-1/2}(-1)^m\frac{1}{\pi\alpha_2}\int_{\pi\alpha_1}^{\pi(\alpha_1+\alpha_2)}\cos\frac{m\sigma}{\alpha_3}\,d\sigma$$
$$= -2\alpha_3(\pi\alpha_2)^{-1}(-1)^m(m)^{-3/2}\sin(m\pi\beta) \qquad (6.4.22)$$

where $\beta = \alpha_1/\alpha_3$, $B_m^{(1)} = -\alpha_2 B_m$, and $B_m^{(2)} = \alpha_1 B_m$.

Although the expressions may seem complicated, the actual calculation of the integral is simple because it is only a Gaussian integral. By performing the integral, we find an explicit form for the vertex function. We first take the integral over $P^{(3)}$, which is trivial because the delta function forces it to be a combination of the momenta for the other two strings. The functional integration over $P^{(1,2)}$ is also easy, because it is also a Gaussian. When we combine terms, we find the compact result for the vertex function in terms of harmonic oscillators [1, 6]:

$$|V_{123}\rangle = \exp\left\{\frac{1}{2}\sum_{r,s=1}^{3}\sum_{m,n=1}^{\infty}\alpha_{-m}^{(r)}N_{mn}^{rs}\alpha_{-n}^{(s)} + \sum_{r=1}^{3}\sum_{m=1}^{\infty}N_{m}^{r}\alpha_{-m}^{(r)}\mathbf{P} + K\mathbf{P}^{2}\right\}|0\rangle \tag{6.4.23}$$

where

$$N_{mn}^{rs} = (C^{-1})_{mn}\delta_{rs} - 2(mn)^{-1/2}(A^{(r)T}\Gamma^{-1}A^{(s)}) \tag{6.4.24}$$

and

$$\begin{aligned} N_{m}^{r} &= -(m)^{-1/2}(A^{(r)T}\Gamma^{-1}B)_{m} \\ K &= -\tfrac{1}{4}B\Gamma^{-1}B \\ \mathbf{P} &= \alpha_{1}p_{2}^{i} - \alpha_{2}p_{1}^{i} \\ \Gamma &= \sum_{r=1}^{3}A^{(r)}A^{(r)T} \end{aligned} \tag{6.4.25}$$

Let us now summarize what we have done. We have postulated that the only interactions between strings are those that are local; i.e., only instantaneous local deformations of the topology of strings are allowed. In this way, we avoided problems with the violation of causality, which historically plagued attempts over the decades to create nonlocal field theories. What is surprising is that this principle alone is sufficient to determine all five string interactions in the light cone gauge.

Notice that the Neumann matrices $N_{nm}^{rs}$ are all determined uniquely from the overlap conditions (6.4.15). Locality and the conservation of momentum alone are sufficient to determine the precise oscillator representation of the vertex.

Next, we wish to check that this reproduces the usual Veneziano formula. There are two things that have to be checked:

(1) We must show that we reproduce the Neumann function for the Veneziano model.
(2) We must show that the Jacobian for the transformation of coordinates from the light cone configuration $\tau$ to the usual upper half $z$-plane is the correct one.

We will now show that we reproduce the usual string amplitudes, demonstrating the equivalence of the first and second quantized interacting theories at the level of perturbation theory.

## §6.5. Neumann Function Method

In Chapter 2 we saw that we could perform the functional integral over a disk with $L$ handles:

$$A_N = \sum_L \int_{T_L} d\mu\, M_{N,L} \exp \sum_{i>j} k_i N(i,j) k_j \tag{6.5.1}$$

where $N(i,j)$ is the Neumann function between the points $i$ and $j$ on the rim of the disk or upper half-plane with handles, and $M_{N,L}$ includes all measure terms.

The disadvantage of this approach, however, is that the functional integral is defined on the disk or the upper half-plane. The string interpretation of the conformal disk is obscure. To make the connection between the string approach and the conformal disk, let us first make the conformal transformation $\rho = \ln z$. This conformal map takes the upper half $z$-plane and maps it into a horizontal strip in the $\rho$-plane that has width $\pi$. This horizontal strip, in turn, can be interpreted as the surface swept out by a single free string. What we want is a generalization of this map to the $N$-point function.

We follow Mandelstam [8] and make the following conformal transformation on the upper half-plane:

$$\rho = \sum_i \alpha_i \log(z - z_i); \qquad \sum_i \alpha_i = 0 \tag{6.5.2}$$

where the parametrization length of each string is $\pi\alpha_i$. This conformal map stretches the upper half-plane into long horizontal strips, which correspond to the motion of splitting strings.

Analytically, we can see qualitatively that this transformation maps the upper half-plane into the proper light cone diagram. Let us start our variable $z$ at positive infinity on the real axis and then slowly move to the left. As we approach one of the singularities on the real axis, $\rho$ moves rapidly to negative infinity on the real axis. But we hop over one of the singularities $z_i$, and then the logarithm picks up an imaginary part:

$$\rho \sim \alpha_i \log(e^{i\pi} z) = i\pi\alpha_i + \alpha_i \log z \tag{6.5.3}$$

Thus, in $\rho$-plane, we jump $\pi\alpha_i$ vertically upward. This corresponds to moving from the bottom of a string of length $\pi\alpha_i$ to the top of the string at infinity. As we keep on moving in the $z$-plane to the left, the point in the $\rho$-plane begins to move to the right (displaced a distance $\pi\alpha_i$ upward).

As we continue to move to the next singularity at $z_{i-1}$ on the real axis, something strange happens. At a certain point, the $\rho$ variable can slow down,

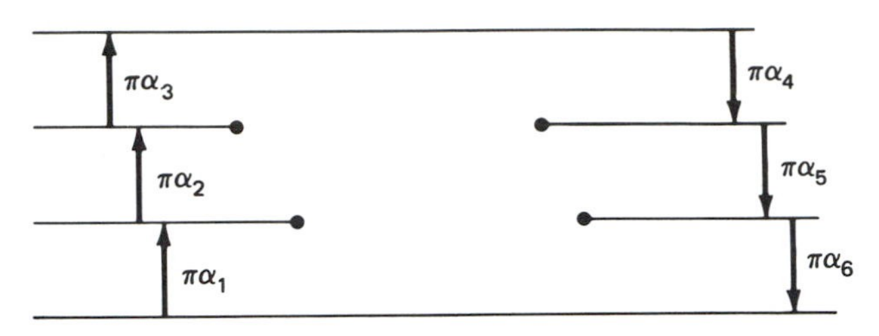


Figure 6.4. Parametrization of the $N$-point function. The incoming (outgoing) legs have positive (negative) parametrization lengths. By changing the lengths of the interior horizontal lines, we create the various field theory graphs that sum up to a dual amplitude.

stop, reverse direction, and begin to move left in the complex plane to negative infinity. This turning point in the $\rho$-plane can be calculated by taking the derivative:

$$\text{Turning point:} \quad \frac{d\rho}{dz} = 0 \tag{6.5.4}$$

and solving for $z$. As we hit $z_{i-1}$ in the $z$-plane, the point in the $\rho$-plane moves to negative infinity. As we hop over the $z_{i-1}$ point, we move another $\pi\alpha_{i-1}$ vertically upward in the $\rho$-plane. In this fashion, by carefully moving on the real axis in the $z$-plane, we sweep out the light cone configuration shown in Fig. 6.4.

Fortunately, we know from (2.5.7) that the Neumann function in the upper half $z$-plane is the sum of two logarithms. Next, we want to power expand this same Neumann function in terms of variables defined on the $\rho$-plane. For the free string, for example, we can take coordinates $\xi$ for the $\tau$ direction and $\eta$ for the $\sigma$ direction. Then

$$z = e^{\zeta} = e^{\xi + i\eta}$$

It is now a straightforward process to power expand the Neumann function in terms of $\xi$ and $\eta$:

$$\begin{aligned} N(z, z') &= \ln|e^{\zeta} - e^{\zeta'}| + \ln|e^{\zeta} - e^{\bar{\zeta}'}| \\ &= -\sum_{n=1}^{\infty} \frac{2}{n} e^{-n|\xi - \xi'|} \cos n\eta \cos n\eta' + 2\max(\xi, \xi') \end{aligned} \tag{6.5.5}$$

The proof that this expansion reproduces the Neumann function is rather simple. First, we notice that the function $\cos n\eta$ has zero $\sigma$ derivative at the endpoint, which is consistent with the fact that $X' = 0$ at the endpoint of a string. Second, by acting on the power expansion by the operator $\nabla^2$, we can show that it vanishes except at the point $z = z'$. Then we can show that the term $2\max(\xi, \xi')$, which selects the maximum of the two, is the correct expression when we take $\xi \to \pm\infty$. By uniqueness, this expression must therefore be correct.

Solving for the Neumann function for the interacting string is done in the same way. First, we make an educated guess that the Neumann function is given by [8]

$$N(\rho, \rho') = -\delta_{r,s} \sum_{n=1}^{\infty} \left(\frac{2}{n}\right) \cos n\eta_r \cos m\eta'_s \exp[-n|\xi_r - \xi'_s|]$$

$$+ \sum_{n,m}' 2N_{mn}^{rs} \cos m\eta_r \cos n\eta'_s + 2\delta_{rs} \max(\xi, \xi')$$

$$- 2\eta_3 \delta_{r,3} - 2\eta_3 \delta_{s,3} + b_{rs} \tag{6.5.6}$$

where the prime means to delete the $n = 0$, $m = 0$ term, and where we can define local coordinates directly on the $r$th string:

$$\rho = \alpha_r(\xi_r + i\eta_r) + \text{const.} \tag{6.5.7}$$

for $r = 1, 2, 3$, and where the constant can be chosen so that the coordinates start at the nearest turning point. The proof that this expansion is correct is also carried out in the same fashion. Notice that the first two terms on the right-hand side of the equation are simply the terms coming from the free string, which guarantees that the equation $\nabla^2 N = 2\pi\delta^2(z - z')$ is satisfied. Second, we note that the rest of the function is nothing but a power expansion in terms of $\cos n\eta_r$ which solves Laplace's equations. Thus, since the Neumann function is unique and the expression solves Poisson's equations, it must be correct.

It is important to notice that we have done nothing new at this point. We have simply written the Neumann function as a Fourier expansion in the various string coordinates. The crucial factor in the above expression is $N_{mn}^{rs}$, which will appear directly in the three-string vertex. These coefficients, in turn, can be explicitly calculated because we know the form of the Neumann function in the upper half-plane, which is the sum of two logarithms. Thus, we can take the various Fourier transforms of this known Neumann function, defined over the upper half-plane, and extract out the Fourier coefficients $N_{mn}$. We reverse the previous equation and now solve for the Neumann coefficients in terms of the Neumann function. If we take the Fourier transform of the Neumann expansion (6.5.5), we arrive at [6]

$$N_{n0}^{rs} = \frac{1}{n} \oint_{x_r} \frac{dz}{2\pi i} \frac{1}{z - x_s} e^{-n\zeta(z)} \qquad (n > 0)$$

$$N_{mn}^{rs} = -\frac{1}{mn(2\pi)^2} \oint_{x_r} dz_r \oint_{x_s} dz_s \frac{e^{-m\zeta_r(z_r) - n\zeta_s(z_s)}}{(z_r - z_s)^2} \tag{6.5.8}$$

($x_1 = 1$, $x_2 = 0$, $x_3 = \infty$) If we insert the Neumann function defined over the upper half-plane into the previous equation and make the proper change of variables, then we have an explicit formula for the Neumann coefficients.

Now that we have an explicit form of the Fourier moments of the Neumann function defined over the light cone diagram, let us consider the case of three strings. The transformation of the upper half-plane to the three-string configuration is simply

$$\rho = \alpha_1 \ln(z - 1) + \alpha_2 \ln z \tag{6.5.9}$$

The turning point for this mapping can be calculated by taking the derivative

$$\frac{d\rho}{dz} = 0 \to z_0 = -\frac{\alpha_2}{\alpha_3} \tag{6.5.10}$$

Thus, the time at which the string splits is equal to

$$\tau_0 = \operatorname{Re} \rho(z = z_0) = \sum_{r=1}^{3} \alpha_r \ln|\alpha_r| \tag{6.5.11}$$

What is remarkable is that the mapping (6.5.9) can actually be inverted, solving for $z$ in terms of $\rho$. This means that we can explicitly calculate all Neumann functions, rather than appealing to abstract expressions such as (6.5.8).

We begin our discussion of calculating the Neumann coefficients by carefully placing coordinates on the various three strings. For example, let us define the coordinate $\zeta_3$ on the third string:

$$\zeta_3 = \frac{\rho}{\alpha_3} + i\pi = \xi_3 + i\eta_3 \tag{6.5.12}$$

By dividing out by $\alpha_3$, we guarantee that the $\eta$ parameter ranges only from 0 to $\pi$. In terms of these new variables, the mapping (6.5.9) can be written as

$$\begin{aligned} -\ln z - \zeta_3 + i\pi &= -\frac{\alpha_1}{\alpha_3} \ln\left(1 - \frac{1}{z}\right) \\ &= -\frac{\alpha_1}{\alpha_3} \ln(1 + e^{\zeta_3} e^{-\zeta_3 + i\pi - \ln z}) \end{aligned} \tag{6.5.13}$$

If we introduce new variables:

$$y \equiv -\zeta_3 + i\pi - \ln z \tag{6.5.14}$$

$$x \equiv e^{\zeta_3} \tag{6.5.15}$$

$$\gamma \equiv -\frac{\alpha_1}{\alpha_3} \tag{6.5.16}$$

then the map (6.5.9) reduces to

$$y = \gamma \ln(1 + x e^{y}) \tag{6.5.17}$$

Fortunately, this equation can be solved explicitly by a power expansion of the form

$$y = \sum_{n=1}^{\infty} \gamma a_n(\gamma) x^n \tag{6.5.18}$$

Inserting (6.5.18) into (6.5.17), we find an explicit expression for $a_n$:

$$a_n = \frac{1}{n!}(n\gamma - 1)\cdots(n\gamma - n + 1) \tag{6.5.19}$$

Inserting back our expressions for $z$ and $\zeta_3$ into (6.5.13), we now find

$$\ln z = -\zeta_3 + i\pi + \sum_{n=1}^{\infty} \frac{\alpha_1}{\alpha_3} a_n\left(-\frac{\alpha_1}{\alpha_3}\right) e^{n\zeta_3} \tag{6.5.20}$$

Now compare this equation to (6.5.6), letting the variable $z'$ go to zero. In that limit, we have

$$\ln|z| = -\xi_3 + \sum_{n=1}^{\infty} N_{n0}^{32} \cos n\eta_3 \tag{6.5.21}$$

Comparing these last two expressions, we see that $N_{n0}^{32}$ is proportional to $a_n$.

The same analysis can be carried out to the other strings, and eventually we can calculate all the Neumann coefficients in terms of $a_n$. The final result is [8]

$$\begin{aligned} \bar{N}_{n,m}^{r,s} &= -\frac{mn}{m\alpha_s + n\alpha_r}\alpha_1\alpha_2\alpha_3 \bar{N}_m^r \bar{N}_n^s \\ \bar{N}_m^r &= \alpha_r^{-1} f_m(-\alpha_{r+1}/\alpha_r)\exp(m\tau_0/\alpha_r) \end{aligned} \tag{6.5.22}$$

where $\gamma_r = -\alpha_{r+1}/\alpha_r$ and

$$\begin{aligned} f_n(\gamma_r) &= (n\gamma_r)^{-1}\binom{n\gamma_r}{n} \\ K &= -\tau_0/(2\alpha_1\alpha_2\alpha_3) \\ \tau_0 &= \sum_{r=1}^{3} \alpha_r \ln|\alpha_r| \\ \mathbf{P} &= \alpha_1 p_2^i - \alpha_2 p_1^i \end{aligned} \tag{6.5.23}$$

At first, we see a vast difference between the functions derived from the second quantized action in (6.4.24) and the Neumann functions in (6.5.22) derived from conformal maps. The first version of the $N$ matrix is defined in terms of products of inverses of overlap integrals. The second version of the $N$ matrices is in terms of Neumann functions on a Riemann surface.

*Miraculously, however, one can show that they are the same* [6, 9]

$$\begin{aligned} N_{nm}^{rs} &= \bar{N}_{nm}^{rs} \\ N_m^r &= \bar{N}_m^r \end{aligned}$$

Thus, although the two derivations seemed at first totally dissimilar, we wind up with an exact equivalence. (The actual proof that the Neumann coefficients found in the field theory (6.4.24) and the coefficients found in the first quantized theory (6.5.22) are the same can proceed in two ways. First, we can use the well-known theorem that the solution to Poisson's equation with known boundary conditions is unique. We can show that both Neumann functions are solutions to the Poisson's equation and are continuous across the $\tau = 0$ configuration when the strings split. Thus, although the two expressions appear vastly different, they are actually the same. There is a second proof, however, which relies on brute force to show the equivalence of these two expressions. The proof requires that we manipulate complicated expressions through the identity (6.5.8) [6]. Details of this calculation, however, are very tedious and will not presented.)

The next step is to prove that the Veneziano model is reproduced by the light cone field theory.

## §6.6. Equivalence of the Scattering Amplitudes

Let us begin our discussion by writing down the $N$-point amplitude in the light cone formalism. The amplitude is a straightforward generalization of the amplitude found in light cone point particle field theory [8]:

$$A_N = \int \sum_{i=2}^{N-1} d\tau_i \int \prod_{r,n,i} dP_{i,n}^{(r)} \, \Psi_{(r)}(P_{i,n}^{(r)}) W \tag{6.6.1}$$

where

$$W = [\det \Delta]^{-12} \prod_r (\alpha_r)^{1/2} e^{-\sum_r (1/2) P_r^- \tau_r}$$
$$\times \exp\left( \tfrac{1}{4} \sum_{rs} \int d\sigma' \, d\sigma'' \, P_r^i(\sigma') N(\sigma', \tau_r; \sigma'', \tau_s) P_s^i(\sigma'') \right) \tag{6.6.2}$$

where $\Psi^{(r)}$ represents an incoming or outgoing state vector of the $r$th external string, $\tau_i$ represent the interaction times, $P^-$ represents the Fourier transform component necessary to convert the expression to the on-shell amplitude, and $P_r(s)$ represents momentum distribution smeared out over the string (which represents an arbitrary collection of higher resonances).

Notice that there are two immediate problems with this amplitude:

(1) The Neumann function occurs only with the transverse momentum excitations. We must show that the complete expression is Lorentz invariant, including the longitudinal momentum factors.
(2) The interaction time variables $\tau_i$ found in the string field theory must be transformed into the usual Koba–Nielsen variables $z_i$. This requires a Jacobian, which in general is quite complicated.

Let us resolve the first problem, the seemingly nonrelativistic form of the amplitude, with the transverse and longitudinal factors occurring in qualitatively different fashion. However, this is an illusion. There is a trick that converts this expression into a Lorentz-invariant integral. Notice that the $\tau$

variable is a solution of Laplace's equation. Thus, because it has the correct boundary conditions, we can write it as a line integral over each external leg at infinity [8]:

$$\tau = \frac{1}{2\pi}\sum_r \int d\sigma'\, N(\sigma, \tau; \sigma', \tau_r) \tag{6.6.3}$$

(This is a rather strange identity, which expresses $\tau$ in terms of expressions containing itself. To prove this equation, simply multiply both sides of (2.5.6) by $\tau'$ and perform integrals over the two-dimensional space. By carefully eliminating terms by integration by parts, we find the above equation.) Thus, we can replace the $\tau$ variable with its Neumann function and write

$$\begin{aligned}\sum_r P_r^- \tau_r &= \frac{1}{2\pi}\sum_{r,s} P_r^- \int d\sigma'\, N(\sigma, \tau_r; \sigma', \tau_s)\\ &= \frac{-1}{\pi^2}\sum_{r,s}\frac{P_r^- p_s^+}{\alpha_r \alpha_s}\int d\sigma \int d\sigma'\, N(\sigma, \tau_r; \sigma', \tau_s)\end{aligned} \tag{6.6.4}$$

By adding this longitudinal contribution to the Neumann function containing $P^- p^+$ with the transverse contribution containing $p_i p_j$, we wind up with a totally Lorentz-invariant Neumann function. Thus, the terms of the form $P^-\tau$ are just the missing pieces necessary to covariantize the terms $p_i p_j$.

Thus, the first problem of establishing the Lorentz invariance of the integrand was a trivial consequence of the uniqueness theorem in electrostatics. The second problem, converting the interaction times $\tau_i$ into the Koba–Nielsen variables, is also a straightforward problem which can be solved using the uniqueness theorem.

We wish to calculate the Jacobian

$$J = \det\left[\frac{\partial \tau_i}{\partial x_j}\right] \tag{6.6.5}$$

where $x_i$ are the usual Veneziano variables on the real axis.

Let us choose the frame $x_1 = 0$, $x_{N-1} = 1$, and $x_N = \infty$. Now compare this Jacobian with the factor

$$\prod_i |c_i|^{1/2} \tag{6.6.6}$$

where

$$\begin{aligned} c_i &= \left.\frac{\partial^2 \rho}{\partial z^2}\right|_{z=x_i}\\ \rho &= \sum_i \alpha_i \ln(x_i - z)\end{aligned} \tag{6.6.7}$$

Notice that the Jacobian and the factor involving the derivatives of the transformation have the same analytic structure. They both have the same singularities when the various $x_i$ touch and have the same boundary conditions. Thus, they must be the same, up to numerical factors! To calculate this overall constant, we are free to take the $x_i$ spaced widely apart, so that

$$\rho(z_i) = \ln x_i \sum_{s \leq i} \alpha_s \tag{6.6.8}$$

so that

$$J = \sum_{i=2}^{N-2} x_i^{-1} \left| \sum_{s \leq i} \alpha_s \right| \tag{6.6.9}$$

(In addition, there are also complicated terms which include the determinant of the Laplacian defined over the four-point configuration and the zero-zero components of the Neumann functions. These terms cancel among each other if we carefully look at their analytic structure and their singularities. The first demonstration of their explicit cancellation was performed in [6] for the four-point function. The cancellation for the arbitrary case, including loops, was performed in [12].) Thus, we have reduced the Jacobian to a trivial factor [10–12], the product of the various Koba–Nielsen variables. Putting everything together, we have

$$\begin{aligned} A_N = & \int \prod_{i,n} \Psi(P_{i,n}^{(r)}) \int d\mu \\ & \times \prod_{i<j} \exp\left( \tfrac{1}{4} \sum_{r,s} \int d\sigma' \, d\sigma'' \, P_\mu^{(r)}(\sigma') N(\sigma', \tau_r; \sigma'', \tau_s) P^{(s)\mu}(\sigma'') \right) \end{aligned} \tag{6.6.10}$$

If we take external tachyons instead of arbitrary external resonance states, we find

$$A_N = \int d\mu \prod_{i<j} |z_i - z_j|^{k_i \cdot k_j} \tag{6.6.11}$$

as before. Thus, the $N$-point amplitude is recovered.

## §6.7. Four-String Interaction

Our previous intuition led us to postulate the existence of a four-string interaction [1], where two strings can combine at their interiors and instantaneously change their local topology. At first, it doesn't appear as if this diagram should exist in the Neumann function method, since the upper half-plane was always mapped into a configuration that was planar. However, the fact that the four-string (and all the other) interaction terms are present in the Veneziano amplitude can be seen if we rigorously examine the region of integration of the Koba–Nielsen variables.

We begin with a four-string mapping and set $x_1 = 1$, $x_2 = \infty$, $x_3 = 0$, and $x_4 = x$. Then the four-string mapping becomes

$$\rho = \alpha_1 \ln(z - 1) + \alpha_3 \ln z + \alpha_4 \ln(z - x) \tag{6.7.1}$$

To find where the mapping is singular, let us set

$$\frac{d\rho}{dz} = 0 \tag{6.7.2}$$

Solving this equation yields the turning points of the transformation:

$$z_{\pm} = \frac{1}{2(1-\gamma_1)}\{1 + (\gamma_2 - \gamma_1)x \pm \Delta^{1/2}\} \tag{6.7.3}$$

where

$$\begin{aligned} \gamma_1 &= -\frac{\alpha_1}{\alpha_1 + \alpha_2} \\ \gamma_2 &= \frac{\alpha_3}{\alpha_1 + \alpha_2} \\ \Delta &= x^2(\gamma_2 - \gamma_1)^2 + 2x(2\gamma_1\gamma_2 - \gamma_1 - \gamma_2) + 1 \end{aligned} \tag{6.7.4}$$

Normally, the two solutions of the turning point equation, given by the $\pm$ in the square root, show that there are two turning points on the Riemann sheet that have the freedom of moving past each other. In the *s*- and *t*-channel diagrams, the world sheets of the strings smoothly transform into each other. However, an interesting thing happens when the imaginary parts of the two interacting points in the $\rho$ plane coincide.

For example, let us study the Feynman graphs for the *t*- and *u*-channel scattering of four particles (see Fig. 6.5). We see that these two graphs cannot smoothly turn into each other. The strings meet either near the top or near the bottom of the diagram, so there is no way of continuously deforming these

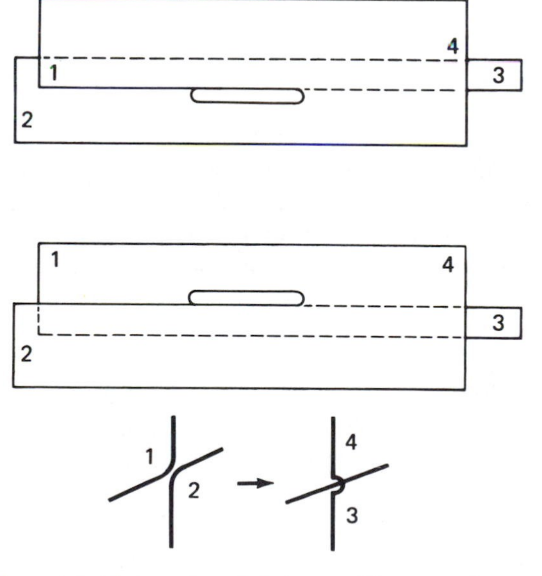


Figure 6.5. The four-string interaction. It is impossible to continuously deform the *t*- and *u*-channel four-string scattering diagrams into each other if we use only three-string vertices. Since the conformal mapping is continuous, this means that there is a missing piece of the integration region, which can be supplied only by postulating a new four-string interaction.

two diagrams into each other. But this is impossible. The conformal map, by definition, was a smooth one that allowed us to go continuously from the $t$ to the $u$ channel and vice versa. *In other words, a piece of the integration region is missing*! To see this, let us calculate precisely when the $t$- and the $u$-channel diagrams meet. We set $\Delta$ equal to zero and then solve for $x$:

$$x_{\pm} = \frac{1}{(\gamma_2 - \gamma_1)^2}\{\gamma_1 + \gamma_2 - 2\gamma_1\gamma_2 \pm 2[\gamma_1(\gamma_1 - 1)\gamma_2(\gamma_2 - 1)]^{1/2}\} \quad (6.7.5)$$

Notice that there are two solutions that give us the $t$- and $u$-channel turning points. The region ($x_+ < x < \infty$) gives us one diagram, and the region ($x_- > x > -\infty$) gives us the other diagram, but what about the region in between?

This is the missing piece. It represents a continuous deformation from one channel graph into the other graph, such that the local topology of the two graphs is only instantaneously deformed into the other (see Fig. 6.6).

In this intermediate region, the local topology of the four strings changes, such that the strings reconnect in a different sequence. Thus, this is graphical proof that the four-string interaction that we postulated earlier is, indeed, part of the Veneziano formula (for the $t$- and $u$-channel graphs). Without this missing piece, the string field theory is actually incomplete and violates conformal invariance and other properties of the $S$-matrix.

The four-string interaction, therefore, requires an extra integration over $d\sigma$. This is like a "zipper" that allows us to locally change the topology of the four strings. At first, one might suspect that this interaction takes place faster than the speed of light. After all, the four-string interaction term takes place instantaneously in time because the $d\sigma$ integration happens instantly. We seem to be violating our postulate of locality, which we originally imposed to yield a causal theory.

The resolution of this puzzle is that the four-string interaction is analogous to the Coulomb interaction term that arises in Yang–Mills theories when quantizing in the Coulomb or light cone gauge. In the Coulomb gauge, the field $A_0$ appears quadratically in $A_0\nabla^2 A_0$ (without any time derivatives) and linearly in the coupling to fermions $A_0\bar{\psi}\gamma^0\psi$. By functionally integrating out over $A_0$, we find

$$\bar{\psi}\gamma^0\psi\nabla^{-2}\bar{\psi}\gamma^0\psi$$

which is the four-fermion Coulomb term. Notice that the operator $\nabla^{-2}$ is an instantaneous operator (i.e., has no time dependence). This term apparently violates relativity, but the $S$-matrix is actually strictly causal even in the

Figure 6.6. Topology of string interactions. The equipotential lines one can draw on a disk with external charges are isomorphic to the interactions of the string. If we have like charges located on opposite sides of the disk, then the equipotential lines collide in the very center and rearrange their topology. This is the four-string interaction. Likewise, all five allowed interactions can easily be displayed in this fashion.

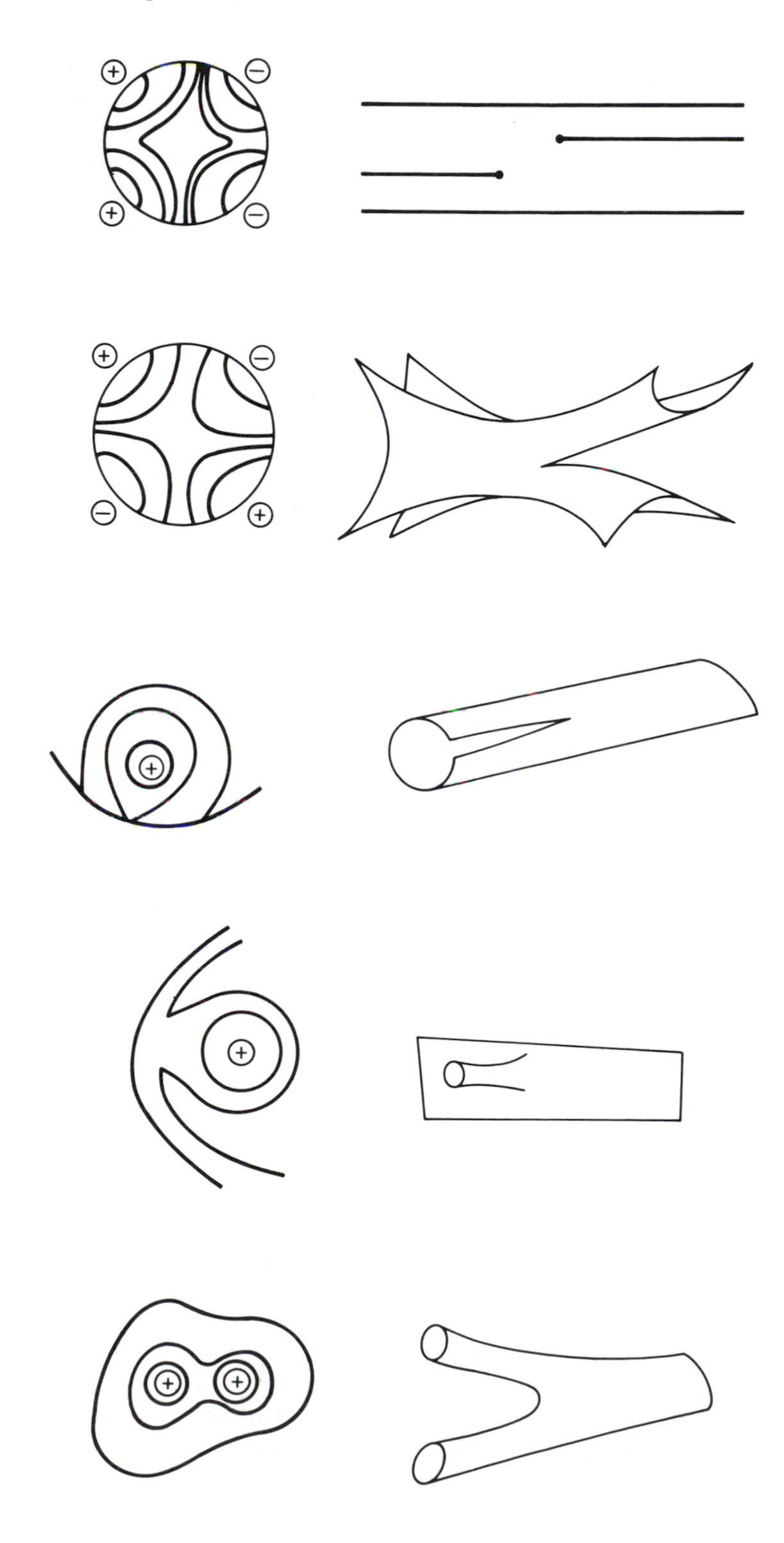

presence of this term. It is an artifact of the gauge fixing, and any apparent violation of causality disappears when we calculate the full $S$-matrix.

The apparent violation of special relativity is only an illusion. Thus, the integration $d\sigma$ over the zipper in the four-string interaction is consistent with special relativity. In fact, in the geometric string field theory, which will be presented in Chapter 8, it can be shown that the covariant four-string interaction term is precisely the Coulomb term found when quantizing nonrelativistically. Thus, there is no conflict with special relativity.

Let us now write down the four-string interaction term for the field theory action. We stress that the four-string interaction term can be guessed simply by postulating locality but is also consistent, as we have seen, with the Veneziano amplitude. The interaction term is [1]

$$S_4 = \int DX_{1234} \int dp_r^+ \delta\left(\sum_{i=1}^{4} \alpha_i\right) \mu \Phi_{p_1^+}^*(X_1) \Phi_{p_2^+}^*(X_2) \Phi_{p_3^+}(X_3) \Phi_{p_4^+}(X_4) \delta_{1234} \tag{6.7.6}$$

where $\mu$ is a measure term and

$$\delta_{1234} = \prod_{\sigma_4} \delta\left(X_4(\sigma_4) - \sum_{i=1}^{2} X_i(\sigma_i)\theta_i\right) \prod_{\sigma_3} \delta\left(X_3(\sigma_3) - \sum_{i=1}^{2} X_i(\sigma_i)\bar{\theta}_i\right)$$

where

$$\begin{aligned}
&\sum_{i=1}^{4} \alpha_i = 0 \\
&\alpha_3 > 0; \qquad \alpha_4 > 0 \\
&\alpha_1 < 0; \qquad \alpha_2 < 0 \\
&\sigma_1 = \pi\alpha_4 - \sigma_4 \quad \text{if } y < \sigma_4 < \pi\alpha_4 \\
&\sigma_2 = \pi|\alpha_1| - \sigma_3 \quad \text{if } 0 < \sigma_3 < x \\
&\sigma_2 = \pi\alpha_3 - \sigma_3 \quad \text{if } x < \sigma_3 < \pi\alpha_3 \\
&\sigma_2 = \pi|\alpha_2| - \sigma_4 \quad \text{if } 0 < \sigma_4 < y \\
&y = x + \pi(\alpha_4 - |\alpha_1|)
\end{aligned} \tag{6.7.7}$$

and

$$\begin{aligned}
\theta_1 &= \theta(\sigma_4 - y) \\
\theta_2 &= \theta(y - \sigma_4) \\
\bar{\theta}_1 &= \theta(\sigma_3 - x) \\
\bar{\theta}_2 &= \theta(x - \sigma_3)
\end{aligned}$$

Although the four-string interaction can be explicitly calculated, it appears hopeless to generalize this to the complete $N$-point amplitude. There appear to be hundreds of possible diagrams that must tediously be checked. However, there is a convenient trick that will reduce this problem enormously.

First, we notice that if external electric charges $q_i$ are placed on a circular disk, then the equipotential lines in the disk are easily drawn. The key observation, however, is that the conformal map takes these equipotential lines and maps them into vertical lines on the light cone diagram. *Thus, the topology of equipotential lines must reproduce the precise topology of interacting open and closed strings.* This remarkable observation reduces a seemingly hopeless task to a relatively simple problem, that of drawing equipotential lines for a disk with external charges.

For example, we know that $\tau = \operatorname{Re} \rho(z) = \sum_i \alpha_i \ln|z - z_i|$. But we also know that the electrostatic potential in two dimensions at a point $\mathbf{r}$ produced by a series of point charges $q_i$ is given by

$$\text{potential} = \sum_i q_i \ln|\mathbf{r} - \mathbf{r}_i|$$

Notice that we can reinterpret $\tau$ as the electrostatic potential created by point charges $\alpha_i$. Therefore, lines of equal potential are precisely lines of equal $\tau$. But strings propagating along the Mandelstam strip are just vertical lines, which correspond to lines of equal $\tau$. *Thus, the equipotential lines on any conformal surface generated by charges $\alpha_i$ correspond to the physical evolution of an interacting string.*

From Fig. 6.6 it is obvious that a four-string vertex must also exist in the light cone formalism. In fact, by analyzing equipotential diagrams, we can show that exactly five distinct interaction terms must be added into the path integral. Notice that closed string interactions emerge from the open string sector as "bound states." We see that the interacting Lagrangian must be the sum of all these five distinct terms.

## §6.8. Superstring Field Theory

It is possible to express both the NS–R and the GS action as a light cone second quantized field theory [13]. New features arise when we do this:

(1) The field functional $\Phi$ is now a functional of the spinor fields as well, which can be represented as the **8** of SO(8) or the $\mathbf{4} + \bar{\mathbf{4}}$ of SU(4).
(2) The theory possesses supersymmetric generators that map interacting bosonic terms of the action into interacting fermionic ones. This will place strong constraints on the possible interactions.
(3) Unlike the bosonic theory, we have to add a specific insertion term at the point at which the strings break. Without this extra insertion term, the theory is neither supersymmetric nor Lorentz invariant.

Let us discuss the GS action in light cone language, because the action is explicitly supersymmetric. Our first quantized light cone action is given in

(3.8.3):

$$S_{\mathrm{lc}} = \frac{-1}{4\pi\alpha'} \int d^2z (\partial_\alpha X^i \partial_\alpha X^i - (2i\alpha')p^+ \bar{\theta}^a \gamma^\alpha \partial_\alpha \theta^a) \tag{6.8.1}$$

where $\theta^{1,2}$ are spinors with eight components in SO(8) space and simultaneously spinors with two components in two dimensions. Recall that in 10 dimensions a Dirac spinor has 32 complex components, a Majorana spinor has 32 real components, a Majorana–Weyl spinor has 16 real components, and in the light cone gauge the spinor has 8 real components transforming under SO(8).

The problem with quantizing this action is that the fermion field is self-conjugate. We have the equation

$$\pi^a = \frac{\delta L}{\delta \dot{\theta}^a} \sim \theta^a \tag{6.8.2}$$

Thus:

$$\begin{aligned} \{\theta^{1a}(\sigma), \theta^{1b}(\sigma')\} &= \{\theta^{2a}(\sigma), \theta^{2b}(\sigma')\} \sim \delta^{ab}\delta(\sigma - \sigma') \\ \{\theta^{1a}(\sigma), \theta^{2b}(\sigma')\} &= 0 \end{aligned} \tag{6.8.3}$$

Notice that this is *not* in canonical form. These fields are self-conjugate. The fermionic fields form a Clifford algebra, while we would prefer to have Grassman states without the delta function on the right-hand side of (6.8.3). The simplest way out of this problem is to divide up the 8 states in the spinor into 4 + 4 states, with one set of 4 states being the conjugate of the other 4 states. One way to do this is to use the SU(4) subgroup of SO(8):

$$\mathrm{SO}(8) \supset \mathrm{SO}(6) \otimes \mathrm{O}(2) = \mathrm{SU}(4) \otimes \mathrm{U}(1) \tag{6.8.4}$$

Under this decomposition, the **8** of SO(8) decomposes into

$$\mathbf{8} = \mathbf{4} \oplus \bar{\mathbf{4}} \tag{6.8.5}$$

Let us now decompose the 8 components of the spinor into

$$\begin{aligned} \theta^{1a} &= (\theta^{1\bar{A}}, \lambda^{1B}) \\ \theta^{2a} &= (\theta^{2\bar{A}}, \lambda^{2B}) \end{aligned} \tag{6.8.6}$$

where $A$ and $B$ range from 1 to 4. We use the notation

$$\begin{aligned} \bar{\mathbf{4}}&: \quad \theta^{\bar{A}} = \theta_A \\ \mathbf{4}&: \quad \lambda^A = \lambda_{\bar{A}} \end{aligned} \tag{6.8.7}$$

With these new definitions, we have the new independent variables $\theta^{\bar{A}}$, which are all mutually anticommuting without any delta functions as in (6.8.3). The desired anticommutation relations with the canonical conjugates are

$$\{\theta^{i,\bar{A}}(\sigma), \lambda^{j,B}(\sigma')\} = \delta(\sigma - \sigma')\delta^{\bar{A}B}\delta^{ij} \tag{6.8.8}$$

Because there are so many stages in the reduction of a Dirac spinor with 32 complex components, let us summarize how we reached only 4 independent states:

$$\begin{aligned} \text{Dirac} &= 32 \text{ complex components} \\ \text{Majorana} &= 32 \text{ real components} \\ \text{Majorana–Weyl} &= 16 \text{ real components} \\ \text{Light cone} &= 8 \text{ real components} \\ \text{Canonical} &= 4 \text{ real components} \end{aligned}$$

Now that we have decomposed the spinors according to the SU(4) subgroup of SO(8), we have to decompose the vectors under SU(4) as well. The vectors $A^I$, $I = 1, 2, \ldots, 8$, will be broken up as follows:

$$A^I B^I = A^i B^i + A^L B^R + A^R B^L \tag{6.8.9}$$

where

$$\begin{aligned} A^R &= 2^{-1/2}(X^7 + iX^8) \\ A^L &= 2^{-1/2}(X^7 - iX^8) \end{aligned} \tag{6.8.10}$$

Our final action for the free theory is given by

$$S = \int D^{16}z[\partial_+ \Psi \partial_- \Psi + \mathrm{Tr}(\partial_+ \Phi \partial_- \Phi)] \tag{6.8.11}$$

where the independent set of integration variables is given by

$$D^{16}z = D^8 X^I \, D^4 \theta^{1\bar{A}} \, D^4 \theta^{2\bar{A}} \tag{6.8.12}$$

and where $\Phi$ are open string fields and $\Psi$ are closed string fields.

Our basic field functional is now given by (where we have placed isospin labels on the field):

$$\Phi^{ab}[X(\sigma), \theta^1(\sigma), \theta^2(\sigma)] = -\Phi^{ba}[X(\pi\alpha - \sigma), \theta^2(\pi\alpha - \sigma), \theta^1(\pi\alpha - \sigma)] \tag{6.8.13}$$

where the $\sigma$ variable is purely symbolic but was added to show how the field transforms under a twist.

Following (6.3.44) for the bosonic case, we can construct canonical quantization relations:

$$\begin{aligned} [\Phi^{ab}(1), \Phi^{cd}(2)] = \frac{1}{2p^+} \delta(\alpha_1 + \alpha_2) \{ & \delta^{ac}\delta^{bc}\Delta^{16}[z_1(\sigma) - z_2(\sigma)] \\ & - \delta^{ad}\delta^{bc}\Delta^{16}[z_1(\sigma) - z_2(\pi|\alpha_2| - \sigma)]\} \end{aligned} \tag{6.8.14}$$

where

$$z = (X^I, \theta^{1\bar{A}}, \theta^{2\bar{A}}) \tag{6.8.15}$$

Now that we have established the free theory of superstrings, let us begin the difficult task of constructing the interacting vertices for superstrings. We will find several complications:

(1) There will be two sets of oscillators, not one, and separate continuity conditions arising from the overlap delta functions. Fortunately, the two sets of oscillators commute with each other and do not mix.
(2) There will be extra terms defined at the joining point of the three strings. In general, extra fields cannot be introduced along the string because they will violate Lorentz invariance and conformal invariance. However, fields can be placed at the precise point where the string breaks. We must be careful, however, because of singularities that exist at that point.
(3) The greatest restriction is supersymmetry, which will completely determine the nature of these insertions at the breaking point.

We begin our discussion of the vertex function by *postulating* the form that it will take. Based on an analogy with the bosonic case, we postulate that the superstring vertex must look like

$$|V\rangle = Z_i \exp[\Delta_0 + \Delta_S]|0\rangle\delta\left(\sum_r \alpha_r\right)\delta(\sum p_r^I)\delta(\sum \alpha_r\theta_r^{\bar{A}}) \tag{6.8.16}$$

where the $Z_i$ fields are insertions at the breaking point and $\Delta_0$ is the usual bosonic term found in (6.4.23):

$$\Delta_0 = \frac{1}{2}\sum_{r,s=1}^{3}\sum_{m,n=1}^{\infty}\alpha_{-m}^{(r)}\bar{N}_{mn}^{rs}\alpha_{-n}^{(s)} + \sum_{r=1}^{3}\sum_{m=1}^{\infty}\bar{N}_m^r\alpha_{-m}^{(r)}\mathbf{P} - \frac{\tau_0}{2\alpha}\mathbf{P}^2 \tag{6.8.17}$$

where $\alpha \equiv \alpha_1\ \alpha_2\ \alpha_3$. We guess, purely by analogy with the bosonic theory, that the fermionic part must be quadratic in the creation operators. Let us adopt the following decomposition:

$$\theta^{1\bar{A}} = \frac{1}{\sqrt{2\alpha}}\sum_n R_n^{\bar{A}} e^{in\sigma/|\alpha|}$$

$$\theta^{2\bar{A}} = \frac{1}{\sqrt{2\alpha}}\sum_n R_n^{\bar{A}} e^{-in\sigma/|\alpha|}$$

$$\lambda^{1A} = \frac{1}{\sqrt{2\pi|\alpha|}}\sum_n R_n^{A} e^{in\sigma/|\alpha|}$$

$$\lambda^{2A} = \frac{1}{\sqrt{2\pi|\alpha|}}\sum_n R_n^{A} e^{-in\sigma/|\alpha|}$$

where

$$\{R_m^A, R_n^{\bar{B}}\} = \alpha\delta_{m+n,0}\delta^{A\bar{B}}$$

$$\Delta_S = \sum_{m,n=1}^{\infty}\sum_{r,s=1}^{3} R_{-m}^{(r)A}U_{mn}^{rs}R_{-n}^{(s)\bar{A}} + \sum_{m=1}^{\infty}\sum_{r=1}^{3} V_m^r R_{-m}^{(r)\bar{A}}\Theta^{\bar{A}} \tag{6.8.18}$$

where the $U$ and $V$ matrices are totally unknown and

$$\Theta^{\bar{A}} = \frac{1}{\alpha_3}(\theta_1^{\bar{A}} - \theta_2^{\bar{A}})$$

There is no justification for this form (6.8.18) other than that it satisfies the basic boundary conditions that we will now apply. We will enforce the conditions

$$\begin{aligned} \sum \varepsilon_r \theta_r^{\bar{A}} &= 0 \\ \sum \varepsilon_r \tilde{\theta}_r^{\bar{A}} &= 0 \end{aligned} \tag{6.8.19}$$

where $\varepsilon_r$ is $+1$ $(-1)$ if the string state is incoming (outgoing) and the tilde represents the second oscillator. These conditions, when written out in Fourier modes, resemble the conditions found for the conservation of momentum. Specifically, we generalize (6.4.17)

$$\begin{aligned} &\sum_{r=1}^{3}\sum_{n=1}^{\infty} \frac{\sqrt{n}}{\alpha_r} A_{mn}^{(r)}(R_n^{(r)\bar{A}} - R_{-n}^{(r)\bar{A}})|V\rangle = 0 \\ &\left\{\sum_{r=1}^{3}\sum_{n=1}^{\infty} \frac{1}{\sqrt{n}} A_{mn}^{(r)}(R_n^{(r)\bar{A}} + R_{-n}^{(r)\bar{A}}) - \sqrt{2}\alpha B_m \Theta^{\bar{A}}\right\}|V\rangle = 0 \end{aligned} \tag{6.8.20}$$

We also enforce the continuity conditions

$$\begin{aligned} \sum \lambda_r^A(\sigma) &= 0 \\ \sum \tilde{\lambda}_r^A(\sigma) &= 0 \end{aligned} \tag{6.8.21}$$

These continuity conditions, in turn, require the following condition on the Fourier modes:

$$\begin{aligned} &\sum_{r=1}^{3}\sum_{n=1}^{\infty} \frac{\sqrt{n}}{\alpha_r} A_{mn}^{(r)}(R_n^{(r)A} - R_{-n}^{(r)A})|V\rangle = 0 \\ &\left\{\sum_{r=1}^{3}\sum_{n=1}^{\infty} \frac{1}{\sqrt{n}} A_{mn}^{(r)}(R_n^{(r)A} + R_{-n}^{(r)A}) + \frac{1}{\sqrt{2}} B_m \frac{\partial}{\partial\Theta^{\bar{A}}}\right\}|V\rangle = 0 \end{aligned} \tag{6.8.22}$$

We now have enough conditions with which to solve for the $U$ and $V$ matrices. The calculation is arduous, but we finally find

$$\begin{aligned} U_{mn}^{rs} &= \frac{m}{\alpha_r}\bar{N}_{mn}^{rs} \\ V_m^r &= -\alpha_1\alpha_2\alpha_3\sqrt{2}\frac{m}{\alpha_r}\bar{N}_m^r \end{aligned} \tag{6.8.23}$$

Next, we wish to construct the supersymmetry operators in the theory, building on our experience with the generators of the free supersymmetric theory. Let the first quantized supersymmetric generators in (3.8.22) be denoted by $q$; then the second quantized supersymmetric generators $Q$ are related to $q$ at the free level by

$$Q_2 = \int_0^\infty \alpha\, d\alpha \int D^{16}Z[\mathrm{Tr}\, \Phi_{-\alpha} q \Phi_\alpha + \Psi_{-\alpha} q \Psi_\alpha] \tag{6.8.24}$$

Notice that the canonical quantization conditions (6.8.14) guarantee that if the $q$'s form a supersymmetric algebra, then the $Q$'s must also. In particular, the first quantized $q$'s obey

$$\begin{aligned} \{q^{-A}, q^{-\bar{B}}\} &= 2h\delta^{A\bar{B}} \\ \{q^{-A}, q^{-B}\} &= \{q^{-\bar{A}}, q^{-\bar{B}}\} = 0 \end{aligned} \tag{6.8.25}$$

Now we wish to construct the second quantized version of these relations. Specifically, the interacting part of the second quantized generator is given by

$$\begin{aligned} Q^{-A} = \cdots + \lambda \int \prod_{r=1}^{3} d\alpha_r D^{16}Z_r \delta\left(\sum_{i=1}^{3} \alpha_i\right) \\ \times \Delta^{16}\left(\sum_{i=1}^{3} \varepsilon_i Z_i\right) \langle\Phi_1|\langle\Phi_2|\langle\Phi_3||Q^{-A}\rangle \end{aligned} \tag{6.8.26}$$

If we now substitute the expression for the second quantized generators into the commutation relations, we have a series of terms that must sum to zero. To the zeroth order in the coupling constant, this is guaranteed because the $q$'s satisfy the relations of supersymmetry. However, the terms that are linear in the coupling constant contain mixing between the free $q$'s and the interacting $|Q^a\rangle$ terms. Specifically, we find

$$\begin{aligned} \sum_{r=1}^{3} q_r^{-A}|Q^{-\bar{B}}\rangle + \sum_{r=1}^{3} q_r^{-\bar{B}}|Q^{-A}\rangle &= 2|H\rangle\delta^{A\bar{B}} \\ \sum_{r=1}^{3} q_r^{-A}|Q^{-B}\rangle + (A \leftrightarrow B) &= 0 \\ \sum_{r=1}^{3} q_r^{-\bar{A}}|Q^{-\bar{B}}\rangle + (\bar{A} \leftrightarrow \bar{B}) &= 0 \end{aligned} \tag{6.8.27}$$

To solve these equations, we *postulate* that the $Q$'s have the general form

$$\begin{aligned} |Q^{-\bar{A}}\rangle &= Y^{\bar{A}}|V\rangle \\ |Q^{-A}\rangle &= k\varepsilon^{ABCD} Y^{\bar{B}} Y^{\bar{C}} Y^{\bar{D}}|V\rangle \end{aligned} \tag{6.8.28}$$

We have now assembled a formidable apparatus, mostly constructed by guesswork and analogies from the bosonic case. Next, we must make one more educated guess, and that is the structure of the functions $Z$ and $Y$ appearing in (6.8.16) and (6.8.28). These last two functions are stringently restricted because they cannot destroy the Fourier matching conditions that we have constructed. Thus, they must either commute or anticommute with the continuity conditions. It turns out that this uniquely specifies $Z$ and $Y$

to be

$$
\begin{aligned}
Z^I &= |2\alpha|^{-1/2}\left(P^I - \alpha \sum_{r,m} \frac{m}{\alpha_r} \bar{N}^r_m \alpha^{(r)I}_{-m}\right) \\
Y^{\bar{A}} &= |\tfrac{1}{2}\alpha|^{1/2}\left(\Theta^{\bar{A}} + \frac{1}{\sqrt{2}} \sum_{r,m} \frac{m}{\alpha_r} \bar{N}^r_m R^{(r)\bar{A}}_{-m}\right)
\end{aligned}
\tag{6.8.29}
$$

Finally, let us turn the crank. We must generate $|H\rangle$ by the commutation relations of supersymmetry. By brute force, we now find that

$$
|H\rangle = \left(2^{-1/2} Z^L - Z^i \rho^i_{\bar{A}\bar{B}} Y^{\bar{A}} Y^{\bar{B}} + \frac{\sqrt{2}}{3} Z^R \varepsilon^{ABCD} Y^{\bar{A}} Y^{\bar{B}} Y^{\bar{C}} Y^{\bar{D}}\right)|V\rangle \tag{6.8.30}
$$

Now that we have a complete expression for the vertex functions that satisfy supersymmetry, let us try to rewrite the $Z$ and $Y$ functions. We noted earlier that they were chosen such that they commuted or anticommuted with the continuity conditions. We suspect, therefore, that they actually vanish except at the breaking point of three strings. Because functions can easily diverge at the breaking point, we must carefully take limits as we approach that point. By careful analysis, we can rewrite $Z$ and $Y$ as

$$
\begin{aligned}
Y^{\bar{A}} &= \lim_{\varepsilon \to 0} (\tfrac{1}{2}\varepsilon)^{1/2} (\theta^{\bar{A}}_1(\pi\alpha_1 - \varepsilon) + \tilde{\theta}^{\bar{A}}_1(\pi\alpha_1 - \varepsilon)) \\
Z^I &= \lim_{\varepsilon \to 0} \sqrt{2\varepsilon\pi}\, p^I_1(\pi\alpha_1 - \varepsilon)
\end{aligned}
\tag{6.8.31}
$$

The crucial thing to note here is that the $Y$'s and the $Z$'s are both local on the string; i.e., they occur only at the point at which the strings join. Thus, we are spared any nonlocalities that would spoil Lorentz invariance.

## §6.9. Summary

Historically, it was conjectured that a field theory of extended objects was not possible for two basic reasons:

(1) Attempts by Yukawa, Heisenberg, and others found that whenever we deviated from locality in quantum mechanics, we violated causality. A vibration at one point of a blob would propagate faster than the speed of light throughout the blob.
(2) The theory would not reproduce the Veneziano model, because cyclic symmetry and many of the properties of the Beta function only hold on-shell, while an action by definition is off-shell.
(3) The theory could not be both Lorentz invariant and unitary off-shell because a quantization program for strings does not exist which is both Lorentz invariant and unitary off-shell.

(4) Likewise, in string theory, a field theory was thought to be impossible because of overcounting introduced through duality. The dual diagrams already have sums over $s$- and $t$-channel poles, and hence adding dual diagrams with poles in each of the various channels would yield severe overcounting, especially at the higher loops.

Fortunately, the field theory of strings evades all these problems. It solves the first problem of nonlocality by introducing a *multilocal* theory. Thus, changes in the string topology only take place locally; i.e., strings can either break or reform only at a single point in the interior or at the endpoints. The vibrations from the break travel at velocities equal to or less than the speed of light (neglecting Coulomb effects in a light cone theory).

String field theory also solves the second problem because it actually breaks some of the properties of the Veneziano model. This is still acceptable, because the theory reproduces the on-shell Veneziano model.

Third, the BRST quantization program is Lorentz invariant, but only at the price of adding ghosts. This breaks unitarity off-shell, but the final $S$-matrix is unitary, so no physical principles are violated.

Lastly, string field theory solves the problem of duality by explicitly breaking duality. Only the sum over all Feynman graphs yields the correct dual diagrams. Thus, at any intermediate stage of the calculation, duality is actually nonexistent.

We have stressed throughout this book that the second quantized formalism allows us to derive the entire model from a single action. In particular, the light cone formalism gives us an interpretation of the Feynman series where the string picture is clear. In the light cone gauge, *we use the simple restriction that all interactions must be local*; i.e., strings can either break or reform only at single points along the string or the endpoints. This uniquely fixes all interaction terms in the action. In particular, we get terms that involve four-string interactions.

We begin by deriving the theory in the same way that Feynman derived the Schrödinger equation from classical mechanics. We use the fundamental relation

$$\begin{aligned}\Delta_{1,2} &= \langle X_1, \tau_1 | X_2, \tau_2 \rangle \\ &= \langle X_1 | e^{-iH\tau} | X_2 \rangle \\ &= \int_{X_1}^{X_2} DX \, e^{i\int_{\tau_1}^{\tau_2} L(\tau)\,d\tau} \\ &= \int D^2\Phi \, \Phi^*(X_1)\Phi(X_2) e^{i\int_{X_1}^{X_2}\int d\tau\, L(\Phi)\, DX}\end{aligned} \tag{6.9.1}$$

where

$$L(\tau) = \frac{1}{2\pi}\int d\sigma(\dot{X}_i^2 - X_i'^2)$$

$$L(\Phi) = \Phi^*(i\partial_\tau - H)\Phi$$

$$H = \frac{\pi}{2} \int d\sigma \left( P_i^2 + \frac{X_i'^2}{\pi^2} \right)$$

We can prove all these relations by simply inserting different sets of intermediate states between the initial and final states:

$$\begin{aligned} &\text{First quantization:} \quad |X\rangle \int DX \langle X| = 1 \\ &\text{Second quantization:} \quad |\Phi\rangle \int D^2\Phi \, e^{-\langle\Phi|\Phi\rangle} \langle\Phi| = 1 \end{aligned} \tag{6.9.2}$$

The field functional $\Phi$ is not a function of $\sigma$. It is a functional of a string variable $X$ that is defined over a region of $\sigma$:

$$\Phi(X) = \langle X|\Phi\rangle = \Phi[X(\sigma_1), X(\sigma_2), X(\sigma_3), \ldots, X(\sigma_N)] \tag{6.9.3}$$

The easiest way to decompose this function is in a basis state of all possible members of the Fock space:

$$|\Phi\rangle = \sum_{\{n\}} \phi_{\{n\}} |\{n\}\rangle \tag{6.9.4}$$

Because $\Phi$ satisfies the string Schrödinger equation, we can power expand the field functional in eigenfunctions of solutions to the string Schrödinger equation:

$$\phi_{p^+,\{n\}}(\tau, X_i) = \int dp_i \, e^{i(\mathbf{p}\cdot\mathbf{x} - E_{\{n\}}\tau)} A_{p^+, p_i, \{n\}} \tag{6.9.5}$$

Notice that $A$ is the creation/annihilation operator for creating or destroying all possible excitations of a string. Thus, it corresponds to an infinite-component field theory. We impose the standard canonical commutation relations, which force us to choose

$$[A_{p^+, p_i, \{n\}}, A^\dagger_{q^+, q_j, \{m\}}] = \delta(p^+ - q^+)\delta(p_i - q_i)\delta_{\{n\},\{m\}} \tag{6.9.6}$$

Now we can power expand the field function $\Phi$ in terms of these eigenfunctions.

$$\begin{aligned} \langle X|\Phi\rangle &= \Phi_{p^+}(X) \\ &= \sum_{\{n\}} \int \prod_i dp_i \, A_{p^+, p_i, \{n\}} H_{\{n\}}(X) e^{-\sum_{i,n} X_{i,n}^2} e^{i(\mathbf{p}\cdot\mathbf{x} - E_{\{n\}}\tau)} \end{aligned} \tag{6.9.7}$$

We can reproduce all the identities found in field theory. In particular, we can show that the Green's function is the matrix element of two fields:

$$\begin{aligned} \Delta_{12} &= \langle\langle 0|\Phi^\dagger_{p^+}(X_1, \tau_1)\Phi_{q^+}(X_2, \tau_2)|0\rangle\rangle \\ &= \delta(p^+ - q^+) \int DX \, e^{i\int L \, d\sigma \, d\tau} \prod_\sigma \delta(\mathbf{X}(\sigma, \tau_1) - \mathbf{X}_1(\sigma)) \\ &\quad \times \prod_\sigma \delta(\mathbf{X}(\sigma, \tau_2) - \mathbf{X}_2(\sigma)) \end{aligned} \tag{6.9.8}$$

Interactions are uniquely determined by our rule that the local topology can only change locally. Thus, our three-string vertex function is given by a delta function:

$$S_3 = \int dp^{+r}\delta\left(\sum_{r=1}^{3} p^{+r}\right)\int DX_{123}\Phi^\dagger(X_3)\Phi^\dagger(X_1)\Phi(X_2)\delta_{123} + \text{h.c.} \quad (6.9.9)$$

Fortunately, it is possible to perform the integration over the strings, which are simple Gaussians. After doing the integration, we find

$$\begin{aligned} |V_{123}\rangle = \exp\Bigg\{&\tfrac{1}{2}\sum_{r,s=1}^{3}\sum_{m,n=1}^{\infty}\alpha_{-m}^{(r)}N_{mn}^{rs}\alpha_{-n}^{(s)} \\ &+\sum_{r=1}^{3}\sum_{m=1}^{\infty}N_m^r\alpha_{-m}^{(r)}\mathbf{P} + K\mathbf{P}^2\Bigg\} \end{aligned} \quad (6.9.10)$$

where

$$N_{mn}^{rs} = (C^{-1})_{mn}\delta_{rs} - 2(mn)^{-1/2}(A^{(r)T}\Gamma^{-1}A^{(s)}) \quad (6.9.11)$$

and

$$\begin{aligned} N_m^r &= -(m)^{-1/2}(A^{(r)T}\Gamma^{-1}B)_m \\ K &= -\tfrac{1}{4}B\Gamma^{-1}B \\ \Gamma &= \sum_{r=1}^{3} A^{(r)}A^{(r)T} \end{aligned} \quad (6.9.12)$$

At this point, we must show that this vertex function can yield the usual Veneziano model. The simplest way to show this is to calculate the Neumann function for the three-string configuration in the first quantized formalism and then compare the results. Our starting point is Mandelstam's mapping:

$$\rho = \alpha_1 \ln(z-1) + \alpha_2 \ln z$$

Since we know the Green's function in the upper half-plane, we just take Fourier components of the Neumann function. The Fourier coefficients look like

$$\begin{aligned} N_{mn}^{rs} &= -\frac{1}{(2\pi)^2}\int_0^{2\pi} d\eta_r\, d\eta_s\, e^{-m\zeta_r - n\zeta_s}\ln(z(\zeta_r) - z(\zeta_s)) \\ &= -\frac{1}{mn(2\pi)^2}\int_{x_r} dz_r \int_{x_s} dz_s\, \frac{e^{-m\zeta_r(z_r) - n\zeta_s(z_s)}}{(z_r - z_s)^2} \end{aligned} \quad (6.9.13)$$

Now we simply insert the Neumann function over the upper half-plane into the above formula. The calculation is straightforward, and we get

$$\begin{aligned} N_{n,m}^{r,s} &= -\frac{mn}{m\alpha_s + n\alpha_r}\alpha_1\alpha_2\alpha_3 \bar{N}_m^r\bar{N}_n^s \\ \bar{N}_m^r &= \alpha_r^{-1}f_m(-\alpha_{r+1}/\alpha_r)\exp(m\tau_0/\alpha_r) \end{aligned} \quad (6.9.14)$$

where

$$\begin{aligned} f_n(\gamma_r) &= (n\gamma_r)^{-1}\binom{n\gamma_r}{n} \\ K &= -\tau_0/(2\alpha_1\alpha_2\alpha_3) \\ \tau_0 &= \sum_{r=1}^{3} \alpha_r \ln|\alpha_r| \\ \mathbf{P} &= \alpha_1 p_2^i - \alpha_2 p_1^i \end{aligned} \tag{6.9.15}$$

Now let us analyze the question of superstrings in the GS light cone formalism. We now have two sets of oscillators, fermionic and bosonic. Our basic strategy is to guess an ansatz for the vertex function and force it to obey the overlap continuity equations. Based on an analogy with the bosonic theory, we guess

$$|V\rangle = Z_i \exp[\Delta_0 + \Delta_S]\,|0\rangle\delta\left(\sum_r \alpha_r\right)\delta(\textstyle\sum p_r^I)\delta(\sum \alpha_r\theta_r^{\bar{A}}) \tag{6.9.16}$$

where

$$\Delta_S = \sum_{m,n=1}^{\infty}\sum_{r,s=1}^{3} R_{-m}^{(r)A} U_{mn}^{rs} R_{-n}^{(s)\bar{A}} + \sum_{m=1}^{\infty}\sum_{r=1}^{3} V_m^r R_{-m}^{(r)\bar{A}} \Theta^{\bar{A}} \tag{6.9.17}$$

In Chapter 3 we showed that the first quantized supersymmetric generators satisfied

$$\begin{aligned} \{q^{-A}, q^{-\bar{B}}\} &= 2h\delta^{A\bar{B}} \\ \{q^{-A}, q^{-B}\} &= \{q^{-\bar{A}}, q^{-\bar{B}}\} = 0 \end{aligned} \tag{6.9.18}$$

Now we must show that the second quantized versions of these supersymmetry generators also satisfy these relations. We take the ansatz

$$\begin{aligned} |Q^{-\bar{A}}\rangle &= Y^{\bar{A}}|V\rangle \\ |Q^{-A}\rangle &= k\varepsilon^{ABCD} Y^{\bar{B}} Y^{\bar{C}} Y^{\bar{D}} |V\rangle \end{aligned} \tag{6.9.19}$$

Remarkably, we can satisfy all conditions placed on the vertex function with the choice

$$\begin{aligned} Y^{\bar{A}} &= \lim_{\varepsilon\to 0} (\tfrac{1}{2}\varepsilon)^{1/2}(\theta_1^{\bar{A}}(\pi\alpha_1 - \varepsilon) + \tilde{\theta}_1^{\bar{A}}(\pi\alpha_1 - \varepsilon)) \\ Z^I &= \lim_{\varepsilon\to 0} \sqrt{2\varepsilon\pi}\, p_1^I(\pi\alpha_1 - \varepsilon) \end{aligned} \tag{6.9.20}$$

It is important to notice about this vertex that the condition of locality is still enforced. The additional insertions with $Y$ and $Z$ take place only at the breaking point of the string, and hence locality is preserved. Thus, our entire treatment of the light cone theory is consistent with our original assumption of locality.

## References

[1] M. Kaku and K. Kikkawa, *Phys. Rev.* **D10**, 1110, 1823 (1974).
[2] M. Kaku, *Introduction to the Field Theory of Strings*, Lewes Superstring Workshop, World Scientific, Singapore, 1985.
[3] M. Kaku, String Field Theory, *Int. J. Mod. Phys.* **A2**, 1 (1987).
[4] P. West, Gauge-Covariant String Field Theory, CERN-TH-4660/86, June 1986.
[5] T. Banks, Gauge Invariant Actions for String Models, SLAC-PUB-3996.
[6] E. Cremmer and J. L. Gervais, *Nucl. Phys.* **B76**, 209 (1974); *Nucl. Phys.* **B90**, 410 (1975); see also M. Ademollo, E. Del Guidice, P. Di Vecchia, and S. Fubini, *Nuovo Cim.* **19A**, 181 (1974).
[7] J. F. L. Hopkinson, R. W. Tucker, and P. A. Collins, *Phys. Rev.* **D12**, 1653 (1975).
[8] S. Mandelstam, *Nucl. Phys.* **B64**, 205 (1973); **B69**, 77 (1974).
[9] M. Green, J. H. Schwarz, and L. Brink, *Nucl. Phys.* **B219**, 437 (1983); **B243**, 475 (1984).
[10] O. Alvarez, *Nucl. Phys.* **B216**, 125 (1983).
[11] H. P. McKean, Jr., and I. M. Singer, *J. Diff. Geom.* **1**, 43 (1967).
[12] S. Mandelstam, in *Unified String Theories* (edited by M. B. Green and D. Gross), World Scientific, Singapore, 1986.
[13] M. B. Green and J. H. Schwarz *Phys. Lett.* **149B**, 117 (1984); *Nucl. Phys.* **B128**, 43 (1983); *Nucl. Phys.* **B243**, 475 (1984).

CHAPTER 7

# BRST Field Theory

## §7.1. Covariant String Field Theory

The great advantage of the light cone field theory, as we saw, was that it was unitary, manifestly ghost-free, and could reproduce the string amplitudes from a single action. There was no need to appeal to intuition in order to construct the unitary $S$-matrix.

The light cone theory, however, was still a broken theory. We would like a covariant description in which all the gauges of the string are in operation. The next step in the development of string field theory is the use of BRST techniques to write down a covariant description of string fields. The power of the BRST formalism is that we can reformulate string field theory in a fully covariant way with the introduction of Faddeev–Popov ghosts.

We will extract the BRST field theory in the same way that Feynman extracted the Schrödinger equation from the classical first quantized theory. We will start with the first quantized theory quantized in the BRST formalism and then extract the field functionals. It must be stressed that the BRST field theory, like the light cone formalism, is still a gauge-fixed field theory. Because we will extract the BRST theory from the gauge-fixed first quantized theory, we will find rather bizarre remnants of the first quantized theory intruding into the second quantized field theory, such as Faddeev–Popov ghosts, ghost counting numbers, parametrization midpoints, and parametrization lengths.

There is also some irony here. Originally, the light cone field theory was introduced to provide a coherent, comprehensive formalism in which to express the entire theory. Unfortunately, the attempts to covariantize the model have produced not one but *two* competing covariant BRST string field theories! These two BRST theories are based on entirely different string topologies, and there does not appear to be any link between them other than the fact that they both can successfully reproduce the Veneziano model.

All these difficulties will be eliminated when we finally go to the geometric version of the field theory in the next chapter.

Previously, we saw that the Gupta–Bleuler formalism, instead of solving the constraints (as in the light cone formalism), applies the constraints directly onto the states:

$$L_n|\phi\rangle = 0; \qquad n > 0 \tag{7.1.1}$$

Normally, one would expect that all 26 components of the theory propagate in the action. However, this is not true if we are careful to include the effect of the constraints being applied to the state vectors. The effect of the previous equation is to kill off the ghost states from the spectrum.

There is yet another way to eliminate the ghost states of the theory, and that is to follow the analogy of gauge theory. Instead of applying these constraints on the Hilbert space, we will now demand that the action be invariant under a transformation of the field $\Phi$ when it rotates into a ghost:

$$\delta|\Phi\rangle = \sum_n L_{-n}|\Lambda_n\rangle \tag{7.1.2}$$

Notice that the state $L_{-n}|\Lambda\rangle$ is a ghost state, as we saw in (2.9.3). We motivate this choice because of an analogy with electromagnetism, where we have the gauge symmetry:

$$\delta A_\mu = \partial_\mu \lambda \tag{7.1.3}$$

The action for the Maxwell field is invariant when we rotate the field $A_\mu$ by a ghost field $\partial_\mu \lambda$. Let us now prove that the string variation actually contains the gauge variation of both the Maxwell field and the linearized gravitational field. As we saw from (6.3.19), the field $|\Phi\rangle$ is the sum over all possible excitations of the string. We power expand the fields into their components:

$$\begin{aligned} |\Phi\rangle &= \phi(x)|0\rangle + A_\mu a_{-1}^\mu|0\rangle + \cdots \\ |\Lambda\rangle &= \lambda(x)|0\rangle + \cdots \\ L_{-1} &= k_1 \cdot a_{-1} + \cdots \end{aligned} \tag{7.1.4}$$

Inserting this power expansion into our gauge transformation of the string field (7.1.2), we find

$$\delta A_\mu a_{-1}^\mu|0\rangle + \cdots = \partial_\mu \lambda a_{-1}^\mu|0\rangle + \cdots$$

Equating terms in the power expansion, we obtain (7.1.3). Thus, we have derived the gauge transformation of the Maxwell field by power expanding the field function [1–3].

Similarly, we can also prove that the closed string sector contains the linearized gravitational field. We now demand

$$\delta|\Psi\rangle = \sum_n L_{-n}|\Lambda_n\rangle + \bar{L}_{-n}|\bar{\Lambda}_n\rangle \tag{7.1.5}$$

where the bar designates the second, independent Hilbert space. Power expanding this expression, we arrive at

$$\begin{aligned}
|\Psi\rangle &= \phi(x)|0\rangle + h_{\mu\nu}a_{-1}^{\mu}\bar{a}_{-1}^{\nu}|0\rangle + \cdots \\
|\Lambda\rangle &= \lambda_{\mu}\bar{a}_{-1}^{\mu}|0\rangle + \cdots \\
|\bar{\Lambda}\rangle &= \lambda_{\mu}a_{-1}^{\mu}|0\rangle + \cdots \\
L_{-1} &= k_{\mu}a_{-1}^{\mu} + \cdots \\
\bar{L}_{-1} &= k_{\mu}\bar{a}_{-1}^{\mu} + \cdots
\end{aligned} \tag{7.1.6}$$

Putting everything together, we obtain

$$\delta h_{\mu\nu}a_{-1}^{\mu}\bar{a}_{-1}^{\nu}|0\rangle + \cdots = \partial_{\mu}\lambda_{\nu}a_{-1}^{\mu}\bar{a}_{-1}^{\nu}|0\rangle + (\mu \leftrightarrow \nu) + \cdots \tag{7.1.7}$$

Equating coefficients, we find

$$\delta h_{\mu\nu} = \partial_{\mu}\lambda_{\nu} + \partial_{\nu}\lambda_{\mu} \tag{7.1.8}$$

Thus, we recover the original variation of the linearized graviton field.

Next, we wish to find an action that is invariant under this gauge transformation (7.1.2).

Let us now repeat the arguments made in the previous chapter concerning how to derive the string field theory from the first quantized formalism. The key step was inserting a complete set of intermediate states at every intermediate point between the two string states. The insertion of the intermediate states is now complicated by the fact that we must preserve the gauge constraints at each step of the calculation. Generalizing (6.3.29), we find that the new set of intermediate states is therefore

$$1 = P|X\rangle \int DX \langle X|P \tag{7.1.9}$$

where $P$ is a projection operator which guarantees that ghost states are eliminated at each intermediate step of the calculation. It satisfies [4]

$$L_n P = PL_{-n} = 0; \qquad P^2 = P$$

Repeating the same steps used in the previous chapter to extract the Lagrangian from the path integral, we now find that the action must be

$$L = \Phi P[L_0 - 1]P\Phi \tag{7.1.10}$$

The crucial fact is that this action possesses the local gauge symmetry (7.1.2). This gauge symmetry is responsible for eliminating the ghosts that appear in the propagator, which in general would propagate all 26 modes if the projection operator did not exist.

It will prove useful to power expand this action and then compare the results with the Maxwell and gravitational actions. To lowest order, we can write the projection operation in the actions as

$$\langle\Phi|(L_0 - \tfrac{1}{2}L_1L_{-1} + \cdots)|\Phi\rangle \tag{7.1.11}$$

Let us now power expand this in terms of the lowest-lying states. The expression now becomes

$$\langle 0| A_\mu a_1^\mu(\Box - k_\nu a_1^\nu k_\rho a_{-1}^\rho) A_\sigma a_{-1}^\sigma |0\rangle \tag{7.1.12}$$

because $L_{-1} \sim k \cdot a_1$, $|\Phi\rangle \sim A_\mu a_{-1}^\mu |0\rangle$, and $L_0 \sim \Box$. When we reduce this expression, it becomes

$$A_\mu(\eta^{\mu\nu}\Box - \partial^\mu \partial^\nu) A_\nu \tag{7.1.13}$$

which, of course, is proportional to the Maxwell action. Thus, to lowest order, the action (7.1.10) contains Maxwell's theory. Similarly, we can repeat all of our steps and show that the closed string action reproduces the linearized graviton action. At the next level, however, there are complications.

By brute force, we can power expand our projection operator to the next levels and we find

$$\begin{aligned} P(L_0 - 1)P = {} & L_0 - 1 - \tfrac{1}{2} L_{-1} L_1 \\ & + L_{-1}^2 (4L_0 + \tfrac{1}{2} D - 9) \Delta L_1^2 \\ & + L_{-1}^2 (6L_0 + 6) \Delta L_2 + \text{h.c.} \\ & - L_{-2}(4L_0 + 2)(2L_0 + 2) \Delta L_2 + \cdots \end{aligned} \tag{7.1.14}$$

where

$$\Delta = 2(16L_0^2 + (2D - 10)L_0 + D)^{-1} \tag{7.1.15}$$

The operator $\Delta$ is *nonlocal* because it contains the inverse of a polynomial in $p^2$. For example, we know that $p^{-2}$ is nonlocal because it equals the Feynman propagator in $x$-space:

$$\Delta(x - y) = \langle x | \frac{1}{\Box} | y \rangle$$

Since $x$ and $y$ are two distinct points, the expression $p^{-2}$ in $x$-space is nonlocal.

Fortunately, there is a way to remove any nonlocality introduced by polynomials in $p^{-2}$ and that is to introduce more auxiliary fields. For example, consider the Lagrangian

$$\psi A \psi + \psi B \phi \tag{7.1.16}$$

where $A$ and $B$ are operators. By eliminating the $\psi$ field (either by calculating its equation of motion and reinserting it back into the Lagrangian, or by functionally integrating over $\psi$ in the path integral), we find the reduced Lagrangian

$$-\frac{1}{4} \phi B A^{-1} B \phi \tag{7.1.17}$$

Notice that if $A$ is a polynomial in $p^2$, then $A^{-1}$ is a nonlocal operator. The point here is that we can trade off nonlocal actions for auxiliary fields. Since these auxiliary fields decouple from the final action (because they are ghosts) they do not affect any of the physics.

Now we wish to construct an explicit form for the full projection operator $P$ for all levels. We can construct $P$ in several ways, such as a power expansion in the $L_n$ operators, or a projector onto each state of level $N$. We will explore the latter possibility.

First, let us define the *Verma module* (see (4.1.42)) as the set of all possible raising operators $L_{-n}$ acting on the vacuum $|0\rangle$ ($\alpha_i < \alpha_{i+1}$):

$$L_{-\{\alpha\}}|0\rangle \equiv L_{-\alpha_1}^{\lambda_1} L_{-\alpha_2}^{\lambda_2} \cdots L_{-\alpha_N}^{\lambda_N}|0\rangle \tag{7.1.18}$$

where $\{\alpha\}$ symbolically represents a vast collection of indices. Then let us power expand the projection operator in terms of $L_{\{\alpha\}}$. We assume that the projection operator $P$ has the following form:

$$P = 1 + \sum_{\alpha,\beta} L_{-\alpha} F_{\alpha\beta}(L_0) L_\beta \tag{7.1.19}$$

where $F_{\alpha\beta}$ are arbitrary functions of $L_0$, and we must remember that we use the symbols $\alpha$ and $\beta$ to represent large collections of indices. If we demand that the operator $P$ vanish when multiplied by the Virasoro generators, then this fixes the $F$'s exactly. Let $P^S$ represent the ghost projection operator. Then

$$\begin{aligned} P &= 1 - P^S \\ P^S &= -\sum_{N=1}^{\infty} \sum_{\substack{\alpha,\beta=1 \\ |\alpha|=|\beta|=N}}^{p(N)} L_{-\alpha} F_{\alpha\beta}^N(L_0) L_\beta \end{aligned} \tag{7.1.20}$$

where we have explicitly written in the sum over different levels $N$. By demanding that the operator vanish when multiplied by $L_n$, this gives us a recursion relation satisfied by the polynomial $F$. Let us define the $F$ matrix as

$$F_{\alpha\gamma}^N(L_0) \equiv \sum_\beta [\delta_{\alpha\beta} - A_{\alpha\beta}^{N-1}(L_0)](S^{-1})_{\beta\gamma}^N(L_0) \tag{7.1.21}$$

where the matrices $S$ and $A$ have yet to be determined. By forcing the operator to have the correct properties, this gives us a recursion relation where the $A$ matrix can be determined iteratively:

$$\begin{aligned} &\sum_{M=1}^{N-1} \sum_{\substack{\alpha,\beta=1 \\ |\alpha|=|\beta|=M}}^{p(M)} L_{-\alpha} F_{\alpha\beta}^M(L_0) L_\beta L_{-\gamma}|R\rangle \\ &\quad \equiv L_{-\delta} A_{\delta\gamma}^{N-1}(L_0)|R\rangle \end{aligned} \tag{7.1.22}$$

Lastly, the matrix $S$ is defined as

$$S_{\alpha\beta}|R\rangle \equiv L_\alpha L_{-\beta}|R\rangle \tag{7.1.23}$$

where $|R\rangle$ is a real state and $|\alpha|$ is equal to $\sum_{i=1}^n i\lambda_i$.

Notice that this expansion is iterative. The $N$th level is determined in terms of all $N$ and lower states. Thus, knowing the projection operator at the first few levels, which we have explicitly written out in (7.1.14), gives us the projection operator at all levels $N$.

This projection operator is nonlocal in $p^2$. This is easily remedied as we can calculate the $F$'s exactly because we know the precise location of all zeros of the determinant.

Fortunately, the determinant of the matrix elements of Verma modules is known exactly. We know that, at the $N$th level, the determinant of the matrix element between $\langle 0|L_\alpha$ and $L_{-\beta}|0\rangle$ is given by

$$\det|\langle 0|L_\alpha L_{-\beta}|0\rangle| = \prod_{p,q} (L_0 - h_{p,q})^{p(N-pq)}$$
$$h_{p,q} = [(mq-(m+1)p)^2 - 1]/4m(m-1) \tag{7.1.24}$$
$$c = 1 - \frac{6}{m(m+1)}$$

where $p$, $q$ are positive integers and their product is less than or equal to $N$. This is a remarkable formula, first written down by Kac [5], which has many ramifications. For example, if the Kac determinant is nonzero, this means that the matrix $S$ introduced in (7.1.23) is invertible. This means that the Verma module forms an irreducible representation of the conformal group. Thus, the Kac determinant is essential in determining the irreducible representations for the string model.

The principal use of this result is to locate all zeros of the determinant and hence the location of all the poles of the projection operator. Inserting (7.1.19) into our action, we find

$$\Phi^* \left[ L_0 - 1 - \sum_{M=1}^{\infty} \sum_{\substack{\alpha,\beta=1 \\ |\alpha|=|\beta|=M}}^{p(M)} L_{-\alpha}\hat{F}^M_{\alpha\beta}L_\beta \right] \Phi \tag{7.1.25}$$

where $F$ is related to $\hat{F}$ by a multiplication by $L_0$. The advantage of this expression is that we know the location of all the singularities of the $\hat{F}$ matrix.

Now let us repeat the analysis of (7.1.16) and (7.1.17), where we introduced auxiliary fields in order to soak up the nonlocal terms in our action. We first decompose the poles within $\hat{F}$:

$$\hat{F}^N_{\alpha\beta}(L_0) = \sum_{l,m} A^{Nlm}_{\alpha\beta}[L_0 - a^{Nl}_{\alpha\beta}]^{-m} + B^N_{\alpha\beta} \tag{7.1.26}$$

We can now introduce auxiliary fields $\rho$ to soak up the extra nonlocal pieces [6] contained within $\hat{F}$:

$$\begin{aligned} &\Phi^* \left[ L_0 - 1 - \sum_{N=1}^{\infty} \sum_{\substack{\alpha,\beta=1 \\ |\alpha|=|\beta|=N}}^{p(N)} L_{-\alpha}B^N_{\alpha\beta}L_\beta \right] \Phi \\ &+ \sum_{N=1}^{\infty} \sum_{\substack{\alpha,\beta=1 \\ |\alpha|=|\beta|=N}}^{p(N)} \rho^{*Nlm}_\alpha (L_0 - a^{Nl}_{\alpha\beta})^m \rho^{Nlm}_\beta \\ &+ \rho^{*Nlm}_\alpha (A^{Nlm}_{\alpha\beta})^{1/2} L_\beta \Phi \\ &+ \Phi^* [L_{-\alpha}(A^{Nlm}_{\alpha\beta})^{1/2}] \rho^{Nlm}_\beta \end{aligned} \tag{7.1.27}$$

The point of this was to show that the action can be written in a form that is explicitly local in $p^2$. The price we paid for locality, however, was introduction of an infinite set of auxiliary fields $\rho_\alpha^{Nlm}$. Our next task is to find a simple way in which to compactly represent this formula.

In order to rearrange the terms in the above action in an elegant fashion, we will need to use the power of symmetry. First, we will introduce the BRST formalism, in which these auxiliary fields can be reinterpreted as being "Faddeev–Popov ghost degrees of freedom." We will see that the auxiliary fields $\rho_\alpha^{Nlm}$ can be represented in terms of states defined on anticommuting ghost fields.

Second, the fact that these auxiliary fields form a Verma module is the first indication that there is an underlying group theoretic origin to the ghost fields. BRST ghost fields, by themselves, are strange objects in a second quantized field theory, but the presence of Verma modules shows that these ghost fields are really representations of a deeper symmetry. This key observation, that the auxiliary fields form a representation space of the conformal group given by a Verma module, will play a crucial role in the development of the geometric formulation of the theory in the next chapter. The geometric formalism of Chapter 8 will allow us to reassemble all the terms appearing in the above action in a compact, group theoretical fashion based on tensor products of irreducible representations of a new infinite-dimensional Lie group. We will show that these ghost fields represent the *tangent space* to a new infinite-dimensional Lie group.

## §7.2. BRST Field Theory

The BRST formalism uses the Faddeev–Popov ghosts to express these auxiliary fields $\rho_\alpha^{Nlm}$ introduced in (7.1.27). These Faddeev–Popov ghosts will kill 2 of the 26 modes, giving us the required 24 physical modes.

Let us review the first quantized BRST string theory. In the Lagrangian (2.1.41), we choose the conformal gauge rewritten in the form $\lambda = 1$; $\rho = 0$. We can write the Faddeev–Popov determinant for the conformal gauge as

$$\Delta_{\mathrm{FP}} \int d\varepsilon\, d\eta\, \delta[\lambda(\varepsilon, \eta) - 1]\delta[\rho(\varepsilon, \eta)] \equiv 1$$

We can, in turn, explicitly calculate this determinant by introducing the Faddeev–Popov ghost field in $\theta$ as in (1.6.22)

$$\Delta_{\mathrm{FP}} = \int d\theta\, dp_\theta \exp - i \int d\sigma\, d\tau\, L(\theta, p_\theta) \tag{7.2.1}$$

where

$$L(\theta, p_\theta) = p_\theta^1(\partial_\sigma + \partial_\tau)\theta^1 + p_\theta^2(\partial_\tau - \partial_\sigma)\theta^2 \tag{7.2.2}$$

Notice that this expression allows us to rewrite the original action of our theory to include the Faddeev–Popov ghosts:

$$L(\sigma, \tau) = P_\mu \dot{X}_\mu - \tfrac{1}{2}[P_\mu^2 + X_\mu'^2 + p_\theta \rho \partial_\sigma \theta] \tag{7.2.3}$$

where $\rho$ is the Pauli spin matrix $\sigma_z$.

As before, this Lagrangian has a local gauge invariance that can be generated on the fields by the application of a nilpotent operator $Q$ [7]. The simplest way to see this is to make a two-dimensional Wick rotation and rewrite the action in a manifestly conformally invariant fashion in terms of the complex variable $z$ and anticommuting fields $b$ and $c$. Rewriting (7.2.2), we find, as in (2.4.4):

$$L = \frac{1}{\pi}\left\{\frac{1}{2}\partial_z X_\mu \partial_{\bar{z}} X_\mu + b\partial_{\bar{z}} c + \bar{b}\partial_z \bar{c}\right\} \tag{7.2.4}$$

As before, we see that this is invariant under

$$\begin{aligned}
\delta X_\mu &= \varepsilon[c\partial_z X_\mu + \bar{c}\partial_{\bar{z}} X_\mu] \\
\delta c &= \varepsilon[c\partial_z c] \\
\delta \bar{c} &= \varepsilon[\bar{c}\partial_{\bar{z}} \bar{c}] \\
\delta b &= \varepsilon[c\partial_z b + 2\partial_z c b - {}^1\!/\!_2 \partial_z X_\mu \partial_z X^\mu] \\
\delta \bar{b} &= \varepsilon[\bar{c}\partial_{\bar{z}} \bar{b} + 2\partial_{\bar{z}} \bar{c} \bar{b} - {}^1\!/\!_2 \partial_{\bar{z}} X_\mu \partial_{\bar{z}} X^\mu]
\end{aligned} \tag{7.2.5}$$

From this variation, we can extract out the nilpotent BRST operator $Q$. However, it is also important to note that, in general, given *any* Lie algebra with commutation relations $[\lambda_m, \lambda_n] = f_{mn}^p \lambda_p$, it is possible to construct a nilpotent operator $Q$ out of anticommuting operators $c_n$ and $b_m$ [8]:

$$Q = \sum_{-\infty}^{\infty} c_{-n}[\lambda_n - \tfrac{1}{2} f_{nm}^p c_{-m} b_p] \tag{7.2.6}$$

For our case, we have two sets of Faddeev–Popov ghosts with

$$\{c_m, b_m\} = \delta_{n,-m} \tag{7.2.7}$$

Thus, our nilpotent BRST operator can be written as

$$\begin{aligned}
Q &= \sum_{n=-\infty}^{\infty} :c_{-n}(L_n + \tfrac{1}{2} L_n^{bc}) \\
&= c_0(L_0 - 1) + \sum_{n=1}^{\infty} [c_{-n} L_n + L_{-n} c_n] \\
&\quad - \tfrac{1}{2} \sum_{n,m=\infty}^{\infty} :c_{-m} c_{-n} b_{n+m}:(m - n)
\end{aligned}$$

where

$$\begin{aligned}
L_n^{bc} &= \sum_{m=\infty}^{\infty} (m + n):b_{n-m} c_m: \\
Q^2 &= 0
\end{aligned} \tag{7.2.8}$$

where the last identity fixes the dimension of space–time to be 26 and the intercept to be equal to 1.

Following the analogy with the point particle case, we can make the transition to the second quantized formalism by taking time slices. This time, however, the basis states $|X\rangle$ are replaced by the complete set of intermediate states $|X, \theta\rangle$ because of the presence of the Faddeev–Popov ghosts. This means that, when we insert an infinite number of intermediate states into our path integral, we must also insert a complete set of intermediate states corresponding to the Faddeev–Popov ghost. Because the $\theta$ variable occurs in the path integral with a $\tau$ derivative, we have no choice but to include it in the complete set of intermediate states when we factorize the functional integral. Thus, our field functional $\Phi$ becomes a function of these ghosts fields. Equations (6.3.29) and (7.1.9) are replaced by

$$\begin{gathered} 1 = |\theta\rangle|X\rangle \int DX\, D\theta \langle\theta|\langle X| \\ \langle\theta|\langle X|\Phi\rangle = \Phi(X, \theta) \end{gathered} \tag{7.2.9}$$

Our Green's functions in (6.3.34) is now modified to read

$$\begin{aligned} \Delta(X_i, \theta_i; X_j, \theta_j) &= \int_{X_i,\theta_i}^{X_j,\theta_j} DX\, DP\, D\theta\, DP_\theta \exp i \int d\sigma\, d\tau\, L(\sigma, \tau) \\ &= \int D^2\Phi\, \Phi(X_i, \theta_i)\Phi^*(X_j, \theta_j) \\ &\quad \times \exp i \int_{X_i\theta_i}^{X_j\theta_j} DX\, D\theta\, L(\Phi) \end{aligned} \tag{7.2.10}$$

The basic field functional is now a function of the Faddeev–Popov ghost fields $\theta$. By making the transition from $|X\rangle|\theta\rangle$ to $|\Phi\rangle$, we find

$$L = \langle\Phi|[L_0 - 1]|\Phi\rangle \tag{7.2.11}$$

where $L_0$ is now a function of both the $X$ and $\theta$ variables. This action was first written down by Siegel [9–14]. If we power expand the BRST field $|\Phi\rangle$ in terms of its ghost modes, we find

$$|\Phi(X)\rangle = \sum_{\{p\}\{q\}} \phi_{\{p\}\{q\}}(X)\{b_{-p}\}\{c_{-q}\}|-\rangle$$

This action explicitly propagates not only the 26 degrees of freedom of the string but also 2 ghost modes, yielding a theory with 24 physical modes. Notice that we are summing over all possible ghost numbers in this expression.

Notice that the above action is totally gauge-fixed. It is also possible, however, to write down another action that contains an explicit gauge degree of freedom based on the nilpotency of the operator $Q$. Let us now replace (7.1.2) with a new gauge transformation. Before, we remarked that the origin of (7.1.2) is that it rotates a string field into a ghost field. In the BRST formalism, however, the physical conditions $L_n|\text{phy}\rangle = 0$ are replaced by $Q|\text{phy}\rangle = 0$. Thus, we suspect that the new BRST gauge invariance is given by

$$\delta|\Psi\rangle = Q|\Lambda\rangle \tag{7.2.12}$$

The state $\langle\Lambda|Q$ does not couple to physical states $|\text{phy}\rangle$, so this confirms our conjecture. Because $Q$ is already nilpotent, our choice for the new action is [15–24]

$$L = \langle\Psi|Q|\Psi\rangle \tag{7.2.13}$$

where we take only the "ghost number $-\frac{1}{2}$ truncation" of the original field:

$$|\Psi\rangle = P_{-1/2}|\Phi\rangle$$

where $P$ is the projection operator that extracts only ghost number $-\frac{1}{2}$. Notice that the new field $|\Psi\rangle$ is a subset or truncation of the original field $|\Phi\rangle$.

If we decompose this action, we find that the lowest excitation of $|\Psi\rangle$ obeys precisely (7.1.10), so this is the desired generalization of our action, where the auxiliary fields that we found there are now regrouped according to the mode expansion found in $|\Psi\rangle$. In summary, we have now compactly reformulated (7.1.27) as (7.2.13).

The equation of motion corresponding to this action is

$$Q|\Psi\rangle = 0 \tag{7.2.14}$$

which reduces to the usual constraints $L_n|\phi\rangle = 0$ for the ground state.

## §7.3. Gauge Fixing

It can also be shown that the previous action (7.2.13) can be gauge-fixed to obtain either the action (7.2.11) or the light cone action (6.3.33). To fix the gauge, it is sometimes helpful to decompose the operator $Q$ according to its zero modes:

$$Q = c_0 K - 2b_0 R + d + \delta \tag{7.3.1}$$

where

$$\begin{aligned}
K &= L_0 - 1 + \sum_{n=1}^{\infty} (nc_{-n}c_n + nb_{-n}b_n) \\
R &= -\sum_{n=1}^{\infty} c_{-n}c_n \\
d &= c_{-n}(L_n + f^{p}_{-m,n} b_{-p} c_m + \tfrac{1}{2} c_{-m} f^{p}_{mn} b_{-p}) \\
\delta &= c_n(L_{-n} + f^{-p}_{m,-n} c_{-m} b_p + \tfrac{1}{2} b_{-p} f^{p}_{-m,-n} c_m)
\end{aligned} \tag{7.3.2}$$

where the $f$'s are the structure constants of the Virasoro algebra. Because $Q$

is nilpotent, these operators must satisfy a large number of simple identities:

$$\begin{gathered} d^2 = \delta^2 = 0 \\ [K, d] = [K, \delta] = [K, R] = 0 \\ [R, d] = [R, \delta] = 0 \\ [d, \delta] = -2RK \end{gathered} \tag{7.3.3}$$

The process of reducing this action down to the usual one begins by writing down the ghost vacuum of the theory.

In Chapter 4 we introduced the "ghost counting number" operator (4.4.20):

$$\sum_{n=1}^{\infty} (c_{-n}b_n + b_{-n}c_n) + \tfrac{1}{2}[c_0, b_0] \tag{7.3.4}$$

This operator counts the number of $c$ modes minus the number of $b$ modes. We can use it to label eigenstates. Because both of the $b$, $c$ ghosts have zero modes, the vacuum of the Faddeev–Popov ghost fields is quite different from the usual unique vacuum of $a_n^\dagger$. The ghost vacuum (as we saw in Chapter 4) is actually degenerate and can have ghost number either $\frac{1}{2}$ or $-\frac{1}{2}$. We can define two vacua:

$$\begin{gathered} c_0|+\rangle = 0; \qquad b_0|-\rangle = 0 \\ |+\rangle = b_0|-\rangle \end{gathered} \tag{7.3.5}$$

There are thus two ghost vacua. Let us fix the ghost number of $\Psi$ to be $-\frac{1}{2}$ and decompose it in terms of the two vacua:

$$|\Psi\rangle = \psi|-\rangle + \phi|+\rangle \tag{7.3.6}$$

Then the equation for action (7.2.13) becomes (using (7.3.1)):

$$\tfrac{1}{2}\langle\psi|K\psi\rangle + \langle\psi|d\phi\rangle + \langle\delta\phi|\psi\rangle + \langle\phi|R\phi\rangle \tag{7.3.7}$$

If we write

$$|\Lambda\rangle = \lambda|-\rangle + \omega|+\rangle \tag{7.3.8}$$

then the variation of the field becomes

$$\delta\psi = (d + \delta)\lambda - 2R\omega \tag{7.3.9}$$

$$\delta\phi = -K\lambda + (d + \delta)\omega \tag{7.3.10}$$

In general, the relations that we have written above can be reexpressed in component language. We can always decompose a field functional into its ghost modes:

$$|\Psi\rangle = \sum_{\{n\},\{m\}}^{\infty} c_{-n_1}c_{-n_2}\cdots c_{-n_N}b_{-m_1}b_{-m_2}\cdots b_{-m_M}\Psi_{n_1,\ldots,n_N;m_1,\ldots,m_M}|0\rangle \tag{7.3.11}$$

Notice that the indices are antisymmetric with each other because of the anticommutation relations between the ghosts. This allows us to write

$$|\Psi\rangle = \sum_{M,N} \binom{N}{M} \Psi_N^M \tag{7.3.12}$$

where $\binom{N}{M}$ represents the product of $N$ antisymmetrized and $M$ anti-symmetrized $c$ and $b$ fields. Notice that this allows us to introduce "forms" defined on anticommuting variables in the same way that differential forms are introduced via anticommuting $dx^\mu$ [10].

We now have the apparatus necessary to perform gauge fixing. Let us first choose the covariant gauge [9]:

$$b_0|\Psi\rangle = 0 \tag{7.3.13}$$

which eliminates half of the fields in $|\Psi\rangle$, that is, $\phi = 0$ in (7.3.6). Now calculate the Faddeev–Popov determinant, which arises from the variation of $b_0|\Psi\rangle = 0$. We find

$$\begin{aligned} L_{\mathrm{FP}} &= \langle\bar{\Lambda}|b_0 Q|\Lambda\rangle \\ &= \langle\bar{\Lambda}|K|\Lambda\rangle \end{aligned} \tag{7.3.14}$$

where $\langle\bar{\Lambda}|$ is a 2-form. The strange feature of this ghost action is that it, in turn, possesses its own gauge invariance:

$$\delta|\Lambda\rangle = Q|\Lambda_1\rangle \tag{7.3.15}$$

Thus, the ghost action itself requires yet another Faddeev–Popov determinant (or else the functional integration over the ghost fields is infinite). The next ghost term coming from the Faddeev–Popov determinant of the ghost field is

$$L_{\mathrm{FP}} = \langle\bar{\Lambda}_2|K|\Lambda_1\rangle \tag{7.3.16}$$

But this action, in turn, also contains its own gauge symmetry, which then requires its own Faddeev–Popov gauge term, and so on. It is clear that there is an infinite tower of "ghosts of ghosts" [10, 12]. This is unavoidable, because each Faddeev–Popov determinant is necessary to eliminate the infinites introduced by the previous set of ghost fields.

Fortunately, it is possible to sum the series. If we start with our action (7.2.13), which has only ghost number $-\frac{1}{2}$, and then add to it this infinite tower of ghost actions, we simply retrieve the action (7.2.11), which is the sum over all possible ghost numbers. Thus, (7.2.11) is the gauge-fixed version of (7.2.13). In other words, a string functional with ghost number $-\frac{1}{2}$ has been transformed, with the addition of the ghosts of ghosts, into a string functional of arbitrary ghost number.

Similarly, it is also possible to use the gauge degrees of freedom within (7.2.13) to reach the light cone gauge [25]. The problem, however, is that the equations of motion must be used explicitly for certain fields to eliminate others. Thus, only on-shell can the BRST formalism reach the light cone gauge, which is a limitation of the formalism. In the geometric formalism, however, we have enough gauge symmetry to eliminate the redundant longitudinal modes in the action in order to obtain the light cone field theory off-shell.

The transition to the light cone gauge is made by observing that the unwanted states, which must be gauged away, can be arranged as the product of all states of the form

$$\{a^+_{-n}a^-_{-m}b_{-p}c_{-q}\}\,|0\rangle \tag{7.3.17}$$

The level number of such a state is given by the sum of the level numbers of each collection of states:

$$n_+ + n_- + n_b + n_c = N \tag{7.3.18}$$

It can be shown that these unwanted states can be eliminated through a variety of mechanisms. For example, for the lowest states $n_+ = n_- = 0, 1$ we find

(a) Most states with $n_c = n_b$; $n_- \geq n_+$ can be eliminated through $\delta\Psi = Q\Lambda$.
(b) States with $n_c = 0$, $n_b = 1$; $n_- \leq n_+$ are Lagrange multipliers.
(c) States not included in (a) with $n_c = 1$, $n_b = 1$; $n_- \geq n_+$ are eliminated by these Lagrange multipliers.

This generalizes to all unwanted states. In general, all these unwanted states can be eliminated by using the gauge invariance of the theory or by using Lagrange multipliers. Only the physical transverse DDF states are left after this elimination.

In summary, we have shown that the action (7.2.13) can be gauge-fixed to yield either the original BRST action (7.2.11) or the light cone action of the previous chapter. We can also show (by eliminating out all higher excitations within $|\Psi\rangle$) that the action reduces back to (7.1.10).

## §7.4. Interactions

In contrast to the light cone field theory, where strings simply split in their interiors, the interacting BRST formalism of Witten [16] relies on the configuration shown in Fig. 7.1. (At first, this configuration seems to violate momentum conservation. However, only in the light cone gauge is the momentum of a string linked to its length. In the covariant approach, the parametrization length is not related to the momentum, so this diagram is allowed.) The length of all strings is set equal. Once again, we can generalize the delta functions appearing in (6.4.4) to include this new configuration:

$$\begin{aligned} S_3 &= \int DX_{123}\, \Phi(X_1)\Phi(X_2)\Phi(X_3)\delta_{123} \\ &= \int DX_{123} \langle\Phi_1|\langle\Phi_2|\langle\Phi_3|X_1\rangle|X_2\rangle|X_3\rangle\delta_{123} \end{aligned} \tag{7.4.1}$$

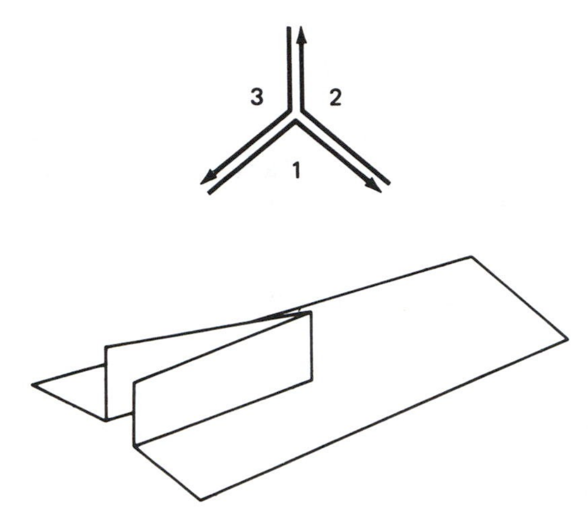


Figure 7.1. Symmetric interaction of BRST string field theory. The surface swept out using this vertex is not flat, as in light cone field theory. There is no need for a four-string interaction.

where

$$\delta_{123} = \prod_{0 \leq \sigma \leq \pi/2} \prod_{r=1}^{3} \delta[X_r(\sigma) - X_{r-1}(\pi - \sigma)] \tag{7.4.2}$$

As in (6.5.2), we must find the conformal map that takes us from the upper half-plane to the configuration being studied in order to construct the vertex in oscillator form [26–32]. Unfortunately, a conformal map of three charges into the upper half-plane (without cuts) does not exist for the BRST vertex because the sum of the changes is usually taken to be zero, while here the sum of the three charges for the symmetric configuration must add up to 3. The solution is to write down electrostatics with six charges adding up to zero charge, and then identify boundaries to simulate the presence of three charges. This requires cutting and pasting together several regions of the complex plane. The map is (see Fig. 7.2)

$$\rho = \sum_{i=-3}^{+3} \alpha_i \ln(z - z_i) = \ln \frac{z^3 - i}{z^3 + i} - i\pi/2 \tag{7.4.3}$$

where

$$\begin{aligned} \alpha_1 = \alpha_2 = \alpha_{-3} &= 1 \\ \alpha_{-1} = \alpha_{-2} = \alpha_3 &= -1 \\ z_1 = e^{i\pi/6}; \qquad z_2 &= e^{i5\pi/6} \\ z_3 = e^{i\pi/2}; \qquad z_{-1} &= -z_1 \\ z_{-2} = -z_2; \qquad z_{-3} &= -z_3 \end{aligned} \tag{7.4.4}$$

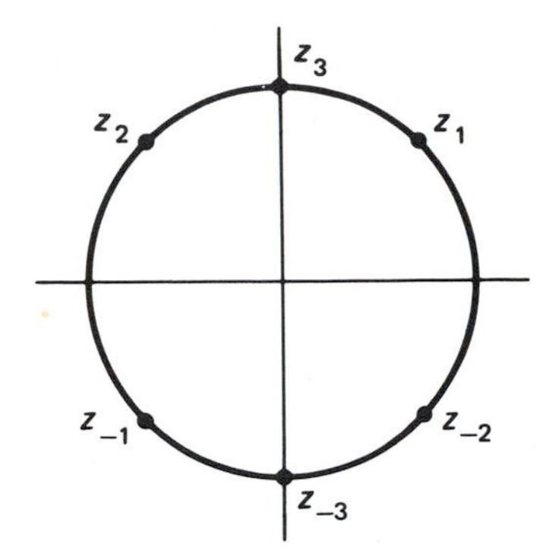


Figure 7.2. Conformal surface of BRST string field theory. This conformal surface can be represented either by taking six charges and resewing the diagram to obtain a surface with three charges or by placing a Riemann cut on the surface.

Fortunately, this map can be inverted, so we can solve for $z$ in terms of $\zeta$, which makes it possible to explicitly construct the Neumann functions. Inverting the map, we find

$$z = z_a\left(\frac{1 + ie^{\zeta_a}}{1 - ie^{\zeta_a}}\right)^{1/3} \tag{7.4.5}$$

where

$$\rho = \zeta_a = \xi_a + i\eta_a \tag{7.4.6}$$

for $a = 1, 2, 3$ and

$$\rho = -(\zeta_a - i\pi) \tag{7.4.7}$$

for $a = -1, -2, -3$. The Fourier coefficients will occur in the combinations

$$\begin{aligned}\left(\frac{1 + ie^{\zeta}}{1 - ie^{\zeta}}\right)^{1/3} &= \sum_{n=2k} A_{2k}e^{n\zeta} + i \sum_{n=2k+1} A_{2k+1}e^{n\zeta} \\ \left(\frac{1 + ie^{\zeta}}{1 - ie^{\zeta}}\right)^{2/3} &= \sum_{n=2k} B_{2k}e^{n\zeta} + i \sum_{n=2k+1} B_{2k+1}e^{n\zeta}\end{aligned} \tag{7.4.8}$$

Putting everything together, we find that the Neumann function appearing in the vertex is

$$\begin{aligned}N_{mn} = &-\tfrac{1}{6}(C + U + \bar{U})1 - \tfrac{1}{6}(C - \tfrac{1}{2}U - \tfrac{1}{2}\bar{U})\begin{pmatrix}0 & 1 & 1\\ 1 & 0 & 1\\ 1 & 1 & 0\end{pmatrix} \\ &- \tfrac{1}{12}i\sqrt{3}(U - \bar{U})\begin{pmatrix}0 & 1 & -1\\ -1 & 0 & 1\\ 1 & -1 & 0\end{pmatrix}\end{aligned} \tag{7.4.9}$$

where

$$
\begin{aligned}
C_{nm} &= 2(-1)^n/n\delta_{n,m} \\
(U + \bar{U})_{nm} &= -2(-1)^n\left[\frac{A_n B_m + B_n A_m}{n+m} + \frac{A_n B_m - B_n A_m}{n-m}\right] \\
(U - \bar{U})_{nm} &= -2i\left[\frac{A_n B_m - B_n A_m}{n+m} + \frac{A_n B_m + B_n A_m}{n-m}\right]
\end{aligned}
\tag{7.4.10}
$$

Written explicitly in terms of operators, this equals

$$
|V_{123}\rangle_X = \exp\left[\tfrac{1}{2}\alpha^r_{-n} N^{rs}_{nm}\alpha^s_{-m} + P^r_0 N^{rs}_{om}\alpha^s_{-m} + \tfrac{1}{2}N_{00}\sum_{r=1}^{3} P^2_{0r}\right]|0_{123}\rangle \tag{7.4.11}
$$

We can, by using the continuity conditions for the ghosts, also calculate the ghost contribution to the vertex function. We simply replace $X$ in (7.4.2) by the $b$ and $c$ ghosts. The ghost vertex is

$$
|V_{123}\rangle_{\mathrm{gh}} = \exp\left(\sum_{n=0}^{\infty}\sum_{m=1}^{\infty} b^r_{-n} X^{rs}_{nm} c^s_{-m}\right)|+++\rangle_{123} \tag{7.4.12}
$$

where the vacuum $|+++\rangle$ is the product of three vacua introduced in (7.3.5). For $s = r$ or $s = r + 2$, the $X$ matrix is given by

$$
X^{rs}_{nm} = -m(N^{rs}_{nm} - N^{r,s+3}_{nm}) \tag{7.4.13}
$$

and for $s = r + 1$

$$
X^{rs}_{nm} = m(N^{rs}_{nm} - N^{r,s+3}_{nm}) \tag{7.4.14}
$$

The final vertex is the product of these two vertices, which are defined in two entirely different spaces:

$$
|V_{123}\rangle = |V_{123}\rangle_X |V_{123}\rangle_{\mathrm{gh}}
$$

By a lengthy calculation, it is possible to show that this vertex satisfies BRST invariance [29]:

$$
[Q_1 + Q_2 + Q_3]|V_{123}\rangle = 0 \tag{7.4.15}
$$

There are several different ways in which to write down this vertex. First, we know from conformal field theory that we can bosonize the anticommuting $b$, $c$ ghost system in terms of a scalar field, which we will call $\phi$. Then the new ghost vertex can be written as

$$
\begin{aligned}
V = \int \prod_{r=1}^{3} DX_r\, D\phi_r e^{i(3/2)\phi(\pi/2)} \\
\times \prod_{0\le\sigma\le(1/2)\pi} \delta(X^r_\mu(\sigma) - X^{r+1}_\mu(\pi-\sigma))\delta(\phi_r(\sigma) - \phi_{r+1}(\pi-\sigma))
\end{aligned}
\tag{7.4.16}
$$

(What is somewhat surprising about this bosonized vertex is the presence of the midpoint insertion. This term does not spoil locality in $\sigma$ because it appears only at $\frac{1}{2}\pi$. Furthermore, it provides the correct ghost number $\frac{3}{2}$ for the multiplication symbol $*$. Because the gauge parameter $\Lambda$ has ghost number $-\frac{3}{2}$, the symbol $*$ must have ghost number $\frac{3}{2}$ so that the product of two gauge parameters $\Lambda_{1,2}$ yields a third gauge parameter $\Lambda_3$ with the same ghost number.)

There is yet another way in which to calculate the symmetric vertex, which depends on using conformal maps with *Riemann cuts* (rather than splicing different regions of the complex plane together). Consider the conformal map

$$\rho(z) = k \log \frac{(z+1)(z-2)(z-\frac{1}{2}) + (z^2 - z + 1)^{3/2}}{z(1-z)} \tag{7.4.17}$$

This map has the usual singularities at $z = 0$, 1, and $\infty$ (which correspond to the three outgoing and ingoing strings at $\pm\infty$). However, what is new is that this conformal map has an explicit Riemann cut that creates a multisheeted $\rho$ plane. This cut is precisely what is necessary to create the surface in the $\rho$ plane that describes the symmetric collision of three strings. The cut extends vertically from $\frac{1}{2} - \frac{1}{2}\sqrt{3}i$ in the lower half-plane to $\frac{1}{2} + \frac{1}{2}\sqrt{3}i$ in the upper half-plane. We can understand the multisheeted $\rho$ plane by tracing our movement along the real $z$-axis in Fig. 7.3.

If we begin at $z = +\infty$ on the positive real $z$-axis and move left, we can choose the constant $k$ such that we trace out a similar line that moves along the positive real $\rho$ axis to negative real infinity. When we reach the point $z = 1$, (at point $B$), we reach negative infinity on the real axis. By hopping over $z = 1$, we suddenly jump $\pi$ units in the vertical direction in the $\rho$ plane. Moving from $z = 1$ to $z = \frac{1}{2}$, we move horizontally right in the $\rho$ plane from $C$ to $D$ until

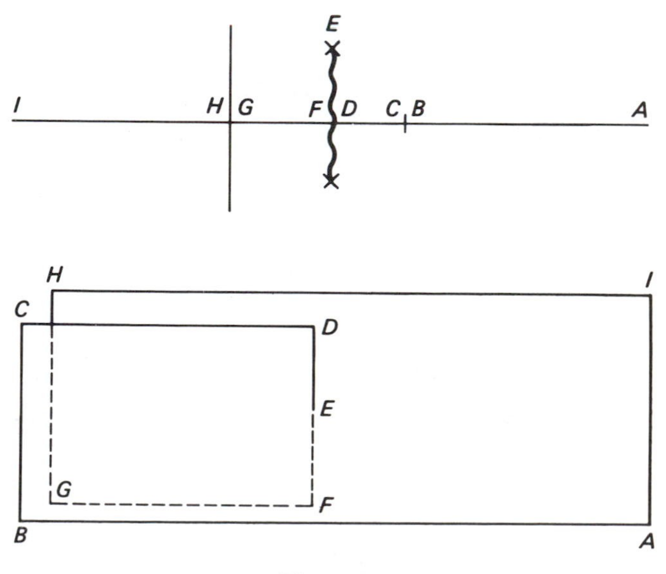


Figure 7.3

we hit the $y$-axis at $\rho = i\pi$. If we now move vertically up the Riemann cut until we hit the point $E$, $z = \frac{1}{2} + \frac{1}{2}i\sqrt{3}$, then we move down the $y$-axis in the $\rho$ plane to $\rho = \frac{1}{2}i\pi$. Going back down the cut to the point $z = \frac{1}{2}$ means that we go from $\rho = \frac{1}{2}i\pi$ to $F$, the origin. Moving from $z = \frac{1}{2}$ to $z = 0$ means moving from $\rho = 0$ to negative real infinity (on the next Riemann sheet). Hopping over $z = 0$ means moving vertically up by $i\pi$ at negative real infinity. Finally, moving from $z = 0$ to $z = -\infty$ means moving from $\rho = i\pi - \infty$ to $\rho = i\pi + \infty$ (from $H$ to $I$).

It is also possible to extract the Neumann functions directly from the map (7.4.17). This alternative approach to the vertex function is presented in Refs. [30–32].

## §7.5. Axiomatic Formulation

If the parametrization midpoint of the string is singled out as a special point, then it becomes possible to define a closed algebra among string field functionals. A string represented by $|A\rangle$ can be joined with a string $|B\rangle$ such that their midpoints exactly coincide. By contracting over their oscillator states, we are left with yet another string of equal length, so we have defined a process similar to multiplication [16]. For example, the abstract notation

$$A * B = C \tag{7.5.1}$$

means concretely

$$\langle A_1 | \langle B_2 | V_{123} \rangle = | C_3 \rangle \tag{7.5.2}$$

where we have contracted over the first and second harmonic oscillators and are left with a string defined over the third harmonic oscillators. Because strings all have equal parametrization lengths, the product rule closes if the parametrization midpoint is singled out as a special point; i.e., the product of two strings of length one is equal to another string of length one. This, in turn, allows us to define a gauge transformation of the field functionals as

$$\delta A = Q\Lambda + A * \Lambda - \Lambda * A \tag{7.5.3}$$

If we define a "curvature" as

$$F = QA + A * A \tag{7.5.4}$$

then we find that

$$\delta F = F * \Lambda - \Lambda * F \tag{7.5.5}$$

Thus, if we can now define an operation called "integration" that preserves:

$$\int A * B = (-1)^{AB} \int B * A \tag{7.5.6}$$

(where we use the anticommutation sign for Grassman odd forms), then we find two invariants, a surface term and the action itself,

$$\int F * F = \text{surface term}$$
$$L = A * QA + \tfrac{2}{3} A * A * A \tag{7.5.7}$$

The remarkable nature of this approach is that we can reduce the essential features of the theory to five "axioms" as in the case of gauge theory:

(1) Existence of a nilpotent derivation

$$Q^2 = 0 \tag{7.5.8}$$

(2) Associativity of the $*$ product

$$[A * B] * C = A * [B * C] \tag{7.5.9}$$

(3) Leibnitz rule

$$Q[A * B] = QA * B + (-1)^A A * QB \tag{7.5.10}$$

(4) The product rule

$$\int A * B = (-1)^{AB} \int B * A \tag{7.5.11}$$

(5) The integration rule

$$\int QA = 0 \tag{7.5.12}$$

where $(-1)^A$ is $-1$ if $A$ is Grassman odd and $+1$ if it is Grassman even.

We define the integration operation as follows:

$$\int \Phi = \int DX \prod_{0 \leq \sigma \leq (1/2)\pi} \delta(X(\sigma) - X(\pi - \sigma))\Phi(X)$$
$$= \langle I | \Phi \rangle \tag{7.5.13}$$

This operation $I$ is a rather strange object. It means that we take a functional of a string $X$, find its midpoint, and then integrate such that we fold the string back on itself around its midpoint. Thus, a functional of a string maps into a $c$-number under this identity operation.

We can also give an explicit representation of this identity operator $I$:

$$I(X) = \langle X | I \rangle = \prod_{n=1,3,5,\ldots} \delta(X_n)$$
$$|I\rangle = e^{-1/2 \sum_{n=0}^{\infty} (-1)^n a_{-n} \cdot a_n} |0\rangle \tag{7.5.14}$$

In addition, the vertex function is invariant under a subgroup of the conformal transformations, i.e., those that do not affect the location of the midpoint. Let us define

$$K_n = L_n - (-1)^n L_{-n} \tag{7.5.15}$$

This operator acts like a derivative on products of forms:

$$K_n(A * B) = K_n A * B + A * K_n B \tag{7.5.16}$$

The integration satisfies the following relation:

$$\int K_n A = \langle A | K_n | I \rangle = 0 \tag{7.5.17}$$

while the vertex function satisfies

$$\sum_{r=1}^{3} K_n^{(r)} | V \rangle = 0 \tag{7.5.18}$$

The advantage of this cohomological approach is that it adheres closely to the cohomological formulation of gauge theory in terms of forms. For Maxwell's equations, we have

$$\begin{aligned} d &= dx^\mu\, \partial_\mu; \qquad d^2 = 0 \\ A &= dx^\mu\, A_\mu^a \tau^a \end{aligned} \tag{7.5.19}$$

We can write this in compact form:

$$\begin{aligned} F &= dA + A * A \\ \delta A &= d\Lambda + A * \Lambda - \Lambda * A \\ \delta F &= F * \Lambda - \Lambda * F \end{aligned} \tag{7.5.20}$$

From these equations, we can write down invariants:

$$\int F * F = \int d[A * dA + \tfrac{2}{3} A * A * A] \tag{7.5.21}$$

The strength of the BRST approach is that we can write down the five axioms and the definitions of curvature forms such that the action is easily shown to be gauge invariant. These elegant axioms sum up a tremendous amount of information, capsuling the string model in five statements.

The disadvantage of the BRST approach, however, is that the origin of these axioms is obscure. These five axioms can, in principle, apply equally well to *any* cohomological system. We have no understanding, therefore, of where these five axioms came from. For example, general relativity theory can be derived, as we have seen, from two principles. These principles, in turn, can be reformulated into the five axioms of cohomology. The five axioms, therefore, are not fundamental, but only a compact and convenient method of writing down the tensor calculus. We seek, therefore, the underlying geometric principles that will allow us to *derive* these five axioms from first principles. The five axioms (i.e., the tensor calculus) should arise naturally from a new infinite-dimensional group.

## §7.6. Proof of Equivalence

Although this is not obvious, this BRST vertex $|V_{123}\rangle$, when the external particles are placed on-shell, is equivalent to any other on-shell vertex function that can be reached by a conformal transformation generated by $L_n$. For example, let us define another vertex function $|\bar{V}_{123}\rangle$ that can be expressed as a conformal transformation of the original vertex:

$$|\bar{V}_{123}\rangle = \prod_{r=1}^{3} \exp\left[\sum_{n\geq 0} \varepsilon_n^r L_{-n}^r\right] |V_{123}\rangle \tag{7.6.1}$$

By acting on this equation with a physical state $\langle \text{phy}|$ that satisfies $\langle \text{phy}| L_{-n} = 0$, we are left with

$$\langle \text{phy}| V_{123}\rangle = \langle \text{phy}| \bar{V}_{123}\rangle \tag{7.6.2}$$

Thus, the two vertex functions have the same on-shell matrix elements. In this way, it has been shown that the vertex function shown here is equivalent on-shell to the covariant oscillator version of the Caneschi–Schwimmer–Veneziano [33, 34] vertex found in the early days of the dual model. (This vertex function was actually derived by taking $N$-point trees and then factorizing the amplitude three times to extract out the three-Reggeon vertex function.) This shows that we must obtain the same $S$-matrix elements by using either vertex. This proves that the three-string vertex found in the BRST formalism yields precisely the same on-shell matrix elements as the old CSV vertex [29, 30], even though the geometry of the BRST vertex is completely different. In fact, one can use this fact to show that all three vertices (the old light cone vertex, the old covariant vertex, and the new BRST vertex) all yield the same $S$-matrix elements.

The proof that the vertex function gives rise to the correct on-shell vertex function of the Veneziano model is quite simple [30–32, 35–36]. We will take matrix elements of both vertex functions with coherent states that are functions of a variable $z$ and then show that the resulting expressions are related to each other by a conformal transformation. We then show that the conformal transformation can be expressed as (7.6.1).

The old CSV vertex is

$$|V_{\text{CSV}}\rangle = \sum_{i=1}^{3} \exp\left(\sum_{n=1}^{\infty} \frac{k_i^\mu \cdot a_{-n,\mu}^{(i+1)}}{\sqrt{n}} + \sum_{n,m=1}^{\infty} a_{-n,\nu}^{(i)} C_{nm} a_{-m}^{(i+1),\nu}\right) |0\rangle_{123} \tag{7.6.3}$$

where the $C$ matrix satisfies the following property:

$$\sum_{m=1}^{\infty} C_{nm} \frac{z^m}{\sqrt{m}} = \frac{(1-z)^n - 1}{\sqrt{z}} \tag{7.6.4}$$

(As a check, we can always hit one of the Reggeon legs with the vacuum, convert a three-oscillator Hilbert space into one oscillator, and reproduce the usual vertex function for the tachyon. Thus, $\langle 0_1 | V_{\text{CSV}} \rangle$ is equal to the usual Veneziano vertex function.)

We will show that this vertex function, up to a transformation generated by the $L_{-n}$, is equal to the symmetric vertex. To show this, let us first analyze the action of a conformal transformation on a coherent state (which was introduced in (2.6.18)):

$$|z, k\rangle = e^{\sum_n k_\mu a^\mu_{-n} z^n} |0\rangle \tag{7.6.5}$$

By a direct calculation, we can show that the Virasoro operator acting on this state yields

$$L_n |z, k\rangle = [(n+1)k^2 z^n + z^{n+1} d_z] |z, k\rangle \tag{7.6.6}$$

This is the transformation of a coherent state when one $L_n$ is applied. Now we wish to apply an arbitrary number of the $L_n$'s to the coherent state:

$$\sum_{n=0}^{\infty} \alpha_n L_n |z, k\rangle = \left[ k^2 f'(z) + f(z) \frac{d}{dz} \right] |z, k\rangle \tag{7.6.7}$$

where we define function $f(z)$ as

$$f(z) = z \sum_{n=0}^{\infty} \alpha_n z^n \tag{7.6.8}$$

Thus, a general conformal transformation generates the following transformation on a coherent state:

$$e^{\sum_{n=0}^{\infty} \alpha_n L_n} |z, k\rangle = \left( \frac{d\bar{z}}{dz} \right)^{k^2} |\bar{z}, k\rangle \tag{7.6.9}$$

where

$$\bar{z} = g^{-1}(g(z) + 1) \tag{7.6.10}$$

and

$$\frac{dg}{dz} = f^{-1}(z) \tag{7.6.11}$$

(7.6.9) should be compared with (4.1.8), which gives the transformation of a field of weight $h$ under the conformal group.

Armed with these results, we can now show the equivalence between the two on-shell vertex functions. We begin by contracting the three-Reggeon vertex function on three arbitrary coherent states [32]:

$$\prod_{i=1}^{3} \langle k_i, z_i | V_{\text{CSV}} \rangle = \prod_{i=1}^{3} \left| z_i - \frac{1}{1 + z_{i+1}} \right|^{k_i \cdot k_{i+1}} |1 - z_i|^{k_i \cdot k_{i-1}} \tag{7.6.12}$$

This can be calculated explicitly from the definition of the $C$ matrices in (7.6.4).

The next step is to contract the symmetric BRST vertex (7.4.11) with three arbitrary coherent states and compare it with the expression given above. We will need the following formulas:

$$\sum_{n,m} N^{12}_{nm} z_1^n z_2^m = \log\left|\bar{z}(z_1) - \frac{1}{1-\bar{z}(z_2)}\right|$$

$$\sum_{n,m} N^{23}_{nm} z_2^n z_3^m = \log\left|\frac{1}{1-\bar{z}(z_2)} - \frac{\bar{z}(z_3)-1}{\bar{z}(z_3)}\right|$$

$$+ \log|z_3| + \log\left|\frac{d\bar{z}}{dz}\right|_{z=0} \tag{7.6.13}$$

$$\sum_{n,m} N^{11}_{nm} z_1^n z_1'^m = \log\left|\frac{\bar{z}(z_1) - \bar{z}(z_1')}{z_1 - z_1'}\right| - \log\left|\frac{d\bar{z}}{dz}\right|_{z=0}$$

$$\sum_{n,m} N^{11}_{nm} z_1^n z_1^m = \log\left|\frac{d\bar{z}}{dz_1}\right| - \log\left|\frac{d\bar{z}}{dz}\right|_{z=0}$$

These identities allow us to construct an explicit form for the matrix element between the coherent state $\prod_{i=1}^3 \langle k_i, z_i|$ and the symmetric BRST vertex function. When we compare this expression to (7.6.12), we find the relationship between the old CSV vertex and the newer symmetric BRST vertex:

$$\prod_{i=1}^{3} \langle k_i, z_i | V_{\text{CSV}}\rangle = \prod_{i=1}^{3} \langle \bar{z}_i(z_i), k_i | V\rangle \prod_{i=1}^{3} \left(\frac{d\bar{z}_i(z_i)}{dz_i}\right)^{k_i^2} \tag{7.6.14}$$

This is the expression that we want. It shows that the matrix element of the CSV vertex and the matrix element of the symmetric BRST vertex are related by a factor raised to the $k^2$ power. Now compare this with (7.6.9), which shows that this factor can be reexpressed in terms of the $L_n$'s. In summary, we now have an explicit solution for (7.6.1) (at least on coherent states), which demonstrates that the CSV vertex and the BRST vertex are related by a conformal transformation and hence have the same on-shell matrix elements.

There is yet another way to show the equivalence of the CSV vertex with the symmetric vertex, which is based on a power expansion of (7.4.17), rather than on using the conformal properties of complex functions. We know that $L_n$ acts like $z^{n+1}d_z$ on a complex function. Let us define

$$W(z) = \frac{-8}{27} \frac{(z+1)(z-2)(z-\frac{1}{2}) - (z^2 - z + 1)^{3/2}}{z(1-z)}$$
$$= 1 + \tfrac{1}{2}z + \tfrac{9}{16}z^2 + \cdots$$

It is always possible to find a series of coefficients $a_n$ such that

$$W(z) = e^{\sum_{n=1}^{\infty} a_n L_n} z$$

In practice, solving for these coefficients might be difficult, but in principle these coefficients are known to any degree of accuracy. For example, $a_1 = \frac{1}{2}$ and $a_2 = \frac{5}{16}$. Then it can be shown that the symmetric vertex and the CSV vertex are related by the following conformal transformation:

$$\langle V | e^{-\ln(3\sqrt{3}/4)\sum_{r=1}^{3}(L_0^r - 1)} = \langle V_{\text{CSV}} | e^{-\sum_{r=1}^{3}\sum_{n=1}^{\infty} a_n L_n^r}$$

Although we have now established the link between the symmetric vertex of Witten and the CSV vertex, we are still facing the problem that there are actually *two* BRST string field theories that do not seem to have any relationship to each other. The other BRST formulation of the interacting theory is based on the old light cone vertex, where strings simply split or rejoin at an interior point. We can also postulate that the three-string vertex function consists of the covariant Dirac delta function [18–24]:

$$|V_{123}\rangle = \int DX_{123}\,|X_1\rangle|X_2\rangle|X_3\rangle \prod_\sigma \delta\left(\sum_{r=1}^{3} \theta_r(\sigma_r)X_r(\sigma_r)\right) \quad (7.6.15)$$

It is important to realize that this new vertex function is precisely the same as the old light cone vertex function (6.4.5), except that the harmonic oscillators are now fully covariant, rather than just transverse, and that we must also multiply it with the ghost vertex. When we multiply these two, we have the "covariantized light cone" vertex.

We can check that the covariantized light cone vertex satisfies BRST invariance:

$$\left\{\sum_{r=1}^{3} Q^{(r)}\right\}|V\rangle = 0 \quad (7.6.16)$$

It is also possible to prove that this covariantized version of the light cone vertex is equivalent to the on-shell CSV three-Reggeon vertex function. We will use a different proof, however, than the one used before. This time, we will show that the continuity or overlap conditions satisfied by the CSV vertex function can be written after a conformal transformation as the continuity condition for the light cone vertex function. Since the continuity or overlap conditions define the vertex, it must follow that the two vertices are related by a conformal transformation.

Let us write the CSV vertex function as

$$|V_{\text{CSV}}\rangle = \exp\left(\sum_{r=1}^{3} \sum_{m=0,n=1}^{\infty} \alpha_{-m}^{\mu(r)}\alpha_{-n,\mu}^{(r+1)}C_{mn}\right)|0\rangle \quad (7.6.17)$$

where

$$C_{mn} = (-1)^n\Gamma(n)\Gamma^{-1}(1+m)\Gamma^{-1}(1+n-m) \quad (7.6.18)$$

We begin with the assumption that we can impose the following continuity condition on this vertex function:

$$(A(z)P_1^\mu[(1-z)^{-1}] + B(z)P_2^\mu[1-z^{-1}] + C(z)P_3^\mu[z])|V_{\text{CSV}}\rangle = 0 \quad (7.6.19)$$

for some unknown values of $A$, $B$, and $C$ and where $P$ is given by

$$P^{\mu(r)} = \sum_{-\infty}^{\infty} \alpha_n^\mu z^{-n} \quad (7.6.20)$$

By commuting the various harmonic oscillators through $|V_{\text{CSV}}\rangle$, we find the following constraints on $A$, $B$, and $C$:

$$-\frac{A}{z} + B - \frac{C}{1-z} = 0 \tag{7.6.21}$$

This still gives us wide latitude in choosing the values of $A$, $B$, and $C$. By carefully looking at where these functions converge, we find that we can set $C = 0$ when $|z| > 1$, $B = 0$ when $|z - 1| < |z|$, and $A = 0$ when $|1 - z| > 1$. For convenience, we will select $B = 0$, which still leaves a certain degree of arbitrariness. Specifically, we choose

$$\begin{aligned} A &= \gamma_2 \frac{z}{1 - \gamma_2 z} \\ B &= 0 \\ C &= -\gamma_2 \frac{1 - z}{1 - \gamma_2 z} \end{aligned} \tag{7.6.22}$$

for some $\gamma_2$.

Our goal is to show the relationship between this vertex and the usual light cone formalism, which maps the upper half-plane into the light cone configuration through

$$\rho = \alpha_1 \ln(z - 1) + \alpha_2 \ln z \tag{7.6.23}$$

As before, in the $\rho$ plane we parametrize the three-string vertex function in the $\sigma$ direction by the parameter $\eta_r$, where

$$\eta_3 = \begin{cases} -\beta_2 \eta_2 + \beta_2 \pi; & 0 \le \eta_3 \le \beta_2 \pi \\ -\beta_1 \eta_1 + \pi; & \beta_2 \pi \le \eta_3 \le \pi \end{cases} \tag{7.6.24}$$

and

$$\begin{aligned} \beta_1 &= \frac{\alpha_1}{|\alpha_3|} \\ \beta_2 &= \frac{\alpha_2}{\alpha_3} \\ \gamma_r &= -\frac{\alpha_{r+1}}{\alpha_r} \\ \delta_r &= e^{i\eta_r} \end{aligned} \tag{7.6.25}$$

Given these definitions, we can rewrite the $A$ and $C$ coefficients as

$$\begin{aligned} A &= \frac{\delta_1}{(1 - z)^{-1}} \frac{d(1 - z)^{-1}}{d\delta_1} (1 - \gamma_1) \\ C &= \frac{\delta_3}{z} \frac{dz}{d\delta_3} \end{aligned} \tag{7.6.26}$$

Now comes the key step. We know that $P$ transforms under the conformal group as follows (see (4.1.7), where we have taken the exponential of the usual variables):

$$P_\mu(z) = \left(\frac{z}{\rho}\frac{d\rho}{dz}\right)^h P_\mu(\rho) \tag{7.6.27}$$

Thus, we can reexpress the continuity equation as

$$\left(\frac{\delta_1}{(1-z)^{-1}}\frac{d(1-z)^{-1}}{d\delta_1}P_1^\mu + \frac{\delta_3}{z}\frac{dz}{d\delta_3}P_3^\mu\right)|V\rangle = 0 \tag{7.6.28}$$

Using the transformation properties of $P$ under the conformal group, we now have

$$(P_1^\mu(\delta_1) + P_2^\mu(\delta_3))|V\rangle = 0 \tag{7.6.29}$$

But $\delta_r = e^{i\eta_r}$. This is the expression we desire. By a conformal transformation, we have created a new vertex that satisfies a different continuity condition on $\eta_r$, which is the $\sigma$ coordinate of three strings. Thus, this new vertex function satisfies precisely the original continuity conditions of the light cone vertex function between strings 1 and 3. Similarly, by taking different combinations we can also show that the continuity conditions between strings 2 and 3 are also satisfied. Thus, we have shown that, by a conformal transformation, we can reexpress the continuity equation for the covariant Veneziano vertex in terms of the continuity equation in the covariantized light cone configuration.

But since conformal transformations are generated by the $L_n$ operators, this means that, on-shell, the covariantized light cone vertex function (7.6.15) is the same as the covariant Veneziano vertex (7.6.17).

At this point, we must explain why there should be two versions of the BRST formalism, one based on the fully symmetric vertex and the other based on the light cone vertex, such that they both have the same on-shell properties. They do not seem to resemble each other in crucial details. The covariantized light cone formalism, for example, has an additional four-string interaction term. This is required because the gauge group for the covariantized light cone theory does not close properly. If we try to adapt (7.5.3) for the light cone configuration, we have problems. The group doesn't close without the addition of the four-string interaction, and even then it only closes on-shell.

The symmetric theory, however, can be shown to be consistent without a four-string interaction. Thus, these two theories differ in fundamental ways.

Although not all the details are clear, it seems that the resolution of the puzzle is that the second field theory, based on the covariantized light cone, is actually an incomplete theory. This because we must integrate over all possible "lengths" of the string $\alpha_r$. In the old light cone theory, this integration was allowed because $\alpha_r$ was really equal to an integration over the momentum of the string. However, in the covariant theory, $\alpha_r$ is a redundant parameter, with no physical meaning. Thus, the integration over $\alpha_r$ produces an infinite number of copies of the same thing. Thus, for example, in the zero slope limit this theory produces an infinite number of Yang–Mills actions:

$$-\tfrac{1}{4}\int_0^\infty d\alpha\, F_{\mu\nu}(\alpha)^2 + \cdots \tag{7.6.30}$$

which is clearly nonsense. This infinite number of copies of the same thing, due to the integration over the fictitious parameter $\alpha$, yields an infinite redundancy in our field theory. Thus, there is a fundamental problem with the second BRST approach.

The resolution of this problem is found in the geometric formulation of the next chapter, where we show that $\alpha$ can be expressed as a genuine *gauge parameter* and hence can be gauge-fixed. However, within the present BRST formalism it is impossible to make $\alpha$ into a genuine gauge parameter. We must vastly increase the number of fields in the theory in order to gauge the "length" of the string, and this is possible only in the geometric formalism (see also [37, 38]).

## §7.7. Closed Strings and Superstrings

Unfortunately, closed strings and superstrings are less developed than the bosonic open string. In the BRST formalism, for example, there exists considerable confusion over how to write down the correct theory. *For closed strings and superstrings, the "ghost counting" comes out all wrong.* In fact, the naive BRST action vanishes exactly because it has the wrong ghost counting number, which necessitates a "consistent truncation" of either $Q$ or the Hilbert space of the "ghost modes." As we shall see, the situation becomes progressively worse and more contrived as we go to more complicated models, until the BRST formalism completely collapses for the fermionic closed string field theory. In some sense, the fact that the open string was so easily expressed in terms of the BRST formalism was a fluke. The harsh reality of the situation is that something important is missing from the BRST formalism.

For the closed string [11, 39, 40], we have two duplicate copies of the Virasoro generators $L_n$ and $\bar{L}_n$. In the BRST formalism, the vacuum therefore possesses double the previous ghost number because $-\frac{1}{2} - \frac{1}{2} = -1$:

$$|0\rangle = |-\rangle|-\rangle; \qquad \langle 0| = \langle -|\langle -| \tag{7.7.1}$$

while $Q$ remains with ghost number 1. This means that the naive action must actually vanish:

$$\langle \Phi|Q|\Phi\rangle = 0 \tag{7.7.2}$$

because the ghost counting does not come out correctly, i.e., $-1 + 1 - 1 \neq 0$. This is clearly undesirable. Various solutions to this ghost counting problem have been proposed, none of them totally satisfactory. For example, we can artificially truncate the zero modes within $Q$, keeping its nilpotency property intact. This means that the zero modes

$$c_0; b_0: \bar{c}_0; \bar{b}_0 \tag{7.7.3}$$

will be arbitrarily removed and replaced by the smaller set

$$\hat{c}_0; \hat{b}_0 \tag{7.7.4}$$

such that the new operator $\hat{Q}$ now decomposes as

$$Q = \hat{c}_0(K + \bar{K}) + d + \delta + \bar{d} + \bar{\delta} - 2\hat{c}_0(R + \bar{R}) \tag{7.7.5}$$

Notice that the old zero modes have been dropped and a smaller set (which resembles the set used for the open string) has been inserted.

In order to have the theory independent of the origin of the $\sigma$ coordinate, we must impose the constraint on the closed string:

$$(K - \bar{K})|\Phi\rangle = 0 \tag{7.7.6}$$

The fact that we have a smaller set of zero modes means that we can use the vacua found earlier for the open string case: $|-\rangle$ and $|+\rangle$.

$$|\Phi\rangle = \phi|-\rangle + \psi|+\rangle \tag{7.7.7}$$

Notice that this trunction was chosen so that we retain the identity

$$\hat{Q}^2 = 0 \tag{7.7.8}$$

The action is

$$\langle\Phi|\hat{Q}|\Phi\rangle \tag{7.7.9}$$

Although the BRST open string theory is straightforward, we begin to see the weaknesses of the BRST approach in the closed string case:

(a) The constraint $(K - \bar{K})|\Phi\rangle = 0$ must be imposed from the outside, without any fundamental understanding. Constraints usually emerge as a result of a symmetry of the action; here, we simply impose it without knowing where it came from.
(b) The ghost states were arbitrarily truncated such that $Q$ remained nilpotent, which seems highly contrived.
(c) This truncation probably does survive in the interacting theory. The dropping of zero modes means we lose the string interpretation of $X_\mu(\sigma)$ and $\psi_\mu(\sigma)$. Locality in $\sigma$ is totally lost, but this was absolutely essential in the construction of the vertex function. Thus, dropping $\sigma$ locality probably destroys any chance of an interacting theory.

Now let us turn to the supersymmetric case, where the situation is much worse. For the Neveu–Schwarz bosons, we actually have no problem with BRST zero mode ghost counting because the commuting ghosts $\gamma$, $\beta$ are half-integral moded, and hence do not change the nature of the vacuum. The BRST vacuum still has ghost number $-\frac{1}{2}$. Thus, the BRST action is still

$$\langle A|Q|A\rangle \tag{7.7.10}$$

However, for the Ramond fermions [11, 41–53], there are severe problems with the integrally moded commuting Faddeev–Popov ghosts. Let $\gamma_0$ be the zero mode of the ghost fermion oscillator. Then there is an infinite number of possible vacua because this ghost is bosonic in nature, not fermionic. Each of this infinite set of vacua is labeled by

$$\gamma_0^n |0\rangle \tag{7.7.11}$$

This means that the field functional must be power expanded into a series of infinite vacua:

$$\Phi(X, \gamma_0) = \sum_n^\infty \phi_n(X)\gamma_0^n \tag{7.7.12}$$

Which vacuum do we use? This is extremely inconvenient, and it shows the limitatons of the BRST approach. (A similar situation occurs in conformal field theory with "picture" changing operators. However, conformal field theory is an on-shell formalism, so all these pictures collapse. Here, the string field theory, by definition, is off-shell, and all these vacua must enter into the theory.)

There are several proposals for truncating the theory in the BRST approach. We can either truncate $Q$, keeping it nilpotent, or truncate the field $|\Phi\rangle$ so we do not have to sum over an infinite number of ghost vacua. Let us begin our discussion by introducing a unified notation by which we can describe the ghost fields for the Neveu–Schwarz and Ramond models simultaneously. Let us define the following [11]:

$$\begin{aligned} A_N^\mu &= (\alpha_n^\mu, \psi_{\dot{n}}^\mu) \\ B_N &= (b_n, \beta_{\dot{n}}) \\ C_N &= (c_n, \gamma_{\dot{n}}) \end{aligned} \tag{7.7.13}$$

where $A_N^\mu$ labels both the usual bosonic string and the anticommuting field $\psi_{\dot{n}}^\mu$, the dot reminds us that the fields can be integrally or half-integrally moded, and $B_N$ and $C_N$ represent the ghost fields for theory. In particular, $b_n$ and $c_n$ are the ghosts corresponding to the bosonic string, and $\beta_{\dot{n}}$ and $\gamma_{\dot{n}}$ label the ghosts for the fermionic field, where the dot again can be either integrally or half-integrally moded. The commutation relations are

$$\begin{aligned} A_M^\mu A_N^\nu - (-1)^{MN} A_N^\nu A_M^\mu &= \eta^{\mu\nu}\delta(M+N) \\ B_M C_N + (-1)^{MN} C_N B_M &= \delta(M+N) \end{aligned} \tag{7.7.14}$$

where $(-1)^{MN}$ is $-1$ if $M$ and $N$ are fermionic and equal to $+1$ otherwise.

If $[L_I, L_J] = F_{IJ}^{-K} L_K$, then the BRST operator, in this notation, is

$$Q = C_{-N}L_N + F_{IJ}^K :C_{-J}C_{-I}B_K: \tag{7.7.15}$$

where $L_N$ is equal to the usual Virasoro operators.

Specifically, for the Ramond string, the operator is

$$Q = c_0 K + \gamma_0 F + d + \delta - 2Rb_0 - 2\hat{R}\beta_0 + \gamma_0^2 b_0 \tag{7.7.16}$$

where

$$\hat{R} = \tfrac{3}{4} M C_{-M} C_{\overline{M}} \tag{7.7.17}$$

and

$$F = F_0 + f_{\overline{M}}^{M} [C^{-\overline{M}} B_M - B_{-M}(-1)^{\overline{M}} C_{\overline{M}}] \tag{7.7.18}$$

where $f_{\overline{M}}^{M}$ equals 2 if $M$ is bosonic and $M/2$ if $M$ is fermionic, the bar interchanges fermionic and bosonic modes, and $d$, $K$, $\delta$, $R$ are equal to their counterparts for the Virasoro case.

$F_0$ is the usual Ramond operator, which satisfies the following commutation relations that define the algebra of Superdiff($S$):

$$\begin{aligned}
\{G_m, G_n\} &= 2L_{m+n} + \tfrac{1}{2}D(m^2 - \tfrac{1}{4})\delta_{m,-n} \\
[L_m, G_n] &= (\tfrac{1}{2}m - n)G_{m+n} \\
[L_m, F_n] &= (\tfrac{1}{2}m - n)F_{m+n} \\
\{F_m, F_n\} &= 2L_{m+n} + \tfrac{1}{2}Dm^2\delta_{m,-n} \\
[L_m, L_n] &= (m-n)L_{m+n} + \frac{D}{8}m^3\delta_{m,-n}
\end{aligned} \tag{7.7.19}$$

Notice that the algebra of Superdiff($S$), which is usually written in terms of physical fields, can be realized on the ghost fields $B_N$ and $C_N$ as well.

The operators $F$, $K$, etc, satisfy, in turn, the following commutation relations:

$$\begin{aligned}
d^2 &= \delta^2 = 0 \\
F^2 &= K \\
2\hat{R} &= [F, R] \\
\{d, \delta\} &= F\{F, R\}
\end{aligned} \tag{7.7.20}$$

These commutation relations, in turn, are sufficient to show that the BRST operator $Q$ is nilpotent.

As we mentioned before, the problem with the Ramond sector is that the ghost mode $\gamma_0$ is integrally moded and satisfies commutation, not anti-commutation, relations, so there is an *infinite number of ghost vacua* described by $\gamma_0^n|0\rangle$ for each $n$. One way out of this problem is to arbitrarily truncate the zero modes of the operator $Q$ while maintaining its nilpotency.

Now we reach the crucial step: as in the case of the bosonic closed string, we will arbitrarily eliminate the bosonic ghost modes $\gamma_0$, $\beta_0$ and introduce fermionic $\hat{b}_0$ and $\hat{c}_0$ such that

$$\begin{aligned}
\hat{b}_0|\hat{0}\rangle &= 0 \\
\langle\hat{0}|\hat{c}_0|\hat{0}\rangle &= 1
\end{aligned} \tag{7.7.21}$$

Let us define a modified $\hat{Q}$ that simply banishes the zero modes $\gamma_0$ and $\beta_0$

from the theory. We define

$$\hat{Q} = \hat{c}_0 F + (-1)^{N_F}(d + \delta) - \hat{b}_0(FR + RF) \tag{7.7.22}$$

where $N_F$ counts the total number of fermion creation operators in a state. Notice that $\hat{Q}^2 = 0$, which was the guiding principle in defining this modified operator.

To define the Ramond states, we define the GSO operator, which in this representation is

$$G = [\hat{b}_0, \hat{c}_0]\Gamma^{11}(-1)^{N_F} \tag{7.7.23}$$

where $N_F$ refers only to nonzero modes and $\Gamma^{11}$ refers to the 10-dimensional chiral Dirac matrix. In this representation, the GSO projection simply eliminates states with the wrong spin statistics.

An arbitrary Ramond superfield can be decomposed in this truncated Fock space as

$$|\Psi\rangle = [|\psi\rangle + \hat{c}_0|\xi\rangle]|\hat{0}\rangle \tag{7.7.24}$$

where $|\psi\rangle$ is a 0-form of chirality $\Gamma^{11} = 1$ and $|\xi\rangle$ is a $G = +1(-1)$-form and

$$\begin{aligned} \hat{b}_0|\hat{0}\rangle &= 0 \\ \langle\hat{0}|\hat{c}_0|\hat{0}\rangle &= 1 \end{aligned} \tag{7.7.25}$$

The truncated action now reads

$$S = \langle\bar{\Psi}|\hat{Q}|\Psi\rangle \tag{7.7.26}$$

where $\langle\bar{\Psi}| = \langle\Psi|\Gamma^0$. The action is invariant under

$$\delta|\Psi\rangle = \hat{Q}|\Lambda\rangle \tag{7.7.27}$$

Substituting our previous decomposition of $|\Psi\rangle$ in (7.7.24) yields

$$S = \langle\bar{\psi}|F|\psi\rangle + \langle\bar{\xi}|FR + RF|\xi\rangle - \langle\bar{\psi}|d + \delta|\xi\rangle - \langle\bar{\xi}|d + \delta|\psi\rangle \tag{7.7.28}$$

The above action has also been found in other ways, such as making a clever choice in the trunction of $|\Psi\rangle$ and preserving the original $Q$. This method also yields the same result.

The situation becomes much worse, however, when we discuss fermionic closed strings. In this case, the constraint $[K - \bar{K}]|\Psi\rangle = 0$ turns into

$$[F - \bar{F}]|\Psi\rangle = 0 \tag{7.7.29}$$

which is a *dynamical* constraint. Hence, we are not able to apply this constraint to the states. Unfortunately, the BRST theory collapses for the fermionic closed string. This demonstrates that the present BRST formalism is incomplete. The closed fermion theory has stretched the BRST formalism to the breaking point, until it no longer produces acceptable results.

We see that the BRST approach becomes increasingly unwieldy as we go to the superstring and the closed string. Unfortunately, many truncation proposals have been made, some of them mutually exclusive. "Truncation," however, is more than a matter of taste; it is a potentially dangerous operation because it probably doesn't generalize to the interacting string. Truncation probably violates locality in $\sigma$, which is essential for the construction of vertices. Thus, the truncations examined so far may actually be inconsistent at the interacting level.

Let us, however, present one more truncation operation [17] that holds promise of being local in $\sigma$ and hence generalizes to interactions. Let us recall the original ghost counting that was performed when we constructed the $*$ and $\int$ operation. The ghost numbers were carefully chosen so that the gauge symmetry closed and the axioms of cohomology (7.5.8) to (7.5.12) were satisfied:

| Object | Ghost Number |
|---|---|
| $\Psi$ | $-\frac{1}{2}$ |
| $Q$ | $+1$ |
| $*$ | $+\frac{3}{2}$ |
| $\Lambda$ | $-\frac{3}{2}$ |
| $\int$ | $-\frac{3}{2}$ |

It's easy to show that the above choice of ghost numbers satisfies the cohomological axioms that we listed before.

Now let us generalize these rules for the open superstring case. We wish to maintain the cohomological axioms and in addition wish to enforce the conditions that two gauge transformations yield a third gauge transformation with the same ghost number.

Given these conditions, we quickly find that the $*$ and the $\int$ that we defined have the wrong ghost numbers. In fact, as we found before, the action is exactly zero. Instead of "truncating" the zero modes, as we did before, *let us use conformal field theory to supply us with the missing ghost numbers.* These insertions from BRST ghost fields will be only at the midpoint, so we do not lose locality in $\sigma$. Thus, this truncation, which truncates only ghost numbers of fields, is superior to the previous truncation, which truncates zero modes.

Several new features must now be added to make this scheme work.

(1) The superfield $\Psi$ is now the sum of an NS field $a$ and an R field $\psi$:

$$\Psi = (a, \psi) \tag{7.7.30}$$

(2) Similarly, the gauge parameter $\Lambda$ is also the sum of two pieces:

$$\Lambda = (\varepsilon, \chi) \tag{7.7.31}$$

(3) The ghost number of $*$ is now equal to $\frac{1}{2}$ (because the conformal ghosts contribute $\frac{3}{2}$, as before in (7.4.16), but now the superconformal ghosts contribute $-\frac{1}{2}$). Thus, we must invent a new multiplication rule for the superstring, which will have ghost number $\frac{3}{2}$ (which we will call $\otimes$).

(4) The integral $\int$ now has ghost number $-\frac{1}{2}$ because of the contribution from the superconformal ghosts. We need to postulate the existence of a new integration $\oint$ which has ghost number $-\frac{3}{2}$.

Now, by trial and error, we can construct the ghost numbers of all of our fields and our operators such that we satisfy closure of the gauge transformation and the previous axioms [17]:

| Object | Ghost Number |
|---|---|
| $a$ | $-\frac{1}{2}$ |
| $\psi$ | $0$ |
| $\varepsilon$ | $-\frac{3}{2}$ |
| $\chi$ | $-1$ |
| $Q$ | $+1$ |
| $*$ | $\frac{1}{2}$ |
| $\otimes$ | $\frac{3}{2}$ |
| $X$ | $1$ |
| $Y$ | $-1$ |
| $\oint$ | $-\frac{3}{2}$ |
| $\int$ | $-\frac{1}{2}$ |

(7.7.32)

where $X$ and $Y$ are still unknown but they have ghost numbers to fill in the missing numbers.

Now that we have made an educated guess that successfully satisfies our original assumptions, we are faced with the task of actually calculating explicit forms for these fields and operators $\otimes$ and $\oint$.

First, it is not a problem to construct BRST fields with these ghost numbers. We simply choose the correct vacuum (from among an infinite number of them) such that it has the correct ghost number. Then we can always apply the ghost number projection operator on each state and obtain a generalized BRST field with the correct ghost number.

More difficult, of course, is defining the operations $\oint$ and $\otimes$.

Since $*$ has only ghost number $\frac{1}{2}$, we need a new conformal operator with ghost number 1 and conformal weight 0. Fortunately, conformal field theory gives us such an operator:

$$X = \{Q, \xi\} \tag{7.7.33}$$

Now, our multiplication rule for gauge parameters must obey

$$\Lambda_1 \otimes \Lambda_2 - (1 \leftrightarrow 2) = \Lambda_3 \tag{7.7.34}$$

With this new $X$ operator, we can satisfy this multiplication rule:

$$\begin{aligned}\Lambda_1 \otimes \Lambda_2 &= (\varepsilon_1, \chi_1) \otimes (\varepsilon_2, \chi_2) \\ &= (X(\varepsilon_1 * \varepsilon_2) + \chi_1 * \chi_2, X(\varepsilon_1 * \chi_2 + \chi_1 * \varepsilon_2))\end{aligned} \tag{7.7.35}$$

By looking up the ghost number of each field and each operator, we find that all ghost numbers are correct. To see this, let us symbolically write down this multiplication rule in terms of ghost numbers. We want to multiply two multiplets, with ghost numbers $(-\frac{3}{2}, -1)$, such that we get back the $(-\frac{3}{2}, -1)$ multiplet. If we carefully write down all the various ghost numbers in this multiplication rule, we find symbolically

$$(-\tfrac{3}{2}, -1) \otimes (-\tfrac{3}{2}, -1) = (1 - \tfrac{3}{2} + \tfrac{1}{2} - \tfrac{3}{2} = -1 + \tfrac{1}{2} - 1, 1 - \tfrac{3}{2} + \tfrac{1}{2} - 1)$$
$$= (-\tfrac{3}{2}, -1) \tag{7.7.36}$$

Thus, we have closure.

Next, we wish to show that $A \otimes \Lambda$, which occurs in the variation of $A$, also has the correct ghost number of $A$. We wish to calculate the ghost numbers of the following operation:

$$(a, \psi) \otimes (\varepsilon, \chi) = (X(a * \varepsilon) + \psi * \chi, X(\psi * \varepsilon + a * \chi)) \tag{7.7.37}$$

By carefully counting the ghost number of each piece, we find

$$(-\tfrac{1}{2}, 0) \otimes (-\tfrac{3}{2}, -1) = (1 - \tfrac{1}{2} + \tfrac{1}{2} - \tfrac{3}{2} = 0 + \tfrac{1}{2} - 1, 1 + 0 + \tfrac{1}{2} - \tfrac{3}{2})$$
$$= (-\tfrac{1}{2}, 0) \tag{7.7.38}$$

so we have closure again.

Finally, we need to construct the new $\oint$ operator that has ghost number $-\frac{3}{2}$. The naive integral operator has only ghost number $-\frac{1}{2}$, so we need an operator that has ghost number $-1$. Once again, conformal field theory supplies us with such an operator:

$$Y(\sigma) = c_- \partial_+ \xi e^{-2\phi} \tag{7.7.39}$$

This operator is called the "inverse picture changing operator" and is, in some sense, the inverse of the operator $X$. We can show

$$\lim_{\sigma_1 \to \sigma_2} X(\sigma_1) Y(\sigma_2) = 1 \tag{7.7.40}$$

so that $X$ and $Y$ are inverses of each other. With this new operator, our new integration operator is defined so that

$$\oint (a, \psi) = \int Y(\tfrac{1}{2}\pi) a \psi \tag{7.7.41}$$

i.e., the integration rule is the same as before except that we insert a factor of $Y(\frac{1}{2}\pi)$ at the midpoint. Notice that this operator is defined only at the midpoint of the string, so that we still maintain a theory local in $\sigma$.

With these new ghost counting operators, we can now show that the following action:

$$S = \oint (A \otimes QA + \tfrac{2}{3} A \otimes A \otimes A) \tag{7.7.42}$$

is invariant under

$$\delta A = QA + A \otimes \Lambda - \Lambda \otimes A \tag{7.7.43}$$

The advantage of this formalism is that it generalizes to the interacting case. All operators are local in $\sigma$, so there is no problem in applying the continuity conditions on interacting strings.

The problem with this approach, however, is that it doesn't generalize to the closed string. Modular invariance is explicitly broken in models of this type. Second, the construction of the model depended more on making shrewd guesses for the ghost numbers than on any new insight into the superstring field theory. There is no underlying motivation for all of this.

Let us now summarize the various difficulties of the BRST formalism, stemming from the fact that it is actually a gauge-fixed formalism:

(1) It is a mystery that there should be two BRST formalisms, both equivalent on-shell, one based on the covariantized light cone and the other on the symmetric vertex.
(2) The meaning of the Faddeev–Popov ghost field, a remnant of gauge fixing in the first quantized conformal gauge, is obscure. Why should a gauge-fixed object play such an important role in the BRST theory?
(3) The multiplication rule $*$ and integration rule $\int$ for strings are simply postulated, with no explanation of their origin.
(4) The meaning of the equation $Q^2 = 0$ is obscure.
(5) The BRST formalism cannot go to the light cone gauge off-shell. As we have seen, we had to use the equations of motion to eliminate the redundant modes to reach the light cone action.
(6) Many truncations are being proposed for the BRST theory, not all of them consistent with the others. Which one is correct? The presence of constraints in the fermionic theory is not only ugly, it is also fatal. These constraints cannot be solved and still maintain the equation of motion; thus the theory collapses completely for the fermionic closed string sector.
(7) Why should the midpoint play such an important role in the theory? Mathematically, the midpoint of a string is not a particularly special point.
(8) Most important, the BRST formalism cannot be derived from first principles. *BRST is not a physical principle.* The BRST method is a formalism for quantizing in a gauge. It is devoid of physical content.

Before concluding this chapter on BRST string field theory, let us say a few words about a promising formulation of string field theory that makes no mention of the background metric and may give clues to the origin of geometry. This new formulation is called the "pregeometrical" approach [54–56] and yields some surprising conclusions.

Let us begin with the action

$$L = \frac{2}{3g^2} \Phi * \Phi * \Phi \tag{7.7.44}$$

At first glance, this action makes no sense. The action makes no mention of the background metric of space–time, which is desirable, but two things seem to be fatally wrong with the action: it has no kinetic term and its equations of motion seem trivial. To study this peculiar action, let us write down the equation of motion:

$$\Phi * \Phi = 0 \tag{7.7.45}$$

which usually has the solution $\Phi = 0$, so the theory is empty in a point particle theory. However, the previous equation is actually shorthand for an infinite set of coupled equations corresponding to an infinite-component field theory, so it is no longer trivial for a string theory. Assume, for the moment, that a classical solution $\Phi_0$ exists for the equations of motion which is nonzero. It must satisfy

$$\Phi_0 * \Phi_0 = 0 \tag{7.7.46}$$

Now power expand the field around this classical solution:

$$\Phi \to \Phi_0 + g\Phi \tag{7.7.47}$$

and reinsert this back into the Lagrangian. Because $\Phi_0$ satisfies the classical equations of motion, we have the new Lagrangian based on a new classical solution:

$$L = 2\Phi * (\Phi_0 * \Phi) + \frac{2g}{3}\Phi^3 \tag{7.7.48}$$

So far, nothing seems new. We still have not made any contact with a physical theory. Now let us make the crucial assumption behind this approach. We assume that an operator $D$ exists that satisfies

$$D\Phi = \Phi_0 * \Phi - (-1)^{\Phi}\Phi * \Phi_0 \tag{7.7.49}$$

If such an operator $D$ exists, then we can show that it is nilpotent. Now insert this expression (7.7.49) back into the Lagrangian, which now becomes

$$L = \Phi * D\Phi + \tfrac{2}{3}g\Phi^3 \tag{7.7.50}$$

If we can make the identification that this $D$ is equal to the usual BRST $Q$, then we have shown that the action is precisely the BRST action with a kinetic term! Thus, the usual BRST action might emerge when we expand about a new classical solution of the theory.

The novel feature of this approach is that nowhere have we made any mention of the background space–time metric. The background occurs only in the kinetic term, not the interacting term. In fact, the choice of the background metric emerges when we expand about a classical solution to the equations of motion. This is why this approach is called the "pregeometrical" theory. In principle, the geometric of space–time should emerge as one among many possible vacua.

In practice, however, we must be careful to check for the consistency of our approach. The key equation was (7.7.49). Under certain assumptions, it appears possible to find solutions for this equation where $D$ satisfies the axioms for the BRST string field theory. In this case, it appears that the usual flat-space BRST string field theory is nothing but one solution of the $\Phi^3$ action. Other solutions of this action presumably will yield BRST theories defined with different classical metrics.

For example, let us choose

$$\Phi_0 = Q_L I \tag{7.7.51}$$

where $Q_L$ is the BRST operator defined only on the left-hand side of the midpoint (for open strings) and $I$ is the identity operator (which equals 1 if left half and right half of the string coincide and zero otherwise). The original BRST $Q$ operator is equal to $Q_L + Q_R$. If we make the substitution in the original action:

$$\Phi = Q_L I + g\Phi \tag{7.7.52}$$

we find that we recover the original BRST action (7.5.7). (The right-hand side of $Q$ emerges in the manipulations because $Q_R I = -Q_L I$.)

Surprisingly, these definitions can be shown to be consistent with the original five axioms in Section 7.5, so that the usual BRST string action seems to be one among many possible solutions to (7.7.45).

So far, it can be shown that flat space is a consistent solution to (7.7.45). It remains to be seen, however, what other kinds of classical backgrounds can be found as solutions to the equations of motion.

## §7.8. Summary

The origin of the covariantized gauge approach was to construct a theory invariant under

$$\delta|\Phi\rangle = \sum_{n=1}^{\infty} L_{-n}|\Lambda_n\rangle \tag{7.8.1}$$

We postulated this invariance based on an analogy with the Yang–Mills theory, which is also invariant under a field that rotates into ghost fields. In fact, if we decompose the previous equation, we will derive precisely the linear part of the variation of the Yang–Mills theory

$$\langle\Phi|P[L_0 - 1]P|\Phi\rangle$$

where $P$ is a projection operator. By a power expansion, we can find the exact solution to this gauge problem:

$$\langle\Phi|(L_0 - \tfrac{1}{2}L_1 L_{-1} + \cdots)|\Phi\rangle \tag{7.8.2}$$

At the free level, this reproduces the usual Maxwell action. The solution can also be written to higher orders by brute force:

$$P(L_0 - 1)P = L_0 - 1 - \tfrac{1}{2}L_{-1}L_1 + L_{-1}^2(4L_0 + \tfrac{1}{2}D - 9)\Delta L_1^2 + L_{-1}^2(6L_0 + 6)\Delta L_2 + \text{h.c.} - L_{-2}(4L_0 + 2)(2L_0 + 2)\Delta L_2 + \cdots \tag{7.8.3}$$

where

$$\Delta = 2(16L_0^2 + (2D - 10)L_0 + D)^{-1} \tag{7.8.4}$$

and $P$ is a projection operator that eliminates ghost fields. Notice that this theory is nonlocal, meaning that we must incorporate auxiliary fields to soak up the nonlocal terms.

The necessity for auxiliary fields already shows up in the BRST quantization approach. Notice that the Faddeev–Popov ghosts propagate in the free theory:

$$p_\theta^1(\partial_\sigma + \partial_\tau)\theta^1 + p_\theta^2(\partial_\tau - \partial_\sigma)\theta^2 \tag{7.8.5}$$

This means that our path integral must be altered to take into account these extra degrees of freedom. The Hilbert space of our field functional must now be expanded:

$$\begin{aligned} 1 &= |\theta\rangle|X\rangle \int DX\, D\theta\langle\theta|\langle X| \\ \langle\theta|\langle X|\Phi\rangle &= \Phi(X, \theta) \end{aligned} \tag{7.8.6}$$

This means that our Green's functions must now be expanded to read

$$\begin{aligned} \Delta(X_i, \theta_i; X_j, \theta_j) &= \int_{X_i,\theta_i}^{X_j,\theta_j} DX\, DP\, D\theta\, DP_\theta \exp i \int d\sigma\, d\tau\, L(\sigma, \tau) \\ &= \int D^2\Phi\, \Phi(X_i, \theta_i)\Phi^*(X_j, \theta_j) \exp i \int_{X_i,\theta_i}^{X_j,\theta_j} DX\, D\theta\, L(\Phi) \end{aligned} \tag{7.8.7}$$

The BRST quantization introduces a new nilpotent operator $Q$ such that

$$Q = c_0(L_0 - 1) + \sum_{n=1}^{\infty} [c_{-n}L_n + L_{-n}c_n] - \tfrac{1}{2} \sum_{n,m=-\infty}^{\infty} :c_{-m}c_{-n}b_{n+m}:(m - n) \tag{7.8.8}$$

$$Q^2 = 0 \tag{7.8.9}$$

Thus, our free action can simply be expressed as

$$\begin{aligned} L &= \langle\Phi|Q|\Phi\rangle \\ \delta|\Phi\rangle &= Q|\Lambda\rangle \end{aligned} \tag{7.8.10}$$

The generalization of this to the interacting theory can also be performed using Neumann function techniques. We need a conformal mapping that will take the upper half-plane into a fully symmetric configuration. One choice is to splice together the mapping of six charges to produce the three-string vertex:

$$\rho = \sum_{i=-3}^{+3} \alpha_i \ln(z - z_i) = \ln\frac{z^3 - i}{z^3 + i} - i\pi/2 \tag{7.8.11}$$

$$\begin{aligned} \alpha_1 &= \alpha_2 = \alpha_{-3} = 1 \\ \alpha_{-1} &= \alpha_{-2} = \alpha_3 = -1 \\ z_1 &= e^{i\pi/6}; \qquad z_2 = e^{i5\pi/6} \\ z_3 &= e^{i\pi/2}; \qquad z_{-1} = -z_1 \\ z_{-2} &= -z_2; \qquad z_{-3} = -z_3 \end{aligned} \tag{7.8.12}$$

This allows us to write down an explicit formula for the vertex function:

$$|V_{123}\rangle = \exp\left[\tfrac{1}{2}\alpha_{-n}^r N_{nm}^{rs} \alpha_{-m}^s + P_0^r N_{om}^{rs} \alpha_{-m}^s + \tfrac{1}{2} N_{00} \sum_{r=1}^{3} P_{0r}^2\right] |0_{123}\rangle \tag{7.8.13}$$

One major advance in the BRST theory is a formalism where everything can be reduced down to the basic assumptions of the theory. It turns out that with five axioms, which are simply postulated with no derivation, we can derive the entire BRST theory:

(1) Existence of a nilpotent derivation

$$Q^2 = 0 \tag{7.8.14}$$

(2) Associativity of the * product

$$[A * B] * C = A * [B * C] \tag{7.8.15}$$

(3) Leibnitz rule

$$Q[A * B] = QA * B + (-1)^A A * QB \tag{7.8.16}$$

(4) The product rule

$$\int A * B = (-1)^{AB} \int B * A \tag{7.8.17}$$

(5) The integration rule

$$\int QA = 0 \tag{7.8.18}$$

With these rules, we can now show that the invariant action is

$$L = \int \Phi * Q\Phi + \tfrac{2}{3}\Phi * \Phi * \Phi \tag{7.8.19}$$

which is a Chern–Simons terms.

Lastly, we can show that the three-string vertex function is equivalent on-shell to the usual Veneziano vertex:

$$|\bar{V}_{123}\rangle = \prod_{r=1}^{3} \exp\left[\sum_{n\geq 0} \varepsilon_n^r L_{-n}^r\right] |V_{123}\rangle \tag{7.8.20}$$

Thus:

$$\langle \text{phy}|V_{123}\rangle = \langle \text{phy}|\bar{V}_{123}\rangle \tag{7.8.21}$$

The embarrassing thing, however, is that we can also write down the covariantized light cone vertex and show that it is also equivalent on-shell to the usual Veneziano vertex. Thus, we are faced with two equivalent BRST vertex functions, both of which produce the same on-shell vertices.

Although the open bosonic string was relatively easy to write down in BRST language, the superstring and closed string have been much less successful in the BRST approach. The problem is that a naive ghost counting of the superstring and closed string actions yields the wrong results:

$$\langle \Phi|Q|\Phi\rangle = 0$$

In order to change the ghost counting, there have been several proposals to "truncate" either the Hilbert space or the $Q$ operator so that the correct ghost counting is achieved. The problem with this approach is that the zero modes are treated differently from the other modes, so that locality in $\sigma$ is probably violated, and hence these approaches probably cannot be generalized to the interacting case.

A more promising approach is to introduce bosonized ghost operators that have the correct missing ghost numbers and insert them at the midpoints where we do not lose $\sigma$ locality. This seems to work well for the open superstring, but it fails for the closed string because selecting out "midpoints" violates modular invariance.

This problem is but a manifestation of a deeper problem. The BRST theory is simply postulated, without any derivation. It is not based on any physical or geometric principles. Thus, it cannot be the final theory. In fact, the presence of two equivalent vertex functions shows that it is, in fact, a gauge-fixed theory. We must now turn to the geometric version of the theory, where everything can be derived from first principles.

## References


[1] M. Kaku, *Nucl. Phys.* **B267**, 125 (1985).

[2] M. Kaku and J. Lykken, in *Symposium on Anomalies, Topology, and Geometry*, World Scientific, Singapore, 1985.

[3] T. Banks and M. Peskin, in *Symposium on Anomalies, Topology, and Geometry*, World Scientific, Singapore, 1985.

[4] R. C. Brower and C. B. Thorn, *Nucl. Phys.* **B31**, 163 (1971).

[5] V. Kac, *Lect. Notes in Phys.* **94**, 441 (1979); see also B. L. Feigin and D. B. Fuchs, *Sov. Math. Dokl.* **27**, 465 (1983).

[6] M. Kaku, *Phys. Lett.* **162B**, 97 (1985).

[7] M. Kato and K. Ogawa, *Nucl. Phys.* **B212**, 443 (1983).
[8] E. S. Fradkin and G. A. Vilkoviskii, *Phys. Lett.* **55B**, 224 (1975).
[9] W. Siegel, *Phys. Lett.* **142B**, 276 (1984); **151B**, 391, 396 (1985).
[10] T. Banks and M. Peskin, *Nucl. Phys.* **B264**, 513 (1986).
[11] T. Banks, D. Friedan, E. Martinec, M. E. Peskin, and C. R. Preitschopf, *Nucl. Phys.* **B274**, 71 (1896).
[12] W. Siegel and B. Zweibach, *Nucl. Phys.* **B263**, 105 (1985).
[13] B. Zweibach, CTP 1308, October 1985.
[14] K. Itoh, T. Kugo, H. Kunitomo, and H. Ooguri, *Prog. Theor. Phys.* **75**, 162 (1986).
[15] D. Pfeffer, P. Ramond, and V. Rogers, *Nucl. Phys.* **B274**, 131 (1986).
[16] E. Witten, *Nucl. Phys.* **B268**, 253 (1986).
[17] E. Witten, *Nucl. Phys.* **B276**, 291 (1986).
[18] A. Neveu and P. C. West, *Phys. Lett.* **165B**, 63 (1985).
[19] A. Neveu, J. H. Schwarz, and P. C. West, *Phys. Lett.* **164B**, 51 (1985).
[20] A. Neveu and P. C. West, *Nucl. Phys.* **B268**, 125 (1986).
[21] A. Neveu, H. Nicolai, and P. C. West, *Nucl. Phys.* **B264**, 173 (1986).
[22] H. Hata, K. Itoh, T. Kugo, H. Kunitomo, and K. Ogawa, *Phys. Lett.* **172B**, 186 (1986); *Phys. Lett.* **172B**, 195 (1986); see also I. Y. Aref'eva and I. V. Volovich, *Theor. Mat. Fiz.* **67**, 484 (1986).
[23] H. Hata, K. Itoh, T. Kugo,H. Kunitomo, and K. Ogawa, *Phys. Rev.* **D34**, 2360 (1986); *Phys. Rev.* **D35**, 3082 (1987); RIFP-674 (1986); RIFP-673 (1986).
[24] H. Hata, K. Itoh, T. Kugo, H. Kunitomo, and K. Ogawa, *Nucl. Phys.* **B283**, 433 (1987); see also N. P. Chang , H. Y. Guo, Z. Qui, and K. Wu, *CCNY-HEP-86/5*, 1986.
[25] M. E. Peskin and C. B. Thorn, *Nucl. Phys.* **B269**, 509 (1986).
[26] S. Giddings, *Nucl. Phys.* **B278**, 242 (1986).
[27] S. Giddings and E. Martinec, *Nucl. Phys.* **B278**, 91 (1986).
[28] S. Giddings, E. Martinec, and E. Witten, *Phys. Lett.* **176B**, 362 (1986).
[29] D. J. Gross and A. Jevicki, *Nucl. Phys.* **B282**, 1 (1987); **B287**, 225 (1987).
[30] S. Samuel, *Phys. Lett.* **181B**, 249, 255 (1986); N. Ohta, *Phys. Rev.* **D34**, 3785 (1986).
[31] E. Cremmer, C. B. Thorn, and A. Schwimmer, *Phys. Lett.* **179B**, 57 (1986).
[32] E. Cremmer, *Quantum Gravity–Integrable and Conformal Invariant Theories Conference*, LPTENS 86/32, September 1986.
[33] L. Caneschi, A. Schwimmer, and G. Veneziano, *Phys. Lett.* **30B**, 351 (1969).
[34] S. Sciuto, *Nuovo Ciminto Lett.* **2**, 411 (1969).
[35] A. R. Bogojevic and A. Jevicki, *Nucl. Phys.* **B287**, 381 (1987).
[36] A. Neveu and P. C. West, *Phys. Lett.* **179B**, 235 (1986).
[37] For an alternative approach to string lengths, see A. Neveu and P. West, *Nucl. Phys.* **B293**, 266 (87).
[38] W. Siegel and B. Zwiebach, *Nucl. Phys.* **B282**, 125 (87).
[39] J. Lykken and S. Raby, *Nucl. Phys.* **B278**, 256 (1986).
[40] S. Sen and R. Holman, *Phys. Rev. Lett.* **58**, 1304 (87).
[41] N. Ohta, *Phys. Rev. Lett.* **56**, 440 (1986); **56**, 1316 (E) (1986); S. P. de Alwis and N. Ohta, *Phys. Lett.* **B174**, 383 (1986); **B188**, 425 (1987); **B200**, 466 (1988).
[42] H. Terao and S. Uehara, *Phys. Lett.* **168B**, 70 (1986); *Phys. Lett.* **B173**, 134 (1986); *Phys. Lett.* **173B**, 409 (1986); *Phys. Lett.* **179B**, 342 (1986).
[43] G. D. Date, M. Gunaydin, M. Pernici, K. Pilch, and P. van Nieuwenhuizen, *Phys. Lett.* **171B**, 182 (1986).
[44] Y. Kazama, A. Neveu, H. Nicolai, and P. C. West, *Nucl. Phys.* **B276**, 366 (1986).
[45] H. Aratyn and A. H. Zimerman, *Nucl. Phys.* **B269**, 349 (1986); *Phys. Lett.* **165B**, 130 (1986); *Phys. Lett.* **166B**, 130 (1986); *Phys. Lett.* **168B**, 75 (1986).
[46] H. Ooguri, *Phys. Lett.* **172B**, 204 (1986).
[47] M. Awada, *Phys. Lett.* **172B**, 32 (1986); **180B**, 45 (1986); *Nucl. Phys.* **B282**, 349 (1987).
[48] A. Ballestrero and E. Maina, *Phys. Lett.* **180B**, 53 (1986).

[49] J. F. Tang and C. J. Zhu, *Phys. Lett.* **180B**, 50 (1986).
[50] K. Suehiro, *Nucl. Phys.* **B296**, 333 (88).
[51] K. Itoh, K. Ogawa, and K. Suehiro, KUNS-846 HE(TH) 86/06.
[52] A. Leclair, *Phys. Lett.* **168B**, 53 (1986).
[53] A. Leclair and J. Distler, *Nucl. Phys.* **B273**, 552 (1986).
[54] A. Friedan, Enrico Fermi Institute Report No. 85-27, 1985; T. Yonea, *Proceedings of the Seventh Workshop on Grand Unification/ICOBAN '86*, Toyama, Japan; E. Witten, unpublished.
[55] H. Hata, K. Itoh, T. Kugo, H. Kunitomo, and K. Ogawa, *Phys. Lett.* **175B**, 138 (1986).
[56] G. T. Horowitz, J. Lykken, R. Rohm, and A. Strominger, *Phys. Rev. Lett.* **57**, 283 (1986). See also K. Kikkawa, M. Maeno, and S. Sawada, *Phys. Lett.* **197B**, 524 (87).

CHAPTER 8

# Geometric String Field Theory

## §8.1. Why Geometry?

String theory, as we have seen, has been historically evolving backward for the past 20 years. This accounts for the fact that we are still searching for the fundamental geometry behind the theory.

By contrast, general relativity was discovered by Einstein, who first tried to explain a physical principle, the *equivalence principle*, and then postulated the geometry behind general covariance. The next step was writing down the unique action that satisfied these principles. Later, the classical theory of curved manifolds was developed, and finally attempts are being made to quantize the theory. Historically, general relativity progressed as follows:

$$\text{Geometry} \rightarrow \text{Action} \rightarrow \text{Classical theory} \rightarrow \text{Quantum theory}$$

Viewed in this fashion, we see that string theory has been evolving backward, starting with the accidental discovery of the Veneziano Born term, the construction of the quantum loops, leading to the classical Nambu–Goto string, then to the light cone action, and finally to attempts at a geometric derivation of the action:

$$\text{Quantum theory} \rightarrow \text{Classical theory} \rightarrow \text{Action} \rightarrow \text{Geometry}$$

Only recently, with the revival of interest in string theory, has there been a concerted effort to try to complete the evolution of the theory. As we have stressed, this is not just an academic question. Ultimately, string theory will succeed or fail on the basis of whether it can select out the correct quantum vacuum from among the tens of thousands of possibilities. Thus, a genuine field theory may be the only framework in which we can solve the pressing problem of the theory, namely the nonperturbative breaking of a 10-dimensional vacuum down to a 4-dimensional one.

There are, in turn, essentially two ways in which one might approach the field theory of strings:

(a) First, one might try to guess the field theory action by extrapolating from the first quantized Nambu action, in the same way that Feynman was able to derive the Schrödinger equation from the classical theory of non-relativistic particles. This is a "bottom-up" strategy that involves making shrewd educated guesses based on features of the first quantized theory. The disadvantage of this approach is that it must break the gauge invariances of the theory (e.g., we must choose the light cone gauge or the BRST conformal gauge). This means that we must arbitrarily postulate the existence of certain fields (e.g., "Faddeev–Popov ghost fields") that occur in a gauge-fixed first quantized approach but appear bizarre and unnatural occurring in a second quantized field theory. Thus, the action appears artificially contrived. In other words, BRST by itself is not a physical principle.
(b) Second, one may try a geometric approach and derive the entire theory by isolating the fundamental *physical principles* behind string theory. This is a "top-down" approach that captures the spirit of Yang–Mills theory and general relativity. In this approach, we start with a *single local gauge group*, based on simple physical principles, and demand that the action be invariant under this group. The major task for this approach is to isolate the gauge group of the string theory and write down its irreducible representations and curvatures and the action itself.

In this chapter, we will explain the most promising candidate for the geometric theory.

We will closely follow the analogy with general relativity or Yang–Mills theory, which can be derived from two simple principles based on an underlying geometry:

**Global Symmetry.** The theory must propagate helicities associated with pure spin-1 $A_\mu^a$ and spin-2 $g_{\mu\nu}$ fields that transform as irreducible representations of SU($N$) and the Lorentz group.

**Local Symmetry.** The action must be locally SU($N$) and generally covariant.

(The first principle is a statement of physics. It identifies the basic representations of the fields and states that they must be pure, i.e., no ghosts and no higher derivatives. We do not want higher derivative ghosts; we are explicitly ruling out higher derivative $R^2$ or $F^4$-type actions. This means that the first principle cannot be included as part of the second one, because local invariance by itself is compatible with higher derivative theories.)

Once we have isolated these two physical principles, the basic strategy we pursue to find the action is as follows:

$$\text{Gauge group} \to \text{Connections} \to \text{Covariant derivatives} \to \text{Curvature tensors} \to \text{Action} \tag{8.1.1}$$

In other words, to derive the action, we first start with two connection fields, one for each group. We construct the connection fields $A^i_\mu$ for the generators $\tau^i$ of SU($N$) and $\omega^{ab}_\mu$ for the generators $\sigma^{ab}$ of the Lorentz group. This allows us to write down the covariant derivatives:

$$\begin{aligned} D_\mu &= \partial_\mu + A^i_\mu \tau^i \\ \nabla_\mu &= \partial_\mu + \omega^{ab}_\mu \sigma^{ab} \end{aligned} \tag{8.1.2}$$

Next, we write down the curvatures for the two groups:

$$\begin{aligned} [D_\mu, D_\nu] &= F^i_{\mu\nu}\tau^i \\ [\nabla_\mu, \nabla_\nu] &= R^{ab}_{\mu\nu}\sigma^{ab} \end{aligned} \tag{8.1.3}$$

And finally, we observe that the tensor calculus for these groups is so restrictive that there is only one action based on these curvatures that contains two derivatives and is both generally covariant and locally SU($N$) invariant:

$$L = -\frac{1}{2\kappa^2}\sqrt{-g}R_{\mu\nu}g^{\mu\nu} - \tfrac{1}{4}\sqrt{-g}F^i_{\mu\nu}F^{\mu\nu i} \tag{8.1.4}$$

What is remarkable is that these two principles, based on group theory alone, are sufficient to determine the unique action. The reason for this is that the tensor calculus (i.e., the rules for multiplying irreducible representations and the measure of integration) is fixed by the gauge group. It does not have to be postulated by hand. Thus, the tensor calculus is so stringent that the unique action is singled out.

Similarly, we follow this basic strategy (8.1.1) found in ordinary point particle gauge theory and in general relativity to find the action of string field theory. We will postulate an entirely new local gauge group for the string, which we call the *unified string group* [1–5], and then proceed to form representations of the group. Following our basic strategy (8.1.1), we first form covariant derivatives, then curvatures, and finally the action itself. *Thus, we find that the reason why string field theory seems so different from conventional gauge theory is that the irreducible representations and tensor calculus of the unified string group are much richer and more complicated than the Lorentz-group and* SU($N$).

The unified string group can, in turn, be further broken up into two smaller local gauge groups, the reparametrization group and the string group:

## The Reparametrization Group

The reparametrization group is the group of transformations on the parametrization of a string:

$$\sigma \to \sigma + \varepsilon(\sigma)$$

It is called Diff($S$)$_-$, where $S$ can apply to either open strings $[0, \pi]$ or closed strings $S_1$. The group Diff($S$)$_-$ is generated only by the odd Virasoro generators $L_n - L_{-n}$ and is a subgroup of Diff($S$) (which is isomorphic to the conformal group and is generated by $L_n$ for all $n$). Diff$(S)_-$ maps physical strings $C$ into themselves, i.e.,

$$\text{Diff}(S)_-: C \to C$$

The key to understanding the reparametrization group is to define the theory totally in *loop space*, i.e., the space of physical, unparametrized space–time strings $\{C\}$. So far throughout this book, we have always treated parametrized strings $X_\mu(\sigma)$ and never unparametrized strings $C$. The essential idea behind the unified string group is to identify a group structure that maps strings into other strings which is totally defined in loop space. The real physics of a geometric theory must necessarily reside in the physical space of unparametrized strings.

Each of the infinitely many string configurations is represented as a single point in loop space. The significance of this is that loop space represents the physical dynamics of interacting strings, stripped clean of extraneous gauge artifacts (such as parametrization midpoints, parametrization lengths, Faddeev–Popov ghosts, and ghost number) that clutter up what is happening in physical space–time.

If $X_\mu(\sigma)$ is a vector that extends from the origin of our coordinate system to a point on the string $C$ labeled by $\sigma$, then we demand that our action be independent of the way we choose a particular parametrization $\{\sigma\}$. Thus, we want our free action $I(C)$ to be independent of any particular parametrization:

$$I(C_{\{\sigma\}}) = I(C_{\{\bar{\sigma}\}}) \tag{8.1.5}$$

In mathematical language, we say that the theory must be invariant under all possible diffeomorphisms of the string $C$. This means that *the "parametrization length" of the string is a pure fiction*:

$$\int_0^l d\sigma = l \tag{8.1.6}$$

In a genuinely covariant formalism, we should be able to change the parametrization length of the string at will. Of course, there is a real physical quantity associated with a string $C$, and that is its actual *physical invariant length*:

$$L = \int_0^l \sqrt{X_\mu'^2(\sigma)}\, d\sigma \tag{8.1.7}$$

Notice that the above expression is invariant under a change in parametrization:

$$\sigma \to \sigma + \varepsilon(\sigma, X) \tag{8.1.8}$$

The distinction between the fictitious parametrization length $l$ and the physical invariant length $L$ will prove to be a crucial factor in the geometric formulation. (All gauge-fixed theories, such as the light cone and BRST string field theories, are based on the fictitious parametrization length. Light cone fields have length $\sim p^+$ and BRST fields have length $\pi$. The geometric theory, we will see, is based on the physical invariant length $L$, measured in centimeters.) We will build our theory on tensors that are invariant under the above transformation rule, so that we can change the parametrization length of the string at any time.

## The String Group

Next, within the unified string group, we also wish to discuss the second symmetry, the group generated by interacting strings. Let us define a *triplet* as a set of three oriented physical strings (of arbitrary parametrization length) $C_1$, $C_2$, $C_3$ that can be arranged as in Fig. 8.1. We say that two strings are "conjugates" of each other if they can appear in the same triplet. (Notice that a string has an infinite number of conjugates. We denote the conjugate of $C$ by $\bar{C}$.) Then the *string group* is the set of maps of $C$ into all its conjugates, i.e.,

$$\text{SG}: C \to \bar{C}$$

These two groups, the reparametrization group and the string group, can in turn be combined into one principle.

**We define the universal string group as the group that maps a string $C$ into itself and all its conjugates.**

$$\text{USG}: \begin{cases} C \to C \\ C \to \bar{C} \end{cases} \tag{8.1.9}$$

When we generalize this bosonic group to include space–time supersymmetry, it becomes the unified string group.

Given this gauge group, we can now summarize the two physical principles from which we will derive all of string field theory [1–5]:

**Global Symmetry.** The fields $A_\sigma^\alpha$ and $e_{\mu\sigma}^{\nu\rho}$ must transform as pure irreducible representations of global Diff($S$).

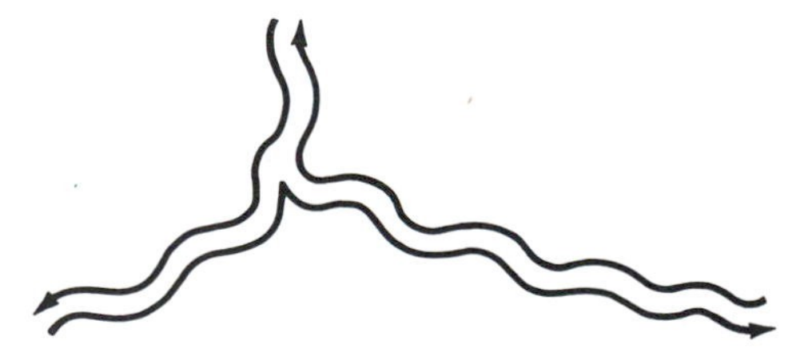

Figure 8.1. Triplet of three strings. Three physical unparametrized strings in loop space form a triplet if they can be arranged as in the diagram. This definition is totally free of parametrization midpoints and parametrization lengths. The joining point of these three strings is arbitrary.

**Local Symmetry.** The theory must be locally invariant under the USG, the group that maps $C$ into itself and $\bar{C}$.

Remarkably, there is only one unique action that satisfies both principles. The advantages of this geometric action are as follows:

(1) The mysterious "ghost sector" of the BRST theory has a simple, elegant interpretation in the geometric theory. We will see that it corresponds to the *tangent space* of the geometric theory.
(2) We can derive both versions of BRST string field theory as gauge-fixed versions of the geometric theory. We call the symmetric BRST theory the "midpoint gauge" and the light cone-like BRST theory the "endpoint gauge." They correspond to different parametrizations of the geometric vertex.
(3) The formalism is defined in loop space and therefore is totally independent of the parametrization or the background classical gravitational metric $g_{\mu\nu}$.
(4) The multiplication and integration rules are fixed strictly by group theory. They do not have to be postulated.

Unfortunately, in the case of general relativity and Yang–Mills theory, one must first construct a formidable mathematical framework to analyze SU($N$) and general covariance. Most of the work involves setting up the group representations and the tensor calculus. Once this is done, writing down the action itself takes only a few lines. The same applies to geometric string field theory. In the next six sections, we will follow our basic strategy (8.1.1) and patiently derive the representations, the connections, and the curvatures for the USG. After that, deriving the action itself in Section 8.7 takes only one page.

## §8.2. The String Group

Let us follow our basic strategy (8.1.1) and begin with a discussion of the string group (SG) by introducing the set of all unparametrized physical space–time strings:

$$\text{Unparametrized physical strings:} \quad \{C\} \tag{8.2.1}$$

It is absolutely essential to notice that these strings are defined to exist independent of any particular parametrization or any background classical gravitational metric. Notice that each string in a given triplet has particular direction or orientation and that the entire triplet configuration is cyclic (or anticyclic). Now let us define

$$f_{C_1C_2C_3} = \begin{cases} +1 & \text{for a triplet} \\ -1 & \text{for an antitriplet} \\ 0 & \text{otherwise} \end{cases} \tag{8.2.2}$$

Notice that these constants are simply 1, −1, 0, depending on whether the strings 1, 2, and 3 form a triplet, antitriplet (with reverse order), or otherwise.

Now let $\tilde{C}$ represent a string with orientation reversed from that of $C$. Then the final form of our structure constants is

$$\text{Structure constants:} \quad \begin{cases} f_{C_1,C_2}^{C_3} = f_{C_1C_2\tilde{C}_3} \\ f_{C_2,C_1}^{C_3} = -f_{C_2C_1\tilde{C}_3} \end{cases} \tag{8.2.3}$$

Notice that these structure constants are antisymmetric in strings 1 and 2.

Next, we introduce an abstract group generator $L_C$ associated with the physical string $C$. The algebra of the string group is defined to be [1, 2]

$$\text{SG:} \quad [L_{C_1}, L_{C_2}] = f_{C_1,C_2}^{C_3} L_{C_3} \tag{8.2.4}$$

(We will always sum over repeated indices.)

The crucial test, of course, is whether this algebra satisfies the Jacobi identities. Let us take the commutator of three of these generators:

$$[L_{C_1}, [L_{C_2}, L_{C_3}]] + \text{permutations} = 0 \tag{8.2.5}$$

Thus, we wish to show

$$f_{[C_1,C_2}^{C_4} f_{C_3],C_4}^{C_5} = 0 \tag{8.2.6}$$

(Notice that the summation over $C_4$ is trivial because there is only one string that is conjugate to both $C_1$ and $C_2$.)

The actual closure of the algebra is easy to prove in Fig. 8.2. Five strings are necessary to show the closure of the algebra. Notice that if we contract over $C_1$ and $C_2$ first, and then over $C_3$, we pick up the factor of −1. However, if we contract over $C_2$ and $C_3$, and then over $C_1$, we get the contribution of +1. Lastly, if we contract over $C_1$ and $C_3$, we get zero because they do not form a triplet. Thus, the sum over the cyclic permutations yields

$$-1 + 1 + 0 = 0$$

It is crucial to note that the structure constants of the group closed on the Jacobi identities without any reference to a parametrization of the string. Each index represents a space–time string $C$, independent of any parametrization or background geometry. Notice that the joining points of four strings do not

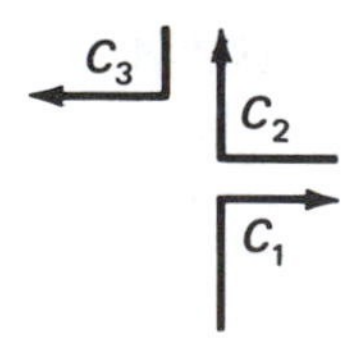


Figure 8.2. Proving the Jacobi identity. Summing over permutations of the triple commutator of three strings vanishes identically. Notice that the joining points of the three strings do not have to match for the Jacobi identity to be satisfied.

have to match, and hence there is no special "midpoint" of a string that is singled out.

Now that we have shown that the algebra closes correctly, let us try to construct some simple representations of the algebra. We will follow closely the derivation of representations of Lie groups presented in the Appendix, where we construct the representations of the orthogonal O($N$). There, we start with the rotations of a vector $x_i$:

$$x_i' = O_{ij} x_j \tag{8.2.7}$$

such that the combination

$$x_i x_i = \text{invariant} \tag{8.2.8}$$

is an invariant. The set of all such matrices $O_{ij}$ that keep $x_i x_i$ invariant forms the group O($N$).

Now, let us repeat these arguments for the string group. We make the transition from O($N$) to the string group by labeling the elements of the representation by the index $C$:

$$x_i \to \phi_C \tag{8.2.9}$$

where $C$ labels all possible space–time string configurations, totally independent of any parametrization of the string. $\phi_C$ plays exactly the same role as $x_i$. As in the case of O($N$), we now wish to make a transformation on this vector with a matrix defined on string states:

$$\phi'_{C_i} = \sum_{C_j} O_{C_i}^{C_j} \phi_{C_j} \tag{8.2.10}$$

Now let us demand that the following combination be invariant under $O$:

$$\phi^C \phi_C = \text{invariant} \tag{8.2.11}$$

with the definition

$$\phi^C = \phi_{\tilde{C}} \tag{8.2.12}$$

where $\tilde{C}$ represents a string $C$ with reversed orientation. Notice that string indices $C$ transform under the string group *covariantly* ($\Lambda$ is a gauge parameter):

$$\delta\phi_{C_i} = f_{C_i, C_j}^{C_k} \Lambda^{C_j} \phi_{C_k} \tag{8.2.13}$$

while the reversed orientation of strings transforms *contravariantly*:

$$\delta\phi^{C_i} = -f_{C_j, C_k}^{C_i} \Lambda^{C_j} \phi^{C_k} \tag{8.2.14}$$

If we insert (8.2.13) and (8.2.14) into (8.2.11), we can show that it is an invariant.

In this representation, an explicit form for the generator is given by

$$L_{C_i} = f_{C_j, C_i}^{C_k} \phi^{C_j} \frac{\delta}{\delta\phi^{C_k}} \tag{8.2.15}$$

The group elements satisfy

$$O^T O = 1 \tag{8.2.16}$$

As in the example of O($N$), where the metric was $\delta_{ij}$, here we see that the metric is

$$\delta_{C_i}^{C_j} = \begin{cases} +1 & \text{if } C_i = \tilde{C}_j \\ 0 & \text{otherwise} \end{cases} \tag{8.2.17}$$

It is easy to check, using (8.2.16), that the delta function transforms as a true mixed tensor under the string group:

$$(O^T)_{C_i}^{C_j} \delta_{C_j}^{C_k} O_{C_k}^{C_l} = \delta_{C_i}^{C_l} \tag{8.2.18}$$

Let us now write the group elements as the exponential of the elements of the algebra:

$$\Omega = e^{\Lambda^C L_C} \tag{8.2.19}$$

where the $\Lambda^C$ are parameters for the group. With this definition of the group elements as operators, we can now write

$$\Omega \phi_{C_i} \Omega^{-1} = O_{C_i}^{C_j} \phi_{C_j} \tag{8.2.20}$$

Let us define the action of the group on the states as

$$[L_{C_i}, \phi_{C_j}] = f_{C_i, C_j}^{C_k} \phi_{C_k} \tag{8.2.21}$$

Plugging (8.2.21) into (8.2.20), we find that we can rederive the infinitesimal transformation:

$$\delta \phi_{C_i} = f_{C_i, C_j}^{C_k} \Lambda^{C_j} \phi_{C_k} \tag{8.2.22}$$

Notice that the group closes correctly under this operation:

$$\Omega_1 \Omega_2 \phi_C \Omega_2^{-1} \Omega_1^{-1} = (O_2 O_1)_C^{C'} \phi_{C'} \tag{8.2.23}$$

This group law can be checked under infinitesimal transformations, in which case we simply rederive the Jacobi identities.

Unlike the orthogonal group, where vectors $x_i$ and the group generators $\lambda_{ij}$ transform differently, we find that the vector $\phi_C$ transforms in the adjoint representation of the group. For example, let the basis states of the string group be represented by the basis vectors

$$|\mathbf{e}^C\rangle \tag{8.2.24}$$

Now define $|\phi\rangle$:

$$|\phi\rangle = \phi_C |\mathbf{e}^C\rangle \tag{8.2.25}$$

Therefore, we have the following invariant:

$$\delta |\phi\rangle = 0 \tag{8.2.26}$$

Thus, we have

$$\begin{cases} \delta \phi_{C_i} = f_{C_i, C_j}^{C_k} \Lambda^{C_j} \phi_{C_k} \\ \delta |\mathbf{e}^{C_i}\rangle = -f_{C_j, C_k}^{C_i} \Lambda^{C_j} |\mathbf{e}^{C_k}\rangle \end{cases} \tag{8.2.27}$$

which preserves the invariant.

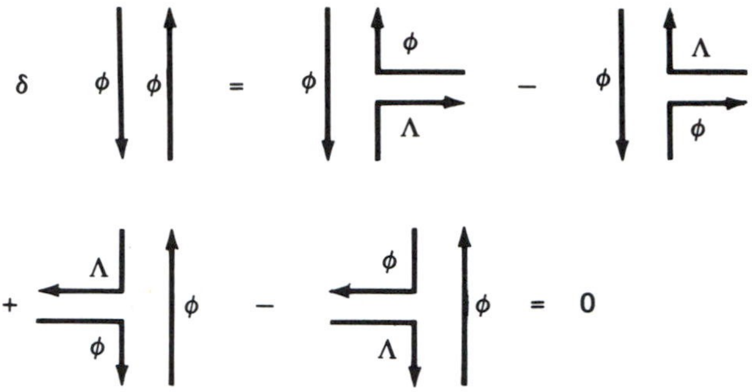

Figure 8.3. Constructing an invariant under the string group. By direct calculation, we can show that the variation of this quadratic term is zero.

Given this set of contravariant basis vectors, we can also define how the generators act on this space:

$$\begin{cases} \langle \mathbf{e}_C | \mathbf{e}^{C'} \rangle = \delta_C^{C'}; \qquad \langle \mathbf{e}_C | \Omega | \mathbf{e}^{C'} \rangle = O_C^{C'} \\ L_{C_i} | \mathbf{e}^{C_j} \rangle = f^{C_j}_{C_k, C_i} | \mathbf{e}^{C_k} \rangle \end{cases} \tag{8.2.28}$$

In this particular representation, the unit operator is given by

$$1 = |\mathbf{e}_C\rangle \langle \mathbf{e}^C|$$

and the generator of the algebra in this new basis can be represented as

$$L_{C_i} = |\mathbf{e}_{C_k}\rangle f^{C_k}_{C_j, C_i} \langle \mathbf{e}^{C_j}|$$

Thus, we have once again the equation

$$\langle \phi | \phi \rangle = \phi^C \phi_C = \phi \times \phi = \text{invariant} \tag{8.2.29}$$

Here, we found it convenient to introduce a new symbol, the multiplication $\times$ of strings. Thus, we have shown that the following combination is invariant under a transformation parametrized by $\Lambda$ (see Fig. 8.3):

$$\delta_\Lambda \langle \phi \times \phi \rangle = 0 \tag{8.2.30}$$

Next, let us look at the question of three-string interactions. Let us define

$$\langle \phi \times \phi \times \phi \rangle = \phi_{C_i} \phi_{C_j} \phi_{C_k} f^{C_i, C_j, C_k} \tag{8.2.31}$$

Notice here that we are summing over all possible triplet combinations $C_i$, $C_j$, $C_k$. We can also show that this combination is invariant under the string group:

$$\delta_\Lambda \langle \phi \times \phi \times \phi \rangle = 0 \tag{8.2.32}$$

(The easiest way to show that the three-string vertex is an invariant is to break up the sum over all possible string interactions and select out just one value of $\Lambda_C$. When we fix the value of the parameter $\Lambda_C$, we find that there are four

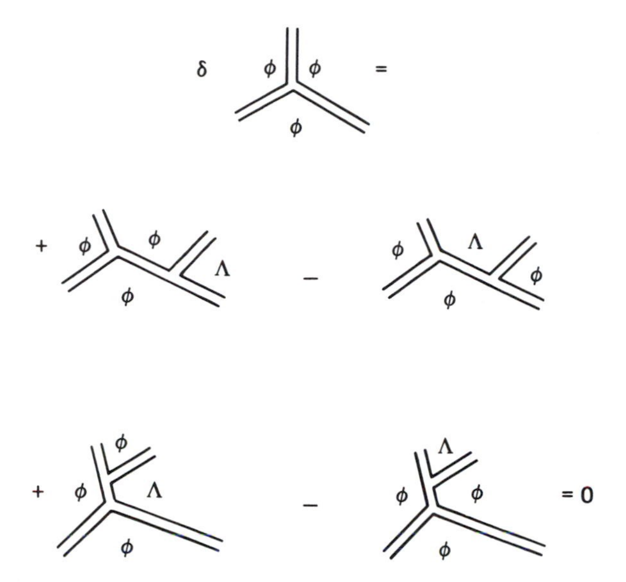


Figure 8.4. Invariance of the cubic interaction. We can show that the cubic interaction is invariant under the string group.

distinct contributions to the variation. In Fig. 8.4, we see that all four diagrams sum up to give zero.)

In summary, we have found that the representations of the string group are quite similar in structure to the representation of O($N$). Furthermore, we have found two invariants $\phi^2$ and $\phi^3$ independent of the parametrization or the background gravitational metric:

$$\begin{cases} \delta_\Lambda \langle \phi \times \phi \rangle = 0 \\ \delta_\Lambda \langle \phi \times \phi \times \phi \rangle = 0 \end{cases} \tag{8.2.33}$$

## §8.3. Unified String Group

Now that we have separately defined the string group SG and Diff($S$), let us put them both together and discuss the universal string group for bosonic strings and its supersymmetric extension, the unified string group.

To write down the generators of the universal string group, we now reintroduce *parametrized* generators of the string group. Let the string $C$ be parametrized by $X_\mu(\sigma)$:

$$C \to X_\mu(\sigma) \tag{8.3.1}$$

Furthermore, this means that $L_C$ becomes

$$L_C \to L_X = L_{X(\sigma_1), X(\sigma_2), X(\sigma_3), \ldots, X(\sigma_N)} \tag{8.3.2}$$

where the string $C$ is now parametrized by the points labled

$$\sigma_1, \sigma_2, \sigma_3, \ldots, \sigma_N$$

and we can take $N$ to be arbitrarily large.

Now that we have introduced a specific parametrization, there is an infinite number of parametrized strings $X$ that correspond to the same physical string $C$. This means that we have an infinite number of parametrized strings within each *equivalence class* of parametrized strings. Within each equivalence class are parametrized strings that parametrize the same physical string. Thus, we have the following decomposition:

$$\frac{\text{USG}}{\text{Diff}(S)_-} = \text{SG} \tag{8.3.3}$$

This means that the universal string group is the semidirect product of the string group and the diffeomorphisms of a string. In general, parametrized strings $X$ and $Y$ belong to the same equivalence class if they describe the same space–time string, i.e.,

$$X \sim Y \quad \text{if } X^\mu(\sigma) = Y^\mu(\sigma) + \varepsilon^\sigma Y'^\mu(\sigma) \tag{8.3.4}$$

and differ by an infinitesimal amount.

We now have to redefine the structure constants of the SG in terms of parametrized strings:

$$[L_X, L_Y] = f_{XY}^Z L_Z \tag{8.3.5}$$

As a first guess, we might postulate

$$f_{XYZ} = \begin{cases} +1 & \text{if } C_{X,Y,Z} \text{ form a triplet} \\ -1 & \text{if they form an antitriplet} \\ 0 & \text{otherwise} \end{cases} \tag{8.3.6}$$

and

$$\begin{cases} f_{XY}^Z = f_{XY\tilde{Z}} \\ f_{YX}^Z = -f_{YX\tilde{Z}} \end{cases} \tag{8.3.7}$$

The above definitions satisfy the Jacobi identities, but there is also an infinite constant created by the summation over all string states within an equivalence class. In other words, we need to define a measure over the integration over these states.

In order to define a measure, let us introduce a vector field $\zeta^\sigma(X)$ that allows us to calculate the difference between two elements within the same equivalence class. If one string is parametrized by $\{\sigma\}$ and another string within the same equivalence class is parametrized by $\{\bar{\sigma}\}$, then

$$\bar{\sigma} = \sigma + \zeta^\sigma(X); \quad \sigma < \pi; \quad \bar{\sigma} < \pi\alpha \tag{8.3.8}$$

Now introduce the geometric vertex:

$$f^{XYZ} = \prod_{r=1}^{3} \prod_{0 \le \sigma \le 1/2\pi} \delta[X_r(\bar{\sigma}_r) - X_{r-1}(\pi\alpha_{r-1} - \bar{\sigma}_r)] \tag{8.3.9}$$

where $\bar{\sigma}$ is defined through (8.3.8). The antisymmetric structure constants $f_{XY}^{Z}$ are introduced via (8.3.7). Notice that there is considerable freedom in defining the structure constant, because we can always make a gauge transformation on the strings that will change the $\zeta^{\sigma}(X)$ field as well. (This freedom is expressed by the fact that the measure of integration is given by a string density: $\int \det|e_{\mu\sigma}^{\nu\rho}|\, DX$, which we will discuss more in detail in Section 8.6. The string density is explicitly reparametrization invariant.)

Let us now write down the algebra of the universal string group [1–5]:

$$\text{USG:} \quad \begin{cases} [L_\sigma, L_\rho] = f_{\sigma\rho}^{\omega} L_\omega \\ [\varepsilon^\sigma L_\sigma, L_X] = \varepsilon^\sigma \partial_{\sigma -} L_X \\ [L_X, L_Y] = f_{XY}^{Z} L_Z \end{cases} \tag{8.3.10}$$

where the structure constants of the Virasoro group are given as

$$f_{\sigma\rho}^{\omega} = \delta(\rho - \sigma)\delta'(\omega - \rho) + 2\delta(\omega - \sigma)\delta'(\rho - \sigma) \tag{8.3.11}$$

and the operator that generates reparametrizations is given by

$$\partial_{\sigma -} = X'_{\mu\sigma} \frac{\delta}{\delta X^{\mu\sigma}}$$

where $\partial_{\sigma -}$ can be Fourier decomposed into $L_n - L_{-n}$.

The first line of (8.3.10) corresponds to the algebra of Diff$(S)_-$, while the last line corresponds to the string group. The middle line shows that the universal string group is the semidirect product between the reparametrization group and the string group.

The field $\phi^C$ is now replaced by $\phi^X$, where

$$\phi^X = \phi[X] = \phi[X(\sigma_1), X(\sigma_2), \ldots, X(\sigma_N)] \tag{8.3.12}$$

The field functional must transform as

$$\delta\phi^X = f_{ZY}^{X} \Lambda^Y \phi^Z$$

This explains the origin of the field functional $\Phi(X)$, eqn. (6.3.3), which we wrote down in the chapter on the light cone field theory. The multilocal field functional $\phi[X]$ is not a scalar but is actually the adjoint representation of the USG.

## §8.4. Representations of the USG

Now that the gauge group is clearly identified, the next step in our basic strategy (8.1.1) is to write down the connection fields, which, of course, requires that we understand the irreducible representations of the group and the constant tensors (Clebsch–Gordon coefficients).

Three irreducible representations of the USG are known: (a) the first is the adjoint representation of the string group, $|\mathbf{e}^C\rangle$, which we have been discussing; (b) the second is called the Verma module **V**, which we first encountered in (4.1.42) and which will be labeled by Greek indices $\alpha$, $\beta$; and (c) the third we shall call the string representation **S**, which contains the adjoint representation **A** of Diff($S$); **S** will be labeled by the continuous indices $\sigma$, $\rho$, $\theta$, which range from 0 to the parametrization length $l$, and its Fourier moments are indexed by Roman letters $m$, $n$ (see Refs. [6–9]).

The key step was made in (7.1.27), when we identified the auxiliary fields as making up the components of a Verma module.

## The V Representation

The representation space for the Verma module **V** is created from the generators $L_n$ of Diff($S$) by forming the universal enveloping algebra. Let us define a *contravariant vector* $|\mathbf{e}^\alpha\rangle$ as

$$|\mathbf{e}^\alpha\rangle \equiv L_{-\alpha_1}^{\lambda_1} L_{-\alpha_2}^{\lambda_2} \cdots L_{-\alpha_n}^{\lambda_n}|0\rangle \equiv L_{-\{\alpha\}}|0\rangle \tag{8.4.1}$$

where $(\alpha_i < \alpha_{i+1})$ and

$$\begin{aligned} L_n|0\rangle &= 0; \qquad n > 0 \\ L_0|0\rangle &= h|0\rangle \end{aligned} \tag{8.4.2}$$

where the index $\alpha$ labels an infinite number of elements, with the number of states at each level $n$ given by $p(n)$, where $p(n)$ is the partition of the integer $n$, and each representation is labeled by $[h, c]$. An arbitrary element $\Omega$ of the full group Diff($S$) (not just the subgroup Diff($S$)$_-$) is written as

$$\Omega = \exp \sum_{n=-\infty}^{\infty} \varepsilon^{-n} L_n = e^{\int d\sigma\, \varepsilon(\sigma) L(\sigma)} = e^{\varepsilon^\sigma L_\sigma} \tag{8.4.3}$$

Integration over the continuous string parameter $\sigma$ is assumed if it occurs twice. This means $\varepsilon^\sigma L_\sigma = \int d\sigma\, \varepsilon(\sigma) L(\sigma)$.

Let us define the action of the group generators $L_n$ on an element of **V**:

$$L_n|\mathbf{e}^\alpha\rangle \equiv f_{n\beta}^{\alpha}|\mathbf{e}^\beta\rangle \tag{8.4.4}$$

Notice that $L_n$ hitting an element of a Verma module in (8.4.1) always creates another element of the same Verma module. These constants $f_{n\beta}^{\alpha}$ are actually Clebsch–Gordon coefficients for the tensor product of the Verma module **V** with the adjoint representation **A** (which includes the generators $L_n$), i.e.,

$$f_{n\beta}^{\alpha}: \mathbf{V} \otimes \mathbf{A} \to \mathbf{V} \tag{8.4.5}$$

This, in turn, allows us to define a *covariant field* $\phi_\alpha$ such that

$$|\phi\rangle = \phi_\alpha|\mathbf{e}^\alpha\rangle \tag{8.4.6}$$

is an invariant:

$$\begin{aligned} \delta|\mathbf{e}^\alpha\rangle &= \varepsilon^{-n} f_{n\beta}^{\alpha}|\mathbf{e}^\beta\rangle \\ \delta\phi_\alpha &= -\varepsilon^{-n} f_{n\alpha}^{\beta}\phi_\beta \end{aligned} \tag{8.4.7}$$

This can be compactly represented as

$$\delta|\phi\rangle = 0$$
$$\Omega|\phi\rangle = |\phi'\rangle \tag{8.4.8}$$

It is also possible to define raising and lowering operators for this representation **V**. Let

$$\langle \mathbf{e}^\alpha| = \langle 0|L_{\{\alpha\}} \tag{8.4.9}$$

represent the adjoint of the ket vector; then let us define the contragradient form

$$S^{\alpha\beta} = \langle \mathbf{e}^\alpha|\mathbf{e}^\beta\rangle \tag{8.4.10}$$

The determinant of $S^{\alpha\beta}$ is called the *Kac determinant* (see (7.1.2)), and when it is nonzero the representation is irreducible and the matrix is invertible. Thus, the operation

$$\phi^\alpha = S^{\alpha\beta}\phi_\beta$$
$$|\mathbf{e}_\alpha\rangle = (S^{-1})_{\alpha\beta}|\mathbf{e}^\beta\rangle \tag{8.4.11}$$

converts a covariant vector into a contravariant one and vice versa. Thus, it is easy to show that $S^{\alpha\beta}$ and its inverse are raising and lowering operators for the representation **V**.

## The S Representation

In addition to **V**, there is also a second representation of Diff($S$) we will call the string representation **S**, which is the conformal representation introduced in (4.1.28) (except it is only a function of $\sigma$, not $\sigma$ and $\tau$). Each irreducible representation is labeled by its "weight" $n$. Each element $\phi^\sigma$ within each representation of weight $n$ is labeled by a continuous string parameter $\sigma$ or by its Fourier components $\phi^n$ [see (4.1.14)]:

$$\delta\phi_\sigma^{[n]} = \varepsilon^\sigma\phi_\sigma'^{[n]} + n\varepsilon'^\sigma\phi_\sigma^{[n]}; \qquad 0 \le \sigma \le l \tag{8.4.12}$$

More compactly,

$$\delta\phi_\sigma^{[n]} = \varepsilon^\rho f_{\sigma,\rho}^{[n]\theta}\phi_\theta^{[n]} \tag{8.4.13}$$

where

$$f_{\sigma\rho}^{[n]\theta} = \delta(\rho - \sigma)\delta'(\theta - \rho) + n\delta(\theta - \sigma)\delta'(\rho - \sigma)$$

(It is important to remember that the adjoint representation **A**, which contains the $L_n$, is a member of the string representation **S** with weight 2.)

These constants are nothing but Clebsch–Gordon coefficients for the following tensor product:

$$f_{\sigma\rho}^{\theta}: \mathbf{S} \otimes \mathbf{A} \to \mathbf{S} \tag{8.4.14}$$

We can multiply two fields of different weights and produce a new irreducible representation with the sum of the two individual weights:

$$\phi^{[n]}\psi^{[m]} = \omega^{[n+m]} \tag{8.4.15}$$

Furthermore, a representation with weight 1 is an invariant under the group Diff($S$) when we integrate over $\sigma$. This identity will prove essential when we are constructing invariant actions:

$$\delta \int d\sigma \, \phi_\sigma^{[1]} = 0 \tag{8.4.16}$$

(Notice that $\phi^\sigma \phi_\sigma = \int d\sigma \, \phi^\sigma \phi_\sigma$ is an invariant, with $\phi_\sigma$ transforming covariantly with weight $n$ and $\phi^\sigma$ transforming contravariantly with weight $1 - n$.)

There is also the important point that the delta function

$$\delta_{\sigma\rho} \tag{8.4.17}$$

is *not* a constant tensor of the group Diff($S$). Thus, we cannot raise and lower indices in this representation. ($\delta_\sigma^\rho$ is a constant tensor of the group, but cannot be used to raise and lower indices.) This small but important fact will place enormous constraints on the construction of the final action (and, in fact, will force us to abandon actions of the form $F^2$ and instead use actions of the Chern–Simons form $FF^*$).

Let us now make an arbitrary reparametrization of the string labeled by $\varepsilon^\sigma$. Then a scalar field $\phi(X)$, using the chain rule, transforms as

$$\begin{aligned} \delta\phi(X) &= \varepsilon^\sigma \partial_{\sigma-} \phi(X) \\ \partial_{\sigma-} &= X'_{\mu\sigma} \frac{\delta}{\delta X^{\mu\sigma}} \end{aligned} \tag{8.4.18}$$

Thus, we can use the fact that the integral of a scalar field is invariant under a global transformation in order to construct actions:

$$\delta \int DX \, L(X) = \int DX \, \varepsilon^\sigma \partial_{\sigma-} L(X) = 0 \tag{8.4.19}$$

Let us now generalize this expression to how elements of **V** or **S** transform under an arbitrary reparametrization of the string:

$$\begin{aligned} \mathbf{V}: \quad U\phi_\alpha U^{-1} &= \phi_\alpha(X + \delta X) + \Lambda_\alpha^\beta \phi_\beta \\ \mathbf{S}: \quad U\phi_\sigma U^{-1} &= \phi_\sigma(X + \delta X) + \Lambda_\sigma^\rho \phi_\rho \end{aligned} \tag{8.4.20}$$

where the $\Lambda$ matrices parametrize the transformation as follows:

$$\begin{aligned} \delta_\alpha^\beta + \Lambda_\alpha^\beta &= \langle \mathbf{e}_\alpha | \Omega | \mathbf{e}^\beta \rangle \\ \Lambda_\sigma^\rho &= \varepsilon^\theta f_{\theta\sigma}^\rho \end{aligned} \tag{8.4.21}$$

Now that we know how to manipulate Verma modules, we should note that the gauge group mentioned earlier can actually be generalized. To each string $C$, we can associate the generator $L_{\alpha,X}$, where $\alpha$ represents a Verma module.

Thus, the algebra (8.3.10) is changed slightly:

$$\begin{aligned}[\varepsilon^\sigma L_\sigma, L_{\alpha,X}] &= \varepsilon^\sigma \partial_{\sigma-} L_{\alpha,X} + \varepsilon^\sigma f_{\sigma\alpha}^\beta L_{\beta,X}\\ [L_{\alpha,X}, L_{\beta,Y}] &= f_{\alpha\beta}^\gamma f_{XY}^Z L_{\gamma,Z}\end{aligned} \tag{8.4.22}$$

Notice that we have enlarge our original gauge group to include transformations on Verma modules. We have now introduced a new Clebsch–Gordon coefficient, the tensor product of Verma modules:

$$f_{\alpha\beta}^\gamma\colon \mathbf{V} \otimes \mathbf{V} \to \mathbf{V} \tag{8.4.23}$$

Remarkably, these Clebsch–Gordon coefficients can be uniquely determined from group theory arguments alone. We define a new vertex $|V\rangle$ such that

$$\langle \mathbf{e}^\alpha | \langle \mathbf{e}^\beta | \langle \mathbf{e}^\gamma | V\rangle = f^{\alpha\beta\gamma} \tag{8.4.24}$$

where $|V\rangle$ satisfies

$$\prod_{r=1}^{3} \prod_{\sigma_r} [L_{\sigma_r}^r - L_{\pi\alpha_r - \sigma_{r-1}}^{r-1}] |V\rangle = 0 \tag{8.4.25}$$

Fortunately, by manipulating how the Virasoro generators reflect when acting on this vertex function, we can show that this identity alone is sufficient to determine all values of $f^{\alpha\beta\gamma}$. This means that (7.4.12) can be derived as a gauge-fixed version of $f^{\alpha\beta\gamma}$ when we use the vertex defined by (8.4.25). Thus, we do not have to postulate the form of the "ghost" vertex function. It emerges automatically as the Clebsch–Gordon coefficient for $\mathbf{V} \otimes \mathbf{V} \to \mathbf{V}$.

The advantage of this new Clebsch–Gordon coefficient is that we can now multiply fields that have Verma module indices. For example,

$$A \times B = C \tag{8.4.26}$$

means, written out explicitly,

$$(A \times B)^{\gamma,Z} = A^{\alpha,X} B^{\beta,Y} f_{\alpha\beta}^\gamma f_{XY}^Z = C^{\gamma,Z}$$

Since our basic fields will transform as Verma modules, we will find this multiplication law extremely useful. In other words, we will define the multiplication symbol $\times$ as

$$\times \equiv f_{\alpha\beta}^\gamma f_{XY}^Z \tag{8.4.27}$$

(When we gauge-fix the theory and choose a parametrization such as the "midpoint gauge," we find that our product rule reduces to the product rule for the $*$ operation introduced in (7.5.1).)

Finally, we want to write down the constant tensors for the USG. For the ordinary Lorentz group in point particle theory, we know that there are three constant tensors, $\delta_\mu^\nu$, the antisymmetric tensor $\varepsilon^{\mu\nu\lambda\kappa}$, and the Dirac matrix $(\gamma^\mu)_{\alpha\beta}$:

$$\text{Constant tensors} = \delta_\mu^\nu;\ e^{\mu\nu\lambda\kappa};\ (\gamma^\mu)_{\alpha\beta}$$

The Dirac matrix is actually the Clebsch–Gordon coefficient found in the tensor product of a spinor and a vector:

$$(\gamma^\mu)_{\alpha\beta} = \langle \mathbf{e}_\alpha | \gamma^\mu | \mathbf{e}_\beta \rangle \tag{8.4.28}$$

where $\langle \mathbf{e}_\alpha |$ are basis states for four-spinors. Surprisingly, the USG has a set of constant tensors quite different from the Lorentz group, which will have an impact on the kinds of invariant actions we can write down. First, the delta functions $\delta^{\sigma\rho}$ and $\delta^{mn}$ are *not* constant tensors. (This can be checked by noticing that neither $\sum_n L_n L_n$ nor $\sum_n L_n L_{-n}$ is invariant under Diff($S$). Furthermore, $\delta_n^m$ and $\delta_\sigma^\rho$ are constant tensors, but they cannot be used to raise or lower indices.) This will greatly complicate finding the action. For example, *it means that there is no counterpart to the $\Box$ operator of Klein–Gordon theory or the square of the curvature tensor in Yang–Mills theory, $F^2$.*

However, the group Diff($S$) does have a Dirac-like constant tensor $(\gamma^\sigma)_{\alpha\beta}$, where the Greek symbols represent Verma modules. Like the Dirac matrix, this constant tensor is found in the Clebsch–Gordon decomposition of the tensor product of a Verma module and the string representation **S**:

$$(\gamma^\sigma)_{\alpha\beta} = \langle \mathbf{e}_\alpha | \gamma^\sigma | \mathbf{e}_\beta \rangle \tag{8.4.29}$$

The simple but powerful observation from group theory means that *the action for string field theory will be more like the Dirac equation than the Klein–Gordon equation.* Thus, $\partial_\sigma^2$ is not invariant, but $\gamma^\sigma \partial_\sigma$ is. We also note that the product of several of these matrices is also a constant tensor: $\varepsilon^{\sigma\rho\cdots} = \gamma^\sigma \gamma^\rho \cdots$. In summary, the action for string field theory will be different from the Klein–Gordon type of action because the constant tensors for Diff($S$) are

$$\text{Constant tensors} = \delta_\sigma^\rho;\ (\gamma^\sigma)_{\alpha\beta};\ (\varepsilon^{\sigma\rho\cdots})_{\alpha\beta}$$

## §8.5. Ghost Sector and the Tangent Space

Up to now, many of the manipulations of the representations of the USG may seem a bit formal, without any practical application to the known string field theories. One goal of this investigation into group theory, however, is to explain some of the rather puzzling aspects of BRST theory. Therefore, let us now show that the group theory of the USG can answer the following questions:

(a) Why is the basic BRST field equal to the "ghost number $-\frac{1}{2}$ truncation" of $\Phi(X, b, c)$?
(b) What is the origin of the "ghost sector" of the BRST theory?
(c) What is the meaning of the identity $Q^2 = 0$?
(d) How can one derive the free action $\langle \Phi | Q | \Phi \rangle$?

All four questions can be answered in terms of group theory.

Let us begin with the basic BRST field:

$$\Phi_{\text{BRST}}(X, b, c) = \mathbf{P}_{-1/2} \sum_{\{p\}\{q\}} \phi_{\{p\}\{q\}}(X)\{b_{-p}\}\{c_{-q}\}|-\rangle \tag{8.5.1}$$

where we sum over all possible ghost excitations, subject to the restriction that the projection operator $\mathbf{P}_{-1/2}$ truncates the field to the "ghost number $-\frac{1}{2}$ sector."

Although this field was postulated from the BRST approach, in the geometric theory it has a rather simple origin. It can be shown that this BRST field is just a Verma module [1], i.e.,

$$|\mathbf{e}^\alpha\rangle \leftrightarrow \mathbf{P}_{-1/2}\{b_{-p}\}\{c_{-q}\}|-\rangle \tag{8.5.2}$$

The proof of this is based on computing the "character" of each space. The character of a space $V$ is a function of $x$ whose Taylor expansion counts the number of states there are at the $n$th level of $V_n$.

$$\operatorname{ch} V = \sum_{n=1}^{\infty} x^n \dim V_n \tag{8.5.3}$$

If the space is an irreducible Verma module, then it can be shown that the dimension of the $n$th level is equal to the partition of the number $n$. (This can be seen from (8.4.1) if we count the number of states at the $n$th level.) Thus, the character of an irreducible Verma module is equal to

$$\begin{aligned} \operatorname{ch} V &= \sum_{n=1}^{\infty} p(n) x^n \\ &= \prod_{n=1} (1 - x^n)^{-1} \end{aligned} \tag{8.5.4}$$

(Notice that this simply reproduces the partition function of the single-loop amplitude.)

It can be shown that the character of an irreducible Verma module and the character of the "ghost number $-\frac{1}{2}$ truncation" of the BRST ghost spectrum are the same to all levels:

$$\operatorname{ch}(|\mathbf{e}^\alpha\rangle) = \operatorname{ch}\left(\mathbf{P}_{-1/2} \prod \{b_{-p}\}\{c_{-q}\}|0\rangle\right)$$

Now that we have shown that a Verma module and the BRST space have the same number of states, the last step is trivial. The final step in the proof is to show that the two Verma modules have the same values of $[h, c]$.

Thus, the BRST field $\Phi(X, b, c)$ is a *reducible* representation of Diff($S$), and the meaning of the "ghost number $-\frac{1}{2}$ truncation" of the BRST is that it is the *irreducible truncation*. We therefore have a simple interpretation of the "truncation" operation that must always be performed in the BRST approach. It corresponds to going from reducible to irreducible representations of the USG.

In the last chapter, on the BRST formalism, we saw that the "truncation" process for the superstring and closed string was a source of great confusion, with several mutually exclusive choices. In the geometric approach, there is only one truncation, the irreducible one.

Next, we will show that the entire "ghost sector" corresponds to the *tangent space* of the USG. Consider, for example, the ordinary four-dimensional Dirac field $\psi^\alpha$. Notice that under a global Lorentz transformation, the Lorentz generator has two parts, the radial ($x$-dependent) and orbital (spinor) parts:

$$M_{\mu\nu} = x^\mu p^\nu - x^\nu p^\mu + (\sigma^{\mu\nu})_{\alpha\beta} \tag{8.5.5}$$

where $\sigma^{\mu\nu}$ is a spinorial representation of the Lorentz generators.

Similarly, global reparametrizations of the string naturally divide up into radial ($X$-dependent) and orbital (Verma module) parts:

$$M_\sigma = L_\sigma^X + L_\sigma^V \tag{8.5.6}$$

where the second operator acts only on Verma modules. The commutators of the complete generator yield

$$[M_\sigma, M_\rho] = f_{\sigma\rho}^{\omega} M_\omega + (D - 26)\cdots \tag{8.5.7}$$

(The complete generator of reparametrizations, which is the sum of the radial and orbital modes, has the correct commutation relations only if $D = 26$. This is the origin of 26 dimensions in the geometric theory.)

Under local transformations, however, the situation becomes more complicated for both the Dirac particle and the string. In particular, *a tangent space is required because there are no finite-dimensional spinor representations of* GL(4). This means we are forced to introduce a tangent space for spinors in general relativity. We have no choice. In other words, the spinor and radial modes of a Dirac spinor $\psi^\alpha$ transform under two quite different local groups:

$$\begin{aligned} \mathrm{O}(3, 1)&: \quad \delta\psi^\alpha = \varepsilon_{\mu\nu}(\sigma^{\mu\nu})_{\alpha\beta}\psi^\beta(x) \\ \mathrm{GL}(4)&: \quad \delta\psi^\alpha(x) = \varepsilon^\mu \partial_\mu \psi^\alpha(x) \end{aligned} \tag{8.5.8}$$

where $\varepsilon^{\mu\nu}$ parametrizes a local Lorentz transformation, and $\varepsilon^\mu$ parametrizes a general coordinate transformation.

A similar situation holds in string field theory. We saw earlier that the $L_n$ generate the group Diff($S$), while the odd generators $L_n - L_{-n}$ generate only the subgroup Diff$(S)_-$:

$$\begin{aligned} &L_n\text{: Diff}(S) \\ &L_n - L_{-n}\text{: Diff}(S)_- \end{aligned} \tag{8.5.9}$$

The essential point is that $X_\mu(\sigma)$ will not support the full Diff($S$); i.e., a representation of Diff($S$) in terms of just the string field does not exist. Under a $\sigma$ reparametrization, a field $\phi(X)$ transforms only under the subgroup generated by $L_n - L_{-n}$:

$$\begin{aligned} \delta\phi(X) &= \int d\sigma\, \varepsilon(\sigma) X'_\mu(\sigma) \frac{\delta}{\delta X_\mu(\sigma)} \phi(X) \\ &= \varepsilon^\sigma \partial_{\sigma-} \phi(X) \end{aligned} \tag{8.5.10}$$

The meaning of this is that the radial and orbital parts of the field $\phi^\alpha(X)$ must transform under two different groups. A Verma module can transform under the full Diff($S$), whereas the radial part cannot. Thus, under a local transformation, the radial and orbital parts transform under two different groups:

$$\begin{aligned} \text{Diff}(S)\colon \quad & \delta\phi^\alpha(X) = \varepsilon^\sigma f^\alpha_{\sigma\beta}\phi^\beta(X) \\ \text{Diff}(S)_-\colon \quad & \delta\phi^\alpha(X) = \varepsilon^\sigma \partial_{\sigma -}\phi^\alpha(X) \end{aligned} \tag{8.5.11}$$

Notice that, at each string coordinate $X$, there is a separate Diff($S$) transformation acting on the Verma module. Thus, Diff($S$) is the symmetry group of the tangent space but not the group of transformations of $X$-dependent variables.

Let us say that a representation of Diff($S$) is called conformal (therefore a Verma module is conformal). Then our fundamental result is: *a tangent space is required because there are no conformal representations of* Diff($S$)$_-$. This explains the origin of the ghost sector. In summary:

Ghost sector $\rightarrow$ Tangent space

Similarly, we can also explain $Q^2 = 0$ from group theory arguments. In ordinary point particle theory, the origin of $d^2 = 0$, where $d = dx^\mu\, \partial_\mu$, is the fact that flat space–time is invariant under parallel displacements, i.e.,

$$[\partial_\mu, \partial_\nu] = 0 \rightarrow d^2 = 0 \tag{8.5.12}$$

To prove an analogous statement for string theory, let us find the counterpart of the derivative operator. First, we notice that our first principle requires global invariance under the full Diff($S$), not just the subgroup Diff($S$)$_-$. However, we notice that

$$\partial_{\mu\sigma} = \frac{\delta}{\delta X^{\mu\sigma}} \tag{8.5.13}$$

is unacceptable because it transforms only under Diff($S$)$_-$ and not the full Diff($S$). Thus, it fails to meet the first principle. The derivative operator with the correct properties is

$$\nabla_{\mu\sigma} = -i\partial_{\mu\sigma} + X'_{\mu\sigma} \tag{8.5.14}$$

Under a transformation generated by $L_n + L_{-n}$, $\partial_{\mu\sigma}$ rotates into $X'_{\mu\sigma}$ and vice versa. Let us now define the globally covariant derivative:

$$\partial_\sigma = \nabla_{\mu\sigma}\nabla_{\mu\sigma} \tag{8.5.15}$$

(where the Fourier components of $\partial_\sigma$ include the *full* set of generators $L_n$).

Let us focus on the most general weight 2 derivative, which is the sum of radial and orbital parts:

$$\nabla_\sigma = \partial_\sigma + aL_\sigma \tag{8.5.16}$$

where $a$ is arbitrary. We now wish to form the commutator $[\nabla_\sigma, \nabla_\rho]$. This means that we have to define the action of $\nabla_\sigma$ on a mixed tensor, such as

$$\nabla_\sigma A^\alpha_\rho = \partial_\sigma A^\alpha_\rho + bf^\theta_{\sigma\rho}A^\alpha_\theta + cf^\beta_{\sigma\alpha}A_{\rho\beta} \tag{8.5.17}$$

If we demand

$$[\nabla_\sigma, \nabla_\rho] = 0 \tag{8.5.18}$$

we find that we must set either $a = -c = -2b = 1$ or $a = -b = c = \frac{1}{2}$. The first choice is ruled out if we wish to eliminate anomaly terms, and the second choice holds only up to terms that vanish on contraction with $\gamma^\sigma$. We will take the second choice. Then it is easy to show

$$\gamma^\sigma \nabla_\sigma = Q \tag{8.5.19}$$

and that $\gamma^\sigma \nabla_\sigma$ is nilpotent.

Thus, the origin of the nilpotent operator $Q$ lies in the vanishing of the commutator of two covariant derivatives $[\nabla_\sigma, \nabla_\rho] = 0$, in the same way that the origin of the nilpotent $d = dx^\mu \partial_\mu$ lies in the vanishing of the commutator of two covariant derivatives $[\partial_\mu, \partial_\nu] = 0$.

Lastly, let us explain the origin of $\langle \Phi | Q | \Phi \rangle$. Because $\delta^{\sigma\rho}$ is not a constant tensor, Klein–Gordon-type actions involving $\Box$ do not exist. (Only Dirac-type actions involving $\gamma^\sigma \nabla_\sigma$ are possible.) Specifically, the following actions are unacceptable:

$$\Phi[L_0 - 1]\Phi; \qquad \partial_{\mu\sigma}\Phi \partial_{\mu\sigma}\Phi; \qquad \partial_\sigma \Phi \partial_\sigma \Phi \tag{8.5.20}$$

They have weights 2, 2, and 4, respectively, while an invariant must have weight 1. In fact, it is easy to see that *it is impossible to construct an invariant action out of scalar string fields alone.* Many attempts have been tried over the years, but group theory alone now tells that such attempts are futile.

The question is sometimes asked: Is the ghost sector really necessary in covariant string field theory? The answer, we now see from group theory, is yes. A tangent space is absolutely necessary because $\delta^{\sigma\rho}$ is not a constant tensor, and Klein–Gordon-type actions are not invariant.

Although scalar fields must be ruled out, it is possible to write down invariant action with higher tensor fields $\phi^\alpha$ and $A^\alpha_\sigma$:

$$\begin{gathered} \langle \phi | \gamma^\sigma \nabla_\sigma | \phi \rangle \\ \langle A_\sigma | \varepsilon^{\sigma\rho\omega} \nabla_\rho | A_\omega \rangle \end{gathered} \tag{8.5.21}$$

These actions, in turn, are invariant under

$$\begin{aligned} \delta|\phi\rangle &= \gamma^\sigma \nabla_\sigma |\Lambda\rangle \\ \delta|A_\sigma\rangle &= \nabla_\sigma |\Lambda\rangle \end{aligned} \tag{8.5.22}$$

We can also show that the first action involving $\phi^\alpha$ is actually a gauge-fixed version of the second action involving $A^\alpha_\sigma$, which is invariant under

$$\delta|A_\sigma\rangle = |\Sigma_\sigma\rangle; \qquad \gamma^\sigma |\Sigma_\sigma\rangle = 0 \tag{8.5.23}$$

This means that we can remove all $\sigma$ dependence in $A^\alpha_\sigma$, i.e., we can replace $|A_\sigma\rangle$ with $|\phi\rangle$, so that the first action emerges when we gauge-fix the second action.

## §8.6. Connections and Covariant Derivatives

Now that we have identified the representations of the USG, we must follow the basic strategy of (8.1.1) and construct the connection fields and covariant derivatives. Once again, we begin by strict analogy to point particle gauge theory.

For Yang–Mills theory and general relativity, we need the connection fields $A_\mu^i$ and $\omega_\mu^{ab}$ in order to gauge SU($N$) and local Lorentz invariance. In string theory, we need gauge fields $A_\sigma^\alpha$ and $\omega_{\mu\sigma,\alpha}^\beta$ in order to gauge both the string group SG and local Diff($S$). The correspondence between point particle connections and string connections is given by

$$\begin{Bmatrix} \mathrm{SU}(N)\colon A_\mu^i \\ \mathrm{O}(3,1)\colon \omega_\mu^{ab} \end{Bmatrix} \to \begin{Bmatrix} \mathrm{SG}\colon A_\sigma^\alpha \\ \mathrm{Diff}(S)_-\colon \omega_{\mu\sigma,\alpha}^\beta \end{Bmatrix}$$

Lastly, we also need the string vierbein, the counterpart of $e_\mu^a$ found in general relativity, which will allow us to multiply and integrate over fields preserving local Diff($S$) invariance. Since the basic string variable has two sets of indices $X_\mu(\sigma) = X^{\mu\sigma}$, the string vierbein necessarily must have four indices:

$$e_\mu^a \to e_{\mu\sigma}^{\nu\rho} \tag{8.6.1}$$

In summary, the minimum set of fields necessary to gauge Yang–Mills theory and general relativity consists of three fields: $A_\mu^a$, $\omega_\mu^{ab}$, $e_\mu^a$, and the minimum set of fields necessary to gauge string field theory also consists of three fields: $A_\sigma^\alpha$, $\omega_{\mu\sigma,\alpha}^\beta$, $e_{\mu\sigma}^{\nu\rho}$.

Now that we have identified our basic set of fields, let us continue to follow the strategy of (8.1.1) and construct covariant derivatives from $\nabla_{\mu\sigma}$ and $\nabla_\sigma$. Notice, first of all, that neither derivative is covariant under the string group. For example, acting on the product of two Verma modules, the derivative acts like

$$\nabla_\sigma[A \times B] = \nabla_\sigma A \times B + A \times \nabla_\sigma B + \Sigma_\sigma \tag{8.6.2}$$

where the product rule is given by (8.4.27) and

$$\gamma^\sigma \Sigma_\sigma = 0 \tag{8.6.3}$$

Thus, the operator $\nabla_\sigma$ acts distributively as a derivation modulo terms that vanish when contracted on $\gamma^\sigma$. (In general, terms in $\Sigma^\sigma$ will contain terms like $\gamma_\sigma Q - k\nabla_\sigma$, where $k$ is the eigenvalue of $\gamma^\sigma\gamma_\sigma$.)

Let us now analyze how derivatives transform under the string group. In general, a field $\phi_\alpha$ transforms under the string group as

$$U\phi^\alpha U^{-1} = \phi^\alpha + [\phi \times \Lambda - \Lambda \times \phi]^\alpha \tag{8.6.4}$$

The point of this discussion is to realize that the derivative of a field $\phi_\alpha$ will not transform correctly under the group, necessitating the introduction of a connection field. We find

$$U\nabla_\sigma \phi U^{-1} = \nabla_\sigma[U\phi U^{-1}] + \phi \times \nabla_\sigma \Lambda - \nabla_\sigma \Lambda \times \phi + \Sigma_\sigma \tag{8.6.5}$$

which does not transform correctly (we have suppressed the indices of **V**).

Let us introduce a connection field $A_\sigma^\alpha$, the counterpart of the Yang–Mills field, which will soak up the unwanted local factors. We define

$$\delta A_\sigma^\alpha = [\nabla_\sigma \Lambda + A_\sigma \times \Lambda - \Lambda \times A_\sigma]^\alpha \tag{8.6.6}$$

where we have suppressed all **V** indices for clarity. This allows us to write

$$D_\sigma = \nabla_\sigma + A_\sigma \tag{8.6.7}$$

which transforms correctly:

$$UD_\sigma \phi U^{-1} = D_\sigma \phi + D_\sigma \phi \times \Lambda - \Lambda \times D_\sigma \phi + \Sigma_\sigma \tag{8.6.8}$$

This is the equation that we desire. It shows that, with the addition of the connection field, we can create derivatives that are covariant under the entire USG.

We should stress that the previous equations were written in a compact notation. For example, in Yang–Mills theory, the covariant derivative, when written out in full, becomes

$$\nabla_\mu = \partial_\mu \delta_{ab} + A_\mu^i (\tau^i)_{ab} \tag{8.6.9}$$

Likewise, in the geometric theory, we must insert the group matrix given in (8.4.27). Thus, our covariant derivative, with all its indices written out, is equal to

$$\nabla_\sigma + A_\sigma = \nabla_{\sigma X} \delta_{\alpha,X}^{\beta,Y} + A_\sigma^{\gamma,Z} (\tau_{\gamma,Z})_{\alpha,X}^{\beta,Y} \tag{8.6.10}$$

where

$$(\tau_{\gamma,Z})_{\alpha,X}^{\beta,Y} = f_{\alpha\gamma}^{\beta} f_{XY}^{Z}$$

So far, we have constructed the covariant derivative $D_\sigma$ associated with the string group. Now, we must construct the covariant derivative $\nabla_{\mu\sigma}$ associated with local reparametrizations Diff($S$) or universal covariance. In order to introduce local Diff($S$) invariance, we will closely follow the example of general relativity and introduce string vierbeins and string connection fields. Let us begin by identifying how $\partial_{\mu\sigma}$ and $dX^{\mu\sigma}$ transform under the universal covariant group:

$$\begin{aligned} \frac{\partial}{\partial X^{\mu\sigma}} &= \frac{\partial \bar{X}^{\nu\rho}}{\partial X^{\mu\sigma}} \frac{\partial}{\partial \bar{X}^{\nu\rho}} \\ dX^{\mu\sigma} &= \frac{\partial X^{\mu\sigma}}{\partial \bar{X}^{\nu\rho}} \, d\bar{X}^{\nu\rho} \end{aligned} \tag{8.6.11}$$

If we let

$$\bar{X}^{\mu\rho} = X^{\mu\sigma} + \varepsilon^\sigma(X) X'_{\mu\sigma} \tag{8.6.12}$$

then we find

$$\begin{aligned} \partial_{\mu\sigma} &= \partial_{\mu\sigma}[\varepsilon^{\rho}(X)X'_{\nu\rho}]\partial_{\nu\rho} \\ dX^{\mu\sigma} &= -\partial_{\nu\rho}[\varepsilon^{\sigma}(X)X'_{\mu\sigma}]\, dX^{\nu\sigma} \end{aligned} \tag{8.6.13}$$

As in general relativity, we will find it convenient to introduce a string vierbein into our theory, in order to make our integration measure invariant, and a connection field into our theory because derivatives of covariant fields are no longer covariant.

As in general relativity, the two principles force us to adopt a unique functional measure for the theory. For string field theory, we are forced to choose a measure in which we integrate once over each physical space–time configuration $C$ and then over all possible $C$-dependent reparametrizations of the curve $C$. To write down such a functional measure, let us define

$$d\mu_X = \det|e^{\nu\rho}_{\mu\sigma}|\, DX \tag{8.6.14}$$

so that it transforms as a string density. If $\Phi(X)$ is a scalar field, then we want

$$\delta \int d\mu_X\, \Phi(X) = 0 \tag{8.6.15}$$

This, in turn, fixes the variation of the vierbein:

$$\delta e^{\nu\rho}_{\mu\sigma} = \partial_{\mu\sigma}[\varepsilon^{\theta}(X)X'_{\lambda\theta}]e^{\nu\rho}_{\lambda\theta} + e^{\nu\omega}_{\mu\sigma} f^{\rho}_{\omega\theta}\varepsilon(X)^{\theta} \tag{8.6.16}$$

A careful analysis of these formulas shows that our choice for the functional measure $d\mu_X$ is the correct one.

We should be careful to point out, however, that the components of the vierbein field are not all independent. This is because it is actually possible to represent the above identities in terms of a smaller vector field $\zeta^{\sigma}$ that transforms under **S**. This is not surprising, because we are not constructing a theory with general coordinate invariance on a string. That would mean

$$\delta X^{\mu\sigma} = \Lambda^{\mu\sigma}(X) \tag{8.6.17}$$

which would no longer be a string theory. The above transformation law means that every point along a string can simply move freely in 26-dimensional space in any direction, thus disintegrating the string. Instead, what we want is to have every point along the string move along the string itself, i.e., move along the tangent vector $X'_{\mu\sigma}$. Thus, universal covariance can be parametrized in terms of a vector field $\zeta^{a}$ rather than the full vierbein. We will use the vierbein formalism, however, because it will provide a compact form for the theory.

The introduction of local Diff($S$) renders noninvariant derivatives like

$$\partial_{\mu\sigma}\phi_{\alpha} \tag{8.6.18}$$

The solution, as in general relativity, is to introduce a connection field. Let us introduce a connection $\omega^{\beta}_{\mu\sigma,\alpha}$ such that the following transform as genuine tensors under Diff($S$):

$$\begin{aligned}\nabla_{\mu\sigma}\phi_\alpha &= \partial_{\mu\sigma}\phi_\alpha + \omega^{\beta}_{\mu\sigma,\alpha}\phi_\beta \\ \nabla_{\mu\sigma}\phi_\rho &= \partial_{\mu\sigma}\phi_\rho + \omega^{\theta}_{\mu\sigma,\rho}\phi_\theta \\ \nabla_{\mu\sigma}\phi_{\nu\rho} &= \partial_{\mu\sigma}\phi_{\nu\rho} + \Gamma^{\lambda\theta}_{\mu\sigma,\nu\rho}\phi_{\lambda\theta}\end{aligned} \tag{8.6.19}$$

If we let the indices $A$, $B$, and $C$ represent either **V** or **S** indices, then we can restore local invariance if we set

$$\begin{aligned}\delta\omega^{B}_{\mu\sigma,A} = &-[\partial_{\mu\sigma}\varepsilon^\rho]f^{B}_{\rho A} + \varepsilon^\rho f^{C}_{\rho A}\omega^{B}_{\mu\sigma,C} \\ &+ \varepsilon^\rho f^{B}_{\rho C}\omega^{C}_{\mu\sigma,A} + \partial_{\mu\sigma}[\varepsilon^\rho X'_{\nu\rho}]\omega^{B}_{\nu\rho,A}\end{aligned} \tag{8.6.20}$$

We can also construct the curvature tensor for Diff($S$), given by

$$R^{\beta}_{\mu\sigma,\nu\rho,\alpha} = [\nabla_{\mu\sigma}, \nabla_{\nu\rho}]^{\beta}_{\alpha} \tag{8.6.21}$$

Notice that we must now generalize our definition (8.5.15) of the derivative:

$$\partial_\sigma = \tfrac{1}{2}(e^{-1})^{\nu\rho}_{\mu\sigma}(e^{-1})^{\lambda\theta}_{\mu\sigma}\nabla_{\nu\rho}\nabla_{\lambda\theta} \tag{8.6.22}$$

Notice also that the contravariant indices of the vierbein transform as part of a linear tangent space that is covariant under the full local Diff($S$), which includes the entire group (not just the elements generated by $\sigma$ reparametrizations). This is the origin of two-dimensional conformal invariance in our theory, which would not be obvious if the theory were only invariant under $\sigma$ reparametrizations. Although the tangent space of the vierbein can transform under the full group Diff($S$), notice that any functional of $X$ is necessarily restricted to transform only under $\sigma$ reparametrizations. (If a functional of $X$ transformed under the entire group Diff($S$), then the theory would describe an on-shell theory, which is undesirable. We avoid this problem by having the tangent space of the vierbein transform under the full Diff($S$).)

We can now summarize the correspondence between our fields and the usual vierbein $e^{\lambda}_{\mu}$, connection $\omega^{ab}_{\mu}$, and gauge field $A^{a}_{\mu}$:

$$\begin{aligned}A^{a}_{\mu} &\to A^{\alpha}_{\sigma} \\ e^{\lambda}_{\mu} &\to e^{\nu\rho}_{\mu\sigma} \\ \omega^{ab}_{\mu} &\to \omega^{\beta}_{\mu\sigma,\alpha}\end{aligned} \tag{8.6.23}$$

The major difference, as we have said, is that universal covariance is parametrized by a vector field $\zeta^a$. In fact, we can show that the covariant derivative of a vierbein is equal to zero and that the curvature tensor defined over the tangent space of Diff($S$) is zero:

$$\begin{aligned}R^{\beta}_{\mu\sigma,\nu\rho,\alpha} &= 0 \\ \nabla_{\mu\sigma}e^{\lambda\theta}_{\nu\rho} &= 0 \\ e^{\nu\rho}_{\mu\sigma} &= \nabla_{\mu\sigma}[X^{\nu\rho} + \zeta^\rho(X)X'_{\nu\rho}] + \cdots\end{aligned} \tag{8.6.24}$$

This shows that the only independent field contained within the vierbein and the connection field is the vector $\zeta^\sigma$. It is tedious, but entirely straightforward, to show that we can eliminate the connection field in terms of the vierbein, and the vierbein itself in terms of the vector field. The symmetry of the vierbein and the connection field can now be derived by setting

$$\delta\zeta^\sigma(X) = \varepsilon^\sigma(X) \tag{8.6.25}$$

This, in turn, allows us to fix the parametrization by eliminating $\zeta^\sigma$ from the theory. By fixing the gauge, we have the freedom to choose

$$e_{\mu\sigma}^{\nu\rho} \to \delta_\mu^\nu \delta_\sigma^\rho \tag{8.6.26}$$

It should be noted that the addition of vierbeins into our theory at first seems to spoil the properties of the displacement operator and the final action will no longer be invariant. However, when we actually calculate these corrections, we find that the extra terms vanish because the curvature tensor is zero and because the covariant derivative of a vierbein is zero. Thus, the important properties of the displacement operator are all preserved.

## §8.7. Geometric Derivation of the Action

The last six sections were devoted to following our basic strategy (8.1.1) and establishing the group theoretical properties of the universal string group (USG). We had to spend a considerable amount of time discussing the tensor product decomposition of irreducible representations of the USG because the mathematicians have not yet investigated the problem. In essence, the previous sections were merely a warm-up for this section. Now that the mathematical preliminaries are over, we can construct the action itself in a few lines. This section marks the heart of the geometric formalism. Similarly in general relativity, it takes a considerable amount of effort to develop the tensor calculus of general covariance, but then the derivation of the action requires only a few lines. There is only one scalar that contains two derivatives and is composed of the metric, and that is the contracted curvature tensor.

Let us construct the curvature out of the covariant derivative $D_\sigma = \nabla_\sigma + A_\sigma$:

$$[D_\sigma, D_\rho] = F_{\sigma\rho} \tag{8.7.1}$$

where we have suppressed indices in $\mathbf{V}$. It transforms under the string group as

$$\delta F_{\sigma\rho} = F_{\sigma\rho} \times \Lambda - \Lambda \times F_{\sigma\rho} + \Sigma_{\sigma\rho} \tag{8.7.2}$$

where the term $\Sigma_{\sigma\rho}$ vanishes when contracted on by $\gamma^\sigma$. Our first choice for an invariant action might be of the form $F^2$, but this is impossible because $\delta^{\sigma\rho}$ is not a constant tensor under the group. Thus, this contraction is impossible.

In fact, our only choice is

$$\det|e|\langle F_{\sigma\rho}|\varepsilon^{\sigma\rho\theta\omega}|F_{\theta\omega}\rangle \tag{8.7.3}$$

where $\varepsilon^{\sigma\rho\theta\omega}$ is the product of four $\gamma^\sigma$ matrices.

This, in turn, can be shown to be a total derivative. If we extract the Chern–Simons topological invariant associated with this term, we find the final form of our action:

$$L = \det|e_{\mu\sigma}^{\nu\rho}|[A_\sigma \times \nabla_\rho A_\theta + \tfrac{2}{3}A_\sigma \times A_\rho \times A_\theta] \tag{8.7.4}$$

where it is understood we contract on Verma modules and the tensor $\varepsilon^{\sigma\rho\theta}$. This action has several local gauge invariances. Our gauge field $A_\sigma^\alpha$ transforms under the universal string group as

$$\begin{aligned} UA_\sigma^\alpha U^{-1} &= A_\sigma^\alpha(X + \delta X) + \Lambda_\beta^\alpha(X)A_\sigma^\beta + \Lambda_\sigma^\rho(X)A_\rho^\alpha \\ &\quad + [\nabla_\sigma\Lambda]^\alpha + [A_\sigma \times \Lambda - \Lambda \times A_\sigma]^\alpha \end{aligned} \tag{8.7.5}$$

The field also has the invariance

$$\begin{aligned} \delta|A_\sigma\rangle &= |\Sigma_\sigma\rangle \\ \gamma^\sigma|\Sigma_\sigma\rangle &= 0 \end{aligned} \tag{8.7.6}$$

Our final action can be written as (8.7.4), which is invariant under (8.7.5) and (8.7.6). Notice that the actual derivation of the action, once the mathematical preliminaries were established, took only a few lines, similar to the situation in general relativity.

Let us now summarize what we have learned from group theory alone. Many of the consequences of the group theory explain the curious features found in the BRST formalism:

(a) $\delta_{\sigma\rho}$ is *not* a constant tensor of Diff($S$), which places severe restrictions on what invariants one can write down for the theory. This means that the Lorentz invariants $\Box$ and $F_{\mu\nu}^2$ have no counterpart in string field theory. This is why the string action differs from the usual actions found in gauge theory. However, the Dirac-like $\gamma^\sigma\nabla_\sigma$ and the Chern–Simons form $FF^*$ are invariant under the USG.
(b) An action invariant under global Diff($S$) cannot be written down with scalar fields alone. This forces us to go to higher representations of the group.
(c) The complete set of irreducible representations of Diff($S$) is not known. However, we know of two, the Verma module **V** and the string representation **S**. In particular, this explains the strange origin of the BRST field $\Phi(X, \theta, \bar{\theta})$, which is now seen to the sum of highly *reducible* representations of Diff($S$).
(d) The mysterious ghost sector of BRST string theory is just the tangent space of the geometric theory. Furthermore, the Faddeev–Popov ghost

field, which enters the BRST formalism through the back door, is easily explained as a Clebsch–Gordon coefficient for $\mathbf{S}^{[-1]} \otimes \mathbf{V} \to \mathbf{V}$. Thus, the geometric formalism explains the strange origin of the ghost field, which performs two functions in the geometric approach, both to span a representation for $\mathbf{V}$ and also to provide a crucial Clebsch–Gordon coefficient that occurs in the free action.

(e) From group theoretic arguments alone, we find two free actions based on the fields $\phi_\alpha$ and $A_\sigma^\alpha$ that are invariant under global Diff($S$):

$$\begin{gathered} \langle \phi | \gamma^\sigma \nabla_\sigma | \phi \rangle \\ \langle A_\sigma | \varepsilon^{\sigma\theta\rho} \nabla_\theta | A_\rho \rangle \end{gathered} \tag{8.7.7}$$

which are invariant under

$$\begin{aligned} \delta | \phi \rangle &= \gamma^\sigma \nabla_\sigma | \Lambda \rangle \\ \delta | A_\sigma \rangle &= \nabla_\sigma | \Lambda \rangle \end{aligned} \tag{8.7.8}$$

(f) We find that the generator of the group Diff($S$), in general, has two contributions:

$$M_\sigma = \partial_\sigma + L_\sigma \tag{8.7.9}$$

By our first principle, we demand that $M_\sigma$ have no central term. This explains why the dimension of space–time is 26.

(g) The geometric formalism explains why $Q^2 = 0$. This is because

$$[\nabla_\sigma, \nabla_\rho] = 0 \tag{8.7.10}$$

or equals zero up to a term that vanishes when contracted on $\gamma^\sigma$. This is similar to the situation in the theory of forms, where the origin of $d^2 = 0$ is $[\partial_\mu, \partial_\nu] = 0$.

(h) The invariant U(1) operator $\gamma^\sigma \gamma_\sigma$ is called the "ghost counting operator" in the BRST formalism. Here, we see that it is an artifact of a particular representation of Diff($S$) and is not an essential feature of the theory. In fact, we can eliminate it entirely by taking abstract formulations of the tensor products on Diff($S$).

(i) The measure of integration, which must be postulated in the BRST formalism, is uniquely determined from the second geometric principle in the geometric formalism:

$$d\mu_X = \det |e_{\mu\sigma}^{\nu\rho}| \prod_\mu \prod_\sigma dX^{\mu\sigma} \tag{8.7.11}$$

(j) The multiplication rule $*$, which must be postulated in the BRST approach, is nothing but a gauge-fixed version of our $\times$ product defined in (8.4.27). In summary, the BRST approach must postulate the integration and multiplication rules. In the geometric approach, the multiplication and integration rules themselves are derived from first principles. This is similar to the situation in general relativity, where general covariance alone is sufficient to determine the multiplication rule for tensors and the integration rules.

(k) The origin of conformal invariance in two dimensions is obscure in the BRST formalism. In our approach, two-dimensional conformal invariance arises because the *tangent space* of the theory is invariant under the full local group Diff($S$), which is generated by the complete set of $L_n$. (Although the tangent space transforms under the full group, the variable $X$ always transforms under the subgroup Diff($S$)$_-$.)

(l) In the BRST formalism, the transition to the light cone field theory can only be made on-shell. However, because we have introduced vierbeins $e_{\mu\sigma}^{\nu\rho}$ and "connection" fields $\omega_{\mu\sigma,\alpha}^{\beta}$ for Diff($S$)$_-$, we have the freedom of choosing the light cone gauge directly in the off-shell Lagrangian.

(m) The fundamental field of the string field theory is the connection field for the string group. Therefore, it must transform as a mixed tensor $A_\sigma^\alpha$, in the same way that the connection field for Yang–Mills theory must transform as $A_\mu^a$ because of group theoretic arguments alone.

(n) The fundamental reason why our string group closes without the benefit of the parametrization midpoint is that physical strings $C$ have arbitrary parametrization lengths. In fact, the two BRST string theories are gauge-fixed versions of the geometric theory. One is called the "midpoint gauge" and the other is called the "endpoint gauge."

Our next task is to fix the gauge and eliminate many of the unwanted fields. The local gauge invariance allows us to eliminate the $\sigma$ parameter within $A_\sigma^\alpha$. We can always choose $\Sigma_\sigma$ such that we replace $A_\sigma^\alpha$ with $\phi^\alpha$:

$$\gamma^\sigma|A_\sigma\rangle = 0 \tag{8.7.12}$$

so we can make the substitution

$$A_\sigma^\alpha \to \phi^\alpha \tag{8.7.13}$$

Our action now reads

$$L = \det|e|[\langle\phi|Q|\phi\rangle + \tfrac{2}{3}\langle\phi|\langle\phi|\langle\phi|V_{123}\rangle] \tag{8.7.14}$$

which is invariant under

$$\delta|\phi\rangle = Q|\Lambda\rangle + \langle\phi|\langle\Lambda|V\rangle - \langle\Lambda|\langle\phi|V\rangle \tag{8.7.15}$$

Finally, we can always choose our field $\zeta^\sigma$ such that each string has the same parametrization length. This means $\det|e| \to 1$.

At this point, it appears that the geometric theory has been reduced to the usual BRST formalism. In fact, we can show that our two axioms, when we fix the gauge covariantly and select out the midpoint, can derive the five axioms of the BRST formalism. Moreover, we can check that the reduced field $|A\rangle$ is equal to the "ghost number $-\frac{1}{2}$" sector of $\Phi(X, \theta, \bar{\theta})$ to all orders.

## §8.8. The Interpolating Gauge

At first, the BRST symmetric vertex, where strings interact via their midpoints, and the light cone-type vertex, where strings interact at their endpoints, seem totally different. But this is because they are both gauge-fixed versions of the geometric theory. As a result, we call the first formalism the "midpoint gauge" and the second formalism the "endpoint gauge."

In the geometric theory, we have the freedom to choose yet another gauge, the "interpolating gauge" which smoothly interpolates between the midpoint gauge and the endpoint gauge. The interpolating gauge is also gauge-fixed, as it is based on a Y-shaped vertex with arbitrary string lengths. When we take the constraint that all parametrization lengths are equal, we have the midpoint gauge. When we take the constraint that one of the legs in the Y-configuration has zero parametrization measure, then we have the endpoint gauge.

It is crucial to note that our geometric vertex (8.3.9) contains *both* the midpoint and the endpoint gauge. As we functionally integrate over $\zeta^r(\sigma)$, we are integrating over all possible parametrizations of the same physical space–time configuration. This means that we are also integrating over both the midpoint and endpoint parametrizations. This is the essential point: *we can parametrize a physical triplet vertex in either the midpoint or the endpoint gauge.* Thus, gauge fixing corresponds to carefully selecting just one copy from each equivalence class of physically distinct string configurations. The only difference between the endpoint and the midpoint gauge is the measure.

To prove this, we will now construct the interpolating vertex explicitly.

The interpolating vertex satisfies the constraint

$$\prod_{r=1}^{3} \prod_{0 \le \sigma_r \le l_r} \{X_\mu^r(\sigma_r) - X_\mu^{r-1}(\pi|\alpha_{r-1}| - \sigma_r)\}|V\rangle = 0 \tag{8.8.1}$$

where:

$$\sum_{r=1}^{3} \alpha_r = 2l_2, \qquad \alpha_3 \le 0, \qquad \alpha_{1,2} \ge 0, \qquad 0 \le \sigma_r \le \pi|\alpha_r| = \pi(l_r + l_{r+1}). \tag{8.8.2}$$

Notice that we recover the midpoint gauge vertex when we set the absolute value of all $\pi\alpha_r$ equal to $\pi$, and we get the endpoint gauge when we fix $\sum_{r=1}^{3} \alpha_r = 0$.

To construct this vertex, we need a conformal mapping that takes us from the upper half $z$-plane to a multisheeted $\rho$-plane. The explicit map is given by [3]

$$\rho(z) = \alpha_1 \log(z-1) + \alpha_2 \log z + \sum_{r=1}^{3} \beta_r \log[(az^2 + bz + c)^{1/2} + a_r z + b_r]$$

$$a_1 = \frac{\alpha_1^2 - \alpha_2^2 + \alpha_3^2}{2\alpha_1}; \qquad a_2 = \frac{-\alpha_1^2 + \alpha_2^2 + \alpha_3^2}{2\alpha_2}; \qquad a_3 = -\alpha_3$$

$$b_1 = \frac{\alpha_1^2 + \alpha_2^2 - \alpha_3^2}{2\alpha_1}; \qquad b_2 = -\alpha_2; \qquad b_3 = \frac{\alpha_1^2 - \alpha_2^2 - \alpha_3^2}{-2\alpha_3}$$

$$a = \alpha_3^2; \qquad b = \alpha_1^2 - \alpha_2^2 - \alpha_3^2; \qquad c = \alpha_2^2; \qquad \beta_r = -\alpha_r \tag{8.8.3}$$

Notice that there is a Riemann cut, which is necessary to have a multisheeted $\rho$ plane. We take the cut from $\rho(z_0)$ to $\rho(\bar{z}_0)$ to be vertical. We take the square root to be $+$ $(-)$ for $z \to \infty$ $(-\infty)$. It is tedious but straightforward to show:

$$\frac{d\rho}{dz} = -\alpha_3 \frac{(z - z_0)^{1/2}(z - \bar{z}_0)^{1/2}}{z(z-1)} \tag{8.8.4}$$

($z_0$ and $\bar{z}_0$ are the roots of $az^2 + bz + c = 0$). Using the techniques pioneered in the light cone theory, it is now possible to extract the Neumann functions for the vertex function from the conformal map. The calculation depends on power expanding the $\rho$ function in terms of coefficients $a_j^{(r,i)}$. We can check that in the light cone limit $l_2 \to 0$ we recover the light cone vertex. In the symmetric limit where $\alpha_r \to \pm 1$, we obtain a new power series for the Neumann functions, which should be equivalent to the midpoint gauge calculation.

Using this interpolating vertex $|V_{\alpha_1\alpha_2\alpha_3}\rangle$, we can now write down the gauge-fixed action in the interpolating gauge:

$$L = \langle\phi|_\alpha Q|\phi\rangle_\alpha + \tfrac{2}{3}\langle\phi|_{\alpha_1}|\langle\phi_{\alpha_2}|\langle\phi_{\alpha_3}||V_{\alpha_1,\alpha_2,\alpha_3}\rangle \tag{8.8.5}$$

where $Q = \gamma^\sigma \nabla_\sigma$. We can always change the parametrization length by the following:

$$U_{\alpha_1,\alpha_2}|\phi\rangle_{\alpha_2} = |\phi\rangle_{\alpha_1}; \qquad U_{\alpha_1,\alpha_2} = e^{\sum_{n=1}^{\infty}\zeta_n(L_n - L_{-n})} \tag{8.8.6}$$

Notice that we are quantizing the theory with a fixed value of $\pi\alpha$. String fields of differing string lengths are all defined in terms of $\pi\alpha$. In other words,

$$\langle\phi(X_{\alpha_1}), \phi(Y_{\alpha_2})\rangle = U_{\alpha_1\alpha_2}\langle\phi(X_{\alpha_2}), \phi(Y_{\alpha_2})\rangle \tag{8.8.7}$$

If we had replaced the $U$ factor appearing in the previous equation with $\delta_{\alpha_1\alpha_2}$, we would have had an infinite number of Hilbert spaces with each string length, which would be undesirable.

The Feynman rules for the interpolating gauge are

$$\text{Feynman rules:} \quad \begin{cases} \text{Vertex} \to |V_{\alpha_1\alpha_2\alpha_3}\rangle \\ \text{Propagator} \to U_{\alpha_i\alpha}D_\alpha U_{\alpha_j\alpha}^{-1} \end{cases} \tag{8.8.8}$$

for arbitrary $\alpha_i$. When explicitly canceling these $U$ factors, we find that we can trivially go back to the midpoint gauge. However, when we remove these $U$ factors and go to the endpoint gauge, there is a remarkable series of identities for the Neumann functions that allow us to *derive the four-string interaction* (6.7.6) [4]. We find that the four-string interaction does not have to be included in the action if we use the vierbein; it is a gauge artifact, the counterpart of the four-fermion instantaneous Coulomb term found in QED. The $t$–$u$ graphs in the midpoint, interpolating, and endpoint gauges are related by a unified string group transformation as follows:

$$
\begin{aligned}
A_{tu} &= \langle V_{\alpha\alpha\alpha}|D_{5,\alpha}(t)|V_{\alpha\alpha\alpha}\rangle + \langle V_{\alpha\alpha\alpha}|D_{7,\alpha}(u)|V_{\alpha\alpha\alpha}\rangle \\
&= \langle V_{\alpha_1\alpha_4\alpha_5}|D_{5,\alpha_5}(t)|V_{\alpha_2\alpha_3\alpha_6}\rangle + \langle V_{\alpha_2\alpha_4\alpha_7}|D_{7,\alpha_7}(u)|V_{\alpha_1\alpha_3\alpha_8}\rangle \\
&\quad + \int_{\alpha_1-|\alpha_4|+2\delta}^{\alpha_1+|\alpha_4|-2\delta} \langle V_{\alpha_1\alpha_4\alpha_X}|V_{\alpha_2\alpha_3\alpha_X}\rangle \mu \, d\alpha_X
\end{aligned}
\tag{8.8.9}
$$

where $|\alpha_3| \geq \alpha_1$, $\alpha_2 \geq |\alpha_4|$ and $-\alpha_5 = \alpha_6 = \alpha_1 - |\alpha_4| + 2\delta$; $-\alpha_7 = \alpha_8 = \alpha_2 - |\alpha_4| + 2\delta$. For the $t$-scattering we have $1 + 4 \to 5 \to 2 + 3$. For $u$-scattering we have $2 + 4 \to 7 \to 1 + 3$. When $\delta = 0$, the last term becomes the usual light cone four-string interaction. When $\delta \neq 0$, then we have the four-string interaction for the interpolating gauge. When $|\alpha_i| = \alpha$ and $\delta = \alpha$, then we have the vanishing of the four-string interaction in the midpoint gauge:

$$
\begin{cases}
\delta = 0 & & \text{endpoint gauge} \\
\delta \neq 0 & & \text{interpolating gauge} \\
\delta = \alpha & (|\alpha_i| = \alpha) & \text{midpoint gauge}
\end{cases}
\tag{8.8.10}
$$

We see, therefore, that the four-string interaction is not a fundamental interaction, but a byproduct of transforming the four-point function from the midpoint gauge to the endpoint gauge. From (8.8.9), we can see that it is, in fact, exactly equal to the square of the interpolating vertex. The same thing happens in QED, where the instantaneous four-fermion Coulomb term is precisely the square of the three-vertex coupling of the $A_0$ field with two fermion fields.

An even more surprising development occurs for the closed string case. Previous formulations for the closed string have all violated modular invariance. Either moduli space is over-counted (because of an unwanted infinite integration over the fictitious string lengths) or is under-counted (because of a missing region of integration if we use a generalization of the Witten vertex). The solution to this problem is to use geometric string field theory.

First, the over-counting problem of moduli space is solved because the fictitious parametrization lengths are gauge parameters. Second, when we go to the midpoint gauge, a new *closed four-string interaction* is created, which precisely fills the missing integration region. This new closed four-string interaction has the topology of a tetrahedron. Each face of the tetrahedron corresponds to one of the four closed string states. It has been checked by an elaborate computer calculation that this new tetrahedron graph fills the missing region of the complex plane. [2].

When we smoothly go to the light cone-like gauge, we find that the tetrahedron graph disappears. When we smoothly change the parametrization lengths to be equal, we find that the tetrahedron graph actually occupies most of the complex plane. In between, the interpolating gauge has a tetrahedron graph which smoothly connects the midpoint and light cone-like gauges. Notice that this is precisely the opposite of the situation found in open string field theory.

Details of this construction can be found in [2].

## §8.9. Closed Strings and Superstrings

In the geometric formalism, we can also set up the local tensor calculus for closed strings based on vierbeins and connections in the same way as for open strings. The new feature for closed strings, however, is that there is no necessity for truncation or for imposing constraints from the outside, which is the principal defect of the BRST approach. The key feature of the geometric approach here is that there is a unique vacuum $|0\rangle$ to the group Diff($S_1$). The ghost states, as we have seen, are nothing but Clebsch–Gordon coefficients for Diff($S$), and therefore *the ghost vacua are not fundamental objects in our geometric approach.* In fact, the particular ghost fields that appear in the BRST formalism are but one of several ways in which to generate the Clebsch–Gordon coefficients. Therefore, the ghost vacua, by themselves, have no intrinsic significance.

The fact that the ghost vacua are not fundamental objects can be seen already at the open string level. Notice that the contragradient matrix

$$\langle \mathbf{e}^\alpha | \mathbf{e}^\beta \rangle = S^{\alpha\beta} \tag{8.9.1}$$

can be reduced out by moving the $L_n$ operators to the right until they annihilate with the vacuum on the right. Of course, a tremendous number of terms are generated in the process, which can be collected into a matrix $M^{\alpha\beta}$, leading to

$$\langle \mathbf{e}^\alpha | \mathbf{e}^\beta \rangle = M^{\alpha\beta} \langle 0|0\rangle \tag{8.9.2}$$

Obviously, we can always choose $\langle 0|0\rangle$ to be equal to one, so that $M^{\alpha\beta}$ equals $S^{\alpha\beta}$. Notice that we have used only pure group theoretical properties of Diff($S$), without making any mention of the "ghost vacua."

Now, let us repeat the same calculation, using an explicit ghost presentation. If we, for example, choose $|-\rangle = |0\rangle$ and $\langle -| = \langle 0|$, then we have

$$\langle \mathbf{e}^\alpha | \mathbf{e}^\beta \rangle = M^{\alpha\beta} \langle -|-\rangle = 0 \tag{8.9.3}$$

which is clearly nonsense. Thus, we cannot naively use the ghost vacuum as a replacement for the unique vacuum of Diff($S$). This example shows, in fact, that Hermitian conjugation does not preserve ghost number. This is because "ghost number counting" is itself not an intrinsic property of Diff($S$) but only a property of a particular representation.

Let us repeat the same calculation used earlier in the open string section to calculate the matrix element $(\gamma^\sigma)_{\alpha\beta}$, except now we will use the ghost coordinates. This time, learning our lesson from the previous example, let us assume that the Hermitian conjugation of $|-\rangle$ is equal to $\langle +|$:

$$\langle \mathbf{e}_\alpha|\gamma^\sigma|\mathbf{e}_\beta\rangle = (\gamma^\sigma)_{\alpha\beta}\langle +|\gamma^0|-\rangle = 0 \tag{8.9.4}$$

Once again, we find that the matrix element collapses if we naively take ghost vacua. Thus, the true vacuum of Diff($S$) need not have any rigorous correspondence to $|-\rangle$ or $|+\rangle$. Once again, we find that ghost counting does not survive Hermitian conjugation.

Now let us apply this knowledge to the closed string sector, where there are four possible vacua:

$$|\pm\rangle|\pm\rangle \tag{8.9.5}$$

In the BRST approach, we are forced to take $|-\rangle|-\rangle$ as the ghost vacua and $\langle -|\langle -|$ as its Hermitian conjugate, which leads us into all sorts of problems with ghost counting. Now let us start over again in the geometric formalism, armed with the fact that ghost vacua and ghost counting are not fundamental features of Diff($S$) but only features of a particular representation of the algebra.

Once again, we start with the universal enveloping algebra:

$$|\mathbf{e}^\alpha\rangle = L_{-\alpha_1}^{\lambda_1}\cdots L_{-\alpha_n}^{\lambda_n}\bar{L}_{-\bar{\alpha}_1}^{\bar{\lambda}_1}\cdots\bar{L}_{-\bar{\alpha}_m}^{\bar{\lambda}_m}|0\rangle \tag{8.9.6}$$

where $|0\rangle$ is the true vacuum of Diff($S_1$). As before, the free propagator is of the form

$$\phi_\alpha\nabla_\sigma\phi_\beta; \qquad \phi_\alpha\bar{\nabla}_\sigma\phi_\beta \tag{8.9.7}$$

where we have a doubling of derivative operators. As usual, we need a Clebsch–Gordon coefficient to form an invariant out of the above expression. We obviously need the matrices

$$\begin{aligned}(\gamma^\sigma)_{\alpha\beta} &= \langle \mathbf{e}_\alpha|\gamma^\sigma|\mathbf{e}_\beta\rangle\\ (\bar{\gamma}^\sigma)_{\alpha\beta} &= \langle \mathbf{e}_\alpha|\bar{\gamma}^\sigma|\mathbf{e}_\beta\rangle\end{aligned} \tag{8.9.8}$$

If we repeat our previous arguments from the open string, we find that we can choose

$$\begin{aligned}|0\rangle &= |-\rangle|-\rangle\\ \langle 0| &= \langle +|\langle -| + \langle -|\langle +|\end{aligned} \tag{8.9.9}$$

Normally, such a choice is not the usual Hermitian conjugation of the ghost vacua. But, as we said earlier, the only criterion is that we produce constant matrices $(\gamma^\sigma)_{\alpha\beta}$ that transform correctly as Clebsch–Gordan coefficients.

For the interacting case, let us now define an oriented triplet of closed strings to be the generalization of Fig. 8.1, so the three strings have the topology of the Greek letter "theta." There are, however, nontrivial complications when we write down the string group for closed strings in loop space. First, because closed strings have no endpoints and can be rotated, anti-triplets cannot exist. (Triplets are actually their own anti-triplets.) Second, anti-symmetric structure constants in loop space do not exist. Only symmetric tensors exist. Third, the Jacobi identities do not close properly on triplets. Although all this may seem disasterous, a solution actually does exist.

Let us define a triplet as three closed strings that can be arranged in a "figure eight" topology (i.e., the light cone-like configuration where the three physical lengths sum to zero). Let $\hat{\imath}_{123}$ be a Grassmann-valued number that is defined at the interaction point of three strings in this figure eight. Let us define the symmetric constant tensor:

$$
\begin{aligned}
f_{C_1C_2C_3} &= \begin{cases} +1(-1) & \text{for triplets if } C_3 = \text{outer (inner) string} \\ 0 & \text{otherwise} \end{cases} \\
\hat{f}^{C_3}_{C_1C_2} &= f_{C_1C_2\tilde{C}_3}\hat{\imath}_{123}
\end{aligned}
\tag{8.9.10}
$$

Now the string algebra and the Jacobi identities are:

$$
\begin{aligned}
\{L_{C_1}, L_{C_2}\} &= \hat{f}^{C_3}_{C_1C_2} L_{C_3} \\
[L_{C_{[1}}, \{L_{C_2}, L_{C_{3]}}\}] &= 0
\end{aligned}
\tag{8.9.11}
$$

Notice the new features of the string algebra. First, the generators are *Grassmann-valued*. Second, the structure constant itself must be Grassmann-valued (so the left and right side of the equation have the same Grassmann number). Third, the algebra is not a traditional Lie algebra at all because of the Grassmann operator defined at the splitting point. By carefully constructing the Jacobi identities for three closed strings (which now have both commutators and anti-commutators in them), we can show, graph for graph, that the resulting sum is precisely zero.

By introducing parametrizations into the algebra, we can show that a standard Lie algebra emerges, and that the mysterious operator at the breaking point actually becomes the "ghost insertion" operator which is found in closed string theories. In [2], we show that this algebra can be gauged, as before, and that the resulting action is again the Chern-Simons form associated with $F\tilde{F}$.

We can only sketch the geometric approach for the open superstring. Once again, we can introduce vierbeins and connection fields, except now they all have supersymmetric partners.

For a physical string $C$, we must now add a supersymmetric partner $F_C$ to the generator of the string group:

$$
(L_C, F_C)
$$

The superstring group (SSG), which is the generalization of the string group (SG), is now defined as

$$
\text{SSG:} \quad \begin{cases} [L_{C_1}, L_{C_2}] = f^{-C_3}_{C_1,C_2} L_{C_3} \\ [L_{C_1}, F_{C_2}] = f^{-C_3}_{C_1C_2} F_{C_3} \\ \{F_{C_1}, F_{C_2}\} = f^{+C_3}_{C_1,C_2} L_{C_3} \end{cases}
\tag{8.9.12}
$$

where the structure constants are

$$f_{\pm C_1 C_2 C_3} = \begin{cases} +1 & \text{for a triplet} \\ \pm 1 & \text{for an antitriplet} \\ 0 & \text{otherwise} \end{cases} \tag{8.9.13}$$

and

$$\begin{aligned} f_{C_1, \tilde{C}_2}^{\pm C_3} &= f_{\pm C_1 C_2 \tilde{C}_3} \\ f_{C_2, \tilde{C}_1}^{\pm C_3} &= \pm f_{\pm C_2 C_1 \tilde{C}_3} \end{aligned} \tag{8.9.14}$$

We can show that the super-Jacobi identities are satisfied. The fundamental field that we wish to quantize is now therefore doubled:

$$\Phi^X L_X + \Psi^X F_X \tag{8.9.15}$$

where $X$ is now a functional of both bosonic and fermionic variables:

$$X = \{X(\sigma_1), X(\sigma_2), \ldots, \psi(\sigma_1), \psi(\sigma_2), \ldots, \psi(\sigma_N)\} \tag{8.9.16}$$

We can also introduce the two irreducible representations **V** and **S**. As before, the crucial fact lies in the existence of the proper Clebsch–Gordon coefficient for a tensor product reduction. For Superdiff($S$), we still have the generalization of the representation **S** and **V**.

We begin with the definition of the Verma modules for Superdiff($S$):

$$|\mathbf{e}^\alpha\rangle = G_{-\alpha_1}^{\lambda_1} G_{-\alpha_2}^{\lambda_2} \cdots G_{-\alpha_N}^{\lambda_N} |0\rangle \tag{8.9.17}$$

Likewise, we can write

$$|\mathbf{f}^\alpha\rangle = F_{-\alpha_1}^{\lambda_1} F_{-\alpha_2}^{\lambda_2} \cdots F_{-\alpha_M}^{\lambda_M} |0\rangle \tag{8.9.18}$$

The particular Clebsch–Gordon coefficient that we wish to evaluate is

$$(\gamma^\sigma)_{\alpha\beta} = \langle \mathbf{f}_\alpha | \gamma^\sigma | \mathbf{f}_\beta \rangle \tag{8.9.19}$$

Once again, we can commute the various $F$'s to the right, given the fact that $\gamma^\sigma$ is an arbitrary contravariant vector. Much as before, we find terms like

$$\langle 0 | \gamma_0 | 0 \rangle \tag{8.9.20}$$

If $\langle 0|$ is the BRST vacuum, then the above expression is equal to zero. However, if $\langle 0|$ is equal to the real highest weight vacuum state of Superdiff($S$), then this does not in general vanish. The essential point here is that, by repeated commutations, we can construct a matrix that transforms like the proper Clebsch–Gordon coefficient under Superdiff($S$) independent of the precise value of the previous matrix element. The advantage of this approach is that we extract the essential Clebsch–Gordon coefficients based on the true vacuum of Superdiff($S$), rather than getting entangled with the infinite number of ghost vacuums that are possible.

Once we have obtained a numerical value for this Clebsch–Gordon coefficient $(\gamma^\sigma)_{\alpha\beta}$, we can use it to contract the following action:

$$\bar{\Psi}_\alpha \nabla_\sigma \Psi_\beta \tag{8.9.21}$$

with the Clebsch–Gordon coefficient to obtain an invariant action. Once again, our action is the Chern–Simons form associated with

$$s \det|e| P\langle A_\sigma | \varepsilon^{\sigma\rho\theta} \nabla_\rho | A_\theta \rangle \tag{8.9.22}$$

Likewise, gauge fixing is possible because the universal superstring group (USSG) is the semidirect product with the Superdiff($S$) group and the superstring group:

$$\frac{\text{USSG}}{\text{Superdiff}(S)} = \text{SSG} \tag{8.9.23}$$

By fixing the parametrization, we can break universal covariance and have a theory that has only the superstring group. In this way, we can retrieve the BRST approach.

Unfortunately, space limitations prevent us from discussing in more detail the closed interacting string and space–time supersymmetry in the geometric formalism. The essential point, however, is that the closed string theory in the geometric formulation is modular invariant, while the BRST closed string theories have difficulty with modular invariance. In addition, when we add space–time supersymmetry to the (bosonic) universal string group, we obtain the unified string group. For a more detailed discussion, the reader should consult Refs. [1, 2].

## §8.10. Summary

In Fig. 8.5, we see the relationship between the various string field theories. The purpose of the geometric formulation is to derive the entire theory from its basic physical assumptions. Remarkably, we have found that all of string field theory, and hence all of string theory itself, can be derived from two simple assumptions:

**Global Invariance.** The fields $A_\sigma^\alpha$ and $e_{\mu\sigma}^{\nu\rho}$ must transform as pure irreducible representations of Diff($S$).

**Local Invariance.** The theory must be locally invariant under the unified string group.

The entire field theory of strings can be derived as the *unique* solution to these two geometric principles. In fact, the first principle alone is powerful enough by itself to uniquely determine the free action. The second principle then determines the complete theory. The search for the action for string field theory therefore reduces to group theory: finding an invariant under the unified string group.

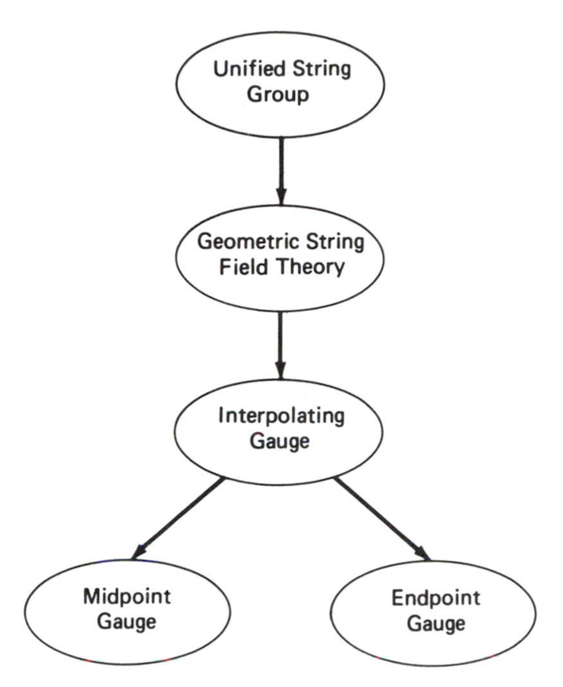


Figure 8.5. Relationship between the various string field theories. Geometric string field theory is defined in physical loop space. By gauge-fixing the geometric string field theory and fixing the parametrization, we obtain the interpolating gauge (where strings have arbitrary parametrization lengths). By choosing different values of the parametrization lengths, we can obtain either the endpoint gauge (light cone-like gauge) or the midpoint gauge.

*Thus, string field theory is the gauge theory of the universal string group* (USG), *in much the same way as Yang–Mills theory is the gauge theory of local* SU($N$). In fact, we now realize that string field theory looks so different from ordinary gauge field theory simply because the irreducible representations and the tensor products of the groups are different.

**We define the universal string group as the group that maps physical string $C$ into themselves and into their conjugates $\bar{C}$:**

$$\begin{aligned} &C \to C \\ &C \to \bar{C} \end{aligned} \tag{8.10.1}$$

and the unified string group as its supersymmetric extension. We can, in turn, extract the generators of the group:

$$\begin{aligned} [L_\sigma, L_\rho] &= f_{\sigma\rho}^{\bar{\sigma}} L_{\bar{\sigma}} \\ [L_\sigma, L_X] &= \partial_{\sigma-} L_X \\ [L_X, L_Y] &= f_{XY}^{Z} L_Z \end{aligned} \tag{8.10.2}$$

$$f_{XYZ} = \prod_{r=1}^{3} \prod_{0 \leq \sigma_r \leq \pi/2} \delta(X_r(\bar{\sigma}_r) - X_{r-1}(\pi\alpha_{r-1} - \bar{\sigma}_r))$$
$$\bar{\sigma} = \sigma + \zeta^\sigma(X) \tag{8.10.3}$$

and

$$f_{XY}^{Z} = f_{XY\tilde{Z}} \tag{8.10.4}$$

To construct the action, we now follow the path pioneered by gauge theories,

$$\text{Gauge group} \to \text{Connections} \to \text{Covariant derivatives}$$
$$\to \text{Curvature tensors} \to \text{Action} \tag{8.10.5}$$

In order to construct the covariant derivatives, we construct two "connections" for each of these two local symmetries, $A_\sigma^\alpha$ for the string group and $\omega_{\mu\sigma}^{\rho}$ for the reparametrization group Diff$(S)_-$. This, in turn, enables us to write down covariant derivatives and the curvature tensors for these two groups:

$$\begin{aligned} D_\sigma &= \partial_\sigma + A_\sigma \\ \nabla_{\mu\sigma} &= \partial_{\mu\sigma} + \omega_{\mu\sigma}^{\rho} f_{\rho,\alpha}^{\beta} \\ F_{\sigma,\rho} &= [D_\sigma, D_\rho] \\ R_{\mu\sigma,\nu\rho,\alpha}^{\beta} &= [\nabla_{\mu\sigma}, \nabla_{\nu\rho}]_\alpha^\beta \end{aligned} \tag{8.10.6}$$

We can now show the striking similarity between the fields occurring in gauge theory and the fields occurring in string field theory:

$$\begin{aligned} A_\mu^i &\to A_\sigma^\alpha \\ e_\mu^a &\to e_{\mu\sigma}^{\nu\rho} \\ \omega_\mu^{ab} &\to \omega_{\mu\sigma,\alpha}^{\beta} \end{aligned} \tag{8.10.7}$$

where $A_\mu^a$ is the Yang–Mills field, $e_\mu^\lambda$ is the vierbein, and $\omega_\mu^{ab}$ is the connection, while $A_\sigma^\alpha$ is the gauge field of the string, $e_{\mu\sigma}^{\nu\rho}$ is a string vierbein, and $\omega_{\mu\sigma,\alpha}^{\beta}$ is the string connection field.

The final action is

$$L = \det|e_{\mu\sigma}^{\nu\rho}|[A_\sigma \times \nabla_\rho A_\theta + \tfrac{2}{3} A_\sigma \times A_\rho \times A_\theta]$$

where we have suppressed **V** indices and the tensor $\varepsilon^{\sigma\rho\theta}$. We can summarize the similarity between ordinary gauge theory and string field theory as follows:

Global Lorentz → Global Diff($S$)

Local SU($N$) → String covariance

General covariance → Universal covariance

Tangent space: Lorentz symmetry → Tangent space: Diff($S$) symmetry

As in general relativity, we see that we can construct only one invariant out of fields transforming under the universal string group which has two derivatives, and that will be the action. Likewise, the multiplication and integration rules for strings do not have to be postulated but can be derived from first principles.

The problem with the bottom-up approach, as we mentioned, is that many features that appear natural in the first quantized theory will appear artificial and contrived when extrapolated into the second quantized approach. This is because the first quantized approach is a gauge-fixed one involving all the remnants of quantum gauge fixing, such as Faddeev–Popov ghosts. Only when we finally discuss the geometric approach will these unnatural objects finally have a beautiful mathematical explanation arising from pure group theory. We will find, in fact, that the Faddeev–Popov ghost fields are nothing but certain Clebsch–Gordon coefficients for tensor products of irreducible representations of the group in the tangent space.

The advantages of the geometric approach are as follows:

(1) The multiplication rule and integration measure are uniquely determined by the tensor calculus. Thus, group theory, and group theory alone, determines how to multiply and integrate over irreducible representations of the group. We do not have to appeal to any "axioms."
(2) The basic BRST field

$$\Phi^{ijklmn\cdots}_{abcdefg\ldots}$$

is no longer a mystery. It is simply an irreducible representation of the USG, a Verma module.
(3) The rather strange features of the BRST theory, such as the ghost sector, can be explained as the tangent space of the geometric theory. Faddeev–Popov ghosts are easily explained in terms of group theory. Ghost counting, on the other hand, is a by-product of taking a particular representation of the theory and is not a fundamental feature at all.
(4) There is no need to place constraints on the theory, which is fatal for the BRST approach. There is no acceptable fermion closed string in the BRST approach because of the constraint problem.
(5) By fixing the gauge, we can go either to the midpoint gauge or to the endpoint gauge.

Yet another advantage to the geometric theory is that the four-string interaction is now seen to be a gauge artifact. It emerges as the counterpart of the instantaneous four-fermion Coulomb term found in QED when we choose the Coulomb gauge. For open strings, the four-string interaction emerges when we transform from the midpoint gauge to the endpoint gauge. For the closed string, it emerges in the opposite way, when we transform from the light cone-like gauge to the midpoint gauge. This new closed four-string interaction has the topology of a tetrahedron, and by computer it can be shown to fill in precisely the missing region of the complex plane necessary to derive the Shapiro-Virasoro model. This solves the problem of writing down a closed string field theory that is compatible with modular invariance [2, 5].

In conclusion, there is still much to be done in completing a geometric theory [1–5, 10–12]. The goal is to isolate the counterpart of the equivalence principle for string theory. We hope that this will allow us to calculate non-perturbatively with the string theory. Already, preliminary nonperturbative results [13, 14]—from instantons, from nonrenormalization theorems for supersymmetric theories, etc.—indicate that nontrivial results should emerge when we go beyond perturbation theory.

Before nonperturbative calculations can be done, however, there is still a wealth of information that can be analyzed from classical solutions to super-string theory. We will now turn to Part III of this book, where we discuss the rather surprising physics that emerges when we reduce the theory to four dimensions.

## References

[1] M. Kaku, Geometric Derivation of String Field Theory from First Principles, I: Curvature Tensors and the Tensor Calculus, HEP-CCNY-14, August 1986; II: Superstrings Without Constraints, CCNY preprint, February 1987.
[2] M. Kaku, Geometric Derivation of String Field Theory from First Principles, III: Closed Strings and Modular Invariance, and III: Space–Time Supersymmetry. See also: M. Kaku and J. Lykken, *Phys. Rev.* **38D**, 30 (1988).
[3] M. Kaku, *Phys. Lett.* **200B**, 22 (88). *Phys. Rev.* **38D**, 3052 (1988).
[4] M. Kaku, Geometric String Field Theory: Deriving String Theory from First Principles, in *Superstrings* (edited by K. T. Mahanthappa and P. Freund), Plenum, New York, 1988.
[5] M. Kaku, Deriving the Four-String Interaction from Geometric String Field Theory, CCNY preprint, 1987.
[6] V. Kac, *Lect. Notes in Phys.* **94**, 441 (1979), Springer-Verlag: *Infinite Dimensional Lie Algebras*, Birkhauser, Boston, 1983.
[7] A. Rocha-Caridi and N. R. Wallach, *Math. Z.* **185**, 1 (1984).
[8] L. Dolan, *Compactified String Theory*, Lectures at the Lewes Workshop on Superstrings, World Scientific, Singapore, 1985.
[9] I. Kaplansky, *Commun. Math. Phys.* **86**, 49 (1982).
[10] For other geometric approaches, which differ significantly from the previous work, see K. Bardakçi, *Nucl. Phys.* **B284**, 334 (1987); **B297**, 583 (1988).
[11] I. Bars and S. Yankielowicz, *Phys. Rev.* **35D**, 3878 (1987).
[12] J. L. Gervais, *Nucl. Phys.* **B276**, 339 (1986); *Nucl. Phys.* **B276**, 349 (1986); LPTENS 86/26; LPTENS 86/29.
[13] M. Dine and N. Seiberg, *Phys. Rev. Lett.* **55**, 366 (1985); *Phys. Lett.* **162B**, 299 (1985).
[14] M. Dine, N. Seiberg, W. G. Wen, and E. Witten, *Nucl. Phys.* **B278**, 769 (1986).

# Part III

# Phenomenology and Model Building

CHAPTER 9

# Anomalies and the Atiyah–Singer Theorem

## §9.1. Beyond GUT Phenomenology

Ideally, we would want a truly unified field theory of all known interactions to satisfy at least two criteria:

(1) It must be based on simple physical assumptions, expressed in terms of a new geometry, which will allow no more than one coupling constant.
(2) It must yield a finite theory of gravity coupled to the minimal $SU(3) \otimes SU(2) \otimes U(1)$ model of particle interactions.

So far in this book, we have only begun to explore the first possibility by demonstrating that a second quantized field theory exists that is based on two physical principles. However, the developments we have presented in string theory have been purely formal. Unless we can make contact with the known experimental data, then the theory, no matter how elegant, must be discarded. The true test of a unified field theory is that it can reproduce the known experimental data at low energies.

The problem, however, is that dimensional breaking from a 10-dimensional theory down to 4 dimensions can occur only nonperturbatively. To any finite order in the perturbation theory, the dimension of space–time seems perfectly stable. In general, field theory is the only reliable formalism in which to perform nonperturbative calculations, because the first quantized formalism is necessarily perturbative. Unfortunately, we do not yet understand how to perform realistic nonperturbative calculations in string theory, mainly because the field theory of strings is still in its infancy. Thus, physicists have not been able to calculate the stability of any classical vacuum solution. Hence we will not deal with the question of quantum stability in Part III of this book and will concentrate exclusively on classical solutions of the equations of motion.

Given this important restriction, it is surprising that initial attempts at exploring the experimental consequences of classical string theory have produced a wealth of new phenomenology that takes us beyond GUTs. In Part III, we will first ask whether the string theory is compatible with the results of the standard GUT theory. Specifically, we will ask whether it can reproduce the GUT theories of SU(5), O(10), or $E_6$. In this respect, the string theory has had some measure of success. We will show in Chapter 11, for example, that the $E_8 \otimes E_8$ heterotic string can easily be broken down classically to $E_6$, which has chiral fermion solutions and an acceptable GUT phenomenology.

However, we must also demand that the string theory go beyond the standard phenomenology of the GUT theory. Specifically, we must ask the following questions of the string model:

(1) Can it explain three generations of chiral fermions?
(2) Can it explain the experimental results on proton decay?
(3) Can it explain the smallness of the electron mass?
(4) Can it explain the vanishing of the cosmological constant after supersymmetry breaking?

Although it is still too early to say, there are indications that the string theory is rich enough in content that it contains mathematical mechanisms which may answer the above questions. Specifically, new topological mechanisms enter into the picture, such that basic phenomenological concepts, such as generation number, are now recast in a new topological language. *Topology is the crucial new mathematical feature that allows us to go beyond standard GUT phenomenology.*

We will begin this chapter with a discussion of the isospin groups that are allowed in string theory. We will find that the properties of the $S$-matrix, such as cyclic symmetry and factorization, provide only the weakest restrictions on the gauge group of the theory. Then we will show that the cancellation of anomalies in string theory places stringent conditions on which gauge groups may be allowed by the theory. Basically, the gauge group of a supersymmetric theory must contain exactly 496 generators, which restricts us to either SO(32) or $E_8 \otimes E_8$.

To understand how the cancellation process takes place, we will find it convenient to review some of the elementary properties of characteristic classes. In particular, new developments in supersymmetry have now made it possible to prove the *Atiyah–Singer index theorem* from a simple Lagrangian. Traditionally, the proof of the Atiyah–Singer theorem has been inaccessible to most physicists because of the intricacies of the mathematical formulation. However, one of the surprising features of supersymmetry is that it gives a relatively simple proof of the Atiyah–Singer index theorem, which we will present at the end of this chapter.

We begin our discussion of phenomenology by introducing isospin into the model via Chan–Paton factors [1]. Ever since the early days of the string theory, it was known that isospin factors can be introduced trivially into the model by a simple multiplicative factor. (In the next chapter, we will discuss a much more sophisticated way to introduce gauge groups via compactification and Kac–Moody algebras.)

The Chan–Paton method produces the scattering amplitude $T$ by simply multiplying the Veneziano Born term $A$ by the trace of isospin matrices, both of which are manifestly cyclically symmetric, and then summing over various permutations of the external legs:

$$T(1, 2, 3, \ldots, N) = \sum_{\text{perm}} \operatorname{Tr}(\lambda_1 \lambda_2 \lambda_3 \cdots \lambda_N) A(1, 2, 3, \ldots, N) \tag{9.1.1}$$

Because the trace is cyclically symmetric regardless of the isospin group, we have no restrictions on the choice of the group itself. Thus, we wish to impose additional constraints that will allow us to eliminate some of the nonphysical choices of the gauge group. We first insert a complete set of intermediate states into the scattering amplitude. We then demand

(1) that the amplitude $T$ factorize explicitly;
(2) that the twisting of external legs yield an eigenvalue of $\pm 1$; and
(3) that the massless Yang–Mills particle on the external legs and in the internal factorized channel belong to the adjoint representation of the gauge group.

Let us begin by imposing the first condition, that the amplitude factorize explicitly. In Fig. 9.1, we divide up the $N$ external particles into two clusters, $a$ and $b$, and let the intermediate line represent the particle $X$. The left-hand piece represents the scattering of particles $a$ into $X$, while the right-hand piece represents the particle $X$ breaking up into particles $b$. Following (5.1.7), we find

$$T(1, 2, \ldots, N) = \frac{1}{s - m_X^2} \sum_X T(a \to X) T(b \to X) + \cdots \tag{9.1.2}$$

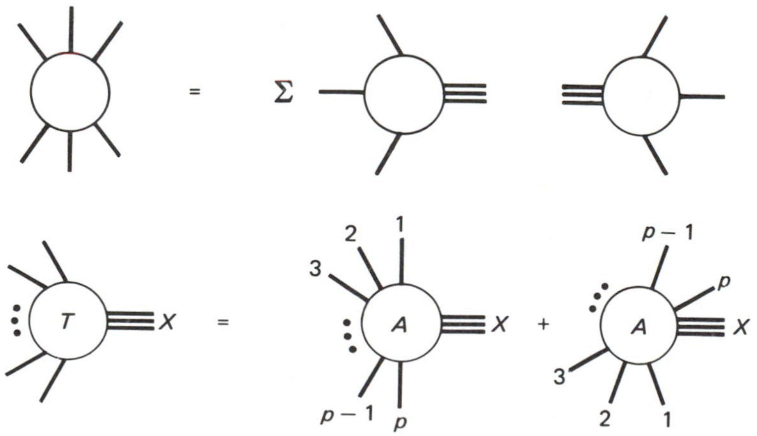

Figure 9.1. Constraints imposed by unitarity. Isospin can be introduced into the model by multiplying the amplitudes by Chan–Paton isospin factors, which must obey stringent conditions arising from unitarity. The factorized scattering amplitude must consist of the sum of Chan–Paton factors where the external lines are arranged in both clockwise and counterclockwise order. ($X$ represents a collection of factorized states.)

Now let us reexpress this factorization formula in terms of the Veneziano amplitudes. Each of the amplitudes obeys

$$A(1, 2, \ldots, N) = \frac{1}{s - m_X^2} \sum_X A(1, 2, \ldots, X) A(X, p, p+1, \ldots, N) \tag{9.1.3}$$

Notice that the Veneziano amplitudes are automatically factorizable and cyclically symmetric, so the factor $T(a \to X)$ must contain two terms, with both cyclic and anticyclic ordering of the external legs. Thus, the product of two $T$'s must contain $2 \times 2$ terms altogether. Let us now write these four terms explicitly:

$$\begin{aligned} T(1, 2, \ldots, N) = \frac{1}{s - m_X^2} \{ & \mathrm{Tr}(\lambda_1 \lambda_2 \cdots \lambda_{p-1} \lambda_p \lambda_{p+1} \cdots \lambda_{N-1} \lambda_N) \\ & \times A(1, 2, \ldots, p-1, p, X)(X, p+1, \ldots, N-1, N) \\ & + \mathrm{Tr}(\lambda_p \lambda_{p-1} \cdots \lambda_2 \lambda_1 \lambda_{p+1} \lambda_{p+2} \cdots \lambda_N) \\ & \times A(p, p-1, \ldots, 2, 1, X) A(X, p+1, p+2, \ldots, N-1, N) \\ & + \mathrm{Tr}(\lambda_1 \lambda_2 \cdots \lambda_p \lambda_N \lambda_{N-1} \cdots \lambda_{p+1}) \\ & \times A(1, 2, \ldots, p, X) A(X, N, \ldots, p+1) \\ & + \mathrm{Tr}(\lambda_p \lambda_{p-1} \cdots \lambda_1 \lambda_N \lambda_{N-1} \cdots \lambda_{p+1}) \\ & \times A(p, \ldots, 1, X) A(X, N, \ldots, p+1) \} \end{aligned} \tag{9.1.4}$$

Next, we wish to simplify this expression. We will now impose our second assumption, that the twist operator corresponds to

$$\Omega = (-1)^{N+1} \tag{9.1.5}$$

Notice that if we twist all external legs, this changes the cyclic ordering into an anticyclic ordering. By assumption, we are considering only the scattering of massless vector particles, which have $N = 0$. Thus, the twist operator picks up a factor of $(-1)$ for each external leg. Thus, the total contribution to the twist operator is

$$A(1, 2, \ldots, p, X) = (-1)^{p+1} A(X, p, p-1, \ldots, 2, 1) \tag{9.1.6}$$

Now let us collect all terms and arrive at some conclusions. After collecting terms,

$$\begin{aligned} & \mathrm{Tr}\{[\lambda_1 \lambda_2 \cdots \lambda_p - (-1)^p \lambda_p \lambda_{p-1} \cdots \lambda_1] \\ & \quad \times [\lambda_{p+1} \lambda_{p+2} \cdots \lambda_N - (-1)^{N-p} \lambda_N \lambda_{N+1} \cdots \lambda_{p+1})]\} \\ & \quad \times A(1, 2, \ldots, X) A(X, p+1, \ldots, N) \end{aligned} \tag{9.1.7}$$

Now let us insert a complete set of isospin matrices $\lambda_\alpha$ in the above equation. We can always do this, so we take

$$\delta_{\alpha\beta} = \mathrm{Tr}(\lambda_\alpha \lambda_\beta) \tag{9.1.8}$$

Then we can write

$$\begin{aligned}
&\mathrm{Tr}\{[(\lambda_1 \cdots \lambda_p - (-1)^p \lambda_p \cdots \lambda_1] \\
&\qquad \times [\lambda_{p+1} \cdots \lambda_N - (-1)^{N-p} \lambda_N \cdots \lambda_{p+1}]\} \\
&\quad = \sum_\alpha \mathrm{Tr}\{[\lambda_1 \cdots \lambda_p - (-1)^p \lambda_p \cdots \lambda_1] \lambda_\alpha\} \\
&\qquad \times \mathrm{Tr}\{\lambda_\alpha [\lambda_{p+1} \cdots \lambda_N - (-1)^{N-p} \lambda_N \cdots \lambda_{p+1}]\}
\end{aligned} \tag{9.1.9}$$

At this point, let us impose our third and final condition. The third criterion becomes the fact that the combination of lambda matrices in the brackets must in turn be part of the algebra of the group:

$$\lambda = \lambda_1 \cdots \lambda_p - (-1)^p \lambda_p \cdots \lambda_1 \tag{9.1.10}$$

Our conclusion is that the Chan–Paton factor, in order to preserve the adjoint representation for the massless vector particle, must have a gauge group for which the above combination of generators is also a generator of the algebra. This necessarily restricts our matrices to Usp($N$), SO($N$), and U($N$). (Actually, U($N$) is also ruled out, but for other reasons. We cannot consistently couple open and closed superstrings with U($N$) because $N = 2$ supergravity cannot couple to $N = 1$ matter multiplets.)

Unfortunately, this analysis does not yield any more constraints, so the model has very little predictive power. We will now look at the question of anomalies, which will fix the gauge group to be either SO(32) or $E_8 \otimes E_8$. As in the case of gauge theories, where the cancellation of anomalies between the quarks and leptons plays a central role in model building, the cancellation of anomalies will play an important rule in fixing the gauge group of the string model.

## §9.2. Anomalies and Feynman Diagrams

Anomalies have a very profound origin [2–4]. They shed light on the deepest dynamics of quantum field theories.

Simply, anomalies arise whenever the symmetries of the classical action do not carry over to the quantum level. The classical symmetries do not survive the process of regularizing the quantum theory. There are two types of anomalies, global and local.

Global anomalies in gauge theories are actually welcomed. For example, a global anomaly in scale invariance in QCD with massless quarks might be responsible for the generation of quark masses. Thus, a breakdown in global scale invariance through anomalies might be the origin of quark masses. Another global anomaly might be responsible for the breakdown of $\mathrm{U}(N)_R \otimes \mathrm{U}(N)_L$ invariance that occurs in quark models of QCD down to $\mathrm{SU}(N) \otimes \mathrm{SU}(N)$, which would be phenomenologically desirable. For the superstring model, however, global anomalies would be undesirable. For example, if global anomalies violated the modular invariance of the multiloop amplitudes, this might have disastrous consequences for the internal consistency and finiteness of the theory. Fortunately, it can be shown that this global anomaly in modular invariance is absent in the string theory [5].

By contrast, local anomalies in gauge theory and in superstring theory must be cancelled at all cost, or else these theories make no sense quantum mechanically. For example, the elimination of chiral anomalies is one of the essential ways in which to build new models of quarks and leptons. In the standard model, the quarks and the leptons have chiralities that just cancel the chiral anomaly. In superstring theory, local anomalies in conformal invariance and chiral symmetry must also be eliminated. The elimination of the conformal anomaly fixes the dimension of space–time and also the fermion content of the theory, while the elimination of chiral anomalies will fix the gauge group of the theory.

We begin our discussion by examining the simplest local anomaly, the chiral anomaly, which occurs because the process of regularization (such as Pauli–Villars or dimensional regularization) does not respect chiral invariance. Specifically, if we have a theory with the invariance

$$\psi \to e^{i\gamma_5 \varepsilon}\psi \tag{9.2.1}$$

then we would expect classically to have a conserved chiral current from (1.9.8):

$$\partial_\mu J^{\mu,5} = \partial_\mu(\bar{\psi}\gamma_5\gamma^\mu\psi) = 0 \tag{9.2.2}$$

However, there are complications due to quantization. The Paul–Villars method, for example, inserts an imaginary massive particle into the theory to make all Feynman diagrams converge:

$$\frac{1}{p^2+m^2} \to \frac{1}{p^2+m^2} - \frac{1}{p^2+M^2} = \frac{M^2-m^2}{(p^2+m^2)(p^2+M^2)} \tag{9.2.3}$$

Propagators, which usually converge as $p^{-2}$, now converge as $p^{-4}$, thus making almost all divergent Feynman diagrams convergent. Once we have obtained a finite $S$-matrix, we set the mass $M$ of the imaginary particle to go to infinity. However, mass terms explicitly break chiral invariance:

$$\delta(\bar{\psi}\psi) \neq 0 \tag{9.2.4}$$

Thus, the Pauli–Villars regularization program does not respect this symmetry, and we expect that the divergence of the axial current is not conserved. *We cannot preserve both conservation of the current and regularization of the Feynman amplitudes at the same time.* But since the regularization of the theory is more important (or else we have no theory), this means that we must sacrifice current conservation.

Similarly, dimensional regularization will also fail to preserve the chiral anomaly. Dimensional regularization assumes that we can analytically continue the Feynman diagrams into complex-dimensional space by the substitution

$$\sum_{\mu=0}^{3} p_\mu^2 \to \sum_{A=0}^{D-1} p_A^2$$
$$\int d^4p \to \int d^Dp \tag{9.2.5}$$

It is easy to generalize the traces over Dirac matrices for vector particle couplings, but this breaks down for axial vector couplings because there is no generalization of $\gamma_5$ in complex space:

$$\gamma_5 \not\to \Gamma_{D+1} \tag{9.2.6}$$

The dimensional regularization program breaks down for chiral fermions because this regularization cannot generalize the $\gamma_5$ matrix to a complex dimension.

We expect, therefore, that the divergence of the axial current is not conserved. In fact, it is the vector–vector–axial vector (VVA) diagram that explicitly breaks the conservation of the axial current. The triangle diagram poses regularization problems because each internal fermion line, which circulates within the triangle diagram, converges only as $p^{-1}$, which is not sufficient to give us a convergent graph and hence has ambiguities. By carefully regularizing the triangle diagram, we wind up with

$$\partial_\mu J^{\mu,5} = \frac{-1}{16\pi^2}\varepsilon^{\mu\nu\sigma\rho}\,\mathrm{Tr}(F_{\mu\nu}F_{\sigma\rho}) \tag{9.2.7}$$

which is a total derivative or a topological term, given by

$$-\frac{1}{4\pi^2}\varepsilon^{\mu\alpha\beta\gamma}\,\mathrm{Tr}(A_\alpha\partial_\beta A_\gamma + \tfrac{2}{3}A_\alpha A_\beta A_\gamma) \tag{9.2.8}$$

This result for the triangle graph was derived in four dimensions. However, the anomaly can also be generalized to $D$-dimensional chiral theories, where the triangle graph is now replaced by a polygon graph. Let us begin by writing down the Feynman rules for external vector and axial vector particles interacting with a spin-$\frac{1}{2}$ fermion. Let us begin with $N-1$ external vector particles and one axial vector particle of external momenta $k_i^\mu$, polarization $\zeta_i^\mu(k)$, and isospin indices $a$, $b$, and $c$ interacting with a fermion of internal momenta $p_i^\mu$. The Feynman rules are

$$\text{propagator} \to \frac{\Gamma\cdot p}{p^2}\delta_{ab}$$
$$\text{vertex (vector)} \to \Gamma\cdot\zeta\lambda_a \tag{9.2.9}$$
$$\text{vertex (axial)} \to \Gamma_{D+1}\Gamma\cdot\zeta\lambda_a$$

We assume that the external particles are on-shell, that the polarization vectors vanish when contracted on external momenta, and that the sum of the external momenta is zero, so that

$$\begin{aligned} k_{i,\mu}^2 &= 0 \\ k_i \cdot \zeta_i(k_i) &= 0 \\ \sum_{i=1}^{N} k_i &= 0 \end{aligned} \tag{9.2.10}$$

Then Feynman's rules for the polygon graph give us

$$G \sim \int d^D p \,\mathrm{Tr}(\lambda_{a_1} \cdots \lambda_{a_N})\, \mathrm{Tr}\left\{ \prod_{i=1}^{N} \frac{\Gamma \cdot p_i \Gamma \cdot \zeta_i}{p_i^2} \tfrac{1}{2}[1 + \Gamma_{D+1}] \right\} \tag{9.2.11}$$

Let us extract the most divergent part of this diagram. We notice that the trace over the term containing the $\Gamma_{D+1}$ can be explicitly carried out:

$$\mathrm{Tr}(\Gamma_{\mu_1} \Gamma_{\mu_2} \cdots \Gamma_{\mu_D} \Gamma_{D+1}) \sim \varepsilon_{\mu_1 \mu_2 \cdots \mu_D} \tag{9.2.12}$$

Notice that the antisymmetric matrix arises only when we take the trace of $D$ $\Gamma$ matrices multiplied by $\Gamma_{D+1}$. Let us recombine the various factors of momenta, which must now be contracted onto the antisymmetric matrix. Notice that the conservation of momentum reduces the number of independent momentum factors by one. In general, the only terms that survive are of the form

$$G \sim \int d^D p \,\mathrm{Tr}(\lambda_{a_1} \cdots \lambda_{a_N}) \prod_{i=1}^{N} \frac{1}{p_i^2} \varepsilon_{\mu_1 \cdots \mu_D} \zeta^{\mu_1} \zeta^{\mu_2} \cdots k^{\mu_D} \tag{9.2.13}$$

Notice that there is now a direct correlation between the number of external lines and the number of dimensions of space–time because the antisymmetric tensor is of rank $D$. Thus, the leading divergence must satisfy the relationship

$$D = 2N - 2 \tag{9.2.14}$$

In four dimensions, this means the triangle graph is divergent. In 10 dimensions, this means the *hexagon graph* is divergent. Notice also that this gives us an explicit expression for the anomaly term. The external momenta $k_{i,\mu}$ can be replaced by derivatives $\partial_\mu$, and the polarization tensor $\zeta^{\mu_i}$ can be replaced by the field $A^\mu$, so that we can replace the contractions over $k_{[\mu}\zeta_{\nu]}$ with a contraction over the Yang–Mills tensor $F_{\mu\nu}$. Thus, after taking the Fourier transform, we find the correspondence

$$\begin{aligned} k_\mu &\to \partial_\mu \\ \zeta_\mu &\to A_\mu \\ k_{[\mu}\zeta_{\nu]} &\to F_{\mu\nu} \end{aligned} \tag{9.2.15}$$

Combining all factors, we find [6–8]

$$\partial_\mu J^{\mu,5} = \frac{i^{(1/2)D}}{2^{D-1} \pi^{(1/2)D} (\frac{1}{2}D)!} \mathrm{Tr}\, \varepsilon_{\mu_1 \cdots \mu_D} F^{\mu_1 \mu_2} \cdots F^{\mu_{D-1} \mu_D} \tag{9.2.16}$$

Similarly, there is an anomaly in the theory of gravity coupled to chiral fermions, which again is proportional to a total derivative or a topological term. If we couple external graviton legs to chiral fermions circulating in the interior of a single-loop graph, we can repeat our previous analysis with Feynman diagrams. The major difference comes from three factors:

(1) We now must include the coupling of fermions to the energy–momentum tensor rather than the chiral current. The graviton coupling to fermions takes place through vierbeins (because GL($N$) has no finite-dimensional spinor representations; see Appendix).
(2) The external polarization vector $\zeta_\mu$ now becomes an external polarization tensor $\zeta_{\mu\nu}$.
(3) The vertex functions now contain higher tensor components.

However, the leading divergent part of the diagram still contains the $\varepsilon_{\mu_1\mu_2\cdots\mu_D}$ tensor contracted onto external momenta $k_i$ and external polarization tensors $\zeta_{\mu\nu}$. Once again, it is trivial to recombine all factors and show that the leading divergence can be reassembled in $x$-space in terms of curvature tensors. For example, the coupling of an axial vector and two energy–momentum tensors in four dimensions yields

$$D_\mu J^{\mu,5} = -\frac{1}{768\pi^2}\varepsilon^{\mu\nu\alpha\beta}R_{\mu\nu\sigma\tau}R^{\sigma\tau}_{\alpha\beta} \tag{9.2.17}$$

In the theory of gravity, there are two well-known total derivative terms, in two and four dimensions. First, there is the celebrated Gauss–Bonnet identity:

$$\chi(M) = \frac{1}{4\pi}\int d^2x\,\sqrt{g}R = 2 - 2g \tag{9.2.18}$$

which we will prove later, where $\chi(M)$ is the Euler number of a two-dimensional manifold $M$, $R$ is the contracted curvature tensor, and $g$ is the genus of a closed manifold (i.e., the number of holes or handles). This identity is possible because the scalar curvature in two dimensions is a total derivative.

In four dimensions, we have the identity

$$\begin{aligned}\varepsilon^{\mu\nu\alpha\beta}\,\mathrm{Tr}(R_{\mu\nu}R_{\alpha\beta}) &= a\sqrt{g}R^2_{\mu\nu\sigma\rho} + b\sqrt{g}R^2_{\mu\nu} + c\sqrt{g}R^2 \\ &= \text{total derivative}\end{aligned} \tag{9.2.19}$$

for specific numbers $a$, $b$, and $c$.

Usually, the integral of a total derivative is zero. However, this is not true if the manifold has a boundary or nontrivial topology. In this case, the integral over a total derivative picks up a term from the boundaries and the topologies. Thus, the integral over the contracted curvature tensor for a closed manifold is related to a linear function of the genus, or the number of holes in the surface, which is a topological number. This is the origin of the relationship between anomalies, total derivatives, and topological numbers.

Thus, both the gauge and gravitational anomalies can be derived from Feynman diagrams. However, for higher dimensions this becomes prohibitively difficult. Indices proliferate very rapidly for higher dimensions. Instead, to generalize these results for higher dimensions, we will derive these expressions from the functional formalism without appealing to Feynman diagrams.

## §9.3. Anomalies in the Functional Formalism

So far, we have been approaching anomalies in bits and pieces. What we would like to have is a systematic method with which to calculate all such total derivative terms for an arbitrary gauge group and arbitrary dimension.

Let us begin with the functional expression for the fermion interacting with the external vector or gravitational particles:

$$\int D\psi\, D\bar{\psi}\, e^{-S} = \int D\psi\, D\bar{\psi} \exp - \left\{ \int d^D x\, \bar{\psi} \tfrac{1}{2}(1 + \Gamma_{D+1}) \not{D} \psi \right\} \tag{9.3.11}$$

where the covariant derivative explicitly represents the coupling of the fermion to the external Yang–Mills and gravitational field:

$$\not{D} = \Gamma^\mu (\partial_\mu + i\lambda^a A_\mu^a + \omega_\mu^{ab} M^{ab}) \tag{9.3.2}$$

where $\Gamma^\mu$ are the gamma matrices in $D$ dimensions multiplied by the vierbein field, and $M^{ab}$ is a generator of the Lorentz group $O(D-1, 1)$.

Notice that the functional integral is quadratic in the fermion fields, so the integral is a Gaussian. Using our results on integration over Grassman variables in the Appendix, we can now perform the integral and obtain a determinant. Then we can reexponentiate the determinant back into the exponential to obtain a term $\Gamma(A, g)$ that is explicitly dependent on the gauge fields. By explicitly performing the Gaussian integration in (9.3.1), we find the following determinant:

$$e^{i\Gamma(A,g)} = \det\{\tfrac{1}{2}(1 + \Gamma_{D+1})\not{D}\} \tag{9.3.3}$$

where $A$ and $g$ are the Yang–Mills and graviton fields. We now use the relation

$$\det M = e^{\mathrm{Tr} \ln M} \tag{9.3.4}$$

(This is most easily proved by making a similarity transformation that diagonalizes the matrix $M$. Then $M$ reduces to a diagonal matrix with its eigenvalues along the diagonal. In this form, the identity is trivial to prove. Finally, we make the reverse similarity transformation to restore the $M$ matrix.) Thus, we can take the log of (9.3.3):

$$\begin{aligned} \Gamma(A, g) &= \ln\{\det[\tfrac{1}{2}(1 + \Gamma_{D+1})\not{D}]\} \\ &= \mathrm{Tr} \ln\{\tfrac{1}{2}(1 + \Gamma_{D+1})\not{D}\} \end{aligned} \tag{9.3.5}$$

Now let us calculate the gauge variation of this field functional. Under the gauge transformation

$$A_\mu \to A_\mu - D_\mu \varepsilon \tag{9.3.6}$$

we find that, after a Taylor expansion in $\varepsilon$, a field functional transforms as

$$\begin{aligned}\Gamma(A'_\mu) &= \Gamma(A_\mu - D_\mu\varepsilon)\\ &= \Gamma(A_\mu) - \int d^Dx\,\mathrm{Tr}\,D_\mu\varepsilon\frac{\delta\Gamma}{\delta A_\mu} + \cdots\\ &= \Gamma(A_\mu) + \int d^Dx\,\mathrm{Tr}\,\varepsilon D_\mu\frac{\delta\Gamma}{\delta A_\mu} + \cdots\\ &= \Gamma(A_\mu) + \int d^Dx\,\mathrm{Tr}\,\varepsilon D_\mu J^{\mu,5} + \cdots \end{aligned} \tag{9.3.7}$$

In the last expression for $\Gamma(A'_\mu)$, we used the fact that the current generated by the gauge transformation is given by

$$J_\mu = \frac{\delta\Gamma(A)}{\delta A_\mu} \tag{9.3.8}$$

We can rewrite the $G$ appearing in (9.2.13) in functional language:

$$G = D_\mu\frac{\delta}{\delta A_\mu}\Gamma(A_\mu) \tag{9.3.9}$$

Then:

$$D_\mu J^\mu = G \tag{9.3.10}$$

*Thus, nonconservation of the axial current means a nonvanishing G.*

The anomaly must also satisfy a self-consistency condition, the Wess–Zumino condition [9]. Notice that the generator of a gauge transformation is equal to

$$\delta_\Lambda = \frac{\delta}{\delta\Gamma} = D_\mu\frac{\delta}{\delta A_\mu} \tag{9.3.11}$$

Then we have a new way of writing $G$:

$$G(\Lambda) = \delta_\Lambda\Gamma(A_\mu) \tag{9.3.12}$$

But we also know that the local gauge generators form a closed algebra. Thus,

$$[\delta_{\Lambda_1}, \delta_{\Lambda_2}] = \delta_{\Lambda_3} \tag{9.3.13}$$

Then this means that the $G$'s must satisfy the constraint

$$\delta_{\Lambda_1}G(\Lambda_2) - (1 \leftrightarrow 2) = G(\Lambda_3) \tag{9.3.14}$$

If we write $G$ as a variation of $\Gamma(A_\mu)$, then it is obvious that the Wess–Zumino consistency condition is satisfied. However, if we write the anomaly term in terms of curvature tensors, then it is highly nontrivial that this self-consistency condition is satisfied. This then provides a powerful check on our computations.

Now that we have a functional formalism in which to discuss anomalies, let us begin a discussion of the mathematics behind anomalies. Specifically, we will show that the anomaly can be written in terms of generalized *characteristic classes* that have been studied by mathematicians. Then we will show the most elegant part of the theory, namely that the integrals over these characteristic classes yield various index theorems. Thus, our strategy is to show

$$\text{Anomalies} \rightarrow \text{Characteristic classes} \rightarrow \text{Index theorems}$$

## §9.4. Anomalies and Characteristic Classes

To study characteristic classes [10, 11], which will give us a systematic analysis of all possible topological terms, we will use the language of forms (see Appendix.) The theory of forms, in some sense, only produces results that can be derived using the usual analytical methods of calculus. However, in higher dimensions the number of indices rapidly proliferates beyond control. Thus, the powerful shorthand notation of the theory of forms allows us to rapidly manipulate tensors of arbitrary rank in any dimensions, which is prohibitive using standard tensor calculus. In Chapter 11 we will see that the theory of forms is the most convenient language for the theory of cohomology and homology.

Armed with the theory of forms, we now construct a series of *characteristic classes* that will allow us to write down by inspection the set of all topological terms for practically any dimension and any group.

Let us now define an *invariant polynomial* that has the property

$$P(\alpha) = P(g^{-1}\alpha g) \tag{9.4.1}$$

where $\alpha$ and $g$ are group matrices. Examples of invariant polynomials are

$$\operatorname{Det}(1 + \alpha); \qquad \operatorname{Tr} e^{\alpha} \tag{9.4.2}$$

Now let $\Omega$ be a curvature two-form:

$$\Omega = d\omega + \omega \wedge \omega \tag{9.4.3}$$

that satisfies the Bianchi identities

$$d\Omega + \omega \wedge \Omega - (-1)^{c_\Omega}\Omega \wedge \omega = 0 \tag{9.4.4}$$

In the Appendix we prove two important statements for invariant polynomials:

$$\begin{aligned} &[1] \qquad dP(\Omega) = 0 \\ &[2] \qquad P(\Omega) = dQ \end{aligned} \tag{9.4.5}$$

for some $Q$. That is, an invariant polynomial based on curvature forms is both closed and exact. The theory of forms allows an explicit construction of the $Q$. For example, in the Appendix we show

$$\begin{aligned} \operatorname{Tr} \Omega^n &= nd \int_0^1 dt\, t^{n-1} \operatorname{Tr}\{A(dA + tA^2)^{n-1}\} \\ &= d\omega_{n-1} \end{aligned} \tag{9.4.6}$$

Thus, the trace of the product of curvature two-forms is exact and closed. This is a generalization in the language of forms of previous identities that were proved by hand. For example, in four dimensions we have (9.2.7) and (9.2.17). For arbitrary dimension we have (9.2.16). This new form $\omega_{n-1}$ is called the *Chern–Simons form*, which we encountered in (9.2.8).

This is important, because earlier we showed that in two and four dimensions there are polynomials in the curvature tensor that are total derivatives. But we were unable to construct these higher topological terms. (9.4.6) summarizes the solution to the problem.

Using the language of forms, we can also find a convenient expression for the Wess–Zumino consistency condition. In $D$-dimensional space, let us formally define a product of curvature forms that we call $I_{D+2}$. (Of course, this is actually zero for $D + 2$ dimensions, but we will put aside this point.) This form, by construction, is exact,

$$I_{D+2} = d\omega_{D+1} \tag{9.4.7}$$

where $\omega_{D+1}$ is a Chern–Simons form. $I_{D+2}$ is also gauge invariant because it consists entirely of curvature forms. The fact that it is both gauge invariant and exact thus leads us to write

$$\begin{aligned} \delta_\Lambda I_{D+1} &= \delta_\Lambda \, d\omega_{D+1} \\ &= d\delta_\Lambda \omega_{D+1} \\ &= d^2 \omega_{D+1} \\ &= 0 \end{aligned}$$

Thus, we see that the gauge variation of the Chern–Simons form is closed. We can write locally

$$\delta_\Lambda \omega_{D+1} = d\omega_D \tag{9.4.8}$$

The purpose of this construction is to find a form for the anomaly $G$ such that it is manifestly a gauge variation of some other form. We now write

$$G = \int_M \omega_D \tag{9.4.9}$$

Notice that we can use Stokes' theorem to show

$$G = \int_M \omega_D = \int_\Sigma d\omega_D = \delta_\Lambda \int_\Sigma \omega_{D+1} \tag{9.4.10}$$

where the surface $M$ and $\Sigma$ are related by $M = \partial\Sigma$.

This is our desired result. We have now shown that the anomaly term $G$ is a gauge variation of another form. Thus, $G$ *automatically* satisfies the Wess–Zumino consistency condition, because

$$\delta_{\Lambda_1} G(\Lambda_2) = \delta_{\Lambda_1} \delta_{\Lambda_2} \int \omega_{D+1} \tag{9.4.11}$$

We will use this construction when we actually cancel the anomalies in the superstring.

Armed with the theory of forms, we can now construct the characteristic polynomials at will.

We will find that there are four major characteristic classes:

(1) Chern classes
(2) Pontrjagin classes
(3) Euler classes
(4) Stiefel–Whitney classes

Let us now restrict the curvature form $\Omega$ to be a member of GL($k$, $C$), i.e., an arbitrary $k \times k$ matrix with complex elements. Let us define the *total Chern form* over this group as

$$c(\Omega) = \det\left(I + \frac{i}{2\pi}\Omega\right) = 1 + c_1(\Omega) + c_2(\Omega) + \cdots \tag{9.4.12}$$

where we have power expanded in powers of $\Omega$. The terms in the expansion, which eventually terminate, are

$$\begin{aligned}
c_0 &= 1 \\
c_1 &= \frac{i}{2\pi}\operatorname{Tr}\Omega \\
c_2 &= \frac{1}{8\pi^2}(\operatorname{Tr}(\Omega \wedge \Omega) - \operatorname{Tr}(\Omega) \wedge \operatorname{Tr}(\Omega)) \\
c_3 &= \frac{i}{48\pi^2}(-2\operatorname{Tr}(\Omega \wedge \Omega \wedge \Omega) + 3\operatorname{Tr}(\Omega \wedge \Omega) \wedge \operatorname{Tr}\Omega \\
&\quad - \operatorname{Tr}\Omega \wedge \operatorname{Tr}\Omega \wedge \operatorname{Tr}\Omega) \\
&\cdots \\
c_n &= \left(\frac{i}{2\pi}\right)^n \det\Omega
\end{aligned} \tag{9.4.13}$$

where

$$dc_j(\Omega) = 0 \tag{9.4.14}$$

Notice that the series eventually ends, because $\Omega$ raised to a high enough power is equal to zero because the $dx^\mu$ are Grassmann-valued. In addition to the Chern classes $c_i$, we can also introduce the *Chern character*:

$$\operatorname{Tr} e^{\Omega} \tag{9.4.15}$$

We can always diagonalize the matrix $\Omega$ so that it has only eigenvalues $x_i$ along the diagonal. Thus, we can always find a matrix $S$ such that

$$S\Omega S^{-1} = (-2\pi i)\begin{pmatrix} x_1 & 0 & 0 & 0 \\ 0 & x_2 & 0 & 0 \\ 0 & 0 & x_3 & 0 \\ 0 & 0 & 0 & \cdots \end{pmatrix} \tag{9.4.16}$$

After diagonalization, the traces and determinants become trivial:

$$c(\Omega) = \prod_{j=1}^{n} (1 + x_j) \tag{9.4.17}$$

Notice that the topological invariant for SU(2) in four dimensions is simply

$$c(M) = \det\left(1 + \frac{1}{4\pi}\lambda_a F^a\right) = 1 + \frac{1}{8\pi^2}\operatorname{Tr}(F \wedge F) \tag{9.4.18}$$

In addition to the Chern classes, there are the closely related *Pontrjagin classes*. Let $\Omega$ be a member of $O(k)$, and then define

$$p(\Omega) = \det\left(I - \frac{1}{2\pi}\Omega\right) = 1 + p_1 + p_2 + \cdots \tag{9.4.19}$$

The *Euler class*, however, is defined slightly differently. It is not based on the determinant, like the Chern and Pontrjagin classes, but on the Pfaffian. Let $\alpha$ be a two-form:

$$\alpha = \tfrac{1}{2}\alpha_{ij}\, dx^i \wedge dx^j$$

Then raise it to the $r$th power:

$$\alpha^r = r!(2\pi)^r e(\alpha)\, dx^1 \wedge dx^2 \wedge \cdots \wedge dx^{2r} \tag{9.4.20}$$

Notice that this is an antisymmetric matrix. We cannot diagonalize an antisymmetric matrix such that its diagonal elements are its eigenvalues $x_i$. However, we can write it in the form

$$\begin{pmatrix} 0 & x_1 & 0 & 0 \\ -x_1 & 0 & 0 & 0 \\ \cdots & \cdots & 0 & x_2 \\ \cdots & \cdots & -x_2 & 0 \\ \cdots & \cdots & \cdots & \cdots \end{pmatrix} \tag{9.4.21}$$

The invariant $e(\alpha)$, which is built out of the antisymmetric matrix $\alpha$, is called the Pfaffian. In terms of these $x$'s, the Pfaffian is equal to

$$e(M) = x_1 x_2 \cdots x_{2r} \tag{9.4.22}$$

We can also conclude that the Pfaffian $e(M)$ is the square root of the highest Pontrjagin class:

$$p(M) = x_1^2 x_2^2 \cdots x_{2r}^2 = e^2(M) \tag{9.4.23}$$

Some examples of the Euler class are

$$\begin{aligned} D = 2: &\qquad e(M) = \frac{1}{2\pi} R_{12} \\ D = 4: &\qquad e(M) = \frac{1}{32\pi^2}\varepsilon_{abcd} R^{ab} \wedge R^{cd} \end{aligned} \tag{9.4.24}$$

where

$$\chi(M) = \int_M e(M) \tag{9.4.25}$$

which are precisely the identities that we mentioned earlier in (9.2.18) and (9.2.19).

Finally, we want to list some of the other characteristic polynomials that are of interest. If $x_i$ labels the eigenvalues of the curvature form $\Omega$ then

$$\begin{aligned}
\text{Todd class} &= \text{td}(M) = \prod \frac{x_i}{1 - e^{-x_i}} \\
\text{Hirzebruch class} &= L(M) = \prod_i \frac{x_i}{\tanh x_i} \\
\text{A roof polynomial} &= \hat{A}(M) = \prod_i \frac{\frac{1}{2}x_i}{\sinh \frac{1}{2}x_i}
\end{aligned} \tag{9.4.26}$$

We will find, for example, that the $\hat{A}$ invariant polynomial plays a crucial role in calculating the anomaly for spin-$\frac{1}{2}$ particles coupled to gravity.

Lastly, we have the *Stiefel–Whitney classes*, which, unfortunately, are not related to any curvature form. Instead, the Stiefel–Whitney class is important in deciding which manifolds are spin manifolds, i.e., which ones can admit spinors. Many manifolds that at first glance seem to admit spinors actually have a problem. We merely quote important features of the Stiefel–Whitney class index. The vanishing of the first Stiefel–Whitney class index means that the manifold is orientable:

$$\omega_1 = 0 \leftrightarrow M \text{ is orientable} \tag{9.4.27}$$

Most important, a manifold $M$ can admit spinors if its second Stiefel–Whitney class index is equal to zero:

$$\omega_1 = \omega_2 = 0 \rightarrow M \text{ is a spin manifold} \tag{9.4.28}$$

## §9.5. Dirac Index

Now that we have listed some of the important properties of characteristic classes, we will show that the integrals over certain characteristic classes yield elegant index theorems. One of the most important will be the *Dirac index*, defined on a spin manifold. In fact, it is possible to reexpress the other invariants in terms of the Dirac index. If we have a spin manifold in $N$-space, then it is possible to define the Dirac equation in this space:

$$\Gamma^A D_A \psi = \not{D}\psi = 0 \tag{9.5.1}$$

where $\Gamma^A$ are the Dirac matrices (multiplied by the vierbein). Let us examine the operator $\not{D}$ given in (9.3.2) and its eigenvalues:

$$i\not{D}\psi_n = \lambda_n \psi_n \tag{9.5.2}$$

The kernel of the operator $\not{D}$ comprises those solutions which are annihilated by the operator, i.e., have zero eigenvalues.

In even-dimensional spaces, the spinor space actually divides in half, depending on the eigenvalue of the operator:

$$\Gamma_{D+1}\psi_{\pm} = \pm\psi_{\pm} \tag{9.5.3}$$

We will call these solutions positive and negative chirality solutions. In general, for nonzero eigenvalue, the chiralities always occur in pairs. This is because the eigenvalue mimics the mass term of the Dirac equation, and massive fermion states cannot be split into two Weyl fermions, which are are always massless. However, for the zero eigenvalue case, the chiralities do not have to occur in pairs. We can have unequal numbers of positive and negative chiral states for zero eigenvalue fermions.

Let us define $n_{\pm}$ to be the number of independent solutions with zero eigenvalue of positive (negative) chirality. Let us also define the *index* of the Dirac operator to be the difference between the numbers of independent zero modes with positive and negative chirality:

$$\text{Index}(\not{D}) = n_+ - n_- \tag{9.5.4}$$

It is crucial to note that this number is a topological number; i.e., it remains invariant under a continuous change in the manifold that does not change the overall topology. The Dirac index is a topological invariant because, after a transformation on the manifold, transitions will occur between zero eigenvalue and nonzero eigenvalue fermion states, *but only pairs of nonzero eigenvalue fermions can enter the zero eigenvalue state or leave it*. This is because, in order to become massive, chiral states must find a partner. Thus, the changes that occur in the index occur only in pairs, and each pair contributes zero to the index. Thus, the index must be a topological invariant (see Fig. 9.2). For example, if we have $N$ pairs of fermions entering the zero eigenvalue state, then

$$\delta\,\text{Index}(\not{D}) = \delta n_+ - \delta n_- = N - N = 0 \tag{9.5.5}$$

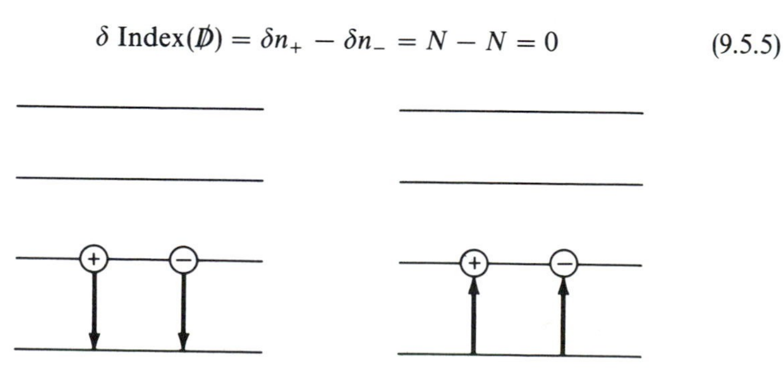


Figure 9.2. Pairwise emission or absorption into the ground state. Fermions can enter or leave the ground state only in pairs with opposite chirality. Thus, the total difference between positive and negative chiral states populating the vacuum always remains the same. This constant is called the Dirac index.

There are several ways in which we can express this index. Since the *kernel* of an operator is defined to be the set of states that it annihilates, we can also write

$$\text{Index}(\not{D}) = \dim \ker(\not{D}_+) - \dim \ker(\not{D}_-) \tag{9.5.6}$$

where we have split up the Dirac operator into positive and negative chiralities.

Another way to write the index formula is

$$\begin{aligned}\text{Index}(\not{D}) &= \sum_n \langle n | \Gamma_{D+1} | n \rangle \\ &= \text{Tr}(\Gamma_{D+1}) \\ &= \sum_+ 1 - \sum_- 1 \\ &= n_+ - n_-\end{aligned} \tag{9.5.7}$$

Notice that each positive chiral state contributes $+1$ to the sum, while each negative chiral state contributes $-1$, so the sum yields the Dirac index.

Although the above formula is quite elegant, it must be carefully regularized. Notice that a careless definition of the index leads to meaningless results:

$$\sum n_+ - \sum n_- = \infty - \infty \tag{9.5.8}$$

Because the cancellations occur level by level, it is incorrect to sum over all positive levels first and then subtract the sum against all negative levels. The sum over all positive or negative levels is infinite by itself. Only the difference is finite and well defined.

The proper way to treat the index is to insert a factor that allows us to define the sum over all positive and negative levels separately (which is potentially divergent) as a function of a parameter. One simple expression for the regularized Dirac index is

$$\text{Index}(\not{D}) = \text{Tr}(\Gamma_{D+1} e^{-\beta D^2}) \tag{9.5.9}$$

where $\beta$ is an arbitrary positive number and $D^2$ is the square of the Dirac operator. Because we are taking the trace, we are free to diagonalize the Dirac operator squared, so we can rewrite the trace in terms of the eigenvalues $\lambda_i$ of $D^2$:

$$\begin{aligned}\text{Index}(\not{D}) &= \sum_i \text{sign}(i) e^{-\beta\lambda_i} \\ \text{sign}(i) &= \pm\end{aligned} \tag{9.5.10}$$

Because the massive eigenvalues come with equal numbers of positive and negative chiral solutions, they cancel because they contribute with different sign$(i)$ and hence do not occur in the sum. The zero eigenvalues $n_{i\pm}$, however, do not necessarily come with equal numbers of positive and negative chiral states, and hence we have once again

$$\begin{aligned}\text{Index}(\not{D}) &= \sum_i \text{sign}(i)e^{-\beta\lambda_i} \\ &= \sum_{\lambda_i=0} [n_{i+} + (-n_{i-})] + \sum_{\lambda_i\neq 0} [e^{-\beta\lambda_i} - e^{-\beta\lambda_i}] \\ &= n_+ - n_- \end{aligned} \tag{9.5.11}$$

In summary, fermions can enter and leave the zero eigenvalue state only in chiral pairs. Thus, the difference of the positive and negative chiralities are not affected by the presence of nonzero eigenvalue fermions (which always occur in pairs, and hence their contribution to the index cancels exactly). Thus, the fermion states with nonzero eigenvalues do not affect the index at all.

This is not an academic question because it bears directly on the anomaly term. For example, we know that the chiral current can be written as

$$J_\mu^{5a} = \bar{\psi}\lambda^a\gamma_5\gamma_\mu\psi \tag{9.5.12}$$

Naively, this is conserved if we use the equations of motion. The covariant divergence (suppressing isospin indices) is equal to

$$\begin{aligned} D_\mu J^{\mu,5} &= \partial_\mu J^{\mu,5} + [A_\mu, J^{\mu 5}] \\ &= \bar{\psi}\gamma_5 \not{D}\psi - (\not{D}\bar{\psi})\gamma_5\psi \end{aligned} \tag{9.5.13}$$

Normally, this is exactly equal to zero because $\gamma^\mu D_\mu \psi = 0$. We must be careful now to include the contribution of the vacuum expectation value, which in general may not vanish. Let us define

$$\begin{aligned} S_F(x, y) &= \langle 0|\bar{\psi}(x)\psi(y)|0\rangle \\ \gamma^\mu D_\mu S_F(x, y) &= \delta(x, y) \end{aligned} \tag{9.5.14}$$

and take the vacuum expectation value of the divergence of the current. Then we find

$$\begin{aligned} D_\mu J^{\mu 5} &= 2\,\text{Tr}(\gamma_5 \not{D} S_F(x, y))_{x=y} \\ &= 2\,\text{Tr}(\gamma_5 \delta(x, x)) \end{aligned} \tag{9.5.15}$$

The last expression, which contains $\delta(x, x)$, makes no sense unless we regularize the integral. This is to be expected because we saw earlier that the triangle graph was divergent and required careful regularization.

One standard method of regularization is called the heat kernel method, and it inserts a converging factor into the expression:

$$D_\mu J^{\mu 5} = \lim_{\tau\to 0} 2\,\text{Tr}\{\gamma_5 e^{-\tau D^2}\delta(x, y)\}_{x=y} \tag{9.5.16}$$

where the trace is taken over spinors, and we have regulated the expression so that it is finite (for positive $\tau$).

But notice that this expression for the anomalous term in the current conservation expression is precisely the value of the Dirac index. Thus, *we have established a direct link between the Dirac index and the nonconservation of the axial current.*

In general, the trace can be explicitly calculated. At the end of this chapter, we will show that supersymmetry yields perhaps the simplest and most elegant proof of the Atiyah–Singer index theorem, which allows us to derive all the known index theorems in one expression.

When we evaluate this trace for an arbitrary spin manifold, we find, using eqns. (9.4.19) and (9.4.26),

$$\begin{aligned}\text{Index}(\not{D}) &= \int_M \hat{A}(M) \\ &= 1 - \frac{1}{24}p_1 + \frac{1}{5760}(7p_1^2 - 4p_2) + \cdots\end{aligned}$$

where the dimension of space–time is a multiple of four. In four dimensions, we have

$$D = 4: \qquad \text{Index}(\not{D}) = -\frac{1}{24}p_1 = \frac{1}{24 \cdot 8\pi^2}\int \text{Tr}(R \wedge R) \tag{9.5.17}$$

For the case of the spinor interacting with a Yang–Mills field in a background metric, the index becomes an integral over the product of two factors, the $A$ roof genus coming from the gravitational part and a term coming from the gauge part:

$$\text{Index}(\not{D}) = \int_M \hat{A}(M)\,\text{ch}(V) \tag{9.5.18}$$

Notice that this formula can exist in even dimensions, not just multiples of four. For $D = 2$ we have

$$\text{Index}(\not{D}) = \int_M c_1(V) = \frac{i}{2\pi}\int \text{Tr}\, F \tag{9.5.19}$$

For $D = 4$ we have

$$\text{Index}(\not{D}) = \frac{\dim V}{24 \cdot 8\pi^2}\int \text{Tr}(R \wedge R) - \frac{1}{8\pi^2}\int \text{Tr}(F \wedge F) \tag{9.5.20}$$

For the case of interest, six dimensions, we have

$$\text{Index}(\not{D}) = \frac{1}{48 \cdot 8\pi^3}\int \text{Tr}(F \wedge F \wedge F) - \frac{1}{8}\int \text{Tr}(F \wedge \text{Tr}\, R \wedge R) \tag{9.5.21}$$

## §9.6. Gravitational and Gauge Anomalies

Now that we have erected a powerful apparatus by which we can construct any invariant polynomial using the method of characteristic classes, let us investigate anomalies in theories where spinors are coupled to gauge and gravitational fields, which were calculated by Alvarez–Gaumé and Witten [12]. Let us begin by reviewing some elementary features of spinors.

In any even dimension $D = 2k$, the Dirac matrices are complex $2^k$-dimensional matrices. In general, these Dirac matrices split into positive and negative chiral representations. In an odd number of dimensions $D = 2k + 1$, the Dirac matrices are also $2^k$-dimensional, and there is only one representation. We do not have negative and positive chiralities in an odd number of dimensions (and hence no anomaly). Therefore, we will restrict our attention to an even number of dimensions.

Let us define the matrix

$$\Gamma_{D+1} = \Gamma_0 \Gamma_1 \cdots \Gamma_{D-1} \tag{9.6.1}$$

Then

$$\begin{aligned} \Gamma_{D+1}^2 &= +1 \quad \text{if } D = 4k + 2 \\ \Gamma_{D+1}^2 &= -1 \quad \text{if } D = 4k \end{aligned} \tag{9.6.2}$$

Thus, $\Gamma_{D+1}$ has eigenvalues $\pm 1$ if $D = 4k + 2$. Thus, under a combined $CPT$ reversal, the chirality of a particular state does not change. Positive chiral states are mapped into positive ones, and the same for negative states. We find

$$CPT(D = 4k + 2): \quad \begin{cases} \psi_+ \leftrightarrow \psi_+ \\ \psi_- \leftrightarrow \psi_- \end{cases} \tag{9.6.3}$$

Thus, it is possible to have unequal numbers of positive and negative chiral states, and hence an anomaly is possible.

However, $\Gamma_{D+1}$ has eigenvalues $\pm i$ if $D = 4k$. Thus, under a $CPT$ reversal, states with $\Gamma_{D+1} = i$ are transformed into states with $\Gamma_{D+1} = -i$. Positive chiral states are converted into negative ones, and vice versa:

$$CPT(D = 4k): \quad \begin{cases} \psi_+ \leftrightarrow \psi_- \\ \psi_- \leftrightarrow \psi_+ \end{cases} \tag{9.6.4}$$

If $D = 4k$, then positive and negative chiralities occur in pairs, and the gravitational interaction will not lead to anomalies.

*In summary, gravitational anomalies are possible only if* $D = 4k + 2$.

Now let us analyze Dirac, Weyl, or Majorana spinors. Weyl fermions can be defined in any even dimension. This is because the operator $\frac{1}{2}(1 + \Gamma_{D+1})$ can be defined in any even dimension. However, it turns out that Majorana spinors can exist only in 2, 3, and 4 (mod 8) dimensions. Furthermore, states that are simultaneously Majorana and Weyl can be defined only in 2 (mod 8) dimensions. We can summarize this:

| Spinor | Dimension |
|---|---|
| Dirac | Any $D$ |
| Weyl | $D$ = even |
| Majorana | $D$ = 2, 3, 4 (mod 8) |
| Majorana–Weyl | $D$ = 2 (mod 8) |

(9.6.5)

Our particular case of interest is 10-dimensional Majorana–Weyl fermions.

In 10 dimensions, three types of internal particle may contribute to the gravitational anomaly:

(1) spin-$\frac{1}{2}$ fermions interacting with $N$ external gravitons,
(2) spin-$\frac{3}{2}$ fermions interacting with $N$ external gravitons, and
(3) antisymmetric fourth-rank tensor fields interacting with $N$ gravitons.

It is obvious why the spinors must contribute to the anomaly. Their propagators converge only as $1/p$, and the usual Pauli–Villars and dimensional regularization methods fail because we choose chiral fields with the $\gamma_5$ factor.

The self-dual fourth-rank antisymmetric tensor is also added because, remarkably, there seems to exist no naive covariant action associated with this tensor. It seems to be well defined in the light cone gauge but not covariantly, so we suspect that it may also have an anomaly.

We begin with a spin-$\frac{1}{2}$ particle coupled to gravitation:

$$S = \int d^D x \, e \, e^{\mu a} \bar{\psi} i \gamma^a D_\mu [\tfrac{1}{2}(1 - \gamma^5)] \psi \tag{9.6.6}$$

Expanding around flat space:

$$e^{\mu a} = \delta^{\mu a} + h^{\mu a} + \cdots \tag{9.6.7}$$

We are interested in the lowest order coupling of the vierbein field with the spinor:

$$\begin{aligned} L_1 &= -\tfrac{1}{4} i h^{\mu\nu} \bar{\psi} \gamma_\mu \partial_\nu (1 - \gamma^5) \psi \\ L_2 &= -\tfrac{1}{16} (h_{\lambda\alpha} \partial_\mu h_{\nu\alpha}) \bar{\psi} \Gamma^{\mu\lambda\nu} \tfrac{1}{2} (1 - \gamma^5) \psi \end{aligned} \tag{9.6.8}$$

Specifically, we are interested in the single-loop diagram with a spin-$\frac{1}{2}$ fermion circulating internally coupled to external graviton legs. The first term contributes a standard three-particle vertex function with a Feynman vertex consisting of one momentum and combinations of polarization tensors. The second term, however, is a four-point Feynman graph, the sea-gull term.

At first, it seems hopeless to calculate the single fermion loop diagram with an arbitrary number of graviton legs, but there are tricks one can use to reduce it to a rather simple problem. Specificially, using simple symmetry arguments, we find that *we can reduce the problem to the scattering of a scalar charged particle interacting with a constant electromagnetic field.* Fortunately, the problem of charged scalars in QED is a well-studied problem with a known solution. Thus, the key to the problem is to reduce the complex problem to one that is already known.

This trick also generalizes for the other two cases. We are also interested in the scattering of internal spin-$\frac{3}{2}$ and antisymmetric tensor fields with external gravitons. Symmetry arguments again allow us to reduce the problem to a much simpler one. The circulating spin-$\frac{3}{2}$ particle can be reduced to an internal vector particle, and the antisymmetric tensor field can be reduced to a spin-$\frac{1}{2}$ particle:

$$\begin{aligned} &\text{spin } \tfrac{1}{2} \text{ and spin } 2 \to \text{spin } 0 \text{ and spin } 1 \\ &\text{spin } \tfrac{3}{2} \text{ and spin } 2 \to \text{spin } 1 \text{ and spin } 1 \\ &\text{antisymmetric tensor and spin } 2 \to \text{spin } \tfrac{1}{2} \text{ and spin } 1 \end{aligned} \tag{9.6.9}$$

By Feynman's rules, it is easy to construct the polygon graph. We know that the amplitude for the scattering of spin-$\frac{1}{2}$ particles can be represented as

$$I_{1/2} = i2^{2k+1}M^2 R(\varepsilon^{(i)}, p^{(j)}) Z(\varepsilon^{(i)}, p^{(j)}) \tag{9.6.10}$$

where $M$ = regulator mass and

$$R(\varepsilon^{(i)}, p^{(j)}) = -\varepsilon^{\mu_1, \ldots, \mu_{4k+2}} p^{(1)}_{\mu_1} \varepsilon^{(1)}_{\mu_2} \cdots p^{(2k+1)}_{\mu_{4k+1}} \varepsilon^{(2k+1)}_{\mu_{4k+2}} \tag{9.6.11}$$

and where the $i$th external graviton's polarization can be given by

$$\varepsilon^{(i)}_{\mu\nu} = \varepsilon^{(i)}_{\mu} \varepsilon^{(i)}_{\nu} \tag{9.6.12}$$

In the above equation, notice that we have the freedom to choose the polarization of the $i$th external graviton to be the product of two spin-1 polarization vectors. This is the key that allows us to rewrite the original scattering amplitude of spin-2 particles in terms of a reduced one with spin-1 particles.

$Z$, fortunately, has now been reduced to *the scattering of a propagating charged scalar particle (of charge $\frac{1}{4}$) interacting with a constant electromagnetic field*, which was computed by Schwinger decades ago [13]. Let the EM field be represented by

$$F_{\mu\nu} = -i \sum_{j=0}^{2k+1} (p^{(j)}_{\mu} \varepsilon^{(j)}_{\nu} - p^{(j)}_{\nu} \varepsilon^{(j)}_{\mu}) \tag{9.6.13}$$

Earlier, in (9.3.5), we used functional techniques to show that the anomaly term was related to the logarithm of a determinant of a propagator. We showed that

$$\Gamma(A_\mu) = \ln \det(\tfrac{1}{2}(1 + \Gamma_{D+1}) \not{D}) \tag{9.6.14}$$

Then $Z$ is

$$\begin{aligned} Z &= \frac{1}{\text{vol}} \ln \det(-D_\mu D^\mu + M^2) \\ &= \frac{-1}{\text{vol}} \int_0^\infty \frac{ds}{s} \operatorname{Tr} e^{s(D_\mu D^\mu)} e^{-sM^2} \end{aligned} \tag{9.6.15}$$

Schwinger proved that

$$\frac{1}{\text{vol}} \operatorname{Tr} e^{sD_\mu D^\mu} = \frac{1}{4\pi} \frac{eB}{\sinh eBs} \tag{9.6.16}$$

where $B$ is the effective magnetic field of the EM field and $e$ is the charge. (We will present a more general proof of this at the end of this chapter.) Although we cannot diagonalize the antisymmetric Maxwell tensor, we can always write it as

$$F_{\mu\nu} = 2\begin{pmatrix} 0 & x_1 & 0 & 0 \\ -x_1 & 0 & 0 & 0 \\ \cdots & \cdots & 0 & x_2 \\ \cdots & \cdots & -x_2 & 0 \\ \cdots & \cdots & \cdots & \cdots \end{pmatrix} \tag{9.6.17}$$

Then the anomaly contribution equals

$$I_{1/2} = -i(2\pi)^{-2k-1} R(\varepsilon^{(i)}, p^{(j)}) \prod_{i=1}^{2k+1} \frac{\frac{1}{2}x_i}{\sinh \frac{1}{2}x_i} \tag{9.6.18}$$

As expected, this is just the index of the Dirac operator, or the integral of $\hat{A}$.

Next, we have to calculate the anomaly for a spin-$\frac{3}{2}$ fermion internal loop coupled to an arbitrary number of external gravitons. Once again, the trick is to use symmetry arguments to reduce the problem to a much simpler one. The action for a spin-$\frac{3}{2}$ particle coupled to gravity is

$$L = -\tfrac{1}{2}\varepsilon^{\mu\nu\lambda\rho}\bar{\psi}_\mu \gamma^5 \gamma_\nu D_\lambda \psi_\rho \tag{9.6.19}$$

Notice that the spinor $\psi_\mu$ has a vector index $\mu$ as well as a spinor index, which we suppress. Again, let us extract only the lowest coupling terms:

$$\begin{aligned} L_1 &= \frac{i}{2} h^{\alpha\beta}\bar{\psi}_\mu \gamma_\alpha \partial_\beta \tfrac{1}{2}(1-\gamma^5)\psi^\mu \\ L_2 &= \frac{i}{16}(h^\alpha_\lambda \partial_\sigma h_{\tau\alpha})\bar{\psi}_\mu \Gamma^{\sigma\lambda\tau}\tfrac{1}{2}(1-\gamma^5)\psi^\mu \\ L_3 &= \frac{i}{2}(\partial_\sigma h_{\alpha\nu} - \partial_\alpha h_{\sigma\nu})\bar{\psi}^\sigma \gamma^\nu \psi^\alpha \end{aligned} \tag{9.6.20}$$

Once again, we can also write the general one-loop amplitude as

$$I_{3/2} = 2^{2k+1} i M^2 R(\varepsilon^{(i)}, p^{(j)}) Z(\varepsilon^{(i)}, p^{(j)}) \tag{9.6.21}$$

where $Z$ represents the scattering of an internal charged vector meson interacting with a constant electric field. Let us write an effective theory of charged vector mesons that reproduces $Z$ to one loop:

$$Z = \text{Tr} \ln H = -\int_0^\infty \frac{ds}{s} \text{Tr}\, e^{-sH} \tag{9.6.22}$$

where

$$H\phi_\mu = -(\partial_\sigma + \tfrac{1}{4}iA_\sigma)^2 \phi_\mu + \tfrac{1}{2}iF_{\mu\nu}\phi^\nu \tag{9.6.23}$$

All determinants can be performed, and we arrive at

$$I_{3/2} = -i(2\pi)^{-2k-1} R(\varepsilon^{(i)}, p^{(j)}) \prod_{i=1}^{2k+1} \frac{\frac{1}{2}x_i}{\sinh \frac{1}{2}x_i} \sum_{j=0}^{2k+1} (2\cosh x_j - 1) \quad (9.6.24)$$

The last anomaly term we wish to calculate is the contribution from an antisymmetric tensor field. When we calculate the anomaly associated with the tensor particle, however, we must be careful because there is no covariant action associated with this particle. Let us begin with an antisymmetric tensor field in $4k+2$ dimensions and its curvature:

$$F_{\mu_1 \cdots \mu_{2k+1}} = \partial_{\mu_1} A_{\mu_2 \cdots \mu_{2k+1}} + \text{cyclic perm.} \quad (9.6.25)$$

If we impose the constraint that it is self-dual:

$$F_{\mu_1 \cdots \mu_{2k+1}} = \frac{1}{(2k+1)!} \varepsilon_{\mu_1 \cdots \mu_{2k+1}}^{\mu_{2k+2} \cdots \mu_{4k+2}} F_{\mu_{2k+2} \cdots \mu_{4k+2}} \quad (9.6.26)$$

then we have problems. By the Bianchi identities, we can show that this relation is equivalent to the equations of motion. In other words, the naive action

$$L \sim F^2_{\mu_1 \cdots \mu_{2k+1}} \quad (9.6.27)$$

propagates *both* self-dual and anti-self-dual states, not just the self-dual ones. Thus, we have a problem. In fact, one can prove that a covariant action that propagates only the self-dual states does not exist [14].

However, we can use a trick to calculate the scattering amplitude. Although there doesn't seem to be a covariant action, the Feynman rules for such a particle can be written down covariantly. The trick is to use a spinor field to mimic the antisymmetric tensor field.

The energy–momentum tensor for this field is

$$T_{\mu\nu} = \frac{1}{(2k)!} F_{\mu\alpha_1 \cdots \alpha_{2k}} F_\nu^{\ \alpha_1 \cdots \alpha_{2k}} - \frac{1}{2(2k+1)!} g_{\mu\nu} F^2_{\alpha_1 \cdots \alpha_{2k+1}} \quad (9.6.28)$$

Fortunately, all the Feynman rules for this antisymmetric $F$ tensor are known, even if its action is not known. Now we will rewrite this antisymmetric tensor in terms of fermion fields by a trick. Let us define

$$\phi_{\alpha\beta} = 2^{-N/4} \sum_{n=0}^{N} \mathrm{Tr}(\Gamma_{\mu_1\mu_2\cdots\mu_n})_{\alpha\beta} F^{\mu_1\mu_2\cdots\mu_n} \quad (9.6.29)$$

We can also invert this:

$$F_{\mu_1\mu_2\cdots\mu_n} = 2^{-N/4} (\Gamma_{\mu_n \cdots \mu_1})_{\beta\alpha} \phi^{\alpha\beta} \quad (9.6.30)$$

The advantage of using this embedding is that fermion fields are much easier to use than antisymmetric tensors. The two-point function is

$$\langle \phi_{\alpha\beta}(q) \phi_{\gamma\delta}(-q) \rangle = (2q^2)^{-1} ((\gamma^5 \gamma^\mu q_\mu)_{\alpha\gamma} (\gamma^5 \gamma^\mu q_\mu)_{\beta\delta} + q^2 \delta_{\alpha\gamma} \delta_{\beta\delta}) \quad (9.6.31)$$

The final amplitude can be written as

$$I_A = \frac{-i}{2} M^2 2^{2k+1} R(\varepsilon^{(i)}, p^{(j)}) Z \tag{9.6.32}$$

where $Z$ represents the coupling of a charged scalar with external photons:

$$\begin{aligned} Z &= \operatorname{Tr} \ln(i\gamma^\mu D_\mu + iM) \\ &= -\tfrac{1}{2} \int_0^\infty ds\, e^{-sM^2} \operatorname{Tr} \exp(-s(-D_\mu D^\mu + i\Gamma^{\mu\nu} F_{\mu\nu})) \end{aligned} \tag{9.6.33}$$

We know, however, that

$$\operatorname{Tr} e^{-is\Gamma_{\mu\nu}F_{\mu\nu}} = 2^{2k+1} \prod_{i=1}^{2k+1} \cosh \tfrac{1}{2} x_i \tag{9.6.34}$$

The final result for the anomaly contribution from the antisymmetric tensor field is

$$I_A = \tfrac{1}{4} i 2^{2k+1} 2\pi^{-2k-1} R(\varepsilon^{(i)}, p^{(j)}) \prod_{i=1}^{2k+1} \frac{\tfrac{1}{2}x_i}{\sinh \tfrac{1}{2} x_i} \cosh \tfrac{1}{2} x_i \tag{9.6.35}$$

Of course, we also have to calculate the anomaly contribution from external gauge fields. The calculation there is almost identical to the one we have done, except now we must have $n$ gauge particles corresponding to $n$ gauge generators.

Let us now put (9.6.18), (9.6.24) and (9.6.35) together, including the contribution from the mixed anomalies coming from the gauge sector. For convenience, we will write the total anomaly contribution in terms of $I_{D+2}$, which we showed earlier in (9.4.7) was a convenient form for the anomaly because we could show that the anomaly term manifestly satisfied the Wess–Zumino consistency condition. We have

$$\begin{aligned} I_{12} = {} & \frac{-1}{720} \operatorname{Tr} F^6 + \frac{1}{24 \cdot 48} \operatorname{Tr} F^4 \operatorname{Tr} R^2 \\ & - \frac{1}{256} \operatorname{Tr} F^2 \left( \frac{1}{45} \operatorname{Tr} R^4 + \frac{1}{36} (\operatorname{Tr} R^2)^2 \right) \\ & + \frac{(n-496)}{64} \left[ \frac{1}{2 \cdot 2835} \operatorname{Tr} R^6 + \frac{1}{4 \cdot 1080} \operatorname{Tr} R^2 \operatorname{Tr} R^4 \right. \\ & \left. + \frac{1}{8 \cdot 1296} (\operatorname{Tr} R^2)^3 \right] + \frac{1}{384} \operatorname{Tr} R^2 \operatorname{Tr} R^4 + \frac{1}{1536} (\operatorname{Tr} R^2)^3 \end{aligned} \tag{9.6.36}$$

where $R$ represents the curvature tensor, and $F$ represents the Yang–Mills tensor. This is our final result. At first glance, it appears as if only a series of miracles can make such a horrible object vanish. However, a series of miracles indeed occurs. Specifically, we need to show that a string model is consistent with the following:

(I) $n = 496$, which kills half the terms in the anomaly;
(II) that the remaining nonzero terms can be regrouped into a factorized form; and
(III) that this reduced, factorized form can be canceled by other terms coming from an effective point particle action $\Delta S$.

Remarkably, all three conditions can be imposed simultaneously.

First, condition (I) is easy to satisfy if we set $n = 496$. Then a large number of terms vanish in (9.6.36). (In the next chapter, we will present the SO(16) $\otimes$ SO(16) model, which manages to eliminate these terms by choosing a different representation for the chiral fields.)

Second, condition (II) is considerably harder to satisfy, but we can show that the remaining terms in (9.6.36) factorize into the product of two terms if Tr $F^6$ can be reexpressed in terms of Tr $F^2$ Tr $F^4$ and (Tr $F^2)^3$, i.e.,

$$\text{Tr } F^6 = \frac{1}{48} \text{Tr } F^2 \wedge \text{Tr } F^4 - \frac{1}{14{,}400} (\text{Tr } F^2)^3 \tag{9.6.37}$$

If this strange-looking equation is satisfied, then we have the further reduction:

$$I_{12} \sim (\text{Tr } R^2 + k \text{ Tr } F^2) X_8 \tag{9.6.38}$$

The anomaly (9.6.36) would vanish if $k = -\frac{1}{30}$ and if

$$X_8 = \tfrac{1}{24} \text{Tr } F^4 - \tfrac{1}{7200} (\text{Tr } F^2)^2 - \tfrac{1}{240} \text{Tr } F^2 \wedge \text{Tr } R^2 + \tfrac{1}{8} \text{Tr } R^4 + \tfrac{1}{32} (\text{Tr } R^2)^2 \tag{9.6.39}$$

At first, it seems remarkable that under any circumstance can we satisfy condition (II) by having these rigid relations hold.

To satisfy condition (II), we must check which Lie groups are compatible with this strange constraint (9.6.37). Notice that the trace over the gauge matrices was defined in the adjoint representation. If we now convert this to traces over the fundamental representation (see Appendix), we will use the symbol "tr."

The proof that groups exist which satisfy (9.6.37) is straightforward. First, we know that, in the fundamental representation of SO($l$), the matrix $F$ can be represented as an antisymmetric $l \times l$ matrix. In the adjoint representation, $F$ can be represented as

$$F_{ab,cd} = \tfrac{1}{2}(F_{ac}\delta_{bd} - F_{bc}\delta_{ad} - F_{ad}\delta_{bc} + F_{bd}\delta_{ac}) \tag{9.6.40}$$

We can insert explicit expressions into the trace calculation of $F^6$, and we find

$$\text{Tr } F^6 = (l - 32) \text{ tr } F^6 + 15 \text{ tr } F^2 \text{ tr } F^4 \tag{9.6.41}$$

In order to satisfy the factorization condition, we must be able to eliminate a number of terms, including the sixth power of the curvature $F^6$. Notice that this is possible if $l = 32$, which has precisely 496 generators. Thus, we are led back again to SO(32).

The proof that $E_8 \otimes E_8$ also satisfies (9.6.37) is a bit more involved. We need to know whether or not the Clebsch–Gordan coefficients of the group enable us to write down independent invariants that allow us to contract four or six $F$'s. Mathematically, we need to know if there are independent Casimir operators of orders four and six. Fortunately, there is a theorem which says that if the homotopy group $\pi_{2n-1}(E_8)$ contains the integers, then there is an independent Casimir operator of order $n$. (Homotopy is a way of generating equivalence classes whose members are continuous maps, rather than spaces or surfaces.) It can be shown that the first homotopy groups satisfying this condition are $\pi_3$ and $\pi_{15}$:

$$\pi_3(\mathrm{E}_8) = \mathbf{Z}; \qquad \pi_{15}(\mathrm{E}_8) = \mathbf{Z}; \qquad \pi_i(\mathrm{E}_8) = 0 \quad \text{for } 3 < i < 15 \tag{9.6.42}$$

Thus, the only independent invariant of interest is of order two, which means that $\mathrm{Tr}\, F^4$ and $\mathrm{Tr}\, F^6$ are not independent and can be rewritten in terms of $\mathrm{Tr}\, F^2$. The last step is to actually calculate the coefficient appearing in $\mathrm{Tr}\, F^6 \sim (\mathrm{Tr}\, F^2)^3$. Since we are only interested in the overall coefficient, we can always calculate this by taking a specific representation of a subgroup within $E_8$, i.e., SO(16), and the adjoint representation 248 can be decomposed, under SO(16), into the sum of $\mathbf{120} \oplus \mathbf{128}$. Thus, (9.6.37) can be proved for $E_8 \otimes E_8$ as well as SO(32). (The equation can also be shown trivially for $\mathrm{U}(1)^{496}$ and $E_8 \otimes \mathrm{U}(1)^{248}$.)

We are not quite finished yet. We still have to satisfy condition (III) and show that there exists an effective low-energy counterterm $\Delta S$ that finally kills the anomaly. If we take naively $D = 10$ chiral supergravity coupled to 496 super-Yang–Mills fields, we find that the $X_8$ term *does not vanish*. Thus, chiral $D = 10$ supergravity must be ruled out as an acceptable quantum theory. However, the superstring theory has new coupling terms in the zero slope limit that can cancel the remaining terms.

As a low-energy approximation, superstrings can have more terms in its action than appear in the $D = 10$ chiral supergravity action. At first, this may seem surprising, because supergravity is the low-energy limit of superstring theory. However, even at the low energy levels, there is a difference: the sum over an infinite number of Reggeon states will, in general, supply us with more Feynman graphs than one would naively expect. (For example, the Yang–Mills theory has an action with up to four fields in its Lagrangian. Yet the first quantized string theory predicts only three-Reggeon couplings. Where is the missing four-particle vertex? It comes from the sum over an infinite number of Reggeon states, which will give us effective four-particle states in the low-energy limit. In general, higher trees and loops give us all the terms necessary to give us the Yang–Mills and gravitation theory.)

In fact, we can show that there exists an effective tree-level term that will cancel the anomaly term. To show this, let us first write the 12-form as

$$I_{12} \sim (\mathrm{Tr}\, R^2 - \mathrm{tr}\, F^2) X_8 \tag{9.6.43}$$

which can be rewritten as

$$I_{12} \sim (d\omega_{3\mathrm{L}} - d\omega_{3\mathrm{Y}}) X_8 \tag{9.6.44}$$

because of the identities

$$\begin{aligned} \mathrm{Tr}\, R^2 &= d\omega_{3\mathrm{L}} \\ \mathrm{tr}\, F^2 &= d\omega_{3\mathrm{Y}} \\ \omega_{3\mathrm{Y}} &= \mathrm{tr}(AF - \tfrac{1}{3}A^3) \\ \omega_{3\mathrm{L}} &= \mathrm{Tr}(\omega R - \tfrac{1}{3}\omega^3) \end{aligned} \tag{9.6.45}$$

We also have

$$\begin{aligned} \delta\omega_{3\mathrm{Y}} &= -d\omega_{2\mathrm{Y}}^1 \\ \delta\omega_{3\mathrm{L}} &= -d\omega_{2\mathrm{L}}^1 \end{aligned} \tag{9.6.46}$$

where the omega terms are Chern–Simons three-forms corresponding to either the Yang–Mills (Y) or the Lorentz (L) connections. We would now like to express the 12-form $I_{12}$ that we have carefully calculated in terms of a new 10-form $\omega_{10}$, so that the Wess–Zumino consistency conditions are satisfied. As in (9.4.10), our strategy is to find the $G$ term by constructing $\omega_{10}$ and $\omega_{11}$, such that

$$\begin{aligned} I_{12} &= d\omega_{11} \\ G &= \delta_\Lambda \int_\Sigma \omega_{11} = \int_\Sigma d\omega_{10} = \int_M \omega_{10} \end{aligned} \tag{9.6.47}$$

It turns out that, working from a 12-form down to a 10-form, there is a certain amount of arbitrariness in the final result. It is not hard to show that, starting from $I_{12}$, we can work down to an 11-form and then a 10-form. We simply quote the final result:

$$\begin{aligned} \omega_{11} = {} & \tfrac{1}{3}(\omega_{3\mathrm{L}} - \omega_{3\mathrm{Y}})X_8 + \tfrac{2}{3}(\mathrm{Tr}\, R^2 - \mathrm{tr}\, F^2)X_7 \\ & + \alpha d((\omega_{3\mathrm{L}} - \omega_{3\mathrm{Y}})X_7) \end{aligned} \tag{9.6.48}$$

where $\alpha$ is an arbitrary constant, and also

$$\omega_{10} = (\tfrac{2}{3} + \alpha)(\mathrm{Tr}\, R^2 - \mathrm{tr}\, F^2)X_6 + (\tfrac{1}{3} - \alpha)(\omega_{2\mathrm{L}} - \omega_{2\mathrm{Y}})X_8 \tag{9.6.49}$$

where we have the following expressions for $X_6$ and $X_7$:

$$\begin{aligned} dX_7 &= X_8 \\ \delta X_7 &= dX_6 \end{aligned} \tag{9.6.50}$$

Combining everything, we now have an expression for $G$:

$$G = (\tfrac{2}{3} + \alpha)\int (\omega_{3\mathrm{L}} - \omega_{3\mathrm{Y}})\, dX_6 + (\tfrac{1}{3} - \alpha)\int (\omega_{2\mathrm{L}} - \omega_{2\mathrm{Y}})X_8 \tag{9.6.51}$$

This is the factor $G$ that we wanted to calculate. Now the question is whether or not the $D = 10$ supergravity coupled to the super-Yang–Mills theory can contribute an anomaly term that cancels $G$. It is not hard, given all these identities, to show that the addition of the following effective action will give a new contribution to the anomaly term that will precisely cancel $G$:

$$\Delta S \sim \int BX_8 - (\tfrac{2}{3} + \alpha) \int (\omega_{3L} - \omega_{3Y}) X_7 \tag{9.6.51}$$

where $\alpha$ is a constant, the $\omega_3$ terms are Chern–Simons terms, $\omega$ is the spin connection, and $B$ is the two-form $B_{MN}$ that appears in the coupling of the supersymmetric Yang–Mills theory with $D = 10$ supergravity (the Chaplain–Manton action). A careful comparison shows that they yield an anomaly-free theory if we choose the variation of the $B$ field as

$$\delta B = \omega_{2L}^1 - \omega_{2Y}^1 \tag{9.6.52}$$

It is important to notice that this is *not* the usual variation of the $B$ field found in supergravity theory. Thus, the Chaplain–Manton supergravity action is riddled with anomalies. This is discouraging, until we realize that in superstring theory we have effective terms arising from loops and the summation over an infinite number of resonance states. Thus, it is possible to have this variation of the $B$ field and still have an acceptable supergravity theory. The next problem, however, is to explicitly show that this cancellation takes place as promised. It is one thing to show that the $B$ field might have the correct variation such that all anomalies cancel. It is another thing to show that it actually happens that way.

Surprisingly, the cancellation of anomalies for the entire superstring turns out to be much simpler than the cancellation of anomalies for the point particle supergravity system!

## §9.7. Anomaly Cancellation in Strings

We saw that a series of miracles happens to cancel all anomalies if we have $n = 496$ and the gauge group is equal to SO(32) or $E_8 \otimes E_8$. These arguments, however, were formulated only in the low-energy approximation and must be redone for the general case. Surprisingly, because the string model treats the tower of resonances in a compact fashion, the calculation for the superstring is much easier than the calculation for the supergravity theory [15].

Let us begin with the Pauli–Villars regularization method, replacing the usual propagator with a massive one:

$$\frac{1}{F_0} \to \frac{1}{F_0 - im} = \frac{F_0 + im}{L_0 + m^2} \tag{9.7.1}$$

Thus, we adopt the philosophy of ordinary field theory, that the anomaly occurs because the Pauli–Villars regularization procedure (or any other regularization program) violates chiral invariance. Here, the addition of the mass term explicitly breaks chiral invariance.

Before regularization, the single-loop hexagon diagram that violates parity looks like

$$T(0) = \int d^{10}p \, \mathrm{Tr}\left(\frac{F_0}{L_0} V(1) \cdots \frac{F_0}{L_0} V(6)\bar{\Gamma}_{11}\right) \tag{9.7.2}$$

where

$$V(i) = V(k_i, \zeta_i, 1) \tag{9.7.3}$$

and where $\frac{1}{2}(1 + \bar{\Gamma}_{11})$ is the projection operator that selects out the "even $G$ parity" states of the NS–R model:

$$\begin{aligned} \bar{\Gamma}_{11} &= \Gamma_{11}\Gamma_d \\ \Gamma_d &= (-1)^{\sum_{n=1}^{\infty} d_{-n} d_n} \end{aligned} \tag{9.7.4}$$

(Without the GSO projection, there is actually no anomaly, because the theory is then parity-conserving.)

The amplitude that we want to calculate is

$$\lim_{m \to \infty} [T(0) - T(m)] \tag{9.7.5}$$

In the limit of large $m$, the dependence on the mass should drop out and leave a finite result. Let us now perform all the traces over the Dirac matrices. Explicitly, the amplitude that is left is

$$G \sim im^2 \varepsilon(\zeta, k) \int d^{10}p \, \mathrm{Tr}\left(\frac{1}{L_0 + m^2} V_0(1) \cdots \frac{1}{L_0 + m^2} V_0(6)\Gamma_d\right) \tag{9.7.6}$$

where

$$\varepsilon(\zeta, k) = \varepsilon_{\mu_1 \mu_2 \cdots \mu_5 \nu_1 \nu_2 \cdots \nu_5} \zeta_1^{\mu_1} \cdots \zeta_5^{\mu_5} k_1^{\nu_1} \cdots k_5^{\nu_5} \tag{9.7.7}$$

Special care must be exercised in taking the limit as the mass goes to infinity. In general, we must add together the contributions from the planar and the nonorientable single-loop graphs. We find

$$\begin{aligned} G_{\mathrm{P}} &\sim \varepsilon(\zeta, k) \int_0^1 \prod_{i=1}^{5} dv_i \, \theta(v_{i+1} - v_i) \langle 0 | V_0(k_1, z_1) \cdots V_0(k_6, z_6) | 0 \rangle \\ G_{\mathrm{NO}} &\sim \varepsilon(\zeta, k) \int_0^2 \prod_{i=1}^{5} dv_i \, \theta(v_{i+1} - v_i) \langle 0 | V_0(k_1, z_1) \cdots V_0(k_6, z_6) | 0 \rangle \end{aligned} \tag{9.7.8}$$

Putting everything together, including the isospin factors coming from the Chan–Paton factors, we have

$$\begin{aligned} G \sim (n + 32l) \, \mathrm{Tr}(\lambda_1 \lambda_2 \cdots \lambda_6) \varepsilon(\zeta, k) \int_0^1 \prod_{i=1}^{5} dv_i \\ \times \theta(v_{i+1} - v_i) \langle 0 | V_0(k_1, z_1) \cdots V_0(k_6, z_6) | 0 \rangle \end{aligned} \tag{9.7.9}$$

To have a vanishing anomaly term, we need only show that $n + 32l$ is equal to zero. It is important to analyze where this $n + 32l$ factor comes from:

(1) $n$ comes from the fact that we have to trace over both the inner and the outer edge of a disk with a hole. The outer trace gives the Chan–Paton factors; the inner trace (which has no external lines) gives the factor $\mathrm{Tr}(1) = n$.
(2) The factor $l$ comes from the isospin group and equals

$$l = \begin{cases} 1 & \mathrm{Usp}(n) \\ 0 & \mathrm{U}(n) \\ -1 & \mathrm{SO}(n) \end{cases} \tag{9.7.10}$$

(3) Most important, the factor 32 comes from several sources. There was a factor of 32 in the Jacobian because the nonorientable graph had an integration from 0 to $\frac{1}{2}$, while the planar graph had an integration from 0 to 1. Furthermore, the integration region was only $\frac{1}{32}$ as big. And finally, there was another factor of 32 coming from the 32 ways in which an odd number of twists can be placed on six internal propagators. Thus, to have a cancellation, we must fix

$$\begin{aligned} l = -1 &\to \text{gauge group is SO}(n) \\ n + 32l = 0 &\to \text{gauge group is SO}(32) \end{aligned} \tag{9.7.11}$$

In summary, we have an anomaly-free theory with Chan-Paton factors if we choose SO(32) as our gauge group.

## §9.8. A Simple Proof of the Atiyah–Singer Index Theorem

Lastly, we want to prove the Atiyah–Singer index theorem [16] for Dirac operators. Because most of the other index theorems can be derived from the Dirac complex, we will only need to prove it for this case. Historically, the Atiyah–Singer theorem has been inaccessible to many physicists because of its mathematical complexity. However, recently physicists have given a remarkably simple proof of theorem using a new approach, the supersymmetric sigma model. Using supersymmetry, the proof of the Atiyah–Singer theorem can be expressed in language familiar to physicists [17, 18]. The new derivation of the theorem is based on the fact that the supersymmetric nonlinear sigma model has a supersymmetric generator that is identical to the Dirac operator, i.e.,

$$Q = \not{D} \tag{9.8.1}$$

To develop the analogy between $Q$ and the Dirac operator, let us begin with a supersymmetric theory with the supersymmetry generator $Q$ and its conjugate. Because the anticommutator of $Q$ with its conjugate yields $p_\mu$, and because $p_0$ is the energy, we have

$$H = Q^*Q \tag{9.8.2}$$

Let us analyze the energy eigenstates of the Hamiltonian:

$$H|E\rangle = E|E\rangle \tag{9.8.3}$$

Thus, if a state has zero energy,

$$Q|E = 0\rangle = 0 \tag{9.8.4}$$

However, if a bosonic state $|E, B\rangle$ or a fermionic state $|E, F\rangle$ has nonvanishing energy $E$, then $Q$ rotates one into the other:

$$\begin{aligned} Q|E, B\rangle &= \sqrt{E}|E, F\rangle \\ Q|E, F\rangle &= -\sqrt{E}|E, B\rangle \end{aligned} \tag{9.8.5}$$

These are powerful statements, because they mean that

(1) The energy is zero or positive, never negative.
(2) The zero energy states do not have to occur in boson–fermion pairs. The zero energy states are supersymmetric by themselves; i.e., they are annihilated by the $Q$.
(3) Nonzero energy states are not annihilated by $Q$, but form a supersymmetric pair of bosons and fermions that turn into each other under the action of $Q$.

Let us now construct the operator $(-1)^F$, where $F$ is the fermion number. For nonzero energy states, there must be an exact pairing of bosons and fermions. However, as we saw, this does not have to be true for zero energy states. Let us construct the Witten index $I$, which counts the number of bosonic zero energy states minus the number of fermionic zero energy states:

$$I = \mathrm{Tr}_{H=0}(-1)^F \tag{9.8.6}$$

Notice that, if states change in energy, the numbers of bosons and fermions entering or leaving the zero energy state must be equal. They must come and go in pairs, because the index is a topological invariant. Thus, we can also write

$$I = \mathrm{Tr}(-1)^F e^{-\tau H} \tag{9.8.7}$$

for *any* $\tau > 0$. Notice that nonzero energy states do not contribute to the trace over states, because they contribute with opposite fermion number. Thus, the trace exists only over zero energy states, as we saw before.

Now let us discuss the index over the Dirac operator. Earlier, in (9.5.4), we saw that this index simply counts the number of zero modes with positive chirality minus the number of zero modes with negative chirality. Thus, by definition, we have

$$\mathrm{Index}(\not{D}) = \mathrm{Tr}_{D^2=0}(\Gamma_{D+1}) \tag{9.8.8}$$

where

$$D^2 = (\not{D})^*(\not{D}) \tag{9.8.9}$$

Notice that fermions can leave or enter the zero mode only in chiral pairs, which is similar to the situation we found earlier with the supersymmetric case. The index can thus be generalized for states with arbitrary eigenvalues of $D^2$ as

$$\text{Index}(\not{D}) = \text{Tr}\{\Gamma_{D+1} e^{-\tau D^2}\} \tag{9.8.10}$$

Our goal is thus to construct a supersymmetric model that will make this correspondence exact. Fortunately, the nonlinear supersymmetric sigma model has this property. Thus, by calculating the supersymmetric index theorem for the supersymmetric sigma model, we will automatically calculate the Atiyah–Singer theorem. Thus, we have the correspondence

| $\sigma$ Model | Spin manifold |
|---|---|
| $Q$ | $\not{D}$ |
| $(-1)^F$ | $\Gamma_{D+1}$ |
| $H$ | $\not{D}^* \not{D}$ |

(9.8.11)

We begin by defining a Lagrangian for a position operator $x^\mu(t)$ and its superpartner $\psi^\mu(t)$, which are functions of the fictitious proper time variable $t$:

$$L = \tfrac{1}{4}\dot{x}^\mu \dot{x}_\mu + \tfrac{1}{4}\psi^\mu \dot{\psi}_\mu \tag{9.8.12}$$

which is invariant under the supersymmetric transformation

$$\begin{aligned} \delta x^\mu &= \varepsilon \psi^\mu \\ \delta \psi^\mu &= -\varepsilon \dot{x}^\mu \end{aligned} \tag{9.8.13}$$

(Compare this with the NS–R action in (3.2.1) when we eliminate all $\sigma$ dependence in the action.) The canonical commutation relations are

$$\begin{aligned} [p_\mu, x^\nu] &= -i\delta_\mu^\nu \\ \{\psi_\mu, \psi^\nu\} &= 2\delta_\mu^\nu \end{aligned} \tag{9.8.14}$$

The supersymmetry generator is thus

$$Q = \psi^\mu (i p_\mu) \tag{9.8.15}$$

where

$$\begin{aligned} \delta(x, \psi) &= [\bar{\varepsilon} Q, (x, \psi)] \\ p_\mu &= \tfrac{1}{2} i \dot{x}_\mu \end{aligned} \tag{9.8.16}$$

Notice that if we let this $Q$ act on an arbitrary space–time spinor state, then $\psi^\mu$ becomes $\gamma^\mu$ and $ip_\mu$ becomes $\partial_\mu$, so that $Q$ becomes $\gamma^\mu \partial_\mu$, which is the Dirac operator.

Our next goal is to add in the gauge and the gravitational fields so that we can calculate index theorems for arbitrary manifolds and gauge groups. Let us introduce a fermion field $\theta$ so that we can define the following fields in superspace:

$$X^\mu = x^\mu + \theta\psi^\mu \tag{9.8.17}$$

In superspace, the supersymmetry operator looks like

$$\begin{aligned} Q &= \theta\partial_t + \partial_\theta \\ D &= \theta\partial_t - \partial_\theta \end{aligned} \tag{9.8.18}$$

such that $D^2 = -\partial_t$. We now introduce the gauge fields

$$g_{\mu\nu}(X); \qquad A_\mu(X) \tag{9.8.19}$$

which are also superfields. Finally, to complete the action, we must also introduce one more superfield that is a gauge object:

$$\begin{aligned} N^a &= \eta^a + \theta\phi^a \\ \bar{N}_a &= \bar{\eta}_a + \theta\phi_a \end{aligned} \tag{9.8.20}$$

Putting everything together, we obtain the supersymmetric action:

$$\begin{aligned} S &= \int_0^1 dt \int d\theta\, L \\ L &= kg_{\mu\nu}(X)\, DX^\mu \partial_t X^\nu - \bar{N} D_A N \end{aligned} \tag{9.8.21}$$

where $k$ is an unimportant one-dimensional "metric" and

$$D_A N = [D + DX^\mu A_\mu(X)]N \tag{9.8.22}$$

To see what this action looks like when we reduce out all the auxiliary fields, we find

$$L = \tfrac{1}{4} g_{\mu\nu}(x)(\dot{x}^\mu \dot{x}^\nu + \psi^\mu \nabla_t \psi^\nu) + \bar{\eta}\nabla_t \eta - \tfrac{1}{2}\bar{\eta} F_{\mu\nu}\psi^\mu\psi^\nu\eta \tag{9.8.23}$$

where

$$\begin{aligned} \nabla_t \psi^\mu &= \partial_t \psi^\mu + \dot{x}^\sigma \Gamma^\mu_{\rho\sigma}\psi^\rho \\ \nabla_t \eta &= (\partial_t + \dot{x}^\mu A_\mu)\eta \end{aligned} \tag{9.8.24}$$

and where $F_{\mu\nu}$ is the Yang–Mills tensor and the $\Gamma^\rho_{\mu\nu}$ are the usual Christoffel symbols defined not in $t$-space but in real $D$-dimensional space–time. We can show that the supersymmetry generator is still the Dirac operator.

The next step is to calculate the supersymmetric index, which, as we know, must also be equal to the Dirac index. The chiral anomaly, as we saw earlier in (9.5.16), can be written in terms of a heat kernel Green's function:

$$\nabla_\mu J^{\mu 5} = 2\,\mathrm{Tr}(\gamma_5 K_\tau(x, x)) \tag{9.8.25}$$

where

$$K_\tau(x, y) = \langle x | e^{-\tau H} | y \rangle \tag{9.8.26}$$

We know from Chapter 1, however, that this Green's function can be rewritten in path integral language as the functional integral over an action. The big difference from the functional integrals found in Chapter 1 is that the eigenfunctions must be periodic in the proper time. This is because the Green's function is a matrix element of $e^{-\tau H}$ rather than $e^{i\tau H}$. Thus, time becomes imaginary (i.e., the functions become periodic). We will use the functional integral

$$\mathrm{Tr}(\Gamma_{D+1}e^{-\tau D^2}) = \int_{\mathrm{PBC}} \prod_t dx(t)\, d\psi(t) \exp - \int_0^\tau L\, dt \tag{9.8.27}$$

where PBC represents periodic boundary conditions. Thus, by calculating this functional integral for the sigma model in one variable $t$, we automatically calculate the Dirac index for the $D$-dimensional manifold! In order to perform this path integral, we must integrate around a classical solution:

$$X_0^\mu = x_0^\mu + \theta\psi_0^\mu \tag{9.8.28}$$

Let us perform the $Dx$ integration first. If we power expand around this solution $x^\mu = x_0^\mu + \delta x^\mu$, we find that the quadratic part of the action includes the term

$$S = \int_0^1 dt\, \delta x(-\partial_t + R)\partial_t\, \delta x + \cdots \tag{9.8.29}$$

where $R$ is a matrix given in terms of the curvature tensor:

$$R_\nu^\mu = \tfrac{1}{2}\psi_0^\alpha\psi_0^\beta R^\mu_{\nu\alpha\beta} \tag{9.8.30}$$

Fortunately, this functional integration is easy. As before, a Gaussian integration of (9.8.29) yields a determinant:

$$\int D\delta x e^{-S} = \det{}^{-1/2}(\partial_t - R) \tag{9.8.31}$$

where the determinant leaves out the constant modes. This determinant can be calculated easily by inserting a complete set of eigenstates into it. However, these eigenstates are periodic in the proper time $t$. Thus, we have the following:

$$\begin{aligned}\det{}^{-1/2}(\partial_t - R) &= \det{}^{-1/2}\left(\prod_{k\neq 0}^{\infty} \langle\phi_k|\partial_t - R|\phi_k\rangle\right)\\ &= \det{}^{-1/2}\left(\prod_{k\neq 0}(R - 2\pi ik)\right)\\ &= \det{}^{-1/2}\{(\tfrac{1}{2}R)^{-1}\sinh\tfrac{1}{2}R\}\end{aligned} \tag{9.8.32}$$

where the product over the integer $k$ occurs because we are inserting periodic functions into the determinant calculation that satisfy

$$\partial_t\phi_k = 2\pi ik\phi_k \tag{9.8.33}$$

and we have used the equation

$$\prod_{n=1}^{\infty}\left(1+\frac{(\frac{1}{2}x)^2}{\pi^2 n^2}\right)=\frac{\sinh\frac{1}{2}x}{\frac{1}{2}x} \tag{9.8.34}$$

Similarly, we can expand around $x_0^\mu$ and $\psi_0^\mu$ and integrate over $\eta$. In this fashion, we obtain the contribution to the integral from the gauge fields. The equation of motion for the $\eta$ field is different from the one found earlier for the $\delta x$ field. In the Heisenberg representation, it is

$$\dot{\eta}+[\bar{\eta}F\eta,\eta]=0 \tag{9.8.35}$$

where

$$F=\tfrac{1}{2}F_{\mu\nu}\psi_0^\mu\psi_0^\nu$$

This means that the evolution of the $\eta$ field is as follows:

$$\bar{\eta}(1)=e^{-\bar{\eta}F\eta}\bar{\eta}(0)e^{\bar{\eta}F\eta} \tag{9.8.36}$$

Thus, the contribution of this field to the integral is a multiplication by the factor $e^F$. Therefore, our final result for the heat kernel is

$$K_\tau(x_0,x_0)\sim e^F\det{}^{-1/2}[(\tfrac{1}{2}R)^{-1}\sinh\tfrac{1}{2}R] \tag{9.8.37}$$

At this point, the curvature for the gauge field and the gravitational field are written in terms of $\psi_0^\mu$. We would like to replace this with an expression in terms of the $dx^\mu$. This is possible because we still have to perform the functional integration over all background $\psi_0^\mu$. Because this integration is Grassman-valued, it selects out only the term proportional to

$$\varepsilon^{\mu_1\cdots\mu_D}\psi_{0,\mu_1}\cdots\psi_{0,\mu_D} \tag{9.8.38}$$

We then integrate the resulting expression over $D$-dimensional space. However, this is equivalent to taking only $D$-dimensional antisymmetric combinations of the curvature forms:

$$\begin{aligned}F&=\tfrac{1}{2}F_{\mu\nu}\,dx^\mu\,dx^\nu\\ R_\nu^\mu&=\tfrac{1}{2}R_{\nu\alpha\beta}^\mu\,dx^\alpha\,dx^\beta\end{aligned} \tag{9.8.39}$$

and then integrating over $D$-space. The integration automatically selects out the proper antisymmetrized product, and so we can freely make the substitution $\psi_0^\mu\to dx^\mu$, because the final expression is the same whether we use one or the other.

Thus, our final expression for the anomaly term is given by

$$\begin{aligned}\mathrm{Index}(\not{D})&=\mathrm{Tr}(-1)^F e^{-\tau D^2}\\ &=\left(\frac{i}{2\pi}\right)^{(1/2)D}\int \mathrm{Tr}\,e^F\det{}^{-1/2}[(\tfrac{1}{2}R)^{-1}\sinh\tfrac{1}{2}R]\end{aligned} \tag{9.8.40}$$

This is the Atiyah–Singer theorem for the Dirac operator on closed, oriented manifolds without boundary (see (9.4.26) and (9.5.18)).

The determinant on the right-hand side can be calculated by diagonalizing the curvature two form and reexpressing it in terms of its eigenvalues $x_i$. Thus, we reproduce the $A$ roof genus given earlier in (9.4.26).

In general, the Atiyah–Singer theorem can be written as the integral over the product of two curvature forms, one for the gravitational part and one for the gauge part. For completeness, let us list the four classical complexes and the various index theorems that result from them:

| Complex | Index | Index theorem |
|---|---|---|
| de Rahm | Gauss–Bonnet | $\text{Index}(d + \delta) = \chi(M) = \int e(M)$ |
| Signature | Hirzebruch | $\tau(M) = \int L(M) \wedge \tilde{\text{ch}}(V)$ |
| Dolbeault | Riemann–Roch | $\text{Index}(\bar{\partial}) = \int \text{td}(M) \wedge \text{ch}(V)$ |
| Spin | A roof genus | $\text{Index}(\not{D}) = \int \hat{A}(M) \wedge \text{ch}(V)$ |

(9.8.41)

where $e$ is the Euler characteristic, td is the Todd characteristic, $\tilde{ch}$ is the Chern characteristic where we double the value of the curvature form, ch is the usual Chern characteristic $e^{\Omega}$, and $\bar{\partial}$ is the complex conjugate of $\partial_z$ when we use complex coordinates $z$, $\bar{z}$ to describe the two-dimensional surface. (We will discuss complex manifolds at length in Chapter 11.)

## §9.9. Summary

We have used Chan–Paton factors and the anomaly cancellation to fix the gauge group of string theory. First, the Chan–Paton factor is simply the trace over various group generators that multiplies the Veneziano Born term:

$$T(1, 2, 3, \ldots, N) = \sum_{\text{perm}} \text{Tr}(\lambda_1 \lambda_2 \lambda_3 \cdots \lambda_N) A(1, 2, 3, \ldots, N) \tag{9.9.1}$$

Unfortunately, the only restriction that unitarity places on the group is that it must be Usp($N$), SO($N$), or U($N$), for arbitrary $N$.

Much more stringent restrictions on the group are found when we demand that the model be anomaly-free. In general, an anomaly occurs whenever a classical symmetry of the Lagrangian does not survive the quantization process. The chiral anomaly arises, for example, because the method of regularization (e.g., either Pauli–Villars or dimensional regularization) must necessarily violate chiral invariance.

In particular, the divergence of the axial current, instead of being zero, is now equal to

$$\partial_\mu J^{\mu,5} = \frac{-1}{16\pi^2}\varepsilon^{\mu\nu\sigma\rho}F_{\mu\nu}F_{\sigma\rho} \tag{9.9.2}$$

which is a total derivative or a topological term, given by

$$J^{\mu 5} = -8\pi^2\varepsilon^{\mu\alpha\beta\gamma}\,\mathrm{Tr}(A_\alpha\partial_\beta A_\gamma + \tfrac{2}{3}A_\alpha A_\beta A_\gamma) \tag{9.9.3}$$

In general, the anomaly in a higher dimension will be proportional to a topological term. For example, using the theory of forms, we can show that the product of the curvature tensor can be written as the divergence of another form. For example, we can show that the trace over curvatures can be written in terms of the divergence of a Chern–Simons form:

$$\mathrm{Tr}\,\Omega^n = nd\int_0^1 dt\, t^{n-1}\,\mathrm{Tr}\{A(dA + tA^2)^{n-1}\} \tag{9.9.4}$$

The study of these invariant polynomials takes us to the theory of characteristic classes. There are four classical characteristic classes. The Chern class can be defined as

$$c(\Omega) = \det\left(I + \frac{i}{2\pi}\Omega\right) = 1 + c_1(\Omega) + c_2(\Omega) + \cdots \tag{9.9.5}$$

The Pontrjagin class is defined for $O(k)$ as:

$$p(\Omega) = \det\left(I - \frac{1}{2\pi}\Omega\right) = 1 + p_1 + p_2 + \cdots \tag{9.9.6}$$

The Euler class is defined in terms of the Pfaffian:

$$\alpha^r = r!(2\pi)^r e(\alpha)\, dx^1 \wedge dx^2 \wedge \cdots \wedge dx^{2r} \tag{9.9.7}$$

Lastly, there is the Stiefel–Whitney class, which cannot be written in terms of curvature forms. However, we will find the Steifel–Whitney class to be important when we are analyzing spin manifolds. In particular, the vanishing of the first two indices yields an orientable spin manifold on which we can define spinors:

$$\begin{aligned} &\omega_1 = 0 \leftrightarrow M \text{ is orientable} \\ &\omega_1 = \omega_2 = 0 \rightarrow M \text{ is a spin manifold} \end{aligned} \tag{9.9.8}$$

The index theorems, in turn, are usually written in terms of the following invariant polynomials:

$$\begin{aligned} \text{Todd class} &= \mathrm{td}(M) = \prod_i \frac{x_i}{1 - e^{-x_i}} \\ \text{Hirzebruch class} &= L(M) = \prod_i \frac{x_i}{\tanh x_i} \\ \text{A roof polynomial} &= \hat{A}(M) = \prod_i \frac{\frac{1}{2}x_i}{\sinh \frac{1}{2}x_i} \end{aligned} \tag{9.9.9}$$

The most useful of the index theorems is the Dirac index theorem, which concerns the number of positive chirality zero eigenvalue solutions to the Dirac equation minus the negative chirality solutions. Using the supersymmetric sigma model, we can show the following:

$$\begin{aligned}\text{Index}(\not{D}) &= \text{Tr}(-1)^F e^{-\tau D^2} \\ &= \tfrac{1}{2}\int d^D x\, \partial_\mu J^{\mu 5} \\ &= \left(\frac{i}{2\pi}\right)^{(1/2)D} \int \text{Tr}\, e^F \det{}^{-1/2}[(\tfrac{1}{2}R)^{-1} \sinh \tfrac{1}{2}R]\end{aligned} \tag{9.9.10}$$

Armed with this theoretical apparatus, we can calculate the gauge and gravitational anomalies found in supergravity and in string theory. In general, the gravitational and gauge anomaly contributions arise because the internal line can be either

(1) a chiral spin-$\frac{1}{2}$ fermion,
(2) a chiral spin-$\frac{3}{2}$ fermion, or
(3) an antisymmetric tensor that has no covariant action.

The calculation of the anomaly contribution is carried out explicitly with Feynman diagrams. The calculation, however, simplifies enormously because we can make certain assumptions about the polarization tensor of the external lines, thus reducing its spin. Thus, the difficult problem of contracting over the various indices is reduced to a much simpler problem of contracting over a lower-spin particle. The final results are

$$\begin{aligned}I_{1/2} &= -i2^{2k+1} R(\varepsilon^{(i)}, p^{(j)})(4\pi)^{2k+1} M^2 \prod_{i=1}^{2k+1} \frac{\frac{1}{2}x_i}{\sinh \frac{1}{2}x_i} \\ I_{3/2} &= -i(2\pi)^{2k+1} R(\varepsilon^{(i)}, p^{(j)}) \prod_{i=1}^{2k+1} \frac{\frac{1}{2}x_i}{\sinh \frac{1}{2}x_i} \sum_{j=0}^{2k+1} (2\cosh x_j - 1) \\ I_A &= \tfrac{1}{4} i 2^{2k+1} 2\pi^{2k+1} R(\varepsilon^{(i)}, p^{(j)}) \prod_{i=1}^{2k+1} \frac{\frac{1}{2}x_i}{\sinh \frac{1}{2}x_i} \cosh \tfrac{1}{2}x_i\end{aligned} \tag{9.9.11}$$

Putting everything together, we find the total anomaly contribution from both gauge and gravitational sectors:

$$\begin{aligned}I_{12} = &\frac{-1}{720}\text{Tr}\, F^6 + \frac{1}{24\cdot 48}\text{Tr}\, F^4\, \text{Tr}\, R^2 \\ &- \frac{1}{256}\text{Tr}\, F^2\left(\frac{1}{45}\text{Tr}\, R^4 + \frac{1}{36}(\text{Tr}\, R^2)^2\right) \\ &+ \frac{(n-496)}{64}\left[\frac{1}{2\cdot 2835}\text{Tr}\, R^6 + \frac{1}{4\cdot 1080}\text{Tr}\, R^2\, \text{Tr}\, R^4\right. \\ &\left.+ \frac{1}{8\cdot 1296}(\text{Tr}\, R^2)^3\right] + \frac{1}{384}\text{Tr}\, R^2\, \text{Tr}\, R^4 + \frac{1}{1536}(\text{Tr}\, R^2)^3\end{aligned} \tag{9.9.12}$$

Miraculously, we can set this to zero if we make a few assumptions. First, we must set $n$ to be 496 (unless we have the $SO(16) \otimes SO(16)$ model, which will be discussed in the next chapter). Then we assume that we can factorize the anomaly into the product of two terms:

$$I_{12} \sim (\mathrm{Tr}\, R^2 + k\, \mathrm{Tr}\, F^2) X_8$$

$$\begin{aligned} X_8 = {} & \frac{1}{24}\mathrm{Tr}\, F^4 - \frac{1}{7200}(\mathrm{Tr}\, F^2)^2 - \frac{1}{240}\mathrm{Tr}\, F^2 \wedge \mathrm{Tr}\, R^2 \\ & + \frac{1}{8}\mathrm{Tr}\, R^4 + \frac{1}{32}(\mathrm{Tr}\, R^2)^2 \end{aligned} \tag{9.9.13}$$

Notice that $D = 10$ supergravity is immediately ruled out because the above identity cannot be satisfied. However, string theory has one big advantage over supergravity. The presence of higher spin fields in superstring theory means that the zero slope limit of the theory does not have to reduce exactly to supergravity. In particular, the couplings of the $B$ field in superstring theory are such that they can, in principle, cancel the terms shown above.

The actual proof, however, that the anomaly term disappears must be carried out with the full single-loop hexagon graph in the string formalism. We use a Pauli–Villars-type regularization on the intermediate lines and then sum over the planar and the nonorientable loops. The anomaly is proportional to

$$G \sim im^2 \varepsilon(\zeta, k) \int d^{10}p\, \mathrm{Tr}\left(\frac{1}{L_0 + m^2} V_0(1) \cdots \frac{1}{L_0 + m^2} V_0(6) \Gamma_d\right) \tag{9.9.14}$$

where

$$\varepsilon(\zeta, k) = \varepsilon_{\mu_1 \mu_2 \cdots \mu_5 \nu_1 \nu_2 \cdots \nu_5} \zeta_1^{\mu_1} \cdots \zeta_5^{\mu_5} k_1^{\nu_1} \cdots k_5^{\nu_5} \tag{9.9.15}$$

and

$$\Gamma_d = (-1)^{\sum_{n=1}^{\infty} d_{-n} d_n} \tag{9.9.16}$$

Finally, adding the planar and nonorientable diagrams together yields

$$\begin{aligned} G \sim (n + 32l)\, \mathrm{Tr}(\lambda_1 \lambda_2 \cdots \lambda_6) \varepsilon(\zeta, k) \int_0^1 \prod_{i=1}^{5} dv_i \\ \times\, \theta(v_{i+1} - v_i) \langle 0 | V_0(k_1, z_1) \cdots V_0(k_6, z_6) | 0 \rangle \end{aligned} \tag{9.9.17}$$

For this term to be zero, we must have $n = 32$ and

$$l = \begin{cases} 1 & \mathrm{Usp}(n) \\ 0 & \mathrm{U}(n) \\ -1 & \mathrm{SO}(n) \end{cases} \tag{9.9.18}$$

Thus, the gauge group must be $O(32)$.

## References

[1] J. E. Paton and H. M. Chan, *Nucl. Phys.* **B10**, 516 (1969).
[2] S. L. Adler, *Phys. Rev.* **177**, 2426 (1969).
[3] J. S. Bell and R. Jackiw, *Nuovo Cimento* **60A**, 47 (1969).
[4] W. A. Bardeen, *Phys. Rev.* **184**, 1848 (1969).
[5] E. Witten, in *Symposium on Anomalies, Geometry, and Topology* (edited by W. A. Bardeen and A. R. White) World Scientific, Singapore, 1985.
[6] P. H. Frampton and T. W. Kephart, *Phys. Rev. Lett.* **50**, 1343 (1983); *Phys. Rev.* **D28**, 1010 (1983).
[7] P. K. Townsend and G. Sierra, *Nucl. Phys.* **B222**, 493 (1983).
[8] B. Zumino, Y. S. Wu, and A. Zee, *Nucl. Phys.* **B239**, 477 (1984).
[9] J. Wess and B. Zumino, *Phys. Lett.* **37B**, 95 (1971).
[10] T. Eguchi, P. B. Gilkey, and A. J. Hanson, *Phys. Rep.* **66**, 213 (1980).
[11] C. Nash and S. Sen, *Topology and Geometry for Physicists*, Academic Press, New York, 1983.
[12] L. Alvarez-Gaumé and E. Witten, *Nucl. Phys.* **B234**, 269 (1983).
[13] J. S. Schwinger, *Phys. Rev.* **82**, 664 (1951).
[14] N. Marcus and J. H. Schwarz, *Phys. Lett.* **115B**, 111 (1982).
[15] M. B. Green and J. H. Schwarz, *Phys. Lett.* **149B**, 117 (1984); **151B**, 21 (1984).
[16] M. F. Atiyah and I. M. Singer, *Ann. Math.* **87**, 485, 546 (1968); **93**, 1, 119, 139 (1971).
[17] L. Alvarez-Gaumé, *Commun. Math. Phys.* **90**, 161 (1983).
[18] D. Friedan and P. Windey, *Nucl. Phys.* **B235** (FS11), 395 (1984).

CHAPTER 10

# Heterotic Strings and Compactification

## §10.1. Compactification

The most pressing problem facing string theory is the question of breaking a 26- or 10-dimensional theory down to a realistic 4-dimensional theory. Until such a dimensional reduction can be made, the theory lacks any real contact with physically measurable quantities.

Until dimensional breaking can be done within the framework of field theory, however, the best we can do is look at classical solutions where spontaneous compactification of the extra dimensions has already taken place. In this chapter we will explore the *heterotic string*, which incorporates the groups $E_8 \otimes E_8$ and Spin(32)/$Z_2$ that arise when we compactify a 26-dimensional space down to 10 dimensions.

As we saw in the previous chapter, the cancellation of anomalies allows for the possibility of either O(32) or $E_8 \otimes E_8$. We saw, however, that the Chan–Paton method does not allow for the possibility of exceptional groups. Thus, we must use yet another method, arising from compactification on a self-dual lattice, to achieve an $E_8 \otimes E_8$ model. Although the heterotic string is a closed string theory, it contains within it the super-Yang–Mills field, which usually emerges in the open string sector for Type I strings.

The heterotic string takes advantage of the deceptively simple identity

$$26 = 10 + 16 \tag{10.1.1}$$

This means that a 26-dimensional string, when compactified down to a 10-dimensional string, leaves 16 extra dimensions that may be placed on the root lattice of $E_8 \otimes E_8$, which is known to yield an anomaly-free theory. This observation was made by Freund [1].

The heterotic string takes advantage of the fact that the closed string has two independently moving sectors, the right-moving and left-moving sectors, depending on whether they propagate as functions of $\sigma + \tau$ or $\sigma - \tau$. The heterotic string makes fundamental use of this splitting. The word "heterosis" implies "hybrid vigor." This means that the asymmetric treatment of the left and right movers creates a hybrid theory considerably more sophisticated than the earlier Type I and II superstrings. It has been shown that it is has no tachyons, no anomalies, and is finite to one loop.

Before we begin a discussion of the heterotic string, however, let us first describe the process of compactification by describing the simplest possible case, that of a scalar particle in a periodic one-dimensional space. This means that we make the identification

$$x = x + 2\pi R \tag{10.1.2}$$

where $R$ is the radius of this space, which is just the real axis divided out by a one-dimensional lattice $\Gamma$ of length $2\pi R$:

$$S_1 = \frac{R_1}{\Gamma} \tag{10.1.3}$$

A field defined in this periodic space must therefore satisfy

$$\phi(x) = \phi(x + 2\pi R) \tag{10.1.4}$$

which means that it can be power expanded in periodic eigenfunctions:

$$\phi(x) = \sum_n \phi_n e^{ipx} \tag{10.1.5}$$

where

$$p = \frac{n}{R} \tag{10.1.6}$$

where $n$ is an arbitrary integer. Thus, we see that *the momentum conjugate to* $x$ *becomes quantized in terms of integers.* This is a feature common to all compactifications.

Now let us generalize this to a particle living in 5-dimensional space–time such that the fifth coordinate has curled up and become periodic. Consider a scalar field that satisfies the massless Klein–Gordon equation:

$$\Box_5 \phi(x_\mu, x_5) = 0 \tag{10.1.7}$$

As before, we can power expand the scalar field in terms of periodic eigenfunctions:

$$\phi(x_\mu, x_5) = \sum_n \phi(x_\mu) e^{ip_5 x_5} \tag{10.1.8}$$

where $p_5 = n/R$. Notice that each eigenfunction can change the effective "mass" of the Klein–Gordon operator:

$$\Box_5 = \Box_4 + \partial_5^2 = \Box_4 - p_5^2 \tag{10.1.9}$$

Several conclusions can be drawn from these simple examples:

(1) The compactification of a dimension creates a quantization of the momentum corresponding to the compactified coordinate. The momentum is labeled by integers.
(2) The mass spectrum in the space–time dimensions that are not compactified is shifted by an effective "mass" term coming from the compactified dimensions.
(3) The radii of the various compactified dimensions can be totally arbitrary. There is considerable freedom in choosing the lattice on which we will compactify the space.
(4) We can power expand the wave function in terms of periodic eigenfunctions of the compactified coordinate. In one dimension it is simply sines or cosines. In higher dimensions we may have spherical harmonics.

Now let us consider the case of compactifying theories of higher spin, such as general relativity, which will generate a Maxwell field out of the fifth dimension.

Historically, the idea of compactification first came from Kaluza [2–4], who wrote a letter to Einstein in 1919 proposing the idea of combining Maxwell's electromagnetic theory with Einstein's general relativity by expanding space–time to five dimensions. Kaluza proposed to write down the metric tensor as

$$\bar{g}_{AB} = \begin{pmatrix} \bar{g}_{\mu\nu} & \bar{g}_{\mu 5} \\ \bar{g}_{5\nu} & \phi \end{pmatrix} \tag{10.1.10}$$

where we define the five-dimensional metric in terms of the Maxwell field $A_\mu$ and the four-dimensional metric tensor $g_{\mu\nu}$:

$$\begin{aligned} \bar{g}_{5\mu} &= \bar{g}_{\mu 5} = \kappa A_\mu \\ \bar{g}_{\mu\nu} &= g_{\mu\nu} + \kappa^2 A_\mu A_\nu \end{aligned} \tag{10.1.11}$$

For the moment, let us assume that the fifth dimension is not experimentally measurable because it has curled up into a very small circle. Thus, the fifth dimension is periodic:

$$x_5 = x_5 + 2\pi R \tag{10.1.12}$$

That is, by traveling a distance $2\pi R$ in the fifth dimension, we arrive at the same starting point. The original five-dimensional Riemannian manifold is now assumed to have split up according to

$$R_5 \to R_4 \times S_1 \tag{10.1.13}$$

Now let us assume that the radius of the fifth dimension is so small that it cannot be measured. Thus, we can set

$$\partial_5 \to 0 \tag{10.1.14}$$

With this added assumption, the equations simplify enormously. The variation of the metric, for example, is usually given by

$$\delta\bar{g}_{\mu\nu} = \partial_\mu \Lambda_\nu + \partial_\nu \Lambda_\mu + \cdots \tag{10.1.15}$$

When the circumference of the fifth dimension is small, we have

$$\delta\bar{g}_{5\mu} = \partial_\mu \Lambda_5 + \cdots \tag{10.1.16}$$

Written in terms of (10.1.11), this reduces to

$$\delta A_\mu = \kappa^{-1} \partial_\mu \Lambda_5 \tag{10.1.17}$$

which is precisely the U(1) gauge variation of the Maxwell field. Since there is only one action that has this U(1) symmetry and has only two derivatives, we thus conclude that *Einstein's theory, written in five dimensions, reduces to Maxwell's theory coupled to a four-dimensional gravitational theory.* For example, we can explicitly calculate some of the Christoffel symbols in this approximation:

$$\Gamma_{5\mu,\nu} = \tfrac{1}{2}\kappa(\partial_\mu A_\nu - \partial_\nu A_\mu) + \cdots = \tfrac{1}{2}\kappa F_{\mu\nu} \tag{10.1.18}$$

We can thus explicitly reduce out the five-dimensional Einstein theory in this approximation, and we find

$$\begin{aligned} L &= \frac{-1}{2k^2}\sqrt{-\bar{g}}\, R_{AB}\bar{g}^{AB} \\ &= \frac{-1}{2k^2}\sqrt{-g}R_{\mu\nu}g^{\mu\nu} - \tfrac{1}{4}\sqrt{-g}F_{\mu\nu}F^{\mu\nu} + \cdots \end{aligned} \tag{10.1.19}$$

Thus, a genuine Maxwell field emerges in four-space when we compactify a five-space theory of gravity.

Similarly, we can generalize this result to higher manifolds, such that an $N$-dimensional manifold peels off a smaller compactified manifold $K_P$ of dimension $P$:

$$R_N \to R_{N-P} \times K_P \tag{10.1.20}$$

When this is done, we can reanalyze the gauge group that results from this breakdown, and we can derive the orthogonal groups or the SU($N$) groups. In this way, we can show that Yang–Mills theory emerges out of higher-dimensional gravitation theory.

Although this Kaluza–Klein formalism elegantly unites both the Yang–Mills and gravitational theories in one framework, there is a severe drawback to this approach that dates back to Kaluza's original proposal, namely, why did the fifth dimension suddenly curl up into a tiny ball? Klein's suggestion that it curled up quantum mechanically into a ball of radius equal to the Planck length was an important one, but it did not answer the difficult problem of how dimensional breaking actually occurs. We are still facing this problem, first raised over 65 years ago, in superstring theory.

Next, we wish to discuss the question of compactification in the framework of string theory [5], which must undergo the reduction from 26 or 10 down to 4 dimensions. Let us first study the compactification of the $i$th coordinate of an open string:

$$X^i(\sigma, \tau) = x^i + 2\alpha' p^i \tau + \sum_{n \neq 0} \frac{1}{n} \alpha_n^i \cos n\sigma e^{-in\tau} \tag{10.1.21}$$

As before, the periodicity of the $i$th coordinate enforces the condition that momentum is quantized according to integer multiples of the integer $M_i$

$$p^i = \frac{M_i}{R_i} \tag{10.1.22}$$

where $R_i$ is the radius of the $i$th compactified coordinate. This is exactly as before. In general, the radii of the compactified dimensions do not have to be the same. As before, we also find that the mass spectrum of the theory is shifted by the presence of the compactified dimensions. By analyzing the Hamiltonian, we see that the masses are given by

$$\alpha' m^2 = \frac{\alpha'}{R^2} \sum_{i=1}^{10-D} M_i^2 + N \tag{10.1.23}$$

where $N$ is the mass operator:

$$N = \sum_{n=1}^{\infty} \sum_{i=1}^{D} \alpha_{-n}^i \alpha_n^i \tag{10.1.24}$$

Thus, the mass spectrum is shifted by the square of an integer, as in (10.1.9).

For the closed string, however, we have an additional contribution to the mass. There is the extra complication that the closed string can loop $N_i$ times completely around the $i$th dimension, much as a rubber band is wound around a cylindrical tube. It is important to notice that this configuration, which contributes a new term to the Hamiltonian, has no counterpart in point particle compactification. We thus have, for the closed string,

$$\begin{aligned} X^i &= x^i + 2\alpha' p^i \tau + 2N_i R\sigma \\ &\quad + (\alpha'/2)^{1/2} \sum_{n \neq 0} \frac{1}{n} (\alpha_n^i e^{-2in(\tau-\sigma)} + \tilde{\alpha}_n^i e^{-2in(\tau+\sigma)}) \end{aligned} \tag{10.1.25}$$

We now have two integers $M_i$ and $N_i$, which represent the effect of compactification for the closed string. The integer $N_i$, in fact, describes the *soliton* state of a string wound around the compactified dimension an integer number of times. Notice that this soliton state is stable because of topology.

The shifted masses are now given by

$$\tfrac{1}{2}\alpha' m^2 = \tfrac{1}{2} \sum_{i=1}^{10-D} (\alpha'^2 M_i^2/R^2 + R^2 N_i^2/\alpha'^2) + N + \tilde{N}$$

where $N$ and $\tilde{N}$ are energy operators for the two different sectors of the closed string.

In this discussion, we have compactified the $10 - D$-dimensional space on a torus. However, this is phenomenologically undesirable, because it leaves us with $N = 4$ supersymmetry and without any chiral fermions. It is more desirable to compactify on more sophisticated surfaces, such as the lattice of a Lie group (see Appendix). We now turn to the heterotic string, where we compactify on the lattice of $E_8 \otimes E_8$ or Spin(32)/$Z_2$.

## §10.2. The Heterotic String

Let us now begin a discussion of the heterotic string of Gross, Harvey, Martinec, and Rohm [6]. We saw in the last chapter that the anomaly cancellation required either O(32) or $E_8 \otimes E_8$. However, we also saw that Chan–Paton factors are not compatible with exceptional groups. Thus, we are forced to use the mechanism of compactification to generate the isospin group $E_8 \otimes E_8$.

The heterotic string is a closed string with the unusual feature that it treats the compactification of the left- and right-moving sectors separately. In the left-moving sector, which is purely bosonic, let us begin with a 26-dimensional string where 16 of the dimensions have been compactified on a lattice. (The 16-dimensional lattice we will eventually choose will be the lattice of $E_8 \otimes E_8$ or Spin(32)/$Z_2$.)

The right movers, on the other hand, are supersymmetric; i.e., they contain the GS Majorana–Weyl fermionic field $S^a(\tau - \sigma)$, where the index $a$ runs from 1 to 8 (in the light cone gauge), and also the space–time string field $X^i$:

| Left movers | Right movers |
|---|---|
| $X^i(\tau + \sigma)$ | $X^i(\tau - \sigma)$ |
| $X^I(\tau + \sigma)$ | $S^a(\tau - \sigma)$ |

where $I$, which labels the directions of the 16-dimensional lattice, runs from 1 to 16, and $i$, which is the space–time index, runs from 1 to $10 - 2 = 8$ in the light cone gauge.

Notice that the space–time string $X^i$ appears in both in the left- and right-moving sectors. Putting everything together, we now have the light cone action for the heterotic string:

$$S = \frac{-1}{4\pi\alpha'} \int d\tau \int_0^\pi d\sigma \left( \partial_a X^i \partial_a X^i + \sum_{I=1}^{16} \partial_a X^I \partial_a X^I + i\bar{S}\gamma^- (\partial_\tau + \partial_\sigma) S \right) \qquad (10.2.1)$$

In this action, we have to enforce the constraints that place the various fields in their proper sector:

$$
\begin{aligned}
(\partial_\tau - \partial_\sigma) X^I &= 0 \\
\gamma^+ S = \tfrac{1}{2}(1 + \gamma_{11}) S &= 0 \\
\gamma^+ &= \frac{1}{\sqrt{2}}(\gamma^0 + \gamma^9)
\end{aligned}
\tag{10.2.2}
$$

This action is explicitly supersymmetric under the variation

$$
\begin{cases}
\delta X^i = (p^+)^{-1/2} \bar{\varepsilon} \gamma^i S \\
\delta S^a = i(p^+)^{-1/2} \gamma_- \gamma_\mu (\partial_\tau - \partial_\sigma) X^\mu \varepsilon
\end{cases}
\tag{10.2.3}
$$

Let us now write down an explicit breakdown of the theory in terms of normal modes (notice that we have slightly changed the normalization of the spinors from that of (3.8.8)):

$$
\begin{aligned}
X^i(\tau - \sigma) &= \tfrac{1}{2} x^i + \tfrac{1}{2} p^i(\tau - \sigma) + \tfrac{1}{2} i \sum_{n=1}^{\infty} \frac{\alpha_n^i}{n} e^{-2in(\tau - \sigma)} \\
X^i(\tau + \sigma) &= \tfrac{1}{2} x^i + \tfrac{1}{2} p^i(\tau + \sigma) + \tfrac{1}{2} i \sum_{n=1}^{\infty} \frac{\tilde{\alpha}_n^i}{n} e^{-2in(\tau + \sigma)} \\
S^a(\tau - \sigma) &= \sum_{n=-\infty}^{\infty} S_n^a e^{-2in(\tau - \sigma)} \\
X^I(\tau + \sigma) &= x^I + p^I(\tau + \sigma) + \tfrac{1}{2} i \sum_{n=1}^{\infty} \frac{\tilde{\alpha}_n^I}{n} e^{-2in(\tau + \sigma)}
\end{aligned}
\tag{10.2.4}
$$

Similarly, we can read off the canonical commutation relations of the theory directly from the action. From these, we can, in turn, write down the canonical commutation relations between the various oscillators:

$$
\begin{aligned}
[x^i, p^j] &= i\delta^{ij} \\
[\alpha_n^i, \alpha_m^j] = [\tilde{\alpha}_n^i, \tilde{\alpha}_m^j] &= n\delta_{n,-m}\delta^{ij} \\
[\alpha_n^i, \tilde{\alpha}_m^j] &= 0 \\
[\tilde{\alpha}_n^I, \tilde{\alpha}_m^J] &= n\delta_{n,-m}\delta^{IJ} \\
\{S_n^a, \bar{S}_m^b\} &= (\gamma^+ h)^{ab} \delta_{n,-m}
\end{aligned}
\tag{10.2.5}
$$

where $h$ is the chirality projection operator $h = \frac{1}{2}(1 + \gamma_{11})$.

(There is one commutator, however, that is rather subtle. Notice that we have constrained the compactified degrees of freedom to be left-moving. Thus, the quantization of these modes must take into account this constraint, or else there will be an inconsistency within the theory. Fortunately, the only modification made by this constraint is to multiply the canonical commutation relation by half:

$$[x^I, p^J] = \tfrac{1}{2} i\delta^{IJ} \tag{10.2.6}$$

This can be explicitly checked by using the Dirac bracket formulation of gauge constraints or by checking that the newly defined commutator is consistent with the constraints.)

Let us define the number operator for each sector:

$$\begin{aligned} N &= \sum_{n=1}^{\infty} (\alpha_{-n}^i \alpha_n^i + \tfrac{1}{2} n \bar{S}_{-n} \gamma^- S_n) \\ \tilde{N} &= \sum_{n=1}^{\infty} (\tilde{\alpha}_{-n}^i \tilde{\alpha}_n^i + \tilde{\alpha}_{-n}^I \tilde{\alpha}_n^I) \end{aligned} \tag{10.2.7}$$

This means that we can categorize the oscillators in both sectors as follows:

| Left movers | Right movers |
|---|---|
| $\tilde{a}^i$ | $a^i$ |
| $\tilde{a}^I$ | $S^a$ |
| $\tilde{N}$ | $N$ |

(10.2.8)

With these definitions, we find, as before, that the mass can be written as

$$\tfrac{1}{4} m^2 = N + (\tilde{N} - 1) + \tfrac{1}{2} \sum_{I=1}^{16} (p^I)^2 \tag{10.2.9}$$

In addition, there is one extra constraint coming from the fact that we have a closed string: the theory should be independent of the origin in $\sigma$ space, as in (2.8.7). Thus, if $U(\theta)$ is the operator that rotates a closed string by $\theta$ in the $\sigma$ parameter, then we must satisfy the following for physical states:

$$U(\theta)|\text{phy}\rangle = |\text{phy}\rangle \tag{10.2.10}$$

For the heterotic string, the rotation operator is

$$U(\theta) = \exp\left\{2i\theta\left(N - \tilde{N} + 1 - \tfrac{1}{2} \sum_{I=1}^{16} (p^I)^2\right)\right\} \tag{10.2.11}$$

This operator, acting on an arbitrary function of $\sigma$, produces the transformation

$$U(\theta) F(\sigma) U^{-1}(\theta) = F(\sigma + \theta) \tag{10.2.12}$$

To ensure that $U(\theta) = 1$ on physical states, we must satisfy the following constraint:

$$N = \tilde{N} - 1 + \tfrac{1}{2} \sum_{I=1}^{16} (p^I)^2 \tag{10.2.13}$$

Up to now, we have not endowed the 16-dimensional lattice with any structure. To fix the theory, we take the space to be

$$T^{16} = \frac{R^{16}}{\Lambda} \tag{10.2.14}$$

where $\Lambda$ represent an as yet unspecified lattice in 16 dimensions.

Let us span this 16-dimensional lattice by basis vectors:

$$e_i^I \tag{10.2.15}$$

To say that we will compactify the 16-dimensional space with a lattice simply means that if we walk in the direction $L^I$ that is specified by one of the lattice vectors, we eventually wind up at the same spot. With this basis, we want the center of mass string coordinate to be periodic along any of these basis vectors:

$$\begin{aligned} X^I &= X^I + 2\pi L^I \\ L^I &= \frac{1}{\sqrt{2}} \sum_{i=1}^{16} n_i e_i^I R_i \end{aligned} \tag{10.2.16}$$

where the $n_i$ are integers and $R_i$ are the radii of the various compactified dimensions. Thus, if we wander off in any of the directions labled by the basis vector, we wind up back at the original position. This places a nontrivial constraint on the momenta. We know from ordinary quantum mechanics that $2p^I$ is the generator of translations in the $I$th direction. Thus, periodicity means the following displacement operator, acting on states, must have an eigenvalue equal to one:

$$e^{i2\pi p^I \cdot L^I} = 1 \tag{10.2.17}$$

This operator simply generates the displacement in (10.2.16) and makes a translation completely around the torus in the $I$th direction, until we wind up at the original spot.

Thus, the canonical momenta must be defined as

$$p^I = \sqrt{2} \sum_{i=1}^{16} m_i e_i^{*I}/R_i \tag{10.2.18}$$

where $m$ is an integer and the $e^*$ define the *dual lattice* $\Lambda^*$ such that

$$\delta_{ij} = \sum_{I=1}^{D} e_i^I e_j^{*I} \tag{10.2.19}$$

At this point we should mention that the lattice over which we compactify the theory is not totally arbitrary. First of all, it must be an *even lattice* because of the constraint (10.2.13) due to demanding $\sigma$ independence of the Hilbert space. Notice that the $U(\theta)$ operator in (10.2.11) can be set equal to one on the Fock space only if $\frac{1}{2}(p^I)^2 =$ integer, i.e.,

$$(p^I)^2 = \text{even integer} \tag{10.2.20}$$

Therefore the winding numbers $L^I$ must lie on an integer, even lattice.

The metric tensor for the lattice corresponding to a Lie group is defined as

$$g_{ij} = \sum_{I=1}^{16} e_i^I e_j^I \tag{10.2.21}$$

we find therefore that the metric must be integer-valued and $g_{ii}$ is even. Second, the lattice must be *self-dual* (i.e., the lattice vectors are equal to the vectors on the dual lattice) because of modular invariance, which will become more apparent as we discuss the first loop diagram in Section 10.6. There are very few even, self-dual lattices. They exist only in $8n$ dimensions. In 8 dimensions, there is only one, $\Gamma_8$, the root lattice of $E_8$. In 16 dimensions, there are only two of them:

$$\Gamma_8 \times \Gamma_8; \qquad \Gamma_{16} \tag{10.2.22}$$

which fixes the lattice to be either $E_8 \otimes E_8$ or $\text{Spin}(32)/Z_2$. Thus, we actually have very little choice in the selection of the gauge group.

The metric for $E_8$ is as follows:

$$\begin{pmatrix} 2 & -1 & & & & & & \\ -1 & 2 & -1 & & & & & \\ & -1 & 2 & -1 & & & & \\ & & -1 & 2 & -1 & & & \\ & & & -1 & 2 & -1 & & -1 \\ & & & & -1 & 2 & -1 & \\ & & & & & -1 & 2 & \\ & & & & -1 & & & 2 \end{pmatrix} \tag{10.2.23}$$

There are several ways in which to describe the root lattice for $E_8$. One of the simplest is as follows. We have 84 root vectors given by

$$\pm\mathbf{e}_i \pm \mathbf{e}_j; \qquad 1 \le i \ne j \le 7 \tag{10.2.24}$$

and 128 root vectors given by

$$\tfrac{1}{2}(\pm\mathbf{e}_1 \pm \mathbf{e}_2 \pm \mathbf{e}_3 \pm \cdots \pm \mathbf{e}_8) \tag{10.2.25}$$

and 28 vectors given by

$$\pm\mathbf{e}_i \pm \mathbf{e}_8 \tag{10.2.26}$$

Including the 8 members of the Cartan subalgebra, that gives us a total of 248 generators, which form the adjoint representation of the group.

Later in this chapter we will also use the concept of *simply laced* groups, i.e., groups that have roots of the same length. The simply laced groups are of the type $A$, $D$, and $E$ (see Appendix).

Let us briefly summarize the kinds of Lie groups we will encounter in our discussion of the heterotic string:

| Lattice | Roots |
|---|---|
| Even | $(p^I)^2 = \text{even}$; ($g_{ii} = \text{even}$) |
| Self-dual | $\Lambda = \Lambda^*$; ($\det g_{ij} = 1$) |
| Simply laced | Same length; ($A$, $D$, $E$) |

These lattices are important because the $U(\theta) = 1$ condition forces us to have even lattices, and modular invariance (as we shall see in Section 10.6) forces us to have self-dual lattices:

$$U(\theta) = 1 \rightarrow \text{Even lattice}$$

$$\text{Modular invariance} \rightarrow \text{Self-dual lattice}$$

In 16 dimensions, the only even, self-dual lattices are the root lattices for $E_8 \otimes E_8$ and Spin(32)/$Z_2$.

## §10.3. Spectrum

Let us now investigate the spectrum of the theory. First, we note that the right movers within the theory are explicitly space–time supersymmetric. The generator of space–time supersymmetry can easily be constructed from the variation of the fields. This, as we saw in earlier chapters, allows us to construct the current associated with the symmetry. We found earlier in (3.8.22) that the light cone supersymmetric operators can be written as

$$Q^a = i\sqrt{p^+}(\gamma^+ S_0)^a + 2i(p^+)^{-1/2} \sum_n (\gamma_i S_{-n})^a \alpha_n^i \tag{10.3.1}$$

which acts on the right movers. We can easily show, given the canonical commutation relations, that

$$\{Q^a, Q^b\} = -2(h\gamma^\mu P_\mu)^{ab} \tag{10.3.2}$$

To analyze the left-moving sector, we notice that isospin is introduced in an entirely different fashion than the Chan–Paton factors. Contrary to the usual Chan–Paton factors, which simply multiply the amplitudes by the trace of a series of generators, the isospin factors emerging from the compactification process astronomically expand the number of isotopically allowed states. Notice that the index $I$ is placed on the harmonic oscillator itself, which greatly proliferates the number of allowed states of the Fock space.

In general, since the two sectors don't really communicate with each other, a member of the Fock space of the heterotic string will be the product of the left- and right-moving sectors:

$$|0\rangle_R \times |0\rangle_L \tag{10.3.3}$$

At the lowest level, let us first investigate the right-moving sector. As before, in (3.8.13) and (3.8.14), the eight states in the fermionic $|a\rangle$ or bosonic $|i\rangle$ ground state of the superstring can be written in terms of each other:

$$\begin{aligned} |a\rangle_R &= \frac{i}{8}(\gamma_i S_0)^a |i\rangle_R \\ |i\rangle_R &= \frac{i}{16}[\bar{S}_0\{\gamma_i, \gamma_+\}]^a |a\rangle_R \end{aligned} \tag{10.3.4}$$

Thus, we have a total of $8 + 8 = 16$ states in the right-moving massless sector.

Now let us investigate the left-moving sector. We have

$$\begin{cases} \tilde{\alpha}_{-1}^j |0\rangle_L & \to 8 \text{ states} \\ \tilde{\alpha}_{-1}^I |0\rangle_L & \to 16 \text{ states} \\ |p^I; (p^I)^2 = 2\rangle & \to 480 \text{ states} \end{cases} \tag{10.3.5}$$

(It is very important to realize that this particular representation of the spectrum breaks manifest $E_8 \otimes E_8$ symmetry. To see this, notice that there are 16 $\alpha^I$ operators, which correspond to the dimension of the root lattice of the group, but these 16 operators do not form a representation of the group. Similarly, the vectors $|p^I; (p^I)^2 = 2\rangle$ correspond to the 480 roots of length two. However, this also does not form a representation of the group. Only when we add 16 and 480 and form 496 states do we finally construct the adjoint representation of the group. At the massless level, it is still possible to recombine the broken states into $E_8 \otimes E_8$ multiplets, but it becomes increasingly prohibitive at higher mass levels. Later, we will give an argument, based on Kac–Moody algebras, demonstrating how we can show that the entire spectrum is, in fact, $E_8 \otimes E_8$ symmetric.)

At the massless level, we obtain the spectrum by multiplying the right and left movers. The total number of states in the lowest level of the heterotic string is thus the product of the left and the right movers: $16 \times 504 = 8064$ states. The breakdown of these states is now given as follows. Of the 8064 states, we find that 128 belong to the $N = 1$, $D = 10$ supergravity (in the light cone):

$$\text{Supergravity} \to \tilde{\alpha}_{-1}^i |0\rangle_L \times |i \text{ or } a\rangle_R \tag{10.3.6}$$

If we break this down into distinct states, we find

$$\begin{aligned} \text{Graviton} &\to \tilde{\alpha}_{-1}^i |0\rangle_L \times |j\rangle_R + (i \leftrightarrow j) \\ \text{Antisym. tensor} &\to \tilde{\alpha}_{-1}^i |0\rangle_L \times |j\rangle_R - (i \leftrightarrow j) \\ \text{Gravitino} &\to \tilde{\alpha}_{-1}^i |0\rangle_L \times |a\rangle_R \end{aligned} \tag{10.3.7}$$

Other states belong to the super-Yang–Mills theory defined on $E_8 \otimes E_8$.

For example, the super-Yang–Mills multiplet $(A_\mu^a, \psi^a)$ is in the adjoint representation of the group, which has 496 states. These 496 isotopic states are given by the 480 + 16 states contained in the left-moving sector:

$$\tilde{\alpha}_{-1}^I|0\rangle_L + |p^I; (p^I)^2 = 2\rangle_L \tag{10.3.8}$$

Thus, the super-Yang–Mills state multiplet can be represented as 496 × 8 states:

$$\text{Super-Yang–Mills} \to [\tilde{\alpha}_{-1}^I|0\rangle_L + |p^I\rangle_L] \times |i \text{ or } a\rangle_R \tag{10.3.9}$$

In summary, we have exactly 10-dimensional supergravity and Yang–Mills fields at $m^2 = 0$.

Notice that we had to combine 480 + 16 to yield an irreducible representation of the group. This means that, in general, the Fock space of the heterotic string does not manifestly recombine into $E_8 \otimes E_8$ multiplets. At the next level, for example, the states form irreducible representations of the algebra only in the last steps of the calculation.

At the next level, where $m^2 = 8$, for the right-moving sector we have 128 bosons and 128 fermions:

$$\begin{aligned} &\text{128 Bosons} \begin{cases} \alpha_{-1}^i|j\rangle_R \\ S_{-1}^a|b\rangle_R \end{cases} \\ &\text{128 Fermions} \begin{cases} \alpha_{-1}^i|a\rangle_R \\ S_{-1}^a|i\rangle_R \end{cases} \end{aligned} \tag{10.3.10}$$

Notice that we have formed bosons out of the tensor product of two fermions in the above equation.

The left-moving sector is considerably more involved, with a total of 73,764 states. We have a total, therefore, of 256 × 73,764 = 18,883,584 states! The scalar states are

$$\text{Scalars} \to \begin{cases} |p^I, (p^I)^2 = 4\rangle_L & \to 61{,}920 \text{ states} \\ \tilde{\alpha}_{-1}^I|p^J, (p^J)^2 = 2\rangle_L & \to 7680 \text{ states} \\ \tilde{\alpha}_{-1}^I\tilde{\alpha}_{-1}^J|0\rangle_L & \to 136 \text{ states} \\ \tilde{\alpha}_{-2}^I|0\rangle_L & \to 16 \text{ states} \end{cases} \tag{10.3.11}$$

for a total of 69,752 scalar states. (To count the states in $|p^I, (p^I)^2 = 4\rangle$, we used the fact that the number of points of length squared equal to $2m$ for integer $m$ on the lattice $\Gamma_{16}$ is 480 times the sum of the seventh powers of the divisors of $m$. Thus, for this states, there are $480(1 + 2^7) = 61{,}920$ states [6].)

At first, it is not obvious at all that these 69,752 scalar states can be rearranged in terms of $E_8 \otimes E_8$ multiplets. However, a careful analysis shows that they can be broken down into the following $E_8 \otimes E_8$ multiplets:

$$(\mathbf{3875}, \mathbf{1}) + (\mathbf{1}, \mathbf{3875}) + (\mathbf{248}, \mathbf{248}) + (\mathbf{248}, \mathbf{1}) + (\mathbf{1}, \mathbf{248}) + (\mathbf{1}, \mathbf{1}) + (\mathbf{1}, \mathbf{1}) \tag{10.3.12}$$

The 3976 vectors in the left-moving sector are arranged as

$$\text{Vectors} \rightarrow \begin{cases} \tilde{\alpha}^i_{-1}|p^I, (p^I)^2 = 2\rangle_L \rightarrow 3840 \text{ states} \\ \tilde{\alpha}^i_{-2}|0\rangle_L \qquad\qquad \rightarrow 8 \text{ states} \\ \tilde{\alpha}^i_{-1}\alpha^I_{-1}|0\rangle_L \qquad\quad \rightarrow 128 \text{ states} \end{cases} \tag{10.3.13}$$

and we have the tensor state

$$\tilde{\alpha}^i_{-1}\tilde{\alpha}^j_{-1}|0\rangle_L \rightarrow 36 \text{ states} \tag{10.3.14}$$

The total number of states in the left-moving sector is thus 73,764.

At higher levels, the degeneracy climbs exponentially as

$$d(M) \sim e^{(2+\sqrt{2})\pi\sqrt{\alpha'}M} \tag{10.3.15}$$

It was not obvious that these states can be reformulated in terms of $E_8 \otimes E_8$ multiplets. At higher levels, it seems like a hopeless task to show that higher masses all can be rearranged into irreducible representations of the group. What we need, of course, is a higher symmetry to prove this to all orders. We will do this when we apply Kac–Moody algebras in Section 10.7.

## §10.4. Covariant and Fermionic Formulations

Although we have analyzed the heterotic string's spectrum in the light cone gauge, we can also write down an explicitly covariant version of the first quantized action. We begin by writing the right-moving supersymmetric sector in covariant form. Let $P_\mu$ and $Q^\alpha$ represent two of the generators of the super-Poincaré group. Let us generalize the operator that gives us translations in the $x$ direction. An element of the supertranslation group is given by

$$h = e^{iX\cdot P + i\theta Q} \tag{10.4.1}$$

where $\theta$ is a 10-dimensional spinor. This operator will shift a function of the coordinate by $X$ and a fermionic coordinate by $\theta$. Now construct ($\alpha = 1, 2$)

$$\Pi_\alpha = h^{-1}\partial_\alpha h = (\partial_\alpha X^\mu - i\bar{\theta}\gamma^\mu\partial_\alpha\theta)P_\mu + \partial_\alpha\theta Q \tag{10.4.2}$$

Now the right-moving GS action can be written as

$$S_R = \int d^2z\,\tfrac{1}{2}e\,\mathrm{Tr}(\Pi_\alpha\Pi_\beta)e^\alpha_a e^{a\beta} + \int d^3\xi\,\varepsilon^{\alpha\beta\gamma}\,\mathrm{Tr}(\Pi_\alpha\Pi_\beta\Pi_\gamma)$$
$$+ \int d^2z\,e\lambda^{++}(e^\alpha_+\Pi_\alpha)^2 \tag{10.4.3}$$

The first term on the right is the usual quadratic term of the GS action. The second term is the nonlinear term of the GS action written as a Wess–Zumino term. It is a three-dimensional integral over a surface whose boundary coincides with the world sheet of the string. The sum of these two terms is an alternative formulation of the GS action. The third term of the action just enforces the right-moving constraint.

Similarly, the left-moving action that contains the isotopic sector can also be written covariantly, but we have a choice of using either *fermions or bosons.* We know from the theory of Lie groups that the generators of an algebra can be written in terms of the product of either bosonic or fermionic fields. Thus, we have

$$S_L = \tfrac{1}{2}i \int d^2z\, e\psi^I \partial_\alpha \psi^I e_+^\alpha \tag{10.4.4}$$

where this time $I = 1$ to 32, which labels the fundamental representation of SO(32). It is important to note that these fermions transform as Lorentz scalars. The $I$ index is an internal index. We can, of course, choose either Ramond (periodic) or Neveu–Schwarz (antiperiodic) boundary conditions on these fermion oscillators.

For the boson representation, we can represent the same left-moving isotopic sector as

$$S_L = \int d^2z\, \tfrac{1}{2}e(e_a^\alpha \partial_\alpha X^I)^2 + e\lambda^{--}(e_-^\alpha\, \partial_\alpha X^I)^2 \tag{10.4.5}$$

where $I = 1, 16$. This, of course, yields the light cone representation we used earlier when we broke Lorentz covariance. (It may seem strange at first to see the left and right sectors labeled by $\pm$, which seems to single out a non-Lorentz-invariant direction in two dimensions. However, the above formulation is still reparametrization invariant because the $\pm$ directions are placed in the tangent space. Thus, two dimensional reparametrization invariance remains intact.)

Now let us write the entire space–time action (minus the isospin part) that contains both left- and right-moving sectors in a covariant fashion:

$$S = \int d^2z\, e\{\tfrac{1}{2}(e_a^\alpha \partial_\alpha X^\mu)^2 - \tfrac{1}{2}i\psi^\mu \rho^- e_-^\alpha\, \partial_\alpha \psi_\mu + \tfrac{1}{2}i(\chi_\alpha \rho^\beta \rho^\alpha \psi^\mu)\partial_\beta X_\mu\} \tag{10.4.6}$$

This action has "$N = \frac{1}{2}$" supersymmetry:

$$\begin{aligned} \delta e_-^\alpha &= i\varepsilon\rho_- \xi^\alpha \\ \delta\chi_\alpha &= -2\nabla_\alpha \varepsilon \\ \delta X^\mu &= i\varepsilon\psi^\mu \\ \delta\psi^\mu &= [\partial_\alpha X^\mu + i\chi_\alpha \psi^\mu \rho^\alpha]\varepsilon \end{aligned} \tag{10.4.7}$$

Now let us, instead of choosing the light cone gauge, choose the conformal gauge. Then the heterotic action reduces to

$$S = \tfrac{1}{2} \int d^2z((\partial_\alpha X^\mu)^2 + i\psi^\mu \rho^- \partial_- \psi_\mu + i\psi^I \rho^+ \partial_+ \psi^I) \tag{10.4.8}$$

The point of all this was to show that a covariant version of the heterotic string exists and that we can use either fermion or bosonic fields to represent the isospin part of the fields. We are not bound to a light cone formulation in terms of boson fields.

It should be noted, however, that even in covariant fashion the heterotic string appears a bit awkward and contrived. Perhaps a future version of the theory will allow a more elegant formulation.

## §10.5. Trees

Trees are constructed in the heterotic string almost exactly the way they were constucted for the light cone superstring, except now the vertices have left and right components and we also must take into account the compactificationon a 16-dimensional lattice.

To describe the supergravity multiplet, we will introduce the spin-$\frac{3}{2}$ field $U^{\mu a}$ to represent the spinor field and $\rho^{\mu\nu}$ to represent the polarization tensor of the graviton. In the on-shell light cone gauge, we will choose the external particles to be massless and transverse:

$$\begin{aligned} &\rho^{+\mu} = \rho^{\mu +} = k_\mu \rho^{\mu\nu} = 0 \\ &U^{+a} = k_\mu U^{\mu a} = \gamma^+ U^{\mu a} = \tfrac{1}{2}(1 - \gamma_{11}) U^{\mu a} = 0 \end{aligned} \tag{10.5.1}$$

In analogy with the superstring light cone vertex developed in Chapter 3, we can construct the vertex functions for the hereotic string based on analogy with the vertices in (3.9.2), except we need an extra insertion of $P^\mu$ to yield more Lorentz indices. In addition, the vertices are *the direct product of left-moving and right-moving vertices.* We now choose for the boson and fermion emission vertices for the supergravity sector:

$$\text{Supergravity:} \quad \begin{cases} V_B = \rho_{\mu\nu}(k) \displaystyle\int_0^\pi d\sigma\, B^\mu \tilde{P}^\nu e^{ik\cdot X} \\ V_F = \displaystyle\int_0^\pi d\sigma\, F^a \tilde{P}_\nu U^{a\nu}(k) e^{ik\cdot X} \end{cases} \tag{10.5.2}$$

where we use the definitions in (3.9.6):

$$\begin{aligned} k^+ &= 0 \\ B^i &= P^i + \tfrac{1}{2} k^j R^{ij} \\ P^i &= \frac{dX^i}{d(\tau - \sigma)} = \tfrac{1}{2} p^i + \sum_{n\neq 0} \alpha_n^i e^{-2in(\tau-\sigma)} \\ R^{ij} &= \tfrac{1}{8} \bar{S} \gamma^{ij-} S \\ \bar{F}^a &= \tfrac{1}{2} i (p^+)^{-1/2} [\bar{S}\gamma \cdot P - \tfrac{1}{6} {:}R^{ij} k^i \bar{S} \gamma^j{:}]^a \\ \tilde{P}^i &= \tfrac{1}{2} p^i + \sum_{n\neq 0} \tilde{\alpha}_n^i e^{-2in(\tau+\sigma)} \end{aligned} \tag{10.5.3}$$

(Notice that we have explicitly chosen the frame where all $k^+$ are set to zero, where we have a large simplification. The calculation for nonzero $k^+$ is actually quite complicated because it requires making a Lorentz transformation with the broken Lorentz generators $M^{-i}$.)

The vertices for the gauge mesons can also be written down. One complication, however, is that the 496 vector mesons in the adjoint representation of the gauge group can be broken down into 16 "neutral" gauge bosons transforming as members of the Cartan subalgebra, and 480 "charged" mesons corresponding to $K^I$ with $(K^I)^2 = 2$. The neutral ones are defined by

$$|i\rangle_R \times \tilde{\alpha}_{-1}^I |0\rangle_L \tag{10.5.4}$$

and the charged ones by

$$|i\rangle_R \times |K^I; (K^I)^2 = 2\rangle_L \tag{10.5.5}$$

Thus, we must have two types of vertices for the gauge bosons. The vertices for the neutral gauge mesons are essentially the same as before, except that we interchange the Lorentz index for an internal one:

$$\text{Neutral gauge mesons:} \quad \begin{cases} V_B^I = \rho_\mu^I(k) \displaystyle\int_0^\pi d\sigma\, B^\mu \hat{P}^I e^{ik_\mu X^\mu} \\ V_F^I = \displaystyle\int_0^\pi d\sigma\, F^a \hat{P}^I U^{aI}(k) e^{ik_\mu X^\mu} \end{cases} \tag{10.5.6}$$

where

$$\hat{P}^I = \frac{dX^I}{d(\tau + \sigma)} = P^I + \sum_{n \neq 0} \tilde{\alpha}_n^I e^{-2in(\tau+\sigma)}$$

Correspondingly, the vertices for the emission of the charged gauge particles, which are functions of an internal momentum $K^I$, can be found:

$$\text{Charged gauge mesons:} \quad \begin{cases} V_B^K = \rho_\mu(k) \displaystyle\int_0^\pi d\sigma\, B^\mu e^{ik_\mu X^\mu} :e^{2iK^I X^I}: C(K) \\ V_F^K = \displaystyle\int_0^\pi d\sigma\, F^a U^a(k) e^{ik_\mu X^\mu} :e^{2iK^I X^I}: C(K) \end{cases} \tag{10.5.7}$$

where the normal ordered part of the vertex comes from the left-moving sector and we will define $C$ later.

One way of checking to see that we have the correct form for the vertex functions is to explicitly operate on them with the supersymmetry operator (10.3.1) to verify the fact that boson vertices turn into fermion vertices and vice versa. However, the proof, which follows the proof given in Section 3.9, is rather involved and will not be presented.

The propagator of the system can also be found easily by generalizing the closed string propagator. We will, however, find it convenient to absorb the $\sigma$ and $\tau$ dependence of the vertex function and put it into the propagator. This can always be done because the $\sigma$ displacement matrix, as we saw earlier, is given by $U(\sigma)$ and the $\tau$ displacement operator is generated by the light cone

Hamiltonian:

$$H = \tfrac{1}{2}p^2 + 2N + 2(\tilde{N} - 1) + \sum_{I=1}^{16} (p^I)^2 \tag{10.5.8}$$

The vertex function, before this extraction, is

$$V(k, \tau) = e^{iH\tau} \int_0^\pi d\sigma \; U(\sigma)\tilde{V}_0 U^\dagger(\sigma) e^{-iH\tau}$$

The vertex function that we want is

$$\tilde{V}_0 = \tilde{V}(\tau = \sigma = 0) \tag{10.5.9}$$

The propagator, which now absorbs the $\sigma$ and $\tau$ integrations, becomes

$$\begin{aligned} \Delta &= \int_0^\infty d\tau \int_0^\pi \frac{d\sigma}{\pi} e^{-H\tau} U(\sigma) \\ &= \int_{|z|<1} d^2 z |z|^{(1/4)p^2 - 2} z^N \bar{z}^{\tilde{N} - 1 + (1/2)\sum_I (p^I)^2} \\ &= \frac{1}{H} \delta\left( N - \tilde{N} + 1 - \tfrac{1}{2} \sum_I (p^I)^2 \right) \end{aligned} \tag{10.5.10}$$

The above propagator is just what one might have expected. The delta function simply enforces the constraint (10.2.13) that makes the states independent of a rotation in $\sigma$, and the poles are given by the inverse of the light cone Hamiltonian.

The $N$-point function can therefore be written as

$$\langle 0, k_1 | \tilde{V}(k_2) \Delta \cdots \Delta \tilde{V}(k_{N-1}) | 0, k_N \rangle \tag{10.5.11}$$

Let us now construct the amplitude for the scattering of four massless gauge bosons with momenta $k_i$, polarization $\rho_i$, and charge $K_i$, where

$$\begin{aligned} \sum_{i=1}^4 k_i^\mu &= 0 \\ \sum_{i=1}^4 K_i &= 0 \\ K_i^2 &= 0 \end{aligned} \tag{10.5.12}$$

The calculation of the four-point function is long but straightforward:

$$\begin{aligned} A_4 &= g^2 K(\rho_i, k_i) \\ &\cdot \varepsilon \frac{\Gamma\left(-1 + \frac{u}{8} + \frac{U}{2}\right) \Gamma\left(-1 + \frac{s}{8} + \frac{S}{2}\right) \Gamma\left(-1 + \frac{t}{8} + \frac{T}{2}\right)}{\Gamma\left(1 - \frac{u}{8}\right) \Gamma\left(1 - \frac{s}{8}\right) \Gamma\left(1 - \frac{t}{8}\right)} \end{aligned} \tag{10.5.13}$$

where

$$\begin{aligned}
K = &-\tfrac{1}{4}[st\,\rho_1\cdot\rho_3\rho_2\cdot\rho_4 + su\,\rho_2\cdot\rho_3\rho_1\cdot\rho_4 + tu\,\rho_1\cdot\rho_2\rho_3\cdot\rho_4] \\
&-\tfrac{1}{2}s[\rho_1\cdot k_4\rho_3\cdot k_2\rho_2\cdot\rho_4 + \rho_2\cdot k_3\rho_4\cdot k_1\rho_1\cdot\rho_3 \\
&+\rho_1\cdot k_3\rho_4\cdot k_2\rho_2\cdot\rho_3 + \rho_2\cdot k_4\rho_3\cdot k_1\rho_1\cdot\rho_4] \\
&-\tfrac{1}{2}t[\rho_2\cdot k_1\rho_4\cdot k_2\rho_3\cdot\rho_1 + \rho_3\cdot k_4\rho_1\cdot k_2\rho_2\cdot\rho_4 \\
&+\rho_2\cdot k_4\rho_1\cdot k_3\rho_3\cdot\rho_4 + \rho_3\cdot k_1\rho_4\cdot k_2\rho_2\cdot\rho_1] \\
&-\tfrac{1}{2}u[\rho_1\cdot k_2\rho_4\cdot k_3\rho_3\cdot\rho_2 + \rho_3\cdot k_4\rho_2\cdot k_1\rho_1\cdot\rho_4 \\
&+\rho_1\cdot k_4\rho_2\cdot k_3\rho_3\cdot\rho_4 + \rho_3\cdot k_2\rho_4\cdot k_1\rho_1\cdot\rho_2]
\end{aligned} \tag{10.5.14}$$

where $\varepsilon$ is a phase factor and

$$\begin{aligned}
s &= -(k_1+k_2)^2 \\
t &= -(k_2+k_3)^2 \\
u &= -(k_1+k_3)^2 \\
s &+ t + u = 0 \\
S &= (K_1+K_2)^2 \\
T &= (K_2+K_3)^2 \\
U &= (K_1+K_3)^2 \\
S &+ T + U = 8
\end{aligned} \tag{10.5.15}$$

## §10.6. Single-Loop Amplitude

The real test of the theory, however, is the calculation of the single-loop diagram [6, 7], where we demand that the theory be finite. It is now a straightforward process to write down the single-loop diagram in terms of these vertices and propagators:

$$A_{\text{loop}} = \operatorname{Tr}(\Delta V(N)\Delta\cdots\Delta V(1)) \tag{10.6.1}$$

where we consider scattering by charged gauge mesons.

Again, the trace calculation is long but straightforward. After the trace is performed, we have

$$\begin{aligned}
A_{\text{loop}} = \bar{\varepsilon}K \int \prod_{i=1}^{4} d^2 z_i |w|^{-2} \left[\frac{-4\pi}{\ln|w|}\right]^5 \bar{w}^{-1} f(\bar{w})^{-24} \\
\times \prod_{1\le i<j\le 4} [\chi(c_{JI}, w)]^{(1/2)k_i\cdot k_j}[\psi(\bar{c}_{ji}, \bar{w})]^{K_i\cdot K_j} L
\end{aligned} \tag{10.6.2}$$

where

$$\chi(z, w) = \exp\left[\frac{\ln^2|z|}{2\ln|w|}\right]\left|\frac{1-z}{\sqrt{z}}\prod_{m=1}^{\infty}\frac{(1-w^m z)(1-w^m/z)}{(1-w^m)^2}\right| \tag{10.6.3}$$

$$\psi(\bar{z}, \bar{w}) = \exp\left[\frac{\ln^2 \bar{z}}{2\ln \bar{w}}\right]\frac{1-\bar{z}}{\sqrt{\bar{z}}}\prod_{m=1}^{\infty}\left(\frac{(1-\bar{w}^m\bar{z})(1-\bar{w}^m/\bar{z})}{(1-\bar{w}^m)^2}\right) \tag{10.6.4}$$

$$L(\bar{w}, \bar{z}_i, K_i) = \sum_{P\in\Lambda}\exp\left[\tfrac{1}{2}\ln\bar{w}\left(P - \sum_{i=1}^{4}\frac{\ln\bar{z}_i}{\ln\bar{w}}Q_i\right)^2\right] \tag{10.6.5}$$

and where

$$\begin{aligned} Q_i &= \sum_{j=1}^{i-1} K_i \\ v_i &= \sum_{j=1}^{i}\frac{\ln z_j}{2\pi i} \\ \tau &= \frac{\ln w}{2\pi i} \\ c_{ji} &= z_i z_{i+1}\cdots z_j \end{aligned} \tag{10.6.6}$$

where the sum over $P \in \Lambda$ represents the sum over all points in the lattice and $K$ is a kinematic factor identical to the one found for the trees in (10.5.14). Because the final result bears so much resemblance to the single-loop superstring amplitude, it is not hard to show that the amplitude is invariant under

$$\begin{cases} v_i \to v_i + 1 \\ v_i \to v_i + \tau \end{cases} \tag{10.6.7}$$

Slightly more difficult is the proof that the integrand is invariant under $\tau \to \tau + 1$ and $\tau \to -\tau^{-1}$, which is necessary to prove modular invariance.

Fortunately, most of the terms in the integrand are identical to those found in the single-loop amplitude in (5.5.1). However, we must check the invariance of the terms that are different, i.e., the factors dependent on the lattice.

As before, the $\Theta$ function, if we make a modular transformation on its arguments, transforms as follows:

$$\Theta_1\left(\frac{\bar{v}}{c\bar{\tau}+d}\middle|\frac{a\bar{\tau}+b}{c\bar{\tau}+d}\right) = \varepsilon(c\bar{\tau}+d)^{1/2}\exp\left(\frac{-i\pi c\bar{v}^2}{c\bar{\tau}+d}\right)\Theta_1(\bar{v}|\bar{\tau}) \tag{10.6.8}$$

where $\varepsilon^8 = 1$. Since the $\chi$ function can be written in terms of $\Theta$ functions, we have

$$\chi\left(\frac{v}{c\tau+d}\middle|\frac{a\tau+b}{c\tau+d}\right) = \frac{1}{|c\tau+d|}\chi(v|\tau)$$

The partition function $f(w)$ also transforms as

$$f(w) = (i/\tau)^{1/2} w^{-1/2} w'^{1/24} f(w')$$
$$\tau = \frac{\ln w}{2\pi i} \to \frac{-1}{\tau} = \frac{-2\pi i}{\ln w'} \tag{10.6.9}$$

The most important transformation is the one for $I$:

$$L(\overline{w}, \overline{z}_i, K_i) = \left(\frac{-2\pi}{\ln \overline{w}}\right)^8 \sum_{P \in \Lambda} \exp\left[\frac{2\pi^2}{\ln \overline{w}}\left(P - \sum_{i=1}^{N} \frac{Q_i \ln \overline{z}_i}{2\pi i}\right)^2 + \frac{1}{2 \ln \overline{w}}\left(\sum_{i=1}^{N} Q_i \ln \overline{z}_i\right)^2\right] \tag{10.6.10}$$

This last identity depended crucially on the lattice being *self-dual*, which is perhaps one of the strongest arguments for this restriction on the lattice. Let us put all these factors together:

$$\overline{w}^{-1} f(w)^{-24} \prod_{1 \le i \le j \le 4} \psi(\overline{v}_{ji} | \tau)^{K_i \cdot K_j} L(\tau, v_i)$$
$$\to (\overline{w}')^{-1} f(\overline{w}')^{-24} \overline{\tau}^{12} \left[\prod_{1 \le i \le j \le 4} \psi\left(\frac{-\overline{v}_{ij}}{\overline{\tau}} \middle| \frac{-1}{\tau}\right) \overline{\tau} \exp\left(\frac{i\pi \overline{v}_{ij}^2}{\overline{\tau}}\right)\right]^{K_i \cdot K_j}$$
$$\times L\left(\frac{-1}{\overline{\tau}}, \frac{-v_i}{\overline{\tau}}\right) (\overline{\tau})^{-8} \exp\left(-\frac{i\pi}{\overline{\tau}}\left(\sum_{i=1}^{4} Q_i \overline{v}_i\right)^2\right) \tag{10.6.11}$$

There are two factors in (10.6.11) that apparently violate modular invariance, the various factors of $\overline{\tau}$ and the exponentials involving $\overline{\tau}$. Fortunately, both sets of terms vanish. The factors of $\overline{\tau}$ cancel because

$$\prod_{1 \le i \le j \le 4} \overline{\tau}^{K_i \cdot K_j} = \overline{\tau}^{-4}$$

This leaves us with $\overline{\tau}^{12-4-8} = 1$. The exponential factors also vanish because

$$\sum_{1 \le i \le j \le 4} \overline{v}_{ij}^2 K_i \cdot K_j = \left(\sum_{i=1}^{4} Q_i \overline{v}_i\right)^2$$

Once we eliminate the factors of $\overline{\tau}$ and the exponentials of $\overline{\tau}$ in (10.6.11), we find that this combination is modular invariant.

The invariance of the amplitude under $v_i \to v_i + 1$ and $v_i \to v_i + \tau$ means we can truncate its region of integration:

$$0 < \operatorname{Im} v_i < \operatorname{Im} \tau$$
$$-\tfrac{1}{2} < \operatorname{Re} v_i < \tfrac{1}{2} \tag{10.6.12}$$

Also, as in the case of the closed single-loop diagram because of invariance under $\tau \to \tau + 1$ and $\tau \to -1/\tau$, we are free to choose a fundamental region:

$$\begin{cases} -\tfrac{1}{2} \le \operatorname{Re} \tau \le \tfrac{1}{2} \\ |\tau| > 1 \end{cases} \tag{10.6.13}$$

Once again, the choice of this fundamental region allows us to avoid the potential singularity at $\tau = 0$, and hence we have a finite one-loop action.

In this discussion, modular invariance was crucial in showing that the amplitude, like the usual superstring, is finite. We can simply choose a fundamental region where no singularities are present. However, the calculation obscures precisely why the theory is modular invariant. Let us analyze the simplest case and isolate the point at which modular invariance comes in. To simplify matters, let us consider the vacuum one-loop amplitude with no external legs.

Let us first define a function $F$ (which appears in in the trace calculation in (10.6.1)):

$$F(\tau, X) = \sum_{L \in \Lambda} e^{-i\pi\tau(L-X)^2} \tag{10.6.14}$$

where we sum over lattice sites, $X$ is an arbitrary 16-dimensional vector in the root lattice space, and each site is labeled by integers $n_i$:

$$L = \sum n_i e_i^I \tag{10.6.15}$$

Notice that $F$ is periodic:

$$F(\tau, X) = F(\tau, X + e_i) \tag{10.6.16}$$

because this shift can be simply absorbed into the integers $n_i$. Let us now write this function in terms of its Fourier transform $\tilde{F}$:

$$F(\tau, X) = \sum_{M \in \Lambda^*} e^{-2iM \cdot X} \tilde{F}(\tau, M) \tag{10.6.17}$$

Notice that, because $F$ is periodic, the $M$'s must necessarily lie on the *dual lattice*:

$$M^I = \sum_{i=1}^{16} m_i e_i^{*I} \tag{10.6.18}$$

Now take the reverse Fourier transform to find $\tilde{F}$ in terms of $F$:

$$\tilde{F}(\tau, M) = \int \frac{d^{16}X}{\sqrt{|g|}} e^{2i\pi M \cdot X} F(\tau, X) \tag{10.6.19}$$

where $\sqrt{|g|}$ is the volume of the torus. Now insert the expression for $F$ into the previous equation. Perform the integration, and we arrive at an explicit form for $\tilde{F}$:

$$\tilde{F}(\tau, M) = \frac{1}{\sqrt{|g|}} \tau^{-8} e^{i\pi M^2/\tau} \tag{10.6.20}$$

Now we would like to insert this identity into the expression that actually occurs in the one-loop vacuum calculation. We must calculate the trace over the Hamiltonian, which contains the piece $(p^I)^2$. Thus, in the trace calculation the function $f$ appears:

$$f(\tau) = \tau^4 \sum_{L \in \Lambda} e^{-i\pi\tau L \cdot L} \tag{10.6.21}$$

From (10.6.14), we have

$$F(\tau, 0) = f(\tau)\tau^{-4} \tag{10.6.22}$$

Let us now substitute the expression for $\tilde{F}$, and we have an expression for $f$:

$$\begin{aligned} f(\tau) &= \frac{1}{\sqrt{|g|}}\left(-\frac{1}{\tau}\right)^4 \sum_{M \in \Lambda^*} e^{i\pi M \cdot M/\tau} \\ &= \frac{1}{\sqrt{|g|}} f^*\left(-\frac{1}{\tau}\right) \end{aligned} \tag{10.6.23}$$

where $f^*$ is nothing but the usual $f$ defined in (10.6.21) over the *dual lattice.* This is the key result.

In summary, we see that *the modular transformation $\tau \to -1/\tau$ has replaced the lattice with the dual lattice.* Thus, for modular invariance we demand that the lattice be self-dual. In fact, this is the origin of the condition that the lattice be self-dual. Here, we see the tight link between self-duality (which eventually restricts us to either $E_8 \otimes E_8$ or Spin(32)/$Z_2$) and modular invariance. Although the original choice of these groups came from the anomaly cancellation, we see that we need precisely these groups for modular invariance and finiteness of the amplitude.

## §10.7. $E_8$ and Kac–Moody Algebras

Earlier, we saw that the spectrum of states was not manifestly $E_8 \otimes E_8$ invariant. Only with difficulty were we able to show that the lowest-lying states could be placed in irreducible representations of the group. Because the number of states rapidly proliferates into the tens of millions, it becomes prohibitive to actually rearrange the states into $E_8 \otimes E_8$ multiplets.

In this section we will use the techniques of Kac–Moody algebras [8–10] developed in Chapter 4 to show, to all orders, that the heterotic string does indeed have a spectrum that is $E_8 \otimes E_8$ symmetric. We will use the *vertex operators* defined in that chapter to generate the representation of a Kac–Moody algebra (which works only if the lattice is even, self-dual, and the algebra is level 1).

The representation that we desire is the Chevalley basis, where the 496 generators are broken up into 16 mutually commuting generators (which form the Cartan subgroup) and the 480 root vectors. Notice that the 16 $p^I$ vectors obey the condition

$$[p^I, p^J] = 0 \tag{10.7.1}$$

Trivially, we can always represent the Cartan subalgebra as simply the mutually commuting vectors $p^I$. More difficult, however, is the construction of the 480 root vectors.

The simplest operator with 480 states is the vertex operator, which we can write in terms of a line integral surrounding the origin:

$$E(K) = \oint \frac{dz}{2\pi i z} V(K, z) C(K) \tag{10.7.2}$$

where the vertex function $V$ is defined over the lattice (rather than over space–time):

$$V(K, z) = :e^{2iK^I X_L^I(z)}: \tag{10.7.3}$$

and where

$$\begin{aligned} (K^I)^2 &= 2 \\ z &= e^{2i(\tau+\sigma)} \end{aligned} \tag{10.7.4}$$

where the cocycle is as yet unspecified. Notice that the vectors $K^I$, by definition, point in 480 directions in the 16-dimensional root lattice space.

We now demand that the 16 elements of the Cartan subalgebra $p^I$ and the 480 elements of root lattice space obey the commutation relations of $E_8 \otimes E_8$. This will, in turn, fix the $C$ operator. We demand

$$[E(K), E(L)] = \begin{cases} \varepsilon(K, L) E(K + L) & \text{if } (K + L)^2 = 2 \\ K^I p^I & \text{if } K + L = 0 \\ 0 & \text{otherwise} \end{cases} \tag{10.7.5}$$

and

$$[p^I, E(K)] = K^I E(K) \tag{10.7.6}$$

where $\varepsilon(K, L)$ are the structure constants of the algebra with values $\pm 1$.

In some sense, we have done nothing yet. We have simply written down the commutations of the algebra of $E_8 \otimes E_8$ in the Chevalley basis, which are well known, and then demanded that our ansatz satisfy them. The nontrivial part is that a solution of these equations actually exists which fixes the value of $C$.

For the moment, let us drop the factor $C$ and see if we can satisfy the commutation relations. It is easy to show

$$V(K, z) V(L, w) = (wz)^{-(1/2)K \cdot L} (z - w)^{K \cdot L} :e^{2iK \cdot X(z) + 2iL \cdot X(w)}: \tag{10.7.7}$$

which is valid for $|w| < |z|$.

From this, we can show

$$E_K E_L - (-1)^{K \cdot L} E_L E_K = \oint \oint \frac{dw}{2\pi i w} \frac{dz}{2\pi i z} \times (z - w)^{K \cdot L} (wz)^{-(1/2)K \cdot L} :e^{2iK \cdot X(z) + 2iL \cdot X(w)}: \tag{10.7.8}$$

where the $w$ integration is performed such that $|z| > |w|$ in the first term and $|z| < |w|$ in the second. By carefully examining the last expression, we find that we have satisfied all the identities of the Chevalley basis except that the statistics are all wrong. Anticommutators arise instead of commutators. That is why we had to introduce the factor $C(K)$. The condition we must impose is

$$C(K)C(L) = \varepsilon(K, L)C(K + L) \tag{10.7.9}$$

The $C(K)$ operator is called a "cocycle" or "twist" and was added into the definition of the generator $E(K)$ in order to get the right statistics. Acting on states of momenta $p^I$, we can show that $C(K)$ has the property

$$C(K)|p\rangle = \varepsilon(K, p)|p\rangle \tag{10.7.10}$$

If we demand that this law be associative, then we have a restriction on these phases. By acting on states successively by the $C$'s and demanding associativity, we easily arrive at

$$\varepsilon(K, L)\varepsilon(K + L, M) = \varepsilon(L, M)\varepsilon(K, L + M) \tag{10.7.11}$$

We call these the two cocycle conditions. Several explicit representations of $C(K)$ and the phase exist. We can always choose

$$\begin{aligned} &\varepsilon(K, L)\varepsilon(L, K) = (-1)^{K\cdot L} \\ &\varepsilon(K, 0) = -\varepsilon(K, -K) = 1 \end{aligned} \tag{10.7.12}$$

In the Appendix, we show that the generators of the Lie algebras can be arranged in terms of the Cartan generators $H_i$ and the eigenvectors $E_\alpha$. We now see the exact correspondence between the operators appearing in the heterotic string and the generators of a Lie group:

| Heterotic | Cartan–Weyl |
|---|---|
| $p^I$ | $H_i$ |
| $E(K)$ | $E_\alpha$ |
| $K^I$ | $\alpha$ |

(10.7.13)

The last step in the proof is to notice that the 16 $p^I$ and the 480 $E(K)$ generate the vertices of the theory (e.g., see (10.5.6), (10.5.7)), so therefore the Fock space consists of these operators acting on the vacuum. But since these generators together make up the 496 generators of $E_8 \otimes E_8$, the entire spectrum itself must be arranged according to irreducible representations of $E_8 \otimes E_8$.

## §10.8. 10D Without Supersymmetry

In the previous sections we saw the crucial importance of modular invariance in establishing the properties of the heterotic string. The logical question to ask is whether it is possible to construct a new string theory that is modular invariant but may have a different gauge group and different properties.

We saw earlier that it is possible to write down the left-moving heterotic sector, which is usually described by the bosonic field $X^I$, in terms of fermion fields. We will now show that, by twisting the boundary conditions on the fermion fields, we can derive yet another 10-dimensional string theory that is modular invariant but breaks supersymmetry explicitly. The gauge group, instead of being $E_8 \otimes E_8$, is now broken down to $O(16) \otimes O(16)$. The advantage of this new formulation is that it is tachyon-free and anomaly-free. (However, it is neither supersymmetric nor finite.) We can view this new string theory as some form of broken heterotic string.

This new string is based on the observation in Chapter 5 that the NS–R theory (without a GSO projection) is not modular invariant or supersymmetric. The choice of NS–R boundary conditions is usually made only in the $\sigma$ coordinate. However, a modular transformation in general will map a region $(\sigma, \tau)$ into $(\tau, \sigma)$. Thus, a modular transformation can interchange the roles of $\sigma$ and $\tau$, so we must be careful about boundary conditions in the $\tau$ coordinate as well. Thus, a naive summing over NS–R states in the closed loop will in general violate modular invariance.

If we review the comments made in Section 5.9 we find that a naive trace only counts the combination

$$\begin{aligned} &\text{NS boson} \to (\text{NS}, \text{NS}) \\ &\text{R fermion} \to (\text{R}, \text{NS}) \end{aligned} \tag{10.8.1}$$

However, modular transformations can permute these boundary conditions. The transformation

$$\tau \to \tau + 1 \tag{10.8.2}$$

changes (NS, NS) into (NS, R). It corresponds to cutting the torus along a line of constant $\tau$ and reconnecting with a $2\pi$ twist. On the other hand,

$$\tau \to -\frac{1}{\tau} \tag{10.8.3}$$

interchanges the roles of $\sigma$ and $\tau$, and thus changes (NS, R) into (R, NS). Thus, what we really want is a sum over all four possible boundary conditions:

$$(\text{NS}, \text{NS}), \quad (\text{NS}, \text{R}), \quad (\text{R}, \text{NS}), \quad (\text{R}, \text{R}) \tag{10.8.4}$$

This combination of boundary conditions is modular invariant because a modular transformation simply interchanges the boundary conditions within these fields.

In order to calculate over the trace with reversed boundary conditions, we can always insert the operator $(-1)^F$, where $F$ is the fermion number, which reverses the $\tau$ boundary condition. Thus, the trace sums over the following configurations:

$$\begin{aligned} \text{Tr}\, x^R &\to (\text{NS}, \text{NS}); (\text{R}, \text{NS}) \\ \text{Tr}(-1)^F x^R &\to (\text{NS}, \text{R}); (\text{R}, \text{R}) \end{aligned} \tag{10.8.5}$$

Thus, the complete sum over all four boundary conditions is given by the sum of these two traces:

$$\mathrm{Tr}([1 + (-1)^F]x^R) \tag{10.8.6}$$

But notice that this operator inserted inside the sum is just the GSO projection operator!

Thus, the GSO projection operator is more than just the extraction of states with the wrong statistics or the removal of the tachyon. We see that the *GSO projection makes the single-loop closed string amplitude modular invariant*. In fact, a generalization of this argument shows that we need both the NS and R sectors to achieve modular invariance. Once periodic or antiperiodic boundary conditions are introduced, we must have all four possibilities included for modular invariance, so a theory with just NS states is probably not modular invariant and possibly not unitary.

Now let us generalize these remarks for the heterotic string. We will now construct a theory that breaks supersymmetry but manages to keep modular invariance intact [11, 12]. Let us begin with a formulation of the theory that has fermions in both the space–time and isospin sectors. We will choose the 10D right-moving heterotic string to be represented by the GS theory and the 16-dimensional isospin sector to be represented by fermion fields (rather than bosons). In this way, we can more easily construct modular-invariant theories by interchanging boundary conditions (the calculation with bosonic fields in the isospin sector is more difficult, since we do not have such a nice interpretation of modular invariance on boson fields).

Let us construct an element $R$ such that $R^2 = 1$, which forms a discrete group $Z_2$. By construction, we will select out only the $R = 1$ subspace of the theory. This will necessarily break $E_8 \otimes E_8$. But by twisting the various boundary conditions, this will allow us to keep modular invariance.

Let us choose $R$ to be a product of two factors, the first being a $2\pi$ rotation in Lorentz space and the second a transformation in the isospin space:

$$R = e^{2\pi i J_{12}}\gamma_\delta \tag{10.8.7}$$

where

$$\gamma_\delta^2 = 1 \tag{10.8.8}$$

and $J_{12}$ is a space–time rotation. Let $\gamma_\delta$ lie in the Cartan subalgebra of the gauge group:

$$\gamma_\delta = e^{2\pi i p^I \delta^I}$$

Notice that this factor simply generates a displacement $2\pi\delta^I$ in the variable $x^I$.

We now demand that the *gauge group commute with this element of* $Z_2$. Of course, this breaks the original gauge group because $R$ contains a member of the group. In fact, the group that commutes with $R$ is

$$\frac{E_8 \otimes E_8}{Z_2} \to \mathrm{O}(16) \otimes \mathrm{O}(16) \tag{10.8.9}$$

Even after selecting out the $R = 1$ sector of the theory, a variety of different types of twisting is still possible, all of them compatible with modular invariance. However, many of them involve tachyons, e.g., the states with $\delta^2 = 1$. The only tachyon-free theory is obtained with the choice

$$\delta^2 = 2 \tag{10.8.10}$$

This choice yields a tachyon-free theory because in the twisted sector the right-handed vacuum at $-\frac{1}{2}$ does not match the left-handed vacuum at $-1 + \delta^2 = 0$. Thus, the constraint $L_0 = \tilde{L}_0$ projects out that tachyon state.

Let us denote the states by (space–time; O(16), O(16)). The fermion string fields $\chi^I$ and $\chi'^I$ transform as

$$(\mathbf{16}, \mathbf{1}) + (\mathbf{1}, \mathbf{16})$$

under $O(16) \otimes O(16)$.

Let us now construct the $R = 1$ subspace of the theory. This means that the following untwisted combinations all appear with the usual $(-1)^F$ factors:

$$\text{Untwisted} \quad \begin{cases} (\text{NS; NS, NS}) \\ (\text{NS; R, R}) \\ (\text{R; NS, R}) \\ (\text{R; R, NS}) \end{cases} \tag{10.8.11}$$

This set, of course, breaks modular invariance. Altogether, we must have $2 \times 2 \times 2 = 8$ combinations represented. In the twisted sector, we must retrieve modular invariance by summing over the remaining set of boundary conditions. Let us now make the transformation $NS \leftrightarrow R$. We take the following combinations with the *opposite* set of states surviving the $(-1)^F$ projection:

$$\text{Twisted} \quad \begin{cases} (\text{R; R, R}) \\ (\text{R; NS, NS}) \\ (\text{NS; R, NS}) \\ (\text{NS; NS, R}) \end{cases} \tag{10.8.12}$$

The partition function for this model can be calculated by summing over untwisted states with $R = 1$, i.e., just those $O(16) \otimes O(16)$ states which survive the projection by $Z_2$. Then we must add in the twisted sector in order to have modular invariance.

The spectrum of the $R = 1$ states can also be easily calculated in this theory. For the massless states, we find that the untwisted sector, with periodic right-moving fermions, yields

(1) $\mathbf{8}_v$ gauge bosons in the $(\mathbf{120}, \mathbf{1}) + (\mathbf{1}, \mathbf{120})$ representation,
(2) $\mathbf{8}_s$ fermions in the $(\mathbf{128}, \mathbf{1}) + (\mathbf{1}, \mathbf{128})$ representation, and
(3) the bosonic part of the supergravity multiplet $(g_{\mu\nu}, B_{\mu\nu}, \phi)$.

Notice that we have only the bosonic part of the supergravity multiplet, which is an indication that the theory is not supersymmetric.

For the twisted sector, with antiperiodic right-moving fermions, we have only the $\mathbf{8}_s$ fermions in the (16, 16) representation.

When we perform the calculation over the partition function by inserting these projections, we find

$$A_{\text{loop}} = \int_F \frac{d^2\tau}{(\text{Im }\tau)^2} \frac{1}{(\text{Im }\tau)^4 \Theta_1'^4} \left\{ \frac{\Theta_2^4(\tau)}{\Theta_2^8(\bar{\tau})} + \frac{\Theta_4^4(\tau)}{\Theta_4^8(\bar{\tau})} - \frac{\Theta_3^4(\tau)}{\Theta_3^8(\bar{\tau})} \right\} \tag{10.8.13}$$

where the theta functions arise in the trace $\text{Tr } e^{i\tau L_0 + i\bar{\tau}\tilde{L}_0}$. The first term comes from the untwisted sector. Notice that it is not modular invariant. The last two come from the twisted sector. Notice that only the combination is modular invariant. Thus, by summing over the various combinations of sectors, we have obtained a theory that is modular invariant. However, as we can see by the absence of the massless gravitino, the model breaks supersymmetry. This also means that there will be dilaton tadpoles that are no longer canceled by supersymmetry, so we expect this theory to have infrared problems.

The fact that this model is anomaly-free is something of a surprise, given that the $n - 496$ term in (9.6.36) played such an important role in selecting out the gauge group. However, the $O(16) \otimes O(16)$ has a different set of anomalous terms. First, the 496 appearing in (9.6.36) was due to the presence of the gravitino, which no longer contributes to this anomaly because it is massive and no longer the partner of the graviton. Second, the contribution of the fermions, which used to be $n$, is now zero. This is because the massless fermions are arranged in the multiplets (**16**, **16**) in the twisted sector with positive chirality and (**1**, **128**) and (**128**, **1**) in the untwisted sector. Thus, the contribution to the anomaly has changed to

$$496 - n \to 16^2 - 128 - 128 + 0 = 0$$

(Although the contribution to the chiral anomaly is now zero, this does not mean that the chiralities themselves cancel. This is because the fermions belong in different representations of the gauge group.) Finally, because the group is an $O(N)$ group, it is easy to evaluate $X_8 = \frac{1}{24} \text{Tr } F^4$ and repeat the steps given earlier.

In summary, the properties of this model are as follows:

(1) Modular invariance (because we added all eight possible contributions, twisted and untwisted, to the single-loop amplitude).
(2) No tachyon (because it does not satisfy the level matching conditions $L_0 = \tilde{L}_0$).
(3) Lack of supersymmetry (because the gravitino spinor is massive).
(4) No anomalies (because the spin-$\frac{1}{2}$ contribution sums to zero and because the gravitino does not contribute).
(5) Not finite (because we cannot use supersymmetry to eliminate infrared divergences).

## §10.9. Lorentzian Lattices

So far, we have been discussing heterotic strings in which one sector is compactified from 26 dimensions down to 10. Then, according to conventional wisdom, we must compactify the 10-dimensional space down to $D$-dimensional space–time. Yet another intriguing possibility is to *directly compactify the* 26- *and* 10-*dimensional spaces down to D dimensions from the very beginning*, bypassing the intermediate stage. This is the approach of Narain [13, 14], which yields gauge groups of rank $26 - D$, which is larger than the $E_8 \otimes E_8$ considered so far.

Let us begin with the two sectors that are 26- and 10-dimensional and now compactify them both down to $D$ space–time dimensions. Then the left-moving sector has $26 - D = p$ dimensions, and the right-moving sector has $10 - D = q$ dimensions. Now parametrize the compactified dimensions as

$$\begin{aligned} X^A &= q^A + 2L^A(\tau + \sigma) + \sum_{n \neq 0} \frac{i}{2n} \alpha_n^A e^{-2in(\tau+\sigma)} \\ X^B &= q^B - 2\tilde{L}^B(\tau - \sigma) + \sum_{n \neq 0} \frac{i}{2n} \alpha_n^B e^{2in(\tau-\sigma)} \end{aligned} \tag{10.9.1}$$

where $A$ ranges from 1 to $p$ and $B$ ranges from 1 to $q$. Notice that we have the relation $p = q + 16$, which guarantees that the number of uncompactified space–time dimensions for both sectors is $D$.

Now impose the constraint that $L_0 - \tilde{L}_0$ annihilates states, so there is no preferred $\sigma$ origin for the closed string. This yields the conditions

$$\tfrac{1}{2}(k^2 - \tilde{k}^2) = \tilde{N} - N + 1 = \text{integer} \tag{10.9.2}$$

The mass relationship is now given by

$$\tfrac{1}{2}m^2 = \tfrac{1}{2}k^2 + \tfrac{1}{2}\tilde{k}^2 + N + \tilde{N} - 1 \tag{10.9.3}$$

Notice that these equations all reduce to the usual heterotic string if we take $p = 16$, $q = 0$.

Now let us check that the final result is modular invariant. This will place a new restriction on the lattice, which at this time is still arbitrary. The single-loop calculation (10.6.2) done for the general case is almost identical to the one for the usual heterotic string. A careful analysis of the terms shows that the generalized case contains the new factor

$$\sum e^{-i\pi\bar{\tau}(L - \sum_{i=1}^{4} K_i \bar{v}_i/\bar{\tau})^2 + i\pi\tau\tilde{L}^2} \tag{10.9.4}$$

Now calculate the change in this factor under the transformation $\tau \to \tau + 1$. It follows that we pick up a new phase factor:

$$e^{-i\pi(L^2 - \tilde{L}^2)} \tag{10.9.5}$$

In order to cancel this term, we are forced to have

$$(L^A)^2 - (\tilde{L}^B)^2 = \text{even integer} \tag{10.9.6}$$

Because of this extra minus sign, the metric on the lattice is "Lorentzian" rather than Euclidean. (For the usual heterotic string, $\tilde{L}$ is zero, so we never see the Lorentzian behavior of the lattice.)

Finally, when we apply the modular transformation $\tau \to -\tau^{-1}$, we find that the lattice must be self-dual. In summary, modular invariance at the one-loop level is preserved if we have an *even, self-dual, Lorentzian lattice.*

Mathematically, it can be shown that such Lorentzian lattices do indeed exist, when we satisfy the condition $p - q = 8n$, for integer $n$. For our case, $n = 2$.

We can also calculate the number of parameters in this lattice. It can be shown that the Lorentzian lattice is unique up to an SO($p$, $q$) transformation. But the mass formulas we have derived are invariant only under SO($p$) and SO($q$). Thus, the total number of parameters is

$$\dim \mathrm{SO}(p, q) - \dim \mathrm{SO}(p) - \dim \mathrm{SO}(q) = pq \tag{10.9.7}$$

Thus, the total number of parameters that characterize the lattice is $p(p - 16)$.

We have considerably expanded the number of groups that are available to us by this method of compactification. The total rank of the group is now $26 - D$. This means that we can have groups as large as SO($52 - 2D$). In four space–time dimensions, this means SO(44). For $D = 4$, we can also have $E_8 \otimes E_6 \otimes E_7 \otimes \mathrm{SU}(2)$, or $E_8 \otimes E_6^2 \otimes G$, where $G$ can be either SO(4) or SU(3). We can also have $E_8 \otimes E_8 \otimes G_{10-d}$, where $G$ must be simply laced.

At this point, we may conclude that Lorentzian lattices are an entirely new way to compactify the heterotic string. Actually, this is not quite right. It turns out that we can still represent the Lorentzian lattice within the conventional string picture of the heterotic string. Consider, for example, the following string action in the presence of background fields:

$$g_{ij}\partial_\alpha X^i \partial^\alpha X^j + \varepsilon^{\alpha\beta} B_{ij}\partial_\alpha X^i \partial_\beta X^j + \varepsilon^{\alpha\beta} A_i^I \partial_\alpha X^i \partial_\beta X^I \tag{10.9.8}$$

where $I$ goes from 1 to 16, $B_{ij}$ is an antisymmetric tensor, and $\varepsilon^{\alpha\beta}$ is an antisymmetric tensor in 2-space. Now assume the $g_{ij}$, $B_{ij}$, and $A_i^I$ fields are approximated by constant background fields. Notice that the total number of parameters in this approach is easily found by counting the number of independent modes within these fields:

$$\begin{Bmatrix} g_{ij} \\ A_i^I \\ B_{ij} \end{Bmatrix} \to \begin{Bmatrix} \frac{1}{2}q(q+1) \\ 16q \\ \frac{1}{2}q(q-1) \end{Bmatrix} \tag{10.9.9}$$

for a total number of $pq$ parameters. Now quantize this system, assuming that $X$ can be approximated by

$$X^i = 2\sigma n^i + q^i(\tau) \tag{10.9.10}$$

Now substitute back this expression into the action, quantizing the system in the presence of these constant background fields. The net effect of these background fields is to add $pq$ new parameters into the compactification process, which is precisely the number of new parameters introduced by the compactification on Lorentzian lattices. It can be shown, furthermore, that the conventional compactification in the presence of these background fields is, in fact, equivalent to compactification on Lorentzian lattices.

In summary, the Lorentzian lattice not only is equivalent to the conventional compactification scheme when we include the presence of background fields but also provides a very convenient framework in which to catalogue an enormously large class of compactifications.

## §10.10. Summary

The anomaly cancellation requires that we have either O(32) or $E_8 \otimes E_8$. However, the Chan–Paton factors are incompatible with exceptional groups. Thus, we have no choice but to consider compactification as the process by which we can generate exceptional groups for the superstring.

The simplest possible compactification process in one dimension requires that we make the identification

$$x = x + 2\pi R \tag{10.10.1}$$

A one-dimensional scalar function must then be expanded in terms of periodic eigenfunctions:

$$\phi(x) = \sum_n \phi_n e^{ipx} \tag{10.10.2}$$

where

$$p = \frac{n}{R} \tag{10.10.3}$$

We see that momentum is quantized by taking periodic boundary conditions.

The heterotic string, by contrast, compactifies 26 dimensions down to 10 dimensions, leaving a 16-dimensional lattice. It is a closed string that separates the right- and left-moving sectors. The right-moving sector is purely 10-dimensional and contains the fermionic superstring and the bosonic string. The left-moving sector was originally 26-dimensional but was compactified down to 10 dimensions, leaving a 16-dimensional theory defined on a lattice and a 10-dimensional bosonic string (which combines with the right-moving bosonic string to complete a bosonic closed string). The final action, in the light cone gauge, is

$$S = \frac{-1}{4\pi\alpha'} \int d\tau \int_0^\pi d\sigma \left( \partial_a X^i \partial_a X^i + \sum_{I=1}^{16} \partial_a X^I \partial_a X^I + i\bar{S}\gamma^-(\partial_\tau + \partial_\sigma) S \right) \tag{10.10.4}$$

where we enforce the constraints

$$\begin{aligned} (\partial_\tau - \partial_\sigma) X^I &= 0 \\ \gamma^+ S = \tfrac{1}{2}(1 + \gamma_{11}) S &= 0 \end{aligned} \tag{10.10.5}$$

This action is explicitly supersymmetric under the variation

$$\begin{cases}\delta X^i = (p^+)^{-1/2}\bar{\varepsilon}\gamma^i S \\ \delta S^a = i(p^+)^{-1/2}\gamma_-\gamma_\mu(\partial_\tau - \partial_\sigma)X^\mu \varepsilon\end{cases} \tag{10.10.6}$$

When we analyze the spectrum of the heterotic string, we must take into account the quantization of the momentum and also the winding of the string around the compactified dimensions. This yields the conditions on the spectrum:

$$\tfrac{1}{4}m^2 = N + (\tilde{N} - 1) + \tfrac{1}{2}\sum_{I=1}^{16}(p^I)^2 \tag{10.10.7}$$

The final constraint on the spectrum arises because we want to make the closed string invariant under an arbitrary $\sigma$ rotation. The operator that generates this rotation is

$$U(\theta) = \exp\left\{2i\theta\left(N - \tilde{N} + 1 - \tfrac{1}{2}\sum_{I=1}^{16}(p^I)^2\right)\right\} \tag{10.10.8}$$

Thus, we demand

$$N = \tilde{N} - 1 + \tfrac{1}{2}\sum_{I=1}^{16}(p^I)^2 \tag{10.10.9}$$

Vertex functions can also be constructed for the heterotic string. They are, in fact, simply left- and right-moving products of the usual superstring vertices that were constructed earlier in Chapter 3. We choose the ansatz for the supergravity multiplet:

$$\begin{aligned} V_B &= \rho_{\mu\nu}(k)\int_0^\pi d\sigma\, B^\mu \tilde{P}^\nu e^{ik\cdot X} \\ V_F &= \int_0^\pi d\sigma\, F^a \tilde{P}_\nu U^{a\nu}(k) e^{ik\cdot X}\end{aligned} \tag{10.10.10}$$

where

$$\begin{aligned} k^+ &= 0 \\ B^i &= P^i + \tfrac{1}{2}k^j R^{ij} \\ P^i &= \frac{dX^i}{d(\tau - \sigma)} = \tfrac{1}{2}p^i + \sum_{n\neq 0}\alpha_n^i e^{-2in(\tau-\sigma)} \\ R^{ij} &= \tfrac{1}{8}\bar{S}\gamma^{ij-}S \\ \bar{F}^a &= \tfrac{1}{2}i(p^+)^{-1/2}[\bar{S}\gamma\cdot P - \tfrac{1}{6}{:}R^{ij}k^i\bar{S}\gamma^j{:}]^a \\ \tilde{P}^i &= \tfrac{1}{2}p^i + \sum_{n\neq 0}\tilde{\alpha}_n^i e^{-2in(\tau+\sigma)}\end{aligned} \tag{10.10.11}$$

With this vertex function and the usual propagator (plus constraint), we can calculate the four-point function for the scattering of four massless gauge bosons:

$$A_4 = g^2 K(\rho_i, k_i) \cdot \varepsilon \frac{\Gamma\left(-1 + \frac{u}{8} + \frac{U}{2}\right)\Gamma\left(-1 + \frac{s}{8} + \frac{S}{2}\right)\Gamma\left(-1 + \frac{t}{8} + \frac{T}{2}\right)}{\Gamma\left(1 - \frac{u}{8}\right)\Gamma\left(1 - \frac{s}{8}\right)\Gamma\left(1 - \frac{t}{8}\right)} \tag{10.10.12}$$

where $K$ is a complicated function of the polarizations and $S$, $T$, and $U$ are defined as

$$\begin{aligned} S &= (K_1 + K_2)^2 \\ T &= (K_2 + K_3)^2 \\ U &= (K_1 + K_3)^2 \\ S + T + U &= 8 \end{aligned} \tag{10.10.13}$$

Similarly, we can define the one-loop amplitude, which can be explicitly shown to be modular invariant. Thus, we have the freedom to eliminate the singularity at $\tau = 0$, which means that the theory is one-loop finite.

When analyzing the spectrum of the theory, we found that it was prohibitively difficult to construct the spectrum in terms of $E_8 \otimes E_8$ multiplets. The proof that the spectrum can be represented in terms of irreducible representations of that group can most easily be constructed using the Kac–Moody algebras. The Kac–Moody generators have the following commutation relation:

$$[T_m^i, T_n^j] = if^{ijl} T_{m+n}^l + km\delta^{ij}\delta_{m,-n} \tag{10.10.14}$$

Notice that this appears to be an ordinary Lie algebra that has been smeared over a circle. We can also rewrite this algebra in terms of the Cartan–Weyl representation, in which the generators look like

$$\begin{aligned} [H_i(\theta), H_j(\theta)] &= 0 \\ [H_i(\theta), E_{\boldsymbol{\alpha}}(\theta')] &= -2\pi\delta(\theta - \theta')\boldsymbol{\alpha}_i E_{\boldsymbol{\alpha}}(\theta) \end{aligned} \tag{10.10.15}$$

and

$$[E_{\boldsymbol{\alpha}}(\theta), E_{-\boldsymbol{\alpha}}(\theta')] = 2\pi\delta(\theta - \theta') \sum_i \boldsymbol{\alpha}_i H_i(\theta) + 2\pi i \delta'(\theta - \theta') \tag{10.10.16}$$

and

$$[E_{\boldsymbol{\alpha}}(\theta), E_{\boldsymbol{\beta}}(\theta')] = \begin{cases} 2\pi\delta(\theta - \theta') E_{\boldsymbol{\alpha}+\boldsymbol{\beta}}(\theta) & \text{if } \boldsymbol{\alpha} + \boldsymbol{\beta} \in \Gamma \\ 0 & \text{otherwise} \end{cases} \tag{10.10.17}$$

Using vertex operators, we can explicitly construct a representation of the Kac–Moody algebra. Thus, since the vertex operators generate the spectrum of the theory, the spectrum itself must be invariant under Kac–Moody transformations. Therefore, even though the spectrum is not manifestly $E_8 \otimes E_8$ symmetric, we have established that it actually is symmetric under that group.

We note that it is possible to construct modular-invariant strings without supersymmetry. If we take the heterotic string, for example, we can construct the operator

$$R = e^{2\pi i J_{12}} \gamma_\delta \tag{10.10.18}$$

which forms the discrete group $Z_2$. It is the product of a space–time rotation and a rotation in isospin space. We will take $\gamma_\delta$ to be a member of the Cartan subalgebra, such that it shifts the coordinate $x^I$ by $2\pi\delta^I$. If $\delta^2$ equals one, the theory has tachyons. But if $\delta^2$ equals two, a tachyon-free theory is possible. The gauge must be smaller than $E_8 \otimes E_8$ or Spin(32)/$Z_2$, because we are projecting onto the $R = 1$ subspace. The remaining gauge group is O(16) $\otimes$ O(16). This theory can be made to be modular invariant if we carefully choose the boundary conditions in both the $\sigma$ and $\tau$ directions for the closed string. Modular transformations will interchange these two boundary conditions, so we must sum over all four possibilities of NS and R boundary conditions.

The resulting model can be shown to be modular invariant, tachyon-free, and anomaly-free. However, it breaks supersymmetry and is not finite.

New compactification schemes more general than the one presented are possible by using a different compactification scheme down to $D$ space–time dimensions. By directly compactifying the left-moving sector from 26 down to $D$ space–time dimensions and compactifying the right-moving sector from 10 down to $D$ dimensions, we can bypass the intermediate step of the $E_8 \otimes E_8$ heterotic string. The new feature of this compactification scheme is that modular invariance restricts us to *Lorentzian* lattices, i.e., lattices where the metric has alternating signs. It can be shown that such Lorentzian lattice are perfectly welldefined. This thus gives us an entirely new class of heterotic-type strings with gauge groups as high as SO($52 - 2D$). Although these models look much different from the standard heterotic string, it can be shown that, with constant background fields $g_{ij}$, $A_i^I$, and $B_{ij}$, we can generate these new classes of models.

## References


[1] P. G. O. Freund, ITP preprint, 1984.
[2] Th. Kaluza, *Sitz. Preuss. Akad. Wiss.* **K1**, 966 (1921).
[3] O. Klein, *Z. Phys.* **37**, 895 (1926).
[4] H. C. Lee, *Introduction to Kaluza–Klein Theories*, World Scientific, Singapore, 1984.
[5] E. Cremmer and J. Scherk, *Nucl. Phys.* **B108**, 409 (1976); **B118**, 61 (1977).
[6] D. J. Gross, J. A. Harvey, E. Martinec, and R. Rohm, *Phys. Rev. Lett.* **54**, 502 (1985); *Nucl. Phys.* **B256**, 253 (1986); **B267**, 75 (1986).
[7] S. Yashikozawa, *Phys. Lett.* **166B**, 135 (1986).
[8] I. B. Frenkel and V. G. Kac, *Invent. Math.* **62**, 23 (1980).
[9] G. Segal, *Commun. Math. Phys.* **80**, 301 (1982).
[10] V. G. Kac, *In finite Dimensional Lie Algebras*, Birkhauser, Boston (1983).
[11] L. Alvarez-Gaumé, P. Ginsparg, G. Moore, and C. Vafa, *Phys. Lett.* **171B**, 155 (1985).
[12] L. Dixon and J. Harvey, *Nucl. Phys.* **B274**, 93 (1986).
[13] K. S. Narain, *Phys. Lett.* **B169**, 41 (1986).
[14] K. S. Narain, M. H. Sarmadi, and E. Witten, *Nucl. Phys.* **B279**, 369 (1987).

CHAPTER 11

# Calabi–Yau Spaces and Orbifolds

## §11.1. Calabi–Yau Spaces

Although the heterotic string is a great advance over the usual formulation of string theory, the question still remains whether we can break the theory down to four dimensions and do rigorous phenomenology. The answer, unfortunately, is no.

At present, our arsenal of techniques is still too primitive to answer the question of whether the theory undergoes spontaneous dimensional breaking. We will have to wait until further developments are made in string field theory or perhaps another formalism before any conclusion can be reached concerning the true vacuum of the theory.

In the absence of such a nonperturbative formulation of the theory, the best we can do is to look for various classical vacua of the theory and see whether reasonable phenomenology can be performed. Surprisingly, it turns out that rather mild restrictions on compactification are sufficient to obtain reasonable phenomenology. Although none of the solutions so far agrees totally with the minimal model $SU(3) \otimes SU(2) \otimes U(1)$, we come remarkably close with just a few assumptions on the classical vacua.

Unfortunately, however, we find an embarrassment of riches. There appear to be hundreds, if not thousands, of possible classical solutions, and it remains to be seen how to choose one vacuum over another. Thus, although naive phenomenology can be performed on the string, we still have to wait for the development of a nonperturbative formalism before any definitive statement can be made concerning the true vacuum of the theory.

In this chapter, we will thus *assume* that the compactification can be performed and will discuss two methods of writing down the classical vacua for the string theory:

(1) Calabi–Yau spaces. We require that $N = 1$ supersymmetry in four dimensions be unbroken. This simple assumption forces us to consider manifolds with a covariantly constant spinor. This, in turn, forces the six-dimensional manifold to be a *Calabi–Yau* manifold [1, 2].
(2) Orbifolds. We compactify on toruses divided by the action of a discrete group. This allows us to break the gauge group and arrive at different low-energy predictions. (Orbifolds are probably special limits of Calabi–Yau spaces, although this is not totally clear.)

Let us begin, however, by taking the zero slope limit of the theory, which reduces to $D = 10$ supergravity coupled to $E_8 \otimes E_8$ super-Yang–Mills theory (see Appendix), and make some reasonable assumptions about the breaking scheme. Candelas, Horowitz, Strominger, and Witten [1, 2] made the following assumptions about the zero slope limit:

(1) That the 10-dimensional universe has compactified down to a product of a 4- and a 6-dimensional universe:

$$M_{10} \to M_4 \times K_6 \tag{11.1.1}$$

where the $M_4$ universe is a maximally symmetric space, i.e.,

$$R_{\mu\nu\alpha\beta} = \frac{R}{12}(g_{\mu\alpha}g_{\nu\beta} - g_{\mu\beta}g_{\nu\alpha}) \tag{11.1.2}$$

and $K$ is compact. (The assumption that the four-dimensional manifold is maximally symmetric restricts the manifold to be either de Sitter, anti-de Sitter, or Minkowski.)
(2) That an $N = 1$ local supersymmetry remains unbroken and has survived the compactification.
(3) That some of the bosonic fields can be set to zero:

$$H = d\phi = 0 \tag{11.1.3}$$

The second assumption about $N = 1$ supersymmetry, in particular, is especially important because it places nontrivial constraints on the structure of the manifold $K^6$. If supersymmetry is unbroken, then the supersymmetry generator $Q$ must vanish on the vacuum $|0\rangle$ (see (9.8.4)). The variation of a fermion field, under supersymmetry, is given by

$$\delta\psi = [\bar{\varepsilon}Q, \psi] \tag{11.1.4}$$

The vacuum expectation value of this equation vanishes if supersymmetry is preserved (since $Q$ annihilates the vacuum):

$$\langle 0|\delta\psi|0\rangle = 0 \tag{11.1.5}$$

But in the classical limit, the variation of a fermionic field and its vacuum expectation value are the same:

$$\delta\psi \sim \langle 0|\delta\psi|0\rangle \tag{11.1.6}$$

*Thus, if $N = 1$ supersymmetry survives compactification, then the variation of the fermion fields must be zero*:

$$\delta\psi = 0 \tag{11.1.7}$$

This, in turn, places nontrivial restrictions on the supersymmetry parameter $\varepsilon$. Originally $\varepsilon$ was an arbitrary spinor field. However, demanding that $N = 1$ supersymmetry be exact means that we will select a subset of the infinitely many possible $\varepsilon$ such that $N = 1$ supersymmetry is preserved.

Now, the variation of the $D = 10$ fermions (see Appendix) is equal to [3]

$$\begin{aligned} \delta\psi_i &= \kappa^{-1} D_i \varepsilon + \frac{\kappa}{32 g^2 \phi} (\Gamma_i^{jkl} - 9\delta_i^j \Gamma^{kl}) \varepsilon H_{jkl} + \cdots \\ \delta\chi^a &= \frac{-1}{4g\sqrt{\phi}} \Gamma^{ij} F_{ij}^a \varepsilon + \cdots \\ \delta\lambda &= \frac{-1}{\sqrt{2\phi}} (\Gamma \cdot \partial\phi)\varepsilon + \frac{\kappa}{8\sqrt{2} g^2 \phi} \Gamma^{ijk} \varepsilon H_{ijk} + \cdots \end{aligned} \tag{11.1.8}$$

where Roman letters $i, j, k, a, b, c$ are six-dimensional indices and we drop higher four-fermion interaction terms. In addition, we have the Bianchi identity, which is satisfied exactly:

$$dH = \operatorname{Tr} R \wedge R - \tfrac{1}{30} \operatorname{Tr} F \wedge F \tag{11.1.9}$$

Now let us impose the second and third conditions, which will place restrictions on our parameter $\varepsilon$. Our variations reduce to

$$\begin{aligned} \delta\psi_i &= \frac{1}{\kappa} D_i \varepsilon = 0 \\ \delta\chi^a &= \Gamma^{ij} F_{ij}^a \varepsilon = 0 \end{aligned} \tag{11.1.10}$$

These are highly nontrivial constraints on the system, especially the first statement that forces $\varepsilon$ to be a *covariantly constant spinor*, i.e., $D_i\varepsilon = 0$. In particular, it places enormous constraints on the spin connection for the covariant derivative and hence the space itself.

The first equation in (11.1.10), for example, states that the parallel transport of the spinor $\varepsilon$ through a distance leaves the spinor invariant. Furthermore, we can perform two such displacements and travel around in a closed path (see Appendix). For example, if we differentiate (11.1.10) again, we obtain the variation of the spinor around a closed path. We find

$$\begin{aligned} D_i \varepsilon = 0 &\to [D_i, D_j]\varepsilon = 0 \\ &\to R_{ijkl} \Gamma^{kl} \varepsilon = 0 \end{aligned} \tag{11.1.11}$$

This implies that the spinor remains unchanged when we make a displacement around a closed path. This, in turn, implies that the manifold $K$ is Ricci flat:

$$R_{ij} = 0 \tag{11.1.12}$$

This last condition, in particular, is important because it says that the metric tensor describes a flat four-dimensional Minkowski space. Thus, de Sitter and anti-de Sitter spaces are ruled out.

Now let us take an arbitrary spinor and make a parallel transport around a closed curve. In the Appendix we see that a spinor after making a closed circuit is given by

$$\varepsilon' \to \varepsilon + \Delta^{mn}[D_m, D_n]\varepsilon \tag{11.1.13}$$

where the area of the closed path is proportional to $\Delta^{mn}$. Thus, we see that a spinor is simply rotated from its original orientation, with the rotation matrix being the commutator of two displacements, which is the curvature tensor:

$$\varepsilon \to U\varepsilon \tag{11.1.14}$$

where

$$U = 1 + \Delta^{mn}[D_m, D_n] + \cdots \tag{11.1.15}$$

Now let us make several consecutive closed paths, each time leaving and returning to the same point. In general, each time we make an arbitrary number of closed paths around the same point, we wind up with the original spinor times a small rotation. Thus, the set of all such rotations forms a group:

$$\varepsilon \to U_2 U_1 \varepsilon = U_3 \varepsilon \tag{11.1.16}$$

This group is called the *holonomy* group.

Now let us apply this result to our special case. For the manifold $K^6$, an arbitrary spinor has a spin connection that is an O(6) gauge field. A spinor transforming under O(6) has $2^3 = 8$ elements. However, we know that

$$\mathrm{O}(6) = \mathrm{SU}(4) \tag{11.1.17}$$

Thus, these eight elements in a spinor of O(6) can also be rearranged according to SU(4):

$$\mathbf{8} = \mathbf{4} \oplus \mathbf{4} \tag{11.1.18}$$

Under SU(4), these two objects transform as spinors of opposite chirality. We can let the spinor $\varepsilon$ have positive chirality. This eliminates half the components, so that it transforms as the **4** of SU(4).

The spinor that we are investigating, however, is not an arbitrary spinor but satisfies the condition

$$D_i \varepsilon = 0 \tag{11.1.19}$$

which means that

$$\varepsilon = U\varepsilon \tag{11.1.20}$$

that is, the spinor remains the same after being transported around a closed path.

The question now is: What is the subgroup of O(6) that leaves invariant the **4** of SU(4)? The answer is a familiar one, taken directly from the theory of Higgs breaking. There, we also want to know the answer to the question: *What is the largest group that will leave a constant spinor or vector invariant?*

To answer this question, note that we can always, by an SU(4) transformation, put $\varepsilon$ in the form

$$\varepsilon \to \varepsilon = \begin{pmatrix} 0 \\ 0 \\ 0 \\ \varepsilon_0 \end{pmatrix} \tag{11.1.21}$$

So far, we have done nothing. We have simply taken an arbitrary spinor and put it into this form by an SU(4) rotation. But now it is obvious that the largest group U that will keep such a spinor invariant is the subgroup of $3 \times 3$ complex matrices within SU(4), that is, SU(3). Notice that we now restrict the $U$ matrix to be block diagonal:

$$U = \begin{pmatrix} \mathrm{SU}(3) & 0 \\ 0 & 1 \end{pmatrix} \tag{11.1.22}$$

We can now trivially satisfy $\varepsilon = U\varepsilon$. It is important to note that our result is quite general. We chose a particular representation of $\varepsilon$ only to show this general result, which works for any representation of $\varepsilon$.

In conclusion, we say that the existence of a covariantly constant spinor forces the holonomy group to reduce from O(6) to SU(3). We say that $K^6$ has SU(3) holonomy.

The logic that we followed can be summarized as

$$N = 1 \text{ SUSY} \to D_i\varepsilon = 0 \to \mathrm{SU}(3) \text{ Holonomy} \tag{11.1.23}$$

This is an important result, but it is rather useless. Very little is known about manifolds with SU(3) holonomy. In fact, none are known explicitly. Therefore, this important piece of information cannot be used for phenomenology. However, there is still hope, because we have not yet exhausted the information that can be derived from our principles.

We have not by any means exhausted all the implications of having a covariantly constant spinor in the theory. We can, for example, always construct the following object out of the spinor field, which transforms as a true tensor on the space:

$$J_j^i = -ig^{ik}\bar{\varepsilon}\Gamma_{kj}\varepsilon \tag{11.1.24}$$

Using a few identities on spinors, we can show that

$$J_m^n J_n^p = -\delta_m^p \tag{11.1.25}$$

Whenever we can construct such a tensor $J$ that maps the space into itself and satisfies $J^2 = -1$, we say that the manifold is *almost complex*. (In two dimensions this tensor statement simply says that there exists a number $i$ such that

$i^2 = -1$, which is trivial.) This tensor is a very interesting object if the spinor $\varepsilon$ is covariantly constant. For example, by differentiation we find

$$D^m J_m^n = 0 \tag{11.1.26}$$

This means that the metric, in addition to being Ricci flat, is *Kähler*. (These terms will be defined shortly.) Lastly, we can also define a one-form:

$$\Gamma = \Gamma_{mp}^q J_q^p \, dx^m \tag{11.1.27}$$

where $\Gamma_{mp}^q$ is the Christoffel symbol. We find that this form satisfies

$$d\Gamma = 0 \tag{11.1.28}$$

which states that the first Chern class vanishes: $c_1 = 0$.

In summary, retracing the logic of our assumptions, we have

$$N = 1 \text{ SUSY} \rightarrow D_i \varepsilon = 0 \rightarrow K \text{ is Ricci flat, Kähler, with vanishing first Chern class} \tag{11.1.29}$$

The great advantage of this new result is that many Ricci-flat, Kähler manifolds with vanishing first Chern class are known. Thus, instead of relying on SU(3) holonomy, which is a dead end, it is much preferable to rely on these types of Kähler manifolds.

One question that remains unresolved is the relationship between these two kinds of manifolds. Fortunately, Calabi conjectured (and Yau later proved) the statement that [4–6]

**Theorem** (*Calabi–Yau*). *A Kähler manifold of vanishing first Chern class always admits a Kähler metric of* SU(3) *holonomy.*

Thus, using the theorem of Calabi–Yau, we can generate (at least in principle) thousands of six-dimensional manifolds for phenomenological purposes.

In order to actually construct such Calabi–Yau manifolds, it is important to first review a few elementary facts about algebraic geometry and cohomology. Let us now make a digression and discuss some simple properties of Kähler and Ricci-flat metrics in the language of cohomology. We will find that many of these results from cohomology theory can be incorporated directly into superstring phenomenology.

## §11.2. Review of de Rahm Cohomology

As stated in the Appendix, the theory of differential forms begins with an operator that is nilpotent:

$$\begin{aligned} d &= dx^\mu \, \partial_\mu \\ d^2 &= 0 \end{aligned} \tag{11.2.1}$$

We say that an $N$-form $\omega$ is *closed* if

$$\text{closed:} \quad d\omega = 0 \tag{11.2.2}$$

Similarly, we say that it is *exact* if there exists an $N - 1$ form $\alpha$ such that

$$\text{exact:} \quad \omega = d\alpha \tag{11.2.3}$$

Notice that the set of exact forms is a subset of the closed forms:

$$\text{exact forms} \subset \text{closed forms} \tag{11.2.4}$$

In three dimensions, these statements can be summarized by the well-known statement that the gradient of a scalar always has zero curl:

$$\mathbf{A} = \nabla \cdot \phi \rightarrow \nabla \times \mathbf{A} = 0 \tag{11.2.5}$$

In fact, the theory of forms allows one to simply reexpress many of the theorems in ordinary tensor calculus in three dimensions.

In Maxwell theory, we say that two field tensors $A_\mu$ and $B_\mu$ describe the same physics if they differ by a total derivative:

$$A_\mu - B_\mu = \partial_\mu \Lambda \rightarrow A_\mu \sim B_\mu \tag{11.2.6}$$

This, of course, is the essence of gauge theory. In this mathematical language, we will say that two forms $\omega$ and $\omega'$ belong to the same equivalence class if they differ by a closed form.

$$\omega - \omega' = d\alpha \rightarrow \omega \sim \omega'$$

In Maxwell theory, we want to construct the set of all inequivalent fields. In the theory of forms, we do this by defining the set $H^p(M)$ to be the set of all closed $p$ forms modulo exact forms, i.e.,

$$H^p(M) = \frac{\text{closed } p \text{ forms}}{\text{exact } p \text{ forms}} \tag{11.2.7}$$

This defines the $p$th *de Rham cohomology group* $H^p(M)$ defined on a manifold $M$.

Notice that the cohomology group counts how many independent $p$ closed forms can be defined on a particular manifold. Cohomology is thus dependent on the *local* structure of the manifold. However, as the name suggests, there is a duality between cohomology and *homology*, which is based on the *global* properties of a manifold:

$$\text{Cohomology} \rightarrow \text{Local properties of a manifold}$$

$$\text{Homology} \rightarrow \text{Global properties of a manifold}$$

To make the relationship between homology and cohomology more precise, let us define how to establish this duality. We begin by writing down the integral over a region $C$ defined on a manifold $M$, where $C$ can be a line, surface, volume, etc.:

$$\int_C \omega$$

We can think of this as a map that takes a $p$ form $\omega$ and marries it with a surface $C$ to produce a real number. Thus, we can view this operation as the "scalar product" of a form with a surface:

$$\langle C|\omega\rangle \equiv \int_C \omega \tag{11.2.8}$$

Now let us write down the familiar Stokes' theorem, generalized to $N$-dimensional space and expressed in the language of forms:

$$\int_C d\omega = \int_{\partial C} \omega \tag{11.2.9}$$

where $C$ is an $N$-dimensional manifold and $\partial C$ is defined as the "boundary" of $C$. Now let us rewrite Stokes' theorem so that it is reexpressed as a scalar product equation:

$$\text{Stokes' theorem: } \langle C|d\omega\rangle = \langle \partial C|\omega\rangle \tag{11.2.10}$$

Thus, we have established that the dual of the cohomology operator $d$ is equal to $\partial$, the boundary operator:

$$d \leftrightarrow \partial \tag{11.2.11}$$

Because of this duality, we suspect that the boundary operator $\partial$ also defines the dual to the cohomology group, which we call homology.

To be precise, let us clarify what we mean by the boundary operator $\partial$ and by the surface $C$ in terms of *simplexes*. Let us define the line segment that extends from a point $p_1$ to $p_2$ as a 1-simplex. The 1-simplex has a definite orientation or direction:

$$\text{line segment} = [p_1, p_2] = -[p_2, p_1] \tag{11.2.12}$$

Let us define a triangle as a 2-simplex, which is labeled by three points or vertices:

$$\text{triangle} = [p_1, p_2, p_3] = -[p_1, p_3, p_2] \tag{11.2.13}$$

Notice that the 2-simplex is cyclic in the three points, but flips sign if the points are arranged anticyclically. Obviously, we can generalize this concept by introducing an $N$-simplex:

$$N\text{-simplex} = [p_1, \dots, p_{N+1}] \tag{11.2.14}$$

(Notice that the vectors formed by taking differences of the points $p_i$ must be linearly independent, or else the simplex collapses to a lower dimension.)

Now let us define the boundary operator that maps $m$-simplexes into

$m-1$-simplexes. For example,

$$\partial_1[p_1, p_2] = [p_1] - [p_2] \tag{11.2.15}$$

where $[p]$ is a point. Thus, we define the "boundary" of a line segment to be the two endpoints. Acting on a triangle, the boundary operator creates line segments:

$$\partial_2[p_1, p_2, p_3] = [p_1, p_2] + [p_2, p_3] + [p_3, p_1] \tag{11.2.16}$$

Thus, the boundary operator acting on a 2-simplex (a triangle) simply breaks it up into its three edges, which are composed of lines or 1-simplexes.

It is easy to check that the boundary operator is nilpotent. For example, for the simple case of triangles we have

$$\begin{aligned}\partial_1\partial_2[p_1, p_2, p_3] &= \partial_1\{[p_1, p_2] + [p_2, p_3] + [p_3, p_1]\} \\ &= [p_1] - [p_2] + [p_2] - [p_3] + [p_3] - [p_1] \\ &= 0\end{aligned} \tag{11.2.17}$$

Thus, we have the important result that the boundary operator is nilpotent:

$$\partial^2 = 0$$

This can be generalized to the arbitrary case. Let us now define how the boundary operator acts on an $N$-simplex:

$$\partial_N[p_1, \ldots, p_{N+1}] = \sum_{i=1}^{N+1} (-1)^i [p_1, \ldots, \hat{p}_i, \ldots, p_{N+1}] \tag{11.2.18}$$

where we omit the points that are hatted. Thus, the boundary operator maps $N$-simplexes into $N-1$-simplexes. It is not difficult to show that

$$\partial_{N-1}\partial_N = 0 \tag{11.2.19}$$

for the general case.

In analogy with the theory of homology, let us define the set of *cycles* $Z$ to be the set of simplexes that obey

$$\partial Z = 0 \tag{11.2.20}$$

For example, take the two-dimensional doughnut or torus. There are several types of cycles that one can draw on this surface. For example, in Chapter 5 we saw that we can draw two types of cycles on the doughnut that cannot be continuously shrunk to a point. However, there is also the cycle that is simply a closed line on the surface of the torus that can be continuously shrunk to a point. We want a method by which we can eliminate the second type of cycle.

Let us define the set of *boundaries* $B$ as the simplexes that can be written as the boundary of some higher simplex $C$:

$$B = \partial C \tag{11.2.21}$$

Notice that the set of boundaries is a subset of the set of cycles:

$$\text{boundaries} \subset \text{cycles} \tag{11.2.22}$$

We say that two simplexes are in the same equivalence class if they differ only by a boundary. Thus, we want to extract the set of all inequivalent simplexes, and we do this by defining the $p$th homology group:

$$H_p(M) = \frac{p - \text{cycles}}{p - \text{boundaries}} \tag{11.2.23}$$

This allows us to eliminate the redundant cycles on the torus that can be continuously shrunk down to a point, keeping only the two independent cycles that encircle the torus. Thus, the concept of homology is a natural one.

## §11.3. Cohomology and Homology

What is the relationship between the homology and cohomology groups for a compact manifold? Because of the duality that exists between them, we can show that they have the same dimension. Let us define the *Betti numbers* to be the dimension of the $H$'s:

$$b^p = \dim H^p(M) = \dim H_p(M) \tag{11.3.1}$$

Let $c_p$ be a set of cycles defined on a compact manifold $M$. Let $\omega_q$ be a set of closed forms defined on $M$. Then construct the matrix $\Omega_{pq}$:

$$\Omega(c_p, \omega_q) = \int_{c_p} \omega_q \tag{11.3.2}$$

The matrix $\Omega$ is called the *period matrix* (see (5.6.31) and (5.11.8)), and it can be shown under quite general conditions that it is an invertible matrix. Thus, it is an $N \times N$ matrix with nonvanishing determinant. But if the period matrix is a square matrix in $N$ dimensions, then the dimension of the closed forms $\omega_q$ is equal to the dimension of the cycles $c_p$, which proves that the Betti numbers are the same for the cohomology and homology groups. This is an extremely important result, because it means that we can use either local (cohomology) or global (homology) properties of a particular manifold to calculate its Betti numbers.

It is also important to realize that the Betti numbers are topological numbers, depending only on the overall topology of a manifold. Thus, any linear combination of the Betti numbers is also a topological number. In particular, the most important one is the Euler number:

$$\chi(M) = \sum_{i=0}^{N} (-1)^i b_i \tag{11.3.3}$$

To understand the properties of these Betti numbers, we will have to introduce a few more operators, including the Laplacian. Let us also introduce the Hodge star operator, which converts a $p$-form into an $N-p$-form:

$$* (dx^{i_1} \wedge dx^{i_2} \wedge \cdots \wedge dx^{i_p})$$
$$= \frac{\sqrt{g}}{(n-p)!} \varepsilon^{i_1 i_2 \ldots i_p}_{i_{p+1} i_{p+2} \ldots i_n} dx^{i_{p+1}} \wedge dx^{i_{p+2}} \wedge \cdots \wedge dx^{i_n} \tag{11.3.4}$$

where $\varepsilon_{ijkl\cdots}$ is the totally antisymmetric tensor in $N$ dimensions. Notice that

$$\begin{aligned} ** \omega_p &= (-1)^{p(N-p)} \omega_p \\ \omega_p \wedge * \omega_q &= \omega_q \wedge * \omega_p \end{aligned} \tag{11.3.5}$$

In $N$-space, the product of a $p$-form and an $N-p$-form yields an $N$-form, which is proportional to the volume $d^N x$. Thus, we can integrate over the product of a $p$ and an $N-p$ form to obtain an ordinary number. Let us now define yet another inner product:

$$(\alpha_p | \beta_p) = \int_M \alpha_p \wedge * \beta_p \tag{11.3.6}$$

Now that we have introduced the definition of the scalar product, this allows us to define the dual $\delta$ of the derivative $d$:

$$(\alpha_p | d\beta_{p-1}) = (\delta\alpha_p | \beta_{p-1}) \tag{11.3.7}$$

Explicitly, the adjoint of $d$ is given by

$$\delta = (-1)^{Np+N+1} * d * \tag{11.3.8}$$

We also have

$$\delta\delta = 0 \tag{11.3.9}$$

Notice that the adjoint reduces by 1 the degree of a differential form, while $d$ raises it.

Let us now define the *Laplacian* as

$$\Delta = (d + \delta)^2 = \delta d + d\delta \tag{11.3.10}$$

Let us define a $p$-form $\omega$ to be *harmonic* if

$$\text{harmonic:} \quad \Delta\omega = 0 \tag{11.3.11}$$

Let us define a $p$-form to be *coclosed* if

$$\text{coclosed:} \quad \delta\omega = 0 \tag{11.3.12}$$

Let us define a $p$-form to be *coexact* if it can be written as

$$\text{coexact:} \quad \omega = \delta\alpha \tag{11.3.13}$$

for some $p + 1$-form $\alpha$.

We now arrive at an important theorem:

**Theorem** (*Hodge*). *On a compact manifold without boundary, any p-form can be uniquely decomposed as the sum of exact, coexact, and harmonic forms, i.e.,*

$$\omega_p = d\alpha_{p-1} + \delta\beta_{p+1} + \gamma_p \tag{11.3.14}$$

*where $\gamma_p$ is a harmonic form.*

This is a powerful result, because we can show that each cohomology class contains only one harmonic form. To see this, first construct the inner product between a form and its Laplacian:

$$(\omega|\Delta\omega) = |\delta\omega|^2 + |d\omega|^2 \tag{11.3.15}$$

Thus, the statement that a form is harmonic (which sets $(\omega|\Delta\omega) = 0$) is equivalent to stating that it is exact and coexact (because $|\delta\omega|^2 = 0$ and $|d\omega|^2 = 0$). In fact, it can be shown that

$$\Delta\omega = 0 \quad \text{iff } \delta\omega = d\omega = 0 \tag{11.3.16}$$

But if $d\omega = 0$, this means that $d\delta\beta = 0$, and so $\delta\beta = 0$. Therefore (11.3.14) reduces to

$$\omega = d\alpha + \gamma \tag{11.3.17}$$

*Thus, every cohomology class contains one harmonic representative.*

The fact that there is one harmonic representative within each equivalence class of exact forms gives us an alternative way to define the Betti numbers. We can also say that the Betti numbers count how many independent harmonic forms there are on the manifold. We have the following equivalent description of the Betti numbers:

$$\text{Betti number} = \begin{cases} \text{Number of independent closed forms} \\ \text{Number of independent cycles} \\ \text{Number of harmonic forms} \end{cases} \tag{11.3.18}$$

Thus, we are free to use either of three equivalent formalisms to calculate the Betti numbers.

The last formulation of Betti numbers, in terms of independent harmonic forms, gives us an alternative definition. The set of harmonic forms can be reexpressed as the kernel of the Laplacian (i.e., the forms that are sent to zero). Thus, we can define the Betti numbers as

$$b_p = \begin{cases} \dim \ker \Delta_p \\ \dim H_p \\ \dim H^p \end{cases} \tag{11.3.19}$$

We will find it useful to study the properties of some of these Betti numbers. First, we can always define the scalar product for an $N$-dimensional manifold:

$$(\omega_p|\omega_{N-p}) = \int_M \omega_p \omega_{N-p} \tag{11.3.20}$$

Consequently, these two spaces contain the same number of independent elements. Thus,

$$b_p = \dim H^p = b_{N-p} = \dim H_{N-p} \tag{11.3.21}$$

This is called *Poincaré duality*. We will usually take $b_0 = 1$, so therefore $b_N = 1$.

Another way to prove Poincaré duality is to notice that if a particular form is harmonic, then it has a dual that is also harmonic:

$$\Delta\omega = 0 \to \Delta * \omega = 0 \tag{11.3.22}$$

But because the number of independent harmonic forms is equal to the Betti number, and because the Hodge star operation converts $p$-forms into $N-p$-forms, we must have Poincaré duality.

Lastly, if we have a product manifold, then the Euler number of the product manifold is the product of the Euler numbers of each manifold:

$$\chi(M \times N) = \chi(M) \times \chi(N) \tag{11.3.23}$$

If we write this out in terms of Betti numbers given in (11.3.3), this becomes

$$b_k(M \times N) = \sum_{p+q=k} b_p(M) b_q(N) \tag{11.3.24}$$

These are called the Kunneth formulas for product manifolds.

Let us now take a few simple surfaces and calculate their Betti numbers:

## (1) Two-Torus

A two-torus can be cut up into cycles in two independent ways. The dimension of the independent one-cycles that one can draw on a torus is thus two. Thus, $b_1 = 2$. Also, by Poincaré duality, we have

$$T_2: \quad \begin{cases} b_0 = b_2 = 1 \\ b_1 = 2 \end{cases} \tag{11.3.25}$$

Thus, the Euler number (11.3.3) for a torus is

$$\chi(T_2) = 1 - 2 + 1 = 0 \tag{11.3.26}$$

## (2) Riemann Surface

As we saw in Chapter 5, there are $2g$ ways in which we can divide up a Riemann surface of genus $g$ into independent cycles. Each hole or handle has two such cycles. Thus, $b_1 = 2g$. We then have

$$\begin{cases} b_0 = b_2 = 1 \\ b_1 = 2g \end{cases} \tag{11.3.27}$$

Thus, the Euler number for a closed Riemann surface is

$$\chi(M) = 1 - 2g + 1 = 2 - 2g \tag{11.3.28}$$

### (3) $N$-Sphere

Cycles on the sphere $S_N$, of course, can always be collapsed down to a point. Hence there are no independent cycles that one can write. Thus,

$$S_N: \quad \begin{cases} b_0 = b_N = 1 \\ b_p = 0 \quad \text{otherwise} \end{cases} \tag{11.3.29}$$

The Euler number is therefore

$$\chi(S_N) = \begin{cases} 0 & \text{if } N \text{ is odd} \\ 2 & \text{if } N \text{ is even} \end{cases} \tag{11.3.30}$$

In particular, given that

$$T_2 = S_1 \times S_1 \tag{11.3.31}$$

we can use the product rule on Euler numbers to show that the Euler number for a torus is

$$\chi(T_2) = \chi(S_1)^2 = 0 \tag{11.3.32}$$

which agrees with our earlier calculation.

### (4) Product of Spheres

Using (11.3.24), we can show that the product $S_3 \times S_3$ has

$$S_3 \times S_3: \quad \begin{cases} b_0 = b_6 = 1 \\ b_3 = 2 \\ b_p = 0 \quad \text{otherwise} \end{cases} \tag{11.3.33}$$

We find that the Euler number is zero, as expected from the fact that the Euler number of the three-sphere is also zero:

$$\chi(S_3 \times S_3) = \chi(S_3) \times \chi(S_3) = 0 \tag{11.3.34}$$

Similarly, we can take the product $S_2 \times S_2 \times S_2$, which has the Betti numbers

$$S_2 \times S_2 \times S_2\colon \quad \begin{cases} b_0 = b_6 = 1 \\ b_2 = b_4 = 3 \\ b_p = 0 \quad \text{otherwise} \end{cases} \tag{11.3.35}$$

We find its Euler number to be

$$\chi(S_2 \times S_2 \times S_2) = \chi(S_2)^3 = 8 \tag{11.3.36}$$

## (5) Four-Torus

We can write the four-torus as

$$T_4 = T_2 \times T_2 \tag{11.3.37}$$

It is a simple exercise to show that the Kunneth relations for product manifolds in (11.3.24) yield

$$T_4\colon \quad \begin{cases} b_0 = b_4 = 1 \\ b_1 = b_3 = 4 \\ b_2 = 6 \end{cases} \tag{11.3.38}$$

Putting these Betti numbers into the Euler number, we find

$$\chi(T_4) = \chi(T_2)\chi(T_2) = 0 \tag{11.3.39}$$

as expected.

## (6) Six-Torus

For the six-torus $T_6$, we have

$$T_6 = T_2 \times T_2 \times T_2 \tag{11.3.40}$$

Therefore we can use the Kunneth relations to solve for the Betti numbers of the product manifold. It is easy to show that

$$T_6\colon \quad \begin{cases} b_0 = b_6 = 1 \\ b_1 = b_5 = 6 \\ b_2 = b_4 = 15 \\ b_3 = 20 \end{cases} \tag{11.3.41}$$

We see that the Euler number is equal to 0:

$$\chi(T_6) = 0 \tag{11.3.42}$$

This is also true because each circle within $T_6$ has Euler number 0. Using these simple rules, we can obviously generate the Betti numbers for $T_N$.

## (7) Real and Complex Projective Space

The space $CP_N$ is formed by taking ordinary complex $N + 1$-space and making the identification

$$z_p = \lambda z_p \tag{11.3.43}$$

for some nonvanishing complex number $\lambda$. This last condition reduces complex $N + 1$-space to complex projective $N$-space. $CP_N$ is a generalization of real projective space $P_N$, which is created by identifying points in real $N + 1$-space by

$$x_p = k x_p \tag{11.3.44}$$

for some nonzero real $k$. We can also construct it by taking the sphere $S_N$ and identifying antipodal points. Some examples of real projective and complex projective space are

$$\begin{aligned} CP_1 &= S_2 \\ P_3 &= \mathrm{SO}(3) = \mathrm{SU}(2)/\mathbf{Z}_2 \\ CP_N &= \frac{\mathrm{SU}(N+1)}{\mathrm{U}(N)} \end{aligned} \tag{11.3.45}$$

Their Betti numbers are given by

$$P_N: \quad \begin{cases} b_0 = 1 \\ b_N = 0 \ (N \text{ even}); \quad = 1 \ (N \text{ odd}) \\ b_i = 0 \quad \text{otherwise} \end{cases} \tag{11.3.46}$$

$$\chi(P_N) = 1 \ (N \text{ even}); \quad = 0 \ (N \text{ odd})$$

and

$$CP_N: \quad \begin{cases} b_{\text{even}} = 1 \\ b_{\text{odd}} = 0 \end{cases}$$

Therefore,

$$\chi(CP_N) = N + 1 \tag{11.3.47}$$

# §11.4. Kähler Manifolds

So far, our discussion has revolved mainly around real manifolds. However, our results generalize quite easily to complex manifolds.

To define an $N$-dimensional complex manifold, we want a $2N$ real manifold and a generalization of the usual definition of a complex number: $z = x + iy$, where $i^2 = -1$. If we have a $2N$-dimensional real space, what we want is a

$2N \times 2N$ tensor field $J_i^j$ that can replace $i$, such that

$$J^2 = -I \tag{11.4.1}$$

as a matrix equation, so that we can define a complex number as

$$\mathbf{z} = \mathbf{x} + J\mathbf{y} \tag{11.4.2}$$

Manifolds that have such a tensor field $J$ are called *almost complex*.

However, we want more than just the existence of the tensor field $J$. We would like to diagonalize it. If we use complex coordinates

$$\begin{aligned} z_a &= x_a + iy_a \\ \bar{z}_a &= z_a - iy_a \end{aligned} \tag{11.4.3}$$

we would like to make a change of coordinates such that

$$J_a^b = i\delta_a^b \tag{11.4.4}$$

The question is: Can we always diagonalize $J$ in a neighborhood of a point $p$ using a coordinate transformation:

$$z'_a = z'_a(z_b) \tag{11.4.5}$$

(which is not a function of $\bar{z}$)? Such a coordinate transformation is called *holomorphic*, which is the generalization of the concept of analytic. For different points $p$, we of course will have to patch them together using different holomorphic transformations until we can cover the entire surface.

If we can always find such holomorphic transformations that can diagonalize $J$ around a neighborhood of any point $p$ of the manifold, we say that manifold has complex structure and is a *complex manifold*.

Thus, we have the analogy

| One dimension | $N$ dimensions |
|---|---|
| $i$ | $J_a^b = i\delta_a^b$ |
| analytic | holomorphic |
| $z = x + iy$ | $\mathbf{z} = \mathbf{x} + J\mathbf{y}$ |

(11.4.6)

(Intuitively, the process of patching together neighborhoods where $J$ is diagonal resembles the "elevator problem" in general relativity. By a general coordinate transformation, we can always locally transform the Christoffel symbols to zero at any specific point, which corresponds to falling freely in an elevator without gravity. In general, we cannot globally transform all the Christoffel symbols away entirely, or else the space is flat, but we can at any point on the manifold enter the elevator frame.)

At first, this definition might seem verbose for a rather intuitive concept. However, certain $2N$-dimensional real manifolds can be shown to fail this criterion. For example the spheres $S_{2N}$ are *not* complex manifolds (except for $S_2$, because $S_2 = CP_1$). Thus, it is not at all obvious that a $2N$-dimensional real manifold can be rewritten as an $N$-dimensional complex manifold.

Now let us discuss differential forms on these complex manifolds. Let us complexify our basis differentials as

$$dz^j = dx^j + i\,dy^j \tag{11.4.7}$$

Now the concept of a $p$-form can be generalized to a $(p, q)$-form:

$$\omega = \omega_{ijk\cdots u;\bar{i}\bar{j}\bar{k}\cdots\bar{u}}\, dz^i \wedge dz^j \cdots dz^u \wedge d\bar{z}^i \wedge d\bar{z}^j \cdots d\bar{z}^u \tag{11.4.8}$$

with $p$ unbarred and $q$ barred indices. We can now define two exterior derivatives:

$$\begin{aligned} \partial\omega &= \frac{\partial}{\partial z^l}\omega_{ijk\ldots;\bar{i}\bar{j}\bar{k}\cdots\bar{u}}\, dz^l \wedge dz^i \cdots \wedge d\bar{z}^u \\ \bar{\partial}\omega &= \frac{\partial}{\partial \bar{z}^l}\, d\bar{z}^l\, \omega_{ijk\cdots;\bar{i}\bar{j}\bar{k}\cdots\bar{u}}\, dz^i \wedge \cdots \wedge d\bar{z}^u \end{aligned} \tag{11.4.9}$$

The properties of these two derivatives are

$$\begin{cases} \partial^2 = \bar{\partial}^2 = 0 \\ \partial\bar{\partial} + \bar{\partial}\partial = 0 \\ d = \frac{1}{2}(\partial + \bar{\partial}) \end{cases} \tag{11.4.10}$$

Every form now takes two indices $(p, q)$ to label it. Thus, we can now define a new cohomology group. Instead of de Rahm cohomology based on $d$, we now have *Dolbeaut cohomology*, based on $\bar{\partial}$. We can define closed and exact for a $(p, q)$-form just as before, except we now use $\bar{\partial}$ instead of $d$:

$$H^{p,q}(M) = \frac{\bar{\partial}\text{ closed }(p, q)\text{-form}}{\bar{\partial}\text{ exact }(p, q)\text{-form}} \tag{11.4.11}$$

As before, we can also construct the adjoint operator and the Laplacian. In the complex case, we actually have two Laplacians:

$$\begin{aligned} \Delta_{\partial} &= \partial\partial^{\dagger} + \partial^{\dagger}\partial \\ \Delta_{\bar{\partial}} &= \bar{\partial}\bar{\partial}^{\dagger} + \bar{\partial}^{\dagger}\bar{\partial} \end{aligned} \tag{11.4.12}$$

We now have the complex version of the Hodge theorem:

**Theorem** (*Hodge*). *Every complex $(p, q)$-form has a unique decomposition*

$$\omega = \alpha + \bar{\partial}\beta + \bar{\partial}^{\dagger}\gamma \tag{11.4.13}$$

*where $\alpha$ is a harmonic form:*

$$\Delta_{\bar{\partial}}\alpha = 0$$

*and $\beta$, $\gamma$ are $(p, q - 1)$- and $(p, q + 1)$- forms, respectively.*

We also have the generalization of the Betti numbers, which are now given by

$$\dim H_{\bar{\partial}}^{p,q}(M) = h^{p,q} \tag{11.4.14}$$

The relation to the Betti numbers is

$$b_n = \sum_{p+q=n} h^{p,q} \tag{11.4.15}$$

We also have, by complex conjugation,

$$h^{p,q} = h^{q,p}$$

and Poincaré duality:

$$h^{p,q} = h^{N-p,N-q} \tag{11.4.16}$$

We now define a *Kähler manifold*. If we have a complex manifold with Hermitian metric $g_{i\bar{j}}$, then we can always construct the form

$$\Omega = g_{i\bar{j}}\, dz^i \wedge d\bar{z}^j \tag{11.4.17}$$

This is called the Kähler form. A complex manifold is called Kähler if

$$d\Omega = 0 \tag{11.4.18}$$

A Kähler manifold has a Kähler form that is closed.

A Kähler manifold, because of the above definition, has a large number of beautiful properties which make it among the most elegant of complex manifolds. We simply list some of its properties:

(1) We can show that the Hermitian metric of a Kähler manifold can be written in terms of a derivative of a single function, the Kähler potential $\phi$:

$$g_{i\bar{j}} = \frac{\partial^2 \phi}{\partial z^i \partial \bar{z}^j} \tag{11.4.19}$$

(2) It is easy to show that a Kähler manifold satisfies, by explicit calculation,

$$2\Delta_d = \Delta_\partial = \Delta_{\bar{\partial}} \tag{11.4.20}$$

This is a powerful identity because it states that the various Laplacians that one can form on a Kähler manifold are all the same. Thus, there is no confusion when using different Laplacians for the Kähler manifold.

(3) If a manifold is to admit a Kähler metric, then the even components of the Betti numbers must be greater than or equal to one, and its odd Betti numbers must be even:

$$\text{Kähler} \to \begin{cases} b_{2p} \geq 1 \\ b_{2n+1} = \text{even} \end{cases} \tag{11.4.21}$$

for integers $p$ and $n$. (This simple criterion rules out a large class of manifolds. For example, this rules out $S_3 \times S_3$ as a manifold that admits a Kähler metric, because $b_2 = 0$. It also rules out $P_N$ but allows $CP_N$.)

(4) If the $J_m^n$ tensor is covariantly constant, then the metric is Kähler. The converse is also true for complex manifolds.

$$\nabla^a J_a^b = 0 \leftrightarrow \text{Hermitian metric is Kähler} \tag{11.4.22}$$

(5) If the torsion form that one constructs on a complex manifold vanishes, then the metric is Kähler.

(6) The only nonvanishing Christoffel symbol of a Kähler manifold is

$$\Gamma_{ab}^c = g^{c\bar{c}} g_{a\bar{c},b} \tag{11.4.23}$$

The only nonvanishing component of the curvature tensor is

$$R_{ab\bar{c}}^d = -\Gamma_{ab,\bar{c}}^d \tag{11.4.24}$$

and other curvature components related by symmetry and complex conjugation. The contracted Ricci tensor becomes

$$R_{a\bar{b}} = -\frac{\partial^2 (\ln \det g)}{\partial z^a \partial \bar{z}^{\bar{b}}} \tag{11.4.25}$$

The two-form

$$R = R_{a\bar{b}}\, dz^a\, d\bar{z}^{\bar{b}} \tag{11.4.26}$$

is closed: $dR = 0$.

(7) As a direct consequence, we can show that Kähler manifolds have $U(N)$ holonomy. In fact, this can be used as an alternative definition of Kähler manifolds. Notice that the holonomy group is the rotation group generated by making closed paths around the manifold, which is a function of

$$[D_a, D_b] \sim R_{ab}^{ij} M_{ij} \tag{11.4.27}$$

where $M_{ij}$ is the rotation matrix for some $2N$-dimensional tangent space group. The coefficients of this rotation, we see, are functions of the antisymmetric Riemann tensor $R_{abcd}$, which can now be viewed as an element in a $2N \times 2N$ antisymmetric rotation matrix. Thus, in general, we have $SO(2N)$ holonomy for a general manifold. However, if the manifold is Kähler, we can show that the restriction on the curvature tensor we found earlier reduces the $SO(N)$ rotation matrix down to $U(N)$, which is a subgroup of $SO(2N)$. The restriction that (11.4.24) is the only nonzero component of the curvature tensor breaks $SO(2N)$ symmetry.

(8) Kähler manifolds, which have $U(N)$ holonomy, can be further restricted if we demand that they have vanishing first Chern class. In this case, $U(N)$ holonomy reduces to $SU(N)$ holonomy. In fact, we have the theorem of Calabi–Yau, which works for all $N$, that a *Kähler manifold of vanishing first Chern class always admits a Kähler metric of* $SU(N)$ *holonomy*. In fact, it can be shown that the vanishing of the first Chern class is equivalent to having a Ricci-flat metric. Thus, we will use these two concepts interchangeably.

It will now be instructive to consider explicit examples of Kahler manifolds.

## (1) Riemann Surface

Any oriented Riemann surface admits a Kähler metric. Because the line element of any Riemann surface can be put in the form:

$$ds^2 = e^\sigma \, dz \, d\bar{z} \tag{11.4.28}$$

then it is Kähler because we can always find a Kähler potential $\phi$ such that:

$$\frac{\partial^2}{\partial z \partial \bar{z}} \phi = e^\sigma \tag{11.4.29}$$

Riemann surfaces are trivially Kähler because any two form on a Riemann surface, including the Kähler form, is closed.

## (2) Complex N Space

Complex N space $C_N$ is trivially Kähler because its standard line element:

$$ds^2 = \sum_i |dz^i|^2 \tag{11.4.30}$$

can always be put into Kähler form.

## (3) Sphere

We note that $S_N$, in general, does not admit a Kähler metric. This is because the even Betti numbers $b_{2p}$ for p equal to a non-zero integer are all usually equal to zero. Hence, they cannot be greater than or equal to one, which was one of the conditions on a Kähler manifold. The exception is the two- sphere $S_2$, which has no even (non-zero) Betti numbers. To show that the two sphere admits a Kähler metric, notice that a two-sphere has the line element:

$$ds^2 = \frac{dx^2 + dy^2}{(1 + x^2 + y^2)^2} \tag{11.4.31}$$

From this, we can write down the Kahler form:

$$\Omega = \tfrac{1}{2} i \frac{dz \wedge d\bar{z}}{(1 + z\bar{z})^2} \tag{11.4.32}$$

if we put it into complex form. This Kähler form is exact, and hence the space admits a Kähler metric. Although $S_2$ is Kähler, one can show that $c_1 \neq 0$, therefore it is not Ricci flat.

We notice that $S_p \times S_q$ is a complex manifold if $p$ and $q$ are odd integers. However, this does not mean that they are Kähler. In fact, $S_3 \times S_3$ and $S_1 \times S_5$ are both complex manifolds, but they are not Kähler. However, $S_3 \times S_3 \times S_3$ is Kähler (but is not Ricci flat).

## (4) Complex Projective $N$-space

To prove that $CP_N$ is Kähler, we note that it is possible to write the metric as a Fubini–Study metric on $CP_N$:

$$\Omega = \tfrac{1}{2} i\partial\bar{\partial} \ln\left(1 + \sum_{i=1}^{N} z^i \bar{z}^i\right) \tag{11.4.33}$$

## (5) Complex Submanifolds of $CP_N$

It is trivial to see that complex submanifolds of $CP_N$ are also Kähler. In fact, we use precisely the metric defined for $CP_N$ for the submanifold, except we must be careful to take only the components of the metric tensor that are tangent to the submanifold. Since the original metric was Kähler, the metric on the submanifold (which is the same metric) must also be Kähler.

## (6) Torus

The two-torus $T_2$ actually has vanishing first Chern class $c_1 = 0$. One can also show that the four-dimensional torus $T_4$ is Kähler. Finally, one can show that the six-torus $T_6$ is both Kähler and Ricci flat. Thus, compactification on the six-torus appears to have the desirable feature that $N = 1$ supersymmetry is preserved. However, the drawback of the six-torus is that too much symmetry is preserved. In fact, $N = 4$ supersymmetry is preserved on $T_6$, making it unacceptable from the point of view of phenomenology.

Let us now summarize some of these results in a table:

| Manifold | Complex | Kähler | R-Kähler | $\chi$ |
|---|---|---|---|---|
| $S_2$ | Y | Y | N | 2 |
| $S_{2n-1}$ | N | N | N | 0 |
| $S_{2n+2}$ | N | N | N | 2 |
| $(S_3)^2$ | Y | N | N | 0 |
| $S_1 \times S_5$ | Y | N | N | 0 |
| $(S_2)^3$ | Y | Y | N | 8 |
| $S_1 \times S_{2n+1}$ | Y | Y | Y | 0 |
| $T_2$ | Y | Y | Y | 0 |
| $T_6$ | Y | Y | Y | 0 |
| $CP_3$ | Y | Y | N | 4 |
| $CP_M$ | Y | Y | — | $M + 1$ |

where R-Kähler means Ricci flat and Kähler, Y (N) means yes (no), $\chi$ equals the Euler number, and $n = 1, 2, 3 \ldots$

In addition, the condition of being Ricci flat places even more restrictions on the Kähler manifold. For example, it can be shown that a Kähler manifold in three complex dimensions has $c_1 = 0$ if and only if there exists a covariantly constant nonvanishing holomorphic three-form $\omega$. This shows, for example, that $S_2 \times S_2 \times S_2$ does not admit a Ricci-flat Kähler metric. (We know that this manifold has $b_3 = 0$. Thus, by definition, there are no harmonic three-forms on this manifold. But this also means that there are no holomorphic three-forms, either. Thus, it cannot admit a Ricci-flat Kähler metric.)

Another simple consequence of this is that any harmonic $(p, 0)$ in three complex dimensions can be multiplied by $\omega$, thus creating a harmonic $(0, 3-p)$-form:

$$c_1 = 0 \rightarrow h^{p,0} = h^{0,3-p} \tag{11.4.34}$$

This identity is true because $\omega$ is covariantly constant and hence acts like a constant under the Laplacian. Thus, a harmonic form remains a harmonic form after multiplication by $\omega$. Also, a $(p, 0)$-form becomes a $(0, 3-p)$-form because we are contracting with the Hermitian metric tensor $g_{a\bar{b}}$ and not $g_{ab}$.

*This, in turn, allows us to eliminate almost all the various Hodge coefficients.* By using the various reflection symmetries and the previous symmetry, we can show that only $h^{1,1}$, $h^{2,1}$, $h^{1,0}$ survive as independent components of a manifold with SU(3) holonomy. Of these, we can also eliminate $h^{1,0}$ for the following reason. We know that the Laplacian can always be expanded out as

$$\Delta_d = -\nabla^2 + \text{curvature terms} \tag{11.4.35}$$

Acting on a one-form, the various curvature tensors reduce to the Ricci tensor. But the Ricci tensor is zero on a Ricci-flat manifold. Thus, a harmonic one-form must also be covariantly constant. This means that the Betti number of harmonic one-forms must be zero: $b_1 = 0$. But it also means that $h^{1,0} = 0$ by (11.4.15). Putting everything together, we can now show, for a Ricci-flat manifold,

$$\chi = \sum_{p,q=n} (-1)^{p+q} h^{p,q} = 2(h^{1,1} - h^{2,1}) \tag{11.4.36}$$

This last identity for a Ricci-flat metric will become extremely important when we discuss the generation problem. It turns out that $h^{1,1}$ and $h^{2,1}$ are related to the number of positive- and negative-chirality fermions one can define on the manifold, so that the above equation states that the generation number is half the Euler number.

$$\text{Generation number} = \tfrac{1}{2}|\chi| \tag{11.4.37}$$

Thus, we have a topological derivation of the generation number!

## §11.5. Embedding the Spin Connection

Armed with some of these elementary results from algebraic topology, let us now return to string phenomenology and apply these results.

Earlier, we saw that the condition of $N = 1$ supersymmetry implied the existence of a covariantly constant spinor, which in turn implied that the six-dimensional manifold $K$ was Kähler, Ricci flat, and had SU(3) holonomy:

$$N = 1 \text{ Supersymmetry} \to \text{Covariantly constant spinor}$$
$$\to \text{Kähler, Ricci flat, SU(3) holonomy}$$

Now we wish to exploit the remaining condition, which is the Bianchi identity (11.1.9):

$$\operatorname{Tr} R \wedge R = \tfrac{1}{30} \operatorname{Tr} F \wedge F \tag{11.5.1}$$

This is actually a strange identity, because we have the Riemann tensor on the left and the Yang–Mills tensor on the right. The equation used to be an exact identity, devoid of any content, but in the presence of our assumptions:

$$d\phi = H = 0 \tag{11.5.2}$$

we find that the equations are actually quite difficult to satisfy. Thus, the preservation of this identity is nontrivial because the system is over-constrained.

One attractive way to satisfy this odd identity is to *set part of the* $E_8 \otimes E_8$ *gauge fields equal to the Riemann spin connections*, which have SU(3) holonomy. This produces a nontrivial link between the spin connection and the Yang–Mills gauge field. We can perform this embedding as follows on the gauge field:

$$A = \begin{pmatrix} 0 & 0 \\ 0 & \omega \end{pmatrix} \tag{11.5.3}$$

where $\omega$ is the spin connection, which occupies part of the gauge field matrix. Thus, to embed the spin connection field into the Yang–Mills gauge field, we need to find a subgroup of the gauge group that contains SU(3). This means, of course, that we are breaking the original gauge symmetry of the Yang–Mills field. The simplest decomposition is

$$E_8 \otimes E_8 \supset \mathrm{SU}(3) \otimes E_6 \otimes E_8 \tag{11.5.4}$$

We must check, however, that the Clebsch–Gordon coefficients are such that the Bianchi identity is satisfied exactly. In particular, we must show that we can obtain the factor of $\frac{1}{30}$ in the Bianchi identity (11.1.9).

We know that the Yang–Mills gauge fields are in the adjoint representation of $E_8$, which has 248 elements. We must now find a breakdown of these 248 elements into representations of $\mathrm{SU}(3) \otimes E_6$, which is always possible to do. We find

$$\mathbf{248} = (\mathbf{3}, \mathbf{27}) \oplus (\overline{\mathbf{3}}, \overline{\mathbf{27}}) \oplus (\mathbf{8}, \mathbf{1}) \oplus (\mathbf{1}, \mathbf{78}) \tag{11.5.5}$$

To see if (11.5.1) is satisfied, let us convert from the adjoint representation of SU(3) to the $\mathbf{3} \oplus \overline{\mathbf{3}}$. If $\lambda$ is a generator of SU(3), then we wish to find the relationship between the trace of this matrix squared in the $\mathbf{8}$ representation and the $\mathbf{3} \oplus \overline{\mathbf{3}}$ representation. The answer is

$$\mathrm{Tr}(\lambda_8^a)^2 = 3\ \mathrm{Tr}(\lambda_{3\oplus\bar{3}}^a)^2 \tag{11.5.6}$$

Let us now focus on the SU(3) content of (11.5.5). Notice that we have 27 sets of fields transforming in the $\mathbf{3} \oplus \overline{\mathbf{3}}$ representation of SU(3), and also 1 set of octets. But the sum of the trace of the squares of the octets, as we saw in (11.5.6), must be multiplied by 3 when we convert to the $\mathbf{3} \oplus \overline{\mathbf{3}}$ representation of the SU(3) matrices. Thus, the total redundancy in the $\mathbf{3} \oplus \overline{\mathbf{3}}$ representation is equal to

$$27 + 3 \times 1 = 30$$

for a total factor of 30. Because of spin embedding, where the curvature tensor is defined even the same space as the Yang–Mills tensor, we can now satisfy (11.5.1) because the missing factor of 30 emerges when we count the specific representations of SU(3) in (11.5.5).

This breakdown to SU(3) $\otimes$ $E_6$ is a good one for phenomenological reasons, because $E_6$ has been extensively explored for GUT model building.

The original $E_8$ group, by contrast, does not have complex representations, which are needed to describe chiral fermions, but $E_6$ does. The **27** multiplet, in fact, is precisely the favored multiplet for fermions for model building with $E_6$. The group $E_6$ is also good from the point of view of low-energy supersymmetry, because the **27** of the fermions can form a supersymmetric multiplet with the **27** of the Higgs.

In summary, we have the phenomenologically acceptable conclusion

$$\mathrm{Tr}\, R \wedge R = \tfrac{1}{30}\, \mathrm{Tr}\, F \wedge F \to \text{Spin embedding} \to \mathbf{27}\ \text{fermions} \tag{11.5.7}$$

Because the fermions are now in the **27** representation, we can ask the question: How many generations are there? The GUT theory, as we saw, was plagued with the problem of generations. There was no reason for assuming more than one generation.

In the superstring picture, the situation is precisely the opposite. We will now show that we get too many generations!

## §11.6. Fermion Generations

One of the most powerful applications of algebraic topology to the phenomenology of string theory is the calculation of the generation number purely from topological considerations.

To calculate the number of generations predicted by the theory, we must first calculate the number of massless particles. The 10-dimensional Klein–Gordon operator for the particles in question becomes, after compactification, the sum of two Klein–Gordon operators in 4 and 6 dimensions:

$$\Box_{10}\psi = (\Box_4 + \Box_6)\psi = 0 \tag{11.6.1}$$

In general, $\Box_6$ will have eigenvalues denoted by $m^2$, that is, $\Box_6\psi_m = m^2\psi_m$, so that our wave equation becomes

$$(\Box_4 + m^2)\psi_m = 0 \tag{11.6.2}$$

We are interested in the massless sector in four dimensions, so we want to keep only the zero eigenvalues of the $\Box_6$ operator. Thus,

$$\Box_4\psi = \Box_6\psi = 0 \tag{11.6.3}$$

This is an important equation because it has two interpretations. First, it means, of course, that the four-dimensional fermions are massless. Second, it also means that $\psi$ is a harmonic form in six dimensions. Therefore the number of massless modes in 4D will be related to the number of harmonic forms that one can write for the 6D manifold. We saw earlier in (11.3.18) that the number of harmonic forms of degree $p$ is equal to the $p$th Betti number. *Thus, topological arguments alone should give us the number of generations*! In summary:

$$m^2 = 0 \to \begin{cases} \psi \text{ is harmonic in six dimensions} \\ \psi \text{ is massless in four dimensions} \end{cases}$$

We expect that the number of generations is a topological number because of the Dirac index theorem. We know that the solutions of the Dirac equation can, in general, have zero modes:

$$i\not{D}\psi = 0 \tag{11.6.4}$$

In fact, the index of this operator is equal to the difference between the positive and negative chiralities of the zero modes:

$$\text{Index}(\not{D}) = n_+ - n_- \tag{11.6.5}$$

The Dirac index is a topological quantity defined on a spin manifold, so we expect that it can be related to the characteristic classes we found earlier. In fact, when discussing SU(3) holonomy, we will find that

$$\text{Index}(\not{D}) = \tfrac{1}{2}|\chi(M)| \tag{11.6.6}$$

But the Dirac index is also equal to the generation number, because we will be considering only fermions of one specific chirality. Thus, the precise relation between the generation number and the Dirac index, or the Euler number, is

$$\text{Generation number} = \tfrac{1}{2}|\chi(M)| \tag{11.6.7}$$

To see this, consider a manifold of SU(3) holonomy, with Betti (Hodge) numbers that have two indices $p$, $q$. The Euler number can be written as

$$\chi(M) = \sum_{p,q}^{3} (-1)^{p+q} h_{p,q} \tag{11.6.8}$$

The multiplicities of the supergravity and Yang–Mills multiplet can be determined by calculating the number of harmonic forms one can write down, which in turn is related to the Betti numbers. We find, if we compare the helicity of supersymmetric pairs with their multiplicity:

$$\begin{array}{ll} (2, \frac{3}{2}) & h_{0,0} \\ (\frac{3}{2}, 1) & h_{0,1} \\ (1, \frac{1}{2}) & (2h_{1,0} + h_{0,1}) \\ (\frac{1}{2}, 0) & (h_{0,0} + h_{1,1} + h_{2,1}) \end{array} \tag{11.6.9}$$

If we analyze the spin-$\frac{1}{2}$ fermion sector, given by $(\frac{1}{2}, 0)$, we find that their multiplicity number equals

$$\text{Fermion multiplicity} = h_{0,0} + h_{1,1} + h_{2,1} \tag{11.6.10}$$

This multiplicity, however, is too large. We want only the subset of this figure corresponding to the **27** and $\overline{\mathbf{27}}$ of the spin-$\frac{1}{2}$ fermions.

Before, in (11.5.5), we found that the **248** of $E_8 \otimes E_8$ can be decomposed into (**27**, **3**) and ($\overline{\mathbf{27}}$, $\overline{\mathbf{3}}$). It can be shown that the multiplicity associated with each of these two representations is equal to

$$\begin{array}{ll} (\mathbf{27}, \mathbf{3}) & h_{2,1} \\ (\overline{\mathbf{27}}, \overline{\mathbf{3}}) & h_{1,1} \end{array} \tag{11.6.11}$$

Thus, the generation number is equal to

$$\#(\mathbf{27}) - \#(\overline{\mathbf{27}}) = h_{2,1} - h_{1,1} \tag{11.6.12}$$

where # represents the multiplicity of the **27** representation. But this number, in turn, is precisely half the absolute value of the Euler number, as we saw in (11.4.36).

The relation between the generation number and the Euler number is rather surprising because there is no reason to believe that any relation should exist between the two. The generation number is a function of the Yang–Mills gauge group $E_8 \otimes E_8$, while the Euler number is a function of the manifold $K_6$. Naively, we do not expect the two to be correlated. But the relationship between the two is established because we performed spin embedding, which breaks the gauge group and produces an intimate relationship between the group manifold $K_6$ and the gauge group. Thus, the essence of this important result is a direct result of compactification and spin embedding.

Next, we would like to calculate the Euler number, and hence the number of generations, for a series of Calabi–Yau metrics.

We showed earlier that a submanifold of the Kähler metric on $CP_N$, because it has the same Hermitian metric $g_{i\bar{j}}$, is also Kähler. Thus, we want to consider the set of submanifolds of $CP_N$ that have vanishing first Chern class $c_1 = 0$. This can be accomplished by placing constraints on the $z$'s by setting certain polynomials to zero.

Consider the submanifold of $CP_4$ obtained by the constraint

$$\sum_{n=1}^{5} z_i^5 = 0 \tag{11.6.13}$$

It can be shown that this manifold has vanishing first Chern class and that $\chi(M)$ is equal to $-200$.

Let

$$Y_{(N;d_1,d_2,\ldots,d_k)} \tag{11.6.14}$$

represent the submanifold obtained by setting $k$ homogeneous polynomials of degrees $d_i$ to zero within $CP_N$. Fortunately, the formula for the total Chern class, not just the first Chern class, of these submanifolds is known. For SU(3) holonomy, the total Chern class equals

$$c = \frac{(1+J)^{k+4}}{\prod_{i=1}^{k}(1+d_i J)} \tag{11.6.15}$$

where $J$ is a certain two-form obtained by normalizing the Kähler form of $CP_{k+3}$. By expanding this formula and then setting the first Chern class $c_1$ equal to zero, we find

$$\sum_{i=1}^{k} d_i = k + 4 \tag{11.6.16}$$

Thus, we find only five possibilities with vanishing first Chern class:

$$\begin{aligned} \chi(Y_{(4;5)}) &= -200 \\ \chi(Y_{(5;2,4)}) &= -176 \\ \chi(Y_{(5;3,3)}) &= -144 \\ \chi(Y_{(6;3,2,2)}) &= -144 \\ \chi(Y_{(7;2,2,2,2)}) &= -128 \end{aligned} \tag{11.6.17}$$

Thus, the generation numbers are unacceptably large! This is phenomenologically undesirable because we know from arguments from nucleosynthesis, cosmology, and asymptotic freedom in QCD that we want very few generations, such as three or four.

Fortunately, we can still reduce the generation number and make it much smaller by considering *nonsimply connected manifolds.* Let us divide the original manifold $M_0$ by a discrete symmetry group $G$ that acts freely on the manifold (i.e., no fixed points), yielding a manifold $M$. If the number of discrete generators of $G$ is $n(G)$, then the Euler number of the original manifold divided by the discrete group $G$ is

$$\chi(M) = \frac{\chi(M_0)}{n(G)} \tag{11.6.18}$$

where

$$M = \frac{M_0}{G} \tag{11.6.19}$$

Let us consider the previous example of the submanifold of $CP_4$ with

$$\sum_{i=1}^{5} z_i^5 = 0 \tag{11.6.20}$$

Notice that this polynomial is invariant under the following symmetry operations:

$$\begin{aligned} (z_1, z_2, \ldots, z_5) &\rightarrow (z_5, z_1, z_2, \ldots, z_4) \\ (z_1, z_2, \ldots, z_5) &\rightarrow (z_1, \alpha z_2, \alpha^2 z_3, \ldots, \alpha^4 z_5) \end{aligned} \tag{11.6.21}$$

where $\alpha$ is a fifth root of unity. The discrete symmetry group is

$$Z_5 \times Z_5 \tag{11.6.22}$$

which has 25 generators. Thus, the Euler number of this new manifold is

$$\chi(M) = \frac{-200}{25} = -8 \tag{11.6.23}$$

which predicts four generations.

Many other types of models are possible with reasonably "low" generation number:

$$\begin{aligned} \chi\left(\frac{Y_{(5;3,3)}}{Z_3 \times Z_3}\right) &= -16 \\ \chi\left(\frac{Y_{(7;2,2,2,2)}}{Z_2 \times Z_2 \times Z_2}\right) &= -16 \\ \chi\left(\frac{Y_{(7;2,2,2,2)}}{X_8}\right) &= -16 \\ \chi\left(\frac{Y_{(4;5)}}{Z_5}\right) &= -40 \end{aligned} \tag{11.6.24}$$

The point here is not that we have obtained the correct phenomenology. The point is that with a clever choice of a discrete group, it is reasonable that we might be able to write down a model with an acceptably low number of generations. We must stress that the very existence of chiral fermions in superstring theory is something of a miracle. In standard Kaluza–Klein theories, for example, there are serious obstructions to constructing a theory with chiral fermions in four dimensions. In fact, an acceptable supersymmetric Kaluza–Klein model with chiral fermions does not exist. Thus, it is quite remarkable that we can obtain any chiral fermions at all with superstrings.

Lastly, the choice of a nonsimply connected manifold may at first be surprising, but it turns out to have other phenomenologically acceptable features.

## §11.7. Wilson Lines

Our previous discussion of embedding the spin connection into the gauge field in (11.5.3) broke the group $E_8$ down to $E_6 \otimes SU(3)$, which has exceptionally good phenomenology because the fermions are in the **27** representation of $E_6$. Our next step is to break $E_6$ further to obtain the minimal group $SU(3) \otimes SU(2) \otimes U(1)$.

The problem here is that most naive methods of breaking $E_6$ down further will necessarily break $N = 1$ supersymmetry as well. However, there is one trick that we can still use on manifolds that are not simply connected.

In general, we know that the path-ordered product of elements of a gauge group around a loop is gauge invariant. We can express this as the Wilson loop:

$$U \sim P \exp \int_C A_\mu \, dx^\mu \tag{11.7.1}$$

where $P$ represents the ordering of each term with respect to the closed path $C$. For small paths, we find that $U$ is proportional to the exponential of the curvature two-form.

Normally, when the curvature $F^a_{\mu\nu}$ vanishes, we expect that the Wilson loop becomes unity. This is because we can always shrink the closed path $C$ to a point. If the area tensor of the small closed path is given by $\Delta^{\mu\nu}$, then the Wilson loop becomes

$$U \to e^{F_{\mu\nu}\Delta^{\mu\nu}} \tag{11.7.2}$$

However, if the path is not simply connected, this argument fails. Thus, we can have vanishing curvature tensors, yet the Wilson loop does not have to be unity [1, 7].

This is ideal for our case, because we are now considering nonsimply connected manifolds. *If $U$ does not equal* 1, *then the group $E_6$ breaks down to the subgroup that commutes with $U$.*

Notice that $E_6$ contains a maximal subgroup:

$$SU(3)_C \otimes SU(3)_L \otimes SU(3)_R \tag{11.7.3}$$

where $C$ is the strong color group and $L$ and $R$ represent left and right weak interactions.

This breaking can be accomplished by choosing one element $U_0$ of $E_6$ that satisfies

$$U_0^n = 1 \tag{11.7.4}$$

This element generates the permutation group $Z_n$. We now want the subgroup of $E_6$ that commutes with $U_0$. In general, this will be a group of rank six. We can, for example, choose a specific element of $E_6$ to be

$$U_0 = (\alpha) \times \begin{pmatrix} \beta & 0 & 0 \\ 0 & \beta & 0 \\ 0 & 0 & \beta^{-2} \end{pmatrix} \times \begin{pmatrix} \gamma & 0 & 0 \\ 0 & \delta & 0 \\ 0 & 0 & \varepsilon \end{pmatrix} \tag{11.7.5}$$

where the matrices represent the elements of the three SU(3)'s within $E_6$, and $\alpha$, $\beta$, $\gamma$, $\delta$ are all $n$th roots of unity. We then have the following breakings for various choices of these elements:

$$\begin{aligned} &\alpha^3 = 1; \qquad \gamma\delta\varepsilon = 1 \to \mathrm{SU}(3)_C \otimes \mathrm{SU}(2)_L \otimes \mathrm{U}(1) \otimes \mathrm{U}(1) \otimes \mathrm{U}(1) \\ &\gamma = \delta \to \mathrm{SU}(3)_C \otimes \mathrm{SU}(2)_L \otimes \mathrm{SU}(2)_R \otimes \mathrm{U}(1) \otimes \mathrm{U}(1) \end{aligned} \tag{11.7.6}$$

By selecting out different elements of $E_6$, we can get different groups. For example,

$$\frac{E_6}{Z_5 \times Z_5} = \mathrm{SU}(3)_C \otimes \mathrm{SU}(2)_L \otimes \mathrm{SU}(2)_R \otimes \mathrm{U}(1) \otimes \mathrm{U}(1) \tag{11.7.7}$$

The explicit form of $U_0$ is

$$U_0 = (1) \times \begin{pmatrix} \alpha^j & 0 & 0 \\ 0 & \alpha^j & 0 \\ 0 & 0 & \alpha^{-2j} \end{pmatrix} \begin{pmatrix} \beta^k & 0 & 0 \\ 0 & \beta^k & 0 \\ 0 & 0 & \beta^{-2k} \end{pmatrix} \tag{11.7.8}$$

Yet another choice is to fix the discrete group to be $Z_3$, in which case

$$\frac{E_6}{Z_3} = \mathrm{SU}(3)_C \otimes \mathrm{SU}(3) \otimes \mathrm{SU}(3) \tag{11.7.9}$$

In general, these solutions that contain the minimal model also contain other gauge interactions. In fact, there are 27 subgroups of $E_6$ that can yield at least the minimal model, and all of these have unwanted U(1) gauge groups that survive the breaking. These, of course, must be further eliminated, probably through another mechanism such as exploiting any "flat" directions in the superpotential.

## §11.8. Orbifolds

Although there are still problems with the Calabi–Yau compactification method, it is rich enough, in principle, to provide qualitatively the mechanisms for breaking the gauge group down to the minimal theory.

In practice, however, Calabi–Yau manifolds are quite difficult to construct and only a few of them are actually known. We would like to have simpler, flat-space solutions to investigate. Unfortunately, the simplest toroidal compactification is unacceptable phenomenologically for, among other reasons, $N = 4$ supersymmetry survives after breaking. If we start with $N = 1$

supersymmetry in 10 dimensions and compactify down to 4 dimensions, we wind up with $N = 4$ supersymmetry. (If we start with $N = 1$ symmetry in 10 dimensions, the 16 supersymmetry generators $Q^\alpha$ after compactification become $Q^{i\beta}$, where $\beta$ is a spinor in four space–time dimensions and $i$ ranges from 1 to 4. Thus, compactification of $N = 1$ supersymmetry down to a smaller number of dimensions always leads to an extra $O(n)$ symmetry.)

However, by constraining the toroidal compactification in a much more rigid fashion, we should be able to reduce $N = 4$ supersymmetry to $N = 1$ supersymmetry. The proposal of Dixon, Harvey, Vafa, and Witten was to compactify on an *orbifold* [8, 9] obtained by taking a manifold and dividing out by a discrete group, such that there are *fixed points* (i.e., points that do not change under the transformation). An orbifold, because of the singularities at these fixed points, is not a manifold, but apparently strings can propagate on these orbifolds without difficulty. The advantage of orbifolds is that they are flat, can break $N = 4$ supersymmetry, can produce chiral fermions, and are easy to generate.

The simplest orbifold is a cone. Simply take the complex plane and make the following identification:

$$z = e^{2\pi i/n} z \tag{11.8.1}$$

for some integer $n$. This divides the complex plane into $n$ equivalent triangular sectors. Notice that this identification divides up the plane with the discrete symmetry group $Z_n$.

Notice that the origin is a fixed point under this transformation; i.e., the origin maps into the origin under this rotation. Now if we were to slice up the complex plane, extract just one of these triangular sectors, and then wrap up this sector according to the above identification, we arrive at a cone. Thus, the cone or orbifold is nothing but a two-dimensional space divided by the action of the discrete group:

$$\text{cone} = \text{orbifold} = \frac{R^2}{Z_n} \tag{11.8.2}$$

The origin, which is a fixed point, now becomes a potential singularity. Thus, a cone is not a manifold. If we follow a path around the origin, the total angle we traverse is not 360 degrees, but 360 degrees divided by $n$.

Let us take another simple example. Let us start with the two-torus defined by making the identification:

$$\begin{aligned} z &= z + 1 \\ z &= z + i \end{aligned} \tag{11.8.3}$$

This divides the complex plane into an infinite number of squares of width 1 whose edges are identified with each other. Now let us create an orbifold out of this two-torus by dividing by $Z_2$, which is generated by reflections:

$$P(z) = -z \tag{11.8.4}$$

If we now construct the surface

$$\frac{T_2}{P} \tag{11.8.5}$$

we have an orbifold. Notice that the reflections leave four points invariant, which are the fixed points:

$$0; \quad \tfrac{1}{2}; \quad \tfrac{1}{2}i; \quad \tfrac{1}{2}(1+i) \tag{11.8.6}$$

Three of these fixed points lie on the edge of the unit square, while one lies within the square.

What does this orbifold look like? Under the action of the reflections, the points within the square are further identified with each other, making the square divide up into smaller squares of width $\frac{1}{2}$. Now imagine two smaller squares, each of width $\frac{1}{2}$. Place them directly on top of each other. Now sew the four edges of these two squares together, forming a closed surface. Topologically, it is the same as the surface of a square beanbag.

Notice that this surface has four singularities, corresponding to the four vertices of the square. If we followed a path around each of these fixed points, the angular deficit would be 180 degrees.

It is possible, however, to convert this orbifold back into a normal manifold and hence calculate its Euler number. Let us cut off each of the four fixed points of $T_2$, divide by $Z_2$, and then resew back a small patch at each of these fixed points. This is called "blowing up" a singularity. In this case, by cutting out these holes, sewing the squares together to form the orbifold, and sewing back four patches, we obtain a manifold topologically equivalent to a sphere.

The Euler number of a disk $d$ is equal to 1 and the Euler number of the two-torus is 0. Therefore, the Euler number of the torus minus four fixed points is equal to

$$\chi(T_2 - 4d) = \chi(T_2) - 4\chi(d) = -4 \tag{11.8.7}$$

If we divide by the action of $Z_2$, then the Euler number of the resulting surface is

$$\chi\left(\frac{T_2 - 4d}{Z_2}\right) = -2 \tag{11.8.8}$$

Finally, by gluing four disks back onto the surface, we must add 4 back to the Euler number:

$$\chi\left(\frac{T_2 - 4d}{Z_2} + 4d\right) = \frac{0-4}{2} + 4 = 2 \tag{11.8.9}$$

This checks with our intuition, because this manifold is equivalent to the two-sphere $S_2$, which has Euler number 2.

Now, we would like to generalize the previous example by taking a more complicated compactified surface in six dimensions. Let us first compactify the complex plane by making the following identification:

$$\begin{cases} z = z + 1 \\ z = z + e^{\pi i/3} \end{cases} \tag{11.8.10}$$

The first transformation divides the complex plane into an infinite number of narrow vertical strips. The action of both transformations divides the complex plane into an infinite number of equilateral triangles. The "fundamental region" $T$ of this space consists of two of these equilateral triangles, back to back.

This space, because it consists of an infinite number of equilateral triangles, is invariant under rotations by 120 degrees:

$$z \to e^{2\pi i/3} z \tag{11.8.11}$$

Thus, this space has $Z_3$ symmetry. Under this rotation by 120 degrees, there are three fixed points, or points that are left invariant:

$$z = 0; \quad \frac{e^{i\pi/6}}{\sqrt{3}}; \quad \frac{2e^{2\pi i/6}}{\sqrt{3}} \tag{11.8.12}$$

Notice that the fundamental region $T$, which consists of two equilateral triangles, contains three fixed points, one at the origin and the other two within the two equilateral triangles.

Now let us perform the folding operation on this space $T$ and obtain an orbifold:

$$Z = \frac{T}{Z_3} \tag{11.8.13}$$

Again, $Z$ is not a manifold because the three fixed points (two of which occur within the region $T$ itself) are potentially singular.

To generalize to the six-dimensional case, we might consider simply taking the trivial product of three of these complex spaces:

$$\bar{Z} = Z \times Z \times Z \tag{11.8.14}$$

which has $3 \times 3 \times 3 = 27$ fixed points.

This space has the great advantage that the $N = 1$ supersymmetry in 10 dimensions can be broken down to $N = 1$ supersymmetry in 4 dimensions. If we compactify on the space:

$$M^4 \times \bar{Z} \tag{11.8.15}$$

we note that the O(10) group is broken down as follows:

$$\mathrm{SO}(10) \supset \mathrm{SO}(4) \otimes \mathrm{SO}(6) \tag{11.8.16}$$

Since SO(6) = SU(4), we can show that the original $Q^\alpha$ of $N = 1$, $D = 10$

supersymmetry is now broken down into four spinors transforming as the components of 4 under SU(4). If we compactify naively on $T_6$, then SU(4) is not broken and each of the four supersymmetry generators in four dimensions transforming as 4 survives. However, if we start with an orbifold by dividing by $Z_3$, we can associate the $Z_3$ as belonging to the SU(3) subgroup of SU(4). Thus, dividing by $Z_3$ necessarily breaks SU(4) symmetry, since there are no three-dimensional representations of SU(4). But only one of the four components within 4 survives the division by $Z_3$, and hence only $N = 1$ supersymmetry survives, which is fortunate.

In summary, orbifolds can be used to break the overall symmetry of the theory because only symmetries that commute with the discrete group survive. This method can be used to break the gauge group as well.

Let $g$ be a specific element of the 10-dimensional gauge group such that

$$g^m = 1 \tag{11.8.17}$$

for some integer $m$. Then we demand that the states of the theory are those that commute with the combined action of

$$g \times Z_m \tag{11.8.18}$$

Thus, we have a mechanism for breaking both gauge and supersymmetry simultaneously.

As an example, we might take $g^3 = 1$, which is an element of SU(3). Thus, demanding that the states of the theory be invariant under the combined operation yields the breaking of $E_8$:

$$E_8 \otimes E_8 \to E_8 \otimes \mathrm{SU}(3) \otimes E_6 \tag{11.8.19}$$

Now let us calculate the Euler number on this surface and hence the number of generations. Once again, we must cut off the 27 fixed points from $T$, then divide by $Z_3$, and then sew back on 27 patches.

The Euler number of the surface $T$ is zero, so the manifold $T$ minus the 27 disks $d$ located at the fixed points has Euler number

$$\chi(T - d) = -27 \tag{11.8.20}$$

If we divide by $Z_3$, the Euler number becomes $-9$. Now we must sew back 27 disks $\bar{d}$. This time, however, each disk has SU(3) holonomy and Euler number 3, so the total Euler number is

$$\chi\left(\frac{T - d}{Z_3}\right) + 27\chi(\bar{d}) = 72 \tag{11.8.21}$$

Thus, we have 36 generations of fermions.

The previous example was only a toy model in six dimensions. It is possible, however, to construct orbifold models that have much fewer generations, as low as two or four.

## §11.9. Four-Dimensional Superstrings

In model building using orbifold compactification to produce four dimensions, stringent constraints must be met, such as modular invariance and the $L_0 = \bar{L}_0$ conditions. These conditions are nontrivial, because modular invariance mixes up the boundary conditions.

For example, when studying modular invariance, we can compactify on the six-torus $T_6$ divided by a discrete point group $P = Z_n$ such that the (twisted) boundary conditions for the orbifold $T_6/P$ are as follows:

$$\begin{aligned} X(\sigma_1 + 2\pi, \sigma_2) &= hX(\sigma_1, \sigma_2) \\ X(\sigma_1, \sigma_2 + 2\pi) &= gX(\sigma_1, \sigma_2) \end{aligned} \tag{11.9.1}$$

where $h$ and $g$ are elements of $P$ of order $n$, that is, $g^n = h^n = 1$. For the usual bosonic or fermionic boundary conditions, we have $g$ and $h$ equal to $\pm 1$. However, when compactifying on orbifolds, this condition must be generalized.

The presence of $g$ and $h$, of course, can break the overall symmetry of the theory, both for space–time and for the internal group. The group that survives the compactification is the subgroup that is unaffected by this process, i.e., the subgroup that commutes with $g$ and $h$.

To study how the one-loop trace is affected by the compactification, let us diagonalize $g$ and $h$ in terms of their eigenvalues, $g, h \to \{e^{2\pi i v_i}\}$, where $v_i = r_i/n$ so that they are of order $n$. For simplicity, we take the following periodicity condition:

$$X(\sigma + 2\pi) = e^{2\pi i v} X(\sigma) \tag{11.9.2}$$

Notice that the Fourier decomposition of the string modes is now shifted. In general, we must now expand the string in terms of a new set of modes:

$$X(\sigma) = \sum_m e^{i(m+v)\sigma} X_m \tag{11.9.3}$$

The presence of the $v$ within the Fourier modes changes the trace calculation of the one-loop amplitude in a subtle but important fashion, which will put nontrivial constraints on the $v_i$.

In the trace calculation, the presence of $v$ enters in the zero point energy. For the usual bosonic string, for example, the unregularized Hamiltonian contains the factor $\frac{1}{2}\sum_n a_{-n}a_n + a_n a_{-n}$. This, of course, has infinite matrix elements and must be normal ordered. Technically speaking, normal ordering the creation and annihilation operators creates an infinite zero point energy given by $\frac{1}{2}\sum_n n$. This infinite zero point energy can be handled in several ways (e.g., requiring the light cone theory to be Lorentz invariant), but it can be shown that this yields the same results as zeta function regularization. The zeta function is defined as

$$\zeta(s) = \sum_{n=1}^{\infty} \frac{1}{n^s} \tag{11.9.4}$$

This function is analytic in $s$, and we can analytically continue the $\zeta$ function to $s = -1$ and show that $\zeta(-1) = -\frac{1}{12}$. This is the desired answer, because $\zeta(-1)$ is the analytic continuation of $\sum_{n=1}^{\infty} n$.

Now use zeta function regularization to compute the contribution coming from the shifted modes. We use the fact that

$$\frac{1}{2} \sum_{n \in Z + a} n = \sum_{n \in Z} (n + a)^{-s}|_{s=-1} = -\frac{1}{24} + \frac{1}{4}a(1 - a) \tag{11.9.5}$$

The $a(1 - a)$ factor is the new contribution to the zero point energy in the trace calculation arising from the shifted modes of the orbifold.

Now consider the heterotic string, where the trace over the single loop yields a factor containing the left- and right-moving energies:

$$e^{2\pi i(\tau E_L - \bar{\tau} E_R)} \tag{11.9.6}$$

Under the transformation $\tau \to \tau + n$, the expression remains invariant if we set

$$n(E_L - E_R) = 0 \bmod 1 \tag{11.9.7}$$

If we label the eigenvalues for the right-moving sector as $e^{2\pi i v_i}$ and the two sets of left-moving eigenvalues as $e^{2\pi i v_{1i}}$ and $e^{2\pi i v_{2i}}$ for the $O(16) \otimes O(16)$ subgroup of $E_8 \otimes E_8$, then from (11.9.5) we have the following zero point contribution to the ground state energy:

$$\begin{aligned} E_R &= \frac{1}{2} \sum_i v_i(v_i - 1) + \frac{1}{12} \\ E_L &= \frac{1}{2} \sum_i v_{1i}(v_{1i} - 1) + \frac{1}{12} + (1 \leftrightarrow 2) \end{aligned} \tag{11.9.8}$$

Since $v = r/n$, we can put everything together until we have the following constraint:

$$\sum_i r_i^2 = \sum_j r_{1j}^2 + \sum_j r_{2j}^2 \tag{11.9.9}$$

which is true mod $n$ for odd $n$ and mod $2n$ for even $n$.

Now consider the effect of these constraints on the compactification discussed earlier with the $Z$-orbifold, i.e., $T_6/Z_3$, so that the point group $P$ is of order 3. This means that $3v_{1i}$ and $3v_{2i}$ can be set to be lattice vectors on $E_8$.

The constraints from modular invariance tell us that $v_{1,2i}^2$ should be $\frac{2}{3}$ times some integer. But since $3v_{1,2i}$ is a lattice vector of $E_8$, we can always fix $v^2$ to be $\frac{2}{9}$ times some integer. Finally, we know that any point in eight-dimensional space containing the $E_8$ lattice is within a distance of 1 from some lattice point. This means that we can always choose $v_{1,2i}^2 \leq 1$.

Modular invariance is so rigid that there are only five solutions consistent with the constraint (11.9.9) on the various $v$'s. Each of the five sets of $v$'s breaks the symmetry group $E_8 \otimes E_8$ down to the subgroup that commutes with the $g$ and $h$ twist factors. It is not hard, given the explicit form for the $v$'s, to calculate what this subgroup is. Because $3v$ is equal to a lattice vector, the subgroup that survives after symmetry breaking is the group that commutes with these lattice vectors. We simply quote the five solutions [8, 9]:

(1)

$$\sum_{i=1}^{8} v_{1i}^2 = \sum_{i=1}^{8} v_{2i}^2 = 0 \tag{11.9.10}$$

This solution has no chiral fermions and hence is unphysical.

(2)

$$\begin{aligned} \sum_{i=1}^{8} v_{1i}^2 &= \tfrac{2}{3} \\ \sum_{i=1}^{8} v_{2i}^2 &= 0 \\ v_{1i} &= (\tfrac{1}{3}, \tfrac{1}{3}, \tfrac{2}{3}, \ldots) \\ v_{2i} &= (0, \ldots, 0) \end{aligned} \tag{11.9.11}$$

This leaves us with the group $E_6 \otimes \mathrm{SU}(3) \otimes E_8$.

(3)

$$\begin{aligned} \sum_{i=1}^{8} v_{1i}^2 &= \tfrac{2}{9} \\ \sum_{i=1}^{8} v_{2i}^2 &= \tfrac{4}{9} \\ v_{1i} &= (\tfrac{1}{3}, \tfrac{1}{3}, \ldots) \\ v_{2i} &= (\tfrac{2}{3}, 0, \ldots) \end{aligned} \tag{11.9.12}$$

This leaves us with the group $E_7 \otimes \mathrm{U}(1) \otimes \mathrm{SO}(14) \otimes \mathrm{U}(1)$.

(4)

$$\begin{aligned} \sum_{i=1}^{8} v_{1i}^2 &= \sum_{i=1}^{8} v_{2i}^2 = \tfrac{2}{3} \\ v_{1i} &= v_{2i} = (\tfrac{1}{3}, \tfrac{1}{3}, \tfrac{2}{3}, \ldots) \end{aligned} \tag{11.9.13}$$

This yields $E_6 \otimes \mathrm{SU}(3) \otimes E_6 \otimes \mathrm{SU}(3)$.

(5)

$$\begin{aligned} \sum_{i=1}^{8} v_{1i}^2 &= \tfrac{8}{9} \\ \sum_{i=1}^{8} v_{2i}^2 &= \tfrac{4}{9} \\ v_{1i} &= (\tfrac{1}{3}, \tfrac{1}{3}, \tfrac{1}{3}, \tfrac{1}{3}, \tfrac{2}{3}, \ldots) \\ v_{2i} &= (\tfrac{2}{3}, 0, \ldots) \end{aligned} \tag{11.9.14}$$

This leaves the symmetry group $\mathrm{SU}(9) \otimes \mathrm{SO}(14) \otimes \mathrm{U}(1)$. (The symbol ... represents a series of zeros.)

The five groups we have constructed so far with orbifolds do not resemble the Standard Model at all. The gauge groups are still too large, and there are too many generations. Models of this type often have 27 generations because we can always construct string fields that are twisted around the 27 fixed points of the orbifold, yielding a redundancy of 27. However, we can further reduce the gauge group and control the number of generations by postulating the existence of background gauge fields (Wilson lines) that correspond to noncontractible loops on the torus. This can, in principle, yield models with only three generations and the gauge group $SU(3) \otimes SU(2) \otimes U(1)^n$. As in the previous discussion, the gauge group that survives will be the subgroup of $E_8 \otimes E_8$ that commutes with the Wilson line. In the case of orbifolds with Wilson lines, the gauge group that survives is the subgroup of $E_8 \otimes E_8$ that commutes with $P = Z_3$ as well as the Wilson line.

For example, define the Wilson integral parametrized by $a_i^I$:

$$\int_i A_\mu^I \, dx^\mu = 2\pi A_\mu^I e_i^\mu = 2\pi a_i^I \tag{11.9.15}$$

where $i = 1$–$6$ and $I = 1$–$16$ and $e_i^I$ define the lattice of the six-torus. Thus, we have now introduced a new vector $a_i^I$ in our constraints that will allow us to construct new solutions. We can show that $3a_i^I$ and $3v^I$ are equal to a lattice vector.

Under a modular transformation $\tau \to -1/\tau$, the $E_8 \otimes E_8$ lattice vectors transform as

$$p^I \to p^I + v^I + n_i a_i^I \tag{11.9.16}$$

where $n_i = 0, \pm 1$. Different values of $n_i$ will yield different twisted sectors. Repeating all the arguments we used earlier, we can also show that the modular transformation $\tau \to \tau + 3$ results in the constraint [10, 11]

$$3(v^I + n_i a_i^I)^2 = 2m \tag{11.9.17}$$

for some integer $m$. This is the desired constraint emerging from modular invariance, which is much less restrictive than (11.9.9).

An extraordinarily large number of solutions are now possible [10, 11] because of the presence of the $a_i^I$ term. However, let us select one such possibility that yields several desirable features, such as three generations. We postulate the following set of values for the $v_i^I$ and $a_i^I$:

$$v^I = \begin{cases} (\frac{1}{3}, \frac{1}{3}, \frac{1}{3}, \frac{1}{3}, \frac{2}{3}, 0, \frac{1}{3}, \frac{1}{3}) \\ (\frac{1}{3}, \frac{1}{3}, 0, 0, 0, \frac{1}{3}, \frac{1}{3}, \frac{2}{3}) \end{cases} \tag{11.9.18}$$

and

$$a_1^I = \begin{cases} (\frac{1}{3}, \frac{1}{3}, \frac{1}{3}, \frac{2}{3}, \frac{1}{3}, 0, 0, 0) \\ (0, 0, 0, 0, 0, \frac{2}{3}, 0, 0) \end{cases} \tag{11.9.19}$$

$$a_3^I = \begin{cases} (0, 0, 0, 0, 0, 0, 0, \frac{2}{3}) \\ (\frac{1}{3}, \frac{1}{3}, \frac{2}{3}, \frac{2}{3}, 0, 0, 0, \frac{1}{3}) \end{cases} \tag{11.9.20}$$

One can check that in the untwisted sector, the subgroup of $E_8$ that commutes with the action of the Wilson line is given by

$$[\mathrm{SU}(3) \otimes \mathrm{SU}(2) \otimes \mathrm{U}(1)^5] \tag{11.9.21}$$

while the other $E_8$ is broken down to $\mathrm{SU}(2) \otimes \mathrm{SU}(2) \otimes \mathrm{U}(1)^6$.

Furthermore, we should note that the number of generations has been reduced to three. (This is because the 27 fixed points can be further divided into three sectors of nine fixed points if there is a Wilson line in one of the three complex dimensions. Adding two Wilson lines creates nine sectors of three fixed points. A careful choice of the $v^I$ and $a_i^I$ will kill all but one sector, leaving us with three generations.) As a further bonus, this model has the desirable property that extra colored triplets, which might mediate fast proton decay, are absent.

The limitation of this model, however, is that there are too many U(1) factors, which is bad for phenomenology. In fact, the Wilson line technique does not reduce the rank of the group (because we are dividing by a discrete group), so we still have gauge groups of rank 8, which is too large. These extra U(1) factors, moreover, must be checked to see if they are anomalous.

So far, none of the solutions obtained with orbifolds [12–25] (or Calabi–Yau manifolds) has exactly the desired low-energy structure. There are still problems getting the right generation number, the right gauge group at low energies, acceptable values for proton decay, etc. The point, however, is that these methods of compactification gives us the ability to construct potentially thousands of solutions that may be compatible with modular invariance.

What is required, of course, is a systematic way to compute *all* four-dimensional physically relevant solutions to the string equations. So far, the bulk of the work has been to postulate a particular compactification scheme and then check for consistency with modular invariance. Recently, there has been considerable work in precisely the other direction, i.e., to starting with modular invariance from the beginning and then searching for all possible compactification schemes compatible with it.

This new program at the outset demands the following set of criteria for any physically relevant model; namely, the model must be (a) tachyon-free, (b) anomaly-free, (c) modular invariant, (d) supersymmetric, and (e) four-dimensional. So far, there are only preliminary results toward this ambitious program, but the early results have been encouraging.

What we want is a way to calculate all coefficients $C$ within the one-loop amplitude (5.9.6) for each spin structure. For example, in Sections 5.9 and 5.11 we analyzed the one-loop and multiloop spin structures and how they changed under modular transformations. Now we want a systematic study of the effect of modular invariance on the multiloop spin structures [26–31]. Let us begin with the expansion of the string amplitude in terms of all possible spin structures:

$$A = \sum_{\mathbf{a},\mathbf{b}} C\begin{bmatrix}\mathbf{a}\\\mathbf{b}\end{bmatrix} A\begin{bmatrix}\mathbf{a}\\\mathbf{b}\end{bmatrix} \tag{11.9.22}$$

where **a** and **b** label various spin structures over the $a$ and $b$ cycles, each element within **a** is equal to either 0 or 1, $C$ is a coefficient that must be determined, and the $A$'s represent the amplitude for each spin structure. For each fermion, there are $2^{2g}$ different spin structures on a Riemann surface of genus $g$.

Let us rewrite the equations of Sections 5.9 and 5.11 in a more systematic fashion. For example, the $\Theta$ functions must satisfy (5.11.15) when we make an Sp$(2g, Z)$ transformation on them. Demanding modular invariance at the one-loop level means that $C$ must satisfy the change in boundary conditions given by (5.9.7) and (5.11.17). In this new notation, we can rewrite these constraints

$$C\begin{bmatrix}\mathbf{a}\\ \mathbf{b}\end{bmatrix} = -e^{(i\pi/8)\sum a_f}C\begin{bmatrix}\mathbf{a}\\ \mathbf{a}+\mathbf{b}+1\end{bmatrix} \tag{11.9.23}$$

and

$$C\begin{bmatrix}\mathbf{a}\\ \mathbf{b}\end{bmatrix} = e^{(i\pi/4)\sum(a_f b_f)}C\begin{bmatrix}\mathbf{b}\\ \mathbf{a}\end{bmatrix} \tag{11.9.24}$$

where the sum $\sum$ is taken over $f$ fermions (taking left movers minus right movers), each $a_f$ takes on value 0 or 1, and vector sums are only mod 2.

When we demand modular invariance at the multiloop level, we must place further restrictions on the $C$'s. For example, unitarity requires the ability to factorize the amplitude describing a genus $g$ surface into the product of genus 1 tori; i.e., by eqn. (5.1.7) we should be able to slice up a genus $g$ surface into different genus 1 surfaces by inserting a complete set of intermediate states. In our new notation, this means

$$C\begin{bmatrix}\mathbf{a}^1 & \cdots & \mathbf{a}^g\\ \mathbf{b}^1 & & \mathbf{b}^g\end{bmatrix} = C\begin{bmatrix}\mathbf{a}^1\\ \mathbf{b}^1\end{bmatrix}\cdots C\begin{bmatrix}\mathbf{a}^g\\ \mathbf{b}^g\end{bmatrix} \tag{11.9.25}$$

As we saw in Chapter 5, Dehn twists can mix up the $(a, b)$ cycles on a genus $g$ surface, so we must also demand (at least at the two-loop level):

$$\delta\delta' e^{(i\pi/4)\sum a_f a'_f}C\begin{bmatrix}\mathbf{a}\\ \mathbf{b}\end{bmatrix}C\begin{bmatrix}\mathbf{a}'\\ \mathbf{b}'\end{bmatrix} = C\begin{bmatrix}\mathbf{a}\\ \mathbf{b}+\mathbf{a}'\end{bmatrix}C\begin{bmatrix}\mathbf{a}'\\ \mathbf{b}'+\mathbf{a}\end{bmatrix} \tag{11.9.26}$$

where $\delta$ equals $+1$ $(-1)$ if the state is a space–time fermion (boson).

What is remarkable is that a rather simple solution of these constraints is possible.

First, let us make a few definitions.

(1) Let us replace the coefficient $C\begin{bmatrix}\mathbf{a}\\ \mathbf{b}\end{bmatrix}$ with the equivalent expression $C(\alpha|\beta)$, where $\alpha$ $(\beta)$ is the set of fermions that are periodic around the $a$ $(b)$ cycle.
(2) Let us introduce a simple "multiplication" rule:

$$\alpha\beta = \alpha \cup \beta - \alpha \cap \beta \tag{11.9.27}$$

This means that $\alpha^2 = \alpha - \alpha = \varnothing$, the empty set. Let $F$ equal the set of all fermions. Then the product $F\alpha$ equals the complement of the set $\alpha$, that is, $F\alpha = F - \alpha$.

(3) Let us introduce $\delta_\alpha$, which equals $+1$ for fermions and $-1$ for bosons.

(4) Also $n(\alpha)$ is the number of left-moving fermions $n_L(\alpha)$ minus the number of right-moving fermions $n_R(\alpha)$.

(5) Let us define $\varepsilon_\alpha = e^{i\pi n(\alpha)/8}$ and the parity operator $(-1)^\alpha$ for a spin structure $\alpha$ that obeys

$$\begin{aligned} (-1)^\alpha \psi &= -\psi(-1)^\alpha \quad \text{if } \psi \text{ belongs to } \alpha \\ (-1)^\alpha \psi &= \psi(-1)^\alpha \qquad \text{if } \psi \text{ doesn't belong to } \alpha \end{aligned} \tag{11.9.28}$$

A spin structure $(\alpha|\beta)$ is compatible with supersymmetry if

$$(-1)^\alpha T_F(-1)^\alpha = \delta_\alpha T_F \tag{11.9.29}$$

where $T_F$ is the generator of supersymmetric transformations in the superconformal group.

(6) Define Y as a collection of sets with the above supersymmetric property (which forms a group closed under our definition of multiplication (11.9.27)). Define $\Xi$ to be a subgroup of Y that satisfies

$$C(\alpha_i|\varnothing) = \delta_{\alpha_i} \tag{11.9.30}$$

where $C(\varnothing|\varnothing) = 1$.

Now that we have our definitions, let us state the constraints arising from modular invariance. Because $n(\alpha)$ is a multiple of eight, we have the constraints on the basis elements $b_i$ of $\Xi$:

$$\begin{aligned} n(b_i) &= 0 \bmod 8 \\ n(b_i \cap b_j) &= 0 \bmod 4 \\ n(b_i \cap b_j \cap b_k \cap b_l) &= 0 \bmod 2 \end{aligned} \tag{11.9.31}$$

as well as the constraints emerging from modular invariance:

$$\begin{aligned} C(\alpha|\beta) &= \begin{cases} \pm 1 & \text{if } \alpha, \beta \in \Xi \\ 0 & \text{otherwise} \end{cases} \\ C(\alpha|\beta) &= e^2_{\alpha\cap\beta} C(\beta|\alpha) \\ C(\alpha|\alpha) &= \varepsilon_F \varepsilon_\alpha C(\alpha|F) \\ C(\alpha|\beta)C(\alpha|\gamma) &= \delta_\alpha C(\alpha|\beta\gamma) \end{aligned} \tag{11.9.32}$$

for all $\alpha$, $\beta$, $\gamma$ within the collection $\Xi$.

Lastly, we have the constraints coming from the conformal anomaly cancellation, which must always be strictly observed. Let us use the fermionic (not bosonic) representation of the lattice compactification. This was used for the heterotic string in eqn. (10.4.4). We must carefully include the fermionic contribution to the conformal anomaly in the super-Virasoro algebra. It is rather remarkable that a representation of the fermionic partner $T_F$ to the energy–momentum tensor $T_B$ can be given strictly in terms of the adjoint representation of fermions:

$$T_F \sim f^{abc}\psi^a\psi^b\psi^c \tag{11.9.33}$$

When we calculate the anomaly arising from this term, we find that the fermionic contribution to the anomaly is $\frac{1}{2}N$, where $N$ is the number of parameters in the group.

In the left-moving sector, the number of compactified dimensions is $26 - D$, where $D$ is the number of space–time dimensions. We represent the compactified sector by fermions in the adjoint representation. In this sector, the three contributions to the anomaly are

$$\text{Left movers:} \quad \begin{Bmatrix} X_\mu \\ b, c \text{ ghosts} \\ \psi^a \end{Bmatrix} \to \begin{Bmatrix} D \\ -26 \\ \frac{1}{2}N \end{Bmatrix} \tag{11.9.34}$$

Since the sum of the three contributions to the anomaly must be zero, we obviously have $N = 2(26 - D)$ fermions in the set $\psi^a$.

Now let us analyze the right movers, where the anomalies must also sum to zero:

$$\text{Right movers:} \quad \begin{Bmatrix} X_\mu, \psi_\mu \\ b, c, \beta, \gamma \\ \psi^a \end{Bmatrix} \to \begin{Bmatrix} 3D/2 \\ -26 + 11 \\ \frac{1}{2}N \end{Bmatrix} \tag{11.9.35}$$

The condition for zero anomaly is therefore $N = 3(10 - D)$. Including the $(D - 2)$ NS–R fermions (in the light cone gauge) contained in $\psi_\mu$, we have $D - 2 + 3(10 - D)$ fermions for the right movers (in the light cone gauge). In summary, we must have the following number of fermions in order to cancel the anomalous term in the Virasoro algebra:

$$\begin{cases} \text{Left movers} \to 2(26 - D) \\ \text{Right movers} \to D - 2 + 3(10 - D) \end{cases} \tag{11.9.36}$$

The anomaly cancellation is automatic if we have this many fermions in the left- and right-moving sectors.

Now that we have explicitly expressed all our constraints, our strategy is as follows:

(a) We first calculate the total number of space–time fermions $\psi$ and internal fermions $\psi^a$ that satisfy (11.9.36), which removes conformal anomalies from the theory. Then we calculate the total number of spin structures that arise from this set, which we call $F$.
(b) We next randomly choose a collection of spin structures as our initial set within $\Xi$ that includes $F$, the complete set.
(c) We then check for closure under multiplication (11.9.27), which generates

the full group $\Xi$ compatible with this initial choice. (By choosing different initial sets for $\Xi$, we can reproduce different compactification schemes.)

(d) We then calculate the square matrix $C(\alpha|\beta)$ defined over the set $\Xi$ that satisfies our constraints. Some arbitrary phases are introduced in this fashion, which allow us to generate more than one solution for each set $\Xi$.)

We will now discuss some of the properties of the solutions to these equations. Remarkably, we find that all consistent solutions necessarily have gravitons, dilatons, and antisymmetric tensors. We also find that the presence of the massless spin-$\frac{3}{2}$ field is sufficient to prove the absence of tachyons and the vanishing of the cosmological constant at the single-loop level. These are encouraging results, and they show there is tremendous self-consistency within these equations that yield phenomenologically desirable results.

Now let us discuss a few specific solutions to these equations. We will first discuss the 10-dimensional Type II theory (without compactification) in the light cone gauge, after we have imposed all the light cone constraints on our fermions. Then the fermions consist of the eight-component right-moving space–time fermions $\alpha_R = \psi^{1-8}$ and the left-moving fermions $\alpha_L = \tilde{\psi}^{1-8}$, and the internal fermions are set to zero, $\psi^a = 0$.

(1) The simplest choice is

$$\Xi = \{\varnothing, F\} \tag{11.9.37}$$

Unfortunately, a careful analysis of the spectrum of this theory shows that it has tachyons and hence is unacceptable.

(2) The choice

$$\Xi = \{\varnothing, \alpha_L, \alpha_R, F\} \tag{11.9.38}$$

has more than one solution, depending on the choice of certain phases within the $C(\alpha|\beta)$ matrix. The two choices reproduce the Type IIA and Type IIB theories.

(3) For the four-dimensional compactified theory, we must choose a different set $\Xi$. In four dimensions, the space–time fermions are represented by the right-moving transverse $\psi^{1-2}$ and left-moving $\tilde{\psi}^{1-2}$. According to (11.9.36), the internal degrees of freedom can also be represented by $18 = 3(10 - D)$ fermions, in which case we have $\lambda^{1-18}$ for the right movers and $\tilde{\lambda}^{1-18}$ for the left movers. We will regroup these 18 internal fermions collectively as $(x^I, y^I, z^I)$, where $I = 1, \ldots, 6$ and represents six SU(2) factors. We then choose the subset $\alpha = (\psi^{1-2}, z^{1-6})$. Then the choice

$$\Xi = \{\varnothing, \alpha, F\alpha, F\} \tag{11.9.39}$$

produces several $N = 4$ supersymmetric four-dimensional theories, with gauge symmetry $\mathrm{SU}(2)^6$, $\mathrm{SU}(4) \otimes \mathrm{SU}(2)$, or $\mathrm{SU}(3) \otimes \mathrm{SO}(5)$.

Now, let us analyze the 10-dimensional heterotic string in the light cone gauge, where we have the transverse right-moving space–time spinors $\alpha_R = \psi^{1-8}$ and the $32 = 2(26 - D)$ internal left-moving spinors $\alpha_L = F\alpha_R = \lambda^{1-32}$.

(1) The choice

$$\Xi = \{\varnothing, F\} \tag{11.9.40}$$

yields an SO(32) theory with tachyons and no massless fermions, which is unphysical.

(2) The choice

$$\Xi = \{\varnothing, \alpha_R, \alpha_L, F\} \tag{11.9.41}$$

yields the usual Spin(32)/$Z_2$ heterotic string.

(3) If we define $\alpha_1 = \{\lambda^{1-16}\}$ and $\alpha_2 = \{\lambda^{17-32}\}$, then the set

$$\Xi = \{\varnothing, \alpha_R, \alpha_1, \alpha_2, \alpha_1\alpha_R, \alpha_2\alpha_R, \alpha_1\alpha_2, F\} \tag{11.9.42}$$

has two possibilities, depending on the choice of phases. One phase choice leads to the standard supersymmetric $E_8 \otimes E_8$ theory, and the other choice leads to the (nonsupersymmetric) SO(16) $\otimes$ SO(16).

(4) After a four-dimensional compactification, we can again represent the right-moving internal fermions as $(x^I, y^I, z^I)$ for $I = 1$–6 and choose $\alpha = (\psi^{1-2}, z^{1-6})$. Then the simplest choice

$$\Xi = \{\varnothing, \alpha, F\alpha, F\} \tag{11.9.43}$$

yields an $N = 4$ supersymmetric theory with SO(44) gauge symmetry. This is one of the solutions found by Narain using even self-dual Lorentzian lattices.

Clearly, there are probably thousands of different sets of $\Xi$ that close under the multiplication rule (11.9.27) and satisfy the modular constraints. This formalism gives us a handle on the problem of constructing a complete classification of all possible compactified solutions compatible with modular invariance.

Before concluding this chapter, it is worth pointing out new directions that compactification may take. We list three interesting topics concerning four-dimensional superstrings:

(1) Asymmetric orbifolds.
(2) "No-go" theorem, which rules out the Type II superstring.
(3) Superstrings from the Nambu–Goto string.

First is the question of "asymmetric orbifolds," which are larger than the set of orbifolds discussed so far. Asymmetric orbifolds [31] may also give us a handle on classifying the thousands of modular invariant solutions to the string equations. Asymmetric orbifolds are string theories in which the left-moving and right-moving degrees of freedom live on different orbifolds. When compactifying 10-dimensional space–time, we have a natural bias toward compactifying symmetrically in the left- and right-moving sectors, but there are consistent modular-invariant asymmetric compactifications in which the six space dimensions are treated asymmetrically.

Although it is difficult to visualize how to twist on asymmetric orbifolds, it can be shown rather simply that modular-invariant solutions exist. For example, let us mod out by a factor of $g$, where $g$ belongs to a discrete group. The partition function for the single-loop diagram must now trace in the $\tau$ direction in the presence of this $g$ factor:

$$\mathrm{Tr}\; g q^{H_L} q^{\bar{H}_R} \tag{11.9.44}$$

The calculation of the trace can be done exactly as before, except now the spaces are different. As before, we find that the condition for modular invariance is level matching:

$$n(E_L - E_R) = 0 \bmod 1 \tag{11.9.45}$$

If the eigenvalues of $g$ are represented by $e^{2\pi i r_i/n}$, this condition gives us restrictions on the sum $\sum_i r_i^2$. The resulting constraints on the $r_i$ are almost identical to the ones found for the symmetric orbifold (11.9.9), except the left-moving and right-moving sectors are now treated differently, depending on the structure of space. Although asymmetric orbifolds are more difficult to work with than symmetric orbifolds, a large class of asymmetric orbifolds can actually be constructed that are compatible with modular invariance. (In practice, the asymmetric orbifold is sometimes embedded in a larger symmetric orbifold, where the calculations are simpler, and then truncated at the end to retrieve the asymmetric orbifold.)

In summary, the advantage of asymmetric orbifolds is that they allow us to classify an enormous class of modular invariant compactifications in four dimensions. This will be of great help toward understanding the complete set of physically relevant superstring compactifications.

The second new direction for string compactification is to explore the use of asymmetric orbifolds to create realistic gauge groups for the Type II superstring. Traditionally, it was thought that the Type II string was unsuitable for phenomenology because it had no Chan–Paton isospin factors. In 10 dimensions, the Type II superstring is a gravitational theory without any Yang–Mills particles at all. Whereas heterotic strings produce too many quasi-realistic gauge groups in four dimensions because they contain 10-dimensional Yang–Mills fields, the Type II string suffers from the opposite problem: it has no 10-dimensional Yang–Mills fields at all. As a consequence, the Type II superstring has been less interesting from a phenomenological point of view.

Contrary to conventional thinking, recent attempts [32, 33] to compactify Type II superstrings have shown a number of isospin groups that may be obtained when reducing the theory to four dimensions. For example, we saw in our discussion of the anomaly cancellation in (11.9.36) that the internal compactified dimensions can be represented fermionically by introducing $3(10 - D) = 18$ fermions $\psi^a$ in the adjoint representation of some group. Thus, we expect that the compactification of Type II superstrings might yield the following gauge groups with 18 generators when reduced to four dimensions:

$$SU(2)^6; \qquad SU(4) \otimes SU(2); \qquad SU(3) \otimes SO(5) \tag{11.9.46}$$

The existence of nontrivial four-dimensional gauge groups for the Type II superstring is a rather surprising result. It shows that superstring theory continues to defy conventional wisdom.

However, it has also been shown recently that, on quite general grounds, *the Type II can never produce the Standard Model containing triplet quarks and doublet leptons* [34]. A negative result that strong would probably not be possible for the heterotic string, which contains a large Yang–Mills gauge group in 10 dimensions. However, because the Type II superstring has no 10-dimensional Yang–Mills gauge group, its potential for phenomenology is less than that of the heterotic string and it is more vulnerable to "no-go" theorems, which might rule it out entirely.

This negative result, which requires only the mildest assumptions about super-Kac–Moody algebras, is remarkable because it apparently eliminates the Type II string completely. The result is independent of precisely how the compactification takes place and is sensitive only to the superconformal anomaly cancellation. Compactifications with triplet quarks without doublet leptons (or vice versa) are still possible, but apparently one cannot have *both* triplet quarks and doublet leptons at the same time.

This no-go theorem is interesting because it shows the power of super-Kac–Moody algebras to rule out an entire class of model building. Because this negative result is sure to be challenged, it is premature to say whether it will stand the test of time, but the details of the proof are interesting in themselves.

Let us begin by constructing the super-Kac–Moody and Virasoro algebras for the compactified model, using the fermionic construction of the isospin sector with the fermion field $\psi^a$. The goal is to calculate the superconformal anomaly for the super-Virasoro generators for various fermion representations and to show that it cannot be canceled for the Standard Model.

Before compactification, the anomaly equals (see (4.2.29))

$$D + \tfrac{1}{2}D - 26 + 11 = \tfrac{3}{2}(D - 10) \tag{11.9.47}$$

where $D$ comes from the bosonic string, $\frac{1}{2}D$ comes from the NS–R fermions, and $-15 = -26 + 11$ comes from the conformal and superconformal ghosts, respectively. After compactification, we must set $D = 4$ and add the contribution to the anomaly coming from the compactified fermions $\psi^a$. Let us call it $c$. Then the anomaly after compactification is

$$4 + \tfrac{1}{2} \times 4 + c - 15 = \tfrac{3}{2}(4 + \hat{c} - 10) \tag{11.9.48}$$

so $c$ must equal 9, or $\hat{c}$ must equal 6.

The first step is to construct the anomaly contribution coming from the compactified $\psi^a$ field. Fortunately, a representation of the super-Kac–Moody generators $J_B^a$ and $J_F^a$ and the super-Virasoro generators $T_B$ and $T_F$ is possible using only the fermion field $\psi^a$. Let $\theta$ represent a Grassmann variable; then the super-Kac–Moody generator can be represented as the superfield $J(z, \theta)$:

$$\begin{aligned} J(z,\theta)^a &= J_F^a + \theta J_B^a \\ J_B^a(z) &= -\frac{i}{2} f^{abc}\psi^b\psi^c \\ J_F^a(z) &= \sqrt{k}\psi^a \end{aligned} \tag{11.9.49}$$

It is now a simple matter to turn the crank and calculate the anomaly term in the Kac–Moody algebra (called the level number $k$; see (4.7.1)) in terms of this representation. We find

$$k = C_2(G) \tag{11.9.50}$$

where $C_2(G)$ is the value of the quadratic Casimir operator for the group $G$. Since we are interested in the Standard Model, we will choose $G$ to be SU($n$), in which case

$$C_2(G) = n \tag{11.9.51}$$

Next, we wish to find the contribution of $\psi^a$ to the super-Virasoro algebra. Fortunately, an explicit representation of the super-Virasoro generator $T(z, \theta)$ is also known in terms of $\psi^a$:

$$\begin{aligned} T(z,\theta) &= T_F + \theta T_B \\ T_B &= -\tfrac{1}{2}\psi^a\partial\psi^a \\ T_F &= \frac{-i}{12\sqrt{k}} f^{abc}\psi^a\psi^b\psi^c \end{aligned} \tag{11.9.52}$$

When we calculate the conformal anomaly for this particular representation of the super-Virasoro algebra, we find that

$$c = \tfrac{1}{2}d(G) \tag{11.9.53}$$

where $d(G)$ is the dimension of the group $G$. In our case,

$$c = \tfrac{1}{2}(n^2 - 1) \tag{11.9.54}$$

for SU($n$). (This result is easy to show. We know from the old NS–R model in $D$ dimensions and from (4.2.29) that fermions contribute $\frac{1}{2}D$ and bosons contribute $D$ to the anomaly. Notice that $T_B$ has precisely the same form as the fermionic part of the NS–R model except that we have $d(G)$ fermions, so we expect a contribution of $c = \frac{1}{2}d(G)$ to the bosonic Virasoro commutators.)

The next step is a bit more complex. We know that, in addition to the above representation, there may be other contributions to the super-Kac–Moody algebra. However, recent results on Kac–Moody algebras [35] tell us that stringent conditions may be placed on their anomalies if we demand unitary representations. For example, let us say that there is a new unknown contribution to the algebra given by a super-Kac–Moody generator $\tilde{J}^a$, whose explicit form we really do not need to know. The new result on Kac–Moody algebras tells us that *all* unitary representations can be obtained in the following way:

$$
\begin{aligned}
J_B^a &= \tilde{J}^a - \frac{i}{2} f^{abc}\psi^a\psi^b\psi^c \\
J_F^a &= \sqrt{k}\psi^a \\
T_B &= -\tfrac{1}{2}\psi^a\partial\psi^a + \frac{1}{2k}:\tilde{J}^a\tilde{J}^a: \\
T_F &= \frac{-i}{12\sqrt{k}} f^{abc}\psi^a\psi^b\psi^c + \frac{1}{2\sqrt{k}}\psi^a\tilde{J}^a
\end{aligned}
\tag{11.9.55}
$$

Assume that $\tilde{J}^a$ contributes a term $\tilde{k}$ to the level number. Then the total level number is the sum of the usual term plus $\tilde{k}$:

$$k = \tilde{k} + C_2(G)$$

Notice that the contribution of $\tilde{J}^a$ to the energy–momentum tensor $T_B$ is in the Sugawara form, i.e., as the square of the Kac–Moody generator $\tilde{J}^a$. This means that the contribution of $\tilde{J}^a$ to the conformal anomaly is known exactly. By direct computation, we can now calculate the commutator between two $T_B$ and hence the total contribution to the conformal anomaly:

$$c = \tfrac{1}{2}d(G) + d(G)\frac{k - C_2(G)}{k} \tag{11.9.56}$$

where the first term comes from the $\psi^a$ field and the second term can be computed because the energy–momentum tensor is in Sugawara form.

Plugging in the values

$$d(G) = n^2 - 1; \qquad C_2(G) = n \tag{11.9.57}$$

we find

$$\hat{c} = (n^2 - 1)\left[1 - \frac{2n}{3k}\right] \tag{11.9.58}$$

which must equal 6.

So far, our results are still too weak; we cannot make any conclusive statements because the value of $k$ is totally unknown. At this point, we introduce the key step in the proof.

The condition that we have *unitary* representations of the super-Kac–Moody algebra places a new constraint that allows us to calculate bounds on $k$ [35]. In particular, we know that the norms of elements of unitary irreducible representations of the algebra must be nonnegative.

In general, irreducible representations of Lie groups are constructed by taking the highest-weight vacuum vector $|h\rangle$ for some representation $h$ and then operating on it by all possible raising operators:

$$J_{-n_1}^{\lambda_1} J_{-n_2}^{\lambda_2} \cdots J_{-n_l}^{\lambda_1}|h\rangle \tag{11.9.59}$$

This creates what is called the universal enveloping algebra. Notice that it forms a representation of the algebra because any transformation of this object by the generators $J_n^p$ simply maps the object back into another member of the enveloping algebra. Now construct the norm of the state $(J_{-1}^a + iJ_{-1}^b)|h\rangle$:

$$\|(J_{-1}^a + iJ_{-1}^b)|h\rangle\|^2 = \langle h|k - 2f^{abc}J_0^c|h\rangle \geq 0 \tag{11.9.60}$$

which must be nonnegative for the representation to be unitary. We have used the commutation relations of the generators to calculate the norm in terms of $k$ and $J_0^c$. We know that $J_0^c$ is nothing but the generator of an ordinary finite-dimensional Lie group, so this matrix element can be explicitly computed for any given heighest-weight state labeled by $h$. Elementary group theory now tells us that the previous equation can be restated as

$$k \geq \lambda_G(h) \tag{11.9.61}$$

where $\lambda_G(h)$ is the number of columns in the Young tableau of the representation $h$. For SU(2), it is just equal to twice the isospin of the state.

We can now summarize the key step in the proof. We must simultaneously satisfy two equations:

$$\hat{c} = (n^2 - 1)\left[1 - \frac{2n}{3k}\right] = 6; \qquad k \geq \lambda_G(h) \tag{11.9.62}$$

Given these restrictions, we can now write down *all possible* four-dimensional gauge groups that are allowed in the Type II superstring:

(a) $SU(2)^6$
(b) $SU(4) \otimes SU(2)$
(c) $SO(5) \otimes SU(3)$
(d) $SO(5) \otimes SU(2) \otimes SU(2)$
(e) $SU(3) \otimes SU(3)$
(f) $G_2$
(g) All proper subgroups of the above

What is even more interesting is that it is a simple matter to calculate $\hat{c}$ for various representations $h$ using the above constraints. We find the following:

(1) If $G = SU(3)$ and we demand triplet quarks, then $\hat{k} \geq 1$ and $\hat{c} \geq 4$:

$$SU(3) \text{ triplets} \to \hat{c} \geq 4 \tag{11.9.63}$$

(2) If $G = SU(2)$ and we demand doublet leptons, then $\tilde{k} \geq 1$ and $\hat{c} \geq \frac{5}{3}$:

$$SU(2) \text{ doublets} \to \hat{c} \geq \tfrac{5}{3} \tag{11.9.64}$$

(3) If $G = U(1)$, then $\hat{c} \geq 1$:

$$U(1) \to \hat{c} \geq 1 \tag{11.9.65}$$

The total contribution to the anomaly from all three groups in the Standard Model is therefore the sum

$$\hat{c} \geq 6\tfrac{2}{3} \tag{11.9.66}$$

which is just $\frac{2}{3}$ above the limit. This is our principal result. Thus, the Type II superstring fails to cancel the conformal anomaly.

The only way out, apparently, is to drop some of the essential parts of the Standard Model (and assume that some of the quarks or leptons emerge as bound states) or drop the Type II string entirely. For example, if we give up doublet leptons, then $\hat{c} = 6$ and the anomaly cancels. We can, therefore, come very close to 6, but not close enough to the Standard Model.

The third new direction for string compactification that we will mention is the ambitious program to derive *all* possible superstring theories from the original $D = 26$ Nambu–Goto string [36–39]. At present, the heterotic string, although it is the leading candidate for a theory of all known forces, appears contrived because of its highly asymmetric nature. If the heterotic string is so fundamental, it appears strange that it should be so inelegant. It is appealing to conjecture that the heterotic string and all other superstring models are simply different compactifications of the original 26-dimensional string.

Three observations make this scheme plausible. First, the process of fermionization and bosonization in two dimensions means that fermions in a superstring model in 10 dimensions may be condensates of 26-dimensional bosons. Thus, the lack of fermions in the 26-dimensional string is not a problem. Supersymmetry, in this picture, emerges by "accident" as we compactify down from 26 to 10 dimensions. Second, all compactifications investigated so far seem superficially to be different truncations of the original bosonic string theory. In particular, the number 26 appears over and over again in superstring theories defined in 10 dimensions. Third, the presence of the tachyon in the 26-dimensional theory (which used to be such a headache in the early years) is now viewed as an asset. It simply means that the naive vacuum is unstable when quantum corrections are calculated, so it is plausible that the bosonic string might prefer to break down to a 10-dimensional theory.

Although this approach has its aesthetic merits, we must be careful to point out the serious problems facing it. First, we have to truncate a large number of particles when we reduce to 10 dimensions. Where do these particles go? Even if we could somehow banish them, they could easily reappear in tree and loop diagrams. Once again, we have to appeal to hand-waving arguments concerning dynamical breaking effects (which are beyond our calculational ability). Second, we have to make the group theory come out correctly. The 10-dimensional fermions of the superstring transform under the same SO(9, 1) Lorentz group as the 10-dimensional bosons. But in truncating from 26 down to 10 dimensions, the excess 16-dimensional bosons do not transform under the SO(9, 1) of the remaining 10-dimensional bosons. These excess 16-dimensional bosons must eventually condense down to the 10-dimensional fermions that do transform under SO(9, 1). Thus, it is not clear how the Lorentz group for the fermions arises when they originally came from the 16-dimensional sector that does not transform under SO(9, 1). For these and other reasons, the appealing idea of compactifying the 26-dimensional string down to the 10-dimensional theory is still a dream.

## §11.10. Summary

We have seen that a few physically reasonable assumptions about the compactification process have led to a wealth of phenomenological predictions. Although no model has yet been proposed that can successfully predict all the known properties of the low-energy particle spectrum, we are encouraged by the qualitative results we have obtained. Let us now specifically isolate the logical sequence by which given assumptions led to certain conclusions.

Our natural choice is to compactify the 10-dimensional space on the six-torus. However, compactifying on this surface transforms $N = 1$ supersymmetry in 10 dimensions into $N = 4$ supersymmetry in 4 dimensions. Thus, we must search for new assumptions that will be more phenomenologically acceptable.

We began with the following assumptions:

(1) The 10-dimensional universe has compactified to

$$M_{10} \to M_4 \times K_6 \tag{11.10.1}$$

where $M_4$ is a maximally symmetric space,

$$R_{\mu\nu\alpha\beta} = \frac{R}{12}(g_{\mu\alpha}g_{\nu\beta} - g_{\mu\beta}g_{\nu\alpha}) \tag{11.10.2}$$

and $K$ is compact.

(2) An $N = 1$ local supersymmetry remains unbroken after the compactification.

(3) Some of the bosonic fields can be set to zero:

$$H = d\phi = 0 \tag{11.10.3}$$

The assumption that $N = 1$ supersymmetry survives the compactification is a stringent assumption. It implies that the variation of the spinors must be zero:

$$\begin{aligned} \delta\psi_A &= \frac{1}{\kappa}D_A\varepsilon = 0 \\ \delta\chi^a &= \Gamma^{ij}F^a_{ij}\varepsilon = 0 \end{aligned} \tag{11.10.4}$$

The existence of a covariantly constant spinor $\varepsilon$ places an enormous number of constraints on the manifold. In particular, if we differentiate the equation once again:

$$\begin{aligned} D_i\varepsilon = 0 &\to [D_j, D_i]\varepsilon = 0 \\ &\to R_{ijkl}\Gamma^{kl}\varepsilon = 0 \end{aligned} \tag{11.10.5}$$

Thus, we can show that the space is Ricci flat. Second, we can show that the space has SU(3) holonomy. We define the holonomy group as the group generated when we take a spinor around successive closed paths around a point. Normally, in six dimensions the holonomy group is SO(6), or equivalently SU(4). But if $\varepsilon$ is covariantly constant, it means that

$$\varepsilon \rightarrow U\varepsilon \tag{11.10.6}$$

By an SU(4) transformation, any spinor can be brought into the form

$$\varepsilon \rightarrow \varepsilon = \begin{pmatrix} 0 \\ 0 \\ 0 \\ \varepsilon_0 \end{pmatrix} \tag{11.10.7}$$

This, in turn, means that the $U$ matrix rotates only the three zero indices in the column matrix. Thus, $U$ must be a member of SU(3), and so the space has SU(3) holonomy.

In general, spaces with SU(3) holonomy are notoriously difficult to deal with. However, we can also show that the space is Kähler and so use a powerful theorem by Calabi–Yau. We can always define the tensor

$$J_j^i = -ig^{ik}\bar{\varepsilon}\Gamma_{kj}\varepsilon \tag{11.10.8}$$

which satisfies $J^2 = -1$, just like $i^2 = -1$ for the usual definition of complex numbers, and so we can use this tensor to define a complex manifold. However, we know that this tensor must be covariantly constant (because $\varepsilon$ is covariantly constant), and thus the space is Kähler.

Fortunately, Kähler manifolds that are Ricci flat (and hence have vanishing first Chern class) can be shown to be equivalent to spaces of SU(3) holonomy and are easy to construct. Thus, we have the logical sequence

$$D_i\varepsilon = 0 \rightarrow K \text{ is Ricci flat, Kähler, with vanishing first Chern class} \tag{11.10.9}$$

The problem, however, is that now we are flooded with potentially thousands of manifolds preserving $N = 1$ supersymmetry!

So far, we have not broken the $E_8 \otimes E_8$ gauge symmetry. Our next step is to notice that the Bianchi identities, which used to be empty identities, now become nontrivial with our assumptions:

$$\text{Tr } R \wedge R = \frac{1}{30} \text{Tr } F \wedge F \tag{11.10.10}$$

This equation is surprisingly difficult to satisfy. On the left, we have the curvature forms over the Riemannian space, and on the right we have the Yang–Mills curvature forms. In order to solve this rather strange-looking equation, we will embed the spin connection into the gauge field, thereby mixing the two manifolds and breaking the original gauge symmetry. Because we have SU(3) holonomy in the Riemann sector, we will have the following breaking in the gauge sector when we mix the two sectors:

$$\text{Bianchi identity} \rightarrow \text{Spin embedding} \rightarrow \text{SU}(3) \otimes E_6 \otimes E_8 \tag{11.10.11}$$

Fortunately, this gives rise to desirable results. We know that $E_8$ by itself does not have chiral representations and hence is not an acceptable candidate for model building, but $E_6$ does. In fact, the decomposition under this breaking of the 248 elements of the adjoint of $E_8$ becomes

$$\mathbf{248} = (\mathbf{3}, \mathbf{27}) \oplus (\bar{\mathbf{3}}, \overline{\mathbf{27}}) \oplus (\mathbf{8}, \mathbf{1}) \oplus (\mathbf{1}, \mathbf{78}) \tag{11.10.12}$$

This is gratifying, because the **27** is the most suitable representation for the quarks and leptions in GUT theories with $E_6$. Thus, we have

$$\operatorname{Tr} R \wedge R = \frac{1}{30} \operatorname{Tr} F \wedge F \rightarrow \text{Spin embedding} \rightarrow \mathbf{27}\ \text{Fermions} \tag{11.10.13}$$

The next question is: How many generations of the **27** do we have? Normally, the generation number has nothing to do with the gauge group. In GUT theories, they are entirely distinct. One can have an arbitrary number of exact copies of fermion generations in GUT theories. However, in string theory the process of spin embedding produces a constraint on the generation number. Because of spin embedding, the gauge fermions are now linked to the $K_6$ manifold. The generation number can be viewed as a topological number, because the difference between positive and negative chiral zero eigenvalue solutions of the Dirac equation is a topological number. Specifically,

$$\text{Generation number} = \tfrac{1}{2}|\chi(M)| \tag{11.10.14}$$

Unfortunately, the Euler number of Ricci-flat Kähler manifolds is usually quite high, but we can always reduce this number drastically by taking a submanifold. For example, we can divide by a discrete group that preserves a certain polynomial in the coordinates. A good example is $CP_4$ divided by $Z_5 \times Z_5$, which is not simply connected. This manifold has four generations.

Next, we want to break the model even further to the Standard Model $SU(3) \otimes SU(2) \otimes U(1)$ without destroying $N = 1$ supersymmetry. Normally, this is quite difficult. Breaking schemes that break the group down to the standard model also break $N = 1$ supersymmetry. One solution to this problem is to use the method of Wilson lines. We saw earlier that the manifolds $K_6$ that we are considering are not simply connected in order to produce low generation number. For nonsimply connected manifolds, the Wilson loop

$$U = P \exp \int_C A_\mu \, dx^\mu \tag{11.10.15}$$

does not necessarily equal 1 even if the curvature tensor vanishes. This is because we cannot always shrink closed loops into points. Thus, we can break $E_6$ symmetry down to a subgroup $G$ by choosing an element of U so that $G$ commutes with all elements of this element. For example, if we take an element of U that is $Z_5 \times Z_5$, we find

$$\frac{E_6}{Z_5 \times Z_5} = SU(3)_C \otimes SU(2)_L \otimes SU(2)_R \otimes U(1) \otimes U(1) \tag{11.10.16}$$

Unfortunately, we see that, in addition to arriving at the standard model, we also have unwanted U(1) groups. Thus,

$$\text{Wilson lines} \rightarrow SU(3) \otimes SU(2) \otimes U(1)^n$$

Orbifolds are another way of compactifying the superstring, which is probably the limiting case of a Calabi–Yau space. Orbifolds are created by taking a torus $T_6$ and dividing by a discrete group $Z_n$, which allows fixed points.

$$\text{Orbifold: } \frac{T_6}{Z_n} \tag{11.10.17}$$

(These fixed points apparently do not spoil the properties of the string model.) Nontrivial constraints are placed on orbifolds by modular invariance, which mixes the boundary conditions nontrivially. The boundary conditions

$$\begin{aligned} X(\sigma_1 + 2\pi, \sigma_2) &= hX(\sigma_1, \sigma_2) \\ X(\sigma_1, \sigma_2 + 2\pi) &= gX(\sigma_1, \sigma_2) \end{aligned} \tag{11.10.18}$$

(where $h$ and $g$ are elements of $Z_n$) can break the symmetries of the string model. The subgroup of the space–time group and the internal group that survives the process is the subgroup that commutes with $g$ and $h$. If we diagonalize $g$ and $h$ to equal the elements $e^{2\pi i v_i}$, where $v_i = r_i/n$ for some integer $n$, then the zero point energy of the Hamiltonian is shifted by an amount proportional to $v_i(v_i - 1)$. Because the trace over the single loop is sensitive to the zero point energy, this means that modular invariance places a restriction on the eigenvalues:

$$\sum_{i=1}^{8} r_i^2 = \sum_{j=1}^{8} r_{1j}^2 + \sum_{j=1}^{8} r_{2j}^2 \tag{11.10.19}$$

where $r_i$ are the eigenvalues from the right-moving sector, and the two sets of eigenvalues $r_{1,2;i}$ are from the left-moving $E_8 \otimes E_8$ sector. If we include Wilson lines, then the group that survives the compactification is the subgroup that commutes with $g$, $h$, and the Wilson line.

Phenomenologically, there are many solutions to this compactification, some with three generations and the group $SU(3) \otimes SU(2) \otimes U(1)^n$, which yields too many U(1) factors. Unfortunately, the method of Wilson lines does not change the rank of the group, so in general we will have too many U(1) factors.

The great advantage of orbifolds over Calabi–Yau spaces is that they are simple, flat, and thousands of them can be explicitly constructed. Unfortunately, both suffer from the problem that there is no systematic way to catalogue these thousands of solutions.

One step in this direction is to use modular invariance and the absence of tachyons and anomalies to systematically derive all possible solutions. We begin with the amplitude written as a sum over spin structures:

$$A = \sum_{\mathbf{a},\mathbf{b}} C\begin{bmatrix}\mathbf{a}\\\mathbf{b}\end{bmatrix} A\begin{bmatrix}\mathbf{a}\\\mathbf{b}\end{bmatrix} \tag{11.10.20}$$

Now we demand that the $C$'s factorize properly, have modular invariance, and yield models that have no tachyons or anomalies. Remarkably, it is possible to solve the constraints on $C$ and obtain solutions. The simple ones reproduce some of the known compactifications, but thousands of other compactifications are still being investigated. This yields the hope that we may be able to exhaust all possibilities and obtain a realistic model.

We list three avenues of research into four-dimensional solutions. First, another way in which to generate large classes of four-dimensional solutions is through asymmetric orbifolds. The heterotic string is an example of an asymmetric compactification, i.e., treating left- and right-moving sectors differently. Asymmetric orbifolds are the largest class of orbifolds found so far and may overlap significantly with the work of others using different types of compactifications. (In addition to asymmetric orbifolds, we should also mention the possibility of non-abelian orbifolds, i.e., orbifolds based on dividing out by non-abelian finite groups such as the crystal groups. The advantage of such constructions is that they help us eliminate some of the unwanted U(1) factors and also yield three and four generations. See [40].)

Second, the Type II superstring is being investigated carefully for its properties under compactification. Especially significant is a "no-go" theorem which states that the Type II string can *never* produce the Standard Model after compactification because the conformal anomaly cannot be eliminated. The contribution of the compactified fermions should yield a factor of 6, but the Standard Model gives $6\frac{2}{3}$. Thus, we either rule out triplet quarks and doublet leptons, or else we rule out Type II strings.

Lastly, there has been effort to construct all known superstrings from the Nambu string. This idea is not as fanciful as one might suspect, because perhaps the excess bosonic modes fermionized into the supersymmetric partners of the 10 $D$ string. However, the problem is now that after compactifying down from 26 to 10 we have created too many particles, which have to be banned somehow from the model. Although the idea is a clever one, there are still serious problems with making it realistic.

## §11.11. Conclusion

We hope that this book has conveyed some of the flavor of superstring models and why they have generated so much excitement within the past few years. Superstrings give us the most promising formalism for the unification of gravity with all known forces. Not only have Calabi–Yau and orbifold solutions to the string theory opened up the possiblity of thousands of four-dimensional models, but also the mathematical structure of the theory has generated intense interest among mathematicians in wide-ranging and diverse areas, such as theta functions, Kac–Moody algebras, super-Riemann surfaces, Kähler manifolds, index theorems, and cohomology. In several areas, string theories have actually generated new mathematical ideas not yet explored by mathematicians.

The biggest drawback to the superstring theory, however, is that particle accelerators able to probe down to the Planck length are prohibitively expensive. Even indirect tests of the superstring theory are probably not possible in the near future. One can take at least two different attitudes toward this fundamental problem.

First, superstring theory may be incorrect but might persist for decades without the decisive experiments necessary to verify or reject the theory. At worst, the theory could waste the efforts of many physicists. However, we should note that even if the theory is incorrect, it has already opened up new methods of handling the divergences of Feynman diagrams. The use of topological arguments to control the divergences of quantum field theory represents a qualitative leap in our understanding of how to construct new field theories of gravity. Also, string theory has already become a permanent area of mathematical research, independent of whether it is physically relevant.

Second, one may take the view that superstring theory is correct but that the main stumbling block is *theoretical*, not experimental. If we could solve for the true vacuum of the theory, we might be able to compare the rigorous predictions of the model with, say, the masses of the quarks and the proton mass. Therefore, we do not have to wait for the development of super-accelerators that will cost more than the combined GNP of the planet. The decisive results will come from the theoreticians, who will be able to make testable predictions about our low-energy universe.

According to the second philosophy, the main problem is not necessarily building larger and larger accelerators; the problem is that we are still unlocking the secrets of the theory nearly two decades after its birth. We are still searching for the counterpart of the equivalence principle on which the entire theory rests. When the theory "died" prematurely in 1974, it was only a partially compete theory. We are still in the process of completing it.

We have, therefore, assembled a list of some of the unsolved problems of the theory in order to stimulate vigorous research in these areas. Let us summarize, therefore, the main unsolved problems for the future:

## Unsolved Problems

(1) The theory must be shown, rigorously and to all orders, to be finite. Hand-waving discussions and plausibility arguments are not enough.
(2) The theory must predict why the cosmological constant is nearly zero *after* supersymmetry is broken.
(3) All classical solutions of the string theory should be catalogued and one must be found that correctly reproduces the Standard Model at low energies.
(4) The field theory of strings must be extended to the nonperturbative domain to calculate the true vacuum of the theory.

(5) The theory of theta functions must be further developed until it yields conclusions concerning the sum of the perturbation series or nonperturbative information concerning the vacuum, possibility within the framework of universal moduli space or Grassmannians.

## References


[1] P. Candelas, G. Horowitz, A. Strominger, and E. Witten, *Nucl. Phys.* **B258**, 46 (1985).
[2] G. Horowitz, in *Unified String Theories* (edited by M. Green and D. Gross), World Scientific, Singapore, 1986.
[3] G. F. Chapline and N. S. Manton, *Phys. Lett.* **120B**, 105 (1983).
[4] E. Calabi, *Algebraic Geometry and Topology: A Symposium in Honor of S. Lefschetz*, Princeton University Press, Princeton, N.J., 1957.
[5] S.-T. Yau, *Proc. Natl. Acad. Sci. U. S. A.* **74**, 1798 (1977).
[6] S.-T. Yau, in *Symposium on Anomalies, Geometry, and Topology* (edited by W. A. Bardeen and A. R. White), World Scientific, Singapore, 1985.
[7] Y. Hosotani, *Phys. Lett.* **126B**, 303 (1983).
[8] L. Dixon, J. Harvey, C. Vafa, and E. Witten, *Nucl. Phys.* **B261**, 678 (1985); **B274**, 286 (1986).
[9] C. Vafa, *Nucl. Phys.* **B273**, 592 (1986).
[10] L. E. Ibanez, H. P. Nilles, and F. Quevedo, *Phys. Lett.* **187B**, 25 (1987).
[11] L. E. Ibanez, J. E. Kim, H. P. Nilles, and F. Quevedo, *Phys. Lett.* **191B**, 282 (1987).
[12] D. X. Li, *Phys. Rev.* **D34** (1986).
[13] D. X. Li, in *Super Field Theories* (edited by H. C. Lee), Plenum, New York.
[14] V. P. Nair, A. Sharpere, A. Strominger, and F. Wilczek, *Nucl. Phys.* **B287**, 402 (1986).
[15] B. R. Greene, K. H. Kirklin, P. J. Miron, and G. G. Ross, *Phys. Lett.* **180B**, 69 (1986); *Nucl. Phys.* **B278**, 667 (1986).
[16] A. Strominger, *Phys. Rev. Lett.* **55**, 2547 (1985).
[17] P. Ginsparg, Harvard preprint HUTP-86/A053, 1986.
[18] S. Karlara and R. N. Mohapatra, *Phys. Rev. D*, LA-UR-86-3954, 1986.
[19] M. Dine, V. Kaplunovsky, M. Mangano, C. Nappi, and N. Sieberg, *Nucl. Phys.* **B259**, 46 (1985).
[20] S. Cecotti, J. P. Deredinger, S. Ferrara, L. Girardello, and M. Roncadelli, *Phys. Lett.* **156B**, 318 (1985).
[21] J. P. Deredinger, L. Ibanez, and H. P. Nilles, *Nucl. Phys.* **B267**, 365 (1986).
[22] J. Breit, B. Ovrut, and G. Segre, *Phys. Lett.* **158B**, 33 (1985).
[23] A. Strominger and E. Witten, *Commun. Math. Phys.* **101**, 341 (1985).
[24] G. Segre, Schladming Lecture notes (1986).
[25] H. Kawai, D. C. Lewellen, and S. H. H. Tye, *Phys. Rev. Lett.* **57**, 1832 (1986); *Phys. Rev.* **D34**, 3794 (1986); *Nucl. Phys.* **B288**, 1 (1987); *Phys. Lett.* **191B**, 63 (1987).
[26] W. Lerche, D. Lust, and A. N. Schellekens, *Nucl. Phys.* **B287**, 477 (1987).
[27] I. Antoniadis, C. Bachas, and C. Kounnas, *Nucl. Phys.* **B289**, 87 (1987).
[28] I. Antoniadis and C. Bachas, CERN-TH-4767/87, 1987.
[29] C. Kounnas, Berkeley preprint UCB-PTh 87/21, 1987.
[30] I. Antoniadis, C. Bachas, C. Kounnas, and P. Windey, *Phys. Lett.* **171B**, 51 (1986).
[31] K. S. Narain, M. H. Sarmadi, and C. Vafa, *Nucl. Phys.* **B288**, 55 (1987).

[32] R. Bluhm, L. Dolan, and P. Goddard, *Nucl. Phys.* **B289**, 364 (1987).
[33] L. Castellani, R. D'Auria, F. Gliozzi, and S. Sciuto, *Phys. Lett.* **168B**, 77 (1986).
[34] L. Dixon, V. Kaplunovsky, and C. Vafa, SLAC-PUB-4282.
[35] V. G. Kac and I. T. Todorov, *Commun. Math. Phys.* **102**, 337 (1985).
[36] P. G. O. Freund, *Phys. Lett.* **151B**, 387 (1985).
[37] A. Casher, F. Englert, H. Nicolai, and A. Taormini, *Phys. Lett.* **162B**, 121 (1985).
[38] F. Englert, H. Nicolai, and A. Schellekens, *Nucl. Phys.* **B274**, 315 (1986).
[39] D. Lust, *Nucl. Phys.* **B292**, 381 (1987).
[40] N. P. Chang and D. X. Li, "Models of Non-Abelian Orbifolds," *CCNY-HEP-*87-15.

# Appendix

Because the mathematics of superstring theory has soared to such dizzying heights, we have included this short appendix to provide the reader with a brief mathematical understanding of some of the concepts introduced in this book. We apologize that we must necessarily sacrifice a certain degree of mathematical rigor if we are to cover a wide range of topics in this appendix. However, the interested reader is advised to consult some of the references listed later for more mathematical details.

## §A.1. A Brief Introduction to Group Theory

A *group* $G$ is a collection of elements $g_i$ such that

(1) There is an identity element $I$.
(2) There is closure under multiplication:

$$g_1 \times g_2 = g_3$$

(3) Every element has an inverse:

$$g_i \times g_i^{-1} = I$$

(4) Multiplication is associative:

$$(g_i \times g_j) \times g_k = g_i \times (g_j \times g_k)$$

Groups come in a variety of forms. Specifically, we have the *discrete groups*, which have a finite number of elements, and the *continuous groups*, such as the *Lie groups*, which have an infinite number of elements. Examples of discrete groups include:

(1) The *Alternating groups*, $Z_n$, based on the set of permutations of $n$ objects.
(2) The 26 *sporadic groups*, which have no regularity. The largest and most interesting of the sporadic groups is the group $F_1$, commonly called the "Monster group," which has

$$2^{46} \cdot 3^{20} \cdot 5^9 \cdot 7^6 \cdot 11^2 \cdot 13^3 \cdot 17 \cdot 19 \cdot 23 \cdot 29 \cdot 31 \cdot 41 \cdot 59 \cdot 71$$

elements.

In this book, however, we mainly encounter the continuous groups, which have an infinite number of elements. The most important of the continuous groups are the *Lie groups*, which come in the following four classical infinite series $A$, $B$, $C$, $D$ when we specialize to the case of compact, real forms

$$\begin{aligned} A_n &= \mathrm{SU}(n+1) \\ B_n &= \mathrm{SO}(2n+1) \\ C_n &= \mathrm{Sp}(2n) \\ D_n &= \mathrm{SO}(2n) \end{aligned} \tag{A.1.1}$$

as well as the exceptional groups:

$$G_2;\ F_4;\ E_6;\ E_7;\ E_8 \tag{A.1.2}$$

of which $E_6$ and $E_8$ are the most interesting from the standpoint of string phenomenology.

Let us give concrete examples of some of these groups by analyzing the set of all real or complex $n \times n$ matrices. Clearly, the set of arbitrary $n \times n$ invertible matrices satisfies the definitions of a group and hence is called the group $\mathrm{GL}(n, R)$ or $\mathrm{GL}(n, C)$. The notation stands for general linear group of $n \times n$ matrices with real or complex elements. If we take the subset of $\mathrm{GL}(n, R)$ or $\mathrm{GL}(n, C)$ with unit determinant, we arrive at $\mathrm{SL}(n, R)$ and $\mathrm{SL}(n, C)$, the group of special linear $n \times n$ matrices with real or complex elements.

## O(*n*)

Now let us take a subgroup of $\mathrm{GL}(n, R)$, the *orthogonal group* $\mathrm{O}(n)$, which consists of all possible $n \times n$ real invertible matrices that are orthogonal:

$$O \times O^T = 1 \tag{A.1.3}$$

This obviously satisfies all four of the conditions for a group. Any orthogonal matrix can be written as the exponential of an antisymmetric matrix:

$$O = e^A \tag{A.1.4}$$

It is easy to see that

$$O^T = e^{A^T} = e^{-A} = O^{-1} \tag{A.1.5}$$

In general, an orthogonal matrix has

$$\tfrac{1}{2}n(n-1)$$

independent elements. Thus, we can always choose a set of $\frac{1}{2}n(n-1)$ linearly independent matrices, called the *generators* $\lambda_i$, such that we can write any element of O as

$$O = e^{\sum_{i=1}^{(1/2)n(n-1)} \rho^i \lambda_i} \tag{A.1.6}$$

The real numbers $\rho^i$ are called the *parameters* of the group, and there are thus $\frac{1}{2}n(n-1)$ parameters in $O(n)$. The number of parameters of a Lie group is called its dimension. The commutator of two of these generators yields another generator:

$$[\lambda_i, \lambda_j] = f_{ij}^k \lambda_k \tag{A.1.7}$$

where the $f$'s are called the *structure constants* of the algebra. Notice that the structure constants determine the algebra completely.

Notice that if we take cyclic combinations of three commutators, we get an exact identity:

$$[\lambda_{[i}, [\lambda_j, \lambda_{k]}]] = 0 \tag{A.1.8}$$

By expanding out these commutators, we find that the combination identically cancels to zero. This is called the *Jacobi* identity and must be satisfied for the group to close properly. By expanding out the Jacobi identity, we now have a constraint among the commutators that must be satisfied, or else the group does not close:

$$f_{[i,j}^l f_{k],l}^m = 0 \tag{A.1.9}$$

Of course, the set of orthogonal matrices closes under multiplication. A more complicated problem is to prove that this particular parametrization of the orthogonal group, with generators and parameters, closes under multiplication. Let us write

$$e^A e^B = e^C \tag{A.1.10}$$

Fortunately, the Baker–Hausdorff theorem shows that $C$ equals $A$ plus $B$ plus all possible multiple commutators of the $A$ and $B$. But since the $A$ and $B$ satisfy the Jacobi identities, the set of all possible multiple commutators of $A$ and $B$ only creates linear combinations of the generators. Thus, the group closes under multiplication.

Notice that the structure constants of the algebra form a representation, called the *adjoint representation*. If we write the structure constants as a matrix:

$$f_{ij}^k = (\lambda^k)_{ij} \tag{A.1.11}$$

Thus, the structure constants themselves form a representation.

We can always choose the commutation relations to be

$$[M^{ab}, M^{cd}] = \delta^{ac}M^{bd} - \delta^{ad}M^{bc} + \delta^{bd}M^{ac} - \delta^{bc}M^{ad} \tag{A.1.12}$$

for the antisymmetric matrix $M^{ab}$.

One convenient representation of the algebra is now given by

$$(M^{ab})_{ij} \sim \delta_i^a\delta_j^b - \delta_i^b\delta_j^a \tag{A.1.13}$$

which, we can show, satisfies the commutation relations of the group.

Let us define a set of $n$ elements $x_i$ that transforms as a *vector* under the group

$$x_i' = O_{ij}x_j \tag{A.1.14}$$

In general, we can also define a *tensor*

$$T_{\mu_1,\mu_2,\mu_3,\ldots,\mu_P} \tag{A.1.15}$$

of rank $P$ that transforms in the same way as the product of $P$ ordinary vectors $x_{\mu_i}$:

$$\bar{T}_{\nu_1,\nu_2,\ldots,\nu_N} = O_{\nu_1,\mu_1}O_{\nu_2,\mu_2}\cdots O_{\nu_P,\mu_P}T_{\mu_1,\mu_2,\ldots,\mu_P} \tag{A.1.16}$$

In addition to the vector and tensor representations of O($N$), we have the *spinor* representation of the group. Let us define the *Clifford algebra*:

$$\{\Gamma^a, \Gamma^b\} = 2\delta^{ab} \tag{A.1.17}$$

Now define a representation of the generators in terms of these Clifford numbers:

$$M^{ab} = \frac{1}{4i}[\Gamma^a, \Gamma^b] \tag{A.1.18}$$

The Clifford numbers themselves transform as vectors:

$$[M^{ab}, \Gamma^c] = i(\delta^{ac}\Gamma^b - \delta^{bc}\Gamma^a) \tag{A.1.19}$$

In general, these Clifford numbers can be represented by $2^n \times 2^n$ matrices

$$(\Gamma^a)_{\alpha\beta} \tag{A.1.20}$$

for the group O($2n$). Therefore, a spinor $\psi_\alpha$ that transforms under O($2n$) has $2^n$ elements and transforms as

$$\psi_\alpha' = (e^{M^{ab}\rho^{ab}})_{\alpha\beta}\psi_\beta \tag{A.1.21}$$

where the $M$'s are written in terms of the Clifford algebra and the $\rho$ variables are parameters.

For the group O($2n + 1$), we need one more element. This missing element is

$$\Gamma^{2n+1} = \Gamma^1\Gamma^2\cdots\Gamma^{2n} \tag{A.1.22}$$

We can easily check that this new element allows us to construct all the $M$ matrices for O($2n + 1$).

Let us now try to construct invariants under the group. Orthogonal transformations preserve the scalar product $x_i x_i$:

$$x_i x_i = \text{invariant} \tag{A.1.23}$$

If $x_i' = O_{ij} x_j$, then

$$x_i' x_i' = x O^T O x = x_i x_i \tag{A.1.24}$$

This invariant can be written

$$x_i \delta_{ij} x_j \tag{A.1.25}$$

where the metric is $\delta_{ij}$. In principle, we could also have a metric with alternating positive and negative signs along the diagonal, $\eta_{ij}$, which would create a parameter space that is noncompact. If we have $N$ positive and $M$ negative elements in $\eta_{ij}$, then the set of matrices that preserve this form is called O($N$, $M$):

$$\begin{aligned} (O^T)_{ij} \eta_{jk} O_{kl} &= \eta_{il} \\ \eta_{ij} &= \varepsilon(i) \delta_{ij} \\ \varepsilon(i) &= \pm 1 \end{aligned} \tag{A.1.26}$$

If all the elements of $\varepsilon$ are positive, this gives the group O($n$). If the signs alternate, then the group is noncompact. Special cases include

| | | |
|---|---|---|
| Projective group | O(2, 1) | |
| Lorentz group | O(3, 1) | |
| de Sitter group | O(4, 1) | (A.1.27) |
| anti-de Sitter group | O(3, 2) | |
| Conformal group | O(4, 2) | |

For example, the de Sitter group can be constructed by taking the generators of O(4, 1) and then writing the fifth component as

$$P^a \sim M^{5a} \tag{A.1.28}$$

Thus, the algebra becomes

$$\text{O(4, 1)}: \quad \begin{cases} [P^a, P^b] = M^{ab} \\ [P^a, M^{bc}] = P^b \eta^{ac} - P^c \eta^{ab} \\ [M^{ab}, M^{cd}] = \eta^{ac} M^{bd} - \cdots \end{cases} \tag{A.1.29}$$

Notice that this is almost the algebra of the Poincaré group. In fact, if we make the substitution

$$P^a \to \pm r P^a \tag{A.1.30}$$

then the only commutator that changes is

$$[P^a, P^b] = \frac{1}{r^2} M^{ab} \tag{A.1.31}$$

$r$ is called the *de Sitter radius*. This means that if we go around a circle in de Sitter space and return to the same spot, we will be rotated by a Lorentz transformation from our original orientation. Notice that if $r$ goes to infinity, we have the Poincaré group. Thus, $r$ corresponds to the radius of a five-dimensional universe such that, if $r$ goes to infinity, it becomes indistinguishable from the flat four-dimensional space of Poincaré. Letting the radius go to infinity is called the *Wigner–Inönü contraction* and will be used extensively in supergravity theories. After the contraction, the de Sitter group becomes the Poincaré group.

## SU(*n*)

The group SU($n$) consists of all possible $n \times n$ complex matrices that have unit determinant and are unitary:

$$UU^\dagger = 1 \tag{A.1.32}$$

The notation stands for special unitary $n \times n$ matrices with complex coefficients. Any unitary matrix can be written as the exponential of a Hermitian matrix $H^\dagger = H$:

$$U = e^{iH} \tag{A.1.33}$$

We can show that

$$U^\dagger = e^{-iH^\dagger} = e^{-iH} = U^{-1} \tag{A.1.34}$$

Let $n$ elements in a complex vector $u_i$ transform linearly under SU($n$):

$$u_i' = U_{ij} u_j \tag{A.1.35}$$

The $n$ complex vectors $u_i$ generate the *fundamental representation* of the group. Then an invariant can be constructed:

$$u_i^* u_i = \text{invariant} \tag{A.1.36}$$

If $u_i' = U_{ij} u_j$, it is easy to check that

$$u_i'^* u_i' = u_i^* (U^\dagger)_{ij} U_{jk} u_k = u_i^* u_i \tag{A.1.37}$$

The metric tensor for the scalar product is again $\delta_{ij}$. If we were to reverse some of the signs in this diagonal matrix, the groups that would preserve this metric are called SU($N$, $M$). An example of this would be the conformal group: SU(2, 2).

Any $n \times n$ complex traceless Hermitian matrix has $n^2 - 1$ independent elements and hence can be written in terms of $n^2 - 1$ linearly independent matrices $\lambda_i$. Thus, any element of SU($n$) can be written as

$$U = e^{i\sum_{i=1}^{n^2-1}\rho_i\lambda_i} \tag{A.1.38}$$

The BakerHausdorff theorem then guarantees that the group closes under this parametrization and that we can write the algebra of the group as

$$[\lambda_i, \lambda_j] = if_{ij}^k\lambda_k \tag{A.1.39}$$

Again, knowledge of the structure constants determines the algebra completely.

We can also construct representations of SU($n$) out of spinors. If we have the group O($2n$), then SU($n$) is a subgroup. If we construct the elements

$$A^j = \tfrac{1}{2}(\Gamma^{2j-1} - i\Gamma^{2j}) \tag{A.1.40}$$

where $\Gamma^{2j}$ are Grassman variables, then the generators of SU($n$) can be written as

$$\lambda^a = \sum_{j,k} A^{\dagger j}(\lambda^a)_{jk}A^k \tag{A.1.41}$$

Thus, we have an explicit representation for the inclusion

$$\mathrm{SU}(n) \subset \mathrm{O}(2n) \tag{A.1.42}$$

## Sp($2n$)

The symplectic groups are defined as the set of $2n \times 2n$ real matrices $S$ that preserve an antisymmetric metric $\eta$:

$$(S^T)_{ij}\eta_{jk}S_{kl} = \eta_{il} \tag{A.1.43}$$

where

$$u_i' = S_{ij}u_j \tag{A.1.44}$$

and

$$\eta_{ij} = \begin{pmatrix} 0 & 1 & 0 & 0 & \cdots \\ -1 & 0 & 0 & 0 & \cdots \\ 0 & 0 & 0 & 1 & \cdots \\ 0 & 0 & -1 & 0 & \cdots \\ \cdots & \cdots & \cdots & \cdots & \cdots \end{pmatrix} \tag{A.1.45}$$

## Accidents

Fortunately, there is a series of "accidents" that allow us to make local isomorphisms between groups. For example, O(2) is locally isomorphic to U(1):

$$\mathrm{O}(2) = U(1) \tag{A.1.46}$$

To see this, we simply note the correspondence between a matrix element of

O(2) and U(1):

$$\begin{pmatrix} \cos\theta & \sin\theta \\ -\sin\theta & \cos\theta \end{pmatrix} \leftrightarrow e^{i\theta} \tag{A.1.47}$$

Thus, they have the multiplication law $\theta_1 + \theta_2 = \theta_3$.

Another accident is

$$\mathrm{O}(3) = \mathrm{SU}(2) \tag{A.1.48}$$

The easiest way to prove this is to note that the Pauli spin matrices $\sigma_i$ are $2 \times 2$ complex matrices with the same commutation relations as the algebra of O(3). Thus,

$$e^{\sum_{i=1}^{3} \theta^i \lambda_i} \leftrightarrow e^{\sum_{i=1}^{3} i\theta^i \sigma_i} \tag{A.1.49}$$

where the matrix on the left is a $3 \times 3$ orthogonal matrix and the one on the right is a unitary matrix.

Another useful accident is

$$\mathrm{O}(4) = \mathrm{SU}(2) \otimes \mathrm{SU}(2) \tag{A.1.50}$$

To prove this, we note that the generators $M^{ij}$ of O(4) can be divided into two sets:

$$A = \{M_{12} + M_{34}, M_{31} + M_{24}, M_{23} + M_{14}\} \tag{A.1.51}$$

and

$$B = \{M_{12} - M_{34}, M_{31} - M_{24}, M_{23} - M_{14}\} \tag{A.1.52}$$

Notice that the $A$ and $B$ matrices separately generate the algebra of O(3) and that

$$[A, B] = 0 \tag{A.1.53}$$

Thus, we can parametrize any element of O(4) such that it splits up into a product of O(3) and another commuting O(3). Thus, we have proved that an element of O(4) can be split up into the product of two elements of a commuting set of SU(2) groups.

Unfortunately, these accidents are the exception, rather than the rule, for Lie groups. We list some of the accidents:

dim = 3

$$\begin{gathered} \mathrm{SU}(2, c) \sim \mathrm{SO}(3, r) \sim \mathrm{Usp}(2) \sim \mathrm{U}(1, q) \sim \mathrm{SL}(1, q) \\ \mathrm{SU}(1, 1; c) \sim \mathrm{SO}(2, 1; r) \sim \mathrm{Sp}(2, r) \sim \mathrm{SL}(2, r) \end{gathered} \tag{A.1.54}$$

dim = 6

$$\begin{gathered} \mathrm{SO}(4, r) \sim \mathrm{SU}(2, c) \otimes \mathrm{SU}(2, c) \\ \mathrm{SO}^*(4) \sim \mathrm{SU}(2, c) \otimes \mathrm{SL}(2, r) \\ \mathrm{SO}(3, 1; r) \sim \mathrm{SL}(2, c) \\ \mathrm{SO}(2, 2; r) \sim \mathrm{SL}(2, r) \otimes \mathrm{SL}(2, r) \end{gathered} \tag{A.1.55}$$

dim = 10

$$\begin{aligned} \mathrm{SO}(5, r) &\sim \mathrm{Usp}(4) \\ \mathrm{SO}(4, 1; r) &\sim \mathrm{Usp}(2, 2) \\ \mathrm{SO}(3, 2; r) &\sim \mathrm{Sp}(4, r) \end{aligned} \tag{A.1.56}$$

dim = 15

$$\begin{aligned} \mathrm{SO}(6, r) &\sim \mathrm{SU}(4, c) \\ \mathrm{SO}(5, 1; r) &\sim \mathrm{SU}^*(4) \sim \mathrm{SL}(2, q) \\ \mathrm{SO}^*(6) &\sim \mathrm{SU}(3, 1; c) \\ \mathrm{SO}(4, 2; r) &\sim \mathrm{SU}(2, 2; c) \\ \mathrm{SO}(3, 3; r) &\sim \mathrm{SL}(4, r) \end{aligned} \tag{A.1.57}$$

For arbitrary $N$, we have

$$\begin{aligned} \mathrm{SL}(N, q) &= \mathrm{SU}^*(2N) \\ \mathrm{U}(N, q) &= \mathrm{Usp}(2N) \\ \mathrm{Sp}(N, q) &= \mathrm{Usp}(2N) \\ \mathrm{O}(N, q) &= \mathrm{SO}^*(2N) \end{aligned} \tag{A.1.58}$$

also, for $N \leq 6$, we have

$$\begin{aligned} \mathrm{Spin}(3) &= \mathrm{SU}(2) \\ \mathrm{Spin}(4) &= \mathrm{SU}(2) \otimes \mathrm{SU}(2) \\ \mathrm{Spin}(5) &= \mathrm{Usp}(4) \\ \mathrm{Spin}(6) &= \mathrm{SU}(4) \end{aligned} \tag{A.1.59}$$

where $\mathrm{SL}(n)$ is the set of all $n \times n$ matrices with unit determinant that can have real, complex, or quaternionic elements $q$. Quaternions are generalizations of complex numbers such that any element can be written as

$$q = \sum_{i=0}^{3} c_i q_i \tag{A.1.60}$$

where the $c$'s are real numbers and

$$\begin{aligned} q_0 &= I \\ q_{1,2,3}^2 &= -I \\ q_1 q_2 &= -q_2 q_1 = q_3 \\ q_2 q_3 &= -q_3 q_2 = q_1 \\ q_3 q_1 &= -q_1 q_3 = q_2 \end{aligned} \tag{A.1.62}$$

## Cartan–Weyl Representation

In general, the exceptional groups do not have such a nice representation in terms of subgroups of GL($n$). We will instead use the methods of Lie and Cartan.

Among the generators $\lambda_i$ of a group, let us select out those elements $H_i$ which are all mutually commuting:

$$[H_i, H_j] = 0 \tag{A.1.63}$$

This is called the *Cartan subalgebra*. The number of elements in the Cartan subalgebra is called the *rank* $r$ of the group. Let us call all the other elements of the algebra $E$. What are the commutation relations between $H$ and the other elements $E$? In general, we cannot have the commutator between $H$ and $E$ yielding another $H$, because this would not satisfy the Jacobi identities. Therefore, the commutator of $H$ and $E$ must yield another $E$. We can always rearrange the rest of the elements $E$ of the algebra so that they become eigenvectors of the $H$'s. We will denote them by

$$E_{\boldsymbol{\alpha}} \tag{A.1.64}$$

where $\boldsymbol{\alpha}$ is called a *root vector* in $r$-dimensional space. In general, these root vectors live in $r$-dimensional space. Notice that if the group has $N$ parameters, then there are $N - r$ elements in $E$. Thus, there are $N - r$ root vectors living in $r$-dimensional space. We can always take linear combinations of the various $E$'s until we have the eigenvalue equation

$$[H_i, E_{\boldsymbol{\alpha}}] = \alpha_i E_{\boldsymbol{\alpha}} \tag{A.1.65}$$

By the Jacobi identities, the other commutation relations become

$$\begin{aligned} [E_{\boldsymbol{\alpha}}, E_{-\boldsymbol{\alpha}}] &= \alpha_i H_i \\ [E_{\boldsymbol{\alpha}}, E_{\boldsymbol{\beta}}] &= N_{\boldsymbol{\alpha},\boldsymbol{\beta}} E_{\boldsymbol{\alpha}+\boldsymbol{\beta}} \end{aligned} \tag{A.1.66}$$

The $N$'s are the structure constants of the group. They are given explicitly by

$$N_{\boldsymbol{\alpha},\boldsymbol{\beta}}^2 = \tfrac{1}{2}n(m+1)\alpha_i\alpha_i \tag{A.1.67}$$

The symmetries of the $N$'s are given by

$$\begin{aligned} N_{\boldsymbol{\alpha},\boldsymbol{\beta}} &= -N_{\boldsymbol{\beta},\boldsymbol{\alpha}} = -N_{-\boldsymbol{\alpha},-\boldsymbol{\beta}} \\ N_{\boldsymbol{\alpha},\boldsymbol{\beta}} &= N_{\boldsymbol{\beta},-\boldsymbol{\alpha}-\boldsymbol{\beta}} = N_{-\boldsymbol{\alpha}-\boldsymbol{\beta},\boldsymbol{\alpha}} \end{aligned} \tag{A.1.68}$$

## Dynkin Diagrams

Let us make a few definitions concerning the roots. Each root vector lives in an $r$-dimensional space. Thus, we can choose a subset of them with $r$ elements such that any other root can be written as a linear combination:

$$\rho = \sum_{i=1}^{r} c_i \alpha_i \tag{A.1.69}$$

A root is called a *positive root* if the first nonzero $c$ in the above equation is positive. A *simple root* is a positive root that cannot be written as the sum of two positive roots. Just as the structure constant specifies the algebra exactly, the same can be said of the *Cartan matrix*, which is an $r \times r$ matrix defined by

$$A_{ij} = 2\frac{\langle \alpha_i, \alpha_j \rangle}{\langle \alpha_j, \alpha_j \rangle} \tag{A.1.70}$$

where the $\alpha_i$ are the $r$ simple roots.

The diagonal elements of the Cartan matrix, by construction, are all equal to 2. But the matrix is not necessarily symmetric. In fact, it can be shown that the only possible values of the off-diagonal matrix elements are 0, $-1$, $-2$, $-3$. Because the Cartan matrix defines the group entirely, some simple properties of the Cartan matrix can be used to define the group graphically. The most convenient of these is the *Dynkin diagram.*

For a group of rank $r$, let us draw $r$ dots. Each simple root thus is represented by a dot. Now connect the $i$th and $j$th dot with $n$ lines, where $n$ is equal to the product of the off-diagonal elements:

$$n = A_{ij} A_{ji} \tag{A.1.71}$$

The resulting diagram is a series of dots connected by a series of multiple lines, called the *Dynkin diagram.* The power of Dynkin diagrams is that they uniquely specify the structure of any Lie group, and thus we can visually distinguish the various Lie groups.

Since the elements of the Cartan matrix are related to the scalar products on the root lattice space, we can also represent $n$ in terms of the angle $\theta$ between two root vectors:

| $n$ | $\theta$ |
|---|---|
| 0 | 90° |
| 1 | 120° |
| 2 | 135° |
| 3 | 150° |

If we specialize to compact, real forms, we have the following Dynkin diagrams (see Fig. A.1).

$$A_n = \mathrm{SU}(n+1)$$

$$\mathbf{e}_i - \mathbf{e}_j; \qquad 1 \le i \ne j \le n+1$$

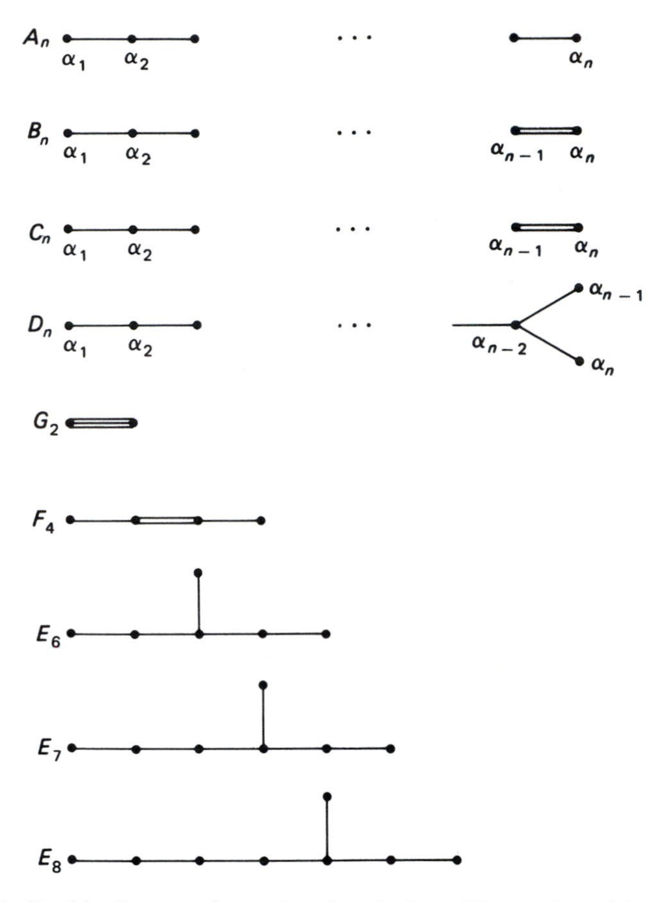


Figure A.1. Dynkin diagrams for various Lie algebras. The number of dots represents the rank of the group. The number of lines connecting the dots depends on the angle between two root vectors. Dynkin diagrams give us a convenient method of visualizing the entire structure of a Lie algebra.

$$B_n = \mathrm{SO}(2n+1)$$

$$\pm \mathbf{e}_i \pm \mathbf{e}_j$$

$$\pm \mathbf{e}_i$$

$$C_n = \mathrm{Sp}(2n)$$

$$\pm \mathbf{e}_i \pm \mathbf{e}_j$$

$$\pm 2\mathbf{e}_i$$

$$D_n = \mathrm{SO}(2n)$$

$$\pm \mathbf{e}_i \pm \mathbf{e}_j$$

$$G_2$$

$$\mathbf{e}_i - \mathbf{e}_j$$

$$\pm(\mathbf{e}_i + \mathbf{e}_j - 2\mathbf{e}_k); \qquad 1 \le i \ne j \ne k \le 3$$

$$F_4$$

$$\pm\mathbf{e}_i \pm \mathbf{e}_j; \qquad 1 \le i \ne j \le 4$$

$$\pm 2\mathbf{e}_i$$

$$\pm\mathbf{e}_1 \pm \mathbf{e}_2 \pm \mathbf{e}_3 \pm \mathbf{e}_4$$

$$E_6$$

$$\pm\mathbf{e}_i \pm \mathbf{e}_j; \qquad 1 \le i \ne j \le 5$$

$$\tfrac{1}{2}(\pm\mathbf{e}_1 \pm \mathbf{e}_2 \pm \cdots \pm \mathbf{e}_5) \pm (2 - \tfrac{5}{4})^{1/2}\mathbf{e}_6$$

where we have an even number of + signs in the last expression.

$$E_7$$

$$\pm\mathbf{e}_i \pm \mathbf{e}_j; \qquad 1 \le i \ne j \le 6$$

$$\pm\sqrt{2}\mathbf{e}_7$$

$$\tfrac{1}{2}(\pm\mathbf{e}_1 \pm \mathbf{e}_2 \pm \cdots \pm \mathbf{e}_6) \pm (2 - \tfrac{6}{4})^{1/2}\mathbf{e}_7$$

where we have an even number of + signs in the last expression.

$$E_8$$

$$\pm\mathbf{e}_i \pm \mathbf{e}_j; \qquad 1 \le i \ne j \le 7$$

$$\tfrac{1}{2}(\pm\mathbf{e}_1 \pm \mathbf{e}_2 \pm \cdots \pm \mathbf{e}_7) \pm \tfrac{1}{2}\mathbf{e}_8$$

where we have an even number of + signs in the last expression.

## §A.2. A Brief Introduction to General Relativity

The most general reparametrization of space–time, a general coordinate transformation, is given by

$$\bar{x}^\mu = \bar{x}^\mu(x) \tag{A.2.1}$$

Under this reparametrization, we use the chain rule to find that the differentials and partial derivatives transform as

$$\begin{aligned} dx^\mu &= \frac{\partial x^\mu}{\partial \bar{x}^\nu}\, d\bar{x}^\nu \\ \frac{\partial}{\partial x^\mu} &= \frac{\partial \bar{x}^\nu}{\partial x^\mu}\frac{\partial}{\partial \bar{x}^\nu} \end{aligned} \tag{A.2.2}$$

We say that the differential $dx^\mu$ transforms *contravariantly* and the derivative $\partial_\mu$ transforms *covariantly*. In direct analogy, we now define vectors that also transform in precisely the same fashion:

$$A'_\mu = \frac{\partial \bar{x}^\nu}{\partial x^\mu} A_\nu$$
$$B'^\mu = \frac{\partial x^\mu}{\partial \bar{x}^\nu} B^\nu \tag{A.2.3}$$

A tensor $T^{\mu\nu\lambda\cdots}_{\alpha\beta\gamma\cdots}$ simply transforms like the product of a series of vectors. The number of indices on a tensor is called the *rank* of the tensor.

It is now easy to show that the contraction of a covariant and a contravariant tensor yields an invariant:

$$A_\mu B^\mu = \text{invariant} \tag{A.2.4}$$

We can show that the partial derivative of a scalar is a genuine vector:

$$\partial_\mu \phi = \frac{\partial \bar{x}^\nu}{\partial x^\mu} \partial_\nu \phi \tag{A.2.5}$$

The fundamental problem of general covariance, however, arises because the partial derivative of a tensor is not itself a tensor. In order to rectify this situation, we are forced to add in another object, called the Christoffel symbol, which converts the derivative into a genuine tensor:

$$\nabla_\mu A_\nu = \partial_\mu A_\nu + \Gamma^\lambda_{\mu\nu} A_\lambda \tag{A.2.6}$$

We demand that

$$(\nabla_\mu A_\nu)' = \frac{\partial \bar{x}^\lambda}{\partial x^\mu} \frac{\partial \bar{x}^\alpha}{\partial x^\nu} \nabla_\lambda A_\alpha \tag{A.2.7}$$

This, in turn, uniquely fixes the transformation properties of the Christoffel symbol, which is *not* a genuine tensor.

We can, of course, now define the covariant derivative of a contravariant tensor:

$$\nabla_\mu A^\nu = \partial_\mu A^\nu - \Gamma^\nu_{\mu\lambda} A^\lambda \tag{A.2.8}$$

as well as the covariant derivative of an arbitrary tensor of rank $r$.

So far, we have placed no restrictions on the Christoffel symbols or even the space–time. Now let us define a metric on this space by defining the invariant distance to be

$$ds^2 = dx^\mu \, g_{\mu\nu} \, dx^\nu \tag{A.2.9}$$

where the $g$ is the metric, which transforms as a genuine second-rank tensor.

Now let us restrict the class of metric we are discussing by defining the covariant derivative of the metric to be zero:

$$\nabla_\mu g_{\nu\lambda} = 0 \tag{A.2.10}$$

Notice that there are

$$D \times \tfrac{1}{2}D(D+1) \tag{A.2.11}$$

equations to be satisfied, which is precisely the number of elements in the Christoffel symbol if we take it to be symmetric in its lower indices. Thus, we can completely solve for the Christoffel symbol in terms of the metric tensor:

$$\begin{aligned} \Gamma^{\alpha}_{\mu\nu} &= g^{\alpha\beta}\Gamma_{\mu\nu,\beta} \\ \Gamma_{\mu\nu,\beta} &= \tfrac{1}{2}(\partial_\mu g_{\nu\beta} + \partial_\nu g_{\mu\beta} - \partial_\beta g_{\mu\nu}) \end{aligned} \tag{A.2.12}$$

Notice that we assumed the Christoffel symbol to be symmetric in its lower indices. In general, this is not true, and the antisymmetric components of the Christoffel symbol are called the *torsion* tensor:

$$T^{\lambda}_{\mu\nu} = \Gamma^{\lambda}_{\mu\nu} - \Gamma^{\lambda}_{\nu\mu} \tag{A.2.13}$$

In flat space, we have the equation

$$[\partial_\mu, \partial_\nu] = 0 \tag{A.2.14}$$

Since the derivative of a field generates parallel displacements, intuitively this equation means that if we parallel transport a vector around a closed curve in flat space, we wind up with the same vector.

In curved space, however, this is not obviously true. The parallel displacement of a vector around a closed path on a sphere, for example, leads to a net rotation of the vector when we have completed the circuit.

The analog of the previous equation can also be found for curved manifolds. We can interpret the covariant derivative as the parallel displacement of a vector and the Christoffel symbol as the amount of derivation from flat space. If we now parallel displace a vector completely around a closed loop, we arrive at

$$[\nabla_\mu, \nabla_\nu]A_\lambda = R^{\alpha}_{\mu\nu\lambda}A_\alpha \tag{A.2.15}$$

where

$$R^{\alpha}_{\mu\nu\lambda} = \partial_\mu \Gamma^{\alpha}_{\nu\lambda} - \partial_\nu \Gamma^{\alpha}_{\mu\lambda} + \Gamma^{\alpha}_{\mu\beta}\Gamma^{\beta}_{\nu\lambda} - \Gamma^{\alpha}_{\nu\beta}\Gamma^{\beta}_{\mu\lambda} \tag{A.2.16}$$

Let us now try to form an action with this formalism. We first note that the volume of integration is not a true scalar:

$$d^D x = \det\left[\frac{\partial x^\mu}{\partial \bar{x}_\nu}\right] d^D \bar{x} \tag{A.2.17}$$

To create an invariant, we must multiply by the square root of the determinant of the metric tensor:

$$\sqrt{-g} = \det\left[\frac{\partial \bar{x}^\mu}{\partial x^\nu}\right]\sqrt{-\bar{g}} \tag{A.2.18}$$

The product of the two is an invariant:

$$\sqrt{-g}\, d^D x = \text{invariant} \tag{A.2.19}$$

Notice that the square root of the metric tensor does not transform as a scalar, because of the Jacobian factor. We say that it transforms as a *density*.

Notice that the curvature tensor has two derivatives. In fact, it can be shown that the contracted curvature tensor

$$R^{\alpha}_{\mu\nu\lambda}\delta^{\mu}_{\alpha}g^{\nu\lambda} = R \tag{A.2.20}$$

is the *only* genuine scalar one can write in terms of metric tensors and Christoffel symbols with two derivatives. Thus, the *only* possible action with two derivatives is

$$S = \frac{-1}{2\kappa^2}\int d^D x\,\sqrt{-g}R \tag{A.2.21}$$

This formalism, however, cannot be generalized to include spinors. If we treat the transformation matrix

$$\frac{\partial x^{\mu}}{\partial \bar{x}^{\nu}} \tag{A.2.22}$$

as an element of GL($D$), we find that there are *no finite-dimensional spinor representations of this group*. Thus, we cannot define spinors with metric tensors alone.

To remedy this situation, we construct a flat *tangent* space at every point on the manifold that possesses O($D$) symmetry. Let us define vectors in the tangent space with Roman indices $a$, $b$, $c$, $d$, .... Let us define the *vierbein* as the matrix that takes us from the $x$-space to the tangent space and vice versa.

$$\begin{aligned} e^a_{\mu}e^a_{\nu} &= g_{\mu\nu} \\ e^{\alpha a} &= g^{\alpha\beta}e^a_{\beta} \\ e^a_{\alpha}e^{\alpha b} &= \delta^{ab} \end{aligned} \tag{A.2.23}$$

We can now define a set of gamma matrices defined over either the tangent or the base space:

$$\begin{aligned} \gamma^a e^{a\mu} &= \gamma^{\mu} \\ \{\gamma^{\mu}, \gamma^{\nu}\} &= -2g^{\mu\nu} \end{aligned} \tag{A.2.24}$$

Thus, the derivative operator on a spinor becomes

$$\gamma^a e^{a\mu}\partial_{\mu} = \gamma^{\mu}\partial_{\mu} = \not{\partial} \tag{A.2.25}$$

With this tangent space, we can now define the covariant derivative of the spinor $\psi$:

$$\nabla_{\mu}\psi = \partial_{\mu}\psi + \omega^{ab}_{\mu}\sigma^{ab}\psi \tag{A.2.26}$$

where $\sigma^{ab}$ is the antisymmetric product of two gamma matrices and $\omega^{ab}_{\mu}$ is called the *spin connection*. Notice that the spin connection is a true tensor in the $\mu$ index. Under a local Lorentz transformation, the field transforms as:

$$\psi \to e^M \psi; \qquad \nabla_\mu \psi \to e^M \nabla_\mu \psi \tag{A.2.27}$$

We can also use the O(3, 1) formulation of general relativity and dispense with Christoffel symbols. We can define

$$\nabla_\mu = \partial_\mu + \omega_\mu^{ab} M^{ab} \tag{A.2.28}$$

where the $M$'s are the generators of the Lorentz group. Then we can form

$$[\nabla_\mu, \nabla_\nu] = R_{\mu\nu}^{ab} M^{ab} \tag{A.2.29}$$

where

$$R_{\mu\nu}^{ab} = \partial_\mu \omega_\nu^{ab} - \partial_\nu \omega_\mu^{ab} + \omega_\mu^{ac} \omega_\nu^{cb} - \omega_\nu^{ac} \omega_\mu^{cb} \tag{A.2.30}$$

Notice that this tensor $R_{\mu\nu}^{ab}$ yields an alternative formulation of the curvature tensor.

We also demand that the covariant derivative of the vierbein be equal to zero:

$$\begin{aligned} \nabla_\mu e_\nu^a &= \partial_\mu e_\nu^a \\ &+ \Gamma_{\mu\nu}^\lambda e_\lambda^a + \omega_\mu^{ab} e_\nu^b = 0 \end{aligned} \tag{A.2.31}$$

If we antisymmetrize this equation in $\mu\nu$, the Christoffel symbol disappears. Notice that the spin connection has

$$D \times \tfrac{1}{2} D(D-1) \tag{A.2.32}$$

components. This is precisely the number of components in the antisymmetrized version of the above equation. Thus, we can solve exactly for the connection in terms of the vierbein. The Christoffel symbol and the vierbein are very complicated expressions of each other.

Given these constrained expressions for the Christoffel symbol and the connection fields, we can now show the relationship between the curvature tensors in the two formalisms:

$$R_{\mu\nu} = R(\Gamma)_{\mu\alpha\nu}^{\alpha} = R(\omega)_{\mu\alpha}^{ab} e_{a\nu} e_b^\alpha \tag{A.2.33}$$

If we take an arbitrary spinor and make a parallel transport around a closed circuit with area $\Delta^{\mu\nu}$, we have

$$\psi \to (1 + \Delta^{\mu\nu} R_{\mu\nu}^{ab} \sigma^{ab}) \psi \tag{A.2.34}$$

Notice that the $\sigma^{ab}$ matrices are the generators of Euclidean Lorentz transformations O($D$). Thus, after a parallel displacement around a closed path, the spinor simply is rotated from its original orientation by an angle proportional to

$$\Delta^{\mu\nu} R_{\mu\nu}^{ab} \tag{A.2.35}$$

Notice that we can make an arbitrary number of closed paths starting from a single point. Each time, the spinor performs a rotation. Notice that this forms a group. In fact, the group is simply $O(D)$, which is called the *holonomy* group.

## §A.3. A Brief Introduction to the Theory of Forms

Let us define a *one-form* $A$ by

$$A = A_\mu \, dx^\mu \tag{A.3.1}$$

where $A_\mu$ is a vector field and the differentials $dx^\mu$ are now anticommuting:

$$\begin{aligned} dx^\mu \wedge dx^\nu &= -dx^\nu \wedge dx^\mu \\ dx^\mu \wedge dx^\mu &= 0 \end{aligned} \tag{A.3.2}$$

Let us define the derivative operator as

$$d = dx^\mu \, \partial_\mu \tag{A.3.3}$$

Notice that because the derivatives commute

$$[\partial_\mu, \partial_\nu] = 0 \tag{A.3.4}$$

so therefore

$$d^2 = 0 \tag{A.3.5}$$

where $d$ is nilpotent.

Let us now define a *two-form*:

$$F = F_{\mu\nu} \, dx^\mu \wedge dx^\nu \tag{A.3.6}$$

Notice that the curvature associated with a vector field is a two-form:

$$\begin{aligned} F = dA &= dx^\mu \, \partial_\mu A_\nu \, dx^\nu \\ &= \tfrac{1}{2}(\partial_\mu A_\nu - \partial_\nu A_\mu) \, dx^\mu \wedge dx^\nu \end{aligned} \tag{A.3.7}$$

Because the $d$ operator is nilpotent, we have

$$dF = d^2 A = 0 \tag{A.3.8}$$

Thus the Bianchi identity for the Maxwell theory, expressed in terms of forms, is nothing but the nilpotency of $d$.

A form $\omega$ is called *closed* if

$$\text{closed:} \quad d\omega = 0 \tag{A.3.9}$$

A form $\omega$ is called *exact* if

$$\text{exact:} \quad \omega = dQ \tag{A.3.10}$$

for some form $Q$. Thus, the curvature form $F$ is exact because it can be written as the divergence of the one form $A$. It is also closed because of the Bianchi identities.

We can also combine this with a local gauge group with generators $\lambda_a$. Let

$$A = A_\mu^a \lambda_a \, dx^\mu \tag{A.3.11}$$

Then the curvature form is

$$F = dA + A \wedge A \tag{A.3.12}$$

Furthermore, the gauge variation of the Yang–Mills field under

$$\Lambda = \Lambda^a \lambda_a \tag{A.3.13}$$

is

$$\delta A = d\Lambda + A \wedge \Lambda - \Lambda \wedge A \tag{A.3.14}$$

Inserting the variation of the field $A$ into the curvature $F$, we find

$$\delta F = F \wedge \Lambda - \Lambda \wedge F \tag{A.3.15}$$

Thus, the variation of the action is zero:

$$\delta \operatorname{Tr}(F^2) = 2 \operatorname{Tr}(F \wedge \Lambda) - 2 \operatorname{Tr}(\Lambda \wedge F) = 0 \tag{A.3.16}$$

Let us now write the anomaly term $F\tilde{F}$ in the language of forms. The divergence of the axial current is also the square of two curvatures, which is also a total derivative. In the language of forms, we find that this is an exact form:

$$\operatorname{Tr}(F \wedge \tilde{F}) = d\omega_3 \tag{A.3.17}$$

where

$$\omega_3 = \operatorname{Tr}(A \, dA + \tfrac{2}{3} A^3) \tag{A.3.18}$$

$\omega_3$ is a three-form which we call a *Chern–Simons* form. In turn, its gauge variation is equal to another form that is also exact:

$$\delta\omega_3 = \operatorname{Tr}(d\Lambda \wedge dA) = d\omega_2 \tag{A.3.19}$$

where

$$\omega_2 = \operatorname{Tr}(\Lambda \wedge dA) \tag{A.3.20}$$

We also note that these identities apply equally well to Yang–Mills theories as to general relativity. For gravity, we have the gauge group O(3, 1). In other words, gravity has two gauge invariances, the general covariance of the coordinates $x$ and the local Lorentz transformations of the tangent space. We will explain this more later.

In general, an *N form* is defined as

$$\omega_N = \omega_{\mu_1 \mu_2 \cdots \mu_N} dx^{\mu_1} \wedge dx^{\mu_2} \wedge \cdots \wedge dx^{\mu_N} \tag{A.3.21}$$

All of the above equations, of course, can be derived without the use of the theory of forms. However, forms give us a powerful shorthand that allows us to manipulate complex mathematical quantities. Notice, for example, that

Stokes' theorem, expressed in the language of forms, now becomes

$$\int_M d\omega = \int_{\partial M} \omega \tag{A.3.22}$$

where $\partial M$ is the boundary of the manifold $M$.

Some simple properties of these forms are

$$\begin{aligned} \omega_p \wedge \omega_q &= (-1)^{pq} \omega_q \wedge \omega_p \\ d(\omega_p \wedge \omega_q) &= d\omega_p \wedge \omega_q + (-1)^p \omega_p \wedge d\omega_q \end{aligned} \tag{A.3.23}$$

Let us also introduce the Hodge star operator, which allows us to take the dual of a $p$-form and convert it to an $n-p$-form in $n$ dimensions:

$$*(dx^{\mu_1} \wedge \cdots \wedge dx^{\mu_p}) = \frac{|g|^{1/2}}{(n-p)!} \varepsilon^{\mu_1 \cdots \mu_p}_{\mu_{p+1} \cdots \mu_n} dx^{\mu_{p+1}} \wedge \cdots \wedge dx^{\mu_n} \tag{A.3.24}$$

Some properties of the star operation are

$$\begin{aligned} ** \omega_p &= (-1)^{p(n-p)} \omega_p \\ \omega_p \wedge *\omega_q &= \omega_q \wedge *\omega_p \end{aligned} \tag{A.3.25}$$

We will devote the next few pages to proving the assertion in Chapter 9 that an invariant polynomial is both closed and exact. We define an invariant polynomial as one that satisfies

$$P(\alpha) = P(g^{-1} \alpha g) \tag{A.3.26}$$

Let us start by defining a homogeneous invariant polynomial of degree $r$, which is dependent on the forms $\alpha_i$:

$$P = P(\alpha_1, \alpha_2, \ldots, \alpha_r) \tag{A.3.27}$$

Let us differentiate this polynomial, being careful to differentiate each form separately within $P$:

$$dP = \sum_{1 \leq i \leq r} (-1)^{d_1 + \cdots + d_{i-1}} P(\alpha_1, \ldots, d\alpha_i, \ldots, \alpha_r) \tag{A.3.28}$$

Each time the derivative $d$ passes over a form $\alpha_i$, it picks up a factor:

$$d\alpha_i = (d\alpha_i) + \alpha_i d(-1)^{d_i} \tag{A.3.29}$$

Now assume that

$$\alpha \to g^{-1} \alpha g \tag{A.3.30}$$

For $g$ close to unity, we can always write

$$\begin{aligned} g &= 1 + \omega \\ \delta\alpha &= -\omega \wedge \alpha + \alpha \wedge \omega \end{aligned} \tag{A.3.31}$$

and let us calculate the variation of a homogeneous polynomial $P$ of degree

$r$ under this shift:

$$\delta P(\alpha_1, \alpha_2, \ldots, \alpha_r) = 0 = \sum_{1 \le i \le r} (-1)^{d_1 + \cdots + d_{i-1}} [P(\alpha_1, \alpha_2, \ldots, \omega \wedge \alpha_i, \ldots, \alpha_r) - (-1)^{d_i} P(\alpha_1, \alpha_2, \ldots, \alpha_i \wedge \omega, \ldots, \alpha_r)] \tag{A.3.32}$$

The trick is now to add the contribution of both $dP$ and $\delta P$:

$$dP + \delta P = \sum_{1 \le i \le r} (-1)^{d_1 + \cdots + d_i} P(\alpha_1, \alpha_2, \ldots, D\alpha_i, \ldots, \alpha_r) \tag{A.3.33}$$

Now set each of the $\alpha_i$ to be a curvature form that obeys the Bianchi identity $D\alpha_i = 0$. Thus, we find

$$dP(\Omega) = 0 \tag{A.3.34}$$

as we claimed.

The second part of the proof is a bit more involved. Let us define

$$\begin{aligned} \Omega &= d\omega + \omega \wedge \omega \\ \Omega' &= d\omega' + \omega' \wedge \omega' \end{aligned} \tag{A.3.35}$$

Our plan is now to show that

$$P(\Omega') - P(\Omega) = dQ \tag{A.3.36}$$

for some form $Q$.

First we want to write a curvature form that allows us to interpolate continuously between $\Omega$ and $\Omega'$. Let

$$\begin{aligned} \omega_t &= \omega + t\eta \\ \eta &= \omega' - \omega \end{aligned} \tag{A.3.37}$$

Notice that the variable $t$ allows us to interpolate between the two forms:

$$\begin{aligned} t = 0 &\to \omega_t = \omega \\ t = 1 &\to \omega_t = \omega' \end{aligned} \tag{A.3.38}$$

It is easy now to find the curvature form that interpolates between the two, as a function of $t$:

$$\Omega_t = d\omega_t + \omega_t \wedge \omega_t \tag{A.3.39}$$

where

$$t = 0 \to \Omega_t = \Omega \tag{A.3.40}$$

Notice that the form varies between $\Omega$ and $\Omega'$ as $t$ varies from zero to one.

Now let us write down the form $q$ such that

$$q(\beta, \alpha) = rP(\beta, \alpha, \alpha \cdots (r - 1 \text{ times}) \cdots \alpha) \tag{A.3.41}$$

where the form $\alpha$ is repeated $r - 1$ times in the invariant polynomial. This form $q$ will be the key to showing that $P$ is an exact form.

By the reasoning given earlier, if we differentiate $q$, we find

$$dq(\eta, \Omega_t) = r\, dP(\eta, \Omega_t \cdots (r-1 \text{ times}) \cdots \Omega_t)$$
$$= q(D\eta, \Omega_t) - r(r-1)tP(\eta, \Omega_t \wedge \eta - \eta \wedge \Omega_t, \Omega_t, \ldots, \Omega_t) \quad \text{(A.3.42)}$$

But we also have the identity, due to the fact that $q$ is an invariant polynomial,

$$2q(\eta \wedge \eta, \Omega_t) + r(r-1)P(\eta, \Omega_t \wedge \eta - \eta \wedge \Omega_t, \ldots, \Omega_t) = 0 \quad \text{(A.3.43)}$$

In the previous equation, we used

$$g = 1 + \eta \quad \text{(A.3.44)}$$

Putting both equations together, we find

$$dq(\eta, \Omega_t) = q(D\eta, \Omega_t) + 2tq(\eta \wedge \eta, \Omega_t) = \frac{dP(\Omega_t)}{dt} \quad \text{(A.3.45)}$$

Thus, we arrive at:

$$\int_0^1 \frac{dP(\Omega_t)}{dt} = P(\Omega') - P(\Omega)$$
$$= d\int_0^1 q(\omega' - \omega, \Omega_t)\, dt = dQ \quad \text{(A.3.46)}$$

We thus obtain our final result:

$$P(\Omega') - P(\Omega) = dQ$$
$$Q = \int_0^1 q(\omega' - \omega, \Omega_t)\, dt + \text{closed forms} \quad \text{(A.3.47)}$$

Thus, an invariant polynomial based on curvature two-forms is both closed and exact.

## §A.4. A Brief Introduction to Supersymmetry

In the late 1960s, physicists tried to construct the master group that would allow a synthesis of an internal symmetry group (like SU(3)) and the Lorentz or Poincaré group. They sought a group $M$ that was a nontrivial union of an internal group $U$ and the Poincaré group:

$$M \supset U \otimes P \quad \text{(A.4.1)}$$

Intense interest was sparked in groups like SU(6, 6). However, the celebrated Coleman–Mandula theorem showed that this program was impossible. There are no unitary finite-dimensional representations of a noncompact group. So either

(1) The group $M$ has continuous masses, or
(2) The group $M$ has an infinite number of particles in each irreducible representation.

Either way, it is a disaster. However, it turns out that *supergroups* or *graded Lie groups* allow for an evasion of the no-go theorem.

The work of Lie and Cartan concerned only continuous simple groups where the parameters $\rho_i$ were real. However, if we allow these parameters to be Grassmann-valued, we can extend the classical groups to the *supergroups*.

Two large infinite classes of groups we will be interested in are the $\mathrm{Osp}(N/M)$ and the $\mathrm{SU}(N/M)$.

Let us begin with the group $\mathrm{O}(N)$, which preserves the invariant:

$$\mathrm{O}(N): \quad x_i x_i = \text{invariant} \tag{A.4.2}$$

and the group $\mathrm{Sp}(M)$, which preserves the form:

$$\mathrm{Sp}(M): \quad \theta_m C_{mn} \theta_n = \text{invariant} \tag{A.4.3}$$

where the $C$ matrices are real antisymmetric matrices because the $\theta_i$ are Grassman-valued. The *orthosymplectic group* is now defined as the group that preserves the sum:

$$\mathrm{Osp}(N/M): \quad x_i x_i + \theta_m C_{mn} \theta_n = \text{invariant} \tag{A.4.4}$$

Notice that the orthosymplectic group obviously contains the product:

$$\mathrm{Osp}(N/M) \supset \mathrm{O}(N) \otimes \mathrm{Sp}(M) \tag{A.4.5}$$

The simplest way to exhibit the matrix representation of this group is to use the block diagonal form:

$$\mathrm{Osp}(N/M) = \begin{Bmatrix} \mathrm{O}(N) & A \\ B & \mathrm{Sp}(M) \end{Bmatrix} \tag{A.4.6}$$

with simple restrictions on the $A$ and $B$ matrices.

Similarly, the superunitary groups can be defined as the groups that preserve the complex form:

$$\begin{gathered} (x^i)^* x^j \delta_{ij} + (\theta^m)^* \theta^n g_{mn} \\ g_{mn} = \pm \delta_{mn} \end{gathered} \tag{A.4.7}$$

The bosonic decomposition of the group is given by

$$\mathrm{SU}(N/M) \supset \mathrm{SU}(N) \otimes \mathrm{SU}(M) \otimes \mathrm{U}(1) \tag{A.4.8}$$

Let us write the generators of $\mathrm{Osp}(1/4)$ as

$$M_A = (P_\mu, M_{\mu\nu}, Q_\alpha) \tag{A.4.9}$$

which have the commutation relations

$$[M_A, M_B]_\pm = f_{AB}^C M_C \tag{A.4.10}$$

Written out explicitly, the commutators involving the supersymmetry generator are

$$\{Q_\alpha, Q_\beta\} = 2(\gamma^\mu C)_{\alpha\beta} P_\mu$$
$$[Q_\alpha, P_\mu] = 0 \tag{A.4.11}$$
$$[Q_\alpha, M_{\mu\nu}] = (\sigma_{\mu\nu})_\alpha^\beta Q_\beta$$

What we want is an explicit representation of these generators, in the same way that

$$P_\mu = -i\partial_\mu \tag{A.4.12}$$

is the generator of translations in $x$-space. Now we must enlarge the concept of space–time to include the supersymmetric partner of the $x$-coordinate. Let us define *superspace* as the space created by the pair

$$x_\mu, \theta_\alpha \tag{A.4.13}$$

where $\theta_\alpha$ is a Grassmann number. Let us define the supersymmetry generator

$$Q_\alpha = \frac{\partial}{\partial\bar{\theta}^\alpha} - i(\gamma^\mu\theta)_\alpha\partial_\mu \tag{A.4.14}$$

where $\theta$ is a Grassmann number. We choose this particular representation because the anticommutator between two such generators yields a displacement, as it should:

$$\{Q_\alpha, Q_\beta\} = -2(\gamma^\mu C)_{\alpha\beta} i\partial_\mu \tag{A.4.15}$$

Notice that $\bar{\varepsilon}Q$ makes the following transformations on superspace:

$$x_\mu \to x_\mu - i\bar{\varepsilon}\gamma_\mu\theta$$
$$\theta_\alpha \to \theta_\alpha + \varepsilon_\alpha \tag{A.4.16}$$

Notice also that we can construct the operator

$$D_\alpha = \frac{\partial}{\partial\bar{\theta}^\alpha} + i(\gamma^\mu\partial_\mu\theta)_\alpha \tag{A.4.17}$$

This anticommutes with the supersymmetry generator:

$$\{Q_\alpha, D_\beta\} = 0 \tag{A.4.18}$$

This is very important because it allows us to place restrictions on the representations of supersymmetry without destroying the symmetry. This permits us to extract the irreducible representations from the reducible ones.

Let us now try to construct invariant actions under supersymmetry. Let us define the *superfield* $V$ as the most general power expansion in this superspace:

$$V(x, \theta) \tag{A.4.19}$$

Then a representation of supersymmetry is given by

$$\delta V(x, \theta) = V(x + \delta x, \theta + \delta\theta) - V(x, \theta) = \bar{\varepsilon}^\alpha Q_\alpha V(x, \theta) \tag{A.4.20}$$

Notice that this definition proves that the product of two superfields is also a superfield:

$$V_1 V_2 = V_3$$

Thus, we can construct a large set of representations of supersymmetry by this simple product rule. Now let us calculate the explicit transformation of the fields. We will sometimes find it useful to break up the 4-component spinor into two 2-component spinors, because of the identity:

$$\mathrm{O}(4) = \mathrm{SU}(2) \otimes \mathrm{SU}(2) \tag{A.4.21}$$

Using indices $A$ and $\dot{A}$, $A = 1, 2$, we will write a Majorana four-spinor in terms of its $\mathrm{SU}(2) \otimes \mathrm{SU}(2)$ content:

$$\begin{aligned} \chi^\alpha &= \begin{pmatrix} \chi^A \\ \bar{\chi}_{\dot{A}} \end{pmatrix} \\ \bar{\chi}_\alpha &= (\chi_A, -\bar{\chi}^{\dot{A}}) \end{aligned} \tag{A.4.22}$$

If we invert this, we find

$$\begin{aligned} \chi^A &= \tfrac{1}{2}(1 + \gamma_5)\chi \\ \bar{\chi}_{\dot{A}} &= \tfrac{1}{2}(1 - \gamma_5)\chi \end{aligned} \tag{A.4.23}$$

and

$$\begin{aligned} C &= \begin{pmatrix} \varepsilon_{AB} & 0 \\ 0 & \varepsilon^{\dot{A}\dot{B}} \end{pmatrix} \\ \varepsilon_{AB} = \varepsilon^{AB} &= -\varepsilon_{\dot{A}\dot{B}} = -\varepsilon^{\dot{A}\dot{B}}, \qquad \varepsilon_{12} = 1 \end{aligned} \tag{A.4.24}$$

In this notation, the covariant derivatives can be written as

$$\begin{aligned} D_A &= \frac{\partial}{\partial \theta^A} - i(\sigma^\mu)_{A\dot{A}}\theta^{\dot{A}}\partial_\mu \\ D_{\dot{A}} &= \frac{\partial}{\partial \theta^{\dot{A}}} - i(\sigma^\mu)_{A\dot{A}}\theta^A\partial_\mu \end{aligned} \tag{A.4.25}$$

where

$$\sigma^\mu = (1, \boldsymbol{\sigma}) \tag{A.4.26}$$

The real vector superfield $V$ can be decomposed as

$$\begin{aligned} V(x, \theta, \bar{\theta}) = C &- i\theta\chi - i\bar{\chi}'\bar{\theta} - \tfrac{1}{2}i\theta^2(M - iN) + \tfrac{1}{2}i\bar{\theta}^2(M + iN) - \theta\sigma^\mu\bar{\theta}A_\mu \\ &+ i\bar{\theta}^2\theta(\lambda - \tfrac{1}{2}i\partial\!\!\!/\bar{\chi}') - i\theta^2\bar{\theta}(\bar{\lambda}' - \tfrac{1}{2}i\partial\!\!\!/\chi) \\ &- \tfrac{1}{2}\theta^2\bar{\theta}^2(D + \tfrac{1}{2}\Box C) \end{aligned} \tag{A.4.27}$$

We can now read off the supersymmetry transformation parametrized by $\zeta$

on these 16 fields:

$$
\begin{aligned}
\delta C &= \bar{\zeta}\gamma_5\chi \\
\delta\chi &= (M + \gamma_5 N)\zeta - i\gamma^\mu(A_\mu + \gamma_5\partial_\mu C)\zeta \\
\delta M &= \bar{\zeta}(\lambda - i\not{\partial}\chi) \\
\delta N &= \bar{\zeta}\gamma_5(\lambda - i\not{\partial}\chi) \\
\delta A_\mu &= i\bar{\zeta}\gamma_\mu\lambda + \bar{\zeta}\partial_\mu\chi \\
\delta\lambda &= -i\sigma^{\mu\nu}\zeta\partial_\mu A_\nu - \gamma_5\zeta D \\
\delta D &= -i\bar{\zeta}\not{\partial}\gamma_5\lambda
\end{aligned}
\tag{A.4.28}
$$

We call it a vector superfield because it contains a vector particle in its representation (not because the superfield itself is a vector field under the Lorentz group). In general, vector fields can be complex and they are *reducible.* To form irreducible representations, we will find it convenient to place constraints on them that do not destroy their supersymmetric nature. The constraints must therefore commute with the supersymmetry generator.

Notice that, because $D_\alpha$ anticommutes with the supersymmetry generator, we can impose this derivative on a superfield and still get a representation of supersymmetry. Let us now try to construct various representations of supersymmetry based on this simple principle. We can impose

$$D_{\dot{A}}\phi = 0 \tag{A.4.29}$$

A superfield that satisfies this constraint is called a *chiral superfield.* Notice that it has half the number of fields of the original superfield but that it still transforms under the group correctly. A chiral superfield has the decomposition

$$\phi(x, \theta) = A + 2\theta\psi - \theta^2 F \tag{A.4.30}$$

The variation of this superfield can also be read off the variation

$$\delta\phi = -i[\phi, \zeta Q + \bar{Q}\bar{\zeta}] \tag{A.4.31}$$

We easily obtain

$$
\begin{aligned}
\delta A &= 2\zeta\psi \\
\delta\psi &= -\zeta F - i\partial_\mu A\sigma^\mu\bar{\zeta} \\
\delta F &= -2i\partial_\mu\psi\sigma^\mu\bar{\zeta}
\end{aligned}
\tag{A.4.32}
$$

We could also try other combinations of constraints, such as

$$D_A\phi = 0 \tag{A.4.33}$$

on a chiral superfield. We find, however, that the combination of the two constraints imposed simultaneously implies that $\phi$ is a constant.

Another constraint might be

$$D^A D_A\phi = 0 \tag{A.4.34}$$

This yields the linear multiplet. (Unfortunately, actions based on this are usually equivalent to actions based on chiral superfields, so we learn nothing new.) Another constraint might be

$$[D_A, D_{\dot{B}}]\phi = 0 \tag{A.4.35}$$

Unfortunately, this yields a constant field. We might also impose

$$D_{\dot{B}} D^A D_A \phi = 0 \tag{A.4.36}$$

This again gets us back to the chiral superfield. Finally, we might also try

$$D_A \bar{D}^2 \phi = 0 \tag{A.4.37}$$

for real $\phi$. This actually yields an entirely new superfield, which we will use to build the Yang–Mills action.

In summary, the only new fields that transform as irreducible representations of supersymmetry are the chiral superfield, the vector superfield, and the Yang–Mills superfield. The other combinations that one might try are either empty or redundant to the original set.

Now let us discuss the problem of forming an invariant action by defining Grassmann integration.

Integration over these Grassmann variables must be carefully defined. Ordinary integration over real numbers, of course, is translation invariant:

$$\int_{-\infty}^{\infty} dx\, \phi(x) = \int_{-\infty}^{\infty} dx\, \phi(x + c) \tag{A.4.38}$$

where $c$ is a real displacement. We would like Grassmann-valued integration to have the same property:

$$\int d\theta\, \phi(\theta) = \int d\theta\, \phi(\theta + c) \tag{A.4.39}$$

If we power expand this function $\phi(\theta)$ in a Taylor series, we have the simple expression

$$\phi(\theta) = a + b\theta \tag{A.4.40}$$

If we define

$$\begin{aligned} I_0 &= \int d\theta \\ I_1 &= \int d\theta\, \theta \end{aligned} \tag{A.4.41}$$

then translation invariance forces us to have

$$\int d\theta\, \phi(\theta) = aI_0 + bI_1 = (a + bc)I_0 + bI_1 \tag{A.4.42}$$

Thus, we must have $I_0 = 0$ and $I_1$ can be normalized to one:

$$\begin{aligned} I_0 &= 0 \\ I_1 &= 1 \end{aligned} \tag{A.4.43}$$

or

$$\int d\theta = 0; \qquad \int d\theta\, \theta = 1 \tag{A.4.44}$$

In other words, we have the curious-looking identity

$$\int d\theta = \frac{\partial}{\partial\theta} \tag{A.4.45}$$

With these identities, we can show that

$$\int \prod_{i=1}^{N} d\theta_i\, d\bar{\theta}_i \exp\left[\sum_{i,j=1}^{N} \bar{\theta}_i A_{ij} \theta_j\right] = \det(A_{ij}) \tag{A.4.46}$$

Thus, in general, invariant actions can be formed (see (A.4.27, 30))

$$\begin{aligned} &\int d^4\theta\, d^4x\, V(x, \theta) \to D \text{ term} \\ &\int d^2\theta\, d^4x\, \phi(x, \theta) \to F \text{ term} \end{aligned} \tag{A.4.47}$$

The first integral only selects out the $D$-term of the superfield. The second integral only selects out the $F$-term of the chiral superfield. In general, we call these "$F$ and $D$" terms. We can check that these are invariant actions:

$$\delta \int d^8x\, V = \int d^8x\, \bar{\varepsilon}^\alpha Q_\alpha V = 0 \tag{A.4.48}$$

This is because the integral of a total derivative, in either $x$ or $\theta$ space, is zero.

Now let us try to write down simple invariant actions based on $F$ and $D$ invariant terms. The simplest invariant action is called the Wess–Zumino model:

$$S = \int d^8x\, \bar{\phi}\phi + \left(\int d^6x \left[\mu\phi + \tfrac{1}{2}m\phi^2 + \frac{\lambda}{3!}\phi^3\right] + \text{h.c.}\right) \tag{A.4.49}$$

Written out in components after performing the $\theta$ integration, it contains

$$S = \int d^4x \{-\tfrac{1}{2}(\partial_\mu A)^2 - \tfrac{1}{2}(\partial_\mu B)^2 - \tfrac{1}{2}\bar{\chi}\gamma^\mu\partial_\mu\chi + \tfrac{1}{2}F^2 + \tfrac{1}{2}G^2\} \tag{A.4.50}$$

Notice that we have now constructed an action with an irreducible representation of supersymmetry with a spin-0 and spin-$\frac{1}{2}$ multiplet: $(\frac{1}{2}, 0)$. To construct the $(1, \frac{1}{2})$ multiplet, we need the following construction for the Maxwell action, given by

$$S = \int d^4x\, d^2\theta\, W^A W_A \tag{A.4.51}$$

where

$$
\begin{aligned}
W_A &= \bar{D}^2 D_A V \\
W_{\dot{A}} &= D^2 D_{\dot{A}} V \\
D_{\dot{A}} W_B &= 0
\end{aligned}
\tag{A.4.52}
$$

where $V$ is a real vector supermultiplet, which transforms as

$$
\delta V = \bar{\Lambda} - \Lambda \tag{A.4.53}
$$

$V$ is real, but $\Lambda$ is chiral: $\Lambda^+ = -\bar{\Lambda}$. Under this transformation, we find that

$$
\delta W_A = 0 \tag{A.4.54}
$$

so the action is trivially invariant under both supersymmetry and U(1) gauge invariance. Notice that the vector supermultiplet contains the Maxwell field $A_\mu$, while the chiral supermultiplet $\Lambda$ contains the gauge parameter $\lambda$. Written out in components, this equals

$$
S = \int d^4x \left(-\tfrac{1}{4}F_{\mu\nu}^2 - \tfrac{1}{2}\bar{\psi}\gamma^\mu\partial_\mu\psi + \tfrac{1}{2}D^2\right) \tag{A.4.55}
$$

which is invariant under

$$
\begin{aligned}
\delta A_\mu &= \bar{\varepsilon}\gamma_\mu\psi \\
\delta\psi &= \left(-\tfrac{1}{2}\sigma^{\mu\nu}F_{\mu\nu} + \gamma_5 D\right)\varepsilon \\
\delta D &= \bar{\varepsilon}\gamma_5\gamma^\mu\partial_\mu\psi
\end{aligned}
\tag{A.4.56}
$$

The next multiplet we wish to investigate is the $(2, \frac{3}{2})$ multiplet. Historically, it was thought that the Rarita–Schwinger theory was fundamentally flawed because it permitted no consistent couplings to other fields. However, physicists neglected to couple the Rarita–Schwinger field with the graviton. The inconsistencies all disappear for this multiplet.

## §A.5. A Brief Introduction to Supergravity

There are at least four ways to formulate supergravity:

(1) Components—this method relies heavily on trial and error. However, it yields the most explicit form of the action.
(2) Curvatures—this method stresses the group theory and the analogy with Yang–Mills theory.
(3) Tensor calculus—this yields precise rules for the multiplication of representations of supersymmetry.
(4) Superspace—this is the most elegant formulation of supergravity; however, it is also the most difficult. The higher $N$ superspace formulations of supergravity are still not known because the torsion constraints are too difficult to solve.

We will concentrate on the method of curvatures because it resembles the Yang–Mills construction we have been using so far.

Since Osp(1/4) has 14 generators, let us define the 14 *connection fields* of Osp(1/4) as

$$h_\mu^A = (e_\mu^a, \omega_\mu^{ab}, \bar{\psi}_\mu^\alpha) \tag{A.5.1}$$

Then the global variation of the connection fields is

$$\delta h_\mu^A = f_{BC}^A \varepsilon^B h_\mu^C \tag{A.5.2}$$

where

$$\varepsilon^A = (\varepsilon^a, \varepsilon^{ab}, \bar{\varepsilon}^\alpha) \tag{A.5.3}$$

The covariant derivative is now given by

$$\begin{aligned} \nabla_\mu &= \partial_\mu + h_\mu^A M_A \\ &= \partial_\mu + e_\mu^a P_a + \omega_\mu^{ab} M_{ab} + \bar{\psi}_{\mu,\alpha} Q^\alpha \qquad (a > b) \end{aligned} \tag{A.5.4}$$

The fields transform under a local gauge transformation as

$$\delta h_\mu^A = \partial_\mu \varepsilon^A + h_\mu^B \varepsilon^C f_{CB}^A \tag{A.5.5}$$

Now form the commutator of two covariant derivatives:

$$[\nabla_\mu, \nabla_\nu] = R_{\mu\nu}^A M_A \tag{A.5.6}$$

where

$$R_{\mu\nu}^A = \partial_\mu h_\nu^A - \partial_\nu h_\mu^A + h_\nu^B h_\mu^C f_{CB}^A \tag{A.5.7}$$

In component form, we have

$$\begin{aligned} R_{\mu\nu}^a(P) &= \partial_\mu e_\nu^a + \omega_\mu^{ab} e_\nu^b - (\mu \leftrightarrow \nu) \\ R_{\mu\nu}^{ab}(M) &= \partial_\mu \omega_\nu^{ab} + \omega_\mu^{ac} \omega_\nu^{cb} - (\mu \leftrightarrow \nu) \\ R_{\mu\nu}^\alpha(Q) &= \partial_\mu \bar{\psi}_\nu^\alpha + \bar{\psi}_\nu \omega_\mu^{ab} \sigma^{ab} - (\mu \leftrightarrow \nu) \end{aligned} \tag{A.5.8}$$

The variation of the curvature can easily be shown to be

$$\delta R_{\mu\nu}^A = R_{\mu\nu}^B \varepsilon^C f_{CB}^A \tag{A.5.9}$$

The action for supergravity is now given by

$$S = \int d^4x\, \varepsilon^{\mu\nu\rho\sigma} \{ R_{\mu\nu}(M)^{ab} R_{\rho\sigma}(M)^{cd} \varepsilon_{abcd} + R_{\mu\nu}(Q)^\alpha R_{\rho\sigma}(Q)^\beta (\gamma_5 C)_{\alpha\beta} \} \tag{A.5.10}$$

If we make a variation of this action, we find that the action is not totally invariant unless we set

$$R_{\mu\nu}(P)^a = 0 \tag{A.5.11}$$

The action is invariant up to the term

$$\delta \omega_a^{\mu\nu} R_{\mu\nu}^a(P) \tag{A.5.12}$$

However, since we imposed this constraint (A.5.11) from the beginning, we find that the action, indeed, is fully invariant under the transformation.

This constraint appears to be highly unnatural until one realizes that it is the same as the vanishing of the covariant derivative of the vierbein in (A.2.31). Thus, we will choose the vierbein to have zero derivative in order to have the final invariance of the action.

The final action is

$$L = -\frac{1}{2\kappa^2} eR - \tfrac{1}{2}\bar{\psi}_\mu \gamma_\nu \gamma^5 D_\sigma \psi_\rho \varepsilon^{\mu\nu\sigma\rho} \tag{A.5.13}$$

Unfortunately, the situation for higher $N$ actions is much less clear. The superspace method still has not been solved for the higher $N$, but the $N = 8$ supergravity has been constructed using a trick: by extending the dimension of space–time to 11, we can construct $N = 1$, $D = 11$ supergravity. Then we compactify in order to reduce down to $N = 8$, $D = 4$ supergravity.

The starting point for constructing the 11-dimensional supergravity is to realize that we need equal numbers of bosonic and fermionic fields. By trial and error, we see that the following choices yield equal numbers of fields:

$$\begin{aligned} e_M^A &= \tfrac{1}{2}9 \times 10 - 1 = 44 \text{ components} \\ \psi_M &= \tfrac{1}{2}(9 \times 32 - 32) = 128 \text{ components} \\ A_{MNP} &= \binom{9}{3} \\ &= 84 \text{ components} \end{aligned} \tag{A.5.14}$$

where the vierbein is transverse and traceless and $\Gamma \cdot \psi = 0$. Then, by brute force, Cremmer, Julia, and Scherk proved that the following action is invariant in 11 dimensions:

$$\begin{aligned} L = {} & -\frac{1}{2\kappa^2} eR - \tfrac{1}{2} e\bar{\psi}_M \Gamma^{MNP} D_N[\tfrac{1}{2}(\omega + \hat{\omega})]\psi_P - \tfrac{1}{48} eF_{MNPQ}^2 \\ & - \frac{\sqrt{2}\kappa}{384} e(\bar{\psi}_M \Gamma^{MNPQRS} \psi_S + 12\bar{\psi}^N \Gamma^{PQ} \psi^R)(F + \hat{F})_{NPQR} \\ & - \frac{\sqrt{2}\kappa}{3456} \varepsilon^{M_1,\ldots,M_{11}} F_{M_1,\ldots,M_4} F_{M_5,\ldots,M_8} A_{M_9,M_{10},M_{11}} \end{aligned} \tag{A.5.15}$$

where

$$\begin{aligned} \delta e_M^A &= \tfrac{1}{2}\kappa\bar{\eta}\Gamma^A \psi_M \\ \delta A_{MNP} &= -\frac{\sqrt{2}}{8}\bar{\eta}\Gamma_{[MN}\psi_{P]} \\ \delta\psi_M &= \kappa^{-1} D_M(\hat{\omega})\eta + \frac{\sqrt{2}}{288}(\Gamma_M^{PQRS} - 8\delta_M^P \Gamma^{QRS})\eta \hat{F}_{PQRS} \end{aligned} \tag{A.5.16}$$

and where

$$\hat{\omega}_{MAB} = \omega_{MAB} + \tfrac{1}{8}\bar{\psi}^P \Gamma_{PMABQ} \psi^Q \tag{A.5.17}$$

where $F_{MNPQ}$ is the curl of $A_{MNP}$ and $\hat{F}_{MNPQ}$ is the supercovariantization of $F_{MNPQ}$, where we choose $\{\Gamma^A, \Gamma^B\} = 2\eta^{AB}$, we antisymmetrize according to $\Gamma^{AB} = \frac{1}{2}(\Gamma^A\Gamma^B - \Gamma^B\Gamma^A)$, and the indices $ABC$ are flat and $MNP$ are curved.

The $N = 1$, $D = 10$ supergravity can be found by truncating the earlier action. The spinor decomposes into a pair of Majorana–Weyl gravitinos and a pair of Majorana–Weyl spin-$\frac{1}{2}$ fermions. The vierbein decomposes into a 10-dimensional vierbein and a scalar field $\phi$, while the antisymmetric tensor field decomposes into a boson field $B_{MN}$. Thus, the set of reduced fields is

$$\{e_M^A; \phi; B_{MN}\}; \qquad \{\psi_M; \lambda\} \tag{A.5.18}$$

Our final action in 10 dimensions is

$$\begin{aligned} e^{-1}L = {} & -\frac{1}{2\kappa^2}R - \tfrac{1}{2}\bar{\psi}_M \Gamma^{MNP} D_N \psi_P - \tfrac{3}{4}\phi^{-3/2} H_{MNP}^2 \\ & -\tfrac{1}{2}\bar{\lambda}\Gamma^M D_M \lambda - \frac{9}{16\kappa^2}(\phi^{-1}\partial_M \phi)^2 - \frac{3\sqrt{2}}{8}\bar{\psi}_M \Gamma^N \Gamma^M \lambda(\phi^{-1}\partial_N \phi) \\ & + \frac{\sqrt{2}\kappa}{16}\phi^{-3/4} H_{NPQ}(\bar{\psi}_M \Gamma^{MNPQR}\psi_R + 6\bar{\psi}^N \Gamma^P \psi^Q \\ & - \sqrt{2}\bar{\psi}_M \Gamma^{NPQ}\Gamma^M \lambda) + \cdots \end{aligned} \tag{A.5.19}$$

where

$$\begin{aligned} \delta e_M^A &= \frac{\kappa}{2}\bar{\eta}\Gamma^A \psi_M \\ \delta\phi &= -\frac{\sqrt{2}\kappa}{3}\phi\bar{\eta}\lambda \\ \delta B_{MN} &= \frac{\sqrt{2}}{4}\phi^{3/4}\left(\bar{\eta}\Gamma_M \psi_N - \bar{\eta}\Gamma_N \psi_M - \frac{\sqrt{2}}{2}\bar{\eta}\Gamma_{MN}\lambda\right) \\ \delta\lambda &= -\frac{3\sqrt{2}}{8}\phi^{-1}(\Gamma\cdot\partial\phi)\eta + \tfrac{1}{8}\phi^{-3/4}\Gamma^{MNP}\eta H_{MNP} + \cdots \\ \delta\psi_M &= \kappa^{-1}D_M \eta + \frac{\sqrt{2}}{32}\phi^{-3/4}(\Gamma_M^{NPQ} - 9\delta_M^N \Gamma^{PQ})\eta H_{NPQ} + \cdots \end{aligned} \tag{A.5.20}$$

and $H = dB$, and where $\cdots$ means four-fermion Fermi-type terms that we are omitting. This is the low-energy limit of a Type IIA string (because the fermions have opposite chiralities from the dimensional reduction). Thus, there are no anomalies from this theory because there is no chiral asymmetry.

The Type IIB theory, however, cannot be derived from dimensional reduction because it has fermions of the same handedness (and hence can have anomalies). Type IIB has no covariant action at all (but has on-shell equations of motion and also a well-defined light cone action).

Next, we would like to couple supergravity to Yang–Mills matter. The super-Yang–Mills multiplet by itself is given by

$$\{A_M^a; \chi^a\} \tag{A.5.21}$$

where $a$ represents the elements of the isospin group. Notice that we have equal numbers of fermions and bosons on-shell. The final action is equal to

$$\begin{aligned} e^{-1}L = {} & e^{-1}L_{SG}(H_{MNP}) - \tfrac{1}{4}\phi^{-3/4}F_{MN}^a F^{MNa} - \tfrac{1}{2}\bar{\chi}^a\Gamma^M(D_M(\hat{\omega})\chi)^a \\ & - \tfrac{1}{8}\kappa\phi^{-3/8}\bar{\chi}^a\Gamma^M\Gamma^{NP}(F_{NP}^a + \hat{F}_{NP}^a)(\psi_M + \tfrac{1}{12}\sqrt{2}\Gamma_M\lambda) \\ & + \tfrac{1}{16}\sqrt{2}\kappa\phi^{-3/4}\bar{\chi}^a\Gamma^{MNP}\chi^a H_{MNP} \\ & - \tfrac{1}{1536}\sqrt{2}\kappa^2\bar{\chi}^a\Gamma_{MNP}\chi^a\bar{\psi}_Q(4\Gamma^{MNP}\Gamma^Q + 3\Gamma^Q\Gamma^{MNP})\lambda \\ & - \tfrac{1}{512}\kappa^2\bar{\chi}^a\Gamma_{MNP}\chi^a\bar{\lambda}\Gamma^{MNP}\lambda - \tfrac{1}{384}\kappa^2\bar{\chi}^a\Gamma_{MNP}\chi^a\bar{\chi}^b\Gamma^{MNP}\chi^b \end{aligned} \tag{A.5.22}$$

where the surprising things is that we must modify the condition $H = dB$ so that

$$H = dB - 2^{-1/2}\kappa\omega_3 \tag{A.5.23}$$

where $\omega_3$ is the Chern–Simons term

$$\omega_3 = \mathrm{Tr}(AF - \tfrac{1}{3}gA^3) \tag{A.5.24}$$

The variation of $B$ under a gauge transformation is now

$$\begin{aligned} \delta B &= 2^{-1/2}\kappa\,\mathrm{Tr}(\Lambda\, dA) \\ \delta H &= 0 \end{aligned} \tag{A.5.25}$$

The action is invariant under

$$\begin{aligned} \delta A_M^a &= \tfrac{1}{2}\phi^{3/8}\bar{\eta}\Gamma_M\chi^a \\ \delta\chi^a &= -\tfrac{1}{4}\phi^{-3/8}\Gamma^{MN}\hat{F}_{MN}^a\eta + \frac{\sqrt{2}}{64}\kappa\{3\bar{\lambda}\chi^a\eta \\ & \quad - \tfrac{3}{2}\bar{\lambda}\Gamma^{MN}\chi^a\Gamma_{MN}\eta - \tfrac{1}{24}\bar{\lambda}\Gamma^{MNPQ}\chi^a\Gamma_{MNPQ}\eta\} \end{aligned} \tag{A.5.26}$$

The transformations of the supergravity fields are the same as before (with modified $H$ field) and the new pieces that must be added to the transformation law are

$$\begin{aligned} \delta'\lambda &= \frac{\sqrt{2}}{432}\kappa\bar{\chi}^a\Gamma^{MNP}\chi^a\Gamma_{MNP}\eta \\ \delta' B_{MN} &= 2^{-1/2}\kappa\phi^{3/8}\bar{\eta}\Gamma_{[M}\chi^a A_{N]}^a \\ \delta'\psi_M &= -\frac{1}{256}\kappa\bar{\chi}^a\Gamma^{NPQ}\chi^a(\Gamma_{MNPQ} - 5g_{MN}\Gamma_{PQ})\eta \end{aligned} \tag{A.5.27}$$

One may ask the obvious question: Are there supergravity theories beyond $N = 8$? The answer is probably no. This is because the supersymmetric generators of the $\mathrm{Osp}(N/4)$ group have spin $\frac{1}{2}$. If we take the maximum helicity state of the graviton and then operate on it with all possible $Q$'s, we find that the series eventually must terminate or else we create particles of spin $\frac{5}{2}$ and 3:

$$Q_\alpha Q_\beta \cdots Q_\zeta |\text{graviton}\rangle \tag{A.5.28}$$

Theories of spin $\frac{5}{2}$ and 3, although they can be constructed as free theories, are thought to be inconsistent when coupled to other particles. Thus, since there are eight half-steps between 2 and $-2$, we find that $N$ can only equal eight in the above series. Thus, O(8) is the largest supergravity theory that does not have spins greater than 2.

## §A.6. Glossary of Terms

**Almost Complex Structure:** A manifold has almost complex structure if there exists a linear map $J$ from the space to itself such that $J^2 = -1$. This is a generalization of the concept of $i$. Manifolds with almost complex structure are even-dimensional and orientable (this rules out $S_{\text{odd}}$ and $P_{2n}$).

**Anomaly:** An anomaly is the failure of a classical symmetry (global or local) to survive the process of quantization. It occurs when the current associated with the symmetry is no longer conserved because of quantum corrections. The most important anomaly in superstring theory is the conformal anomaly, which fixes the dimension of space–time to be either 26 or 10. Also, the vanishing of the chiral anomaly forces us to go to groups like $E_8 \otimes E_8$ and O(32). The string theory can also be shown to be free of global anomalies that might spoil modular invariance, which is a global symmetry.

**Automorphic Function:** An automorphic function is one that remains invariant under the action of a projective transformation: $\psi(z) = \psi[P(z)]$. Automorphic functions appear as the integrand of the $N$-loop string amplitude.

**Betti Number:** The $p$th Betti number is the number of independent harmonic $p$-forms for a given real manifold. It is also equal to the dimension of the $p$th cohomology or $p$th homology group for that surface. For a two-torus, the first Betti number simply counts the number of independent one-cycles. Because a two-torus has only two independent cycles, the first Betti number is equal to 2. Betti numbers are important because they are topological numbers, giving us a convenient way in which to classify topologically equivalent surfaces. If two surfaces are topologically equivalent, they have the same Betti numbers. The important topological indices, like the Euler index, are composed out of the Betti numbers.

**Bianchi Identity:** The Bianchi identity is an identity on curvature two-forms that is a result of the Jacobi identities on covariant derivatives:

$$[D_{[\mu}, [D_\nu, D_{\lambda]}]] = 0$$

Written out explicitly, it is

$$D_{[\alpha} R_{\beta\gamma]\delta\varepsilon} = 0$$

Because it is only an identity, no new physical information is contained within the Bianchi identities.

**Bosonization:** Bosonization is the process by which fermions are created out of bosons in two dimensions, i.e., $\psi = :e^{\phi(z)}:$. Historically, this was thought to be impossible. The intuitive reason why this is possible only in two dimensions is that the Lorentz group has only one generator in two dimensions and the concept of "spin" becomes trivial. Thus, the actual content of bosonization is the formation of anticommuting variables from commuting ones. In conformal field theory, bosonization plays an important part in creating a truly satisfactory fermion vertex function. This is accomplished because bosonization yields an explicit form for the irreducible representations of the Kac–Moody algebra for SO(10).

**BRST Transformation:** The Becchi–Rouet–Stora–Tyupin transformation is a transformation on gauge fields and their Faddeev–Popov ghosts. It is a global symmetry, so no new constraints can arise from it. The symmetry is generated by the BRST charge $Q$, such that $Q^2 = 0$. For the string, this fixes $D = 26$ and intercept equal to 1.

**Calabi–Yau Manifold:** Calabi conjectured, and Yau later proved, that a compact Kähler manifold of vanishing first Chern class always admits a Kähler metric of SU(3) holonomy. Such manifolds are called Calabi–Yau manifolds, and they form classical vacua for the six-dimensional string manifold after dimensional breaking. They are derived from the assumption that $N = 1$ supersymmetry survives the compactification, which implies the existence of a covariantly constant spinor.

**Chan–Paton Factor:** The Chan–Paton factor is a multiplicative factor in front of the Veneziano amplitude which introduces isospin indices on the strings. It is the simplest method of introducing isospin in superstring theory consistent with duality. We simply multiply the $N$-point amplitudes by the trace of isospin matrices, which is cyclically symmetric. By demanding factorization and that the Yang–Mills particle be in the adjoint representation of the group, we fix the group to be U($N$), O($N$), or Usp($N$). Chan–Paton factors are needed for the O(32) string, but the heterotic string introduces isospin via an entirely new mechanism, compactification.

**Chern Character:** The Chern character of a manifold $M$ with curvature form $\Omega$ is given by

$$\mathrm{ch}(M) = \mathrm{Tr}\, e^{(i/2\pi)\Omega}$$

**Chern Class:** The Chern class $c(M)$ of a curvature form $\Omega$ is equal to

$$c(M) = \det\left(1 + \frac{i}{2\pi}\Omega\right)$$

**Chern–Simons Form:** The $N$th Chern class, because it is closed and exact, can be written as $\omega_N = d\omega_{N-1}$. The Chern–Simons form is $\omega_{N-1}$. The Chern–Simons form occurs in formulations of gauge theories, especially the anomaly.

**Closed Form:** A closed form satisfies the identity $d\omega = 0$. For a manifold, the dimension of closed $p$-forms (modulo exact form) is the $p$th Betti number for that manifold.

**Compactification:** This is the process by which we take a manifold $R^n$ and divide by a lattice. The simplest compactification was introduced by Kaluza, who compactified a five-dimensional manifold down to four-dimensional space–time by making the fifth dimension periodic. Depending on the lattice, after compactification the manifold has a symmetry associated with it. Thus, isospin may be introduced into the superstring model by compactification rather than by Chan–Paton factors. Compactification allows us to break the ten-dimensional string down to a four-dimensional theory by curling up the unwanted six dimensions. Compactification, therefore, is the key to performing meaningful four-dimensional phenomenology on the string. The problem with compactification, which has persisted for the last 70 years since Kaluza first proposed compactification of a fifth dimension, is how to select which of the many possible classical solutions the theory really prefers. At present, tens of thousands of classical solutions are now possible with orbifolds and Calabi-Yau spaces. Only nonperturbative formulations (such as string field theory) will answer which of these possible vacua is the correct quantum one.

**Complex Manifold:** Roughly speaking, a complex manifold of dimension $N$ is a real manifold of dimension $2N$ if we can always find complex coordinates with holomorphic (i.e., analytic) transition functions. Thus, not every real manifold of $2N$ dimensions is a complex manifold.

**Complex Projective Space:** Complex projective space $CP_N$ is a complex $N + 1$-dimensional space where all points $z_i$ are identified with $\lambda z_i$ for some nonzero complex number. It is both compact and Kähler. It is equal to the sphere $S_{2N+1}$ with the identification

$$CP_N = \frac{S_{2N+1}}{\mathrm{U}(1)}$$

**Conformal Anomaly:** The conformal anomaly is the failure of the quantized string action to be conformally invariant. It appears most clearly in the failure of the Polyakov action to be invariant under a scale transformation: $g_{ab} \to \sigma g_{ab}$. New noninvariant terms are generated by this transformation, which vanish if (a) we fix the dimension of space–time to be 26 or 10 or (b) we allow the scalar particle to be part of the action, which generates Liouville theory (which is not discussed in this book).

**Conformal Field Theory:** Conformal field theory is a formalism by which we can calculate all correlation functions $\langle \phi(z)\phi(z') \rangle$ of string theory if we know the short-distance behavior of the product of the various $\phi(z)$. Since the amplitudes are correlation functions, this allows us to calculate all bosonic and fermionic amplitudes using conformal arguments alone. (It is not a field theory in the sense of second quantization.)

**Conformal Gauge:** The conformal gauge is a gauge that fixes the two-dimensional metric tensor to be proportional to $\delta_{ab}$. This gauge choice can be taken classically only for Riemann surfaces of genus zero. When we go to higher genuses or if we quantize the system, there are complications; e.g., the metric depends on $3N - 3$ complex Teichmüller parameters and the conformal anomaly spoils conformal invariance except when $D = 26$.

**Conformal Ghosts:** Conformal ghosts are Faddeev–Popov ghosts that arise when taking the conformal gauge. The BRST operator $Q$ is composed of these ghost fields. These conformal ghost fields are one of the features of the "new" string that did not exist in the "old" string theory of the early 1970s.

**Conformal Group:** The conformal group in four dimensions is equal to O(4, 2) or SU(2, 2). It has 15 generators, which correspond to 4 translations $P^a$, 6 Lorentz generators $M^{ab}$, 4 proper conformal boosts $K^a$, and one scale transformation or dilatation $D$. In two dimensions, however, the conformal group has an infinite number of generators, given by the Virasoro $L_n$. It is equivalent to Diff$(S_1)$.

**Conformal Weight:** The conformal weight labels irreducible representations of the conformal group. Under a conformal transformation, a field $\phi(z)$ transforms with conformal weight $h$ if

$$\phi(z) \to \phi(\bar{z}) \left( \frac{d\bar{z}}{dz} \right)^h$$

If a field has weight one, its line integral is invariant under a conformal transformation. In constructing string models, we require that the vertex function have weight one so that the Virasoro gauge conditions apply. The string variable $X_\mu(z)$ has conformal weight zero. The Virasoro generators have conformal weight 2. The $b$, $c$ ghosts of conformal field theory have conformal weight 2 and $-1$, respectively.

**Connection Form:** A connection form is a one-form $\omega$ defined over a Lie algebra that is added to the partial derivative to produce a genuine covariant derivative:

$$D_\mu = \partial_\mu + \omega_\mu$$

The connection is a mixed tensor. It is a vector in space–time but is also a member of the Lie algebra. In the language of fiber bundles, the *base space* is ordinary space–time while the *fiber space* is that associated with a Lie algebra. The connection form allows us to "connect" the base with the fiber as we make displacements in the base manifold. For SU($N$), the connection is just the usual Yang–Mills field:

$$\omega = A = A_\mu^a \, dx^\mu \, \lambda_a$$

For the Lorentz group, the connection is called the spin connection:

$$\omega = \omega_\mu^{ab} \, dx^\mu \, M^{ab}$$

where $M$ is a matrix representation of the generators of the Lorentz group.

**Cosmological Constant:** The cosmological constant is the term $\lambda\sqrt{-g}$ that appears in the action for general relativity in addition to the curvature tensor. One of the pressing problems of cosmology is to explain why it is such an exceedingly small number without "fine-tuning" one's equations by hand. Supersymmetry is sufficient to cancel the cosmological constant, but eventually supersymmetry must be broken, so supersymmetry alone cannot solve this problem. This is one of the important problems facing any quantum theory of gravity, such as superstring theory.

**Curvature Form:** In the theory of differentiable forms, the form

$$d\omega + \omega \wedge \omega$$

is called the curvature two-form if $\omega$ is a connection one-form defined over a Lie algebra. If we choose the tangent space to be the Lorentz group, then we can write the curvature two-form as follows:

$$R_b^a = d\omega_b^a + \omega_c^a \wedge \omega_b^c = \tfrac{1}{2}R_{bcd}^a e^c \wedge e^d$$

where $a$, $b$, $c$ represent Lorentz indices.

**DDF States:** The Del Guidice–Di Vecchi–Fubini states are real transverse oscillator states defined in 24 dimensions. They span the physical Hilbert space. They are equivalent to the physical oscillators of the string model after light cone quantization.

**Dehn Twist:** A Dehn twist is a global diffeomorphism on a Riemann surface. It is equivalent to cutting a torus along one of its cycles, twisting one of its open ends by $2\pi$, and then resealing the cut. The resulting diffeomorphism is global and cannot be reached by making continuous deformations of the identity. The set of Dehn twists forms a discrete group, the mapping class group, which is identical to the modular group.

**De Rahm Cohomology:** The $p$th de Rahm cohomology group is the set of $p$-forms that are closed modulo exact forms. The dimension of the $p$th de Rahm cohomology group for a manifold is equal to the $p$th Betti number for that manifold.

**Dirac Index:** The Dirac index is the difference between the positive and negative chirality solutions of the Dirac equation with zero eigenvalue for a given manifold:

$$\text{Index}(\not{D}) = n_+ - n_-$$

With proper regularization, it can also be represented as

$$\text{Index}(\not{D}) = \text{Tr}\ \Gamma_{D+1}$$

If a gauge and a spin connection can be defined over a manifold $M$, then

$$\text{Index}(\not{D}) = \int_M \text{ch}(V)\hat{A}(M)$$

**Dolbeault Cohomology Group:** For a complex manifold, this cohomology group is the set of $p$-forms that are closed under the operator $\bar{\partial}$ modulo the forms that are exact under this operator. The index associated with Dolbeault cohomology is the Riemann–Roch theorem.

**Duality:** Duality is the property that a scattering amplitude can be expressed as the sum over either $s$-channel or $t$-channel poles but not the sum over both, i.e.,

$$A(s, t) = \sum_n \frac{A_n(t)}{\alpha(s) - n} = \sum_n \frac{A_n(s)}{\alpha(t) - n}$$

This is in contrast to standard field theory, where we sum over both sets of poles. For this reason, a field theory of strings was thought to be impossible.

**Euler Index:** The Euler number is defined as the sum of all positive Betti numbers minus the sum of negative Betti numbers:

$$\chi(M) = \sum_{i=1}^{N} (-1)^i b_i$$

It is perhaps the most important topological number in the theory of cohomology. It is equal to the integral over the Euler characteristic through the Gauss–Bonnet theorem in two dimensions. In superstring theory, after compactification on a Calabi–Yau space and spin embedding, the generation number is equal to half the absolute value of the Euler number.

**Exact Form:** A $p$-form $\omega_p$ is said to be exact if there is a $p-1$-form $\omega_{p-1}$ such that $\omega_p = d\omega_{p-1}$.

**Exceptional Groups:** Exceptional groups are Lie groups that do not fit in the usual classification of O($N$), SU($N$), or Sp($N$). They are $F_4$, $G_2$, $E_6$, $E_7$, $E_8$. From the point of view of phenomenology, $E_6$ is the favored exceptional group because it has complex representations for the quarks and leptons if they belong to the **27**.

**Faddeev–Popov Determinant:** The Faddeev–Popov determinant is the measure term that arises in the path integral when we fix the gauge. If we choose $F(A) = 0$ to be the gauge, and if the gauge transformation is parametrized by $\Lambda$, then the Faddeev–Popov determinant is

$$\det \left. \frac{\delta F(A^\Lambda)}{\delta \Lambda} \right|_{F=0}$$

This determinant can be exponentiated into the action, yielding the term

$$L \sim \bar{\eta} \frac{\delta F(A^\Lambda)}{\delta \Lambda} \eta$$

where the $\eta$ are called Faddeev–Popov ghosts.

**First Quantization:** First quantization is the quantization of the coordinates. If $p$ is the canonical conjugate of $x$, then $[p, x] = -i$. The first quantized approach treats the coordinates of a point or string (rather than a field) as the fundamental object for the action. The first quantized approach, in order to add in interactions, must sum over the topologies of Feynman diagrams. Thus, the first quantized formalism is necessarily perturbative.

**Gauss–Bonnet Theorem:** The Gauss–Bonnet theorem states that the Euler number is equal to the integral over the Euler characteristic in 2 dimensions:

$$\chi(M) = \int_M e(M)$$

The Gauss–Bonnet theorem is the index theorem associated with the derivation $d$ of de Rahm cohomology in two dimension. For a Riemann surface, it is equal to $2 - g$, where $g$ is the genus.

**Genus:** The genus of a compact Riemann surface is equal to the number of holes or handles. Thus, a torus has genus 1.

**Ghost:** A ghost is a particle with negative metric. It thus propagates in the Green's function with the wrong sign. Ghosts are associated with the negative sign appearing in the Lorentz metric. A physical theory must be totally free of ghosts (or have ghosts cancel against other ghosts in the $S$-matrix). If a theory has ghosts that are part of its physical spectrum, then the theory allows negative probabilities and is hence physically unacceptable. The physical space of the string theory, it has been shown, is ghost-free because of the Ward identities generated by the Virasoro algebra. Historically, string amplitudes were constructed by explicitly inserting a complicated projection operator along each propagator, which extracted the ghost states from theory by hand. Unfortunately, the projection operator was quite complicated, especially for fermion states. The remarkable property of Faddeev-Popov ghost states is that they allow a consistent Lorentz covariant quantization (BRST) that cancels against these ghosts, making the calculation of fermion string amplitudes relatively easy. This forms the basis of conformal field theory.

**Grassmann Number:** A Grassmann-valued number $\theta_i$ anticommutes with all other Grassmann-valued numbers and commutes with ordinary real or complex numbers.

**GS Model:** The Green–Schwarz model, unlike the Neveu–Schwarz–Ramond model, is based on space–time Majorana–Weyl spinors in 10 dimensions. It is manifestly space–time supersymmetric. However, it is difficult to work with because covariant quantization is highly nonlinear. In the light cone gauge, however, the GS model is a convenient way in which to calculate multifermion processes.

**GSO Projection:** The Gliozzi–Scherk–Olive projection is a truncation of the Neveu–Schwarz–Ramond model that selects out the even $G$-parity sector, which renders the model space–time supersymmetric and tachyon-free. At the one-loop level, it is equivalent to modular invariance. It is also equivalent to summing over all possible spin structures for the one-loop amplitude.

**Harmonic Form:** A real form is said to be harmonic if it is annihilated by the Laplacian. By the Hodge decomposition theorem, there is at most one harmonic $p$-form for every $p$th cohomology group. Thus, the $p$th Betti number is equal to the number of harmonic $p$-forms.

**Heterotic String:** The heterotic string (after the Greek word for "hybrid vigor") is the leading candidate for a theory of all known forces. It is based on closed strings, but it treats the left- and right-moving sectors separately. The right-moving sector is the 10-dimensional superstring, while the left-moving sector is the 26-dimensional Nambu–Goto string that has been compactified down to 10-dimensional space–time, leaving a 16-dimensional surface that has been compactified on the lattice of either $E_8 \otimes E_8$ or Spin(32)/$Z_2$. At the first loop, it is anomaly-free and finite.

**Holomorphic:** A function $f$ on complex $n$ space $\mathbf{C}^n$ is holomorphic if

$$\frac{\partial f}{\partial \bar{z}_k} = \bar{\partial} f = 0$$

for all $k$. This is a generalization of the concept of analytic function for complex $n$-space.

**Holomorphic Factorization:** Holomorphic factorization states that the measure of the multiloop closed string amplitude is the square of the absolute value of a certain holomorphic form, divided by a function of the determinant of the period matrix. Intuitively, it means that the left- and right-moving modes (except for zero modes) contribute equally to the measure. Holomorphic factorization may eventually make it possible to write down the measure for the multiloop integrand by inspection using simple invariance arguments. The proof of the theorem works only in 26 dimensions because of the presence of an anomaly called the "analytic anomaly."

**Holonomy Group:** The holonomy group is the group generated when we take a spinor $\psi$ and transport it around a closed path on a manifold. The spinor transforms as $U_1 U_2 \cdots U_N \psi$ upon successive transport around closed paths $1, 2, \ldots, N$. In $N$ dimensions, the holonomy group will be a subgroup of $O(N)$.

**Homology Group:** The homology group of a manifold is equal to the set of cycles divided by the set of boundaries. Homology is a global property of manifolds that allows us to distinguish topologically inequivalent manifolds. What is remarkable is the relationship between this global concept and the local concept of de Rahm cohomology; i.e., the dimension of the $p$th homology group is equal to the dimension of the $p$th de Rahm cohomology group.

**Hyperbolic:** A projective transformation is hyperbolic if its multiplier is real, positive, and not equal to one. We can always choose the invariant points so that the multiplier is less than one. Hyperbolic projective transformations figure prominently in the formulation of the $N$-loop amplitude.

**Kac Determinant:** The Kac determinant is the determinant of the matrix

$$\langle \alpha | \beta \rangle$$

where $|\alpha\rangle$ is a Verma module. The Kac determinant has been explicitly evaluated by Kac for all Verma modules of the two-dimensional conformal group. If the determinant is nonzero for all elements of a Verma module, then the Verma module forms an irreducible representation of the group. Verma modules are important because they give us nontrivial information concerning the tower of resonances appearing in the string model. In the geometric string theory, the basic string fields are Verma modules (which explains the origin of the strange Faddeev–Popov ghost appearing in BRST string field theory).

**Kac–Moody Algebra:** A Kac–Moody algebra is an infinite-dimensional Lie algebra. The algebra is defined by

$$[T_m^i, T_n^j] = f^{ijk} T_{m+n}^k + c\delta^{ij}\delta_{m,-n}$$

where $f$'s are the structure constants of an ordinary Lie algebra and $c$ is an undetermined central term. By treating the index $n$ as the Fourier transform of a continuous function defined from 0 to $2\pi$, we can smear the Kac–Moody generators around a circle. The Virasoro algebra is not a Kac–Moody algebra, but it can form the semidirect product with the Kac–Moody algebra:

$$[L_m, T_n^i] = -nT_{m+n}^i$$

Kac–Moody algebras are useful for several reasons. First, the string states that appear after compactification on the root lattice of some Lie algebra can all be shown to transform under the Kac–Moody generalization of that Lie algebra. Thus, the Kac–Moody algebra based on exceptional groups is essential to understanding the heterotic string. Second, the various fermionic and bosonic fields of conformal field theory all transform under the SO(10) Kac–Moody algebra. Thus, conformal field theory and Kac–Moody algebras are intimately linked.

**Kähler Form:** The Kähler form is

$$\Omega = g_{i\bar{j}}\, dz^i \wedge d\bar{z}^j$$

for a Hermitian metric.

**Kähler Manifold:** A complex manifold is Kähler if it has a Hermitian metric and its Kähler form is closed, i.e.,

$$d\Omega = 0$$

A number of elegant theorems can be proved for Kähler manifolds; e.g., the various Laplacians one can write for complex manifolds are all the same.

**Kähler Potential:** For a Kähler manifold, it can be shown that the Hermitian metric can be written in terms of a single function called the Kähler potential $\phi$:

$$g_{i\bar{j}} = \frac{\partial^2 \phi}{\partial z^i \partial \bar{z}^j}$$

**Koba–Nielsen Variable:** Koba–Nielsen variables are parameters defined on a circle that allow us to parametrize the $N$-point Veneziano amplitude. Koba–Nielsen variables are the most versatile and commonly used coordinates for the dual model. They allow us to show manifest cyclic symmetry.

**Invariant Points:** The invariant points of a projective transformation are those that remain the same under the group action: $P(z) = z$. There are two invariant points for a projective transformation:

$$x^{(1)} = P^{\infty}(z)$$

$$x^{(2)} = P^{-\infty}(z)$$

where $z$ is arbitrary. We can parametrize any real projective transformation by its two invariant points and its multiplier. If the transformation is multiplicative, i.e., $P(z) = wz$, then the two invariant points are 0 and $\infty$ and its multiplier is $w$. (This is the parametrization of the single-loop amplitude.)

**Lattice:** A lattice $\Gamma$ is a set of points of the form $p^I = \pi \sum n_i e_i^I$, where the $n$'s are integers and $e_i^I$ is a set of independent vectors in $N$-dimensional space. Lattices occur in string theory when we compactify from 26 or 10 down to 4 dimensions. One method is to compactify on the torus defined by Euclidean space divided by the lattice, i.e., $R^N/\Gamma$. Not every lattice can be associated with the lattice of a Lie algebra. For example, the Leech lattice in 24 dimensions is not the lattice of any Lie algebra. We say the lattice is even if $(p^I)^2 =$ even. We say it is self-dual if the lattice is equal to its dual. The dual lattice is defined by the independent vectors $e_i^{*I}$ such that

$$\sum_{I=1}^{N} e_i^{*I} e_j^I = \delta_{ij}$$

In 16 dimensions, the only even self-dual lattices are those associated with the root lattices of $E_8 \otimes E_8$ and Spin(32)/$Z_2$. We say the root lattice of a Lie algebra is simply laced if it has roots of equal length. The only simply laced Lie groups are $A$, $D$, and $E$.

**Majorana Spinor:** A Majorana spinor is a purely real representation of the Dirac matrices. Thus, particles are their own antiparticles. Majorana spinors can be defined only in 2, 3, 4 (mod 8) dimensions.

**Majorana–Weyl Spinor:** A Majorana–Weyl spinor is simultaneously Weyl and Majorana, i.e., real and with definite chirality. It exists only in 2 (mod 8) dimensions. Majorana–Weyl fermions are found in the Green–Schwarz model.

**Manifold:** A space $M$ is a manifold if it can be covered by patches $U_i$ if each patch is a subspace of $n$-dimensional real space $\mathbf{R}^n$ or $n$-dimensional complex space $\mathbf{C}^n$. Interacting string theories are defined over manifolds (i.e., Riemann surfaces) but interacting point particle theories are not. Feynman graphs do not constitute a manifold because their local topology is not that of $\mathbf{R}^n$.

**Mapping Class Group:** The mapping class group is the group of diffeomorphisms of a Riemann surface divided by the diffeomorphisms that are connected to the identity:

$$MCG = \frac{\text{Diff}(M)}{\text{Diff}_0(M)}$$

It is equivalent to Teichmüller space divided by the moduli space. It is equivalent to the set of global diffeomorphisms on the Riemann surface (e.g., including Dehn twists). We must divide out by the MCG if we are to eliminate overcounting in the closed string amplitude due to modular invariance. The mapping class group is identical to the modular group.

**Modular Group:** The modular group for the torus is SL(2, $Z$), i.e., the set of $2 \times 2$ real matrices with integer elements with unit determinant. It is the invariance group of the integrand of the $N$-loop closed string amplitude. If a torus is parametrized as a square in $(\sigma, \tau)$ space whose opposite sides are identified, then the modular group is responsible for interchanging the roles of $\sigma$ and $\tau$. Modular invariance is crucially linked to supersymmetry and the finiteness of the multiloop amplitude.

**Moduli Space:** For a closed Riemann surface, moduli space is defined as the set of all constant-curvature metrics divided by all possible diffeomorphisms. For a manifold of genus $N$ greater than two, it has dimension $6N - 6$ (the number of independent parameters necessary to parametrize the surface). It is the space over which we will integrate the Polyakov action. It is closely related to Teichmüller space, differing only by the set of global diffeomorphisms or Dehn twists.

**Multiplier:** A projective transformation SL(2, $R$) can always be diagonalized and placed in the form $P(z) = wz$. The multiplicative constant $w$ is called the multiplier. In the single-loop superstring amplitude, the multiplier of the first loop's projective transformation appears in the partition function.

**Neumann Function:** The Neumann function is the Green's function for a Riemann surface such that the normal derivative on the boundary is equal to zero. We require this boundary condition to preserve the fact that $X' = 0$ on the edge of the open string. The exponential of the Neumann function for a Riemann surface of genus $N$ appears in the superstring amplitude for $N$-loop scattering.

**NS–R Model:** The Neveu-Schwarz–Ramond model (after the GSO projection) is the simplest superstring theory. It consists of the string $X_\mu(\sigma)$ and its supersymmetric partner, an anticommuting vector field $\psi_\mu(\sigma)$. Depending on the boundary conditions on the $\psi_\mu$, we can describe either fermions or bosons. The model is 2D supersymmetric, but 10D space–time supersymmetry is very obscure. By contrast, the Green–Schwarz action is fully supersymmetric in 10 dimensions (but difficult to quantize covariantly).

**Off-Shell:** Off-shell means that a particle does not satisfy $p_\mu^2 = -m^2$ and hence is virtual. Off-shell particles or fields appear in Green's functions, the intermediate states of Feynman diagrams, and the action.

**On-Shell:** On-shell means that a particle satisfies the equation $p_\mu^2 = -m^2$ and hence is physical, not virtual. The properties of the Veneziano amplitude, such as cyclic symmetry, hold only on-shell. The $S$-matrix is necessarily on-shell, whereas the Lagrangian and Green's functions are off-shell. For the string, the on-shell condition is

$$(L_0 - 1)|\phi\rangle = 0$$

**Orbifold:** An orbifold is the surface created by taking a torus and dividing by a discrete group. Orbifolds usually have fixed points, i.e., points that are invariant under the action of the discrete group, and hence they are not manifolds. A cone, for example, is an orbifold created by taking two-dimensional flat space and dividing it by $Z_n$ such that the origin is a fixed point. When the heterotic string is compactified on an orbifold, the group $E_8 \otimes E_8$ is usually broken down to the subgroup that commutes with the action of this discrete group. These fixed points apparently do not spoil the physical properties of the resulting theory. An orbifold is probably a special limit of a Calabi–Yau space.

**Orientable:** A manifold $M$ that is made up of patches (open sets) $M_\alpha$ is said to be orientable if $\det(\phi_\alpha \cdot \phi_\beta^{-1}) > 0$, where $M_\alpha$ and $M_\beta$ overlap and where $\phi_\alpha$ are transformations that map the patch into $\mathbf{R}^n$. If the determinant is negative, then the orientation of the manifold has changed, and the manifold is non-orientable (e.g., Möbius strip or Klein bottle). We can construct nonorientable and orientable surfaces with higher genus by taking a two-dimensional surface with two holes in it and connecting the rims of these two holes by identifying points on the rims in either cyclic or anticyclic fashion.

**Orthosymplectic Group:** The orthosymplectic groups $\mathrm{Osp}(N/M)$ are the graded Lie groups that leave invariant the following:

$$x_i x_i + \theta_m C_{mn} \theta_n$$

where the $\theta$ are Grassmann-valued. It can be broken down to

$$\mathrm{Osp}(N/M) \supset \mathrm{O}(N) \otimes \mathrm{Sp}(M)$$

Supergravity is the gauge theory of $\mathrm{Osp}(N/4)$.

**Partition Function:** The partition function is

$$\prod_{n=1}^{\infty} (1 - w^n)^{-D}$$

If $D = 1$, then the coefficient of $w^n$ is the partition of the integer $n$. This function not only determines the number of states in the string model at the $n$th level but also governs the divergence of the single-loop amplitude.

**Polyakov Action:** The Polyakov Lagrangian is

$$L = \frac{1}{2\pi} \sqrt{g} g^{ab} \partial_a X^\mu \partial_b X_\mu$$

It is classically reparametrization and Weyl invariant. When quantized, Weyl invariance is broken unless $D = 26$. Classically, it is equivalent to the Nambu–Goto action. However, at the quantum level the Polyakov action contains specific information concerning the metrics over which we sum to produce multiloop surfaces, while for the Nambu–Goto action this must be added by hand.

**Pontrjagin Class:** The Pontrjagin class of a manifold with curvature lying in the Lie algebra of O($n$) is

$$p(\Omega) = \det\left(1 - \frac{1}{2\pi}\Omega\right)$$

**Projective Transformation:** A projective transformation is given by

$$P(z) = \frac{az + b}{cz + d}$$

where $ad - bc = 1$. This generates the group SL(2, $R$), i.e., the set of $2 \times 2$ real matrices with unit determinant. An $N$-loop string amplitude has $N$ projective transformations associated with it.

**Real Projective Space:** Real projective space is the set of lines in $\mathbf{R}^{n+1}$ passing through the origin. It is equivalent to the sphere $S_n$ in $\mathbf{R}^{n+1}$ where we identify antipodal points (i.e., two unit vectors $\mathbf{x}$, $\mathbf{x}'$ determine the same line in $\mathbf{R}^{n+1}$ if $\mathbf{x} = \lambda \mathbf{x}'$). Notice that

$$P_3(\mathbf{R}) = \mathrm{SO}(3) = \frac{\mathrm{SU}(2)}{Z_2}$$

The Euler number of $P_n(\mathbf{R})$ is $n + 1$. It is orientable if $n$ is odd, nonorientable if $n$ is even.

**Regge Pole:** A Regge pole is a singularity in the $S$-matrix in complex angular momentum space. In $S$-matrix theory, the angular momentum and energy of a scattering can become complex numbers. If $\alpha(s)$ is the angular momentum as a function of energy squared, then the scattering amplitude will exhibit a Regge pole in this variable. Each Regge pole corresponds to a resonance. The string theory sums over an infinite number of Regge poles or resonances.

**Regge Trajectory:** A Regge trajectory is the line formed when resonances of the $S$-matrix are plotted on a graph, with the energy squared on the $x$-axis and the angular momentum on the $y$-axis. In string theory, the resonances form an infinite set of points that can be connected to form linearly rising parallel curves. The curve farthest to the left is called the leading trajectory. When the leading trajectory crosses the $y$-axis, this is called the "intercept." The intercept for the string model is fixed at 1. The slope of these trajectories is called the Regge slope $\alpha'$.

**Ricci Flat:** A metric $g_{\mu\nu}$ is Ricci flat if

$$R_{\mu\nu} = 0$$

**Riemann-Roch Index Theorem:** The Riemann-Roch index theorem is the index theorem corresponding to the operator $\partial_{\bar{z}}$ on a Riemann surface. It can be expressed in many forms, but for our purposes it states that the dimension of the space of quadratic differentials minus the dimension of the conformal Killing vectors is proportional to $1 - g$ for a genus $g$ surface. For the ghost system, we can say that the number of independent zero modes of the $c$ ghost minus the number of zero modes of the $b$ ghost equals $3(g - 1)$. The Riemann-Roch theorem is thus useful in establishing that the quadratic differentials span a space of $3g - 3$ dimensions, i.e., moduli space. For superstrings, the super Riemann-Roch theorem allows us to calculate the dimension of super moduli space.

**Riemann Surface:** A Riemann surface is a complex two-dimensional manifold. The string sweeps out a Riemann manifold with holes as it moves in space–time. Hence, the perturbation theory of strings is based on the theory of holomorphic functions defined on Riemann surfaces of genus $g$.

**Second Quantization:** Second quantization is the quantization of fields (not co-ordinates) defined over space–time. If the field $\phi(x_1)$ has a canonical conjugate $\pi(x_2)$, then second quantization postulates

$$[\pi(x_2), \phi(x_1)]_{t_1=t_2} = -i\delta^{(3)}(x_1 - x_2)$$

Second quantization forms the basis of quantum field theory.

**Spin Connection:** The spin connection $\omega_\mu^{ab}$ appears in the covariant derivative of a spinor:

$$D_\mu \psi = (\partial_\mu + \omega_\mu^{ab} M^{ab})\psi$$

where $M^{ab}$ is the generator of the Lorentz group in matrix form. With the addition of the spin connection, the covariant derivative becomes a true tensor.

**Spin Embedding:** Spin embedding is placing the spin connection, which transforms under O(6) in the tangent space, into the gauge connection, which transforms under the gauge group. Spin embedding is the simplest way to satisfy the Bianchi identities when we compactify on Calabi–Yau spaces. It also allows us to break the gauge group. If the spin connection has SU(3) holonomy (in order to perserve a covariantly constant spinor) then embedding it into the gauge group of the heterotic string breaks the gauge group down to $E_6$ via the decomposition:

$$E_8 \otimes E_8 \supset \mathrm{SU}(3) \otimes E_6 \otimes E_8$$

**Spin Manifold:** A spin manifold is one that admits spinors; i.e., it is possible to define spinors and the Dirac equation on the manifold. Many smooth manifolds are not spin manifolds because a spin manifold must have vanishing first and second Stiefel–Whitney indices. Every orientable manifold in two and three dimensions is a spin manifold.

**Spin(*N*):** Spin($N$) is the simply connected covering group of the compact doubly connected group SO($N$), e.g.,

$$\mathrm{Spin}(3) = \mathrm{SU}(2)$$

**Spin Structure:** Spin structure is the set of all possible boundary conditions that a fermion may take on a spin manifold of genus $g$. In the functional integral over the single-loop diagram in the NS–R model, we must integrate over all spin structures, i.e., the four possible combinations of boundary conditions in the $(\sigma, \tau)$ directions: (NS, R), (NS, NS), (R, NS), (R, R), in order to perserve modular invariance. For a surface of genus $g$, there are $2^{2g}$ spin structures.

**Spurious State:** A spurious state is one that does not couple to physical state: $\langle R|S\rangle = 0$. In 26 dimensions, spurious states with negative norm all decouple from the string theory.

**Superconformal:** The superconformal group in four dimensions is equal to SU(2, 2/$N$). It is a graded Lie group that has the decomposition

$$\mathrm{SU}(2, 2/N) \supset \mathrm{SU}(2, 2) \otimes \mathrm{U}(N)$$

The gauging of this group yields superconformal gravity. In two dimensions, however, the superconformal algebra is given by the super-Virasoro or NS–R algebra.

**Superfield:** A superfield is a function of both space–time and a supersymmetric Grassman-valued coordinate: $\phi(x, \theta)$. In four dimensions, if the Grassman coordinates are Majorana, then there are 16 independent fields contained within this superfield. It forms a representation of supersymmetry:

$$\delta\phi = \bar{\varepsilon}Q\phi$$

Usually, a superfield is reducible, so we are allowed to place constraints on the superfield. The superfield remains a representation of supersymmetry if the operators enforcing the constraint commute with the supersymmetric generator.

**Symplectic:** The symplectic group is the group that leaves invariant the following:

$$\theta_m C_{mn} \theta_n$$

where the $\theta$ are Grassman-valued. The most important symplectic group for supergravity theory is Sp(4), which is locally isomorphic to the de Sitter group.

**Tangent Space:** At every point on a curved manifold $M$, we can erect a flat coordinate system called the tangent space. It is particularly useful in defining spinors on curved manifolds, because spinors can be constructed for the Lorentz group in $N$ dimensions but not for GL($N$). Thus, spinors are defined only in the tangent space.

**Teichmüller Parameter:** Teichmüller parameters are a set of $6N - 6$ conformally distinct numbers necessary to parametrize a closed Riemann surface of genus $N$.

**Teichmüller Space:** Teichmüller space is the space of Riemann surfaces of constant curvature divided out by diffeomorphisms that can be connected to the identity. The dimension of this space is $6N - 6$ for a Riemann surface of genus $N$; that is, $6N - 6$ Teichmüller parameters parametrize a closed Riemann surface of genus $N$. They occur explicitly as integration variables of the multiloop diagram. Teichmüller space is actually too "large" for the functional integral. We must still divide out by the global diffeomorphisms of the mapping class group (modular group), which are generated by Dehn twists.

**Vacuum State:** The vacuum state is the lowest state in a Hilbert space. It corresponds to a classical solution of the equations of motion before quantum corrections are introduced. Thus, in string theory, classical solutions defined on Calabi–Yau or orbifold spaces correspond to the vacuum state of the full quantum theory. The fundamental problem is to calculate which vacua are unstable and which will decay to the "true" vacuum. Unfortunately, perturbation theory, the main tool developed for solving quantum field theory, is not sufficient to determine which vacua are unstable in string theory.

**Verma Module:** A Verma module is the set of states created from a vacuum vector $|h\rangle$ with eigenvalue $h$ when acted on by all possible combinations of raising operators $L_{-n}$ $(\alpha_i < \alpha_{i+1})$:

$$L_{-\alpha_1}^{\lambda_1} L_{-\alpha_2}^{\lambda_2} \cdots L_{-\alpha_N}^{\lambda_N} |h\rangle$$

A Verma module is characterized by two numbers, the eigenvalue $h$ of $L_0$ and the central term $c$. If the Kac determinant is nonzero for all members of the module, then it forms an irreducible representation of the conformal group.

**Vierbein:** In $N$-dimensional space, the vierbein is an $N \times N$ real matrix $e_\mu^a$ that transforms as a first-rank tensor under general coordinate transformations in the index $\mu$ and under local Lorentz transformations in the index $a$. Its square is equal to the metric tensor:

$$e_\mu^a e_\nu^a = g_{\mu\nu}$$

Vierbeins are absolutely necessary to define spinors in general relativity (because there are no finite-dimensional spinor representations of GL($N$)).

**Virasoro Algebra:** The Virasoro algebra is an infinite-dimensional algebra whose generators obey

$$[L_n, L_m] = (n - m)L_{n+m} + \frac{c}{12}(n^3 - n)\delta_{n,-m}$$

If $c = 0$, we have the Witt algebra, which defines reparametrization of the circle generated by $z^{n+1}d_z$. The Virasoro algebra is often designated Vect($S_1$). The group generated by this algebra is Diff($S_1$), the group of diffeomorphisms of the circle. The Virasoro algebra is not a Kac–Moody algebra, but one can take the semidirect product of the two. Furthermore, through the Sugawara construction, one may form the Virasoro generators by taking the square of the Kac–Moody generators. The Virasoro algebra is absolutely essential to understanding the string theory for several reasons. First, the Virasoro generators arise naturally when we calculate the moments of the energy-momentum tensor for the string. Second, their commutation relations correspond to the reparametrization and conformal symmetry of the string. Third, the Virasoro operators generate conformal transformations on the two-dimensional world sheet when they operate on a string amplitude. Fourth, their positive moments annihilate physical states, which is essential in the proof that the string theory is ghost-free. Fifth, the BRST operator Q is composed out the Virasoro generators.

**Weyl Spinor:** A Weyl spinor is a representation of the Dirac matrices with definite chirality; i.e., the eigenvalue of $\Gamma_{D+1}$ is either plus or minus one. Weyl spinors can be defined only in even dimensions.

**Weyl Transformation:** A Weyl transformation is a scale transformation.

**Wilson Lines:** The gauge-invariant Wilson loop is the line integral

$$U = Pe^{i\int_c dz A}$$

where we take path ordering around a closed curve $C$, and $A$ is the connection one-form for a Lie algebra. If a manifold is simply connected (i.e., if a line drawn on the surface can always be contracted smoothly to a point), then the vanishing of $A$ means that $U$ equals one. However, if the manifold is not simply connected, then $U$ need not equal one even if $A$ vanishes. Thus $E_6$ can be broken down to the subgroup that commutes with all elements of $U$. This is called symmetry breaking via Wilson lines.

**World Sheet:** The world sheet is the two-dimensional Riemann manifold swept out by the motion of the string.

***$Z_N$*:** This is the discrete permutation group, whose elements all satisfy $g^N = 1$.

## §A.7. Notation

For the purpose of clarity, we have deliberately dropped the normalization factors appearing in the functional path integrals and the $N$-point amplitude $A_N$. This should not be a problem because they can be easily restored by the reader.

In our units, the Planck length and Planck mass are equal to:

$$L = [\hbar G c^{-3}] = 1.6 \times 10^{-33} \text{ cm}$$

$$M = [\hbar c G^{-1}] = 2.2 \times 10^{-5} \text{ gm} = 1.2 \times 10^{19} \text{ GeV/c}^2$$

We set $c$ and $\hbar = 1$.

We use the space-time Lorentz metric $(-, +, +, \ldots, +)$. On the world sheet, we use the two-dimensional metric $(-, +)$, where the first (second) index refers to $\tau(\sigma)$. Our two-dimensional matrices are:

$$\rho^0 = \begin{pmatrix} 0 & -i \\ i & 0 \end{pmatrix}, \qquad \rho^1 = \begin{pmatrix} 0 & i \\ i & 0 \end{pmatrix}$$

where:

$$\bar{\psi} = \psi \rho^0$$

In general, we will use $k_\mu$ to represent the physical momentum and $p_\mu$ to represent the operator whose eigenvalue is the momentum. However, as in the literature, we will often drop this distinction and use both interchangeably.

We use the following choices for the Regge slope:

$$\alpha' = 1/2 \text{ for open strings}$$

$$\alpha' = 1/4 \text{ for closed strings}$$

Whenever possible, we use Greek letters to denote curved space indices.

We use $\mu$, $\nu$ to denote 26- and 10-dimensional Lorentz indices in curved space. We use Greek letters $\alpha$, $\beta$ to denote two-dimensional curved world sheet indices. Flat space indices, either in Lorentz or in two-space, are usually in Roman letters $a$, $b$, $c$. When the context is clear, we will sometimes use the metric $(+, +, \ldots, +)$ for the flat tangent space.

We choose our gamma matrices such that:

$$\{\Gamma_A, \Gamma_B\} = -2g_{AB}$$
$$\Gamma_{D+1} = \Gamma_0 \Gamma_1 \cdots \Gamma_{D-1}$$

In an even number of dimensions, they are normalized so that:

$$(\Gamma_{D+1})^2 = +1 \quad \text{if} \quad D = 4k + 2$$
$$(\Gamma_{D+1})^2 = -1 \quad \text{if} \quad D = 4k$$

When the gamma matrices appear with more than one index, we take the sum of all anti-symmetric combinations of these indices. For example, we normalize the gamma matrices so that:

$$\Gamma^{AB} = \tfrac{1}{2}[\Gamma^A \Gamma^B - \Gamma^B \Gamma^A]$$

We define the light cone gamma matrices as:

$$\Gamma^+ = \frac{1}{\sqrt{2}}(\Gamma^0 + \Gamma^{D-1})$$
$$\Gamma^- = \frac{1}{\sqrt{2}}(\Gamma^0 - \Gamma^{D-1})$$

When specializing to the case of 10 or 4 dimensions, we will often just use the symbol $\gamma^\mu$.

When we make the reduction down to SO(8) for the light cone formalism, we will use the direct products of $2 \times 2$ Pauli matrices:

$$\gamma^i = \begin{pmatrix} 0 & \gamma^i_{a\dot{a}} \\ \gamma^i_{\dot{b}b} & 0 \end{pmatrix}$$
$$\gamma^1 = -i\tau_2 \otimes \tau_2 \otimes \tau_2$$
$$\gamma^2 = i\mathbb{1} \otimes \tau_1 \otimes \tau_2$$
$$\gamma^3 = i\mathbb{1} \otimes \tau_3 \otimes \tau_2$$
$$\gamma^4 = i\tau_1 \otimes \tau_2 \otimes \mathbb{1}$$
$$\gamma^5 = i\tau_3 \otimes \tau_2 \otimes \mathbb{1}$$
$$\gamma^6 = i\tau_2 \otimes \mathbb{1} \otimes \tau_1$$
$$\gamma^7 = i\tau_2 \otimes \mathbb{1} \otimes \tau_3$$
$$\gamma^8 = \mathbb{1} \otimes \mathbb{1} \otimes \mathbb{1}$$
$$\gamma^{ij}_{ab} = \tfrac{1}{2}(\gamma^i_{a\dot{a}}\gamma^j_{\dot{a}b} - \gamma^j_{a\dot{a}}\gamma^i_{\dot{a}b})$$

where $\tau_i$ are the Pauli matrices, and each $\gamma$ matrix is a direct product of three $2 \times 2$ blocks.

## References

For an introduction to group theory, see

[1] R. Gilmore, *Lie Groups, Lie Algebra, and Some of Their Representations*, Wiley–Interscience, New York, 1974.
[2] N. Jacobson, *Lie Algebras*. Dover, New York, 1962.
[3] H. Georgi, *Lie Algebras in Particle Physics*, Benjamin/Cummings, Reading, Mass., 1982.
[4] R. N. Cahn, *Semi-Simple Lie Algebras and Their Representations*, Benjamin/Cummings, Reading, Mass., 1984.

For an introduction to general relativity, see

[1] C. W. Misner, K. S. Thorne, and J. A. Wheeler, *Gravitation*, Freeman, San Francisco, 1973.
[2] S. Weinberg, *Gravitation and Cosmology*, Wiley, New York, 1972.
[3] S. W. Hawking and G. F. R. Ellis, *The Large Scale Structure of Space–Time*, Cambridge Univ. Press, Cambridge, 1973.
[4] R. Adler, M. Basin, and M. Schiffer, *Introduction to General Relativity*, McGraw–Hill, New York, 1965.
[5] M. Carmeli, *Group Theory and General Relativity*, McGraw–Hill, New York, 1977.

For an introduction to the theory of forms, see

[1] T. Eguchi, P. B. Gilkey, and A. J. Hanson, *Phys. Rep.* **66**, 213 (1980).
[2] C. Nash and S. Sen, *Topology and Geometry for Physicists*, Academic Press, New York, 1983.
[3] C. von Westenholtz, *Differential Forms in Mathematical Physics*, North-Holland, Amsterdam, 1978.
[4] P. B. Gilkey, *Invariance Theory, the Heat Equation, and the Atiyah–Singer Index Theorem*, Publish or Perish, Wilmington, Del.,
[5] S. I. Goldberg, *Curvature and Homology*, Dover, New York, 1962.

For an introduction to supersymmetry and supergravity, see

[1] P. van Nieuwenhuizen, *Phys. Rep.* **68C**, 189 (1981).
[2] S. J. Gates, M. Grisaru, M. Rocek, and W. Siegel, *Superspace or One Thousand and One Lessons in Supersymmetry* Benjamin/Cummings, Reading, Mass., 1983.
[3] M. Jacob, ed., *Supersymmetry and Supergravity*, North-Holland, Amsterdam, 1986.
[4] P. West, *Introduction to Supersymmetry and Supergravity*, World Scientific, Singapore, 1986.
[5] J. Wess and J. Bagger, *Introduction to Supersymmetry*, Princeton Univ. Press, Princeton, N.J., 1983.
[6] R. N. Mohapatra, *Unification and Supersymmetry*, Springer-Verlag, New York, 1986.

# Index

**A**
Action, first quantized
  bosonic point particle 25
  bosonic string 52
  conformal field theory 150, 156, 174
  first order form 28, 58
  ghost action 150, 156–8, 174
  Hamiltonian 28, 58
  heterotic string 422, 448
  second order form 28, 58
  superparticle 101
  superstring (conformal gauge) 103
  superstring (GS) 124, 136
  superstring (NS-R) 116, 134, 150
Action, second quantized
  BRST string field theory 298, 301
  geometric string field theory 357
  light cone field theory 257, 261, 276
  point particle 40, 249
  pre-geometry 323
  superstring 279, 322
Almost complex structure 468, 544
$A_n$ 512
Anomaly
  cancellation in strings 404–6
  characteristic classes 390
  chiral 384–6
  conformal anomaly 216–9
  gravitational 394–6
  hexagon diagrams 382, 415
  O(16) × O(16) string 445
$\hat{A}$-roof genus 390, 413
Atiyah–Singer Theorem 376, 406–12
Automorphic function
  multiloop amplitude 202–5
  single loop amplitude 183, 544

**B**
Betti number 461, 463, 544
Bianchi identity 399, 454, 475, 528, 544
$B_n$ 512
Bosonization 141, 162, 167–8, 545
BRST, first quantized
  conformal field theory 158
  current 158–9
  bosonic string 69, 96
  fermionic string 119–20
  point particle 30–3
  transformation 545
BRST, second quantized
  axiomatic formulation 307, 327
  closed 315–6
  point particle 32–4
  pre-geometry 324–5
  problems 323
  superstrings 316–23

**C**
Calabi–Yau manifold 453, 545
Cartan subalgebra 166, 433, 520
Cartan–Weyl representation 166–7, 520
Chan–Paton factor 377–9, 412, 545

Characteristic classes 386
Chern class 388, 413, 545
Chern–Simons form 387, 543
Chiral invariance 380–3
Closed form 458
Closed strings
  field theory of closed strings 315–7, 364–6
  modular invariance 196
  multiloop amplitude 208–9
  single loop amplitude 195
  tree amplitude 86–9
$C_n$ 512
Cocycle 167, 170
Coherent state method 81, 195
Cohomology
  de Rahm 457–60, 548
  Dolbeault 469, 548
  Hodge theorem 228, 469
Compactification 418, 546
Complex maifold 546
Conformal anomaly 216–9
Conformal field theory
  fermion vertex 163–4, 174–5
  multiloop 232–4
  operator product expansion 144
  spin fields 141, 153–6, 167–8
  superconformal ghosts 156
Conformal gauge
  multiloops 210–2
  single loop 60, 95, 547
Conformal ghost 70, 119, 156–63, 547
Conformal group 515
Conformal Killing vector 213
Conformal weight 142, 156–63, 174, 345, 547
Connection form 547
Cosmological constant problem 508, 548
$CP_n$ 467, 473
Currents 41, 104, 117, 158, 380–2, 385, 393
Curvature form 356, 548
Curvature tensor 6, 13, 453, 525

**D**
DDF state 93, 548
Dehn twist 197, 212, 226–7, 548
de Rahm cohomology 457–8, 548
de Sitter group 515
Differential forms 457–9, 462–3
Dilaton 191
Dirac index 391–3, 414, 548
Divergences
  bosonic single loop 189–91
  closed bosonic loop 196–8
  fermionic single loop 198–9
  multiloop 202, 209, 224
  Selberg zeta function 224, 239
$D_n$ 512
Dolbeault operator 469, 548
Duality
  dual amplitudes 72, 80, 549
  incompatibility with string field theory 247
Dynkin diagram 520–523

**E**
$E_6$ 523
$E_7$ 523
$E_8$
  anomaly cancellation 402
  Dynkin diagram 523
$E_6$ subgroup 481, 475
  heterotic string
  Kac–Moody 439–41
  metric 426
  modular group 438–9
Energy-momentum tensor
  conformal field theory 144, 147, 150–1
  ghost 157
  point particle field theory 42
  superstring 104–5, 116
  Virasoro algebra 57
Euler beta function 15, 79
Euler index 549
Euler number 461, 477
Exact form 458, 549
Exceptional groups 549

**F**
$F_{1-2}$ formalism 111–2, 135–6
$F_4$ 523
Faddeev–Popov quantization
  conformal field theory 146, 156, 160
  multiloop 214, 238
  point particle 30–2
  string field theory 295–7
  superconformal ghost 119
Fermion vertex, first quantized
  conformal field theory 163-4, 174-5
  GS 132
  NS-R 124

Fermion vertex, second quantized
BRST 321
light cone 280–2
Feynman diagrams
point particle 4, 10, 51
string 50–1
First quantization, bosonic string
BRST 69–71, 96
Gupta–Bleuler 95
light cone 66–9, 95
problems 245
First quantization, point particle
BRST 30, 34
Coulomb 28–9
Gupta–Bleuler 29
First quantization, superstring
BRST 119
Gupta–Bleuler 117
light cone 118
Fock space 61
Forms, theory of 528–32
Four dimensional superstrings 491–6
Fubini–Study metric 473

**G**

$G_2$ 523
Gauss–Bonnet theorem 161, 383, 550
General relativity 523–8
Generation number
Calabi–Yau manifolds 480
Euler number 477–80
orbifold phenomenology 490
Genus 550
Geometric string field theory
action 358
connection fields 353
endpoint gauge 362
four string interaction 362–3
ghost sector 348
interpolating gauge 361
midpoint gauge 362
string group 336–40
universal string group 335, 342
Ghost number
current 159
second quantized superstrings 320–1
Ghosts
as tangent space 350–2
decoupling 89–93
Faddeev–Popov 32–3, 120, 159, 320–1, 550
G-parity 121–4
Grand unification theories
GUT 8, 11, 376
hierarchy problem 9
Grassman
integration 537
number 32–33, 537, 550
Grassmanian 225–7, 233–5, 240
GS model, first quantization 192, 550
GS model, second quantization 277–83
GSO projection
G-parity 121–4, 550
modular invariance 221, 442–4
Gupta–Bleuler quantization
bosonic strings 95
point particles 29, 43, 46
superstrings 117

**H**

Hamiltonian
first quantized 28, 58
second quantized 257
Harmonic form 462, 550
Harmonic oscillators 38, 61–63, 77–9, 184–5
Heterotic string
covariant form 430–2
divergences 435–9
$E_8 \otimes E_8$ 426, 439–441
modular invariance 439
single loop amplitude 435–7
spectrum 427–30
supersymmetry 427
trees 432–5
Hirota equations 233
Hirzebruch class 390, 412–3
Hodge theorem 228, 469
Holomorphic factorization 225, 230–2, 239, 551
Holomorphic function 551
Holonomy group 551
SU(3) 456
SU($N$) 471
Homology 458
Homology cycle 206, 226
Hyperbolic 201, 551

**I**

Index theorems
Atiyah–Singer 376, 406–12
Dirac 391–3, 414, 548
Gauss–Bonnet 161, 383, 550
Hirzebruch 412
Riemann–Roch 161, 412

Infrared divergences 189–91, 196–8, 198–9
Internal symmetries
  Chan–Paton factors 377–9, 412, 545
  via compactification 425–7
Invariant points 200

**K**
Kac determinant 551
Kac–Moody algebra
  commutation relations 552
  conformal field theory 169
  $E_8 \times E_8$ 439–41, 450
  SO(10) 169–72
  super Kac–Moody 499
Kadomtsev–Petviasvili hierarchy 233–4
Kähler manifold 467, 470, 552
Kähler potential 470, 552
Kaluza–Klein theory 419
Klein bottle 180
Koba–Nielsen variable 552
Kunneth formula 464

**L**
Laplacian 462, 470
Lattice
  even 425–7, 553
  Lorentzian 446–7
  modular invariance 439
  self-dual 425–7, 437
Lie algebra
  accidents 518–9
  Dynkin diagrams 520–3
  Kac–Moody 169, 439–41, 450, 552
  types 512
Light cone, first quantized
  bosonic string 66–9, 95
  superstring 126
Light cone, second quantized
  action 257, 261, 276
  four string interaction 272–7
  supersymmetry 277–283
  vertex 260–2
Lorentz group in light cone gauge 68–9
Lorentzian lattice 446–8

**M**
Majorana spinor 535, 553
Majorana–Weyl spinor 125–6, 278–9, 553
Mandelstam map 265
Manifold
  almost complex 468, 544
  complex 468
  Kähler 467, 470, 552
  Riemann surface 199, 209, 472, 557
Mapping class group 211, 553
Mass-shell conditions 64, 71
Modular group
  heterotic string 437
  multiloop amplitude 552
  O(16) ⊗ O(16) 445
  single loop amplitude 196
  SL(2, $Z$) 196
Moduli space 210–11, 225, 554
Multiloop amplitudes
  closed string 206-9
  constant curvature metrics 199
  Dehn twists 197, 212, 227, 548
  divergences 236
  Grassmanians 225–7, 233–5, 240
  modular group 211, 225, 437, 445, 552
  open bosonic 199
  Schottky groups 199, 203
  Schottky problem 226
  Selberg zeta function 224, 239
  theta functions 183–4, 187, 195, 200, 221 228–9, 233–4, 239–40
Multiplier 201, 223, 554

**N**
Neumann coefficients 267, 554
No-ghost theorem 89–93
Non-orientable 180, 189, 405
NS-R model 106–9, 116, 134–5, 150, 317

**O**
O(6) 455
Off-shell 245–6, 554
O($N$) 512–6
One loop amplitude
  bosonic 187
  closed 195
  heterotic 437
  slope renormalization 190
  superstring 192
On-shell 245–6, 554
Orbifold
  asymmetric 496–7
  generation number three 490
  modular invariance 487
  Wilson lines 490
  $Z$-orbifold 485

Osp(1/2) 114
Osp(8/4) 12
Osp($N/M$) 11, 12, 555

**P**
Partition function 186, 189, 203
Path integral
  anomaly 384
  multiloops 181, 209
  point particle 18–25
  trees 71–7
  string field theory 222
Period matrix
  cohomology 461
  multiloop 228
  Schottky problem 228
  single loop 207
Pictures
  $F_1$ and $F_2$ pictures 111–2, 135–6
  conformal field theory 160
  picture changing operators 321–2
Planck length 10
Poincaré duality 464
Polyakov action 57, 209, 555
Pontryagin class 389, 555
Pre-geometry 324–5
Prime form 233, 240
Projective transformation 555

**Q**
Quadratic differential 213
Quantum gravity 10, 15
Quaternion 518–20

**R**
Ramond sector 105–9, 134–5, 317
Regge behavior 80
Regge pole 556
Regge slope 55
Regge trajectory 55, 62, 556
Ricci flat 454, 457, 473, 556
Riemann-Roch theorem
  conformal field theory 161, 556
  index theorem 412
  moduli space 211, 225, 554
  supermoduli space 222
Riemann surface 199, 209, 472, 556
Riemann tensor 525

**S**
Schottky group 199, 203, 238
Schottky problem 225, 231–5, 240
Second quantization
  advantage over first quantization 34, 44
  BRST field theory 289–328
  geometric string field theory 331–372
  light cone field theory 252–87
  point particle 34–44
Selberg zeta function 223–4, 239
Shapiro–Virasoro model
  amplitude 86
  geometric string field theory 364–6
  modular invariance 196
  multiloop 209
  problems with BRST string field theory 315–6
  single loop 195
Sigma model 408–10
Simplex 459–60
Singularities
  hexagon graph 382, 415
  multiloop 204–5, 209, 224
  single loop 189–91, 196–8, 198–9
SL(2, $R$) 84, 148, 183
SL(2, $Z$) 196
SO($N$) 512
SO(8) 278, 561
SO(10) 166, 169, 173, 175
SO(16) $\otimes$ SO(16) string 127, 441–5
SO(32) 402
SO(44) 447
Sp($N$) 512, 517, 522
Sp($2g$, $Z$) 226–7
Spectrum
  bosonic string 62
  closed string 64
  GS superstring 129
  heterotic string 428–30
  NS-R superstring 108, 134–5
Spin($N$) 519, 557
Spin(32) 417
Spin connection 557
Spin embedding 474–5, 557
Spin field 141, 153–6, 167–8
Spin manifold 390, 557
Spin structure
  single loop 220
  multiloop 220
Spinor
  covariantly constant 455, 504
  Dirac 395
  Majorana 125, 395
  Majorana–Weyl 395
  Weyl 395
Spurious states 65, 557

Stiefel–Whitney class 388, 390, 413
String field theory
  BRST 289–328
  closed string 315, 364
  derivation from group theory 336
  duality 247
  endpoint gauge 360
  equivalence of 309
  four string interaction 272–7, 362
  geometric 331–372
  interpolating gauge 360–3
  light cone 252–87
  midpoint gauge 360
  objections to string field theory 246, 283
  superstrings 277–83, 320–2, 366–8
  Witten's 306–7, 320–2
Strings, types
  four dimensional 491–6
  heterotic 427–32
  $O(16) \otimes O(16)$ 127, 441–5
  Type I, $II_{ab}$ 128–30
Sugawara form 171
$SU(N)$ 516–7
$SU(N/M)$ 533, 558
SU(6, 6) 532
Superconformal ghosts 119, 156, 163, 320–1
Superconformal group 107, 120, 135, 318, 557
Supercurrent 104–5
Superfield
  chiral 536
  vector 535
Supergravity
  $N = 1$ 11, 432, 540–1
  $N = 8$ 12, 541–2, 544
  problems with finiteness 12
  problems with phenomenology 12
  Yang–Mills 543
Supermoduli space 161, 222
Superspace 534
Supersymmetry
  2D world sheet 102
  10D space time 125
  conformal field theory 172–3
  geometric string field theory 366–7
  heterotic string 427
  light cone gauge 131, 137
  local 2D supersymmetry 115–6, 134
  string field theory 280–3
Super Yang–Mills 539, 543
Szego kernel 234

**T**
Tachyon 61–2, 112, 191
Tangent space 350–2, 526
Teichmüller parameter 202, 211–3, 238, 558
Teichmüller space 211–3, 558
Theta function
  heterotic string 436
  multiloop 200, 228–9, 233–4, 239–40
  single loop 183–4, 187, 195
  spin structure 221, 229, 240
Todd class 390, 412
Trees
  bosonic 71–7
  conformal field theory 168
  heterotic 434
  supersymmetric 109–115, 133
Twist operator 85
Type I, $II_{ab}$ 128–30, 180, 237, 542–3
Type II no-go theorem 496–502

**U**
Unitarity
  multiloop 177
  second quantization 35, 244, 245–6
  single-loop 177, 181
$Usp(N)$ 379, 512

**V**
Veneziano amplitude 79
Verma module
  conformal field theory 148, 559
  covariant string field theory 293
  geometric string field theory 344–5
Vierbein 526, 559
Virasoro algebra 56, 145–6, 171, 499, 559

**W**
Wess–Zumino action 538
Wess–Zumino condition 387, 400
Weyl scaling 116, 210, 560
Weyl spinor 560
Wilson lines 481, 505, 560
World sheet 560

**Y**
Yang–Mills theory 8, 308, 528–9

**Z**
Zero mode problems 317–20
Zero norm state 65–6
Zeta function regularization 487–8